# THE IMPACTS OF CLIMATE CHANGE

# THE IMPACTS OF CLIMATE CHANGE

## A Comprehensive Study of Physical, Biophysical, Social, and Political Issues

*Edited by*

TREVOR M. LETCHER

*Laurel House, Stratton on the Fosse, Bath, United Kingdom*

ELSEVIER

Elsevier
Radarweg 29, PO Box 211, 1000 AE Amsterdam, Netherlands
The Boulevard, Langford Lane, Kidlington, Oxford OX5 1GB, United Kingdom
50 Hampshire Street, 5th Floor, Cambridge, MA 02139, United States



**Library of Congress Cataloging-in-Publication Data**
A catalog record for this book is available from the Library of Congress

**British Library Cataloguing-in-Publication Data**
A catalogue record for this book is available from the British Library

ISBN: 978-0-12-822373-4

For information on all Elsevier publications
visit our website at https://www.elsevier.com/books-and-journals

*Publisher:* Candice Janco
*Acquisitions Editor:* Marisa LaFleur
*Editorial Project Manager:* Naomi Robertson
*Production Project Manager:* Sruthi Satheesh
*Cover Designer:* Christian Bilbow

Typeset by SPi Global, India

# Contents

## 5. Effect of climate change on marine ecosystems

Phillip Williamson and Valeria A. Guinder



## 6. Natural disasters linked to climate change

Raktima Dey and Sophie C. Lewis



## 7. Climate change and microbes

Stanley Maloy



## 8. Effects of climate change on food production (fishing)

Heike K. Lotze, Andrea Bryndum-Buchholz, and Daniel G. Boyce



## 9. Effect of climate change on food production (animal products)

Haorui Wu and Florence Etienne



## 10. Emerging typology and framing of climate-resilient agriculture in South Asia

Rajesh S. Kumar, Shilpi Kundu, Bishwajit Kundu, N.K. Binu, and M. Shaji



# C

# Social impacts

## 11. Social issues related to climate change and food production (crops)

Thandi F. Khumalo

# D

## Political impacts

### 19. Security implications of climate change: The climate-conflict nexus

Elisabeth Lio Rosvold



### 20. Climate change governance: Responding to an existential crisis

Heike Schroeder and Yuka Kobayashi



### 21. Justice and climate change

Steve Vanderheiden



### 22. Climate change and the law

John F. McEldowney



### 23. The ethics of measuring climate change impacts

Kian Mintz-Woo



### 24. Climate change and refugees

John F. McEldowney and Julie L. Drolet

# Contributors

**N.K. Binu** College of Forestry, Kerala Agricultural University, Thrissur, India

**Daniel G. Boyce** Ocean Frontier Institute, Dalhousie University, Halifax, NS, Canada

**Andrea Bryndum-Buchholz** Department of Biology, Dalhousie University, Halifax, NS, Canada

**Raktima Dey** Fenner School of Environment and Society, Australian National University, Canberra, ACT, Australia

**Julie L. Drolet** Faculty of Social Work, University of Calgary, Edmonton, AB, Canada

**Tobias Emilsson** Department of Landscape Architecture, Planning and Management, Swedish University of Agricultural Sciences, Alnarp, Sweden

**Florence Etienne** Independent Researcher, Vancouver, BC, Canada

**Valeria A. Guinder** Argentine Institute of Oceanography, National Scientific and Technical Research Council, Bahía Blanca, Argentina

**Rhosanna Jenkins** Tyndall Centre for Climate Change Research, University of East Anglia, Norwich, United Kingdom

**Thandi F. Khumalo** Department of Sociology and Social Work, University of Eswatini, Kwaluseni Campus, Kwaluseni, Eswatini

**Yuka Kobayashi** Department of Politics and International Studies, SOAS, London, United Kingdom

**Adam D. Krauss** Traub Lieberman Straus & Shrewsberry LLP, Hawthorne, NY, United States

**Rajesh S. Kumar** Indian Forest Service (IFS), New Delhi, India

**Bishwajit Kundu** Bangladesh Jute Research Institute, Dhaka, Bangladesh

**Shilpi Kundu** Cities Research Institute & School of Environment and Science, Griffith University, Brisbane, QLD, Australia; Sher-e-Bangla Agricultural University, Dhaka, Bangladesh

**Trevor M. Letcher** Laurel House, Stratton on the Fosse, Bath, United Kingdom

**Sophie C. Lewis** School of Science, University of New South Wales, Canberra, ACT, Australia

**Heike K. Lotze** Department of Biology, Dalhousie University, Halifax, NS, Canada

**Daniel P. Loucks** Cornell University, Ithaca, NY, United States

**Stanley Maloy** San Diego State University, San Diego, CA, United States

**John F. McEldowney** School of Law, University of Warwick, Coventry, United Kingdom

**Kian Mintz-Woo** University Center for Human Values and Princeton School of Public and International Affairs, Princeton University, Princeton, NJ, United States; Department of Philosophy and Environmental Research Institute, University College Cork, Cork, Ireland

**David Mond** Mathematics Institute, University of Warwick, Coventry, United Kingdom

**Jane O'Sullivan** School of Agriculture and Food Sciences, University of Queensland, St Lucia Campus, Brisbane, QLD, Australia

**Jeff Price** Tyndall Centre for Climate Change Research, University of East Anglia, Norwich, United Kingdom

**Juliana Reu Junqueira** Environmental Planning Programme, School of Social Sciences, University of Waikato, Hamilton, New Zealand

**Elisabeth Lio Rosvold** Department of Peace and Conflict Research, Uppsala University, Uppsala; Department of Economic History and International Relations, Stockholm University, Stockholm, Sweden

**S. Santamaria-Aguilar** Coastal Risks and Sea-Level Rise Research Group, Institute of Geography, Christian-Albrechts University, Kiel, Germany

**Heike Schroeder** School of International Development; Tyndall Centre for Climate Change Research, University of East Anglia, Norwich, United Kingdom

**Silvia Serrao-Neumann** Environmental Planning Programme, School of Social Sciences, University of Waikato, Hamilton, New Zealand; Cities Research Institute, Griffith University, Brisbane, QLD, Australia

**Maria Shahgedanova** Department of Geography and Environmental Sciences, University of Reading, Reading, United Kingdom

**M. Shaji** College of Forestry, Kerala Agricultural University, Thrissur, India

**A.T. Vafeidis** Coastal Risks and Sea-Level Rise Research Group, Institute of Geography, Christian-Albrechts University, Kiel, Germany

**Steve Vanderheiden** Department of Political Science, University of Colorado at Boulder, Colorado, CO, United States

**Rachel Warren** Tyndall Centre for Climate Change Research, University of East Anglia, Norwich, United Kingdom

**Iain White** Environmental Planning Programme, School of Social Sciences, University of Waikato, Hamilton, New Zealand

**Phillip Williamson** University of East Anglia, Norwich, United Kingdom

**C. Wolff** Coastal Risks and Sea-Level Rise Research Group, Institute of Geography, Christian-Albrechts University, Kiel, Germany

**Haorui Wu** School of Social Work, Dalhousie University, Halifax, NS, Canada

# Preface

The evidence that our climate is warming is overwhelming. This evidence comes not only from land and sea surface temperature records but also from indicators such as the coverage of Arctic sea ice—all of which, and much more, is discussed in this book and in related books: *Climate Change 3rd edition* (Letcher, 2020) and *Managing Global Warming* (Letcher, 2019). Most scientists in the world now accept that anthropogenic activities and specifically the emissions of greenhouse gases are responsible for the major part of the observed warming. May 9, 2013, was an auspicious day for the warming of the planet, when it was reported by both the National Oceanic and Atmospheric Administration (NOAA) and the Scripps Institute of Oceanography that the daily mean concentration of $CO_2$ in the atmosphere at Mauna Loa laboratory exceeded 400 ppm (400 $\mu$mol $mol^{-1}$ or $400 \times 10^{-6}$) for the first time in millions of years. In June 2020, it was 417 ppm, with the rate of increase accelerating each year. The fundamental aim of this book is to alert the public to these impacts so that adaptations can be made to a world of increasing global temperature. It is also a clarion call to do something about global warming and urgently reduce our dependence on fossil fuels and embrace renewable forms of energy. This book focuses mainly on the social and political impacts of climate change.

We are regularly bombarded in the media by the evidence of the physical impacts of climate change; hurricanes, tornadoes, flooding, wash-aways, record high temperatures, melting sea ice, glaciers and ice sheets, unpredicted and exceptional weather patterns, acidic oceans, dying coral beds, and fast increasing concentrations of $CO_2$ in the atmosphere. It is most likely that the target of keeping global temperatures below 1.5°C above the preindustrial age will be breached and that we will have to accept a much warmer world and all that means.

Projections of our global warming indicate that the temperature will exceed the 2°C global average regarded by many scientists as the upper limit in temperature within the next 50 years. If we do not take action to halt this rise in temperature, we must expect the serious consequences of extreme weather: droughts, floods, winds, and storms. The book is a urgent appeal to humans to take immediate action to reduce the amount of $CO_2$ that we are pumping into the atmosphere, which arguably can best be accomplished by reducing our dependency on fossil fuels. We must strive to stop burning coal and oil in our power stations with the ultimate aim of keeping most of the fossil fuel in the ground and find new, renewable ways of producing electricity and propelling our vehicles.

The book contains 24 chapters and is divided into 4 sections:

- INTRODUCTION
- PHYSICAL and BIOPHYSICAL IMPACTS
- SOCIAL IMPACTS
- POLITICAL IMPACTS

The audience we hope to reach are: policy makers in local and central governments;

students, teachers, researchers, professors, scientists, engineers, and managers working in fields related to climate change and future energy options; editors and newspaper reporters responsible for informing the public; and the general public who need to be aware of the impending disasters that a warmer Earth will bring. An introduction is provided at the beginning of each chapter for those interested in a brief synopsis, and copious references are provided for those wishing to study each chapter topic in greater detail.

Many of the authors were not involved in recent assessments of the IPCC, and here they present fresh evaluations of the evidence testifying to a problem that was described by Sir David King as the most severe calamity our civilization has yet to face (David, 2008).

The IPCC assessments have produced two basic conclusions: firstly, that the current climate changes are unequivocal, and secondly, that this is largely because of the emission of greenhouse gases resulting from human activity. This book reinforces these two conclusions and the chapters on "Indicators of Climate Change" and on the "Possible Causes of Climate Change" are particularly relevant. Furthermore, the section on "Modeling of Climate Change" further supports these conclusions through simulations of past climate changes and projections of future climate.

The International System of Quantities (SI units) has been used throughout the book, and where necessary other units are given in parentheses. Furthermore, the authors have rigorously adhered to the IUPAC notation and spelling of physical quantities.

This book has the advantage that the chapters have each been written by world-class experts working in their respective fields. As a result, this volume presents a balanced picture across the whole spectrum of climate change. Furthermore, the authors are from both the developing and developed countries, thus giving a worldwide perspective of looming climatic problems. The 12 countries represented are: Australia, Bangladesh, Canada, Germany, Ireland, India, New Zealand, South Africa, Swaziland, Sweden, The United Kingdom, and the United States of America.

The success of the book ultimately rests with the 34 authors and co-authors. As editor, I would like to thank all of them for their cooperation and their highly valued, willing, and enthusiastic contributions. I would also like to thank my wife for her patience while I wrote and edited this volume. Finally, my thanks are due to Naomi Robertson of Elsevier whose expertise steered this book to its publication.

*Trevor M. Letcher*
Laurel House, Stratton on the Fosse,
Bath, United Kingdom

## References

David, K.S., 2008. In: Letcher, T.M. (Ed.), Foreword to Future Energy: Improved, Sustainable and Clean Options for Our Planet, first ed. Elsevier, Oxford, ISBN: 978-0-08-054808-1.

Letcher, T.M. (Ed.), 2020. Climate Change: Observed Impacts on Planet Earth, third ed. Elsevier, New York, USA, ISBN: 978-0-12-821575-3.

Letcher, T.M. (Ed.), 2019. Managing Global Warming: An Interface of Technical and Human Issues. Elsevier, Cambridge, MA, USA, ISBN: 978-0-12-814104-5.

SECTION A

# Introduction

# Introduction

CHAPTER

# 1

# Why discuss the impacts of climate change?

*Trevor M. Letcher*

**Laurel House, Stratton on the Fosse, Bath, United Kingdom**

OUTLINE



## 1 Introduction

The world is entering an unprecedented time of global warming which is affecting our climate on which we depend for our very existence. Global warming is causing changes in rain and snow patterns; rising sea levels; increased severity and frequency of droughts, wildfires, storms, tornadoes, and hurricanes; high temperatures and heatwaves and changes to our social fabric and political structures. Global warming is the most important calamitous change our civilization has ever had to face. In another publication *Climate Change 2nd edition* (Letcher, 2015), the physical and biological effects of rising global temperatures were discussed but little was made of the effects on society and on human life. This book puts that to right. These impacts which are now blatantly obvious become more and more important


https://doi.org/10.1016/B978-0-12-822373-4.00020-3

with each passing year and are poised to change our lives and those of our children and their children forever. We must plan our future with these changes in mind. This is the *raison d'etre* for this volume.

Before reading the chapters in this book, it is important that we look at the origins and the physics and chemistry of global warming and let the science tell us just how serious a position our ecosystem and our society is in. The temperature and climate of our planet has been more or less constant for the best part of a million years and it is under this regime of climate that our ecosystem and indeed human life evolved. Any significant deviation from this equilibrium will have a devastating effect on both the ecosystem and on human life. We are fast reaching this stage.

The fundamental mechanism leading to the warming of our planet is the greenhouse effect. This initial warming effect is followed by certain feedback mechanisms (e.g., evaporation of water from the oceans, the reduction in albedo effect on polar ice sheets) which exacerbates the situation leading to further global warming and perhaps, in the not too distant future, a run-a-way global warming catastrophe. Understanding the causes of global warming and the present situation give reader a background to appreciating the different impacts climate change is having on our society. This must indeed educate and galvanize the reader to do something about reducing the onset of a catastrophic collapse of our society and the way we live.

## 2 The greenhouse effect

Much of what follows in this section has been discussed in Chapter 1 of *Managing Global Warming* (Letcher, 2019). It is pertinent to include it here at the beginning of *The Impacts of Climate Change.* The concept of the greenhouse effect goes back to the 1820s, when Joseph Fourier suggested that some component of the earth's atmosphere was responsible for the temperature at the surface of the earth. He was researching the origins of ancient glaciers and the ice sheets that once covered much of Europe (Fourier, 1824). Decades later, Tyndall followed up the Fourier's suggestion, and used an apparatus designed by Macedonio Melloni to show that $CO_2$ was able to absorb a much greater amount of heat than other gases. This fitted in with Fourier's concept and pointed to $CO_2$ as the component in the atmosphere that Fourier was looking for. The Melloni apparatus was called a thermomultiplier, and was reported in 1831 (Nobili and Melloni, 1831; Sella, 2018). Tyndall's results were published in references (Tyndall, 1861, 1863). As a result, Tyndall can be named as the discoverer of the $CO_2$ greenhouse gas effect.

Linking $CO_2$ in the atmosphere to the burning of fossil fuels was to be the last link in the chain in understanding the reasons for the ice ages and also our own climate change. In the 1890s, Svante Arrhenius, an electrochemist, calculated that by reducing the amount of $CO_2$ in the atmosphere by half, the temperature of Europe would be lowered by about 4–5°C. This would bring it in line with ice age temperatures. This idea would only answer the question of why the ice age formed and then retreated, if there were large changes in atmospheric composition and in particular, changes in $CO_2$ concentration. At much the same time, also in Sweden, a geologist, Arvid Högbom, had estimated that $CO_2$ from volcanic eruptions, together with the ocean uptake of $CO_2$, could explain how the $CO_2$ concentrations in the atmosphere could change and hence provide some explanation for the ice ages. Along the way Högbom stumbled on a strange and new idea that the $CO_2$ emitted from industrial coal burning factories might influence the atmospheric $CO_2$ concentration. He did indeed find that human activities were contributing $CO_2$ to the atmosphere at a rate comparable to the natural geochemical processes. The increase was small compared to what was already in the atmosphere, but if continued, it would influence the climate. Arrhenius took up this concept, and his calculations are published (Arrhenius, 1896). Arrhenius concluded that the emissions from human industry might someday bring on global warming. Hence, Arrhenius's name is forever linked to the greenhouse theory of global warming. However, thanks must also go to those who paved the way—Fourier, Melloni, Tyndall, Högbom, and probably many others.

Arrhenius's calculations were at first dismissed as unimportant or at worst faulty. A similar fate was met by G.S Callendar who, in 1938, made the point that $CO_2$ levels were indeed climbing (https://www.rmets.org/sites/default/files/qjcallender38.pdf). It was only in the 1960s, after C D Keeling measured the $CO_2$ concentration in the atmosphere and showed that it was rising rapidly, that scientists woke up to the fact that global warming was real and that anthropogenic activity was to blame.

Water vapor is an even more effective greenhouse gas than $CO_2$. Furthermore, its concentration in the atmosphere is very much higher than that of $CO_2$ (of the order of a hundred times higher), and $H_2O$ contributes over 60% of the global warming effect. The amount of water vapor in the atmosphere is controlled by the temperature. An increase in the $CO_2$ concentration in the atmosphere results in a relatively small increase of the global temperature but that change is enough to increase the amount of water vapor in the air, through evaporation from the oceans. It is this feedback mechanism that has the greatest influence on global temperature. In a sense, paradoxically, the concentration of $CO_2$ acts as a regulator for the amount of water vapor in the atmosphere and is thus the determining factor in the equilibrium temperature of the earth. Without $CO_2$ in the atmosphere, the temperature of the earth would be very much cooler than it is today; in fact, 33°C cooler.

The amount of solar energy shining on the earth (with wavelengths ranging from 0.3 to 5 μm) is vast. It heats our atmosphere and everything on the Earth and provides the energy for our climate and ecosystem. At night, much of this heat energy is radiated back into space but at different wavelengths, which are in the infrared range from 4 to 50 μm (earthguide.ucsd.edu/virtualmuseum/climatechange1/02_3.shtml). The frequencies of the heat radiating from a body is dependent of the temperature of the body (Planck's Law of blackbody radiation). This energy, leaving the Earth, heats the greenhouse gas molecules (such as $H_2O$, $CO_2$, $CH_4$, etc.) in the atmosphere. The explanation is as follows: using $CO_2$ and $H_2O$ as examples, this heating process takes place because the radiated IR frequency is in sync (resonates) with the natural frequency of the carbon-oxygen bond of $CO_2$ (4.26 μm being the asymmetric stretching vibration mode and 14.99 μm being the bending vibration mode) and the oxygen-hydrogen bond of $H_2O$. The increased vibration of the bonds effectively heats the $CO_2$ and $H_2O$ molecules. These heated molecules then pass the heat to the other molecules in the atmosphere ($N_2$, $O_2$) and this keeps the earth at an equitable temperature. The vibrating frequencies of the O—O bond in oxygen and the N—N bond in nitrogen molecules are very different from the radiation frequencies and so are unaffected by the radiation leaving the Earth at night.

## 3 Global warming

The scientific evidence that global warming is largely because of the rising $CO_2$ levels in the atmosphere is overwhelming and, furthermore, that the rising $CO_2$ concentration is because of human activities. Every scientific society and every research organization working in the field of climate change accepts this view. The atmospheric $CO_2$ concentration has increased from 280 ppm (280 ppm or 280 molecules per million molecules) (https://link.springer.com/article/10.1007/BF02423528) before the industrial revolution to 417 ppm (observed at Mauna Loa Observatory on May 2020) (https://www.co2.earth/daily-co2), and it is this increase of almost 50% that has triggered the present increase in global temperature.

The total amount of $CO_2$ in the atmosphere and its concentration value are the most dependable measurements we have for the progress of global warming. In 1960, the *rate* of increase of $CO_2$ (as measured at Mauna Loa, in Hawaii) was less than 1 ppm per year. It is now 2.4 ppm per year (https://link.springer.com/article/10.1007/BF02423528). It is this rate of change that is the best indicator of any progress we are making in reducing global warming. At the moment there is no sign that this is happening, in fact the reverse is true. Even if we stopped burning fossil fuel, the $CO_2$ levels will take a long time to decrease as the lifetime of $CO_2$ in the upper atmosphere is of the order of hundreds of years.

The most compelling evidence that the increase in $CO_2$ is the most likely cause of global warming can be seen in the related graphs of $CO_2$ concentration in the atmosphere and the global average temperature as functions of time over the past many decades (see Figs. 1 and 2). The $CO_2$ increase is mirrored by an increase in the relative increase in average global temperatures over the past 60 years.

A question which needs answering is this: we know that the $CO_2$ level in the atmosphere is rising steadily: but is the $CO_2$ increase because of human activity? The evidence that it is indeed because of human activity is based on the relative ratios of carbon isotopes. The relative amount of $^{13}C$ in the atmosphere has been declining and that is because the ratio of $^{13}C$ in fossil fuel-derived $CO_2$ is significantly lower that the $CO_2$ produced from present-day decaying plants (http://www.realclimate.org/index.php/archives/2004/12/how-do-we-know-that-recent-cosub2sub-increases-are-due-to-human-activities-updated/).

From the properties of each of the greenhouse gases (such as the wavelengths of energy), scientists can calculate how much of each gas contributes to global warming. The results show that $CO_2$ is responsible for about 20% of the earth's greenhouse effect, water vapor between 60% and 80% (https://www.acs.org/content/acs/en/climatescience/.../its-water-vapor-not-the-co2.html or https://www.nasa.gov/topics/earth/features/vapor_warming.html). It is, however, $CO_2$ that is the driver and trigger of global warming. The rest is caused by minor greenhouse gases such as methane and chlorinated hydrocarbons. The relative concentrations of the major greenhouse gases emitted by human activity in the USA are: $CO_2$ 81%; $CH_4$ 10%, and $N_2O$ 7% (https://www.epa.gov/ghgemissions/overview-greenhouse-gases).

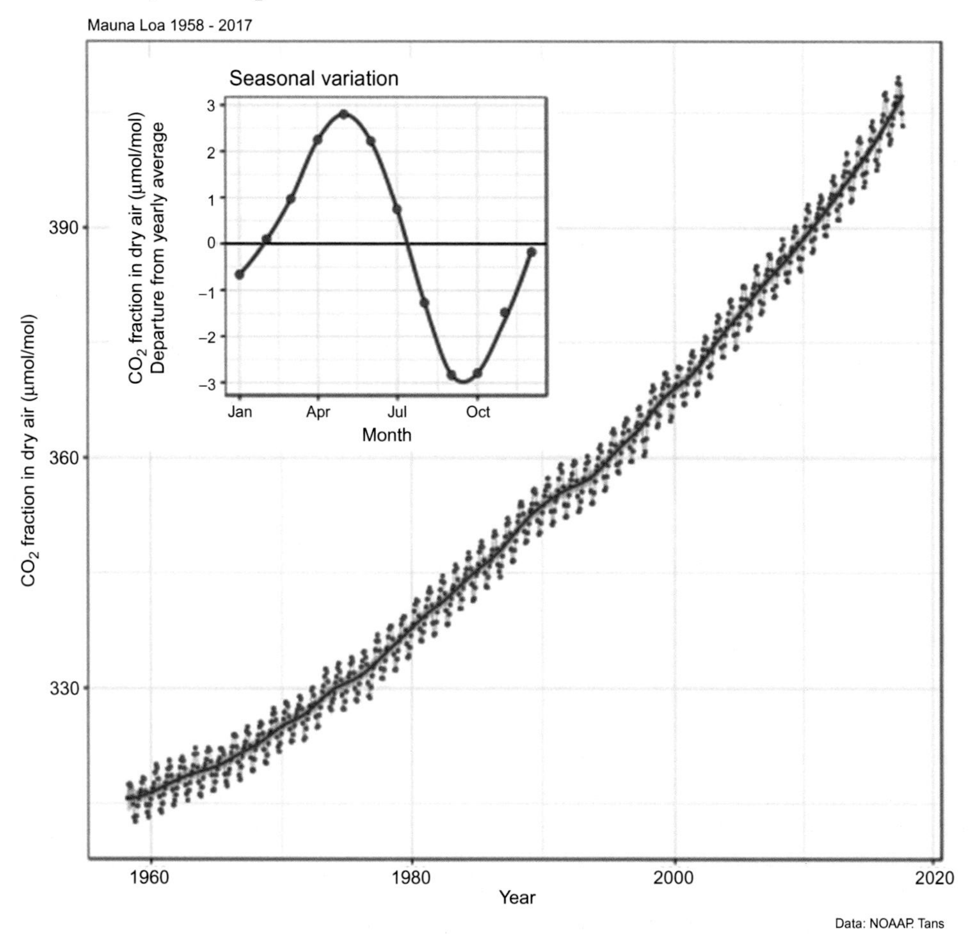


**FIG. 1** The increase of $CO_2$ concentration over the past 60 years. *https://en.wikipedia.org/wiki/Keeling_Curve Data from Dr. Pieter Tans, NOAA/ESRL and Dr. Ralph Keeling, Scripps Institution of Oceanography.*

It is perhaps of interest to note that it is not possible to obtain absolute proof that it is $CO_2$ which is largely responsible for global warming because we cannot do the definitive experiment of suddenly stopping the use of fossil fuels. And even if we could do this experiment it would take decades to obtain a definite conclusion because of the long-life $CO_2$ has in the atmosphere (https://www.epa.gov/ghgemissions/global-greenhouse-gas-emissions-dat).

Most of the anthropogenic $CO_2$ entering the atmosphere comes from fossil fuels. The relative fraction of energy produced by fossil fuels has remained at over 86% over the past decade as is illustrated by worldwide primary energy consumption listed in Table 1. However, the quantity of fossil fuel extracted from the earth has increased significantly over the past 11 years, as seen in Table 2. This is reflected in the steadily increasing amount of $CO_2$ entering the atmosphere. However, between 2015 and 2016, world oil production increased by only

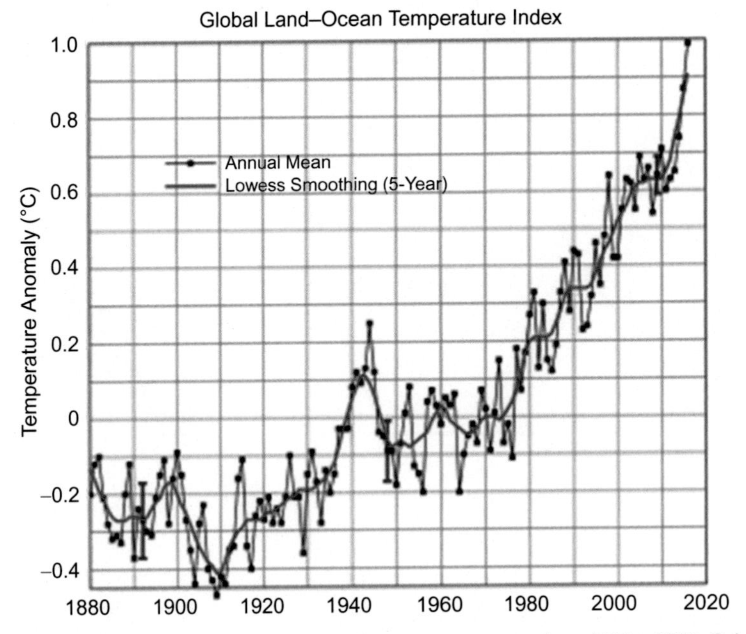


**FIG. 2** The relative increase in the world's average surface air temperature from 1880 to 2009. *Original data produced by ASA's Goddard Institute for Space Studies (http://data.giss.nasa.gov/gistemp/graphs/).*

**TABLE 1** Worldwide primary energy consumption percentages from 2005 to 2015 (https://www.worldenergy.org/wp-content/uploads/2016/10/World-Energy-Resources-Full-report-2016.10.03.pdf).

| | Energy/% | | |
|---|---|---|---|
| | **2005** | **2010** | **2015** |
| Oil | 36 | 33.5 | 32.9 |
| Coal | 28.6 | 29.8 | 29.2 |
| Natural gas | 22.9 | 23.7 | 23.9 |
| Nuclear | 5.7 | 5.1 | 4.4 |
| Hydropower | 6.1 | 6.4 | 6.8 |

**TABLE 2** The quantities of oil, coal, and natural gas mined pumped over the decade 2005–15.

| | **2005** | **2010** | **2016** |
|---|---|---|---|
| Oil volume/($10^6$ barrels) per day (https://www.bp.com/en/global/corporate/energy-economics/statistical-review-of-world-energy/oil/oil-production.html) | 81,908 | 83,251 | 92,150 |
| Coal/($10^6$ t oil equivalent) per year (https://www.bp.com/en/global/corporate/energy-economics/statistical-review-of-world-energy/coal/coal-production.html) | 3039 | 3633 | 3656 |
| Natural Gas ($10^9$ $m^3$) per year (https://www.bp.com/en/global/corporate/energy-economics/statistical-review-of-world-energy/natural-gas/natural-gas-production.html) | 2773 | 3192 | 3551 |

0.4%, *world* coal production fell by 6.2%, and natural gas increased by only 0.3%. This is the first sign that fossil fuel usage is slowing down.

The Global Carbon Project (GCP) (http://www.globalcarbonproject.org/) has reported that emissions in 2015 from burning fossil fuels and also from industry (especially cement production) account for 91% $CO_2$ caused by human activity with 9% from land use changes. In 2015, the GCP has reported that $9.9 \times 10^9$ t of carbon in the form of $CO_2$ from burning fossil fuels entered the atmosphere. Nevertheless, the GCP felt that there were signs that the emission of $CO_2$ from human activity was indeed showing signs of peaking.

## 4 Feedback mechanisms to further increase the heating of the planet

There are many $CO_2$ feedback mechanisms at play and five of them are summarized below.

The water vapor feedback mechanism has been discussed above and it is this feedback mechanism that has the greatest influence on global temperature.

The melting of ice contributes to another feedback mechanism. When ice melts, land or open water takes its place. Both land and open water are on average less reflective than ice and thus absorb more solar radiation. This causes more warming, which in turn causes more melting, and this cycle continues.

The oceans are a storehouse for $CO_2$ but the amount it stores is limited by the solubility of $CO_2$ in seawater. This solubility is dependent on the temperature. Global warming results in a warmer sea and a lowering of the $CO_2$ solubility, resulting in some $CO_2$, leaving the oceans and entering the atmosphere, which in turn increases global warming and so on...

Another feedback mechanism is at play in the peat bogs and permafrost regions of the world, such as in Siberia and in Greenland. Rising global temperatures are melting the permafrost and will in time release vast quantities of methane gas ($CH_4$). This gas is over 25 times more effective than $CO_2$ as a greenhouse gas.

Yet another feedback mechanism involves methane clathrates, a form of water ice that contains methane within its crystalline structure. Extremely large deposits of it have been found under the sediments on ocean floors. An increase in temperature breaks the crystal structure releasing the caged methane. Rising sea temperatures could cause a sudden release of vast amounts of methane from such clathrates resulting in a runaway global warming events.

## 5 Other possible causes of climate change

In spite of the evidence presented above, there has been much debate as to whether our present global warming and climate change could in fact be because of effects other than atmospheric gases. These include: the variation in the sun's energy; volcanic activity; changes in the earth's orbital characteristics, including the Malankovitch cycles; cosmic ray effects; and atmospheric aerosols. These have all been discussed in chapters in Climate Change (Letcher, 2015) by world experts and the consensus of opinion is that none of them could not possibly be responsible for our present climate change (Stenchikov, 2016). The conclusion of scientists around the world is summed up by Macott et al. who wrote "The earth's climate is complex and responds to multiple forcings, including $CO_2$ and solar insolation. Both of those have changed very slowly over the past 11,000 years. But in the last 100 years, the increase in $CO_2$ through increased emissions from human activities has been significant. It is the only variable that can best explain the rapid increase in global temperatures" (Shakun et al., 2012).

Overall, human sources of $CO_2$ are much smaller than the natural emissions from animals, plants, decaying animals, vegetation, and volcanoes (https://earthobservatory.nasa.gov/Features/CarbonCycle/page5.php). However, human activity has upset the balance in a cycle that has existed for thousands of years. The amount of carbon on the earth and in the atmosphere is fixed but it is in a dynamic and equilibrium cycle, moving between living and nonliving things, and changing into different carbon compounds such as carbon dioxide in the air and in the oceans, solid carbonate rock, and the living cells of plants and animals. In the first step of the cycle, plants take up $CO_2$ from the air through the process of photosynthesis, and release oxygen. The $CO_2$ is then converted into living cells. In the next step, the animals eat the plants, and the carbon in the plant cells are used to build animal tissue and cells. Animals also breathe in oxygen and exhale carbon dioxide which in turn enter the atmosphere and the cycle continues. Dead plants and dead animals decompose and carbon is either released as $CH_4$ and $CO_2$ or stored in the soil. Superimposed on this cycle is the exchange of $CO_2$ between the atmosphere and the oceans. These processes were in near perfect equilibrium before the industrial revolution (https://earthobservatory.nasa.gov/Features/CarbonCycle/page5.php) (Denman et al., 2007).

The evidence that global warming is altering our climate is very well documented, and almost no day goes by without more evidence for climate change. The indicators include: more extreme weather events in the future; melting of Arctic sea ice; Antarctic Sea changes; land ice behavior (including glaciers and ice sheets); weather pattern changes; bird ecology changes (including migration); mammal and insect ecology changes and biodiversity loss; sea life and coral reef changes; marine diversity and intertidal indicators; plant ecology and plant pathogen changes; rising sea levels; and ocean acidification (Letcher, 2016).

## 6 Urgent action is required

There is a growing threat of environmental collapse in the future, as a result of changes in our present climate. We are beginning to see this with extreme weather events such as flooding, droughts and water crises, high winds, runaway fires, wash-aways, and mud flows from land denuded of its natural rain soaking properties, high seas in coastal areas, together with rising sea levels, to mention a few. One consequence of climate change is the migration of insects and animals to more hospitable climates. A more frightening involuntary mass migration has already begun: of humans from lands unable to support the growing of crops and from areas where rising sea levels are beginning to threaten livelihood. It is not only natural disasters that are a cause for concern but also man-made disasters which result indirectly from global warming that are a cause for concern. These include: the reduced ability of land to soak up rain water as a result of land clearances and urban development resulting in flooding; chemical pollution in the form of pesticides, endocrine disruptors and hormonally active agents used on farms to increase yields; nuclear disasters through extreme weather;

land-use decisions for agriculture; oil fires, coal mine fires and even tyre fires which add their own contribution to rising $CO_2$ levels (https://www.ecotricity.co.uk/our-green-energy/energy-independence/the-end-of-fossil-fuels).

Most world governments have accepted the assessment of the United Nations Framework Convention on Climate Change (UNFCCC) that a 2°C rise in mean global temperature above the preindustrial level must be the maximum limit. In order to meet this objective, studies generally indicate the need for global greenhouse gas emissions to peak before 2020 with a substantial reduction in emissions thereafter.

We need to reduce the amount of $CO_2$ entering the atmosphere and if possible, we should find ways of removing some of the $CO_2$ presently in the atmosphere. Present day $CO_2$ levels in the atmosphere exceed the natural equilibrium of dissolved $CO_2$ in the oceans and with the $CO_2$ uptake by biota on land. Unfortunately, this rising nonequilibrium amount of $CO_2$ remains in the air for a very long time. The reason is that $CO_2$, unlike other greenhouse gases such as $CH_4$, is very un-reactive. It does not naturally react with most chemicals and in thermodynamic terms, it has a very high Gibbs Energy of Formation. In order to bring about a reaction of $CO_2$ with another chemical, a significant amount of energy must be given to the system (e.g., heat energy). This is also the reason why it is so difficult to get rid of waste $CO_2$ from chemical reactions (e.g., cement manufacture, or even from burning fossil fuels) and why it is rarely used as a chemical feedstock in industry.

There are still large reserves of coal oil and gas in the earth. These convenient sources of energy are not only easy to use for heating and for producing energy, but exist in a stored form which allows them to be used at any time in the future. It has been estimated (https://www.ecotricity.co.uk/our-green-energy/energy-independence/the-end-of-fossil-fuels) that globally, we currently consume the equivalent of over 11 billion tonnes of oil from fossil fuels every year. Crude oil reserves are vanishing at a rate of more than 4 billion tonnes a year—so if we carry on as we are, our known oil deposits could run out in just over 53 years. If we increase gas production to fill the energy gap left by oil, our known gas reserves only give us just 52 years left. Although it's often claimed that we have enough coal to last hundreds of years, this doesn't take into account the need for increased production if we run out of oil and gas. If we step up production to make up for depleted oil and gas reserves, our known coal deposits could be gone in 150 years. Another set of estimates have been given by British Petroleum (BP) in 2018. The figures were a little less optimistic. Their estimation of the time left for fossil fuel as a result of present-day usage was predicted to be: oil will end in 30 years, gas in 40 years and coal in 70 years (https://mahb.stanford.edu/library-item/fossil-fuels-run/). Our future mindset must however not be seduced by the convenient properties of fossil fuel, but for the sake of the planet, the reserves must stay forever below ground and nonfossil fuel sources of energy should be embraced.

What is also required is the need for the world to replace growth in the financial sector with sustainability for the future of the society and the world's ecosystem. We cannot carry on as we are and perhaps the present COVID-19 pandemic has given us time to rethink our lives and follow the advice of Riccardo Mastini in reference (http://unevenearth.org/2020/02/a-post-growth-green-new-deal/). "To summarize, from a postgrowth perspective a Green New Deal must pursue three distinct but interrelated goals: decreasing energy and material use, decommodifying the basic necessities of life, and democratizing economic production.

Any Green New Deal proposal that does not address head-on the drivers of economic growth is doomed to fall short of the challenge of steering away from the worst scenarios of ecological breakdown." This is also the sentiment of Jason Hickel who wrote in reference (Hickel, 2020), "The world has finally awoken to the reality of climate breakdown and ecological collapse. Now we must face up to its primary cause. Capitalism demands perpetual expansion, which is devastating the living world. There is only one solution that will lead to meaningful and immediate change: and that is degrowth."

## 7 Our present situation

This past year, 2019, was again one of the hottest on record (https://climate.nasa.gov/news/2945/nasa-noaa-analyses-reveal-2019-second-warmest-year-on-record/). According to independent analyses by NASA and the National Oceanic and Atmospheric Administration (NOAA), Earth's average global surface temperature in 2019 was the second warmest since modern record-keeping began in 1880. Globally, average temperature in 2019 was second only to that of 2016 and continued the planet's long-term warming trend: the past 5 years have been the warmest of the last 140 years. This past year was 0.98°C warmer than the 1951–80 mean, according to scientists at NASA's Goddard Institute for Space Studies (GISS) in New York (https://climate.nasa.gov/news/2945/nasa-noaa-analyses-reveal-2019-second-warmest-year-on-record/). It shows that we are not doing enough to reduce the amount of $CO_2$ in the atmosphere. The only way to reduce global warming is to reduce the amount of $CO_2$ we are pumping into the air, and if possible, removing $CO_2$ from the atmosphere.

At present, less than 20% of all energy sources are either renewable (wind, solar, hydropower, biomass tide, and geothermal) or nuclear. Replacing fossil fuel to reduce significantly our $CO_2$ emissions is going to be a mammoth task.

The world is not replacing fossil fuel with renewable forms of energy fast enough. This was emphasized in 2019 by Spencer Dale, Chief Economist at BP who stated: "There is a growing mismatch between societal demands for action on climate change and the actual pace of progress, with energy demand and carbon emissions growing at their fastest rate for years. The world is on an unsustainable path" (https://www.bp.com/en/global/corporate/news-and-insights/press-releases/bp-statistical-review-of-world-energy-2019.html).

It is possible for solar energy to power the world. In less than 80 min, the solar equivalent energy of the total world energy use for a year, strikes the Earth; this implies that in theory the sun could power the world 7000 times.

It has been estimated, that in 2015, human activities contributed $36.8 \times 10^9$ t of $CO_2$ through burning coal and other fossil fuels, cement production, deforestation, and other landscape changes (https://www.carbonbrief.org/analysis-global-fossil-fuel-emissions-up-zero-point-six-per-cent-in-2019-due-to-china#:~:text=Emissions%20from%20fossil%20fuel%20and,Global%20Carbon%20Project%20(GCP)). It has also been estimated that since the Industrial Revolution, over $2000 \times 10^9$ t of $CO_2$ have been added to the atmosphere. Human activities emit 60 or more times the amount of carbon dioxide released by volcanoes each year (https://www.climate.gov/news-features/climate-qa/which-emits-more-carbon-dioxide-volcanoes-or-human-activities).

The population of the world is increasing and so is the need for more energy with a greater demand for more electricity. The world population (it is now $7.6 \times 10^9$ according to the latest 2018 United Nation estimate) is expected to reach $9 \times 10^9$ in 2050. It is increasing at a rate of 1.09% per year at the moment (2018) down from 1.14% $yr^{-1}$ in 2016 and down from the peak in 1963 of 2.2% $yr^{-1}$. The expected rate of growth in energy demand over the next decade is greater than the growth rate of the population; this is largely because of the increase demand for electricity in developing countries. Electricity generation is expected to increase from $25 \times 10^{12}$ kWh in 2017 to $31.2 \times 10^{12}$ kWh in 2030 an increase of almost 2% per year (https://www.statista.com/statistics/238610/projected-world-electricity-generation-by-energy-source/).

At the moment, coal is still the largest producer of electricity worldwide, and is not expected to be overtaken by renewables until 2040. The relative breakdown of electricity producers and future predictions is given in Table 3. It illustrates the energy dilemma of our time—the positive and encouraging increase in the deployment of renewable forms of energy is masked by the increasing overall energy needs of the world and that increase is still being met by further increases in fossil fuel usage. The present and future world electricity generation is dominated by the burning of fossil fuels (over 60%) and the prediction for 2040 is not much better (58%). It is no doubt driven by a number of forces including: the relative

**TABLE 3** Breakdown of electricity production worldwide and a prediction over the next two decades (https://www.statista.com › Energy & Environmental Services › Electricity).

| | 2012 | 2020 | 2030 | 2040 |
|---|---|---|---|---|
| | Electricity production energy/($10^{12}$ kWh) | | | |
| Oil | 1.06 | 0.86 | 0.62 | 0.56 |
| Nuclear | 2.34 | 3.05 | 3.95 | 4.50 |
| Renewables | 4.73 | 6.87 | 8.68 | 10.63 |
| Natural gas | 4.83 | 5.20 | 7.47 | 10.10 |
| Coal | 8.60 | 9.73 | 10.12 | 10.62 |

**TABLE 4** Worldwide source of $CO_2$ (mostly fossil fuel) emissions, 2018 (https://www.epa.gov/ghgemissions/sources-greenhouse-gas-emissions).

| | Carbon dioxide emissions (%) |
|---|---|
| Electricity | 27 |
| Transport | 28 |
| Industrial (including cement manufacture) | 22 |
| Residential (heating, wood fires) | 12 |
| Agriculture | 11 |

economics of fossil fuels versus renewable energy; the massive inertia linked to status quo situations; and the fear of things new as opposed to well-tried technologies.

Electricity production is not the only producer of $CO_2$ in our atmosphere. The various sectors responsible for $CO_2$ generated as a result of human activity is given in Table 4.

If there is the necessary political will to do so, we can replace the fossil fuel-derived electricity with renewable forms of energy, or nuclear energy or hydropower. However, we do have a problem with replacing transport fuel. We could 1 day have electric cars replace petrol vehicles and possibly even diesel vehicles, but replacing fossil fuel for air travel and sea travel is difficult if not impossible. Furthermore, some industrial processes such as cement manufacture, involving the heating of $CaCO_3$ resulting in the waste product, $CO_2$, are also problematic. Attempts at replacing petrol in transport with renewable fuel-derived from biomass (sugar cane as done in Brazil or corn as done in the US for petrol, and palm oil in Malaysia for biodiesel) has had some success but the overall contribution has been relatively small (https://www.iea.org/etp/tracking2017/transportbiofuels/). In 2016, the biofuels contributed 4% to the world's transport fuels. The US, Brazil, and Malaysia are the world leaders in biofuels.

All of this does indicate that the world is not on top of solving the global warming problem, in spite of the steady increase in the deployment of renewable forms of energy. The changeover from fossil fuel to renewables is just too slow. It is predicted that renewables will increase their share of electricity production from 21.9% in 2012 to 29.2% in 2040 (less than 0.3% per year) (see Table 1). We will have to work very much harder to replace fossil fuel as the main driving force of our energy industry.

One slight glimmer on the horizon is the fact that natural gas, methane, (including shale gas) is better for the planet than burning coal and in many countries, coal is being replace by natural gas. The reason why natural gas is better than coal is that the amount of $CO_2$ produced from burning $CH_4$ per unit of energy ($50\,g\,MJ^{-1}$) is less than it is for coal ($92\,g\,MJ^{-1}$) and moreover coal burning produces particulates. Of course, the burning of $CH_4$ still produces $CO_2$:

$$CH_4 + 2O_2 = CO_2 + 2H_2O$$

## 8 Global warming, climate change, and the new pandemic—COVID-19

The coronavirus pandemic has been linked to climate change issues. There seems to be little doubt that there is a link between population density, human encroachment on natural areas, and zoonotic disease transmission (https://www.eco-business.com/opinion/covid-19-is-a-product-of-our-unhealthy-relationship-with-animals-and-the-environment/). Climate change through drought, flooding, rising sea levels, unpredictable weather conditions is slowly reducing the arable land in many parts of the world forcing people to move into areas close to wildlife populations that humans had not previously been in close contact with. The disruption of pristine forests driven by logging, mining, the need to find new places to live, the spread of urban development, and population growth is bringing people into closer contact with animal species. In the case of COVID-19, the contact was most likely with bats. As Jane Goodall says, "COVID-19 is a product of our unhealthy relationship with animals and the environment and that our exploitation of animals and the environment has contributed to pandemics, including the COVID-19 crisis. Wildlife trafficking, factory farming, and the destruction of habitats are drivers of zoonotic diseases" (https://www.eco-business.com/opinion/covid-19-is-a-product-of-our-unhealthy-relationship-with-animals-and-the-environment/). With global warming on the increase, we can expect that climate change will further impact on the spread of infectious diseases through animals as it has in the past with rabies, the plague, Ebola, SARS, MERS, and ZIKA to mention but a few zoonotic diseases.

## 9 How global warming affects society

Global warming affects societies in many ways. Here below are a few examples:

- Reduces the area available for farming and for human occupation through droughts, floods, and climate change resulting in food shortages.
- Sea level rises result in a loss of housing and farmland which in turn involves human migration, and expensive new housing/buildings.
- Risk of life increases and insurance premiums rise—all insurers suffer.
- Need rapid development of renewable energy to replace fossil fuel.
- Pressure on industry to improve efficiencies, leading to more expensive products.
- Health suffers in many ways. For example, malaria becomes more widespread as a small rise in temperature results in a large vectoral capacity of mosquitoes (development of anopheles is shorter, bites by females increases as gonadotrophic cycle is shortened, incubation period of plasmodium decreases).
- Heat waves kill. It has been estimated that the heatwave in Europe, in 2003, with temperatures of over 45°C, killed 70,000 people. In France, the number of heatwaves has doubled over the past 40 years and is expected to double again by 2050.
- Economics—everything costs more—electric car, electricity, imported food, insurance, need more air conditioning.
- Wild fires—loss of homes, loss of habitat for animals and wild animals.
- Rise in extreme weather patterns (flooding, gales, droughts, too hot, too cold...) reduce time for working.
- Flooding low lying areas of the world causes a loss of infrastructure and housing.
- Increased tropical typhoons and hurricanes leading to loss of life, housing and occupations.
- Droughts—causes of starvation, food shortages, loss of occupation, start of wars—catalyst for unrest in Syria, civil war and human migration, refugees, droughts are reputed to be the most expensive weather-related disasters;
- Mental anguish as a result.

## 10 Conclusions

We believe that we do understand the underpinnings of global warming and that greenhouse gases and $CO_2$ in particular are the root causes. We know that renewable forms of energy and possibly nuclear energy MUST replace fossil fuel where possible and that this must be done soon. There are limits to what can be replaced as we unfortunately depend on fossil fuels for transport both now and in the foreseeable future. Other areas such as electricity production using fossil fuel can and should be phased out.

Much depends on governments around the world having the will and energy to drive a nonfossil fuel policy. It should be a basic moral decision for the sake of future generations. Governments should not be driven by short-range decisions that benefit the few (including parliamentarians and also shareholders in fossil fueled industries). They should be bold and brave enough to create a legacy for our children and our children's children.

This book is about the impact of climate change on the planet's ecosystem and human beings. The indicators of climate change on the weather patterns, sea levels, wind systems, ocean currents, animal, bird and insect ecology, sea life, corals, marine and intertidal ecology, plant ecology and plant pathogens have been dealt with in detail and at great length in *Climate Change 2nd edition*. The major part of the book relates to the social and political issues as impacted by climate change. The social impacts of climate change stem from: food production problems; population movements and expansion; economics related to ecological changes; societal adaption, attitudes and pressures on urban life. There is no precedent for coping with these issues. We must however note the changes and make plans with new ways to deal with them. This is a global problem and it is hoped that countries and governments will come together to find solutions to new ways of life as they have for the COVID-19 pandemic. The section on political aspects focuses on security, governance, justice, the law and ethics as related to climate change; all of which involve issues that are new and need new solutions and preferably solutions that are universal and can be adopted worldwide. Again, we must look at international cooperation for finding solutions; this is bought home to us in the final chapter on climate refugees. This is a very new problem and can only be solved with collaboration from all governments.

## References

Arrhenius, S., 1896. On the influence of carbonic acid in the air upon the temperature of the ground. Philiso. Mag. J. Sci. 41, 237–276.

Denman, K.L., Brasseur, G., Chidthaisong, A., Ciais, P., Cox, P.M., Dickinson, R.E., Hauglustaine, D., Heinze, C., Holland, E., Jacob, D., Lohmann, U., Ramachandran, S., da Silva Dias, P.L., Wofsy, S.C., Zhang, X., 2007. Couplings between changes in the climate system and biogeochemistry. In: Climate Change 2007: The Physical Science Basis. Contribution of Working Group I to the Fourth Assessment Report of the Intergovernmental Panel on Climate Change. Cambridge University Press, Cambridge, United Kingdom and New York, NY, USA.

Fourier, J., 1824. Remarques Générales sur les Températures Du Globe Terrestre et des Espaces Planétaires. Ann. Chim. Phys. 27, 136–167.

Hickel, J., 2020. Less is More: How Degrowth Will Save the World. William Heinemann, Milton Keynes, UK. https://www.penguin.co.uk/books/1119823/less-is-more/9781785152498.html.

Letcher, T.M. (Ed.), 2015. Climate Change: Observed Impacts on Planet Earth. second ed. Elsevier, New York. ISBN 978-0-444-63524-2.

Letcher, T.M. … (Ed.), 2016. Climate Change: Observed Impacts on Planet Earth. second ed. Elsevier, pp. 21–Oxford Chapters 2–21.

Letcher, T.M., 2019. Managing Global Warming: An Interface of Technical and Human Issues. Elsevier, Cambridge, MA ISBN: 978-0-12-814104-5.

Nobili, L., Melloni, M., 1831. Le Thermo-multiplicateur. Ann. Chim. (Phys.) 48, 198–199.

Sella, A., 2018. Melloni's thermomultiplier. Chem. World 15, 70.

Shakun, J.D., Clark, P.U., He, F., Marcott, S.A., Mix, A.C., Liu, Z., Otto-Bliesner, B., Schmittner, A., Bard, E., 2012. Global warming preceded by increasing carbon dioxide concentrations during the last deglaciation. Nature 484, 49–54.
Stenchikov, G., 2016. The role of volcanic activity in climate and global change. In: Letcher, T.M. (Ed.), Climate Change, Observed Impacts on Planet Earth. second ed. Elsevier, Oxford, pp. 419–448.
Tyndall, J., 1861. On the absorption and radiation of heat by gases and vapours.... Philos. Mag. Ser. 4. 22, 169–94, 273–85.
Tyndall, J., 1863. On radiation through the earth's atmosphere. Philos. Mag. Ser. 4 25, 200–206.

## Further reading

Vallero, D.A., Letcher, T.M. (Eds.), 2013. Unraveling Environmental Disasters. Elsevier, Oxford (Chapters 8, 12, 13, 14).

CHAPTER

# 2

# Impacts of climate change on economies, ecosystems, energy, environments, and human equity: A systems perspective

*Daniel P. Loucks*

**Cornell University, Ithaca, NY, United States**

OUTLINE




https://doi.org/10.1016/B978-0-12-822373-4.00016-1

# 1 Introduction

Global climate change and the mitigation of its adverse impacts is arguably one of the most critical issues facing humanity today. It has many dimensions and addressing them requires the insights from many disciplines—science, economics, society, governance, and ethics. The impacts of global warming will be around for decades, if not centuries. The discharges of gases from our extraction, processing, and use of fossil fuels are the major causes of the relatively recent rapid increase in global warming. Some of these gases, especially carbon dioxide, can remain in the atmosphere for hundreds of years. So even if all such gas emissions stopped, global warming and its adverse impacts will continue to be felt and affect us and our descendants. While we cannot stop global warming, we can control its rate of change.

The adverse impacts of global warming are felt in our economic and energy sectors, in the quality of our environment and in the functioning and health of our ecosystems. And we humans are part of our ecosystems. The pandemic caused by COVID-19 is also affecting our economy and our health, but because it has happened relatively quickly, and because its impact on our physical health is pronounced, dramatic, and undebatable, it got our attention and willingness to take measures to reduce its spread and find a cure. Not so for managing the causes of our warming climate. The adverse impacts resulting from the increase in global warming are multiple, major, and long term. They are changing the world we live in as well as affecting our health, but at rates that so far have not, with few exceptions, generated the political will needed to control it.

The aim of this chapter is to take a systems view of how climate change is impacting us and our planet and in turn how what we consume, and then discard or discharge into our environment, is impacting our climate. Focusing on these linkages and feedbacks among the various climatic, economic, ecologic, energy, and environmental components of this system may help us identify more comprehensive and effective approaches to addressing what the Oxford Dictionaries has called a climate emergency—"a situation in which urgent action is required to reduce or halt climate change and avoid potentially irreversible environmental damage resulting from it" (Ripple et al., 2020).

The section that follows first briefly reviews the causes of global warming and the physical impacts being observed today resulting from increasing temperatures. The next four sections focus on how global warming is impacting, and being impacted by, the world's ecosystems, economies, energy sectors, and environments. The chapter ends with a discussion of how these impacts are inequitably distributed among the world's countries and communities. The schematic in Fig. 1 illustrates the interdependencies among each of these system components.

# 2 Climate change

Changes in the earth's climate have been happening relatively slowly. Over the last 650,000 years, ice glaciers have advanced and retreated, with the most recent one ending some 11,700 years ago. These changes are attributed to variations in Earth's orbit and the resulting

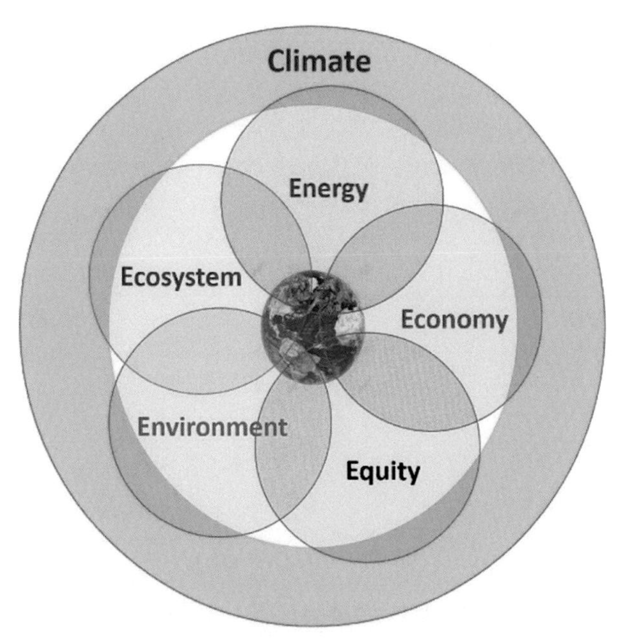


**FIG. 1** The interdependent climate-economic-ecosystem- energy-environment-equity nexus.

changes in the amount of solar energy reaching the Earth. Since the mid-20th century, the major cause of the relatively faster rate of warming is different. Rather than geophysical, it is anthropological. Scientists now consider the current rate of increase in global warming to be the result of human activities. Since the industrial revolution, we humans have learned we have the ability to dramatically alter the climate of our planet and at the global scale. Current increases in average global temperatures, as illustrated in Fig. 2, are now occurring roughly 10 times faster than the average rate of warming after the last ice age (NASA, 2020a).

In the 1860s, physicist John Tyndall (shown in Fig. 3) recognized the impact of carbon dioxide, methane, and nitrous oxides gases on global temperatures. Their discharge into the atmosphere result from natural processes such as plant and animal respiration and volcano eruptions and through a variety of human activities. These activities include deforestation, land use changes, burning fossil fuels. Decomposition of wastes, the use of commercial and organic fertilizers, domestic livestock, soil and rice cultivation, manure management, nitric acid production, and biomass burning.

Some 30 years after Tyndall's discovery, Swedish scientist Svante Arrhenius published a paper explaining how changes in the levels of carbon dioxide in the atmosphere could increase the Earth's surface temperature (Graham, 2000). As such, carbon dioxide acts like a shield or greenhouse, preventing much of the sun's reflected radiations from exiting the atmosphere. Indeed, as shown in Figs. 4–6, warming rates have increased, especially after the industrial revolution. Humans have increased atmospheric $CO_2$ concentration by more than a third since the Industrial Revolution began. The agricultural, industrial, and transportation activities that support our economies have raised atmospheric carbon dioxide levels from preindustrial revolution 280–412 ppm.

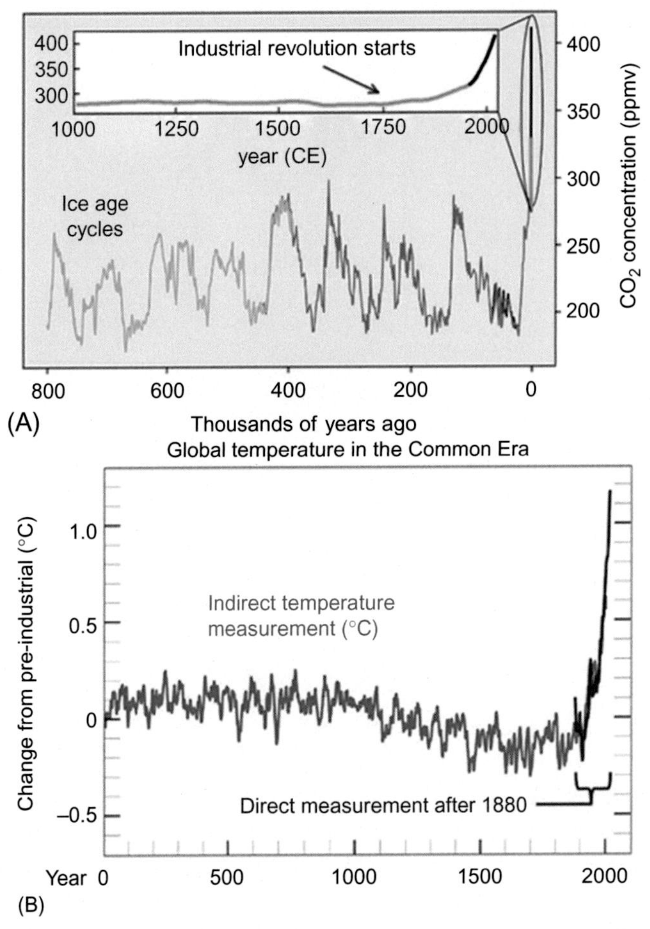


**FIG. 2** (A) $CO_2$ concentrations over the last 800,000 years as measured from ice cores (blue/green) and directly (black). (B) Global surface temperature reconstruction over the last millennia using proxy data from tree rings, coral reefs, and ice cores, are shown in blue. Observational data is from 1880 to 2019. *(A) Graph by Femke Nijsse [CC BY-SA 3.0, https://commons.wikimedia.org/w/index.php?curid=69480542]. (B) Graph by Efbrazil [CC BY-SA 4.0, https://commons.wikimedia.org/w/index.php?curid=87410053].*

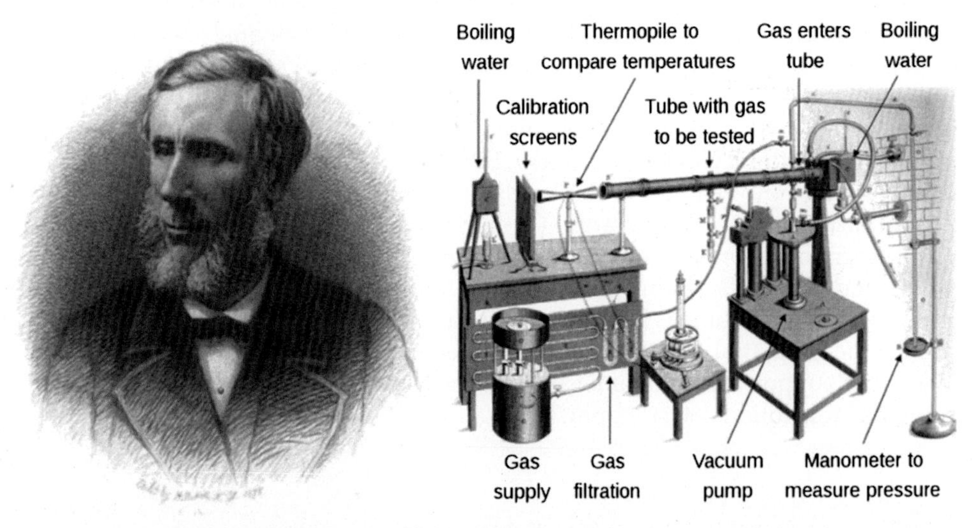


**FIG. 3** Physicist John Tyndall and his apparatus to investigate the heat-trapping properties of various gases. *[https://en.wikipedia.org/wiki/John_Tyndall].*

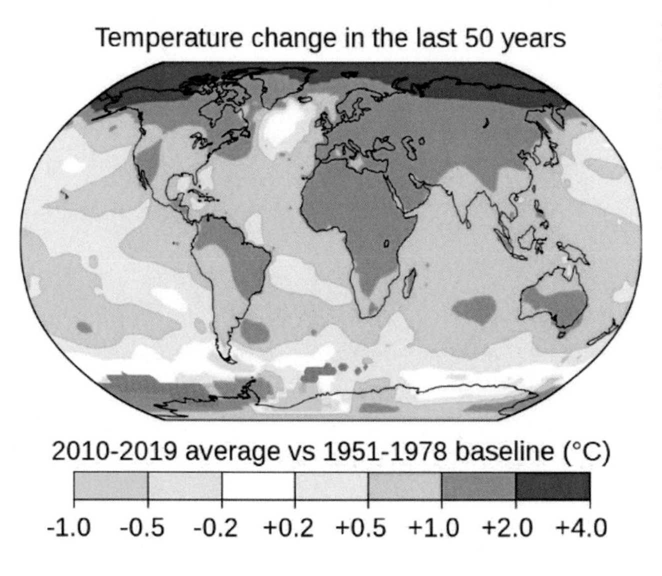


**FIG. 4** Average global temperatures from 2010 to 2019 compared to a baseline average from 1951 to 1978. *Source: https://data.giss.nasa.gov/gistemp/maps/index_v4.html [Public Domain https://en.wikipedia.org/wiki/Global_warming].*

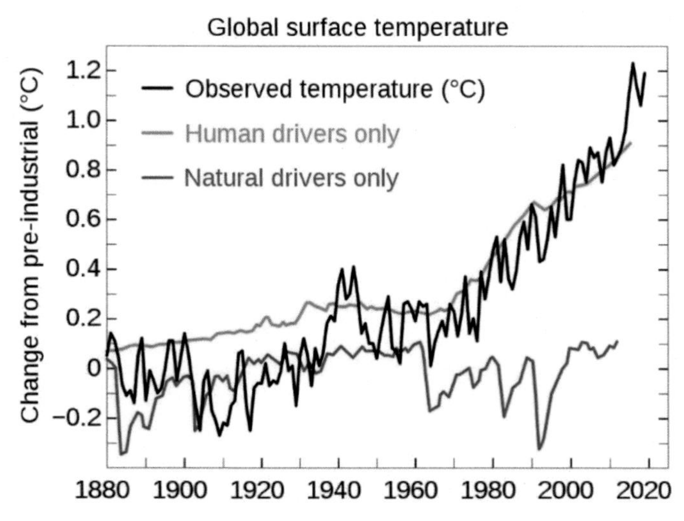


**FIG. 5** Observed temperature from NASA vs the 1850–1900 average used by the IPCC as a preindustrial baseline. The primary driver for increased global temperatures in the industrial era is human activity, with natural forces adding variability. *Graph by Efbrazil [CC BY-SA 4.0, https://commons.wikimedia.org/w/index.php?curid=87373456, https://en.wikipedia.org/wiki/Global_warming].*

As the adverse impacts from global warming became more evident, an increasing number of scientists and a few politicians began issuing urgent calls to reduce the rate of global warming. Today, much of scientific community, and informed general public, admit to the fact that humans are responsible for the increasing temperatures and that there are adverse impacts occurring throughout the world as a result of it (Leiserowitz et al., 2019). Yet the debate continues on when and just how to address this issue in an effective and equitable manner (Hansen et al., 2006, 2012; Gore, 2019; Martinich and Crimmins, 2019).

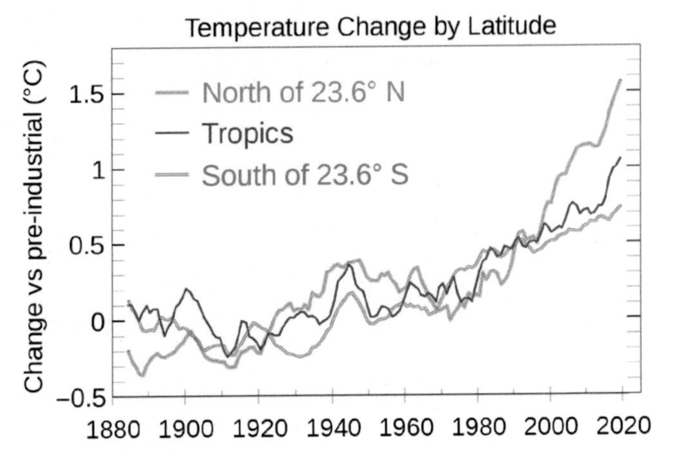


**FIG. 6** Three latitude bands that respectively cover 30%, 40%, and 30% of the global surface area show mutually distinct temperature growth patterns in recent decades. *Graph by RCraig [CC BY-SA 4.0, https://commons.wikimedia.org/w/index.php?curid=88086817].*

Most of the greenhouse gases we humans have discharged into the atmosphere have occurred in the past 35 years. What physical impacts have we observed over these past 35 years?

So far, our hotter atmosphere has resulted in:

- A warming Earth. As temperatures increase, what have been once-in-20-year extreme heat days may occur every 2 or 3 years on average. The 6 warmest years on record have occurred since 2014. In 2016, the months from January through September, with the exception of June, were the warmest on record (NASA, 2020c). The 10 warmest years of record (140 years) have occurred since 2005. The 6 warmest years are the 6 most recent years (IPCC, 2018).
- Warming oceans. Over 90% of the Earth's warming that has happened during the past 50 years has occurred in the oceans (NASA, 2020c). Hotter oceans contribute to rising sea levels, ocean heat waves and coral bleaching, more intense storms, changes in marine life, and melting of ocean-terminating glaciers and ice sheets around Greenland and Antarctica (Dahlman and Lindsey, 2020). Last year, the oceans were warmer than any time since measurements began over 60 years ago (Berwyn, 2020; Cheng et al., 2020).
- Shrinking Ice Sheets. The Greenland ice sheet has lost an average of 286 billion tons of ice per year between 1993 and 2016. The Antarctica ice sheet lost about 127 billion tons of ice per year during the same time period. The rate of Antarctica ice mass loss has tripled in the last decade. The Arctic Ocean is expected to become essentially ice free in summer before mid-century (NASA, 2020b).
- Glacial retreat. Most of the world's glaciers are retreating, including those in Africa and Alaska, the Alps, Andes, Himalayas, and the Rocky Mountains (WGMS, 2020; Pelto, 2015). Melting glaciers and ice sheets are the biggest cause of sea level rise in recent decades. Glacier loss is a serious threat to ecological and human water supplies in many parts of the world.

- Decreased snow cover. Satellite observations in the Northern Hemisphere over the past five decades show spring snow cover melting earlier and decreasing in extent (National Snow and Ice Data Center, 2019; NASA, 2013). The earlier decrease in snow cover increases surface temperatures and durations of growing seasons.
- Sea level rise. A warming of the oceans and the partially melting of the glaciers and other ice increases sea levels. Ocean water expands as it warms, contributing further to sea level rise. Global sea level rose about 20 cm in the last century. The rate of increase in the last two decades was twice that of the last century and this rate is increasing. Storm surges and high tides together with sea level rise and land subsidence increases the extent and frequency of flooding in many regions. Because of the delayed response to any change in global warming, sea levels are expected to rise well beyond this century (Nerem, et al., 2018; NASA, 2020c).
- Loss of Artic Sea Ice. Both the extent and thickness of Arctic sea ice has rapidly declined over the last several decades. The Arctic Ocean is expected to become essentially ice free in summer before mid-century (SNIDE, 2019; NASA, 2020c).
- More extreme hydrological and meteorological events. Since the middle of the previous century, the occurrence of record high temperature and rainfall intensity events has been increasing. The frequency of record low temperature events has been decreasing. Since the early 1980s the intensity, frequency, and duration of the hurricanes have increased. As the oceans continue to warm, hurricane-associated storm intensity and rainfall rates are expected to increase.
- Ocean acidification. Since the beginning of the Industrial Revolution, the acidity of surface ocean waters has increased by about 30%. This increase is the result of humans emitting more carbon dioxide into the atmosphere and hence more being absorbed into the oceans. The rate of carbon dioxide absorbed by the upper layer of the oceans is increasing by about 2 billion tons per year (NOAA, 2020b; NASA, 2020c).

Depending on the extent to which greenhouse gas emissions are reduced over the next century we could sustain what remains of the Earth's land mass and glaciers or, if current emission rates continue, achieve temperatures warmer than what can sustain life as we know it. Various scenarios, called representative concentration pathways (RPCs), are shown in Figs. 7 and 8 (NAS, 2020). RCP8.5 refers to the concentration of carbon that delivers global warming at an average of 8.5 $W/m^2$ across the planet.

Reports prepared by the Intergovernmental Panel on Climate Change (IPCC) tell us that we are on track for an eventual 3°C increase in average global warming. Their recommended goal is a maximum of 1.5°C but staying below 1.5°C will require "rapid, far-reaching, and unprecedented changes in all aspects of society." Even a half of a degree above that recommended level may make the difference between a world with coral reefs and Arctic summer sea ice, and a world without them. To meet a goal of 1.5°C warming will require reductions in greenhouse gas emissions to 45% below 2010 levels by 2030. And as stated previously, even if all such emissions ceased today, the global temperature will continue to increase for decades, because of the cumulative effects in the atmosphere and oceans (USGCRP, 2017; IPCC, 2019).

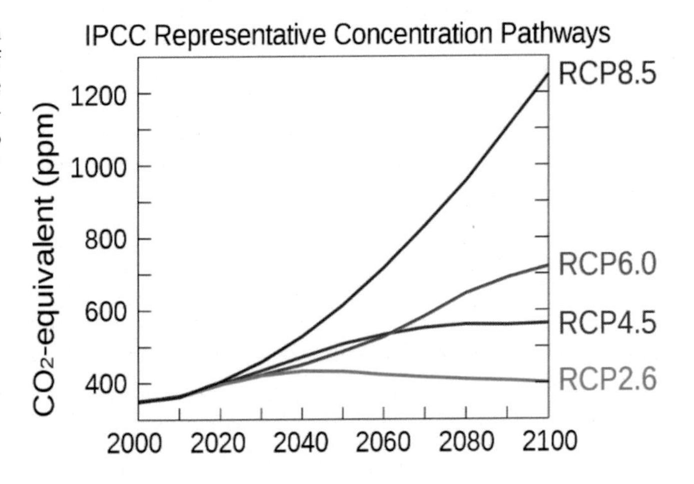


**FIG. 7** Representative Concentration Pathways in terms of the concentration of carbon in the atmosphere at any date. The RCP 8.5 pathway delivers a temperature increase of about 4.3°C by 2100, relative to preindustrial temperatures, https://climatenexus.org/climate-change-news/rcp-8-5-business-as-usual-or-a-worst-case-scenario.

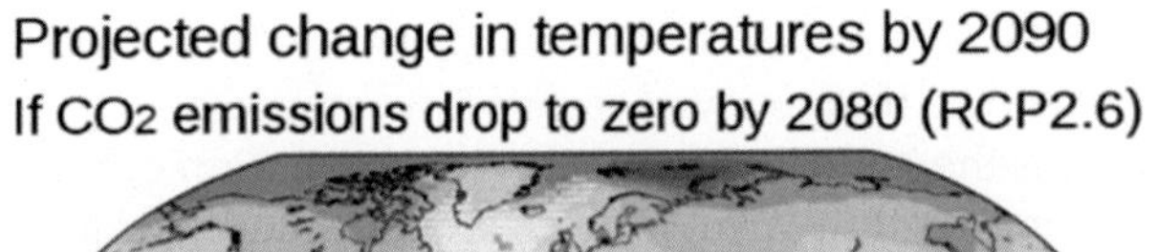


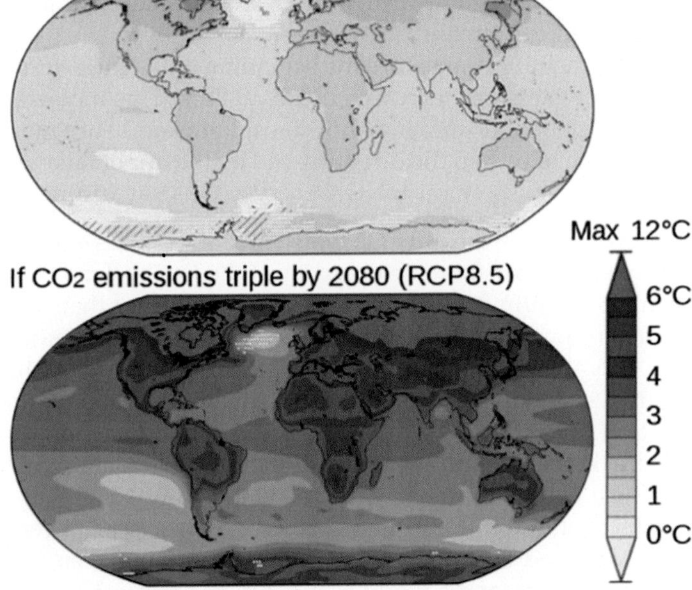


**FIG. 8** Average of climate model temperature projections for 2081–2100 relative to 1986–2005, under low and high emission scenarios, www.wikiwand.com/en/Regional_effects_of_global_warming.

Half of a degree doesn't seem much, but this is an average increase over the entire globe. Making up those averages are the extremes that result in more frequent and intense heat waves as shown in Fig. 9, more damaging storms from both wind and rain, and higher oceans, as shown in Fig. 10. Life in this world will be affected differently, depending on where it is located. Species least able to adapt will face the greatest risks of extinction.

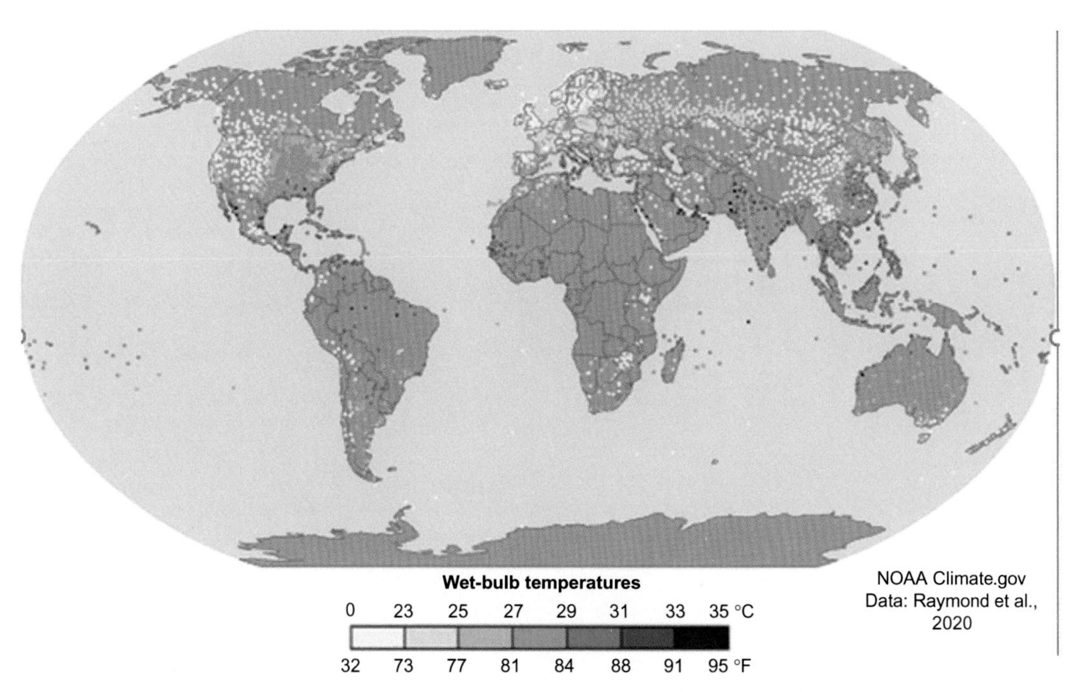


**FIG. 9** Locations that experienced extreme heat and humidity levels briefly (hottest 0.1% of daily maximum wet-bulb temperatures) from 1979 to 2017. Darker colors show more severe combinations of heat and humidity. Some areas have already experienced conditions at or near humans' survivability limit of 35°C (95°F) (Raymond et al., 2020).

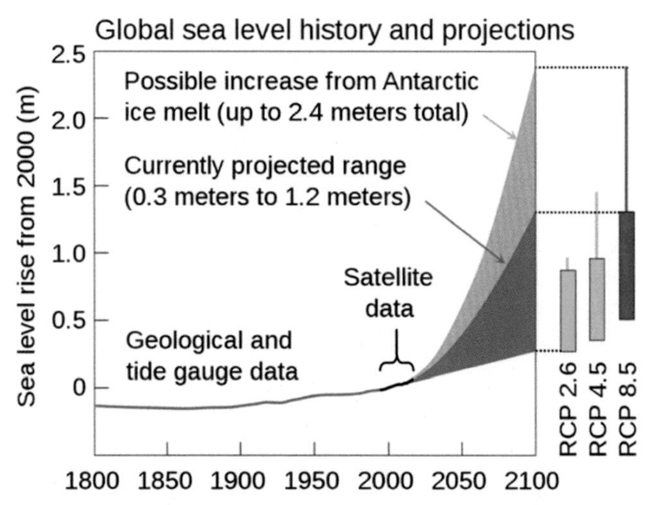


**FIG. 10** Observed and projected changes in global mean sea level for 1800–2100. The boxes on the right show the very likely ranges in sea level rise by 2100 (relative to 2000) corresponding to three different representative concentration pathway scenarios. The lines above the boxes show possible increases based on newer research of the potential contribution to sea level rise from Antarctic ice melt. *Data from the U.S. Global Research Program for the Fourth National Climate Assessment, https://nca2018.globalchange.gov/chapter/appendix-3#fig-A3-1.*

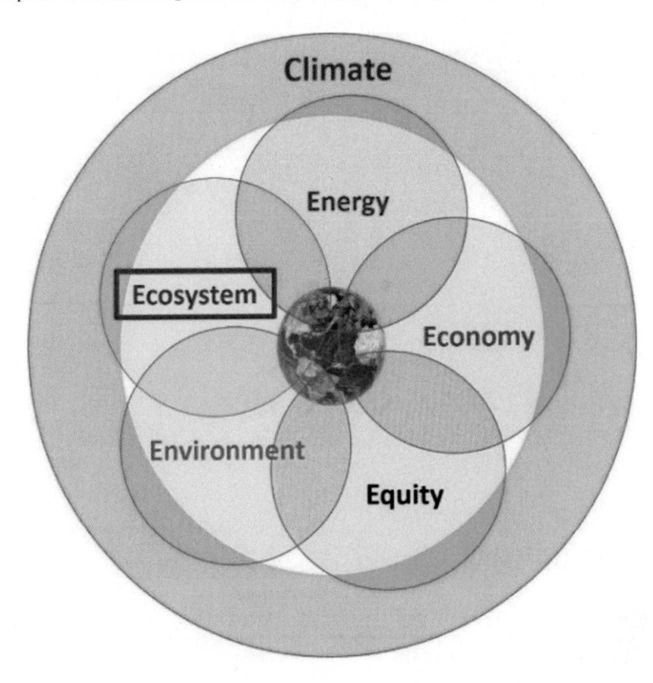


## 3 Ecosystem impacts

Ecosystems are habitats that contain life. They range from wildlands free of human influence, to built environments totally designed and managed by humans. They include fresh and salt water bodies, wetlands, estuaries, forests, peatlands, terrestrial-coastal systems (e.g., mangroves and salt marshes), deserts, agricultural lands, and cities. The warming climate is impacting all of these habitats in multiple ways. On land, increasing temperatures are increasing precipitation variability and the probability of extreme dry and wet events leading to increasing physiological and hydrological stresses and the risks of wildfire. In the ocean, an increased occurrence of heatwaves and long-term trends of acidification increase the physiological stress experienced by many organisms. Interactions with other anthropogenic or natural stressors such as poaching, overfishing, invasive species, habitat fragmentation, direct degradation, or other measures that alter species populations, all tend to impact the sustainability of existing ecosystems and their resilience to climate change. Not only the warming itself but also the consequences of warming such as rising sea levels and more frequent and extreme flooding and droughts, and fires, also result in changes in plant and animal composition and in predator-prey and food chain relationships.

The resulting changes in these habitats can alter the products and services provided by them, such as food, fuel, timber, water, clean air, and medicines as well as their aesthetic or cultural value. Other examples include reefs and barrier islands that protect coastal developments from storm surges, wetlands that absorb floodwaters, and cyclical wildfires that clear excess forest debris and reduce the risk of larger more damaging fires (NAS, 2019). Shifts in ecological conditions from warming can support the spread of pathogens, parasites, and diseases that could impose serious effects on human health and income. For example, the oyster parasite, *Perkinsus marinus*, is causing large oyster die-offs. This parasite has extended its range northward along the Atlantic coast of the US because of increasing winter temperatures (Marquis et al., 2020).

Warming can cause both animal and plant species to migrate to habitats more suited to their survival. Patterns of animal and plant migration, breeding, pest avoidance, and food availability can be affected in different ways. Different species differ in their ability to adjust and adapt. For many species, the climate's temperature influences where they live and key stages of their life cycle. As winters become shorter and milder, the timing of these life cycle events can change.

As temperatures increase, the habitat ranges of many species are shifting to cooler elevations. For some species, it means movement into less hospitable habitat, increased competition, and/or range reduction. For some others it has led to local extinctions. For example, boreal forests are invading tundra, reducing habitat for the many unique species that depend on the tundra, such as caribou, arctic foxes, and snowy owls. Other observed changes include a shift in the range boundaries of shrub lands and broadleaf and conifer forests. As rivers and streams warm, warm water fish are expanding into areas previously inhabited by coldwater species. Coldwater fish, including many highly valued species, are losing their habitat. Decreases in the duration and extent of Artic sea ice has led to declines in ice algae that are eaten by zooplankton, which are then eaten by Arctic cod, an important food source for many marine mammals, including seals. Seals are eaten by polar bears. Thus, the loss of sea ice can ultimately affect survival of polar bears (Malhi et al., 2020).

## 3.1 Agricultural ecosystems

Agricultural ecosystems cover almost 40% of Earth's land area. About 11% is cultivated, and approximately 27% is permanent pasture. Agricultural ecosystems provide a range of services including food production, carbon sequestration, nutrient recycling, and climate regulation (Andrén and Kätterer, 2019).

Agricultural crop, livestock, and fisheries production are dependent on the climate. Increases in the frequency and severity of droughts and floods pose challenges for farmers, ranchers, and fishers and threaten food safety as well as food production security. Warming water temperatures result in changes in the habitat ranges of many fish and shellfish species. Livestock may be at risk, both directly from heat stress and indirectly from reduced quality of their food supply. Warmer water temperatures attract invasive species and alter the ranges or lifecycles of certain fish species.

Climate change can affect food security at the global, regional, and local levels. Projected increases in temperatures, changes in precipitation patterns, changes in extreme weather events, and reductions in water availability may all result in reduced agricultural productivity, interrupt food delivery, and contribute to food spoilage and contamination (Brown et al., 2015).

## 3.2 Urban ecosystems

The impacts of global warming pose a series of interrelated challenges to those living in densely populated cities. Cities depend on a functioning infrastructure, including water and sewage systems, roads, bridges, subways and trains, power plants and transmission grids, buildings, telecommunications, and information technology, to mention a few. All are essential for the overall well-being of residents. This complex interdependent infrastructure can be vulnerable to rising sea levels, storm surges, heatwaves, and extreme weather events that are the result of or exacerbated by a warming climate (Hunt and Watkiss, 2011).

Urban dwellers are particularly vulnerable to disruptions in essential infrastructure services, in part because many of these infrastructure components are reliant on each other. For example, a failure in the electrical grid can affect water treatment, transportation services, and public health and safety.

Climate changes affect the built, natural, and social infrastructure of cities, from storm drains to urban waterways to the capacity of emergency responders. Climate change increases the risk, frequency, and intensity of certain extreme events like intense heatwaves, heavy downpours, flooding from intense precipitation and coastal storm surges, and disease incidence related to temperature and precipitation. As climate change impacts increase, climate-related events may tax the capacity of the existing infrastructure to contain or control them. This can result in economic damages and even adversely impact the health of significant numbers of people living in cities or suburbs. Also at risk from climate change are historic properties as well as cultural resources and archeological sites.

The vulnerability of urban dwellers multiplies when the effects of climate change interact with preexisting urban stressors, such as deteriorating infrastructure, housing and recreational developments along coastal beaches, and areas of intense poverty and high population density (Dawson, et al., 2009; De Sherbinin et al., 2007; Rosenzweig, et al., 2011).

## 3.3 Forest ecosystems

Forests provide many services including clean air and water, recreation, wildlife habitat, carbon storage, climate regulation, and a variety of forest products (CCSP, 2008). Climate influences the structure, functioning, and health of forests through changes in its temperature, rainfall, and other factors. A warming climate may increase many of the threats to forests, such as pest outbreaks, disease, fires, residential development, and drought.

The climate-induced modifications of forest wildfire frequency and intensity, outbreaks of insects and pathogens, and extreme meteorological events such as high winds and ice storms may be more important than the direct impact of higher temperatures and elevated $CO_2$, both of which may increase growth. Increased temperatures alter the timing of snowmelt, affecting seasonal runoff and streamflow regimes. Although many tree species are resilient to some degree of drought, increases in temperature could result in more tree mortality than experienced in the past. In addition, droughts increase the risks of wildfires, since dry forest land cover provides fuel to fires. Droughts also reduce trees' ability to produce sap, which in turn decreases their resistance to destructive insects such as pine beetles.

Increases in frequency of extreme events, such as strong winds, winter ice storms, droughts, etc. can adversely impact commercial forestry by increasing costs for road and facility maintenance and by tree damage through branch or trunk breakage, crown loss, or complete stand destruction.

These disturbances can reduce forest productivity and change the distribution of tree species. Disturbances can interact with one another so as to increase risks to forests. For example, drought can weaken trees and make a forest more susceptible to wildfire or insect outbreaks. Similarly, wildfire can make a forest more susceptible to pest damage. Fires can also contribute to climate change, since they can cause rapid, large releases of carbon dioxide to the atmosphere (USEPA, 2016; Kirilenko and Sedjo, 2007).

Mangrove forests play a vital role in storing carbon dioxide and protecting coastal communities from storms and erosion. These trees create barriers against destructive storm surges, reduce erosion, and shelter wildlife. Their roots can serve as a nursery for fish, crustaceans, and shellfish. If greenhouse gas emissions continue unabated these forests will not be able to move inland fast enough to escape rising sea levels and could disappear over the next three decades. Sea level rise coupled with land development has already destroyed over one-fifth of the world's mangrove forests over the past four decades (Calma, 2020).

## 3.4 Aquatic ecosystems

Climate change is a significant threat to the species composition and the functioning of aquatic ecosystems. Global temperatures of the magnitude projected over the next 100 years will pose substantial risks for inland freshwater ecosystems (lakes, streams, rivers, and wetlands) and coastal wetlands, and adversely affect numerous critical services they provide to human populations.

Projected increases in mean temperature are expected to greatly disrupt present patterns of plant and animal distributions in freshwater ecosystems and coastal wetlands. For example, in the continental United States, cold-water fish like trout and salmon are projected to disappear from large portions of their current geographic range if warming causes water temperatures to exceed their thermal tolerance limits. Fish species that prefer warmer water, such as largemouth bass and carp, will potentially expand their ranges as surface waters warm. Species migration to more suitable habitats may occur if such migration is possible. If not, there is an increased likelihood of species extinction and loss of biodiversity.

Changes in the hydrological characteristics of aquatic systems can impact their biological properties. The species composition, reproduction, and productivity of many aquatic species is sensitive to changes in the frequency, duration, and timing of extreme precipitation events, including floods and droughts, as well as to changes in stream flow and quality regimes. These changes can be caused by changes in the seasonal timing of snowmelt as well as by human activities.

The productivity of inland freshwater and coastal wetland ecosystems also will be significantly altered by increases in water temperatures. Warming is expected to melt permafrost areas, allowing shallow summer groundwater tables to drop; the subsequent drying of wetlands will increase the risk of catastrophic peat fires and the release of substantial quantities of carbon dioxide and possibly methane into the atmosphere.

In addition to its independent effects, temperature changes can act synergistically with changes in the seasonal timing of runoff to freshwater and coastal systems. Expected summertime reductions in runoff and elevated temperatures will decrease water quality as well as quantity. The loss of winter snowpack will reduce a major source of groundwater recharge and summer runoff, resulting in reductions of water levels in streams, rivers, lakes, and wetlands during the growing season (Lane et al., 2015).

While warmer waters are typically more biological productive, the particular species that flourish in warmer waters may be undesirable or even harmful. For example, the blooms of "nuisance" algae and accompanying fish kills can occur in many lakes and along coast lines during warm, nutrient-rich periods. Large fish predators that require cool water may be lost from smaller lakes as surface water temperatures warm, and this may indirectly cause more algae blooms, reduce water quality, and pose potential health problems.

Reducing the likelihood of significant adverse impacts to aquatic ecosystems as a result of global warming will depend on human activities that reduce other sources of ecosystem stress. These include maintaining riparian forests, reducing nutrient loading, restoring damaged ecosystems, minimizing groundwater withdrawal, and siting, designing and operating any new reservoirs in ways that minimize the risks of adverse ecosystem effects.

Admittedly, many uncertainties still exist regarding how regional climates will change and how complex aquatic ecological systems will respond. But it is certain that a warming climate will alter ecosystem productivity and species composition, and threaten the goods and services these systems provide to humans (Adams, 2011; Poff et al., 2002).

## 3.5 Grassland ecosystems

Grasslands are major contributors to food production, aquifer recharge, pollination, and recreational opportunities, among other services benefiting humans. While typically defined as lands on which the existing plant cover is dominated by grasses, natural grasslands are diverse communities of grasses, forbs, and nonvascular and woody plants, possibly containing wetlands, that provide critical wildlife habitat. Disturbances such as fire and grazing contribute to sustaining grasslands. Some temperate savannas transition into grasslands, and are maintained as grasslands principally by human-caused disturbance (Loarie et al., 2009; Bagne et al., 2012).

The relatively flat terrain of grasslands can increase their vulnerability to climate change impacts, since species must migrate long distances to compensate for temperature shifts. This is in contrast to mountainous terrain, where temperatures and habitat conditions can change over relatively short distances because of steep elevation changes. Warmer temperatures bring greater evaporation and alter rainfall patterns, which can decrease aquifer recharge and alter water-dependent habitats.

Extreme weather conditions can increase the frequency of droughts, floods, fires, and hurricanes all of which can impact grassland ecosystems. Droughts typically lead to soil erosion and aquifer depletion and more frequent fires, which can reduce encroachment of woody plants into grasslands. Fires are a natural element in grassland ecosystems, but too many and too extreme fires will adversely impact plant and animal life over very large areas.

Similarly, floods can recharge aquifers, transport nutrients, and enhance the habitat of wildlife species, but high intensity run-off events can decrease retention of organic matter and the numbers of grassland organisms. Insect outbreaks in response to warmer temperature can contribute to the transition of forests and woodlands to grasslands (Ford, 2002; Morgan et al., 2008).

Warmer temperatures increase evaporation which can lead to increased water and soil salinity. Sea level rise can inundate coastal grasslands with salt water and increase erosion. These changes can impact bird populations that breed or winter in, or migrate through, grassland habitats.

## 3.6 Desert ecosystems

Global warming is predicted to increase the extent of desert areas, which already cover a quarter of Earth. Earth's largest hot desert, the Sahara in northern Africa, is getting bigger. It is advancing south into more tropical terrain in Sudan and Chad, turning green vegetation dry and soil once used for farming into barren ground. It is happening during the African summer, when there is usually more rain. But reduced precipitation has allowed the boundaries of the desert to expand even during the wet season. This expansion is not unique to the Sahara. Climate changes are causing other hot deserts to expand as well.

Rising temperatures and reduced precipitation are affecting many plant and animal species inhabiting desert ecosystems. From Arizona to Australia, hotter, longer, and more frequent heatwaves are adversely impacting desert birds. And in the hot, sandy soils of South Africa and Namibia, dwarf succulents are predicted to disappear in 50 years (McDermott, 2016). Biological soil crusts are vital biotic components of desert ecosystems. They help maintain soil stability and carbon and nitrogen levels, and serve as habitats for microorganisms. The abundance of moss, surface cover, and biomass are expected to sharply decrease if continued warming coupled with reduced precipitation occurs. Since the diversity and biomass of crustal communities rely on moss, their reductions will result in a structural and functional change in crustal communities and an imbalance of soil water in deserts. This will lead to detrimental effects on the stability and sustainability of a desert's ecology (Chen, 2018).

Ecosystems are not only impacted by our climate but also impact our climate. They play an active role in the carbon cycle, the hydrologic cycle, and other biogeochemical cycles. If sustainably managed, ecosystems can be a major source of human resilience and can support the adaptation of human societies to environmental change.

Climate change is ongoing, and within the next few decades, societies and ecosystems will either be committed to a substantially warmer world or major actions will have been taken to limit warming. Ecosystems play a major role in both of these scenarios. Extensive and connected ecosystems, species and genetic diversity, trophic intactness, and habitat heterogeneity can buffer the impacts of climate change. Ecosystem management and restoration can play an important role in climate change mitigation and societal adaptation, but will only provide benefits if deployed in conjunction with a reduction in the causes of global warming: fossil fuel extraction, processing, and consumption emissions (NAS, 2020).

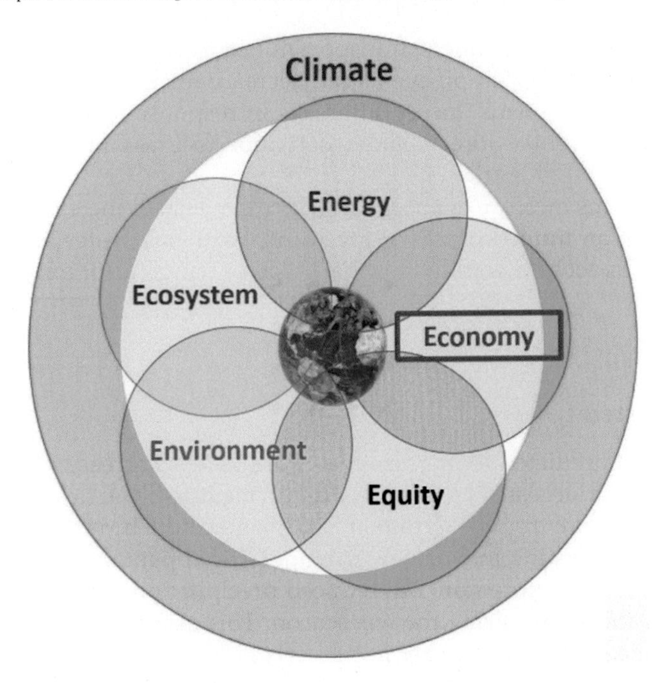


## 4 Economic impacts

The economic impacts of global warming are many and diverse. Whether these impacts are economically beneficial or detrimental, small or large, depends on who is being impacted and how. Crop production is usually adversely impacted by droughts and floods and possibly beneficially impacted because of carbon dioxide fertilization and longer growing seasons. Individuals living in warming urban areas may be spending more energy to cool their buildings in the longer summer seasons but less to warm them in the shorter winter seasons. Mitigating or adapting to the increasing risks of infectious disease, wildfires, sea level rise, and saltwater intrusion all involve costs. Climate change has and will continue to affect the world's economies, both directly and indirectly in many ways.

Over the past four decades, it has been widely accepted that climate change is, on balance, a negative externality. Economists have been proposing various pricing and taxing policies as economic incentives to reduce greenhouse gas emissions. Fig. 11 identifies the relative amount of greenhouse gas emissions from various economic sectors. Despite a general agreement on the overall benefits of reducing these emissions, the debate continues among economists about how to quantify climate change costs and benefits and how to manage them (Blanc and Reilly, 2017).

A recent study examined how climate change could affect 22 different sectors of the U.S. economy if global temperatures reach 2.8°C and 4.5°C. above preindustrial levels by 2100. The study concluded that if the higher-temperature scenario prevails, climate change impacts on

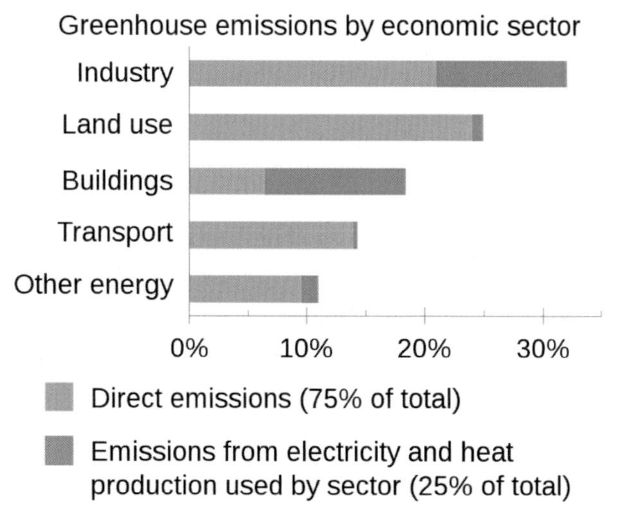


**FIG. 11** Emissions produced by different economic sectors, including emissions generated by electricity and heat production used by the sectors, according to the IPCC AR5 report. *Bar Plot by Efbrazil [CC BY-SA 4.0, https://commons.wikimedia.org/w/index.php?curid=87864359].*

these 22 sectors could cost the U.S. $520 billion each year. If kept to 2.8°C, the annual cost would be about $300 billion. Recognizing the uncertainty associated with these estimates, what is not uncertain is that large economic losses will result from a warming climate. Just since 2016, climate disaster damages have exceeded $415 billion in North America, much of that because of wildfires and hurricanes. While it's not yet possible to directly link climate change to hurricanes or wildfires, warmer temperatures are known to enhance their intensity and destructiveness (Martinich and Crimmins, 2019). Such information, along with the costs of measures taken to reduce these damages, are needed to carry out any benefit cost analyses, and to examine distributional and equity impacts.

Even if no actual physical damage occurs in a flood prone area, for example, just the fact that it could impacts the value of buildings in that area. Buildings can lose some of their value before the water actually arrives, once people realize that eventually the water could arrive. (As the saying goes, it is not a question of if, but when.) This example illustrates the interconnection between physical and social aspects of damage estimation.

## 4.1 Agriculture

Agriculture is clearly vulnerable to climate change. Regions where agriculture dominates the economy are very vulnerable to the effects of global warming on crop productivity, livestock health, and the economic vitality of rural communities. Many commodity crops such as corn, soybean, wheat, rice, cotton, and oats do not grow well above certain temperature thresholds. While some regions may see conditions conducive to expanded or alternative crop productivity over the next few decades, overall, crop yields are expected to decline as temperatures increase. Increases in extreme heat leading to livestock heat stress can result in economic losses for livestock producers. Add to this the increased costs of irrigation water, and increases in soil erosion, wildfires, and disease and pest outbreaks.

Climate change is also expected to lead to changes in the availability and prices of many agricultural products across the world. As farmers adapt to changing conditions, their costs will likely increase and be passed along to consumers (Blanc and Reilly, 2017; Walthall et al., 2012).

## 4.2 Business and industry

Industry is a main contributor to climate change and is adversely impacted by it. Industrial activity, from logistics, operations, marketing, sales, and after-sales service, can generate emissions. The simple ratio of profits to total emissions in the value chain can be a measure of its climate impact. Restrictions in emissions or cap-and-trade policies are likely to be imposed on industry if they are not already. A company's "carbon exposure" is the impact of reducing carbon emissions on profits. Like other risks, carbon exposure carries opportunities as well as challenges: forestry companies, for example, may find that depending on economic incentive programs, removing carbon dioxide from the air by planting trees may be as profitable as cutting them down and producing paper or plywood (Economist, 2015).

A company's emissions can be under its direct control or result from the activities of its suppliers and/or customers. Both types are possible targets for reduction. Once managers understand their firm's overall carbon exposure and the emissions impact of specific activities, they can better identify more cost-effective emission reduction plans.

Changing temperature and weather patterns can affect a business's production or operating costs as well as the demand for its products or services. Either can affect the profitability of a business. For example, a property insurers' own carbon emissions may be low but its carbon exposure may be high if it is insuring or reinsuring coastal real estate that is threatened, and thus losing value, by rising sea levels. Similarly, the carbon emissions associated with oil come not only from its production and refinement but also from its use. Restrictions or taxes placed on emissions reduce the demand for and hence income derived from oil products.

The frequency and intensity of extreme weather events can adversely impact industries, commerce, disrupt transport and supply chains and damage associated infrastructure. Droughts can lead to higher water prices, which affects the cost of raw materials and production of goods and services, including energy, much of which depends on water. As the climate warms, some industrial products could become obsolete or lose some of their market value, such as skiing slopes and associated equipment in areas that no longer have snow. Owners having "stranded assets" whose assets are decreasing in value also include owners of fossil fuels that are being left in the ground for lack of demand, and owners of real estate along rivers or coasts that risks being flooded by storms and sea level rise made more likely because of a warming climate.

A survey in 2018 found that, unless they took preemptive measures, 215 of the world's 500 biggest companies could lose an estimated one trillion US dollars within 5 years because of climate change. For example, Alphabet (Google's parent company) indicated they will likely have to deal with rising cooling costs for its data centers. Hitachi Ltd.'s suppliers in Southeast Asia stated they are vulnerable to increased rainfall and flooding. Some companies have already been impacted by climate change-related losses. Western Digital Technologies, maker of hard disks, suffered flood damage in 2011 that disrupted its production (CDP, 2019).

PG&E, the utility supplying electrical energy to much of the northwest US, became liable for fire damages and had to file for bankruptcy after its power lines sparked California's deadliest wildfire. And General Electric cost its investors $193 billion between 2015 and 2018 because it overestimated the demand for natural gas and underestimated the rate of growth in the supply of less expensive renewable energy.

The movement away from fossil fuels is impacting banks and investment firms that have financial relationships with the fossil fuel industry. Overall, oil and gas producers have lost $400 billion in market value over the last 4 years. ExxonMobil, one of the world's most valuable public company in 2012, saw its stock value drop to 9-year lows in June 2020. The public, including activists on college campuses and contributors to pension funds, is increasingly calling for outright divestment of fossil fuel industries (Egan, 2020).

Given that fossil fuels are nonrenewable, we once worried that at some point all known reserves would be exhausted. What's clear now is that there is far more coal, oil, and gas remaining in the ground than we can safely afford to consume. Studies suggest companies would have to forgo consuming over 80% of known reserves if emissions of greenhouse gases were capped to meet a 1.5°C (2.7°F) global warming target. Meeting a 2° target would still mean writing off nearly 60% of known reserves.

## 4.3 Infrastructure

As global warming increases, an increasing proportion of the world's infrastructure will likely need to be rebuilt or repaired. Much of the world's flood protection infrastructure is at risk of failing to protect from flooding events of the magnitudes expected in a warmer future. Permafrost melting is destroying much of the public infrastructure in the artic regions (Melvin et al., 2017; Duncombe, 2020). Sea level rise is already threatening entire islands in the Pacific, as well as housing, roads, railways, military bases, communication systems, and portions of airports on the coasts and along rivers. Communication systems consist of fiber optic cable as well as data centers, traffic exchanges, and termination points—the lifeblood of the global information network, much of which is buried along highways and coastlines. While the cables are water resistant, they typically are not waterproof. Threats to the internet infrastructure could have substantial economic implications for individuals and businesses (Neumann and Price, 2009; Henson, 2011).

## 4.4 Human health and productivity

Productivity decreases when temperatures increase. This is true of humans as well as economic entities. Decreases in productivity have economic consequences. A 2014 Rhodium Group study predicts the largest climate change-related economic losses in the U.S. will be from lost labor productivity. Local warming, caused by the urban heat island effect, can significantly impact productivity as well as costs of cooling if air conditioning is available. The added effects of urban heat island warming could double the economic losses expected from climate change (Houser et al., 2014; Estrada, et al., 2017).

If global temperatures rise 4.5°C by 2090, more people will die in cities because of extreme heat. Increasing temperatures increases the risk of waterborne and foodborne diseases and allergies, and of insect-spread diseases like Zika, West Nile, dengue, and Lyme diseases. Extreme heat can also exacerbate mental health issues that become economic ones as well. Often among the most economically vulnerable populations, such as the elderly, children, and low-income communities are expected to be most affected by these health impacts.

## 4.5 Tourism

Tourism is an economic sector that strongly depends on the climate. Warm periods motivate many to head to sites where there is water. Cold seasons motivate many to head to sites where it is either warmer or where skiing and ice skating is available. As cold seasons shorten because of the warming of the climate, tourism income is decreasing in regions whose economies count on snow and ice. For example, rapid warming in the Adirondack Mountains in northeastern US could end the outdoor winter recreational industry, which makes up 30% of the local economy (ONYSC, 2017).

In summer seasons, as the air temperatures increase so do the water temperatures. This can impact water quality such as from more frequent and more intense toxic algae blooms, thus curtailing recreational water activities and freshwater fishing and the income derived from them. More frequent and severe wildfires can worsen air quality and discourage tourism in otherwise scenic or cultural sites. Sea level rise can submerge small islands and coastal areas, while deforestation and its destructive impacts on biodiversity can detract from their value as ecosystem tourist destinations. Each of these likely impacts from increased warming will impact the tourism industry and the economies of the respective regions.

Fig. 12 provides a general view of the economic impact global warming will have on the economies of various countries.

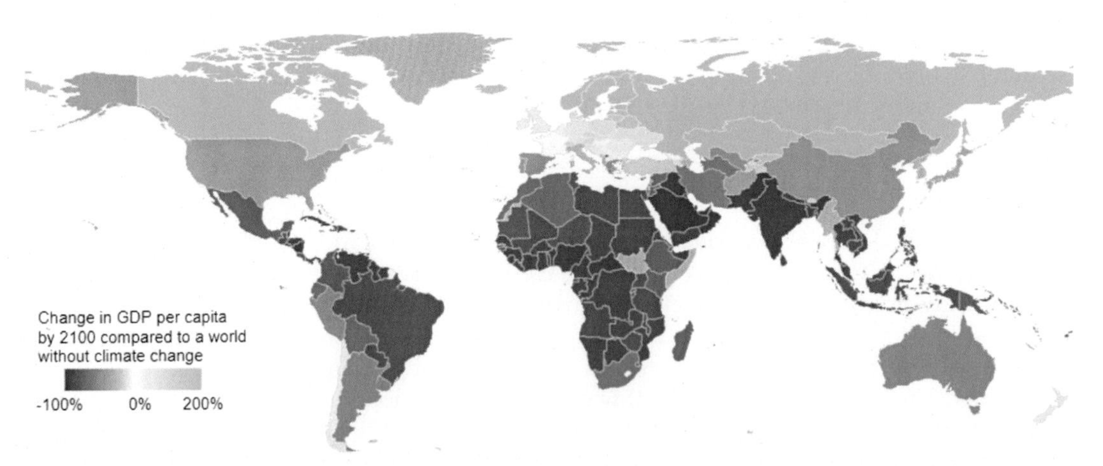


**FIG. 12** Estimates of how climate change will affect GDP per capita by 2100 assuming the high RCP8.5 emissions scenario as calculated in Burke et al. (2015).

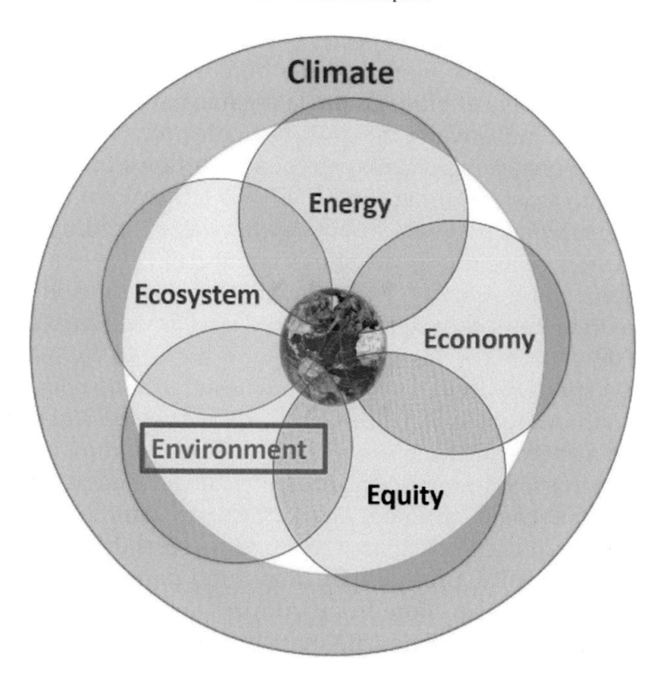


## 5 Environmental impacts

The health of the Earth's ecological systems depends on the quality of our air, soil, and water resources. Pollutants in the emissions discharged from the extraction, processing and burning of fossil fuels end up degrading the quality of our air, land, and water resources. The quality of our air, land, and water is also impacted by events exacerbated by global warming, such as floods, droughts, wind storms, sea level rise, and salt water intrusion. Flooding and runoff carry pollutants picked up from land surfaces into water bodies. Warmer waters promote biological processes that can degrade the quality of the water such as dissolved oxygen depletion and promotion of algal blooms. Drought-induced soil moisture and flow reductions can increase salinity concentrations on the land, in the water, and even in the air during wind storms over dry areas.

Emissions from the global transportation systems are major contributors of pollutants affecting air quality, climate change, and public health. Tailpipe emissions from the gasoline and diesel vehicles contributes to respiratory and cardiovascular diseases. Globally, gasoline and diesel vehicle emissions are considered to cause over 120,000 premature deaths annually. There is substantial regional variability in premature death rates as expected, especially those caused by diesel emissions (Fann et al., 2016; Huang et al., 2020).

Runoff containing nutrients, pesticides, or hydrocarbon pollutants threaten the health of humans and other species in aquatic and terrestrial ecosystems. Flood waters can contaminate drinking waters and increase the spread of waterborne pathogens. Runoff from agricultural lands can elevate the concentrations of nutrients, chemicals, and pesticides in streams, rivers and lakes and in the fish and other species that are in them. Resulting algal blooms and fish kills are evident in many of the world's water bodies, and along coasts.

The discharge into waterways of biodegradable organic material from excessive runoff reduces the dissolved oxygen in the water that many aquatic pollutants into organisms, including fish, need to survive. Intense rainstorms can cause overflows from wastewater treatment plants, sewers, hazardous waste sites, agricultural lands and animal feeding operations, carrying pollutants into nearby lakes, rivers, or other waterways. Flooding of industrial areas or agricultural chemical storage locations can cause chemicals to move into nearby waters and watersheds, degrading water quality and even contaminating some residential areas. Droughts can also result in the accumulation of salt on land as well as in any remaining water (Whitehead et al., 2009).

Greenhouse gases, especially those from automobiles in urban areas, affect the quality of the air both indoors and outdoors. Air pollutants such as ground-level ozone ($O_3$) and fine particulate matter and airborne allergens (aeroallergens) are common in many of the world's industrial and urban regions. Summer heat exacerbates the formation of ground-level ozone, a major component of photochemical smog. This surface pollutant differs from the protective layer of ozone in the upper atmosphere that needs to be preserved. Changing patterns of temperatures, cloud cover, humidity, precipitation, and wind patterns can influence the extent and duration of smog that can also come from wildfires and dust storms.

Wildfires are a major source of particulate matter, Climate change has already led to an increased frequency of large wildfires, as well as longer durations of individual wildfires and wildfire seasons. The areas burned by wildfires are expected to increase over this century because of global warming (Stavros et al., 2014).

Likewise, the spatial extents of droughts are projected to increase as a result of global warming. Dust blowing off dry areas can be an important constituent of particulate matter in both outdoor and indoor environments. Dust can contain biological particles, including pollen and bacterial and fungal spores. Pathogenic fungi and bacteria found in dust both indoors and outdoors can be allergenic (D'Amato et al., 2011; Sheffield et al., 2011).

Poor air quality, whether outdoors or indoors, can negatively affect human respiratory and cardiovascular systems. Outdoor ground-level ozone and particle pollution can have a range of adverse effects on human health, including premature deaths (IASS, 2020).

As is the case for ozone, atmospheric particulate matter (PM) concentrations depend on emissions and on meteorology. Particulate matter can include sulfate, nitrate, ammonium, organic carbon, elemental carbon, sea salt, and dust. These particles (also known as aerosols) can either be directly emitted or can be formed in the atmosphere from gas-phase precursors. Particulate matter smaller than 2.5 μm in diameter ($PM_{2.5}$) is associated with serious chronic and acute health effects, including lung cancer, chronic obstructive pulmonary disease, cardiovascular disease, and asthma.

A changing climate can also influence the level, distribution, and seasonal timing of aeroallergens. Higher pollen concentrations and longer pollen seasons can increase allergic reactions. Changes in the climate may also increase the amounts and concentration of pollutants generated indoors, such as mold and volatile organic compounds (NOAA, 2020a).

As this chapter is being written, the world is experiencing a pandemic. The restrictions imposed on economic activities to reduce the spread of the disease have resulted in reductions of greenhouse gas emissions, as shown in Fig. 13. The result has been decreasing nitrogen dioxide concentrations by an average of 40% over Chinese cities and by 20%–38% percent over Western Europe and the United States in early 2020 as compared to those recorded for the same time the previous year. In addition to nitrogen dioxide, particulate matter (particles smaller than 2.5 μm) decreased by 35% in northern China (AGU, 2020).

These observations give a glimpse into how reductions in greenhouse gas emissions might impact air quality if emissions reductions become a reality in the future.

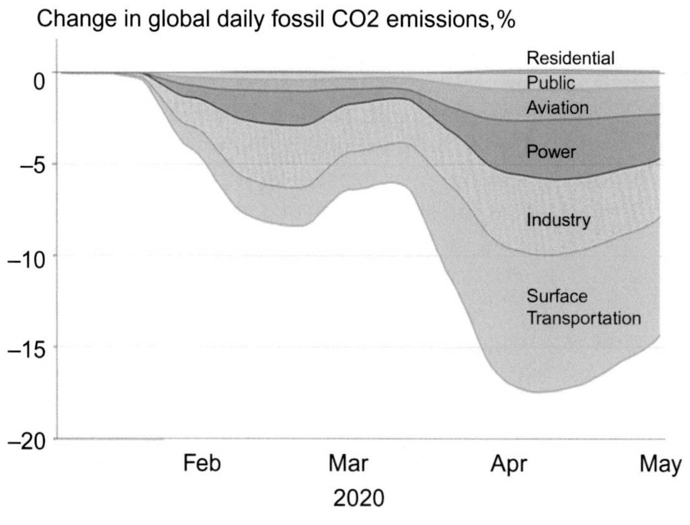


**FIG. 13** Temporary reduction in daily global $CO_2$ emissions during the COVID-19 forced confinement (Le Quéré et al., 2020).

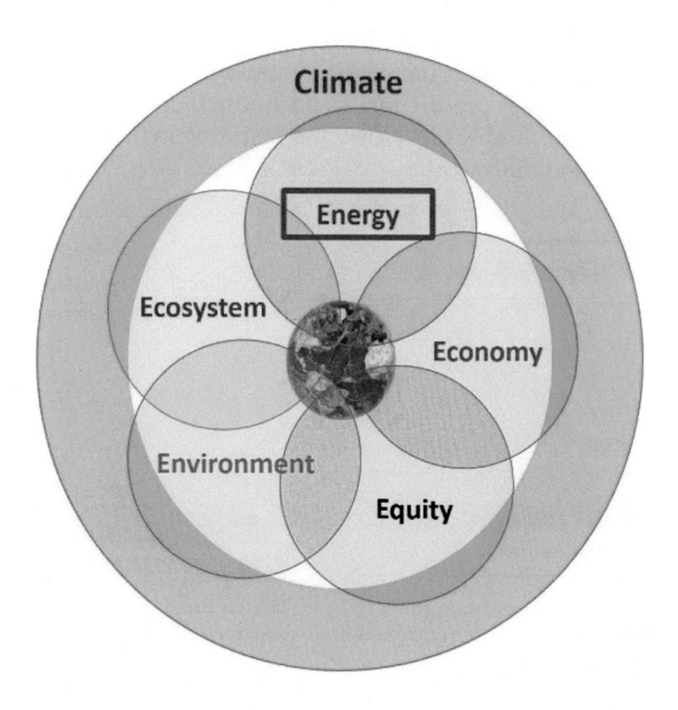

## 6 Energy impacts

As the world's population grows, so does its demand for energy, even on a per capita basis, along with its greenhouse gas emissions, as shown in Fig. 14.

Since before the industrial revolution, humans have relied on fossil fuels to meet much of their energy demands. Especially since the industrial revolution this has resulted in the substantial increase of greenhouse gas emissions that have caused the accelerating increase in global warming. Global warming in turn is impacting the energy production sector. Increasing temperatures will likely increase energy used for cooling and decrease its use for heating.

Increased air and water temperatures reduce thermal power plant cooling efficiencies. Restrictions in water supplies coupled with increasing temperatures of ambient air and water used for cooling increases the likelihood of exceeding water thermal intake or effluent limits, and increases the risk of partial or full shutdowns of generation facilities. As ambient air temperatures increase, the capacities of electricity transmission and distribution systems decrease. Heat reduces their current carrying capacities and operating efficiencies. They also face increasing risks of physical damage from more intense and frequent storm events and wildfires (Zamuda et al. 2013).

Some of these effects, such as higher temperatures of ambient cooling water, are projected to occur worldwide. Other effects may vary more by region, and the vulnerabilities faced by the energy sectors may differ. However, regional variation does not imply regional isolation as energy systems have become increasingly interconnected. Compounding factors may create additional challenges. For example, combinations of persistent drought, extreme heat events, and wildfire may create short-term peaks in demand and diminish means of responding to that changing demand (DOE, 2013).

In some regions and seasons, reduced inflows to hydropower reservoirs, possibly because of increased diversions upstream to meet increased heat-induced domestic, industrial and agricultural demands, coupled with increased evaporation losses, reduces the amounts of water available for hydropower production. In addition, increasingly intense and more frequent storm events, flooding, and sea level rise, are resulting in increasing damages to energy production and transmission infrastructure (CCSP, 2007).

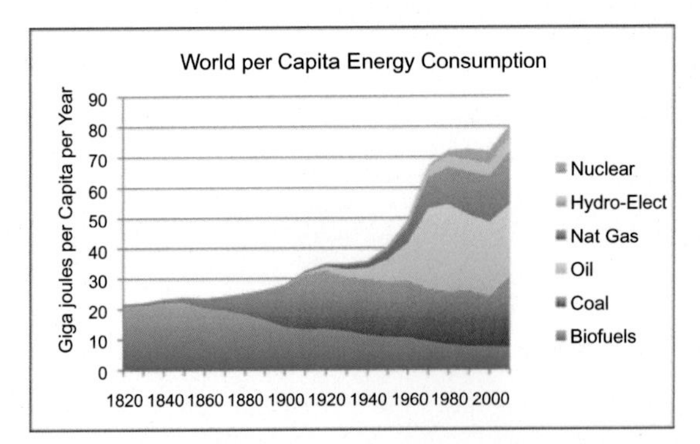


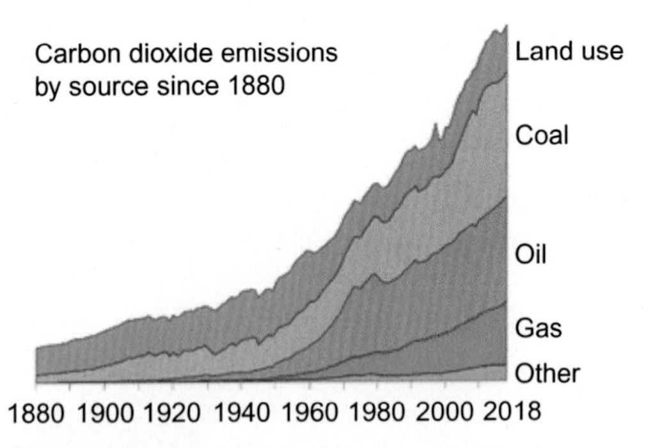


**FIG. 14** Energy use has exceeded population growth over the last few hundred years, most coming, along with the relative increasing amounts of carbon dioxide emissions, from fossil fuels (Burke et al., 2015). *Graph by Efbrazil 2020.*

Oil and gas production, including unconventional oil and gas production (which constitutes an expanding share of the world's energy supply) is becoming increasingly vulnerable to increasing periods of water shortages caused by a warming climate and the volumes of water required for enhanced oil recovery, hydraulic fracturing, and refining.

Hydropower, bioenergy, and concentrating solar power are at risk from changing precipitation patterns, increasing frequency and intensity of droughts, and increasing temperatures. Hydropower generation has doubled over the last three decades and is projected to double again by 2050, even in the face of hydrologic uncertainties and ecologic challenges (Schaeffer et al., 2012).

Fuel transport by rail and barge is susceptible to increased interruption and delay during more frequent periods of drought and flooding that affect water levels in rivers and ports.

Onshore oil and gas operations in Arctic regions are vulnerable to thawing permafrost, which can damage infrastructure and restrict seasonal access, while offshore operations could benefit from longer sea ice-free seasons.

Any measures or policies taken to reduce the vulnerability of energy production and transmission facilities to the impacts of global warming will involve both land and water. All are interdependent, as shown in Table 1.

**TABLE 1** The energy, land and water nexus.

| Resource system interaction | Illustrative components involved |
|---|---|
| Water needed for energy | Energy resource extraction<br>Fuel processing<br>Thermal power plant cooling<br>Carbon capture and storage (CCS) |
| Water needed for land | Agriculture<br>Industrial, municipal, commercial, and residential uses<br>Natural ecosystems |
| Energy needed for water | Water extraction<br>Water transport<br>Water treatment |
| Energy needed for land | Resource extraction and conversion<br>Agriculture<br>Transportation<br>Industrial, municipal, commercial, and residential uses |
| Land needed for energy | Energy resource extraction<br>Energy infrastructure, including dams/reservoirs, mines/wells, power plants, solar and wind farms, power lines, pipelines, and refineries<br>Bioenergy cropland<br>CCS |
| Land needed for water | Water capture and watershed<br>Ground cover vegetation |

Pacific Northwest National Laboratory, http://www.pnnl.gov/main/publications/external/technical_reports/PNNL-21185.pdf.

As changes in climate impact water resources, they will also impact energy resource extraction, fuel processing, thermal power plant cooling, and carbon capture and storage. As water is needed for energy, energy is needed for water extraction, transport, and treatment. Water is also needed for land-based agricultural, industrial, municipal, commercial, and residential uses and for the functioning of natural ecosystems.

As changes in climate impact land, they will also impact energy resource extraction, energy related infrastructure including dams/reservoirs, mines/wells, power plants, solar and wind farms, power lines, pipelines, and refineries. As land is needed for energy, energy is needed for land-based fuel resource extraction and conversion, transportation, and agriculture, industrial, municipal, commercial, and residential uses. Land resources are also needed for water storage and control such as watershed catchment areas and aquifers and surface water infrastructure including reservoirs, levies, distribution networks, and storm water drainage.

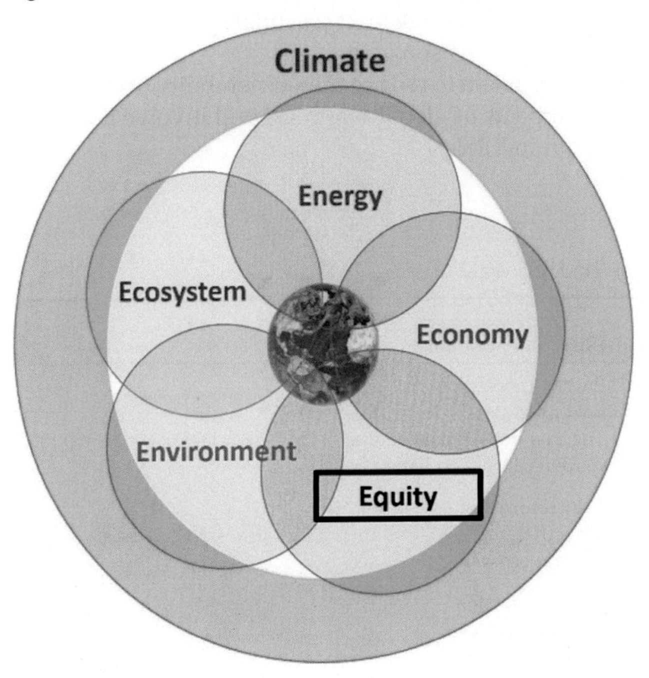


## 7 Equity impacts

Climate change and its physical, economic, and social impacts are affecting the well-being of people throughout much of the world. Also, throughout much of the world those communities most vulnerable to climate change impacts tend to be the least prepared to manage and adapt to them. Children, the elderly, minorities and indigenous peoples, and those with preexisting health conditions are especially vulnerable to climate-driven disruptions such as flooding, dangerous heat, drought, and poor air quality. This is particularly true if they have low incomes or limited means to address the cumulative impacts of climate change.

The health risks and adverse impacts of climate change disproportionately affect the well-being of low-income and minority communities. The same physical, social, and economical environments that are associated with poor health outcomes for low-income and minority communities also increase their exposure and vulnerability to the health impacts of climate change. These communities are often historically disenfranchised and lack the political and economic power and voice to ensure that their perspectives, needs, and ideas are fully considered. This contributes to health inequities and constrains the ability of these communities from building climate resilience and to contributing fully to climate change solutions. Addressing climate change and health inequities requires transformational change in our systems and communities. Many climate solutions offer health benefits and opportunities to promote greater equity. But to assure that all have equal opportunities for good health requires that we maintain a healthy planet. We cannot have healthy people without healthy places for them to live and work.

What applies to poorer communities also applies to poorer countries (see Fig. 12). Poorer countries tend to be the hotter ones. Poorer and hotter countries are notably more vulnerable to climate change than richer and cooler ones. The negative impacts of climate change will be borne by developing economies more than developed ones. Poorer countries are more exposed to the weather because of the important role of agriculture and water resources in the economy. In contrast, richer countries have a larger share of their economic activities in manufacturing and services, which are typically less vulnerable to variations of weather and hence climate change. Poorer countries often lack access to the technology and institutions that can help protect against the weather (e.g., air conditioning, malaria medicine, crop insurance). They may also lack the ability, and sometimes the political will, to mobilize the resources for large-scale infrastructure—for example irrigation systems and coastal protection.

The smallest and least developed countries often bear the major burdens of climate change, while contributing the least to its cause. Industrialized nations are responsible for the majority of greenhouse gas emissions causing global warming. From 1850 to 2011, the United States, European Union, China, Russian Federation, and Japan emitted two-thirds of the global $CO_2$ emissions. The World Health Organization estimates that 99% of the disease burden from climate change occurs in developing countries, 88% of which occurs in children under five. From January 1980 through July 2013, there were 2.52 million deaths globally because of climate-related disasters, 51% of which occurred in the 49 least developed countries (WHO, 2018).

Research suggests that the drivers behind climate injustice and health disparities are fundamentally the same—social inequities, institutional power, and the need for broader systems changes in our health systems, transportation infrastructure, and the production and distribution of energy (White-Newsome et al., 2018). These causes inequities are often the same throughout the world: Choices regarding energy, transportation, land use, housing, food and agriculture, and socioeconomic systems are key shapers of community living conditions. The institutions largely responsible for constructing and managing these systems influence, and are influenced by social inequities such as class and race. Because climate, health, and equity are linked, smart public policies must draw on all applicable disciplines, perspectives, and resources if they are to end this inequity that global warming is making increasingly evident (Wahowiak, 2018).

## 8 Conclusion

Any attempt to apply systems thinking or systems analysis approaches to study a complex issue forces one to decide what components and linkages to include in, and what to consider exogenous to, the system. This chapter has focused on the linkages between the warming climate and various components of our world that this warming trend is impacting. These aspects or components are the Earth's ecosystems, the Earth's air, land and water environments, our economy and energy sectors, and human equity issues arising from, and being made more evident by, global warming. In general, as each component either contributes or adapts to a warmer world it impacts each of the others, and in turn is impacted by them. Describing many of these interactions among components of this system leaves it to others to model it for the purposes of identifying comprehensive policies for reducing global warming and all its adverse impacts.

Evidences that our Earth is warming, and with adverse impacts at increasing rates, are there for any individual to see. Those who deny that this accelerating rate of warming is not going to change the planet and the quality and well-being of life on it, is just admitting they find it convenient to ignore an inconvenient truth. Life and the state of the world we inhabit today will not be sustainable at the increasing rate of global warming currently occurring. As mentioned in the introduction, we are witnessing a "climate emergency" (Leiserowitz et al., 2019; Ripple et al., 2020).

With warmer temperatures come reductions in the area and volume of glaciers and sea ice. The top half-mile of most major ocean waters is getting warmer, causing damage to coral reefs, threatening marine ecosystems, and disrupting global fisheries. Melting glaciers and permafrost are leading to hazardous landslides and avalanches with accompanying losses of economic and aesthetic values.

Just as air temperatures are getting warmer, so are water temperatures. This leads to more intense precipitation and storms. Ocean surface water temperatures have been increasing over the past three decades, adding to the observed changes in marine life, increases in algal blooms, and to the destructive potential of tropical cyclones and hurricanes (Berwyn, 2020; Chapra et al., 2017).

Sea level rise resulting from melting ice covers and warming oceans also threatens freshwater supplies and ecosystem services such as natural water filtration, as well as places coastal areas and infrastructure at greater risks of erosion and damage.

A warming climate impacts the quality of our environment, our food supplies, our susceptibility to various diseases and other health treats, the services provided by our ecosystems, as well as our economic productivity. Most of these impacts are and will continue to be felt disproportionately by the poorer of us.

As expressed by the Intergovernmental Panel on Climate Change (NASA, 2020a; IPCC, 2013):

> "Taken as a whole, the range of published evidence indicates that the net damage costs of climate change are likely to be significant and to increase over time."

It is up to all of us at local levels to decide what to do about it, and its inequities, and then do it. Given the speed and extent of the responses to control our ongoing pandemic, it seems we can if we have the political will to try.

## References

Adams, S.B., 2011. Climate Change and Warmwater Aquatic Fauna. U.S. Department of Agriculture, Forest Service, Climate Change Resource Center. http://www.fs.fed.us/ccrc/topics/climate-change-and-warmwater-faunage.

AGU, 2020. COVID-19 Lockdowns Significantly Impacting Global Air Quality. American Geophysical Union. Science Daily www.sciencedaily.com/releases/2020/05/200511124444.htm.

Andrén, O., Kätterer, T., 2019. Encyclopedia of Ecology., ISBN: 978-0-444-64130-4. https://www.sciencedirect.com/topics/earth-and-planetary-sciences/agricultural-ecosystem#:~:text=of%20Ecology.

Bagne, K., Ford, P., Reeves, M., 2012. Grasslands. U.S. Department of Agriculture, Forest Service, Climate Change Resource Center. www.fs.usda.gov/ccrc/topics/grasslands/.

Berwyn, B., 2020. Ocean Warming is Speeding Up, With Devastating Consequences. Study Shows https://insideclimatenews.org/news/14012020/ocean-heat-2019-warmest-year-argo-hurricanes-corals-marine-animals-heatwaves.

Blanc, E., Reilly, J., 2017. Approaches to assessing climate change impacts on agriculture: an overview of the debate. Rev. Environ. Econ. Policy 11, 247–257. https://doi.org/10.1093/reep/rex011.

Brown, M.E., Antle, J.M., Backlund, P., Carr, E.R., Easterling, W.E., Walsh, M.K., Ammann, C., Attavanich, W., Barrett, C.B., Bellemare, M.F., Dancheck, V., Funk, C., Grace, K., Ingram, J.S.I., Jiang, H., Maletta, H., Mata, T., Murray, A., Ngugi, M., Ojima, D., O'Neill, B., Tebaldi, C., 2015. Climate Change, Global Food Security, and the U.S. Food System. 146 pp. Available online at: http://www.usda.gov/oce/climate_change/FoodSecurity2015Assessment/FullAssessment.pdf.

Burke, M., Hsiang, S., Miguel, E., 2015. Global non-linear effect of temperature on economic production. Nature 527, 235–239. https://doi.org/10.1038/nature15725.

Calma, J., 2020. Mangrove Forests are Crucial to Protecting Coastlines. https://www.theverge.com/2020/6/4/21280580/sea-level-rise-mangroves-climate-change-2050.

CCSP, 2008. The effects of climate change on agriculture, land resources, water resources, and biodiversity in the United States. Chapter 3, In: Land Resources: Forest and Arid Lands. A Report by the U.S. Climate Change Science Program and the Subcommittee on Global Change Research. P. Backlund, A. Janetos, D. Schimel, J. Hatfield, K. Boote, P. Fay, L. Hahn, C. Izaurralde, B.A. Kimball, T. Mader, J. Morgan, D. Ort, W. Polley, A. Thomson, D. Wolfe, M. Ryan, S. Archer, R. Birdsey, C. Dahm, L. Heath, J. Hicke, D. Hollinger, T. Huxman, G. Okin, R. Oren, J. Randerson, W. Schlesinger, D. Lettenmaier, D. Major, L. Poff, S. Running, L. Hansen, D. Inouye, B.P. Kelly, L Meyerson, B. Peterson, R. Shaw, U.S. Environmental Protection Agency, Washington, DC, USA.

CCSP (U.S. Climate Change Science Program), 2007. Effects of climate change on energy production and distribution in the United States. In: Bull, S.R., Bilello, D.E., Ekmann, J., Sale, M.J., Schmalzer, D.K. (Eds.), Effects of Climate Change on Energy Production and Use in the United States, Synthesis and Assessment Product 4.5. U.S. Climate Change Science Program, Washington, DC, pp. 8–44.

CDP, 2019. https://www.cdp.net/en/articles/media/worlds-biggest-companies-face-1-trillion-in-climate-change-risks.

Chapra, S.C., et al., 2017. Climate change impacts on harmful algal blooms in U.S. freshwaters: a screening-level assessment. Environ. Sci. Technol. 51, 8933–8943.

Chen, N., 2018. Climate Change to Impact Desert Ecosystems. Chinese Academy of Sciences. http://english.cas.cn/newsroom/archive/news_archive/nu2018/201808/t20180817_196187.shtml.

Cheng, L., Abraham, J., Zhu, J., et al., 2020. Record-setting ocean warmth continued in 2019. Adv. Atmos. Sci. 37, 137–142. https://doi.org/10.1007/s00376-020-9283-7.

D'Amato, G., et al., 2011. Climate change, migration, and allergic respiratory diseases: an update for the allergist. World Allergy Organ. J. 4, 121–125. https://doi.org/10.1097/WOX.0b013e3182260a57. Detail.

Dahlman, L.A., Lindsey, R., 2020. Climate Change: Ocean Heat Content. February 13, 2020 https://www.climate.gov/author/luann-dahlman-and-rebecca-lindsey. https://www.climate.gov/news-features/understanding-climate/climate-change-ocean-heat-content.

Dawson, J.P., Racherla, P.N., Lynn, B.H., Adams, P.J., Pandis, S.N., 2009. Impacts of climate change on regional and urban air quality in the eastern United States: role of meteorology. J. Geophys. Res. Atmos. 114. https://doi.org/10.1029/2008JD009849.

De Sherbinin, A., Schiller, A., Pulsipher, A., 2007. The vulnerability of global cities to climate hazards. Environ. Urban. 19, 39–64. https://journals.sagepub.com/doi/abs/10.1177/0956247807076725.

DOE, 2013. US Energy Sector Vulnerabilities to Climate Change and Extreme Weather. DOE PI 0013, July https://www.energy.gov/sites/prod/files/2013/07/f2/20130710-Energy-Sector-Vulnerabilities-Report.pdf.
Duncombe, J., 2020. The ticking time bomb of Arctic permafrost. Eos 101. https://doi.org/10.1029/2020EO146107.
Economist, 2015. Recognising the Value at Risk From Climate Change. Economist Intelligence Unit. July http://www.economistinsights.com/sites/default/files/The%20cost%20of%20inaction.pdf.
Egan, M., 2020. Exxon's Market Value has Crumbled by $184 billon. CNN Business. February 5 https://www.cnn.com/2020/02/05/business/exxonmobil-oil-stock/index.html.
Estrada, F., Botzen, W., Tol, R., 2017. A global economic assessment of city policies to reduce climate change impacts. Nat. Clim. Change 7, 403–406. https://doi.org/10.1038/nclimate3301. https://www.nature.com/articles/nclimate3301.
Fann, N., Brennan, T., Dolwick, P., Gamble, J.L., Ilacqua, V., Kalb, L., Nolte, C.G., Ziska, T.L., 2016. Impacts of Climate Change on Human Health in the United States: A Scientific Assessment. U.S. Global Change Research Program, Washington, DC, https://doi.org/10.7930/J0GQ6VP6. https://health2016.globalchange.gov/air-quality-impacts. (Chapter 3).
Ford, P.L., 2002. Grasslands and savannas. In: Encyclopedia of Life Support Systems. UNESCO, p. 20. http://www.eolss.net/Sample-Chapters/C12/E1-01-06-06.pdf.
Gore, 2019. https://www.nytimes.com/2019/09/20/opinion/al-gore-climate-change.html.
Graham, S., 2000. Svante Arrhenius (1859–1927). https://earthobservatory.nasa.gov/features/Arrhenius.
Hansen, J., Sato, M., Ruedy, R., Lo, K., Lea, D.W., Medina-Elizade, M., 2006. Global temperature change. PNAS 103 (39), 14288–14293. https://doi.org/10.1073/pnas.0606291103.
Hansen, J., Sato, M., Ruedy, R., 2012. Perception of climate change. PNAS 109 (37), E2415–E2423. https://doi.org/10.1073/pnas.1205276109.
Henson, R., NRC, 2011. Warmıng World Impacts by Degree, Based on the National Research Council Report, Climate Stabilization Targets: Emissions, Concentrations, and Impacts Over Decades to Millennia (2011). http://dels.nas.edu/resources/static-assets/materials-based-on-reports/booklets/warming_world_final.pdf.
Houser, T., et al., 2014. American Climate Prospectus: Economic Risks in the United States (Rhodium Group, 2014). https://rhg.com/research/american-climate-prospectus-economic-risks-in-the-united-states/.
Huang, Y., Unger, N., Harper, K., Heyes, C., 2020. Global Climate and Human Health Effects of the Gasoline and Diesel Vehicle Fleets. GeoHealth, https://doi.org/10.1029/2019GH000240. March.
Hunt, A., Watkiss, P., 2011. Climate change impacts and adaptation in cities: a review of the literature. Clim. Change 104 (1), 13–49. https://doi.org/10.1007/s10584-010-9975.
IASS, 2020. Air Pollution and Climate Change. Institute for Advanced Sustainability Studies, Potsdam, Germany. https://www.iass-potsdam.de/en/output/dossiers/air-pollution-and-climate-change.
IPCC, 2013. Summary for policymakers. In: Stocker, T.F., Qin, D., Plattner, G.-K., Tignor, M., Allen, S.K., Boschung, J., Midgley, P.M. (Eds.), Climate Change 2013: The Physical Science Basis. Contribution of Working Group I to the Fifth Assessment Report of the Intergovernmental Panel on Climate Change. Cambridge University Press, Cambridge, United Kingdom and New York, NY, USA. http://www.climatechange2013.org/images/report/WG1.
IPCC, 2018. Summary for Policymakers. https://www.ipcc.ch/site/assets/uploads/2018/02/ipcc_wg3_ar5_summary-for-policymakers.pdf.
IPCC, 2019. Summary for Policymakers of IPCC Special Report On Global Warming of 1.5°C Approved by Governments. https://www.ipcc.ch/sr15/.
Kirilenko, A.P., Sedjo, R.A., 2007. Climate change impacts on forestry. PNAS 104 (50), 19697–19702. https://doi.org/10.1073/pnas.0701424104. https://www.pnas.org/content/104/50/19697.
Lane, D. et al. Climate change impacts on freshwater fish, coral reefs, and related ecosystem services in the United States. Clim. Change 131, 143–157 (2015).
Leiserowitz, A., Maibach, E., Rosenthal, S., Kotcher, J., Bergquist, P., Ballew, M., Goldberg, M., Gustafson, A., 2019. Climate change in the American mind: April 2019. Yale University and George Mason University. Yale Program on Climate Change Communication, New Haven, CT, https://doi.org/10.17605/OSF.IO/CJ2NS.
Le Quéré, C., Jackson, R.B., Jones, M.W., Smith, A.J.P., Abernethy, S., Andrew, R.M., De-Gol, A.J., Willis, D.R., Shan, Y., Canadell, J.G., Friedlingstein, P., Creutzig, F., Peters, G.P., 2020. Temporary reduction in daily global $CO_2$ emissions during the COVID-19 forced confinement. Nat. Clim. Change. https://doi.org/10.1038/s41558-020-0797-x.
Loarie, S., Duffy, P., Hamilton, H., et al., 2009. The velocity of climate change. Nature 462, 1052–1055. https://doi.org/10.1038/nature08649. https://www.nature.com/articles/nature08649#citeas.

Malhi, Y., Franklin, J., Seddon, N., Solan, M., Turner, M.G., Field, C.B., Knowlton, N., 2020. Climate change and ecosystems: threats, opportunities and solutions. Phil. Trans. R. Soc. B 375, 20190104. https://doi.org/10.1098/rstb.2019.0104.

Marquis, Nicholas D., Theodore J. Bishop, Nicholas R. Record, Peter D. Countway, José A. Fernández Robledo, 2020, A qPCR-Based Survey of Haplosporidium nelsoni and Perkinsus spp. in the Eastern Oyster, Crassostrea virginica in Maine, USA Pathogens, 9(4), 256; https://doi.org/10.3390/pathogens9040256.

Martinich, J., Crimmins, A., 2019. Climate damages and adaptation potential across diverse sectors of the United States. Nat. Clim. Change 9, 397–404. https://doi.org/10.1038/s41558-019-0444-6. https://www.nature.com/articles/s41558-019-0444-6#citeas.

McDermott, A., 2016. Climate Change May Be as Hard on Lizards as on Polar Bears, The Atlantic. May 23 https://www.theatlantic.com/science/archive/2016/05/climate-change-deserts/483896/.

Melvin, A.M., Larsen, P., Boehlert, B., Neumann, J.E., Chinowsky, P., Espinet, X., Martinich, J., Baumann, M.S., Rennels, L., Bothner, A., Nicolsky, D.J., Marchenko, S.S., 2017. Climate change damages to Alaska public infrastructure and the economics of proactive adaptation. PNAS 114 (2), E122–E131. First published December 27, 2016 https://doi.org/10.1073/pnas.1611056113.

Morgan, J.A., Derner, J.D., Milchunas, D.G., Pendall, E., 2008. Management Implications of Global Change for Great Plains Rangelands. https://www.ars.usda.gov/ARSUserFiles/1354/221.Morganetal.Rangelands2008.pdf.

NASA, 2020a. The Effects of Climate Change. June 26 https://climate.nasa.gov/effects/.

NASA, 2020b. Ice Melt Linked to Accelerated Regional Freshwater Depletion. https://www.jpl.nasa.gov/news/news.php?feature=7669.

NASA, 2020c. Climate Change: How Do We Know? https://climate.nasa.gov/evidence/#:~:text=Warming.

NASA, 2013. The Virtues of Pure Snow. https://earthobservatory.nasa.gov/images/81959/the-virtues-of-pure-snow. https://eoimages.gsfc.nasa.gov/images/imagerecords/81000/81959/P1030604_lrg.jpg.

NAS, 2020. Climate Change: Evidence and Causes: Update 2020. The National Academies Press, Washington, DC, https://doi.org/10.17226/25733.

NAS, 2019. Climate Change and Ecosystems. The National Academies Press, Washington, DC, https://doi.org/10.17226/25504. https://www.ipcc.ch/report/srccl/.

National Snow and Ice Data Center, 2019. https://nsidc.org/cryosphere/sotc/snow_extent.html.

Nerem, R.S., Beckley, B.D., Fasullo, J.T., Hamlington, B.D., Masters, D., Mitchum, G.T., 2018. Climate-change-driven accelerated sea-level rise detected in the altimeter era. PNAS. https://doi.org/10.1073/pnas.1717312115.

Neumann, J.E., Price, J.C., 2009. Adapting to Climate Change. The Public Policy Response. Public Infrastructure. RFF Report http://www.rff.org/rff/documents/RFF-Rpt-Adaptation-NeumannPrice.pdf.

NOAA, 2020a. The Impact of Weather and Climate Extremes on Air and Water Quality. https://www.ncdc.noaa.gov/news/impact-weather-and-climate-extremes-air-and-water-quality.

NOAA, 2020b. What is Ocean Acidification? http://www.pmel.noaa.gov/co2/story/What+is+Ocean+Acidification%3F.

ONYSC, 2017. Special Report: North Country Region Economic Profile. Office of New York State Comptroller. October 2017. https://www.osc.state.ny.us/sites/default/files/local-government/documents/pdf/2019-01/north-country.pdf.

Pelto, M., 2015. 31 years of observations on Retreating Columbia Glacier, Washington. https://blogs.agu.org/fromaglaciersperspective/2015/04/07/31-years-of-observations-on-retreating-columbia-glacier-washington/.

Poff, N.L.R., Brinson, M.M., Day, J.W., Jr., 2002. Aquatic Ecosystems and Global Climate Change, Potential Impacts on Inland Freshwater and Coastal Wetland Ecosystems in the United States. Pew Center on Global Climate Change. January. https://www.c2es.org/site/assets/uploads/2002/01/aquatic.pdf. https://www.c2es.org/document/aquatic-ecosystems-and-global-climate-change/.

Raymond, C., R.M. Horton, J. Zscheischler, O. Martius, A. AghaKouchak, J. Balch, S.G. Bowen, S.J. Camargo, J. Hess, K. Kornhuber, M. Oppenheimer, A.C. Ruane, T. Wahl, and K. White, 2020: Understanding and managing connected extreme events. Nat. Clim. Change, 10, no. 7, 611-621, https://doi.org/10.1038/s41558-020-0790-4.

Ripple, William & Wolf, Christopher & Newsome, Thomas & Barnard, Phoebe & Moomaw, William & Gutiérrez Cárdenas, Paul David Alfonso. (2020). World scientists' warning of a climate emergency. BioScience. 70. 8–12.

Ripple, W.J., Wolf, C., Newsome, T.M., Barnard, P., Moomaw, W.R., 2020. Warning of a climate emergency. BioScience 70 (1), 8–12. https://doi.org/10.1093/biosci/biz088.

Rosenzweig, C., Solecki, W.D., Hammer, S.A., Mehrotra, S., 2011. Climate Change and Cities: First Assessment Report of the Urban Climate Change Research Network. Cambridge University Press. https://pubs.giss.nasa.gov/abs/ro00210r.html.

Schaeffer, R., Szklo, A.S., de Lucena, A.F.P., Borba, B.S.M.C., Nogueira, L.P.P., Fleming, F.P., Troccoli, A., Harrison, M., Boulahya, M.S., et al., 2012. Energy sector vulnerability to climate change: a review. Energy. https://doi.org/10.1016/j.energy.2011.11.056.

Sheffield, P.E., Weinberger, K.R., Kinney, P.L., 2011. Climate change, aeroallergens, and pediatric allergic disease. Mt. Sinai J. Med. 78, 78–84. https://doi.org/10.1002/msj.20232.

SNIDE, 2019. Sea Ice. https://nsidc.org/cryosphere/sotc/sea_ice.html.

Stavros, E.N., McKenzie, D., Larkin, N., 2014. The climate-wildfire-air quality system: interactions and feedbacks across spatial and temporal scales. Wiley Interdiscipl. Rev.: Clim. Change 5, 719–733. https://doi.org/10.1002/wcc.303.

USEPA, 2016. Climate Impacts on Forests. https://19january2017snapshot.epa.gov/climate-impacts/climate-impacts-forests.

USGCRP, 2017. In: Wuebbles, D.J., Fahey, D.W., Hibbard, K.A., Dokken, D.J., Stewart, B.C., Maycock, T.K. (Eds.), Climate Science Special Report: Fourth National Climate Assessment. vol. I. U.S. Global Change Research Program, Washington, DC, USA, https://doi.org/10.7930/J0J964J6. 470 pp.

Wahowiak, L., 2018. Climate change, health equity 'inextricably linked': vulnerable populations most at risk from harmful effects. Nation's Health 48 (7), 1–10. http://thenationshealth.aphapublications.org/content/48/7/1.1#:~:text=Climate%20change%20has%20the%20worst,%2C%20learn%2C%20work%20and%20play.

Walthall, C.L., Hatfield, J., Backlund, P., Lengnick, L., Marshall, E., Walsh, M., Adkins, S., Aillery, M., Ainsworth, E. A., Ammann, C., Anderson, C.J., Bartomeus, I., Baumgard, L.H., Booker, F., Bradley, B., Blumenthal, D.M., Bunce, J., Burkey, K., Dabney, S.M., Delgado, J.A., Dukes, J., Funk, A., Garrett, K., Glenn, M., Grantz, D.A., Goodrich, D., Hu, S., Izaurralde, R.C., Jones, R.A.C., Kim, S.-H., Leaky, A.D.B., Lewers, K., Mader, T.L., McClung, A., Morgan, J., Muth, D.J., Nearing, M., Oosterhuis, D.M., Ort, D., Parmesan, C., Pettigrew, W.T., Polley, W., Rader, R., Rice, C., Rivington, M., Rosskopf, E., Salas, W.A., Sollenberger, L.E., Srygley, R., Stöckle, C., Takle, E.S., Timlin, D., White, J. W., Winfree, R., Wright-Morton, L., Ziska, L.H., 2012. Climate Change and Agriculture in the United States: Effects and Adaptation. USDA Technical Bulletin 1935. Washington, DC. 186 pp. https://www.usda.gov/oce/climate_change/effects_2012/CC%20and%20Agriculture%20Report%20(02-04-2013)b.pdf.

WGMS, 2020. World Glacier Monitoring Service. https://wgms.ch/.

Whitehead, P.G., Wilby, R.L., Battarbee, R.W., Kernan, M., Wade, A.J., 2009. A review of the potential impacts of climate change on surface water quality. Hydrol. Sci. J. 54 (1), 101–123. https://doi.org/10.1623/hysj.54.1.101.

White-Newsome, J., Meadows, P., Kabel, C., 2018. Bridging climate, health, and equity: a growing imperative. Am. J. Publ. Health 108 (Suppl. 2). https://doi.org/10.2105/AJPH.2017.304133.

WHO, 2018. Climate Change and Heath. February https://www.who.int/news-room/fact-sheets/detail/climate-change-and-health.

SECTION B

# Physical impacts

CHAPTER

# 3

# Climate change and melting glaciers


*Maria Shahgedanova*

**Department of Geography and Environmental Sciences, University of Reading, Reading, United Kingdom**


OUTLINE



## 1 Introduction

About 10% of the Earth's land surface or $15 \times 10^6$ km$^2$ is at present covered by glacier ice including ice sheets of Antarctica and Greenland, ice caps, and glaciers. About 91% and 8% of glacial ice is contained in the Antarctic and Greenland ice sheets, respectively. Ice caps, defined as bodies of ice with area of less than 50,000 km$^2$ and covering the underlying land completely, occur primarily in the polar and subpolar regions, for example, in the Canadian and Russian Arctic, Svalbard, and Iceland. Glaciers, which are often referred to as "small glaciers" irrespective of their size to distinguish them from ice sheets and ice caps, are located in polar and subpolar regions and in the mountains and span the whole range of elevations. In the polar and subpolar regions, they terminate in the ocean (tidewater glaciers) and in the High Mountain Asia (HMA) and in the Andes, descend from over 7000 m above sea level




*The Impacts of Climate Change*
**https://doi.org/10.1016/B978-0-12-822373-4.00007-0**

(a.s.l.). Estimates of the total area of glaciers and ice caps (including those in Greenland and Antarctica, but excluding the ice sheets) vary between $680 \times 10^3$ km$^2$ and $785 \times 10^3$ km$^2$ (Radić and Hock, 2010). Total glacier volume, based on the total area of $741 \times 10^3$ km$^2$, is estimated as $(241 \pm 29) \times 10^3$ km$^3$ (Radić and Hock, 2010). Farinotti et al. (2019) provides a lower estimate of $(158 \pm 41) \times 10^3$ km$^3$ which is equivalent to $0.32 \pm 0.08$ m of global mean sea level (GMSL) change.

About 69% of the world's fresh water is stored as ice. Glacial melt (including ice sheets and small glaciers) is responsible for about 50% of the observed sea level rise (Church et al., 2011; Oppenheimer et al., 2019). Glaciers deliver a range of ecosystem services (Huss et al., 2017). They are important sources of water (Viviroli and Weingartner, 2004; Viviroli et al., 2020) and in regions, where mountains are surrounded by arid plains, such as Central Asia and the Andes, glacial meltwater sustains irrigation without which sufficient food production may be impossible (Kaser et al., 2010). Healthy glaciers prevent the formation and expansion of glacier lakes and such hazards as glacier lake outburst floods (GLOF), mudflows, and landslides (Haeberli and Whiteman, 2015). It is estimated that in 2010, almost 10% (671 million people) of the global population lived in high mountain regions at a distance of less than 100 km from mountain glaciers and this population is expected to grow to 736–844 million (Hock et al., 2019). Those living in the mountains and residing further downstream depend on the ecosystem services provided by glaciers. Last but not least, glaciers have a high aesthetic value and represent some of the most spectacular landscapes in the world.

Glacial ice, in all its forms, is extremely vulnerable to climatic warming because its very existence depends on subzero temperatures. In the Arctic, summer temperatures, relevant to glacier melt, increased by 0.5–1.0°C in 2000–09 in comparison with 1960–2009 (Serreze and Barry, 2011). There is evidence that in the mountains, especially in the lower latitudes, climate is warming at a higher rate than on the plains (Pepin et al., 2015). It is not surprising that shrinking ice caps and glaciers have become symbols of the adverse effects of climate change.

In this chapter, the observed changes in the state of glaciers and ice sheets are reviewed together with factors initiating and controlling these changes. Indicators of glacier change and methods of assessments are discussed. The contribution of ice sheets and glaciers to sea level change is addressed.

## 2 Mass balance of glaciers and ice sheets

The state of glaciers, ice caps, and ice sheets depends on their mass balance (budget) defined as a difference between mass gain and mass loss in a given time period expressed either in metres of water equivalent (m w.e.) or in units of mass per year (Cuffey and Paterson, 2010). Glaciers which receive more (less) mass than they lose, have positive (negative) mass balance. They advance and thicken (retreat and thin). Glaciers, gaining and losing approximately the same mass have zero mass balance and if this persists over time, they are said to be in a state of equilibrium and their termini neither advance nor retreat. Mass gain occurs by accumulation of snow from direct precipitation, windblown snow, or avalanches on glacier surface. Net accumulation occurs in the upper part of glaciers, in the accumulation zone, where accumulated snow is transformed into firn and then ice which over years flows downslope. For glaciers terminating on land, mass loss (ablation) is dominated by melt in the lower part of glaciers but can also occur via sublimation while basal melting is usually a smaller term. For temperate glaciers, liquid precipitation and meltwater are considered as runoff. Sublimation is particularly important for tropical glaciers. For example, at Kilimanjaro, ice loss occurs almost entirely through sublimation (Kaser et al., 2004; Mölg et al., 2008).

Difference between accumulation and ablation on the surface forms surface mass balance (SMB). The elevation, at which net accumulation equals net ablation is known as the equilibrium line altitude (ELA). If a glacier is receding, its ELA progressively moves up, making the accumulation area ratio (AAR) smaller. In colder areas (either geographical or at higher elevations), liquid precipitation and meltwater percolate into ice and refreeze, resulting in internal accumulation which, together with SMB, forms climatic mass balance. Confusingly, in ice sheet studies, climatic mass balance can be referred to as SMB to distinguish it from ice discharge (van den Broeke et al., 2017).

For marine-terminating glaciers (as well as for those terminating in lakes) and ice sheets, front calving and melting underneath floating ice shelves are important components of ice loss and they are often aggregated as ice discharge or dynamic ice loss (D) across the grounding line, a zone of transition from grounded ice to freely floating ice (e.g., ice sheet to ice shelf). Knowledge of its position is important because ocean warming around the grounding line can cause rapid melt, increase in ice velocity, and ice discharge to the ocean (Rignot, 1996). If SMB equals D, ice mass is in balance providing that basal melt is negligible (Cuffey and Paterson, 2010).

Mass balance is controlled primarily, but not exclusively, by air temperature of the ablation season and by precipitation. The length of the ablation season varies from a few weeks in the polar regions to all year in the tropics. Increasing air temperature can lead to both more intense melt and an extension of the melt season increasing mass loss. Thus, the progressively more negative mass balance of Glacier de Sarennes in the French Alps during the 1949–2007 period was caused by summer warming with the lengthening of the ablation season and higher ablation rates explaining 65% and 35% of the increase in glacier melt, respectively (Thibert et al., 2013). Precipitation during the accumulation season controls mass gain. For example, a decline in precipitation in the 1970–80s in Central Asia significantly contributed to glacier mass loss in the Tien Shan (Shahgedanova et al., 2018). The occurrence of solid precipitation during the ablation season is also important because it interrupts and reduces ablation by changing the albedo (reflectivity) of the glacier surface. For this reason, the high-elevation glaciers of Kilimanjaro are more sensitive to changes in precipitation than temperature (Mölg et al., 2008). Atmospheric humidity controls sublimation which is particularly important for tropical glaciers (Vuille et al., 2008; Rabatel et al., 2013). Cloudiness controls both incoming long-wave and short-wave radiation. Melt at the highest elevations in Greenland is caused by the long-wave radiation emitted by clouds (van den Broeke et al., 2017). Short-wave radiation is a particularly important source of energy for melt in combination with surface albedo. The transition from fresh snow to wet, aged snow, and ice and the presence of light absorbing impurities (LAI) on the glacier surface strongly influence melt particularly when solar radiation values are high. Changes in sea surface temperature (SST) and heat supplied to glacier tongues by the ocean affect mass loss from marine-terminating glaciers.

The response of glaciers to climatic perturbations is not uniform because glaciers have different SMB sensitivities to climatic forcing and different response times. Sensitivity is the change in mass balance per unit change in a climatic variable such as air temperature or precipitation. Sensitivity depends on climate (as a rule, maritime glaciers have higher sensitivity than continental glaciers because they have high mass turnover), slope, and morphology of glacier. The response time is the time which is required for a glacier to adjust to a change in mass balance. Small cirque glaciers react to changes in climate and mass balance almost without delay; short and steep glaciers reach equilibrium in a few years while larger valley glaciers have a response time of several decades (Haeberli, 1995).

Mass balance is used to calculate glacier and ice sheet contributions to GMSL change. However, change in mass and contribution to GMSL are not identical. Changes in floating ice shelves do not cause any significant changes in GMSL directly although there is a negligible effect because of changes in the salinity of ocean water. However, ice shelves buttress marine ice sheets and their demise can cause marine ice-sheet instability resulting in a discharge of land ice into the ocean. Therefore, for ice sheets and marine-terminating glaciers, change in mass balance and contribution to GMSL are different because they contain floating ice and because of the effects of a migrating grounding line. A more detailed explanation is provided by Bamber et al. (2018). There is also a difference in the case of glaciers in landlocked (endorheic) basins, for example, in the Pamir and Tien Shan in Central Asia, because water, which they lose, may never directly reach the ocean, contributing to GMSL only indirectly through land hydrology processes (Brun et al., 2017). Meltwater can be withdrawn for human use or stored in glacial lakes (Haeberli and Linsbauer, 2013). For example, in Patagonia, terrestrial storage in the lakes of the Northern Patagonian Icefield is as high as 10% of its glacier melt contribution to sea level rise (Loriaux and Casassa, 2013).

## 3 Observing glacier change

### 3.1 Measuring mass balance

Glacier mass balance is measured either by glaciological (stake) or geodetic methods (Cogley, 2009; Zemp et al., 2009, 2015). Both annual mass balance rates and cumulative mass balance, showing glacier mass in a given time period relative to an earlier one, are used. The glaciological method involves establishing a grid of ablation stakes across a glacier, measuring accumulation and ablation with reference to the stakes several times per year, and interpolating measurements across the glacier surface. These measurements are accompanied by snow density measurements (to calculate water equivalent) and supported by meteorological observations. The surveys are time-consuming, physically demanding, and occasionally dangerous. They have limited coverage but provide the longest mass balance series dating back to the late 19th (the Rhone glacier)—early 20th (the Silvretta glacier) centuries (Zemp et al., 2009). Glaciological mass balance measurements are coordinated by the World Glacier Monitoring Service (WGMS). In the 1946–2005 period, results were available from 228 glaciers around the world but only 12 provided time series dating back to 1960 or earlier (Zemp et al., 2009). For climate change assessments, ongoing mass balance series with over 30 observation years are used from the so-called reference or benchmark glaciers. Fig. 1 shows two of these glaciers, Tuyuksu in the Tien Shan and Djankuat in the Caucasus. Fig. 2 shows cumulative mass balance of the Tuyuksu glacier which has the longest uninterrupted data series in Central Asia.

FIG. 1 The WGMS reference glaciers: Tuyuksu in the Tien Shan (left-hand panel) and Djankuat in the Caucasus (right-hand panel).

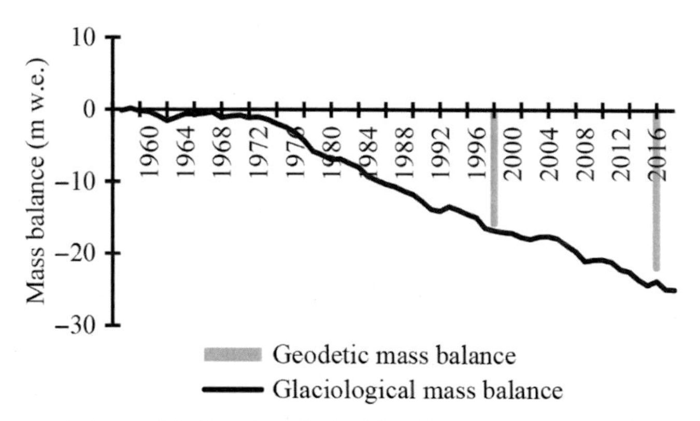


FIG. 2 Cumulative mass balance of the Tuyuksu glacier, Tien Shan, measured using the glaciological method and geodetic mass balance for the 1958–98 and 1958–2016 periods (Kapitsa et al., 2020). Using both methods helps to verify data quality: A close match between the two types of measurements confirms data reliability. Note that cumulative mass balance has been persistently negative since the 1970s indicating that Tuyuksu is losing mass.

The geodetic method is based on repeated measurements of glacier surface elevation using consecutive digital elevation models (DEM) developed from field surveys, aerial, and spaceborne data (Cogley, 2009; Marzeion et al., 2017). Satellite radar (ERS-1, ENVISat, CryoSat 2) and laser (ICESat) altimetry is used to measure changing surface elevation of ice sheets, ice caps and, in the case of ICESat, larger glaciers (Bamber et al., 2018). The Shuttle Radar Topography Mission (SRTM) provides elevation data collected between 60°N and 54°S in 2000. Photogrammetric methods using repeat airborne and optical satellite imagery such as ASTER (e.g., Brun et al., 2017) and submetre Pléiades (e.g., Berthier et al., 2014; Kutuzov et al., 2019; Kapitsa et al., 2020) enable development of higher-resolution DEMs from spaceborne data for smaller mountain glaciers worldwide. Fig. 3 shows changes in the surface elevation of the Tuyuksu glacier between 1958 and 2016 derived from a ground-based geodetic survey

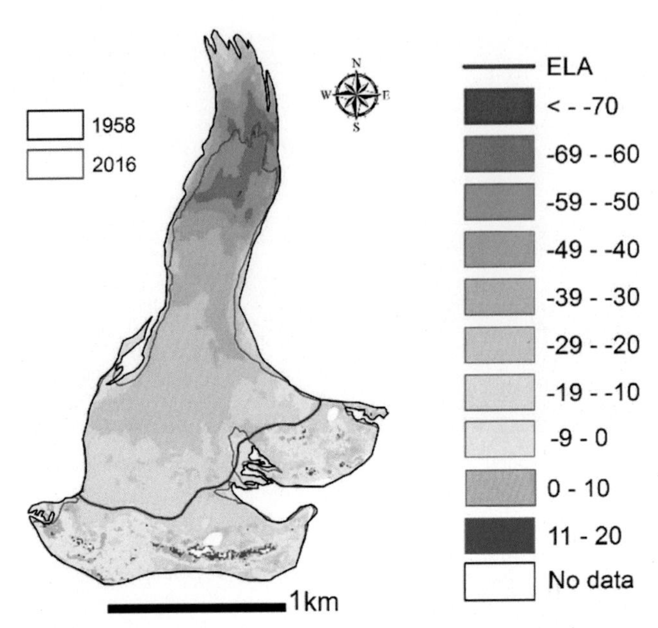


**FIG. 3** Changes in the surface elevation of the Tuyuksu glacier between 1958 and 2016, derived from a ground-based geodetic survey and the Pléiades imagery (Kapitsa et al., 2020). These data were used to derive the geodetic mass balance of the Tuyuksu glacier (Fig. 2). Note that at the lower part of the glacier, surface lowering exceeded 60 m and about 23% of the glacier area has been lost.

and the Pléiades imagery. The USA declassified spy Corona and Hexagon imagery, particularly plentiful over the territory of the former Soviet Union, enables development of historical DEM from the 1970s (e.g., Pieczonka and Bolch, 2015). Using these data, changes in glacier volume are measured and converted to mass balance using glacier-averaged density (Huss, 2013). Thus, the geodetic mass balance of the Tuyuksu glacier (Fig. 2) was derived from changes in surface elevation (Fig. 3). The geodetic method expands geographical coverage of mass balance measurements and extends them to inaccessible parts of glaciers. The use of spaceborne data enabled global assessments, particularly measurements of mass balance of ice caps and ice sheets, and this method is widely used in calculations of glacier contribution to sea level rise (Marzeion et al., 2017).

Gravimetric method of measuring changes in glacier mass at the global scale became available since the launch of the Gravity Recovery and Climate Experiment (GRACE) twin satellites in summer 2002 (Wahr et al., 2004). Changes in gravity are mainly caused by changes in the mass of surface water and, therefore, gravity anomalies provide information about its redistribution. GRACE has a coarse resolution of 300 km and is used predominantly to measure changes over ice sheets (Velicogna, 2009; Bamber et al., 2018) or ice caps and larger polar glaciers (Jacob et al., 2012; Box et al., 2018). Attempts were made to use GRACE to assess changes in mass of mountain glaciers (Jacob et al., 2012) but with limited success because the coarse resolution resulted in a very high uncertainty. Changes in regional hydrology, particularly water withdrawal for irrigation in Central Asia, affected the redistribution of mass and the quality of measurements.

All sources of data and methods are valuable and their complementary use enables validation of data sets, accuracy assessments, and inter-comparison of measurements leading to a wider coverage and an improved quality of observations. For example, the Pléiades Glacier Observatory was established by the WGMS in collaboration with the French Space Agency (CNES) to expand mass balance monitoring from individual benchmark glaciers to regional scale through the application of the high-resolution Pléiades satellite DEM validated with ground control data (see, e.g., Fig. 2 as well as Berthier et al., 2014; Kutuzov et al., 2019; Kapitsa et al., 2020) and to fill in gaps in observations (e.g., Barandun et al., 2018). Another example is the Ice Sheet Mass Balance Inter-comparison Exercise (IMBIE) project who combined assessments of changes in the mass balance of the Antarctic and Greenland Ice sheets derived from satellite altimetry, gravimetry, and the so-called mass balance input-output method, comparing modeled SMB with satellite-derived ice flow (D), to reduce uncertainty of these complex assessments and improve our understanding of glacier contribution to sea level change (The IMBIE Team, 2018, 2020).

## 3.2 Measuring changes in glacier length and area

Another way to assess glacier change is to measure changes in glacier length (position of glacier termini or front variations) or area. Glacier length data sets go back the furthest. This is because direct field observations of the position of glacier tongues were the earliest type of measurement to be taken and because it is relatively easy to map the past locations of glacier fronts on satellite images by identifying the Little Ice Age (LIA) moraines which usually mark the maximum extent of glaciers over the last millennium. There are numerous regional studies. For example, Stokes et al. (2006) mapped changes in glacier length in the Caucasus Mountains from their LIA extent at multiple intervals using Landsat and Corona imagery while Solomina et al. (2016a) used lichenometry to date terminal moraines and tree-ring climatic reconstructions to extend the record to the end of the 16th century. They attributed changes in glacier length to changes in temperature. Lichenomtery and radiocarbon dating of moraines were used to develop detailed chronologies of glacier fluctuations during the LIA in the tropical Andes (Rabatel et al., 2013). A global data set, containing over 470 lengths with the longest dating back to 1535, was developed by Leclercq et al. (2014). Selected records from this data set are shown in Fig. 4. Solomina et al. (2016b) developed a global data set detailing glacier length change over the last 2000 years. Several world-wide data sets were used in assessment of impacts of climate change on glaciers and to model glacier contribution to sea level rise (Oerlemans et al., 2007; Oerlemans, 2010; Leclercq et al., 2011). WGMS has a collection of glacier frontal variations containing about 40,000 observations of about 2000 glaciers extending back to the 16th century.

A global initiative Global Land Ice Monitoring from Space (GLIMS) was launched to monitor changes in glacier area worldwide using optical satellite imagery (Raup et al., 2007; Bhambri and Bolch, 2009; Kargel et al., 2014). A globally complete inventory of glacier outlines—the Randolph Glacier Inventory (RGI)—was produced (Pfeffer et al., 2014) and its latest version, released in 2017, includes 215,547 glaciers (RGI Consortium, 2017). Numerous regional studies report changes in glacier area from the 1980s using first Landsat and later ASTER imagery and extend assessments back in time by using aerial photography, Corona and Hexagon satellite imagery, and by mapping positions of glacial moraines using modern

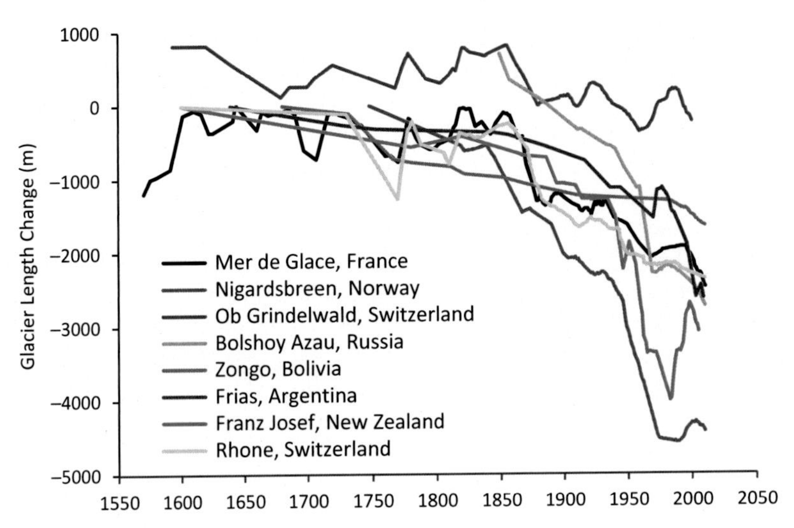


**FIG. 4** Examples of the long-term records of glacier length change (m) from different parts of the world. *Data are from Leclercq, P.W., Oerlemans, J., Basagic, H.J., Bushueva, I., Cook, A.J., Le Bris, R., 2014. A data set of worldwide glacier length fluctuations. Cryosphere 8, 659–672. doi:10.5194/tc-8-659-2014.*

satellite images. For example, Kutuzov and Shahgedanova (2009) and Shahgedanova et al. (2012) used these multiple sources to quantify changes in glacier area at multiple time steps in the Tien Shan and in the Polar Urals, respectively. Repeat terrestrial photographs taken from the same vantage point, at the same time of year but years apart are used to assess changes in individual glaciers providing important historical perspective. Repeated images of the MGU glacier in the Polar Urals are shown in Fig. 5. Excellent archives of repeated photographs are available from the National Snow and Ice Data Center (NSIDC) and WGMS.

Assessments of changes in glacier area over multiple time steps are useful because they provide information about trends through time (i.e., acceleration of glacier wastage which can be attributed to changes in climate), elevational dependencies, and enable comparisons between regions and different types of glaciers. When making these comparisons, however, one should keep in mind that absolute changes in glacier length and area depend on length and area itself and relative changes produce a more coherent global pattern. Accurate glacier outlines improve calculations of geodetic mass balance, constrain spaceborne gravimetric measurements and provide input to various glaciological and hydrological models. Information about area change is important in assessments of glacial hazards: glacial lakes (and potential GLOF) develop following glacier retreat and knowledge about glacier area and its change helps to model the development of existing and formation of future glacier lakes (e.g., Kapitsa et al., 2017).

## 4 Observed changes in the state of ice sheets and glaciers

Since the end of the 19th century glaciers and ice caps have been thinning and retreating worldwide in response to climatic warming (with exception of a small number of regions) and this trend intensified in the second half of the 20th century. Recently, several estimates of

**FIG. 5** Repeated ground-based photographs of the MGU glacier, Polar Urals. See Shahgedanova et al. (2012) for discussion of glacier change in this region. *Photo from 2008 is by G. Nosenko, Institute of Geography, Russian Academy of Science.*

global-scale glacier mass budget, based on different methods, were published (Church et al., 2011; Gardner et al., 2013; Bamber et al., 2018; Wouters et al., 2019; Zemp et al., 2019; IPCC, 2019). Both remote sensing, in situ measurements, and all indicators of glacier change confirm that there is a globally coherent pattern of negative mass budget and glacier recession (Zemp et al., 2015, 2019). Despite significant decadal and regional variations, the recent IPCC Special Report on the Ocean and Cryosphere in a Changing Climate (SROCC) stated that there is very high confidence in this conclusion (Hock et al., 2019).

Reliable assessments of changes in mass of the Antarctic ice sheet (AIS) and Greenland ice sheet (GrIS) became available in the satellite era when both were repeatedly measured using altimetry, interferometry, and gravimetry (Bamber et al., 2018). This is a very short period to account for response time and to separate signals of climate change from impacts of climatic variability (Wouters et al., 2013). Modeling of SMB and glacial isostatic adjustment, analyses of ice cores and examination of geological evidence are used to put the recent changes in perspective (Hanna et al., 2013). There have been more than 150 assessments of mass loss from the AIS and numerous assessments for the GrIS and inevitably, the estimates refer to different time periods, use different methods and are characterized by a large spread of values (Meredith et al., 2019). Numerous peripheral glaciers and ice caps surround both the AIS and the GrIS margins and changes in their mass balance can be considered together or separately from the ice sheets (see, e.g., Hock et al., 2019; Meredith et al., 2019). For Greenland, they account for a significant proportion of changes in ice mass (Bolch et al., 2013). Recently, independently derived estimates, based on the uniform assessment periods, were collated by the IMBIE project for the AIS (1992–2017) (The IMBIE Team, 2018) and the GrIS (1992–2018) (The IMBIE Team, 2020). These and other assessments are reviewed by Meredith et al. (2019).

## 4.1 Antarctic ice sheet

The AIS contains enough water to raise the global sea level by 58m if it melts (The IMBIE Team, 2018). Assessments of changes in mass of the AIS focus on three distinctly different regions: the East Antarctic ice sheet (EAIS), which covers about 85% of the total AIS area, the West Antarctic ice sheet (WAIS), and the Antarctic Peninsula (AP). The EAIS mostly rests on landmass although there are marine sectors with limited unstable areas. Its central part, the East Antarctic Plateau has an average elevation of about 3000m a.s.l. and is the coldest place of Earth where the absolute minimum temperature of −89°C was registered at the Vostok station. Rising air temperatures, therefore, do not have a direct effect on mass balance as they remain below freezing, but they lead to higher atmospheric moisture-holding capacity and precipitation. Due to the very low temperatures, precipitation is also extremely low over the Eastern Antarctic Plateau, which accounts for most of the EAIS, and its prevailing type is the so-called diamond dust which forms in the very cold atmosphere. Precipitation caused by weather systems is rare but an order of magnitude higher. Accumulation is very low, less than 50 mm $a^{-1}$, and exhibits strong interannual variability. An increase in precipitation and snow accumulation was observed since the beginning of the 20th century and attributed to the observed atmospheric warming, although a shorter-term negative trend observed since 1979 should be noted (Medley and Thomas, 2019). Longer precipitation and accumulation records, derived from ice cores, do not show spatially coherent trends over the EAIS (Thomas et al., 2017).

The WAIS is predominantly a marine ice sheet grounded below the sea level with the seabed deepening inland. Due to this configuration, the WAIS is considerably more unstable than the EAIS (Bamber et al., 2018). Similarly to the WAIS, the AP also has regions satisfying instability criteria (ibid.) but its geography and climatic conditions are distinctly different. Both the WAIS and particularly the AP are high-accumulation areas with annual precipitation around 500 mm $a^{-1}$ in the AP. Snow accumulation has increased since the beginning of the 20th century, following an earlier decline over the AP (Goodwin et al., 2016; Thomas et al., 2017; Medley and Thomas, 2019). Here, variability in precipitation and accumulation is linked to variability in atmospheric circulation associated with the El Nino-Southern Oscillation (ENSO) and the Pacific Decadal Oscillation (PDO) (Goodwin et al., 2016). The reduction in the extent of sea ice in the Bellingshausen Sea, observed in the 20th century, also plays an important part in increasing the availability of surface-level moisture which leads to an increase in accumulation (Thomas et al., 2017). Over the WAIS, changes in snow accumulation were balanced between the eastern and western sectors (Medley and Thomas, 2019).

Complex systems of tributary glaciers and ice streams transport ice from the interior regions to the ocean (Rignot et al., 2011). Ice shelves buttress 90% of the AIS outflow (Meredith et al., 2019). The unstable configuration around the WAIS and the AP enables rapid grounding line retreat, acceleration of ice flow, and mass loss to the ocean (Bamber et al., 2009; Wouters et al., 2015). These processes accelerate when the buttressing effect of ice shelves is weakened by their thinning. Ice shelves are vulnerable to atmospheric warming at their surface and to changing ocean temperatures at their bases. Acceleration of ice shelf thinning has been observed around the WAIS and the AP since the mid-1990s and particularly after 2003 (Mouginot et al., 2014; Pritchard et al., 2012; Paolo et al., 2015). In the Amundsen and Bellingshausen regions, some ice shelves have lost up to 18% of their thickness in less than two decades. At the same time, the total volume of the EAIS ice shelves increased followed by a moderate loss after 2003 (Paolo et al., 2015).

Therefore, climatic warming may increase snowfall and accumulation, especially in the continental interiors, but enhance glacier discharge at the coast where rising air and ocean temperatures erode the buttressing ice shelves. Assessments agree that since 1992, the EAIS (whose area is larger by an order of magnitude than the WAIS) has been close to balance or slightly gaining mass because of increasing precipitation while the smaller WAIS and AP have been losing mass through an increase in ice flow (dynamic thinning) at a greater rate (Bamber et al., 2018; Meredith et al., 2019; The IMBIE Team, 2018). The EAIS mass balance was estimated as $5 \pm 46\,\text{Gt}\,\text{a}^{-1}$ in the 1992–2017 period by The IMBIE Team (2018) (Table 1). Other studies agree that there was no clear trend in the EAIS mass balance over the period of satellite records, and stress interannual variability and very large uncertainties in measurements, which exceed the central estimate. The flow of glaciers, draining the EAIS, has remained largely constant (Gardner et al., 2018).

All assessments agree that ice loss from the AIS was dominated by acceleration and rapid dynamic thinning of the WAIS outlet glaciers driven by the melting of ice shelves and the loss of the buttressing effect (Mouginot et al., 2014; Meredith et al., 2019). The WAIS ice loss increased from $53 \pm 29$ to $159 \pm 26\,\text{Gt}\,\text{a}^{-1}$ between the first and final 5 years of the 1992–2017 period with an average rate of $-94 \pm 27\,\text{Gt}\,\text{a}^{-1}$ (The IMBIE Team, 2018) (Table 1). The strongest loss occurred in the Amundsen Sea sector where acceleration of glacier flow resulted in a 77% increase in ice discharge since the 1970s (Mouginot et al., 2014). In the AP, the average rate of

**TABLE 1** Mass balance of the Antarctic ice sheets ($\text{Gt}\,\text{a}^{-1}$).

| Region | 1992–1997 | 1997–2002 | 2002–2007 | 2007–2012 | 2012–2017 | 1992–2017 |
|---|---|---|---|---|---|---|
| EAIS | $11 \pm 58$ | $8 \pm 56$ | $12 \pm 43$ | $23 \pm 38$ | $-28 \pm 30$ | $5 \pm 46$ |
| WAIS | $-53 \pm 29$ | $-41 \pm 28$ | $-65 \pm 27$ | $-148 \pm 27$ | $-159 \pm 26$ | $-94 \pm 27$ |
| AP | $-7 \pm 13$ | $-6 \pm 13$ | $-20 \pm 15$ | $-35 \pm 17$ | $-33 \pm 16$ | $-20 \pm 15$ |
| AIS | $-49 \pm 67$ | $-38 \pm 64$ | $-73 \pm 53$ | $-160 \pm 50$ | $-219 \pm 43$ | $-109 \pm 56$ |

*Data are from The IMBIE Team, 2018. Mass balance of the Antarctic ice sheet from 1992 to 2017. Nature 558, 219–222. https://doi.org/10.1038/s41586-018-0179-y. See Bamber, J.L., Westaway, R.M., Marzeion, B., Wouters, B., 2018. The land ice contribution to sea level during the satellite era. Environ. Res. Lett. 13, 063008. https://doi.org/10.1088/1748-9326/aac2f0 and Meredith, M., Sommerkorn, M., Cassotta, S., Derksen, C., Ekaykin, A., Hollowed, A., Kofinas, G., Mackintosh, A., Melbourne-Thomas, J., Muelbert, M.M.C., Ottersen, G., Pritchard, H., Schuur, E.A.G., 2019. Polar Regions. In: Pörtner, H.-O., Roberts, D.C., Masson-Delmotte, V., Zhai, P., Tignor, M., Poloczanska, E., Mintenbeck, K., Alegría, A., Nicolai, M., Okem, A., Petzold, J., Rama, B., Weyer, N.M. (Eds.), IPCC Special Report on the Ocean and Cryosphere in a Changing Climate. IPCC, Geneva for a wider range of assessments.*

ice loss was 20±15 Gt $a^{-1}$ with an increase of about 15 Gt $a^{-1}$ since 2000. Of 860 marine-terminating glaciers located in the AP, 90% retreated from their 1940s positions (Cook et al., 2014) confirming that negative trend in mass balance extends beyond the satellite record. Dynamic thinning in the Amundsen Sea Embayment and the western AP accounted for 88% of the increase in the AIS mass loss from 2008 to 2015 (Gardner et al., 2018). These losses were partially offset by changes in precipitation and accumulation. However, mass loss from the AIS observed between 1992 and 2017, amounting to 2720±1390 Gt of ice (The IMBIE Team, 2018), accounted for 70% of the century-long snow accumulation gain of 3815±1105 Gt (Medley and Thomas, 2019).

## 4.2 Greenland ice sheet

The GrIS occupies about $1.71 \times 10^6$ $km^2$ and if its total volume of $2850 \times 10^3$ $km^3$ is to melt, the GMSL will rise by 7.4 m (The IMBIE Team, 2020). Here, changes in air temperature directly control melt (which extended to the summit in 2012), and, together with variability in snow accumulation, SMB. The relative contributions of SMB and D vary in time and spatially (Meredith et al., 2019; Mouginot et al., 2019; The IMBIE Team, 2020).

The observed air temperature increase in the Arctic and sub-Arctic regions, amplified by the reduced sea-ice coverage (the Arctic or Polar Amplification), is among the largest in the world (Serreze and Barry, 2011). Analysis of ice cores suggested that the 2000–10 decade was the warmest in about 2000 years (Vinther et al., 2009; Masson-Delmotte et al., 2012). However, quantifying its impacts on the evolution of the GrIS mass balance on a centennial time scale and longer remains challenging because of data availability and quality (Khan et al., 2015; van den Broeke et al., 2017). Our understanding of changes in the GrIS mass, elevation, and area (including peripheral glaciers) improved when spaceborne (Section 3.2) and airborne (Koenig et al., 2016; Lewis et al., 2017) data became available in the 1990s. Airborne surveys in particular helped to resolve large discrepancies in surface accumulation data. The availability of meteorological data used in modeling and mass balance reconstructions improved from the 1950s with emergence of the meteorological reanalysis data sets. Climatic and mass balance reconstructions for earlier years relied on data from coastal stations and ice cores used in combination with regional climate and mass balance models. Observational data were too sparse to capture spatial variability in temperature and especially in precipitation in high-accumulation areas leading to uncertainties which obscured climatic signals. Aerial photography and historical maps were used to reconstruct changes in glacier area and surface elevation. Kjeldsen et al. (2015) used aerial imagery from the 1980s to map moraines and reconstruct the maximum extent of the GrIS at the end of the 19th century and build its only historical DEM.

Studies of the 19th–early 20th century changes in the GrIS climate and mass balance agree on the direction of the general trend but discrepancies between absolute values and decadal anomalies are large before the 1950s. Thus Box (2013) estimated that summer temperature increased by 1.6°C over Greenland between 1840 and 2010. Temperature reconstruction by Hanna et al. (2011) for 1870–2010 suggested a smaller increase because in their record, temperatures prior to the 1940s were considerably higher. Both records were in agreement from the 1950s confirming that temperatures in the 1990–2000s were the highest over the GrIS since records began. Box (2013) suggested that accumulation increased over the GrIS from the mid-20th century in comparison with the 1840–1940 period and reconstruction by Fettweis et al. (2017) for 1900–2015 agrees. Hanna et al. (2011) reported an opposite trend. Both, Hanna et al. (2011) and Box (2013) showed a negative anomaly in SMB in the 1960s attributing it to low accumulation. These and other studies agree that the GrIS gained mass in the 1970–80s and overall, the 1960–90s was a period of relative stability. van den Broeke et al. (2016, 2017) calculated SMB as $400 \pm 70\,\text{Gt}\,\text{a}^{-1}$ in 1961–90 and the 1996 value of ice discharge as $410 \pm 20\,\text{Gt}\,\text{a}^{-1}$ showing that ice flow adjusted to mass balance forcing and that the GrIS had near zero mass balance within the bounds of uncertainty.

In the early 2000s, the GrIS experienced a shift to negative annual and cumulative mass balance (Hanna et al., 2013; Khan et al., 2015; Box et al., 2018; Bamber et al., 2018; Mouginot et al., 2019). Kjeldsen et al. (2015) estimated mass loss from the GrIS as $75.1 \pm 29.4\,\text{Gt}\,\text{a}^{-1}$ in the 1900–83 period, $73.8 \pm 40.5\,\text{Gt}\,\text{a}^{-1}$ in the 1983–2003 period, and $186.4 \pm 18.9\,\text{Gt}\,\text{a}^{-1}$ between 2003 and 2010.

Satellite and mass balance modeling data showed a strongly negative trend in SMB between 1992 and 2018 (Table 2), with a strong increase in melt and runoff (van den Broeke et al., 2016; Bamber et al., 2018; The IMBIE Team, 2020) in line with extensive surface warming (Hall et al., 2013; McGrath et al., 2013; Orsi et al., 2017), aided by an increasing fraction of liquid precipitation (Box et al., 2018). Box (2013) estimated that meltwater production increased by 59% between 1840 and 2010. In the 1994–2013 period, runoff from the GrIS occurred at a level unprecedented over at least 350 years, being 33% and 50% higher than the 20th century and the 18th century means, respectively (Trusel et al., 2018).

After 2002, mass balance became significantly more negative than in the previous years including both, SMB and dynamic discharge (Bamber et al., 2018; The IMBIE Team, 2020) (Table 2). The IMBIE Team (2020) estimated that SMB (and indirect firn processes) and dynamic discharge contributed to mass loss in approximately equal measures between 1992 and 2018 while in earlier periods, about 60% of the mass loss was through ice discharge and 40% was from surface ablation (van den Broeke et al., 2016; Bamber et al., 2018; Mouginot et al., 2019). Due to a very strong surface melt, observed in the first decade of the 21st century and peaking in 2012, the input of these processes in recent imbalance is the other way round: change in surface melt was mainly responsible for the observed imbalance (Bamber et al., 2018). Ice discharge also increased following the acceleration of large outlet glaciers in the

**TABLE 2** Mass balance of the Greenland Ice Sheet ($\text{Gt}\,\text{a}^{-1}$).

| Component | 1992–1997 | 1997–2002 | 2002–2007 | 2007–2012 | 2012–2017 | 1992–2018 |
|---|---|---|---|---|---|---|
| SMB | $26 \pm 35$ | $-15 \pm 36$ | $-78 \pm 36$ | $-193 \pm 37$ | $-139 \pm 38$ | $-76 \pm 16$ |
| Dynamic | $-52 \pm 44$ | $-29 \pm 50$ | $-96 \pm 47$ | $-82 \pm 46$ | $-105 \pm 47$ | $-75 \pm 21$ |
| Total | $-26 \pm 27$ | $-44 \pm 35$ | $-174 \pm 30$ | $-275 \pm 28$ | $-244 \pm 28$ | $-150 \pm 13$ |

*Data are from The IMBIE Team, 2020. Mass balance of the Greenland ice sheet from 1992 to 2018. Nature 579, 233–239. https://doi.org/10.1038/s41586-019-1855-2. See Bamber, J.L., Westaway, R.M., Marzeion, B., Wouters, B., 2018. The land ice contribution to sea level during the satellite era. Environ. Res. Lett. 13, 063008. https://doi.org/10.1088/1748-9326/aac2f0 and Meredith, M., Sommerkorn, M., Cassotta, S., Derksen, C., Ekaykin, A., Hollowed, A., Kofinas, G., Mackintosh, A., Melbourne-Thomas, J., Muelbert, M.M.C., Ottersen, G., Pritchard, H., Schuur, E.A.G., 2019. Polar Regions. In: Poörtner, H.-O., Roberts, D.C., Masson-Delmotte, V., Zhai, P., Tignor, M., Poloczanska, E., Mintenbeck, K., Alegría, A., Nicolai, M., Okem, A., Petzold, J., Rama, B., Weyer, N.M. (Eds.), IPCC Special Report on the Ocean and Cryosphere in a Changing Climate. IPCC, Geneva for a wider range of assessments.*

southeast and northwest and ubiquitous retreat of glacier calving fronts at an increasing rate (King et al., 2020). While only several large glaciers accounted for half of the dynamically driven ice loss from the GrIS in the 2000s, many more contribute now (Mouginot et al., 2019). Overall, between 1992 and 2018, the GrIS lost $1964 \pm 565$ and $1938 \pm 541$ Gt of ice because of meteorological processes and due to the dynamic imbalance of glaciers, respectively (The IMBIE Team, 2020). Both atmospheric and ocean forcing control dynamic mass loss from the GrIS, and the declining suppression effect of glacier calving by sea ice is important too (O'Leary and Christoffersen, 2013; Khan et al., 2015; Meredith et al., 2019).

While the observed increase in mass loss is attributed to rising temperatures, attribution of the observed atmospheric warming is less certain. Natural climatic variability plays an important part in controlling the GrIS mass balance. Thus, in the first decade of the 21st century, frequent occurrence of the negative phase of the North Atlantic Oscillation (NAO) led to enhanced advection of warm air masses, high pressure and increased insolation, and a reduction in solid precipitation over the GrIS in summer (Masson-Delmotte et al., 2012; Fettweis et al., 2013; Bevis et al., 2019). About 70% of the 1993–2012 summer warming is attributed to the NAO (Fettweis et al., 2013). These factors enhanced melt leading to more negative SMB particularly along the south-western margin of the GrIS which has greater area of lower-elevation ice surface per unit length, gentler slope, and where surface melting occurs much farther inland (Khan et al., 2015). Record melt was registered in 2012 when it extended to the summit of the GrIS. However, the negative trend in SMB slowed down between 2013 and 2018 in response to changing NAO phase which brought about lower summer temperatures and higher precipitation (Bevis et al., 2019).

Various positive feedbacks amplify melt at the GrIS. A very important feedback is linked to the darkening of the surface of the ice sheet. More intensive summer melt is further amplified by an increase in the spatial extent and duration of wet snow and areas of bare ice which have lower albedo (Box et al., 2018; Ryan et al., 2019). Algal growth and decay and, to a lesser extent, deposition of dust from the ice sheet periphery and black carbon, transported from industrial regions, on snow and ice reduce albedo, increasing melt further (van den Broeke et al., 2017). Only about a half of surface melt runs off the GrIS. A considerable portion is stored in snow and firn aquifers which spread to higher elevations as air temperature increases. The potential of these aquifers to buffer runoff is reduced by firn densification and formation of ice layers which prevent drainage and promote the formation of water ponds on the surface. This also reduces albedo enhancing melt further (Steger et al., 2017).

The GrIS appears to be significantly out of equilibrium (van den Broeke et al., 2016, 2017) and the observed increase in ice discharge alone is sufficient for transition to a new state of persistent negative mass balance (King et al., 2020). There is a considerable difference between a very long time of 4200 years, which would be required for the GrIS mass balance to adjust to a hypothetic increase in accumulation under the warmer climate, and mass loss through surface melt and calving, which occurs on the time scale of decades, significantly outpacing potential gains (van den Broeke et al., 2016). When the ice sheet has lost its contact with the ocean ($D = 0$) and SMB becomes persistently negative resulting in negative total mass balance, the GrIS has reached a "tipping point" beyond which it will not be able to recover. At the current rate of ice loss, the tipping point could be reached by the mid-21st century although in the absence of reliable long-term data uncertainty surrounding this scenario is high (van den Broeke et al., 2016).

## 4.3 Glaciers and ice caps

A period of glacier advance, which started with the onset of the LIA in the 15th century and peaked in the 17th century, ended at the turn of the 20th century (Solomina et al., 2016b). In the 20th century and particularly in its second half, and in the 21st century the overwhelming pattern of glacier behavior was that of loss of mass, retreat, and thinning in response to climatic warming. The combined glaciological and geodetic measurements showed that globally, glacier mass change amounted to $-9625 \pm 7975$ Gt during the 1961–2016 period (Zemp et al., 2019). Increase in air temperature is primarily responsible for the observed glacier wastage (Hock et al., 2019). However, there were considerable spatial and temporal variations in the rates of glacier mass loss (and much more rarely, gain) (Winkler et al., 2010) resulting from multiple and often interdependent factors creating complex patterns of glacier response to climate change.

### *4.3.1 Factors affecting regional variability in glacier behavior and their response to climate change*

The geography of glaciers is varied—from the poles to the tropics—and geographical location affects vulnerability of glaciers to climate change. Geographical location controls exposure to positive temperatures and solar radiation. Much of the Antarctic is not exposed to positive temperatures and ice loss is not directly affected by its increase. By contrast, glaciers in warmer regions are vulnerable to even a small temperature increase. Examples include the glaciers of southern Europe, for example, the Pyrenees where glaciers remain only on the highest peaks (González Trueba et al., 2008), tropical Andes (Vuille et al., 2008; Rabatel et al., 2013), and the Papua province of Indonesia where the last remaining glaciers in the West Pacific were melting rapidly since the 1960s in response to rising temperatures augmented by ENSO events (Permana et al., 2019). Geographical location controls amount and seasonality of precipitation. The former is higher in maritime regions where it can compensate for glacier melt. The latter predetermines whether glaciers are of winter or summer accumulation type. For example, glaciers in the European Alps, Scandinavia, and the Caucasus lose mass in summer and accumulate snow between September and May. On the glaciers of much of the Tien Shan, Pamir, and in the monsoonal regions of the Himalayas precipitation peaks in summer at the same time with melt (summer accumulation type). Fraction of liquid precipitation, which depends on temperature, and interruptions of melt by solid precipitation are important factors controlling mass balance of these glaciers. For these reasons, glaciers affected by monsoon may be more vulnerable to climatic warming than the winter-accumulation glaciers (Fujita, 2008). The Arctic glaciers are affected too because their ablation season is short and summer snowfalls can reduce melt very significantly.

Elevation controls temperature and precipitation regimes, and their relation to glacier area. Glaciers positioned at higher elevations have a larger AAR which sustains accumulation and transfer of ice downstream. Such glaciers are able to dynamically respond to climate change by retreating to higher elevations and, therefore, they are less vulnerable. By contrast, glaciers located at lower elevations have lost large parts of their accumulation area with ELA approaching the uppermost glacier elevation. Difference between values of cumulative mass balance of glaciers with higher and lower maximum elevations was clearly demonstrated by the region-wide mass balance assessment in the European Alps (Davaze et al., 2020).

For example, the Careser glacier in the Italian Alps (one of the WGMS reference glaciers) has lost its accumulation zone and is in the state of decay, which is intensified by decreasing surface albedo and increased input of heat from the exposed rocks (Carturan et al., 2013). The same is observed on many lower-elevation glaciers in other regions of the European Alps (Paul et al., 2007) and in the lower-elevation mountain ranges located in very different climates, for example, in the Kodar Mountains in north-eastern Siberia (Shahgedanova et al., 2011; Stokes et al., 2013).

Since size and hypsometry of glaciers (and, consequently, glacier type) control their response time and sensitivity to climatic perturbations, it is always important to take into account the predominant glacier parameters when comparing rates of glacier change in different regions. Due to higher sensitivity and shorter response time, small glaciers currently often account for larger proportions of regional area and mass loss. For example, in the Swiss and Lombardy Alps, in the last two decades of the 20th century glaciers with individual areas less than 1 $km^2$, accounting for only 15% and 30% of the glacierized area, contributed 40% and 58% of total area loss, respectively (Paul et al., 2004; Citterio et al., 2007). This is also why some smaller, steeper, fast-response glaciers in the Southern Alps of New Zealand, such as the Franz Josef glacier, advanced slightly at the beginning of the 21st century in response to decadal-scale climatic variability (Fig. 4) while large glaciers did not (Winkler et al., 2010; Chinn et al., 2005). This general pattern, however, can be disrupted by other factors. For example, due to local topography, small cirque glaciers had lower wastage rates than larger glaciers in the Polar Urals because they were shaded by cirque walls while larger valley glaciers were exposed to solar radiation (Shahgedanova et al., 2012).

Presence of debris cover on glacier surface affects melt rates. A very thin debris layer (up to about 2 cm) enhances glacier melt by reducing the albedo of the surface, however, thicker debris insulate glacier surface and reduce melt (Benn et al., 2012). A thick debris cover can alter the ablation gradient pushing higher melt rates from the well-insulated glacier tongues to the middle of the ablation zone (ibid.). About 7% of the total area of mountain glaciers is covered by debris. It is present on 44% of Earth's glaciers and prominent on 15% and, importantly, it is most abundant at lower elevations where melt rates are higher (Herreid and Pellicciotti, 2020). Fig. 6 shows an example of a debris-covered glacier. Over half of supra-glacial debris cover is concentrated in Alaska and in the Karakorum-Hindukush region. It is widespread in the Himalayas and in Central Asia. The presence of debris cover on the Himalayan glaciers contributes to their inhomogeneous response to climate change (Bolch et al., 2012). A comparison of retreat rates of the debris-free and debris-covered monsoon-influenced glaciers in the Himalayas showed that 65% of the former were retreating while the latter had stable fronts (Scherler et al., 2011). There are numerous studies showing lower retreat rates of the debris-covered glaciers in different parts of the world (e.g., Winkler et al., 2010; Chinn et al., 2005). However, it should be noted that downwasting rather than retreat may be a predominant way in which the debris-covered glaciers respond to climatic warming and this signal and a difference to the response of the debris-free glaciers are not always detected especially over short periods of time (Kääb et al., 2012). Another complicating factor is the formation of supra-glacial and moraine-dammed lakes on the debris-covered glaciers which, once formed, increase melt (Benn et al., 2012). At present, the mechanisms through which debris cover affects glacier melt are known but its contribution to glacier response is not quantified sufficiently well (Pellicciotti et al., 2015).

**FIG. 6** The Shkhelda glacier in the northern Caucasus has extensive debris cover.

The LAI—black carbon (BC) and desert dust—enhance glacier melt. BC, a product of incomplete combustion in industrial processes, wildfires, and domestic biofuel use, is by far the most efficient type of aerosol absorbing solar radiation, changing albedo of glacier surface, and warming the atmosphere (Flanner et al., 2007; Bond et al., 2013; Khan et al., 2017). The Fifth IPCC report estimated the at-surface forcing of BC as +0.04 (0.02–0.09) $W m^{-2}$. BC can be transported over long distances and deposited as far as on the Antarctic glaciers, in the Arctic and Alaska, and was detected on the surface of glaciers and in ice cores in the Himalayas and Tibet, Andes, Rockies and European Alps. In the polar regions of the Northern Hemisphere, the highest concentrations are found in the Russian Arctic, particularly in its western sector where there are more settlements and large industrial facilities, and also in the Canadian Arctic and Alaska (Doherty et al., 2010; Dou and Xiao, 2016). Among the mountain regions, the Himalayas and Tibetan plateau stand out because of the widespread use of coal by power stations in India, China, and Iran and a widespread use of biofuel by households in India and Nepal (Menon et al., 2010). Deposition of BC on glaciers peaks in winter and spring. The latter is climatologically important because it can cause earlier onset and more intensive snow melt on glacier surface (Bond et al., 2013). Observations in the Khumbu Valley in Nepal showed that snow melt can start three weeks earlier if concentrations of BC on snow reach 300 ppb in comparison with clean snow (Jacobi et al., 2015). Moreover, the spring maximum in deposition of BC coincides with a peak in dust storms in the Middle East (Hennen et al., 2019) and Central Asia

(Nobakht et al., 2019) and deposition of desert dust on glaciers. Although the effect of dust on snow melt is substantially lower than that of BC, large dust storms events can generate strong radiative forcing and enhance melt (Painter et al., 2012; Gautam et al., 2013). In Fig. 1, mineral dust deposited on the Djankuat glacier can be recognized by its characteristic brownish color. Desert dust serves as a vector for transportation of various substances including BC. Therefore, a combined effect of BC and dust and their interactions with other impurities on glacier surface (cryoconite, algae), changes in snow grain size and, importantly, presence of debris cover (which can negate their impacts) should be considered. These processes are complex. As with debris cover, it is possible to quantify their effects on melt of a glacier but it is difficult to assess their contribution to glacier wastage overall. What is known is that deposition of BC and dust explains some of the variability in glacier change in the Arctic and in the Himalayas. It can be expected to increase if wildfires and dust storms become more frequent with climate change.

Whether glaciers terminate on land or in water affects their dynamics and rates of change. The formation of lakes at glacier tongues usually results in an acceleration of their retreat. For example, Yang et al. (2020) showed that in the Kenai Peninsula in Alaska, glaciers terminating in lakes lost about 30% more mass than glaciers terminating on land (and also nearly three times as much as tidewater glaciers) between 1986 and 2016. However, tidewater glaciers terminating in the ocean exhibit much more complex behavior which is not controlled by climate change alone. Tidewater glaciers go through a cycle whereby slow advance, which can continue for many centuries, is followed by stable position of glacier tongues and by rapid retreat which can occur within a century or less or even at surging speed (Cuffey and Paterson, 2010). The Columbia Glacier in south-eastern Alaska which flows from the elevation of 3050 m a.s.l. to the inlet leading to the Prince William Sound is an example. The glacier advanced between the 15th and 19th century but started retreating in the 1980s losing about 1 km of its length per year (Pfeffer, 2003). This does not imply that changes in the extent of tidewater glaciers are not controlled or triggered by changes in climate. Indeed, a consistent pattern of retreat of outlet glaciers was observed across the entire Atlantic sector of the Arctic from Greenland to Novaya Zemlya between 1992 and 2010 (Carr et al., 2017). However, tidewater glaciers respond to the climate-driven changes in ocean heat advection, sea ice conditions, atmospheric warming, and changes in precipitation in a complex way. Even within the same region they can be out of sync with land-terminating and other tidewater glaciers, and exhibit more varied responses than land-terminating glaciers. Carr et al. (2014) analyzed changes in the position of glacier tongues in the north of Novaya Zemlya between 1992 and 2010 and found that while 90% of all studied glaciers retreated, the observed changes could not be directly attributed to atmospheric warming; correlations with sea ice conditions and SST were significant but not strong; and fjord geometry (a nonclimatic factor) was the most important control over variability between retreat rates of individual glaciers. These conclusions were re-iterated in a wider study of the Atlantic Arctic (Carr et al., 2017). McNabb and Hock (2014) examined changes in the positions of 50 tidewater glacier termini between 1948 and 2012 in Alaska and found that while 60% of the glaciers retreated overall, there was no coherent pattern of their retreats and advances and it was not clear whether retreats were triggered by increasing SST. Cook et al. (2019) investigated changes in more than 300 marine-terminating glaciers between 1958 and 2015 in the Canadian Arctic Archipelago, reporting gradual retreat before 2000 and a fivefold increase in retreat rates up to 2015. The latter was attributed to atmospheric warming driving retreat in the same way as that of land-terminating glaciers, in contrast to other regions where the delivery of ocean heat was a stronger control over the retreat of marine-terminating glaciers.

Last but not least, interannual variability in atmospheric circulation is an important control over short-term glacier fluctuations which can enhance or temporarily obscure the long-term change. Decadal variability in mass balance of glaciers in the Andes, particularly in Bolivia, Ecuador, Columbia and to a lesser extent in Peru, is governed by SST variability in the Pacific. Here glaciers respond strongly to the ENSO cycle with El Niño years featuring a strongly negative mass balance and La Niña events producing a nearly balanced or even slightly positive mass balance (Vuille et al., 2008; Rabatel et al., 2013). The NAO controls variability in glacier mass balance in western Scandinavia. Rapid glacier advances during the positive NAO phases, which result in abundant winter precipitation in this region, were recorded throughout the LIA (Nesje and Dahl, 2003) and in the 20th century, especially in the 1990s when maritime glaciers in Norway experienced a mass gain and advanced (Nesje et al., 2000; Chinn et al., 2005; Paul and Andreassen, 2009). In the Southern Alps of New Zealand, glaciers with short response time advanced between the 1980s and 2000 (Fig. 4) in response to change in the Interdecadal Pacific Oscillation which occurred in 1976/77 and continued until 1998 (Chinn et al., 2005).

### 4.3.2 Changes in mass and extent of glaciers

Despite regional variations and exceptions, there is a coherent global pattern of glacier retreat. Glacier length records from all continents and almost all latitudes show that glaciers receded worldwide since the start of the 20th century (Leclercq et al., 2014) (Fig. 4). Numerous regional analyses of glacier area change and their compilations show glacier shrinkage especially since the second half of the 20th century (Pfeffer et al., 2014; Kargel et al., 2014). Glacier mass change reconstructions for the 20th century also agree on glacier mass loss rates (Leclercq et al., 2011; Marzeion et al., 2015). Zemp et al. (2019) estimated an increase in mean global glacier mass loss by approximately 30% between 1986 and 2005 and 2006 and 2015.

The best data are available for 2006–15 and Table 3 summarizes glacier mass balance of mountain and polar glaciers for this common assessment period (Hock et al., 2019; Meredith et al., 2019). Due to the large ice extent, the highest glacier mass loss occurred in Alaska, the Greenland periphery and the Canadian Arctic and in the mountain regions, in the Southern Andes (including the Patagonian ice fields) and the HMA. A comparison of the specific mass balance (loss per unit area) shows that they were comparable in the Arctic and in the mountains in 2006–15 at $-500 \pm 70\,\mathrm{kg\,m^{-2}\,a^{-1}}$ and $-490 \pm 100\,\mathrm{kg\,m^{-2}\,a^{-1}}$, respectively (Hock et al., 2019). In the Arctic, glaciers of the southern sector of the Canadian Arctic had the most negative specific mass balance at $-800 \pm 220\,\mathrm{kg\,m^{-2}\,a^{-1}}$ followed by Alaska, while in Svalbard it was $-270 \pm 170\,\mathrm{kg\,m^{-2}\,a^{-1}}$ (Hock et al., 2019). Among the mountain regions, the most negative averages (less than $-850\,\mathrm{kg\,m^{-2}\,a^{-1}}$) were observed in the Southern Andes, Caucasus, European Alps, and Pyrenees. The European Alps were losing mass at a rate of $-910 \pm 70\,\mathrm{kg\,m^{-2}\,a^{-1}}$ in 2006–15 (Table 3) and this trend is a part of longer, widespread glacier wastage across the entire European Alps (Sommer et al., 2020) (Fig. 4). The least negative mass budget characterized the High Mountain Asia where it averaged $-140 \pm 110\,\mathrm{kg\,m^{-2}\,a^{-1}}$.

On a multidecadal time scale, the largest cumulative mass loss was observed from Alaska accounting for about a third of the total change of $-8305 \pm 5115$ Gt from all glaciers, excluding the peripheries of Antractica and Greenland between 1961 and 2016 (Zemp et al., 2019). All other assessments agree that Alaska is a hotspot of glacier change (Jacob et al., 2012;

**TABLE 3** Regional estimates of glacier mass budget in the common assessment period of 2006–15 by the IPCC Special Report on Ocean and Cryosphere in a Changing Climate (SROCC) (Hock et al., 2019). Regional glacier areas and volumes are from RGI Consortium (2017) and Farinotti et al. (2019). Regions of assessment are as in RGI Consortium (2017). "Arctic Total" values include Alaska, Iceland, and Scandinavia which are also included in "Mountains Total".

| Mass budget | Area | Volume | Mass budget | | |
|---|---|---|---|---|---|
| Region | $km^2$ | mm SLE | $Gt a^{-1}$ | $kg m^{-2} a^{-1}$ | $mm SLE a^{-1}$ |
| Antarctic periphery | 132,867 | 69.4±18 | −11±108 | −90±860 | 0.03±0.3 |
| Greenland periphery | 89,717 | 33.6±8.7 | −47±16 | −570±200 | 0.13±0.04 |
| Arctic Canada North | 105,111 | 64.8±16.8 | −39±8 | −380±80 | 0.11±0.02 |
| Arctic Canada South | 40,888 | 20.5±5.3 | −33±9 | −800±220 | 0.09±0.03 |
| Western Canada and USA | 14,524 | 2. 6±0.7 | −8±13 | −500±910 | 0.02±0.04 |
| Alaska | 86,725 | 43.3±11.2 | −60±16 | −700±180 | 0.17±0.04 |
| Iceland | 11,060 | 9.1±2.4 | −7±3 | −690±260 | 0.02±0.01 |
| Svalbard | 33,959 | 17.3±4.5 | −9±5 | −270±170 | 0.02±0.01 |
| Russian Arctic | 51,592 | 32.0±8.3 | −15±12 | −300±270 | 0.04±0.03 |
| High Mountain Asia | 97,605 | 16.9±2.7 | −14±11 | −150±110 | 0.04±0.03 |
| Southern Andes | 29,429 | 12.8±3.3 | −25±4 | −860±170 | 0.07±0.01 |
| Central Europe | 2092 | 0.3±0.1 | −2±0 | −910±70 | 0.01±0.00 |
| Scandinavia | 2949 | 0.7±0.2 | −2±1 | −660±270 | 0.01±0.00 |
| Caucasus and Middle East | 1307 | 0.2±0.0 | −1±1 | −880±570 | 0.00±0.00 |
| Northern Asia | 2410 | 0.3±0.1 | −1±1 | −400±310 | 0.00±0.00 |
| New Zealand | 1162 | 0.2±0.0 | −1±1 | −590±1140 | 0.00±0.00 |
| Low latitudes | 23,41 | 0.2±0.1 | −1±1 | −590±580 | 0.00±0.00 |
| Arctic total | 422,000 | 221±25 | −213±29 | −500±70 | 0.59±0.08 |
| Mountains total | 251,604 | 87±15 | −123±24 | −490±100 | 0.34±0.07 |
| Global total | 705,739 | 324±84 | −278±113 | −390±160 | 0.77±0.31 |

*Data are from Hock, R., Rasul, G., Adler, C., Cáceres, B., Gruber, S., Hirabayashi, Y., Jackson, M., Kääb, A., Kang, S., Kutuzov, S., Milner, A., Molau, U., Morin, S., Orlove, B., Steltzer, H., 2019. High Mountain Areas. In: Poörtner, H.-O., Roberts, D.C., Masson-Delmotte, V., Zhai, P., Tignor, M., Poloczanska, E., Mintenbeck, K., Alegría, A.,Nicolai, M., Okem, A., Petzold, J., Rama, B., Weyer, N.M. (Eds.), IPCC Special Report on the Ocean and Cryosphere in a Changing Climate. IPCC, Geneva. Also see the reference for other assessments.*

Gardner et al., 2013; Box et al., 2018). Most of the ice loss occurred in the southern coastal part of the region and surface melt strongly dominated over tidewater glacier retreat (Larsen et al., 2015). The largest reduction in specific cumulative mass balance was observed in the Southern Andes resulting in total mass loss of 1200 Gt, which is less than in Alaska where the total glacierized area is three times higher (Table 3). Other assessments of glacier change in the Southern Andes show that the loss of glacier area has continued since the LIA with the Northern and Southern Patagonian Icefields losing over 100 $km^2$ and 500 $km^2$ of glacierized area since their maximum extent in 1870 and 1650, respectively (Glasser et al., 2011). The centennial rates of mass loss were an order of magnitude lower than in the second half of the 20th century.

In the HMA, very strong spatial variations in the response of glaciers to climate change were observed (Azam et al., 2018). Brun et al. (2017) reported contrasting regional trends in 2000–16 varying from the strongly negative mass balance in Nyainqentanglha, Tibet ($-620 \pm 230\,kg\,m^{-2}\,a^{-1}$) through moderately negative in the Himalayas (ranging from $-420 \pm 200\,kg\,m^{-2}\,a^{-1}$ in Bhutan to $-330 \pm 200\,kg\,m^{-2}\,a^{-1}$ in eastern Nepal) to slightly positive in the Kunlun Mountains ($+140 \pm 80\,kg\,m^{-2}\,a^{-1}$). Glaciers in the western and central Pamir exhibited low mass losses and in the eastern Pamir they had balanced budgets (Brun et al., 2017; Bolch et al., 2019). Detected originally in the Karakoram, where regional mass budgets were nearly balanced since the 1970s (Azam et al., 2018), this unusual pattern was termed the Karakoram Anomaly although at present the highest increases in ice mass are registered over the eastern part of the Karakoram and the western Kunlun. Various explanations were suggested ranging from the widespread occurrence of surging glaciers and extensive debris cover to climatic controls and even anthropogenic influence (Azam et al., 2018; Farinotti et al., 2020). Archer and Fowler (2004) identified negative trends in summer and annual temperatures, and an increase in summer, winter and annual precipitation over the Karakoram in 1961–2000 while Treydte et al. (2006) showed that the 20th century was the wettest in the last millennium. These changes are consistent with increase in mass balance. Attributions of the observed trends in precipitation include the impact of the NAO, the strengthening of the westerly jet stream, the main mechanism of moisture delivery (Archer and Caldeira, 2008; Cannon et al., 2014; Norris et al., 2019), and interactions with the Karakoram/West Tibet Vortex (Forsythe et al., 2017; Li et al., 2018) while regional cooling in the 1960–80 period appears to be influenced by the weakening monsoon (ibid.). Another suggested explanation is a dramatic expansion of irrigation in China (Cook et al., 2015) which caused an increase in evaporation and cloud cover and, consequently, decrease in solar radiation and more frequent summer snowfalls over the western Kunlun and Pamir (Bashir et al., 2017; Norris et al., 2019). This hypothesis is plausible but requires further investigation. Sakai and Fujita (2017) argued that not only climatic trends but also the higher mass balance sensitivity of regional glaciers to climatic forcing is behind the observed mass gain. Given the importance of the state of regional glaciers to water supply and glacier-related hazards, it is important to understand the mechanisms creating the Karakoram-Kunlun-Pamir anomaly.

## 5 Contribution to global mean sea level change

Perhaps the main impact of the observed glacier melt and projected future glacier change is their contribution to GMSL rise. It is estimated that up to 187 million people could be displaced by a GMSL rise of 1 m and the aggressive climate change scenarios in the future (Nicholls et al., 2011) but costs to coastal population, infrastructure, agricultural land and ecosystems can be immense even under more restrained scenarios (Oppenheimer et al., 2019). Observed changes in sea level are derived from a global network of tide gauges (with some records starting as early as the 1700s) and, at present, from satellite altimetry (Church et al., 2011). Contributions from different sources are calculated. Under the warming climate,

GMSL rise is caused by the expanding volume of the ocean due to lower density of warmer water (thermal expansion), increase in the mass of the ocean caused by the loss of land ice, and net changes in other terrestrial water reservoirs, for example, ground water depletion (Church et al., 2011; Oppenheimer et al., 2019).

There are several assessments of glacier melt contribution to GMSL rise. Cogley (2009) and Zemp et al. (2019) used glaciological and geodetic records of mass balance to calculate its input for 2000–05 and 1961–2016, respectively. Oerlemans et al. (2007) and Leclercq et al. (2011) used data on glacier front variations as proxies for volume change to reconstruct input of glaciers to GMSL rise in 1850–2000 and 1800–2005, respectively. Marzeion et al. (2012) reconstructed glacier input to GMSL rise for 1900–2005 from meteorological records calibrated with mass balance data and this work was extended by Marzeion et al. (2015). Gardner et al. (2013) used ICESat altimetry to assess input of glaciers to GMSL rise in 2003–09. There were substantial variations between the earlier assessments because of uncertainties associated with upscaling of mass balance records, which are not uniformly distributed throughout the world, and modeling (e.g., important processes such as calving were not explicitly included). Uncertainties in assessments based on glacier length resulted from the fact that change in glacier length is a function of length itself and that glaciers can lose mass through thinning while their tongues remain nearly stagnant. Rapid extension of geodetic mass balance assessments from spaceborne data alleviated many of these problems (Zemp et al., 2019).

Reliable assessments of contributions of ice sheets to GMSL rise became available with frequent satellite observations in the 1990s with further improvements in data quality in the 2000s, when GRACE data became available. Multimethod assessments were produced by Gardner et al. (2013), Bamber et al. (2018), Wouters et al. (2019), and IMBIE (The IMBIE Team, 2018, 2020). Estimations of ice sheet contribution to GMSL in the 20th century were attempted based on the combined use of modeling and tide gauge records (Hay et al., 2015) and geodetic reconstruction of the GrIS since 1900 by Kjeldsen et al. (2015).

**TABLE 4** Global mean sea level budget derived from observations (mm $a^{-1}$). Numbers in parentheses are uncertainties ranging from 5% to 95%.

| Source | 1901–1990 | 1970–2015 | 1993–2015 | 2006–2015 |
|---|---|---|---|---|
| Thermal expansion | | 0.89 (0.84–0.94) | 1.36 (0.96–1.76) | 1.40 (1.08–1.72) |
| Glaciers | 0.49 (0.34–0.64) | 0.46 (0.21–0.72) | 0.56 (0.34–0.78) | 0.61 (0.53–0.69) |
| GrIS and peripheral glaciers | 0.40 (0.23–0.57) | | 0.46 (0.21–0.71) | 0.77 (0.72–0.82) |
| AIS and peripheral glaciers | | | 0.29 (0.11–0.47) | 0.43 (0.34–0.52) |
| Land water storage | −0.12 | −0.07 | 0.09 | −0.21 (−0.36–0.06) |
| Total | | | 2.76 (2.21–3.31) | 3.00 (2.62–3.38) |
| Tide gauge and altimetry records | 1.38 (0.81–1.95) | 2.06 (1.77–2.34) | 3.16 (2.79–3.53) | 3.58 (3.10–4.06) |

*Data are from Oppenheimer, M., Glavovic, B.C., Hinkel, J., van de Wal, R., Magnan, A.K., Abd-Elgawad, A., Cai, R., CifuentesJara, M., DeConto, R.M., Ghosh, T., Hay, J., Isla, F., Marzeion, B., Meyssignac, B., Sebesvari, Z., 2019. Sea level rise and implications for low-lying islands, coasts and communities. In: Poörtner, H.-O., Roberts, D.C., Masson-Delmotte, V., Zhai, P., Tignor, M., Poloczanska, E., Mintenbeck, K., Alegría, A., Nicolai, M., Okem, A., Petzold, J., Rama, B., Weyer , N.M. (Eds.), IPCC Special Report on the Ocean and Cryosphere in a Changing Climate. IPCC, Geneva. See their Table 4.1 and accompanying text for a detailed explanation of how different components of GMSL were obtained.*

Observed changes in and contribution to GMSL change were summarized by Oppenheimer et al. (2019) for four time periods according to data availability (Table 4). The 1993–2015 and 2006–15 periods are short and might be affected by climatic variability. The 2006–15 assessments are based on improved data coverage and methods.

In 1993–2015, thermal expansion and contribution from glaciers were the largest contributors to the observed GMSL rise, accounting for 43% and 20%, respectively, while Greenland and Antarctica contributed 15% and 9%. In 2006–15, the observed sea level rise accelerated and the combined input of land ice became the main contributor over thermal expansion with glaciers, Greenland and Antarctica contributing 20%, 21%, and 12%, respectively. Although Antarctica contains eight times more ice above flotation than Greenland, the latter currently provides the largest input in GMSL rise. Glacier melt in the Canadian Arctic, Alaska, and the Southern Andes were the largest contributors among the Arctic and mountain glaciers (Table 3). Box et al. (2018) estimated that over a longer time period of 1971–2013, Alaska's contribution to GMSL rise was 1.8 times higher than that of the Canadian Arctic because of the earlier onset of continuing ice loss. Overall, the Arctic is a very significant contributor to GMSL rise.

# 6 Synthesis and outlook

Assessments of climate and glacier change agree that accelerated wastage of glaciers was observed in the 20th century and increasing loss of ice mass occurred from the GrIS, the AP, and the WAIS in the last few decades. The observed loss of ice can be confidently attributed to the continuing climatic warming. Due to the observed imbalance between the current glacier mass and climate, a large loss of glacier area and mass are expected in the mountain and polar regions in the future (Hock et al., 2019). Future changes in the extent of glaciers and ice sheets and their contributions to GMSL change depend on which representative concentration pathway (RCP) or emission scenario is followed. However, sea level rise is expected to be faster at the end of the 21st century under all scenarios including those compatible with achieving the long-term temperature goal set out in the Paris Agreement.

Global glacier mass loss between 2015 and 2100 is expected to reach 18% (with a range of 11%–25%) for the least aggressive RCP2.6 scenario and 36% (26%–47%) for the most aggressive RCP8.5 scenario. This corresponds to 94 (69–119) mm and 200 (156–240) mm of sea level rise, respectively. Mountain ranges, dominated by small glaciers, such as the European Alps, Pyrenees, Caucasus, North Asia and tropical regions, may lose more than 80% of their ice cover, however, the Arctic region and Antarctic periphery are expected to be the main contributors to GMSL rise (Hock et al., 2019). While the direction of the global trend is clear, the magnitude and timing of ice loss are subject to large uncertainties. These originate from the uncertainty in climate scenarios but also from many unknowns in glaciological models, such as the omission of mass loss by iceberg calving, subaqueous melt processes and instability mechanisms, from most global studies.

A likely range of the GrIS contribution to GMSL rise by the end of the 21st century is 4–10 cm under RCP2.6 and 7–21 cm under RCP8.5 (Oppenheimer et al., 2019). Surface melt and runoff are expected to dominate over ice discharge. Greenland's bedrock topography, whereby there is limited contact between thick ice and the ocean, is expected to limit the potential pace of GMSL rise from GrIS (ibid.). Evolution of the AIS is characterized by deep uncertainties. At the upper end of RCP8.5, Antarctica can contribute up to 28 cm of sea level rise by the end of the 21st century and the instability of the WAIS and its peripheral marine-terminating glaciers is expected to make the most significant impact. Potential loss of ice from the EAIS could make a major contribution to sea level in the future because of its vast extent (Oppenheimer et al., 2019).

Glacier wastage can affect regional water availability especially in the arid regions where glaciers are important sources of water, such as Central Asia and the Andes (Huss and Hock, 2018). Streamflow in glacier-fed rivers shows distinct seasonality with a pronounced maximum in the melt season which can coincide either with the wet season (e.g., monsoon-dominated regions where changes in glacier runoff make less impact on water availability) or with the dry season (e.g., Hindukush, Karakoram, Central Asia where impacts can be significant). At present, runoff is increasing in many glacierized catchments because water is released from long-term storage (glacier ice) until maximum or so-called peak water is reached. Afterwards, runoff will decline because the diminished glacier area will not be able to support high flow rates. Knowledge about the timing of peak water as well as reliable estimations of future changes in runoff are important for the development of adaptation strategies. Many regional assessments exist, for example, Juen et al. (2007) for the Cordillera Blanca, Farinotti et al. (2012) for the Swiss Alps, Immerzeel et al. (2012) for the Nepalese Himalayas, Lutz et al. (2014) for the HMA, Ragettli et al. (2016) for the Himalayas and the Andes, Duethmann et al. (2015, 2016), and Shahgedanova et al. (2018, 2020) for the Tien Shan, and many others. These studies and a global assessment by Huss and Hock (2018) showed that peak water is expected in the middle of the 21st century in the HMA (with an earlier peak in the outer ranges of the Tien Shan and later in the Indus headwaters) but in the Alps and the Andes, it has passed or is expected in the next decade. Changes in glacier runoff can have many consequences beyond water availability, for example, too much water can lead to the formation of glacier lakes and GLOF, too little to deterioration of water quality. Many of these impacts are reviwed by Huss et al. (2017) and Hock et al. (2019).

The global cryosphere is declining and this trend is expected to continue. The global community of scientists is working on assessments of the past, and prediction of the future, changes and their impacts, but a different intiative is under way. The Ice Memory Project under the auspices of UNESCO is building the first world archive of hundreds of ice cores in the ice cave at the Antarctica's Concordia research station, where mean annual temperature is about −54 °C, to preserve ice for generations of future researchers.

## References

Archer, C.L., Caldeira, K., 2008. Historical trends in the jet streams. Geophys. Res. Lett. 35, L08803. https://doi.org/10.1029/2008GL033614.

Archer, D.R., Fowler, H.J., 2004. Spatial and temporal variations in precipitation in the Upper Indus Basin, global teleconnections and hydrological implications. Hydrol. Earth Syst. Sci. 8, 47–61. https://doi.org/10.5194/hess-8-47-2004.

Azam, M.F., Wagnon, P., Berthier, E., Vincent, C., Fujita, K., Kargel, J.S., 2018. Review of the status and mass changes of Himalayan-Karakoram glaciers. J. Glaciol. 64 (243), 61–74. https://doi.org/10.1017/jog.2017.86.

Bamber, J.L., Riva, R.E.M., Vermeersen, B.L.A., LeBrocq, A.M., 2009. Reassessment of the potential sea-level rise from a collapse of the West Antarctic ice sheet. Science 324, 901–903. https://doi.org/10.1126/science.1169335.

Bamber, J.L., Westaway, R.M., Marzeion, B., Wouters, B., 2018. The land ice contribution to sea level during the satellite era. Environ. Res. Lett. 13, 063008. https://doi.org/10.1088/1748-9326/aac2f0.

Barandun, M., Huss, M., Usubaliev, R., Azisov, E., Berthier, E., Kääb, A., Bolch, T., Hoelzle, M., 2018. Multi-decadal mass balance series of three Kyrgyz glaciers inferred from modelling constrained with repeated snow line observations. Cryosphere 12, 1899–1919. https://doi.org/10.5194/tc-12-1899-2018.

Bashir, F., Zeng, X., Gupta, H., Hazenberg, P., 2017. A Hydrometeorological perspective on the Karakoram anomaly using unique valley-based synoptic weather observations. Geophys. Res. Lett. 44, 10,470–10,478. https://doi.org/10.1002/2017GL075284.

Benn, D.I., Bolch, T., Hands, K., Gulley, J., Luckman, A., Nicholson, L.I., Quincey, D., Thompson, S., Toumi, R., Wiseman, S., 2012. Response of debris-covered glaciers in the Mount Everest region to recent warming, and implications for outburst flood hazards. Earth Sci. Rev. 114, 156–174. https://doi.org/10.1016/j.earscirev.2012.03.008.

Berthier, E., Vincent, C., Magnússon, E., Gunnlaugsson, P., Pitte, P., Le Meur, E., Masiokas, M., Ruiz, L., Pálsson, F., Belart, J.M.C., Wagnon, P., 2014. Glacier topography and elevation changes derived from Pléiades sub-meter stereo images. Cryosphere 8, 2275–2291. https://doi.org/10.5194/tc-8-2275-2014.

Bevis, M., Harig, C., Khan, S.A., Brown, A., Simons, F.J., Willis, M., Fettweis, X., Van Den Broeke, M.R., Madsen, F.B., Kendrick, E., Caccamise, D.J., Van Dam, T., Knudsen, P., Nylen, T., 2019. Accelerating changes in ice mass within Greenland, and the ice sheet's sensitivity to atmospheric forcing. Proc. Natl. Acad. Sci. U. S. A. 116, 1934–1939. https://doi.org/10.1073/pnas.1806562116.

Bhambri, R., Bolch, T., 2009. Glacier mapping: a review with special reference to the Indian Himalayas. Prog. Phys. Geogr. 33, 672–704. https://doi.org/10.1177/0309133309348112.

Bolch, T., Kulkarni, A., Kääb, A., Huggel, C., Paul, F., Cogley, J.G., Frey, H., Kargel, J.S., Fujita, K., Scheel, M., Bajracharya, S., Stoffel, M., 2012. The state and fate of Himalayan glaciers. Science 336, 310–314. https://doi.org/10.1126/science.1215828.

Bolch, T., Sandberg Sørensen, L., Simonsen, S.B., Mölg, N., Machguth, H., Rastner, P., Paul, F., 2013. Mass loss of Greenland's glaciers and ice caps 2003-2008 revealed from ICESat laser altimetry data. Geophys. Res. Lett. 40, 875–881. https://doi.org/10.1002/grl.50270.

Bolch, T., Shea, J.M., Liu, S., Azam, F.M., Gao, Y., Gruber, S., Immerzeel, W.W., Kulkarni, A., Li, H., Tahir, A.A., Zhang, G., Zhang, Y., 2019. Status and change of the cryosphere in the extended Hindu Kush Himalaya region. In: The Hindu Kush Himalaya Assessment. Springer International Publishing, Cham, pp. 209–255. https://doi.org/10.1007/978-3-319-92288-1_7.

Bond, T.C., Doherty, S.J., Fahey, D.W., Forster, P.M., Berntsen, T., DeAngelo, B.J., Flanner, M.G., Ghan, S., Kärcher, B., Koch, D., Kinne, S., Kondo, Y., Quinn, P.K., Sarofim, M.C., Schultz, M.G., Schulz, M., Venkataraman, C., Zhang, H., Zhang, S., Bellouin, N., Guttikunda, S.K., Hopke, P.K., Jacobson, M.Z., Kaiser, J.W., Klimont, Z., Lohmann, U., Schwarz, J.P., Shindell, D., Storelvmo, T., Warren, S.G., Zender, C.S., 2013. Bounding the role of black carbon in the climate system: a scientific assessment. J. Geophys. Res. Atmos. 118, 5380–5552. https://doi.org/10.1002/jgrd.50171.

Box, J.E., 2013. Greenland ice sheet mass balance reconstruction. Part II. Surface mass balance (1840–2010). J. Clim. 26, 6974–6989. https://doi.org/10.1175/JCLI-D-12-00518.1.

Box, J.E., Colgan, W.T., Wouters, B., Burgess, D.O., O'Neel, S., Thomson, L.I., Mernild, S.H., 2018. Global Sea-level contribution from Arctic land ice: 1971–2017. Environ. Res. Lett. 13, 125012. https://doi.org/10.1088/1748-9326/aaf2ed.

Brun, F., Berthier, E., Wagnon, P., Kääb, A., Treichler, D., 2017. A spatially resolved estimate of High Mountain Asia glacier mass balances, 2000–2016. Nat. Geosci. 10, 668–673. https://doi.org/10.1038/NGEO2999.A.

Cannon, F., Carvalho, L.M.V., Jones, C., Bookhagen, B., 2014. Multi-annual variations in winter westerly disturbance activity affecting the Himalaya. Clim. Dyn. 44, 441–455. https://doi.org/10.1007/s00382-014-2248-8.

Carr, J.R., Stokes, C., Vieli, A., 2014. Recent retreat of major outlet glaciers on Novaya Zemlya, Russian Arctic, influenced by fjord geometry and sea-ice conditions. J. Glaciol. 60, 155–170. https://doi.org/10.3189/2014JoG13J122.

Carr, J.R., Stokes, C.R., Vieli, A., 2017. Threefold increase in marine-terminating outlet glacier retreat rates across the Atlantic Arctic: 1992–2010. Ann. Glaciol. 58, 72–91. https://doi.org/10.1017/aog.2017.3.

Carturan, L., Baroni, C., Becker, M., Bellin, A., Cainelli, O., Carton, A., Casarotto, C., Dalla Fontana, G., Godio, A., Martinelli, T., Salvatore, M.C., Seppi, R., 2013. Decay of a long-term monitored glacier: Careser glacier (Ortles-Cevedale, European Alps). Cryosphere 7, 1819–1838. https://doi.org/10.5194/tc-7-1819-2013.

Chinn, T., Winkler, S., Salinger, M.J., Haakensen, N., 2005. Recent glacier advances in Norway and New Zealand: a comparison of their glaciological and meteorological causes. Geogr. Ann. Ser. A Phys. Geogr. 87, 141–157. https://doi.org/10.1111/j.0435-3676.2005.00249.x.

Church, J.A., White, N.J., Konikow, L.F., Domingues, C.M., Cogley, J.G., Rignot, E., Gregory, J.M., van den Broeke, M.R., Monaghan, A.J., Velicogna, I., 2011. Revisiting the Earth's sea-level and energy budgets from 1961 to 2008. Geophys. Res. Lett. 38. https://doi.org/10.1029/2011GL048794.

Citterio, M., Diolaiuti, G., Smiraglia, C., D'agata, C., Carnielli, T., Stella, G., Siletto, G.B., 2007. The fluctuations of italian glaciers during the last century: a contribution to knowledge about alpine glacier changes. Geogr. Ann. Ser. A Phys. Geogr. 89, 167–184. https://doi.org/10.1111/j.1468-0459.2007.00316.x.

Cogley, J.G., 2009. Geodetic and direct mass-balance measurements: comparison and joint analysis. Ann. Glaciol. 50, 96–100. https://doi.org/10.3189/172756409787769744.

Cook, A.J., Copland, L., Noël, B.P.Y., Stokes, C.R., Bentley, M.J., Sharp, M.J., Bingham, R.G., van den Broeke, M.R., 2019. Atmospheric forcing of rapid marine-terminating glacier retreat in the Canadian Arctic archipelago. Sci. Adv. 5, eaau8507. https://doi.org/10.1126/sciadv.aau8507.

Cook, A.J., Vaughan, D.G., Luckman, A.J., Murray, T., 2014. A new Antarctic peninsula glacier basin inventory and observed area changes since the 1940s. Antarct. Sci. 26, 614–624. https://doi.org/10.1017/S0954102014000200.

Cook, B.I., Shukla, S.P., Puma, M.J., Nazarenko, L.S., 2015. Irrigation as an historical climate forcing. Clim. Dyn. 44, 1715–1730. https://doi.org/10.1007/s00382-014-2204-7.

Cuffey, K.M., Paterson, W.S.B., 2010. The Physics of Glaciers. Elsevier, Oxford.

Davaze, L., Rabatel, A., Dufour, A., Hugonnet, R., Arnaud, Y., 2020. Region-wide annual glacier surface mass balance for the European Alps from 2000 to 2016. Front. Earth Sci. 8, 1–14. https://doi.org/10.3389/feart.2020.00149.

Doherty, S.J., Warren, S.G., Grenfell, T.C., Clarke, A.D., Brandt, R.E., 2010. Light-absorbing impurities in Arctic snow. Atmos. Chem. Phys. 10, 11647–11680. https://doi.org/10.5194/acp-10-11647-2010.

Dou, T.-F., Xiao, C.-D., 2016. An overview of black carbon deposition and its radiative forcing over the Arctic. Adv. Clim. Chang. Res. 7, 115–122. https://doi.org/10.1016/j.accre.2016.10.003.

Duethmann, D., Bolch, T., Farinotti, D., Kriegel, D., Vorogushyn, S., Merz, B., Pieczonka, T., Jiang, T., Su, B., Güntner, A., 2015. Attribution of streamflow trends in snow and glacier melt-dominated catchments of the Tarim River, Central Asia. Water Resour. Res. 51, 4727–4750. https://doi.org/10.1002/2014WR016716.

Duethmann, D., Menz, C., Jiang, T., Vorogushyn, S., 2016. Projections for headwater catchments of the Tarim River reveal glacier retreat and decreasing surface water availability but uncertainties are large. Environ. Res. Lett. 11, 1–13. https://doi.org/10.1088/1748-9326/11/5/054024.

Farinotti, D., Huss, M., Fürst, J.J., Landmann, J., Machguth, H., Maussion, F., Pandit, A., 2019. A consensus estimate for the ice thickness distribution of all glaciers on Earth. Nat. Geosci. 12. https://doi.org/10.1038/s41561-019-0300-3.

Farinotti, D., Immerzeel, W.W., de Kok, R.J., Quincey, D.J., Dehecq, A., 2020. Manifestations and mechanisms of the Karakoram glacier anomaly. Nat. Geosci. 13, 8–16. https://doi.org/10.1038/s41561-019-0513-5.

Farinotti, D., Usselmann, S., Huss, M., Bauder, A., Funk, M., 2012. Runoff evolution in the Swiss Alps: projections for selected high-alpine catchments based on ENSEMBLES scenarios. Hydrol. Process. 26, 1909–1924. https://doi.org/10.1002/hyp.8276.

Fettweis, X., Box, J.E., Agosta, C., Amory, C., Kittel, C., Lang, C., Van As, D., Machguth, H., Gallée, H., 2017. Reconstructions of the 1900-2015 Greenland ice sheet surface mass balance using the regional climate MAR model. Cryosphere 11, 1015–1033. https://doi.org/10.5194/tc-11-1015-2017.

Fettweis, X., Hanna, E., Lang, C., Belleflamme, A., Erpicum, M., Gallée, H., 2013. Brief communication important role of the mid-tropospheric atmospheric circulation in the recent surface melt increase over the Greenland ice sheet. Cryosphere 7, 241–248. https://doi.org/10.5194/tc-7-241-2013.

Flanner, M.G., Zender, C.S., Randerson, J.T., Rasch, P.J., 2007. Present-day climate forcing and response from black carbon in snow. J. Geophys. Res. 112, 1–17. https://doi.org/10.1029/2006JD008003.

Forsythe, N., Fowler, H.J., Li, X.-F., Blenkinsop, S., Pritchard, D., 2017. Karakoram temperature and glacial melt driven by regional atmospheric circulation variability. Nat. Clim. Chang. 7, 664–670. https://doi.org/10.1038/nclimate3361.

Fujita, K., 2008. Effect of precipitation seasonality on climatic sensitivity of glacier mass balance. Earth Planet. Sci. Lett. 276, 14–19. https://doi.org/10.1016/j.epsl.2008.08.028.

Gardner, A.S., Moholdt, G., Cogley, J.G., Wouters, B., Arendt, A.A., Wahr, J., Berthier, E., Hock, R., Pfeffer, W.T., Kaser, G., Ligtenberg, S.R.M., Bolch, T., Sharp, M.J., Hagen, J.O., van den Broeke, M.R., Paul, F., 2013.

A reconciled estimate of glacier contributions to sea level rise: 2003 to 2009. Science 340, 852–857. https://doi.org/10.1126/science.1234532.

Gardner, A.S., Moholdt, G., Scambos, T., Fahnstock, M., Ligtenberg, S., van den Broeke, M., Nilsson, J., 2018. Increased West Antarctic and unchanged East Antarctic ice discharge over the last 7 years. Cryosphere 12, 521–547. https://doi.org/10.5194/tc-12-521-2018.

Gautam, R., Hsu, N.C., Lau, W.K., Yasunari, T.J., 2013. Satellite observations of desert dust-induced Himalayan snow darkening. Geophys. Res. Lett. 40, 988–993. https://doi.org/10.1002/GRL.50226.

Glasser, N.F., Harrison, S., Jansson, K.N., Anderson, K., Cowley, A., 2011. Global sea-level contribution from the Patagonian icefields since the little ice age maximum. Nat. Geosci. 4, 303–307. https://doi.org/10.1038/ngeo1122.

González Trueba, J.J., Moreno, R.M., Martínez de Pisón, E., Serrano, E., 2008. 'Little Ice Age' glaciation and current glaciers in the Iberian Peninsula. The Holocene 18, 551–568. https://doi.org/10.1177/0959683608089209.

Goodwin, B.P., Mosley-Thompson, E., Wilson, A.B., Porter, S.E., Roxana Sierra-Hernandez, M., 2016. Accumulation variability in the Antarctic peninsula: the role of large-scale atmospheric oscillations and their interactions. J. Clim. 29, 2579–2596. https://doi.org/10.1175/JCLI-D-15-0354.1.

Haeberli, W., 1995. Glacier fluctuations and climate change detection. Geogr. Fis. Din. Quat. 191–199.

Haeberli, W., Linsbauer, A., 2013. Global glacier volumes and sea level—small but systematic effects of ice below the surface of the ocean and of new local lakes on land. Cryosphere 7, 817–821. https://doi.org/10.5194/tc-7-817-2013.

Haeberli, W., Whiteman, C. (Eds.), 2015. Snow and Ice-Related Hazards, Risks, and Disasters. Elsevier, Oxford.

Hall, D.K., Comiso, J.C., DiGirolamo, N.E., Shuman, C.A., Box, J.E., Koenig, L.S., 2013. Variability in the surface temperature and melt extent of the Greenland ice sheet from MODIS. Geophys. Res. Lett. 40, 2114–2120. https://doi.org/10.1002/grl.50240.

Hanna, E., Huybrechts, P., Cappelen, J., Steffen, K., Bales, R.C., Burgess, E., McConnell, J.R., Peder Steffensen, J., Van den Broeke, M., Wake, L., Bigg, G., Griffiths, M., Savas, D., 2011. Greenland ice sheet surface mass balance 1870 to 2010 based on twentieth century reanalysis, and links with global climate forcing. J. Geophys. Res. Atmos. 116. https://doi.org/10.1029/2011JD016387.

Hanna, E., Navarro, F.J., Pattyn, F., Domingues, C.M., Fettweis, X., Ivins, E.R., Nicholls, R.J., Ritz, C., Smith, B., Tulaczyk, S., Whitehouse, P.L., Zwally, H.J., 2013. Ice-sheet mass balance and climate change. Nature 498, 51–59. https://doi.org/10.1038/nature12238.

Hay, C.C., Morrow, E., Kopp, R.E., Mitrovica, J.X., 2015. Probabilistic reanalysis of twentieth-century sea-level rise. Nature 517, 481–484. https://doi.org/10.1038/nature14093.

Hennen, M., White, K., Shahgedanova, M., 2019. An assessment of SEVIRI imagery at various temporal resolutions and the effect on accurate dust emission mapping. Remote Sens. 11. https://doi.org/10.3390/rs11080965.

Herreid, S., Pellicciotti, F., 2020. The state of rock debris covering Earth's glaciers. Nat. Geosci. 13. https://doi.org/10.1038/s41561-020-0615-0.

Hock, R., Rasul, G., Adler, C., Cáceres, B., Gruber, S., Hirabayashi, Y., Jackson, M., Kääb, A., Kang, S., Kutuzov, S., Milner, A., Molau, U., Morin, S., Orlove, B., Steltzer, H., 2019. High Mountain Areas. In: Poörtner, H.-O., Roberts, D.C., Masson-Delmotte, V., Zhai, P., Tignor, M., Poloczanska, E., Mintenbeck, K., Alegría, A., Nicolai, M., Okem, A., Petzold, J., Rama, B., Weyer, N.M. (Eds.), IPCC Special Report on the Ocean and Cryosphere in a Changing Climate. IPCC, Geneva.

Huss, M., 2013. Density assumptions for converting geodetic glacier volume change to mass change. Cryosphere 7, 877–887. https://doi.org/10.5194/tc-7-877-2013.

Huss, M., Bookhagen, B., Huggel, C., Jacobsen, D., Bradley, R.S., Clague, J.J., Vuille, M., Buytaert, W., Cayan, D.R., Greenwood, G., Mark, B.G., Milner, A.M., Weingartner, R., Winder, M., 2017. Toward mountains without permanent snow and ice. Earth's Future 5, 418–435. https://doi.org/10.1002/eft2.207.

Huss, M., Hock, R., 2018. Global-scale hydrological response to future glacier mass loss. Nat. Clim. Chang. 8, 135–140. https://doi.org/10.1038/s41558-017-0049-x.

Immerzeel, W.W., van Beek, L.P.H., Konz, M., Shrestha, A.B., Bierkens, M.F.P., 2012. Hydrological response to climate change in a glacierized catchment in the Himalayas. Clim. Chang. 110, 721–736. https://doi.org/10.1007/s10584-011-0143-4.

IPCC, 2019. IPCC Special Report on the Ocean and Cryosphere in a Changing Climate. IPCC, Geneva.

Jacob, T., Wahr, J., Pfeffer, W.T., Swenson, S., 2012. Recent contributions of glaciers and ice caps to sea level rise. Nature 482, 514–518. https://doi.org/10.1038/nature10847.

Jacobi, H.-W., Lim, S., Ménégoz, M., Ginot, P., Laj, P., Bonasoni, P., Stocchi, P., Marinoni, A., Arnaud, Y., 2015. Black carbon in snow in the upper Himalayan Khumbu Valley, Nepal: observations and modeling of the impact on snow albedo, melting, and radiative forcing. Cryosphere 9, 1685–1699. https://doi.org/10.5194/tc-9-1685-2015.

Juen, I., Kaser, G., Georges, C., 2007. Modelling observed and future runoff from a glacierized tropical catchment (Cordillera Blanca, Perú). Glob. Planet. Chang. 59, 37–48. https://doi.org/10.1016/j.gloplacha.2006.11.038.

Kääb, A., Berthier, E., Nuth, C., Gardelle, J., Arnaud, Y., 2012. Contrasting patterns of early twenty-first-century glacier mass change in the Himalayas. Nature 488, 495–498. https://doi.org/10.1038/nature11324.

Kapitsa, V., Shahgedanova, M., MacHguth, H., Severskiy, I., Medeu, A., 2017. Assessment of evolution and risks of glacier lake outbursts in the Djungarskiy Alatau, Central Asia, using Landsat imagery and glacier bed topography modelling. Nat. Hazards Earth Syst. Sci. 17. https://doi.org/10.5194/nhess-17-1837-2017.

Kapitsa, V., Shahgedanova, M., Severskiy, I., Kasatkin, N., White, K., Usmanova, Z., 2020. Assessment of changes in mass balance of the Tuyuksu Group of Glaciers, northern Tien Shan, between 1958 and 2016 using ground-based observations and Pléiades satellite imagery. Front. Earth Sci. 8. https://doi.org/10.3389/feart.2020.00259.

Kargel, J.S., Leonard, G.J., Bishop, M.P., Kaab, A., Raup, B., 2014. Global Land Ice Measurements From Space. Springer-Praxis.

Kaser, G., Großhauser, M., Marzeion, B., Barry, R.G., 2010. Contribution potential of glaciers to water availability in different climate regimes. Proc. Natl. Acad. Sci. U. S. A. 107, 21300–21305. https://doi.org/10.1073/pnas.

Kaser, G., Hardy, D.R., Mölg, T., Bradley, R.S., Hyera, T.M., 2004. Modern glacier retreat on Kilimanjaro as evidence of climate change: observations and facts. Int. J. Climatol. 24, 329–339. https://doi.org/10.1002/joc.1008.

Khan, A.L., Wagner, S., Jaffe, R., Xian, P., Williams, M., Armstrong, R., McKnight, D., 2017. Dissolved black carbon in the global cryosphere: concentrations and chemical signatures. Geophys. Res. Lett. 44, 6226–6234. https://doi.org/10.1002/2017GL073485.

Khan, S.A., Aschwanden, A., Bjørk, A.A., Wahr, J., Kjeldsen, K.K., Kjær, K.H., 2015. Greenland ice sheet mass balance: a review. Rep. Prog. Phys. 78, 046801. https://doi.org/10.1088/0034-4885/78/4/046801.

King, M.D., Howat, I.M., Candela, S.G., Noh, M.J., Jeong, S., Noël, B.P.Y., van den Broeke, M.R., Wouters, B., Negrete, A., 2020. Dynamic ice loss from the Greenland ice sheet driven by sustained glacier retreat. Commun. Earth Environ. 1, 1. https://doi.org/10.1038/s43247-020-0001-2.

Kjeldsen, K.K., Korsgaard, N.J., Bjørk, A.A., Khan, S.A., Box, J.E., Funder, S., Larsen, N.K., Bamber, J.L., Colgan, W., van den Broeke, M., Siggaard-Andersen, M.-L., Nuth, C., Schomacker, A., Andresen, C.S., Willerslev, E., Kjær, K.H., 2015. Spatial and temporal distribution of mass loss from the Greenland ice sheet since AD 1900. Nature 528, 396–400. https://doi.org/10.1038/nature16183.

Koenig, L.S., Ivanoff, A., Alexander, P.M., MacGregor, J.A., Fettweis, X., Panzer, B., Paden, J.D., Forster, R.R., Das, I., McConnell, J.R., Tedesco, M., Leuschen, C., Gogineni, P., 2016. Annual Greenland accumulation rates (2009–2012) from airborne snow radar. Cryosphere 10, 1739–1752. https://doi.org/10.5194/tc-10-1739-2016.

Kutuzov, S., Lavrentiev, I., Smirnov, A., Nosenko, G., 2019. Volume changes of Elbrus glaciers from 1997 to 2017. Front. Earth Sci. 7, 153. https://doi.org/10.3389/feart.2019.00153.

Kutuzov, S., Shahgedanova, M., 2009. Glacier retreat and climatic variability in the eastern Terskey-Alatoo, inner Tien Shan between the middle of the 19th century and beginning of the 21st century. Glob. Planet. Chang. 69. https://doi.org/10.1016/j.gloplacha.2009.07.001.

Larsen, C.F., Burgess, E., Arendt, A.A., O'Neel, S., Johnson, A.J., Kienholz, C., 2015. Surface melt dominates Alaska glacier mass balance. Geophys. Res. Lett. 42, 5902–5908. https://doi.org/10.1002/2015GL064349.

Leclercq, P.W., Oerlemans, J., Basagic, H.J., Bushueva, I., Cook, A.J., Le Bris, R., 2014. A data set of worldwide glacier length fluctuations. Cryosphere 8, 659–672. https://doi.org/10.5194/tc-8-659-2014.

Leclercq, P.W., Oerlemans, J., Cogley, J.G., 2011. Estimating the glacier contribution to sea-level rise for the period 1800–2005. Surv. Geophys. 32, 519–535. https://doi.org/10.1007/s10712-011-9121-7.

Lewis, G., Osterberg, E., Hawley, R., Whitmore, B., Marshall, H.P., Box, J., 2017. Regional Greenland accumulation variability from operation IceBridge airborne accumulation radar. Cryosphere 11, 773–788. https://doi.org/10.5194/tc-11-773-2017.

Li, X.F., Fowler, H.J., Forsythe, N., Blenkinsop, S., Pritchard, D., 2018. The Karakoram/Western Tibetan vortex: seasonal and year-to-year variability. Clim. Dyn. 51, 3883–3906. https://doi.org/10.1007/s00382-018-4118-2.

Loriaux, T., Casassa, G., 2013. Evolution of glacial lakes from the northern Patagonia icefield and terrestrial water storage in a sea-level rise context. Glob. Planet. Chang. 102, 33–40. https://doi.org/10.1016/j.gloplacha.2012.12.012.

Lutz, A.F., Immerzeel, W.W., Shrestha, A.B., Bierkens, M.F.P., 2014. Consistent increase in high Asia's runoff due to increasing glacier melt and precipitation. Nat. Clim. Chang. 4. https://doi.org/10.1038/NCLIMATE2237.

Marzeion, B., Champollion, N., Haeberli, W., Langley, K., Leclercq, P., Paul, F., 2017. Observation-based estimates of global glacier mass change and its contribution to sea-level change. Surv. Geophys. 38, 105–130. https://doi.org/10.1007/s10712-016-9394-y.

Marzeion, B., Jarosch, A.H., Hofer, M., 2012. Past and future sea-level change from the surface mass balance of glaciers. Cryosphere 6, 1295–1322. https://doi.org/10.5194/tc-6-1295-2012.

Marzeion, B., Leclercq, P.W., Cogley, J.G., Jarosch, A.H., 2015. Brief communication: global reconstructions of glacier mass change during the 20th century are consistent. Cryosphere 9, 2399–2404. https://doi.org/10.5194/tc-9-2399-2015.

Masson-Delmotte, V., Swingedouw, D., Landais, A., Seidenkrantz, M.-S., Gauthier, E., Bichet, V., Massa, C., Perren, B., Jomelli, V., Adalgeirsdottir, G., Hesselbjerg Christensen, J., Arneborg, J., Bhatt, U., Walker, D.A., Elberling, B., Gillet-Chaulet, F., Ritz, C., Gallée, H., van den Broeke, M., Fettweis, X., de Vernal, A., Vinther, B., 2012. Greenland climate change: from the past to the future. Wiley Interdiscip. Rev. Clim. Chang. 3, 427–449. https://doi.org/10.1002/wcc.186.

McGrath, D., Colgan, W., Bayou, N., Muto, A., Steffen, K., 2013. Recent warming at summit, Greenland: global context and implications. Geophys. Res. Lett. 40, 2091–2096. https://doi.org/10.1002/grl.50456.

McNabb, R.W., Hock, R., 2014. Alaska tidewater glacier terminus positions, 1948–2012. J. Geophys. Res. Earth Surf. 119, 153–167. https://doi.org/10.1002/2013JF002915.

Medley, B., Thomas, E.R., 2019. Increased snowfall over the Antarctic ice sheet mitigated twentieth-century sea-level rise. Nat. Clim. Chang. 9, 34–39. https://doi.org/10.1038/s41558-018-0356-x.

Menon, S., Koch, D., Beig, G., Sahu, S., Fasullo, J., Orlikowski, D., 2010. Black carbon aerosols and the third polar ice cap. Atmos. Chem. Phys. 10, 4559–4571. https://doi.org/10.5194/acp-10-4559-2010.

Meredith, M., Sommerkorn, M., Cassotta, S., Derksen, C., Ekaykin, A., Hollowed, A., Kofinas, G., Mackintosh, A., Melbourne-Thomas, J., Muelbert, M.M.C., Ottersen, G., Pritchard, H., Schuur, E.A.G., 2019. Polar Regions. In: - Poörtner, H.-O., Roberts, D.C., Masson-Delmotte, V., Zhai, P., Tignor, M., Poloczanska, E., Mintenbeck, K., Alegría, A., Nicolai, M., Okem, A., Petzold, J., Rama, B., Weyer, N.M. (Eds.), IPCC Special Report on the Ocean and Cryosphere in a Changing Climate. IPCC, Geneva.

Mölg, T., Cullen, N.J., Hardy, D.R., Kaser, G., Klok, L., 2008. Mass balance of a slope glacier on Kilimanjaro and its sensitivity to climate. Int. J. Climatol. 28, 881–892. https://doi.org/10.1002/joc.1589.

Mouginot, J., Rignot, E., Bjørk, A.A., van den Broeke, M., Millan, R., Morlighem, M., Noël, B., Scheuchl, B., Wood, M., 2019. Forty-six years of Greenland ice sheet mass balance from 1972 to 2018. Proc. Natl. Acad. Sci. 116, 9239–9244. https://doi.org/10.1073/pnas.1904242116.

Mouginot, J., Rignot, E., Scheuchl, B., 2014. Sustained increase in ice discharge from the Amundsen sea embayment, West Antarctica, from 1973 to 2013. Geophys. Res. Lett. 41, 1576–1584. https://doi.org/10.1002/2013GL059069.

Nesje, A., Dahl, S.O., 2003. The 'Little Ice Age'—only temperature? The Holocene 13, 139–145. https://doi.org/10.1191/0959683603hl603fa.

Nesje, A., Lie, O., Dahl, S.O., 2000. Is the North Atlantic oscillation reflected in Scandinavian glacier mass balance records? J. Quat. Sci. 15, 587–601. https://doi.org/10.1002/1099-1417(200009)15:6<587::AID-JQS533>3.0.CO;2-2.

Nicholls, R.J., Marinova, N., Lowe, J.A., Brown, S., Vellinga, P., de Gusmão, D., Hinkel, J., Tol, R.S.J., 2011. Sea-level rise and its possible impacts given a 'beyond 4°C world' in the twenty-first century. Philos. Trans. R. Soc. A Math. Phys. Eng. Sci. 369, 161–181. https://doi.org/10.1098/rsta.2010.0291.

Nobakht, M., Shahgedanova, M., White, K., 2019. New inventory of dust sources in Central Asia derived from the daily MODIS imagery. E3S Web Conf. 99, 01001. https://doi.org/10.1051/e3sconf/20199901001.

Norris, J., Carvalho, L.M.V., Jones, C., Cannon, F., 2019. Deciphering the contrasting climatic trends between the central Himalaya and Karakoram with 36 years of WRF simulations. Clim. Dyn. 52, 159–180. https://doi.org/10.1007/s00382-018-4133-3.

O'Leary, M., Christoffersen, P., 2013. Calving on tidewater glaciers amplified by submarine frontal melting. Cryosphere 7, 119–128. https://doi.org/10.5194/tc-7-119-2013.

Oerlemans, J., 2010. Extracting a climate signal from 169 glacier records. Science 308 (5722), 675–677. https://doi.org/10.1126/science.1107046.

Oerlemans, J., Dyurgerov, M., van de Wal, R.S.W., 2007. Reconstructing the glacier contribution to sea-level rise back to 1850. Cryosphere 1, 59–65. https://doi.org/10.5194/tc-1-59-2007.

Oppenheimer, M., Glavovic, B.C., Hinkel, J., van de Wal, R., Magnan, A.K., Abd-Elgawad, A., Cai, R., CifuentesJara, M., DeConto, R.M., Ghosh, T., Hay, J., Isla, F., Marzeion, B., Meyssignac, B., Sebesvari, Z., 2019. Sea level rise and implications for low-lying islands, coasts and communities. In: Poörtner, H.-O., Roberts, D.C., Masson-Delmotte, V., Zhai, P., Tignor, M., Poloczanska, E., Mintenbeck, K., Alegría, A.,

Nicolai, M., Okem, A., Petzold, J., Rama, B., Weyer, N.M. (Eds.), IPCC Special Report on the Ocean and Cryosphere in a Changing Climate. IPCC, Geneva.

Orsi, A.J., Kawamura, K., Masson-Delmotte, V., Fettweis, X., Box, J.E., Dahl-Jensen, D., Clow, G.D., Landais, A., Severinghaus, J.P., 2017. The recent warming trend in North Greenland. Geophys. Res. Lett. 44, 6235–6243. https://doi.org/10.1002/2016GL072212.

Painter, T.H., Bryant, A.C., Skiles, S.M., 2012. Radiative forcing by light absorbing impurities in snow from MODIS surface reflectance data. Geophys. Res. Lett. 39. https://doi.org/10.1029/2012GL052457.

Paolo, F.S., Fricker, H.A., Padman, L., 2015. Volume loss from Antarctic ice shelves is accelerating. Science 348, 327–331. https://doi.org/10.1126/science.aaa0940.

Paul, F., Andreassen, L.M., 2009. A new glacier inventory for the Svartisen region, Norway, from Landsat ETM + data: challenges and change assessment. J. Glaciol. 55, 607–618.

Paul, F., Kääb, A., Haeberli, W., 2007. Recent glacier changes in the Alps observed by satellite: consequences for future monitoring strategies. Glob. Planet. Chang. 56, 111–122. https://doi.org/10.1016/j.gloplacha.2006.07.007.

Paul, F., Kääb, A., Maisch, M., Kellenberger, T., Haeberli, W., 2004. Rapid disintegration of alpine glaciers observed with satellite data. Geophys. Res. Lett. 31. https://doi.org/10.1029/2004GL020816.

Pellicciotti, F., Stephan, C., Miles, E., Herreid, S., Immerzeel, W.W., Bolch, T., 2015. Mass-balance changes of the debris-covered glaciers in the Langtang Himal, Nepal, from 1974 to 1999. J. Glaciol. 61, 373–386. https://doi.org/10.3189/2015JoG13J237.

Pepin, N., Bradley, R.S., Diaz, H.F., Baraer, M., Caceres, E.B., Forsythe, N., Fowler, H., Greenwood, G., Hashmi, M.Z., Liu, X.D., Miller, J.R., Ning, L., Ohmura, A., Palazzi, E., Rangwala, I., Schöner, W., Severskiy, I., Shahgedanova, M., Wang, M.B., Williamson, S.N., Yang, D.Q., 2015. Elevation-dependent warming in mountain regions of the world. Nat. Clim. Chang. 5, 424–430. https://doi.org/10.1038/nclimate2563.

Permana, D.S., Thompson, L.G., Mosley-Thompson, E., Davis, M.E., Lin, P.-N., Nicolas, J.P., Bolzan, J.F., Bird, B.W., Mikhalenko, V.N., Gabrielli, P., Zagorodnov, V., Mountain, K.R., Schotterer, U., Hanggoro, W., Habibie, M.N., Kaize, Y., Gunawan, D., Setyadi, G., Susanto, R.D., Fernández, A., Mark, B.G., 2019. Disappearance of the last tropical glaciers in the Western Pacific warm Pool (Papua, Indonesia) appears imminent. Proc. Natl. Acad. Sci. 116, 26382–26388. https://doi.org/10.1073/pnas.1822037116.

Pfeffer, W.T., 2003. Tidewater glaciers move at their own pace. Nature 426, 602. https://doi.org/10.1038/426602a.

Pfeffer, W.T., Arendt, A.A., Bliss, A., Bolch, T., Cogley, J.G., Gardner, A.S., Hagen, J.-O., Hock, R., Kaser, G., Kienholz, C., Miles, E.S., Moholdt, G., Mölg, N., Paul, F., Radić, V., Rastner, P., Raup, B.H., Rich, J., Sharp, M.J., 2014. The Randolph glacier inventory: a globally complete inventory of glaciers. J. Glaciol. 60, 537–552. https://doi.org/10.3189/2014JoG13J176.

Pieczonka, T., Bolch, T., 2015. Region-wide glacier mass budgets and area changes for the central Tien Shan between ~1975 and 1999 using hexagon KH-9 imagery. Glob. Planet. Chang. 128, 1–13. https://doi.org/10.1016/j.gloplacha.2014.11.014.

Pritchard, H.D., Ligtenberg, S.R.M., Fricker, H.A., Vaughan, D.G., Van Den Broeke, M.R., Padman, L., 2012. Antarctic ice-sheet loss driven by basal melting of ice shelves. Nature 484, 502–505. https://doi.org/10.1038/nature10968.

Rabatel, A., Francou, B., Soruco, A., Gomez, J., Cáceres, B., Ceballos, J.L., Basantes, R., Vuille, M., Sicart, J.-E., Huggel, C., Scheel, M., Lejeune, Y., Arnaud, Y., Collet, M., Condom, T., Consoli, G., Favier, V., Jomelli, V., Galarraga, R., Ginot, P., Maisincho, L., Mendoza, J., Ménégoz, M., Ramirez, E., Ribstein, P., Suarez, W., Villacis, M., Wagnon, P., 2013. Current state of glaciers in the tropical Andes: a multi-century perspective on glacier evolution and climate change. Cryosphere 7, 81–102. https://doi.org/10.5194/tc-7-81-2013.

Radić, V., Hock, R., 2010. Regional and global volumes of glaciers derived from statistical upscaling of glacier inventory data. J. Geophys. Res. 115, F01010. https://doi.org/10.1029/2009JF001373.

Ragettli, S., Immerzeel, W.W., Pellicciotti, F., 2016. Contrasting climate change impact on river flows from high-altitude catchments in the Himalayan and Andes Mountains. Proc. Natl. Acad. Sci. U. S. A. 113, 9222–9227. https://doi.org/10.1073/pnas.1606526113.

Raup, B., Ka, A., Kargel, J.S., Bishop, M.P., Hamilton, G., Lee, E., Paul, F., Rau, F., Soltesz, D., Jodha, S., Khalsa, S., Beedle, M., Helm, C., 2007. Remote sensing and GIS technology in the Global Land Ice Measurements from Space (GLIMS) Project. Comput. Geosci. 33, 104–125. https://doi.org/10.1016/j.cageo.2006.05.015.

RGI Consortium, 2017. Randolph Glacier Inventory—A Dataset of Global Glacier Outlines: Version 6.0R: Technical Report. Global Land Ice Measurements from Space, Colorado, USAhttps://doi.org/10.7265/N5-RGI-60.

Rignot, E., 1996. Tidal motion, ice velocity and melt rate of Petermann Gletscher, Greenland, measured from radar interferometry. J. Glaciol. 42, 476–485. https://doi.org/10.1017/S0022143000003464.

Rignot, E., Mouginot, J., Scheuchl, B., 2011. Ice flow of the Antarctic ice sheet. Science 333, 1427–1430. https://doi.org/10.1126/science.1208336.

Ryan, J.C., Smith, L.C., van As, D., Cooley, S.W., Cooper, M.G., Pitcher, L.H., Hubbard, A., 2019. Greenland ice sheet surface melt amplified by snowline migration and bare ice exposure. Sci. Adv. 5, eaav3738. https://doi.org/10.1126/sciadv.aav3738.

Sakai, A., Fujita, K., 2017. Contrasting glacier responses to recent climate change in high-mountain Asia. Sci. Rep. 7, 1–8. https://doi.org/10.1038/s41598-017-14256-5.

Scherler, D., Bookhagen, B., Strecker, M.R., 2011. Spatially variable response of Himalayan glaciers to climate change affected by debris cover. Nat. Geosci. 4, 156–159. https://doi.org/10.1038/ngeo1068.

Serreze, M.C., Barry, R.G., 2011. Processes and impacts of Arctic amplification: a research synthesis. Glob. Planet. Chang. 77, 85–96. https://doi.org/10.1016/j.gloplacha.2011.03.004.

Shahgedanova, M., Afzal, M., Hagg, W., Kapitsa, V., Kasatkin, N., Mayr, E., Rybak, O., Saidaliyeva, Z., Severskiy, I., Usmanova, Z., Wade, A., Yaitskaya, N., Zhumabayev, D., 2020. Emptying water towers? Impacts of future climate and glacier change on river discharge in the northern Tien Shan, Central Asia. Water 12 (3), 627. https://doi.org/10.3390/w12030627.

Shahgedanova, M., Afzal, M., Severskiy, I., Usmanova, Z., Saidaliyeva, Z., Kapitsa, V., Kasatkin, N., Dolgikh, S., 2018. Changes in the mountain river discharge in the northern Tien Shan since the mid-20th century: results from the analysis of a homogeneous daily streamflow data set from seven catchments. J. Hydrol. 564, 1133–1152. https://doi.org/10.1016/j.jhydrol.2018.08.001.

Shahgedanova, M., Nosenko, G., Bushueva, I., Ivanov, M., 2012. Changes in area and geodetic mass balance of small glaciers, polar Urals, Russia, 1950–2008. J. Glaciol. 58, 953–964. https://doi.org/10.3189/2012JoG11J233.

Shahgedanova, M., Popovnin, V., Aleynikov, A., Stokes, C.R., 2011. Geodetic mass balance of Azarova glacier, Kodar mountains, eastern siberia, and its links to observed and projected climatic change. Ann. Glaciol. 52 (58), 129–137. https://doi.org/10.3189/172756411797252275.

Solomina, O., Bushueva, I., Dolgova, E., Jomelli, V., Alexandrin, M., Mikhalenko, V., Matskovsky, V., 2016a. Glacier variations in the northern Caucasus compared to climatic reconstructions over the past millennium. Glob. Planet. Chang. 140, 28–58. https://doi.org/10.1016/j.gloplacha.2016.02.008.

Solomina, O.N., Bradley, R.S., Jomelli, V., Geirsdottir, A., Kaufman, D.S., Koch, J., McKay, N.P., Masiokas, M., Miller, G., Nesje, A., Nicolussi, K., Owen, L.A., Putnam, A.E., Wanner, H., Wiles, G., Yang, B., 2016b. Glacier fluctuations during the past 2000 years. Quat. Sci. Rev. 149, 61–90. https://doi.org/10.1016/j.quascirev.2016.04.008.

Sommer, C., Malz, P., Seehaus, T.C., Lippl, S., Zemp, M., Braun, M.H., 2020. Rapid glacier retreat and downwasting throughout the European Alps in the early 21st century. Nat. Commun. 11, 3209. https://doi.org/10.1038/s41467-020-16818-0.

Steger, C.R., Reijmer, C.H., van den Broeke, M.R., Wever, N., Forster, R.R., Koenig, L.S., Kuipers Munneke, P., Lehning, M., Lhermitte, S., Ligtenberg, S.R.M., Miège, C., Noël, B.P.Y., 2017. Firn meltwater retention on the Greenland ice sheet: a model comparison. Front. Earth Sci. 5. https://doi.org/10.3389/feart.2017.00003.

Stokes, C.R., Gurney, S.D., Shahgedanova, M., Popovnin, V., 2006. Late-20th-century changes in glacier extent in the Caucasus Mountains, Russia/Georgia. J. Glaciol. 52 (176), 99–109. https://doi.org/10.3189/172756506781828827.

Stokes, C.R., Shahgedanova, M., Evans, I.S., Popovnin, V.V., 2013. Accelerated loss of alpine glaciers in the Kodar Mountains, south-eastern Siberia. Glob. Planet. Change. 101https://doi.org/10.1016/j.gloplacha.2012.12.010.

The IMBIE Team, 2020. Mass balance of the Greenland ice sheet from 1992 to 2018. Nature 579, 233–239. https://doi.org/10.1038/s41586-019-1855-2.

The IMBIE Team, 2018. Mass balance of the Antarctic ice sheet from 1992 to 2017. Nature 558, 219–222. https://doi.org/10.1038/s41586-018-0179-y.

Thibert, E., Eckert, N., Vincent, C., 2013. Climatic drivers of seasonal glacier mass balances: an analysis of 6 decades at glacier de Sarennes (French Alps). Cryosphere 7, 47–66. https://doi.org/10.5194/tc-7-47-2013.

Thomas, E.R., Melchior Van Wessem, J., Roberts, J., Isaksson, E., Schlosser, E., Fudge, T.J., Vallelonga, P., Medley, B., Lenaerts, J., Bertler, N., Van Den Broeke, M.R., Dixon, D.A., Frezzotti, M., Stenni, B., Curran, M., Ekaykin, A.A., 2017. Regional Antarctic snow accumulation over the past 1000 years. Clim. Past 13, 1491–1513. https://doi.org/10.5194/cp-13-1491-2017.

Treydte, K.S., Schleser, G.H., Helle, G., Frank, D.C., Winiger, M., Haug, G.H., Esper, J., 2006. The twentieth century was the wettest period in northern Pakistan over the past millennium. Nature 440, 1179–1182. https://doi.org/10.1038/nature04743.

Trusel, L.D., Das, S.B., Osman, M.B., Evans, M.J., Smith, B.E., Fettweis, X., McConnell, J.R., Noël, B.P.Y., van den Broeke, M.R., 2018. Nonlinear rise in Greenland runoff in response to post-industrial Arctic warming. Nature 564, 104–108. https://doi.org/10.1038/s41586-018-0752-4.

van den Broeke, M., Box, J., Fettweis, X., Hanna, E., Noël, B., Tedesco, M., van As, D., van de Berg, W.J., van Kampenhout, L., 2017. Greenland ice sheet surface mass loss: recent developments in observation and modeling. Curr. Clim. Chang. Rep. 3, 345–356. https://doi.org/10.1007/s40641-017-0084-8.

van den Broeke, M.R., Enderlin, E.M., Howat, I.M., Kuipers Munneke, P., Noël, B.P.Y., Jan Van De Berg, W., Van Meijgaard, E., Wouters, B., 2016. On the recent contribution of the Greenland ice sheet to sea level change. Cryosphere 10, 1933–1946. https://doi.org/10.5194/tc-10-1933-2016.

Velicogna, I., 2009. Increasing rates of ice mass loss from the Greenland and Antarctic ice sheets revealed by GRACE. Geophys. Res. Lett. 36, 2–5. https://doi.org/10.1029/2009GL040222.

Vinther, B.M., Buchardt, S.L., Clausen, H.B., Dahl-Jensen, D., Johnsen, S.J., Fisher, D.A., Koerner, R.M., Raynaud, D., Lipenkov, V., Andersen, K.K., Blunier, T., Rasmussen, S.O., Steffensen, J.P., Svensson, A.M., 2009. Holocene thinning of the Greenland ice sheet. Nature 461, 385–388. https://doi.org/10.1038/nature08355.

Viviroli, D., Kummu, M., Meybeck, M., Kallio, M., Wada, Y., 2020. Increasing dependence of lowland populations on mountain water resources. Nat. Sustain. 3, 917–928. https://doi.org/10.1038/s41893-020-0559-9.

Viviroli, D., Weingartner, R., 2004. The hydrological significance of mountains: from regional to global scale. Hydrol. Earth Syst. Sci. 8, 1017–1030. https://doi.org/10.5194/hess-8-1017-2004.

Vuille, M., Francou, B., Wagnon, P., Juen, I., Kaser, G., Mark, B.G., Bradley, R.S., 2008. Climate change and tropical Andean glaciers: past, present and future. Earth Sci. Rev. 89, 79–96. https://doi.org/10.1016/j.earscirev.2008.04.002.

Wahr, J., Swenson, S., Zlotnicki, V., Velicogna, I., May, A., 2004. Time-variable gravity from GRACE: first results. Geophys. Res. Lett. 31, 20–23. https://doi.org/10.1029/2004GL019779.

Winkler, S., Chinn, T., Gärtner-Roer, I., Nussbaumer, S.U., Zemp, M., Zumbühl, H.J., 2010. An introduction to mountain glaciers as climate indicators with spatial and temporal diversity. Erdkunde 64, 97–118. https://doi.org/10.3112/erdkunde.2010.02.01.

Wouters, B., Bamber, J.L., van den Broeke, M.R., Lenaerts, J.T.M., Sasgen, I., 2013. Limits in detecting acceleration of ice sheet mass loss due to climate variability. Nat. Geosci. 6, 613–616. https://doi.org/10.1038/ngeo1874.

Wouters, B., Gardner, A.S., Moholdt, G., 2019. Global glacier mass loss during the GRACE satellite Mission (2002-2016). Front. Earth Sci. 7, 1–11. https://doi.org/10.3389/feart.2019.00096.

Wouters, B., Martin-Espanol, A., Helm, V., Flament, T., van Wessem, J.M., Ligtenberg, S.R.M., van den Broeke, M.R., Bamber, J.L., 2015. Dynamic thinning of glaciers on the southern Antarctic peninsula. Science 348, 899–903. https://doi.org/10.1126/science.aaa5727.

Yang, R., Hock, R., Kang, S., Shangguan, D., Guo, W., 2020. Glacier mass and area changes on the Kenai peninsula, Alaska, 1986–2016. J. Glaciol. 66 (258), 603–607. https://doi.org/10.1017/jog.2020.32.

Zemp, M., Frey, H., Gärtner-Roer, I., Nussbaumer, S.U., Hoelzle, M., Paul, F., Haeberli, W., Denzinger, F., Ahlstrøm, A.P., Anderson, B., Bajracharya, S., Baroni, C., Braun, L.N., Cáceres, B.E., Casassa, G., Cobos, G., Dávila, L.R., Delgado Granados, H., Demuth, M.N., Espizua, L., Fischer, A., Fujita, K., Gadek, B., Ghazanfar, A., Ove Hagen, J., Holmlund, P., Karimi, N., Li, Z., Pelto, M., Pitte, P., Popovnin, V.V., Portocarrero, C.A., Prinz, R., Sangewar, C.V., Severskiy, I., Sigurđsson, O., Soruco, A., Usubaliev, R., Vincent, C., 2015. Historically unprecedented global glacier decline in the early 21st century. J. Glaciol. 61 (228), 745–762. https://doi.org/10.3189/2015JoG15J017.

Zemp, M., Hoelzle, M., Haeberli, W., 2009. Six decades of glacier mass-balance observations: a review of the worldwide monitoring network. Ann. Glaciol. 50, 101–111. https://doi.org/10.3189/172756409787769591.

Zemp, M., Huss, M., Thibert, E., Eckert, N., McNabb, R., Huber, J., Barandun, M., Machguth, H., Nussbaumer, S.U., Gärtner-Roer, I., Thomson, L., Paul, F., Maussion, F., Kutuzov, S., Cogley, J.G., 2019. Global glacier mass changes and their contributions to sea-level rise from 1961 to 2016. Nature 568, 382–386. https://doi.org/10.1038/s41586-019-1071-0.

CHAPTER

# 4

# Climate change and terrestrial biodiversity

*Rachel Warren, Jeff Price, and Rhosanna Jenkins*

**Tyndall Centre for Climate Change Research, University of East Anglia, Norwich, United Kingdom**

OUTLINE



*The Impacts of Climate Change*
**https://doi.org/10.1016/B978-0-12-822373-4.00025-2**

# 1 Introduction

The Earth's regional climate determines the geographical location of tropical forests, deserts, tundra, temperate woodland, and all other ecosystems. It also constrains the geographical range of many individual species. Hence, it is not surprising that the 1°C global warming that has so far occurred has already affected biodiversity with the effects of climate change on biodiversity and ecosystems already detected on every continent. At the species level, this includes observed changes in species' geographical ranges, changes in the timing of seasonal events (called phenology), and population declines. The first species extinction, attributable to anthropogenic climate change has already been observed. At the ecosystem level, some ecosystems have begun to transform, and the frequency of fires has increased in some areas. The effects are not uniform around the world—with particularly large effects detected already in the Arctic for example. With further climate change, much greater effects are projected upon species and ecosystems, with large scale species losses expected from many areas, resulting in the potential collapse of ecosystem functioning and a cascade of species extinctions. As well as the Arctic ecosystem, Mediterranean-type ecosystems, and biodiversity hotspots are particularly at risk.

Humans depend on nature in ways that many people are unaware of. Biodiversity performs natural processes resulting in the pollination of many of our crops, the breakdown of much of our waste, the purification of air and water, and the maintenance of the Earth's geochemical cycles. This last process includes the sequestration of around a half of the carbon dioxide that we have emitted into the atmosphere within terrestrial and marine ecosystems. Natural systems provide timber, food and fiber, and they often buffer settlements against the impacts of extreme weather events such as floods and storm surges. They regulate soil erosion and disease outbreaks, while providing health, recreational, and spiritual benefits. Thus, if terrestrial ecosystem functioning is diminished it places human civilization at greater and greater risk (see Chapter 5 for a discussion of marine ecosystems, with which terrestrial systems obviously interact at coastal zones).

The management of land is one necessary part of the solution to the climate change problem. Poor land management results in an escalation of climate change impacts on biodiversity. For example, to reduce (mitigate) the amount of global climate change, energy could be produced from energy crops. However, in this case if land is used on a large scale to grow energy crops in place of cropland or protected ecosystems, this can have deleterious effects on biodiversity and ecosystems that may be comparable with those caused by climate change itself. There are win-win solutions: ecosystems themselves may be restored in order to increase the sequestration of carbon by the biosphere, and carefully tailored use of energy crops and agricultural practices, together with changes in diet, can reduce the land use footprint of our society. In this way, careful land management can synergistically address the goals of the United Nations (UN) Framework Convention on Climate Change, the UN Convention on Biological Diversity, and the UN Convention to Combat Desertification (CCD).

# 2 Major shifts in biomes

## 2.1 Species, ecosystems, and biomes

Earth's biodiversity exists in close-knit ecosystems where the existence of each species is dependent on the existence of many other species variously providing them with food, shelter from climatic conditions or predators, a background against which they are camouflaged, and so on. Ecologically speaking, species interact with other species in a complex web. These interactions may be the interaction of predator and prey, or of host and parasite; of plant and pollinator, or of plant and seed disperser. Interactions between pairs of species may be mutually beneficial, or benefit only one of the two species at the expense of the other. It is this complex web that forms the ecosystem, often maintaining the suitable conditions for the species within it to survive. For example, by capturing energy from the sun through photosynthesis, maintaining the quality of air and water, by providing cooling shade, retaining water in the soil, and breaking down dead organisms so that the recycled material and energy can be used again by new plants and animals. Interactions may exist between almost any two (or more) species within an ecosystem—thus vertebrates interact with plants or insects, as well as other vertebrates, insects interact with vertebrates, plants, and other insects, and so on. Within the complex web, some species may be strongly dependent on their interactions with one or a few others and are known as specialists: they are unlikely to survive if those other species are not present. Other species are generalists and can survive through interactions with large numbers of different species. These generalist species tend to be more resilient to environmental change, which might result in the loss of some species from an ecosystem. When ecosystems become isolated climatically from other ecosystems, such as on mountaintops, species can become uniquely adapted to that particular environment, evolving slowly into a new species, and a biodiversity hotspot may form over millions of years. Such hotspots are also particularly vulnerable to environmental change.

While there are many types of different ecosystems, they fall generally into broad categories: forest, grassland, desert, tundra, and wetlands. Within each of these categories, subcategories are often considered, for example, forest may be deciduous or coniferous, tundra may be woody or not. The geographical boundaries of each of these categories and subcategories are constrained on large scales by the climate. This, in turn, means their presence is often correlated with their latitude on the earth, and with their altitude above sea level. Hence, "arctic" tundra is found on mountaintops as well as in the Arctic—because the climatic conditions there are similar to those found in the Arctic. These broad categories of ecosystems, which consist of a closely interacting system of plants and animals, are called biomes. Their location on earth is determined by the present climate of the earth and its regional variations over space.

The location of biomes on earth is also determined by the variation in climate over time as well as space. Therefore, if the earth's climate changes over time, it is also expected that the position of the earth's biomes will shift over time. Over geological time, climate has changed slowly in the past taking biomes with it. The UK has had tropical and even desert-like climates. Palm trees and crocodiles have flourished. However, when this happened, the land that makes up the UK was actually south of the equator. Over geological time there have been periods of "hothouse" worlds and periods of "icehouse" worlds. These transitions were often accompanied by mass extinction events. These events occurred over millions of years, so at a much slower rate than is currently occurring. However, the rate of present global warming is unprecedented in the past 10,000 years. The fastest regional warming to have occurred naturally, as far as we know, may have occurred near the end of the last ice age when the climate quickly flipped from one state to another. The advancing ice depauperated the diversity of ecological communities in Europe, where species were obliged to retreat south from the advancing ice, and were impeded by the barrier of the Alps, as compared with N America where the Rockies and Appalachians run N-S and hence such barriers were much less prevalent.

## 2.2 Biomes, the terrestrial biosphere, and the carbon cycle

The location of forests, grasslands, and deserts are all determined by regional climate. For example, deserts are all located where falling air dries the surrounding area. These areas are tied to the Earth's atmosphere, the amount of incoming solar radiation and the rotation of the Earth.

Dynamic general vegetation models (DGVMs) treat the earth's surface as if it were thousands of interconnected grid cells. The models take into account the local rainfall, the temperature, and sunshine amount; as well as the soil types and topography of regions. They simulate the local hydrology (that is, how water flows between the squares, down mountains and to the sea, or is absorbed by vegetation and held in the soil). This enables them to explain the location of current biomes on earth, as well as projecting where they may move to as climate changes. While the location of major biomes are largely fixed by climate, each biome is an assemblage of species with individual climatic requirements. Hence, as climate changes, the species composition of the biome can begin to change, while the overall biome can persist. With further climate change, the biome can transform into another one over time.

DGVMs also simulate parts of the biogeochemical cycles, including the carbon cycle. Just as water flows between different parts of the earth (i.e., between the oceans, the clouds, wetlands, groundwater, and back to the oceans), so do other chemicals on earth. One of the most important of these cycles is the carbon cycle that describes the natural flow of carbon from the rocks into the atmosphere via volcanic processes, from the atmosphere into plants, and from plants into soil and hence rock once more. These plants are not only terrestrial plants, but also the quadrillions of small creatures in the ocean known as "phytoplankton." When carbon dioxide is captured by plants through photosynthesis it produces energy in the form of sugars which can be used by the plant to build stems, roots, flowers, and seeds. This is called "primary production," thus plants are "primary producers" because they capture energy from the sun upon which all other life on earth, including human civilization, depends. This process also releases oxygen. Although plants also respire, consuming oxygen and releasing carbon dioxide, as they grow the net effect is one of removal of carbon dioxide from the atmosphere, often termed "carbon sequestration."

Many environmental problems are caused by humans' interference in these natural cycles. This book is being written because of human interference in the carbon cycle, whereby the carbon captured by plants in the Jurassic period (and earlier) and fossilized in coal and oil is being released into the atmosphere millions of years earlier than it would naturally be, because humans are burning these fossil fuels and releasing the carbon dioxide into the atmosphere. This means that the concentration of carbon dioxide in the atmosphere is much higher than it would naturally be, and so the natural greenhouse effect is artificially enhanced by humans, causing global warming. It is important to note that in the absence of a natural greenhouse effect, the Earth would be a snowball and there would likely be no life. The natural greenhouse effect comes from various gases, in normal concentrations, to trap heat and maintain a reasonably steady climate, geologically speaking. The additional "fossil" carbon released acts to push the Earth's climate to be more similar to that occurring when the organisms that eventually make up the oil and gas died.

Terrestrial primary production (plant growth) underpins many ecosystem services, including carbon cycling, food, and timber supplies. Net primary production (NPP) refers to the net amount of carbon captured by plants through photosynthesis. Primary production varies between regions and throughout the year. Tropical forests have a high productivity throughout the year whereas boreal forests have high productivity in the summer months (June and July) but a low productivity in the winter. NPP also varies year to year, in response to climate variation and major volcanic eruptions (Settele et al., 2014). Increases in net primary productivity relative to the preindustrial era have already been observed (Settele et al., 2014). However, these changes cannot be confidently attributed to climate change.

The total of life on earth is known as "the biosphere" and it is fortunate that the biosphere itself has sequestered about half of the carbon dioxide that we have emitted: that is the plants have been able to absorb about half of the extra amount. If that were not so, the other half would have warmed the earth by around another degree leading to warming of 2°C, so we have plants to thank that climate change is not already much larger.

Carbon dioxide ($CO_2$) is a fundamental resource for photosynthesis. Higher levels of carbon dioxide in the atmosphere can lead to higher rates of photosynthesis in plants, which is known as $CO_2$ fertilization. Elevated atmospheric $CO_2$ leads to increased short-term $CO_2$ uptake in plants (Jia et al., 2019). However, this does not uniformly lead to increased plant growth. Plant growth is affected by a number of other factors including soil nutrients, soil water, and light levels. $CO_2$ fertilization also affects water use efficiency in plants, allowing plants to use less water, which may alleviate the effects of droughts. Furthermore, plants grown under elevated $CO_2$ have also been shown to have altered floral traits, such as nectar composition (Hoover et al., 2012) and pollen protein concentration (Ziska et al., 2016), which has important implications for species that rely on them for food.

## 2.3 Projected future biome shifts

As noted in the introduction, without significant and immediate efforts to reduce the emissions of greenhouse gases by mankind (principally, carbon dioxide), the earth's climate is set to continue to change. Whilst the UNFCCC Paris Agreement seeks to limit warming to "well below 2°C" and to "pursue efforts" to limit warming to 1.5°C above preindustrial levels, the Agreement is being implemented by offered contributions to emission reduction made by each country, to be implemented by 2030. These offered contributions are currently sufficient only to limit global warming to perhaps 3°C, with more action needed both before and after 2030 to achieve the goals of the Agreement. If the country commitments are not implemented, then greater climate change might occur. Also, the calculation that the current contributions will limit warming to about 3°C are based on scientists' best estimate of the earth's climate sensitivity—that is how sensitive the planet is to the additional warming or "forcing" as it is scientifically termed, caused by increases in greenhouse gas emissions. There are large uncertainties in this estimate, and if the true value is higher than the best estimate (but still within the range that scientists think is possible) then the amount of climate change could also be larger than 3°C. Also, climate models do not fully account for all possible feedbacks that could occur, as scientific knowledge of these feedbacks, although extensive, is still unavoidably imperfect. Feedback processes are those which accelerate climate change. For example, when Antarctic or Greenland ice melts, less sunlight is reflected, increasing warming in those areas; or, when fire frequency is increased by climate change, the carbon locked up in the forests is lost to the atmosphere, making more carbon dioxide and increasing the greenhouse effect. Taken together, this means that global temperatures could rise by anything between 1.5°C and 4°C, or even more, relative to preindustrial times, by 2100 (Fig. 1).

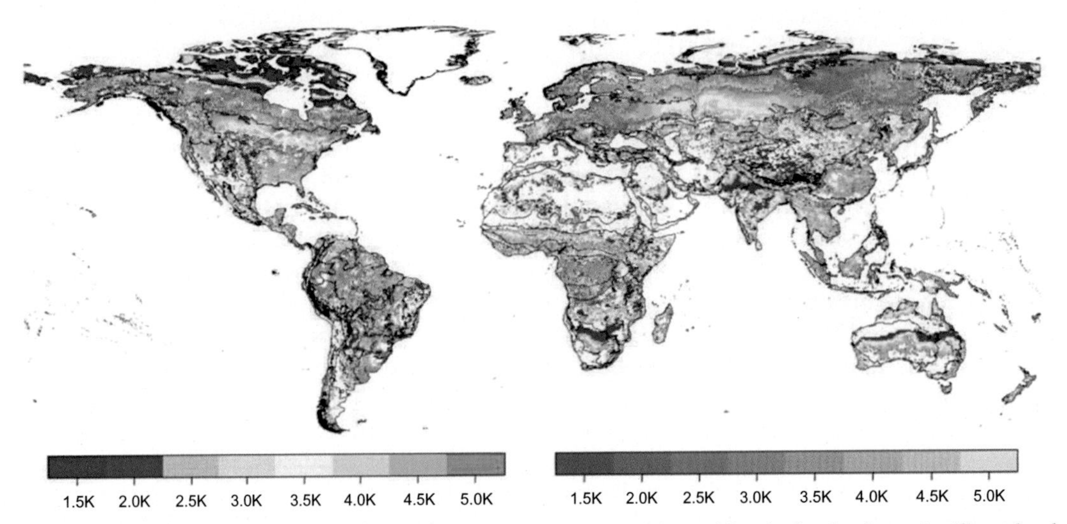


**FIG. 1** Threshold level of global temperature anomaly above preindustrial levels that leads to significant local changes in terrestrial ecosystems. Regions with severe (colored) or moderate (grayish) ecosystem transformation; delineation refers to the 90 biogeographic regions. All values denote changes found in >50% of the simulations. Regions colored in dark red are projected to undergo severe transformation under a global warming of 1.5°C while those colored in light red do so at 2°C; other colors are used when there is no severe transformation unless global warming exceeds 2°C. *Source: Reproduced with permission from Hoegh-Guldberg, O., Jacob, D., Taylor, M., et al., 2018. Impacts of 1.5° C global warming on natural and human systems. In: Masson-Delmotte, V.,* et al. *(Eds.), Global Warming of 1.5°C. An IPCC special report on the impacts of global warming of 1.5°C above pre-industrial levels and related global greenhouse gas emission pathways, in the context of strengthening the global response to the threat of climate change, sustainable development, and efforts to eradicate poverty. World Meteorological Organization, Geneva, Switzerland.*

# 3 Major loss of species geographical ranges

## 3.1 Relationship between geographical range and present-day climate

The vast majority of terrestrial species are not ubiquitous but are found only within a specific geographic range. Thus, most of the trees found in North American woodlands differ from those found in European ones, and again from those found in tropical forests. While the presence or absence of a species on small scales is strongly influenced by soil type, habitat type, topography, and competition with other species; on large scales, a major constraint on the presence/absence of a species is the physical climatic environment. The limitation imposed by climate can be mediated directly either on the organism itself or its ability to reproduce, for example if it becomes too cold, a species may be unable to generate enough energy from its food to survive; or if it becomes too hot and dry, a species may simply over-heat or its offspring will not survive. Alternatively, the effect can be indirect, for example a species' favored food plant may disappear if the climate becomes unsuitable, or a new predator may appear if the climate becomes more suitable for it. Finally, the ability of one species to outcompete another in a particular place can be affected by the climate.

Some species may buffer themselves against temporarily unfavorable climates by hibernating or estivating, while yet others take refuge in holes or burrows. On large scales, analysts have found statistical relationships between species presence and climatic conditions, mediated by so-called "bio-climatic variables" such as the rainfall in the driest month, the temperature of the hottest or coldest month, and so on (Phillips and Dudík, 2008). As a result, if the climate changes, it may become unsuitable for a species that is presently there; or it may become newly suitable for a species that is presently not there. Hence, species ranges may be expected to move across the earth's surface as climate change. Observed range changes are discussed in Section 3.2, and projected ones in Section 3.3. This is what scientists know occurred during previous climate changes, especially around the ice ages, most species shifted ranges to keep up with their climatic niche. These past changes occurred over much longer time periods so the velocity of change has been much lower than that projected for the future. The current projected rate of climate change is thought to exceed the rate at which most species can move to keep up. Thus, the future will see many novel climates, with novel habitat assemblages.

## 3.2 Observations of changes in species range

In 2003, global temperatures had warmed by 0.6°C, and a metaanalysis of 143 studies of observed changes in the geographical ranges of many species in diverse plant and animal taxa revealed that 80% of the shifts were in the direction that would be expected if driven by climate change (Root et al., 2003). Since then, further evidence has accrued showing that climate change is already affecting species ranges. More mobile species, such as birds (Lee and Barnard, 2016; Milne et al., 2015), bats (Wu, 2016) and some insects (Parmesan et al., 1999; DeVictor et al., 2012; Soroye et al., 2020) are shifting their distributions polewards, tracking their preferred conditions as the climate changes. Chen et al. (2011) conducted a metaanalysis and found that distributions of species have shifted to higher elevations at a median rate of 11 meters per decade, and to higher latitudes at a median rate of nearly 17 km per decade.

Rushing et al. (2020) compared observed range shifts of resident and migratory birds in eastern North America and found that migrants may be particularly vulnerable to climate change. Significantly, their results also showed that, on average, observed shifts in range of neither resident nor migratory birds had kept pace with temperature changes. Range contractions have been seen in some species that have not exhibited altitudinal and latitudinal shifts. In freshwater ecosystems, range contraction has been seen in cold-water species (Settele et al., 2014). In general, plants appear to be lagging behind the changes in climate as they lack the capability to rapidly disperse. However, range shifts can be difficult to detect in some groups of species, particularly those with low dispersal rates.

In montane regions, many range-restricted species have already shifted their ranges upslope. This was observed in birds in the Peruvian mountains (Freeman et al., 2018) and around Mount Kilimanjaro in Tanzania (Dulle et al., 2016) where a greater abundance of birds at higher elevations on Mount Kilimanjaro in 2011 than seen in 1991 was attributed to a reduction in temperature gradient with elevation. Altitudinal shifts have also been observed in insects, including beetles in Northern Ecuador (Moret et al., 2016) and bees in Spain (Ploquin et al., 2013).

There is also evidence of climate change induced range expansion in many invasive species in both terrestrial and freshwater ecosystems. The spread of invasive species is due to several factors, including land use change and the ability to confidently attribute this movement to climate change is low in most instances (Settele et al., 2014). Generally, freshwater species have less opportunity to disperse than terrestrial species. However, some changes are already being observed. In the Colorado River, a significant range expansion of nonnative trout was recorded between 1965 and 2004 (Roberts et al., 2017) linked to a reduction in snowmelt runoff flows. These invasive species invaded native trout populations at a rate of around 8% per decade and are projected to be a significant factor in extinction risk of the native species in the future as the climate warms further and nonnative trout move further upstream.

## 3.3 Future projections of changes in species ranges

As the climate continues to change, range changes are expected to become more pronounced. One may simplistically imagine that the "climatically suitable geographic range" of a species is like a "shadow" moving across the surface of the earth as viewed from a hypothetical satellite. As the climate changes, the "shadow" (which is the species' climate envelope) moves and changes shape according to the effect that global warming has on the regional climate and limited by other land uses. The "shadow" tends to move polewards and to higher elevations, but may "fall off" coastlines or "disappear off the top of" mountains. Hence, the area of land where climate is suitable for a species often shrinks as climate changes, although it can sometimes increase. The faster the "shadow" moves, the harder it is for species to keep up; when they cannot keep up, only some of the suitable climate space is occupied. The area of the shadow that is actually occupied decreases the faster the climate is changing. This also contributes to the shrinking of geographical ranges with climate change. Note that the "shadow" may become split by an impassable barrier, such as a sea channel or a large city.

Note that the ability or not to keep up is often termed "dispersal", but this is a misnomer because that is an ecological term that is usually applied to single individuals. The ability of an individual to disperse a great distance can be great—for example consider certain types of seeds or birds—but entire populations cannot move that fast—hence, it is the geographic speed of the whole population that has been measured in Section 3.2.

Scientists have used the combination of the observed present-day geographic range of species and the observed climate to understand how climate constrains present day species ranges (Section 3.1). Combining these statistical relationships with projections of future climate change allows scientists to project future geographic range changes in thousands of species, including or excluding the ability of the more mobile species populations to track a changing climate at the rates observed (Section 3.2). Quantifications of projected range losses of particular taxa in particular geographical locations including European, Mexican and South African plants, African mammals, European and North American birds, and Australian rainforests (Midgley et al., 2002; Peterson et al., 2002; Williams et al., 2003; Thuiller et al., 2005; Thuiller et al., 2006), created growing concern about the potential for climate change increase to extinction risk in general (Thomas et al., 2004) especially in biodiversity hotspots (Malcolm et al., 2006, and see section 7.5). A recent metaanalysis of 115,000 terrestrial species (Warren et al., 2018a) found that at 3.2°C global warming above pre-industrial levels 49% (range 31%–66%) of insects, 26% (range 16%–40%) of vertebrates, and 44% (29%–63%) of plants are projected to lose more than half of their geographic range due to anthropogenic climate change, even when the ability of mobile populations of birds, bats and butterflies to move is taken into account (Fig. 2A–C). The ranges in the projections reflect the varying projections of regional climate change associated with a given level of global warming that emerge from the use of various different global climate models.

A complementary approach to the climate envelope modeling method is a trait-based analysis, in which sensitivity, exposure and adaptive capacity of species is considered (Foden et al., 2013). This approach considers traits such as habitat specialization, environmental tolerance, interspecific interactions, rarity, dispersal ability, and genetic diversity. It found similarly large proportions of species at risk, including 2323–4890 bird species (24%–50%), and 1368–2740 amphibian species (22%–44%). Hence, two independent methods of analysis point to the same general finding: a large proportion of species are projected to be at risk from climate change. Urban (2015) performed a metaanalysis of 131 studies modeling the potential impacts of climate change on extinction risks using a range of different modeling techniques. While Urban found variability among individual studies, he also found that the different techniques largely yielded similar projections of extinction risk. The differences in estimates of extinction risk stemmed largely from differing assumptions of thresholds for extinction risk.

## 3.4 Implications for ecosystem services

When species ranges decline, they tend to shrink to "refugia." These locations tend to be similar across species and are often located in mountainous areas. In refugia, many of the species originally present can persist, whereas outside of them species disappear. Although other species may move in to replace these, the overall decline in species ranges means that the number moving in to an area is generally less than the number moving out, and species richness declines. Species loss undermines ecosystem functioning, and it is this functioning

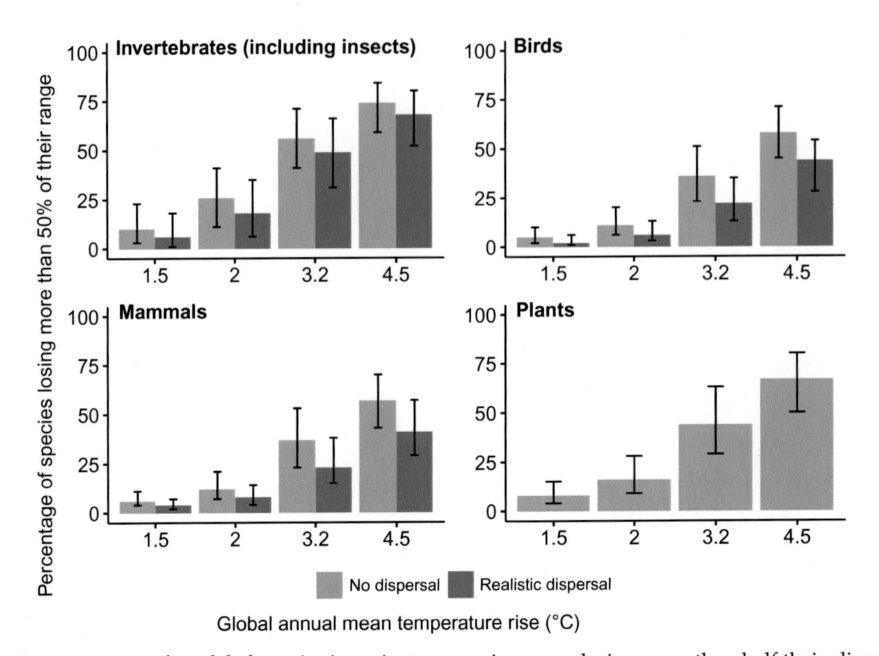


**FIG. 2** The proportion of modeled species in major taxonomic groups losing more than half their climatically determined geographic range by 2100 at specific levels of global warming. Projections are shown for of invertebrates (including insects, 31,536 species), birds and mammals (9735 species), and plants (73,224 species). Colors: Including (blue) and excluding (orange) realistic dispersal. Data are presented as the mean projection across 21 alternative climate model patterns with error bars indicating the 10%–90% range. *Source: These data are published in an alternative format in Warren, R., Price, J., Graham, E., Forstenhaeusler, N., Vanderwal, J., 2018a. The projected effect on insects, vertebrates and plants of limiting global warming to 1.5°C rather than 2°C. Science 360, 791–795.*

that provides ecosystem services (Gaston and Fuller, 2008). While there may be some redundancy in an ecosystem, in terms of more than one species performing a similar function, there are many specialists, for example co-dependent relationships between predators and prey, or pollinators and plants. Some species may perform a "keystone" function without which the whole ecosystem cannot function. A well-known example is the California sea otter which eats sea urchins: when it does not, the urchins eat the kelp to the extent that the kelp forest declines. Species richness loss associated with geographic range shrinkage is projected to be greatest in sub-Saharan Africa, the Amazon, Australia, North Africa, Central Asia, and Southeastern Europe (Warren et al., 2013).

The metaanalysis mentioned in the previous paragraph focuses on widespread and common species, meaning that the range shrinkage will have a strong impact on ecosystem services currently provided. Second, the projected loss of insects would be expected to seriously impact upon ecosystem functioning and services. In particular, the insect taxa that are most critical for crop pollination were found to be at high risk due to climate change (Warren et al., 2018a). Not only do insects perform key functions in ecosystems such as pollination, detrivory, and nutrient cycling, but they are also at the base of the food chain, providing food for organisms at higher trophic levels such as mammals and birds. So, while the models may simulate that a species can remain in a particular location, the analysis does not factor in whether species they may depend upon, such as particular plant or insect species, can remain. It is therefore considered that the projections underestimate the possible loss of ecosystem functioning due to the potential for disrupted ecological interactions including predator-prey, plant-pollinator, or other species-species interactions including parasitism and mutualism. In the UK and the Netherlands, for example, observed declines in pollinators have already been associated with declines in insect pollinated plants (Biesmeijer et al., 2006).

# 4 Changes in phenology

## 4.1 How climate affects phenology

Phenology refers to the timings of cyclical or seasonal biological events, such as migrations, egg laying, flowering, and hibernation. Many species of plants and animals have life cycle events that are influenced by climatic factors. For some, it is accumulated heat following winter that triggers these events, while other species are influenced by reaching a threshold temperature or even the onset of rainfall. Alternations to phenology are some of the least understood consequences of climate change, but it has been argued that it is also one of the most sensitive biological indicators of climate change (Calinger et al., 2013).

Generally, climate change will advance the timing of spring seasonal events for the majority of species, which is already well documented (Cohen et al., 2018; Parmesan, 2006; Parmesan and Yohe, 2003; Root et al., 2003). Shifts in phenology not only affect the individual species but also interactions among them. For instance, the timing of fruit on trees will impact the species that depend on these food sources. In some cases, different species within an ecosystem will experience shifts in phenology at different rates from other species that they interact with. This can lead to phenological mismatches (or asynchrony), such as a de-coupling of predator-prey or plant-pollinator interactions (Samplonius et al., 2020). Mismatches have already been witnessed between many species of birds and their insect food sources (Both et al., 2006; Pearce-Higgins et al., 2010; Reed et al., 2013). Evidence has shown that the phenology of primary producers and consumers has advanced more than secondary consumers (Thackeray et al., 2010). It has also been shown that species in strong mutualistic relationships (relationships that benefit both species) tend to suffer less from mismatch than species that do not exhibit a close relationship with particular other species, possibly due to species in mutually beneficial relationships having adapted together (Renner and Zohner, 2018).

Biological events in some species are triggered by the onset of cooler, rather than warmer, conditions so these would be delayed with greater climate change. For instance, delays in the first arrival of sardines to their spawning grounds off the coast of South Africa has been associated with the polewards shift in sea surface temperature isotherms (Fitchett et al., 2019). Furthermore, there is less of a consensus on how climate change is affecting end of season (autumn) phenology but most studies have reported delays in leaf colouring and leaf fall (Doi and Takahashi, 2008; Estrella and Menzel, 2006).

Phenological changes in vegetation can also lead to feedbacks to the climate system (Jia et al., 2019; Richardson et al., 2013). Feedbacks are processes that have been altered by climate change which in turn diminish (negative feedback) or amplify (positive feedback) climate change. Feedbacks of vegetation to the climate system that are influenced by phenology include albedo (the proportion of solar radiation that is reflected), water and energy fluxes and photosynthesis and $CO_2$ fluxes. However, knowledge of these potential effects of feedbacks is still incomplete.

## 4.2 Observed changes in phenology

There is substantial evidence that the timing of seasonal activities in both plants and animals are already changing as a result of recent warming. In the Northern Hemisphere, spring advancement has been occurring at a rate of 2.8 + 0.35 days per decade (Settele et al., 2014). Root et al. (2003) conducted a metaanalysis of 143 studies covering nearly 700 species and found that earlier timing of spring events ranged from 24 days earlier to 6.3 days later. A study by Thackeray et al. (2016) demonstrated that, at a UK-wide scale, phenological climate sensitivity varies greatly between species and between groups due to interacting factors such as population structure and resource availability. In the UK, the phenology of fish and insects was found to be particularly sensitive to climate changes. Both groups were projected to be particularly sensitive to further warming. Changes to phenology can also vary between different elevation zones (Illan et al., 2012; Saracco et al., 2019).

In plants, phenological changes have been observed across a wide range of taxa (Hoegh-Guldberg et al., 2018). Earlier flowering, usually as a consequence of warmer temperatures in the preceding months, has been observed in many species. For instance, Mohandass et al. (2015) found earlier flowering dates of three Alpine ginger plants in the Himalayas as a result of higher temperatures between 1913 and 2011. Similarly, Singh et al. (2016) investigated 8 native species of trees in the Western Himalayas and found the advancement in growing season of an average of 1.6 days per year between 1985 and 2015. This study also found that leaf drop was initiated earlier. In South Africa, Grab and Craparo (2011) found that higher temperatures and precipitation changes were linked to the earlier blooming of apple and pear trees between 1973 and 2009. Earlier seasonal greening-up of vegetation has been detected by satellites.

There are fewer long-term records of phenological changes in animals but there is still widespread evidence of changes in recent decades. Evidence has shown that some butterflies, moths, and wild bees have become active earlier in the season (Bartomeus et al., 2011; Roy and Sparks, 2000). Earlier egg laying was observed in birds across the world, including the Arctic (Grabowski et al., 2013), the UK (Pearce-Higgins et al., 2005) and in North America (Shipley et al., 2020). However, the latter study also showed that offspring that hatched earlier faced inclement weather events twice as often as they would have done in the 1970s. An advancement in the breeding season of an invasive rodent on the sub-Antarctic Marion Island from 1979 to 2011 was linked to the number of days without precipitation during the nonbreeding season (McClelland et al., 2018). Earlier breeding had a huge impact, increasing the population of mice by 530%.

Similarly, changes in reproductive phenology as a result of recent climate change have been observed in freshwater ecosystems. Krabbenhoft et al. (2014) investigated 8 native freshwater fish species in the Rio Grande, New Mexico, and found that the onset of spawning and emergence of young was 4–28 days earlier in 2008–10 compared to 1995. This was linked to lower precipitation in the later years. This was also seen in rivers in the Tibetan Plateau, where earlier breeding and a later onset of winter resulted in altered population dynamics of fish and other species in the ecosystem (Tao et al., 2018). In the UK, fish showed strong phenological advances with warming but freshwater phytoplankton did not exhibit a clear relationship with higher temperatures (Thackeray et al., 2016). The same study showed that crustaceans will be particularly vulnerable to phenological shifts by the 2050s.

### 4.3 Implications for ecosystem functioning and services

There is limited research into how changes in phenology and potential phenological mismatch may affect whole ecosystem functioning. However, some argue that phenological mismatches will alter nutrient cycling, carbon cycling, and pollination (Beard et al., 2019; Kudo and Ida, 2013). Pollination in particular could have large-scale impacts as nearly 90% of all plant species depend on pollination by animals. Analyses have shown that, to date, phenology of bees and plants are advancing at similar rates and suggest a complete decoupling of plant-pollinator interactions at the ecosystem scale is unlikely (Bartomeus et al., 2011; Rafferty and Ives, 2011). However, changes to pollination would have huge implications for provisioning ecosystem services, particularly food production. Burkle et al. (2013) found that phenological shifts explained 14%–44% of the plant-pollinator mismatches in Illinois (USA) over 120 years. As noted by Morton and Rafferty (2017), it is possible that, if a plant-pollinator interaction is disrupted, a new interaction may take its place. Limiting warming to 1.5°C may reduce the advancement of spring phenology enough to reduce the risks to ecosystem functioning (Hoegh-Guldberg et al., 2018).

Changes to breeding seasons in some invasive species may lead to population explosions, such as that seen in Marion Island, discussed above (McClelland et al., 2018). These changes had significant implications for the ecosystem as the invasive mice have decimated insect populations (their main food source).

Through changes in primary production, phenology can influence carbon cycling in ecosystems. Longer growing seasons have the potential to increase carbon sequestration. Phenological changes in plants can also affect the water cycle by altering evapotranspiration and runoff but further research is needed to fully understand these effects (Richardson et al., 2013).

## 5 Changes in extreme weather

### 5.1 How extreme weather effects ecosystems and species

As well as changes to the average long-term climate, biodiversity and ecosystems will be affected by extreme weather events, including heavy rainfall and heatwaves. Extreme events are, by definition, outside the normal conditions experienced by species and may affect some species more than changes to the average climate. Extreme weather events are already increasing in intensity, frequency, and duration with climate change. Extreme weather events affect biodiversity through direct mortality or reduced reproductive success, and while some impacts are immediate, others may play out on longer timescales, with effects sometimes being seen in subsequent year(s) . Maxwell et al. (2019) reviewed studies of ecological responses to past extreme events. About 57% of responses in these studies were classified as negative, with local extirpation seen following flood events or tropical storms. Other studies documented that invasive species could benefit from extreme events. During extreme years, population crashes are more common than population explosions (Palmer et al., 2017). The risks associated with extreme events are projected to be significantly greater at higher levels of warming (Hoegh-Guldberg et al., 2018).

Heatwaves can have profound effects on ecosystems and cause mortality among a range of species. Extreme heat during overwintering periods has been shown to lead to local extinction events in butterflies (McDermott-Long et al., 2016). Greater mortality in birds has been linked to heat-related stressors such as acute dehydration (Wingfield et al., 2017).

There is high confidence that extreme precipitation and floods are increasing due to climate change (Hoegh-Guldberg et al., 2018). The projected increase in extreme floods in the Amazon has been linked to a reduction in the habitat available for juvenile caimans (Herrera et al., 2015). At the coast, more severe coastal storms and cyclones will have destructive impacts on coastal and ocean ecosystems.

Climate change will also lead to reductions in some extreme events associated with cold, such as cold waves, extreme winters, and ice storms. Extreme low temperatures can reduce body condition and result in reproductive failure with cold waves coincide with the breeding season (Wingfield et al., 2017).

## 5.2 Changes in fire regimes

Climate is a major control on fire regimes in many ecosystems (Bowman et al., 2009). Climate change will affect fire extent, severity, and frequency, as well as lengthening the fire season in many regions. Fire activity is influenced by a range of factors. Higher temperatures and lower rainfall lead to drier vegetation, which acts as fuel for fires. Warmer temperatures are also expected to lead to an increase in lightning strikes (Romps et al., 2014). Other climate-driven changes in vegetation, such as the shift to increased woody biomass projected in several regions including sub-Saharan African savanna and the Arctic tundra, could amplify changes in fire regimes.

Despite the difficulty of attributing changes in fire regimes to anthropogenic climate change because of the influence of other factors, such as land use change and fire suppression management practices, strong evidence is now accruing to show that climate change is already increasing the global and regional risk of wildfire (Abatzoglou et al., 2019; Clarke et al., 2020) in areas as diverse as the Western United States (e.g., Khorshidi et al., 2020; Park-Williams et al., 2019), Alaska (Partain et al., 2016), Sweden (Krikken et al., 2019), Canada (Kirchmeier-Young et al., 2017), Iberia (e.g., Ruffault et al., 2020), Australia (e.g., Boer et al., 2020), the Amazon (e.g., Le Page et al., 2017), and NE China (Zhao et al., 2020), while in Siberia unprecedented fire regimes are being observed (Feurdean et al., 2020). For example, a study of the Mediterranean region found that nearly half of the long-term increases in fire weather and fire-related drought conditions has been caused by anthropogenic climate change (Barbero et al., 2020). In Australia, climate related warming and drying created combustible conditions in months leading up to the damaging fire season of 2016

(Hoffmann et al., 2019), while high temperatures preceded the unprecedented 2018 fire season in Queensland (Lewis et al., 2019).

Much of the mid-latitudes are projected to become warmer and drier because of climate change, including southern Australia, the Mediterranean, Southern Africa, and the western United States, and increased fire frequency has been projected to follow (e.g., Dowdy et al., 2019; Ruffault et al., 2020; Turco et al., 2018). Changes to fire regimes in these regions could lead to large shifts in tree species composition (Cassell et al., 2019) and hence implications for biodiversity. Similarly changes in fire regimes have been projected for boreal regions (de Groot et al., 2013).

## 5.3 Implications

Extreme events will have impacts on ecosystem services and functioning in a variety of ways. All extreme events have the potential to disrupt food webs. Lesser kestrel nestlings in the Mediterranean were found to have a 12% less chance of successfully fledging in drought years, which was linked to a decline in prey species (Marcelino et al., 2020). Extreme events such as heatwaves leading to mass mortality in insects, birds, and bats would have huge implications for pollination services for both wild plants and crops. Heatwaves and drought affect mortality in freshwater fish. Fish provide important provisioning ecosystem services (e.g., food), carbon removal, and important cultural and recreational ecosystem services.

Although fire is a naturally occurring essential part of some ecosystems, the greater occurrence of fires caused by climate change will reduce the ability of ecosystems to recover from the disturbance. Recovery time for different ecosystem services postfire varies considerably. Some services, such as carbon storage, are not fully returned until the vegetation at the site is mature again (Stephenson et al., 2014). Increased fire in some ecosystems could affect the provision of goods obtained from ecosystems such as food, fuelwood, and timber. All extreme events affecting plant growth would influence carbon cycling.

Floods not only lead to habitat loss but also greater soil erosion and changes in nutrient cycling, including increasing the discharge of nutrients into waterways. Flooding also causes huge disturbance to freshwater ecosystems and the ecosystem services they provide, including food and drinking water. However, not all effects of flooding are negative. Floods can recharge groundwater and wetlands and increase soil fertility. The nature of the impacts depends on the size of the flood and the conditions of the ecosystem before the event. Extreme floods are more likely to result in a loss of ecosystem services (Talbot et al., 2018).

Maxwell et al. (2019) reviewed studies that monitored species or ecosystem recovery following extreme events and found 38% of these studies found species or ecosystems did not fully recover. 62% of studies showed full or partial recovery, but this recovery took up to 10 years to reach predisturbance levels.

# 6 Miscellaneous mechanisms

In addition to shifting their ranges or exhibiting changes in phenology, biodiversity may be affected by climate change in other ways. The pace of anthropogenic climate change will likely be too fast for many species to adapt through new genetic mutations. However, some species may exhibit rapid genetic or evolutionary responses to climate change. There is evidence of longer wings or larger thoraces in some insects, which allow them to disperse further (Parmesan, 2006). Myers et al. (2017) reviewed the literature on observed and projected effects of climate change on freshwater fish species and concluded that the limited information on documented evolutionary responses was a key knowledge gap which should be addressed by future research. Changes to metabolic rates of fish have been seen in ocean ecosystems as a consequence of warming waters (Carozza et al., 2019).

Furthermore, changes in climate may lead to changes in animal body size (Sheridan and Bickford, 2011). This has been witnessed in some amphibians (Caruso et al., 2014), mammals (Yom-Tov and Yom-Tov, 2004), birds (Yom-Tov and Yom-Tov, 2006) and marine fish (Todd et al., 2008). In laboratory experiments, Greenspan et al. (2020) linked stunted tadpole growth to changes in environmental bacteria and microbial recruitment, which occurred as a result of warmer temperatures. In areas where plant growth (primary productivity) is projected to decrease, declines in organism size could be linked to lower resource availability. However, there are exceptions to this trend. Some species at high latitudes have increased body size due to longer growing or feeding seasons associated with warmer temperatures (Sheridan and Bickford, 2011).

In addition, some species have already started to show temperature-related changes in the sex ratio (Kallimanis, 2010). Temperature-related sex determination applies to the majority of reptiles and many fish. Several studies have shown that changes to sand temperature is affecting the sex ratios for nesting turtles in the Caribbean, with warmer temperatures leading to a greater proportion of female hatchlings (Laloë et al., 2016; Tanner et al., 2019).

## 7 Geographical implications

### 7.1 Arctic

Over recent decades, the high Arctic has warmed more than the global average (Hoegh-Guldberg et al., 2018). The effects of warming on biodiversity in the Arctic is likely to be particularly significant and many, such as range shifts, melting permafrost, and earlier snowmelt and sea ice break up, are already being observed. Arctic ecosystems are strongly influenced by cryosphere processes such as snow and ice-melt. Changes to snow and ice alter habitats and the species that depend on these habitats or processes within them (e.g., snowmelt). The most profound effects of climate change on Arctic biodiversity will be as a result of the loss of ice. Phenological responses are already being witnessed in Arctic vegetation, which show earlier fruiting and flowering (Panchen and Gorelick, 2017) and birds, which have shown earlier egg laying as a result of earlier snowmelt (Grabowski et al., 2013; Wingfield et al., 2017). However, shorebirds that breed in the Arctic are dependent on short seasonal bursts of invertebrate abundance. This leaves them vulnerable to potential phenological mismatch if the birds' arrival does not correspond to the emergence of insects. Plant flowering advanced by up to 20 days in one decade in some Arctic areas (CAFF, 2013).

Higher temperatures are causing degradation of permafrost in the Arctic tundra. This, along with increased fire probability, could help woody species establish in tundra areas, leading to a northward shift in the boreal coniferous forest and shrinking of the tundra ecosystem, which is known as "Arctic squeeze" (Kaplan and New, 2006). Limiting warming to 1.5°C would prevent the thawing of an estimated 1.5–2.5 million $km^2$ of permafrost (Hoegh-Guldberg et al., 2018).

Overall biodiversity is projected to increase in the Arctic as many animal species currently found in the sub-Arctic move northwards, tracking their preferred climatic conditions. However, native Arctic species are likely to become increasingly threatened. A global analysis by Foden et al. (2013) found that the proportion of bird species threatened was highest in the Arctic and Southern Ocean. Unlike in other areas, where they may be able to move north to track their preferred conditions, species adapted to the high Arctic will not be able to shift their ranges as the climate warms (MacDonald, 2010). Therefore, many Arctic species may go extinct as a result of warming, particularly those inhabiting islands far away from the main continental land mass.

## 7.2 Tropics

There are many different habitats within the tropics including: tropical rainforest, cloud forest, wetlands, dry forest, desert, and even alpine habitat on the highest mountains. Some effects have already been observed, including upslope range shifts and local extirpations (Moret et al., 2016; Freeman et al., 2018).

Tropical forests are the most biodiverse ecosystems in the world and they provide a diverse range of ecosystem services both at the local and global level. Hannah et al. (2020) modeled changes to biodiversity and found that, in tropical regions, species richness could decline by over 60% under both RCP 2.6 (low emissions) and RCP 8.5 (high emissions) by the 2070s. Their analysis showed that the biodiversity of the Asia tropics is at a slightly greater risk than that of the Afrotropics or Neotropics (Central and South America). Reductions in species richness and elevated extinction risks have also been reported by smaller-scale studies. For instance, a 28%–36% decrease in the area suitable for bats in the Amazon by the 2070s was projected (Costa et al., 2017). Bats are important pollinators in tropical regions so changes to their distribution could have profound effects on the ecosystem.

The Amazon forest has been shown to be close to its climatic limits, making it particularly vulnerable to climate change. The Amazon contains more than a fifth of the species on Earth, so changes here could have global consequences. Over the recent decades, there has been an increase in the length and intensity of the dry season, including several droughts, as well as more extreme rainfall in the wet season. Compositional shifts in Amazon forests have already been observed, with a shift towards more drought-tolerant species (Esquivel-Muelbert et al., 2019). Fire frequency is also likely to increase in the Amazon region (Brando et al., 2020). Most wet, tropical forests in the Amazon basin have a low susceptibility to fire but changes to the ecosystem, such as through severe drought or logging, will increase susceptibility to burning.

Phenology in tropical regions is often linked to precipitation and moisture, so tropical forests on different continents may have different responses to climate change depending on the precipitation projections for the area (Richardson et al., 2013). To date, studies of phenological mismatch in tropical forests have shown few wider, cascading effects on the ecosystem as a whole (Kishimoto-Yamada et al., 2009; Sakai and Kitajima, 2019).

Evidence of the effects of climate change on tropical freshwater species is more limited. However, tropical freshwater species, including invertebrates, have been shown to have a narrower thermal tolerance range compared to subtropical and temperate species and so will be especially vulnerable to increases in water temperature (Polato et al., 2018).

## 7.3 Mediterranean

Mediterranean ecosystems are found in the mid-latitudes on all continents, including the Mediterranean Basin, California, Central Chile, the Cape Region of South Africa, and Southwestern and South Australia. The biodiversity in these regions is adapted to warm, dry summers and mild, wet winters. Mediterranean terrestrial and freshwater ecosystems are already affected by summer drought and projections are for reduced precipitation and more intense drought in the future. The phenology of Mediterranean shrub species is sensitive to drought conditions (Richardson et al., 2013). These drier conditions will also contribute to an increase in fire frequency. As discussed in Section 4.2, wildfire risk is already increasing in European and North American Mediterranean ecosystems and is projected to increase further in the future (Lozano et al., 2017). Limiting warming to 2°C could prevent catastrophic heatwaves in Mediterranean ecosystems (Hoegh-Guldberg et al., 2018).

Range shifts are already being seen, particularly at higher elevations. Many modeling studies project that Mediterranean species will face range shifts or increased risk of extinction with further warming. For example, Ahmadi et al. (2019) examined the extinction risk of endemic vipers due to climate change. They projected a reduction in range size of 59–97% for individual species by the 2070s with severe climate change (RCP8.5). Casazza et al. (2014) modeled changes to range-restricted plants in the Mediterranean, projecting a range reduction of around 25% for 20 species by the 2020s. The Mediterranean basin itself was also projected to be a hotspot for terrestrial invasive plant species under both low and high emissions scenarios (Wan et al., 2016), which will further threaten native biodiversity.

## 7.4 Sub-Saharan Africa

Sub-Saharan Africa encompasses a wide range of ecosystems, from tropical forests to arid and semiarid lands. Africa has an extremely rich biodiversity, with high numbers of endemic species, and is also projected to be particularly sensitive to the effects of climate change (Niang et al., 2014). Projections of the impact of future climate change on ecosystems and biodiversity vary for much of Africa, due to contrasting simulations of precipitation trends. This uncertainty over the future of Africa's climate has implications for adaptation and biodiversity protection.

Despite these uncertainties, reductions in species range and potential local extinctions have been projected for a wide range of taxa (Niang et al., 2014). African ecosystems will also be impacted by biome shifts from savanna to woody vegetation and changes to fire regimes (Martens et al., 2021; Midgley and Bond, 2015). However, to date, the pattern of change has been inconsistent across Africa. Increases in woody vegetation are being observed in central, eastern, and southern Africa (Buitenwerf et al., 2012; Wigley et al., 2010), whereas reductions have been noted in West Africa (Gonzalez et al., 2010).

Changes to water availability will have a profound effect on Africa's ecosystems and biodiversity as its ecosystems are particularly water limited (Midgley and Bond, 2015). Lakes in the African Rift Valley are already suffering from drought (Dudgeon et al., 2006), which will likely be exacerbated by further climate change. By the 2050s, 80% of Africa's freshwater fish species are likely to experience conditions substantially different from the conditions they experience at present (Thieme et al., 2010). Reductions in flow are projected to threaten the survival of 9% of freshwater dependent fish and birds in Africa. Freshwater species will also face increased water temperatures. Elevated water temperatures associated with climate changes have already been seen in East African lakes (Olaka et al., 2010).

## 7.5 Biodiversity hotspots

Biodiversity is unevenly distributed across the world. Areas that have a particularly rich biodiversity and many endemic species are often known as biodiversity hotspots. These hotspots often overlap with areas that experienced the most stable climatic conditions during the Pleistocene (Brown et al., 2020). As the conditions were relatively stable, more species were able to persist in these hotspots than other areas of the globe. These areas acted as refugia for species (see Section 3.4 for a discussion of refugia), where the long-term climatic stability supported high biodiversity. Anthropogenic climate change is occurring at a faster rate than previous global change and areas that were refugia in the past are unlikely to be so in the future. Today, many species found within biodiversity hotspots are endemic. Endemic species are more sensitive to climate change as they will not be easily able to respond to climate change by moving to new areas (Chichorro et al., 2019; Dirnböck et al., 2011; Enquist et al., 2019). Therefore, biodiversity hotspots are particularly vulnerable to future climate change.

There have been several attempts to define biodiversity hotspots. Myers et al. (2000) put forward 36 terrestrial biodiversity hotspots. Habel et al. (2019) assessed the impact of climate change on endemic plants within these hotspots. Their results showed that hotspots on the African continent and smaller hotspots suffered the highest change in climate space, meaning that the species there could experience novel conditions. Bellard et al. (2014) predicted that hotspots would experience an average 31% loss of current climatic conditions by the 2080s. This change in climate would negatively impact an average of 25% of endemic species per hotspot. Malcolm et al. (2006) assessed the climate change impact on 25 biodiversity hotspots. They found that the Cape Floristic Region, Caribbean, Indo-Burma, Mediterranean Basin, Southwest Australia, and Tropical Andes were likely to be particularly vulnerable to climate change. These studies highlighted the risk that climate change poses to island biodiversity hotspots. The Caribbean Islands, Madagascar and the Indian Ocean Islands, the Philippines and Sri Lanka could lose all their endemic plants with continued warming (Habel et al., 2019). Islands are often centres of endemicity and are also at a greater risk from invasive species (Kier et al., 2009).

Olson and Dinerstein (2002) created the "Global 200," which are 238 ecoregions around the globe determined as having unique and irreplaceable ecosystems or species. Out of these ecoregions 142 are terrestrial, 53 freshwater, and 43 marine. These areas have been consolidated by World Wildlife Fund (WWF) into 35 "Priority Places" for conservation. An assessment of the impacts of climate change on these Priority Places found that higher levels of warming would increase biodiversity loss (Warren et al., 2018b). With high levels of warming (4.5°C), the area of refugia within Priority Places is projected to be limited to 33% when species are able to disperse and to only 18% if they are not able to track their preferred climatic conditions. Limiting warming to 2°C increases the area of refugia within Priority Places, with the Eastern Himalayas biodiversity hotspot seeing the greater benefit of mitigation.

There has been significantly less research into the effects of climate change on the biodiversity of freshwater biodiversity hotspots compared to terrestrial biodiversity hotspots. Studies of individual freshwater biodiversity hotspots have projected habitat and range loss (Herrera et al., 2015) and reductions in species richness (James et al., 2017).

## 8 Synthesis

### 8.1 Burning embers

The Intergovernmental Panel on Climate Change (IPCC) is tasked by the United Framework Convention on Climate Change (UNFCCC) to regularly assess the peer reviewed literature and summarize it in a series of reports that are designed to inform the UNFCCC policy making process in which world governments participate. The UNFCCC Article 2 (https://unfccc.int/resource/docs/convkp/conveng.pdf) states that the "ultimate objective of this Convention ...is the stabilization of greenhouse gas concentrations in the atmosphere at a level that would prevent dangerous anthropogenic interference with the climate system.... Such a level should be achieved within a time frame sufficient to allow ecosystems to adapt naturally to climate change, to ensure that food production is not threatened and to enable economic development to proceed in a sustainable manner." Owing to the difficulty of defining dangerous anthropogenic climate change, the selection of the appropriate level needed to be informed by some way of clearly indicating how climate change risk accrues with warming. In the IPCC reports, the information about risks due to global warming is typically synthesized in a number of different ways, including the Summary for Policy Makers. However, a particularly important graphic which communicates these risks in a single diagram has been created in several successive reports. This is affectionately known as the "Burning Embers" diagram because it consists of several parallel bars that change color from white to yellow to red as one looks along the bar: the distance along the bar (or the *y*-axis of the graphic) indicates the level of global warming since preindustrial times. The diagram splits different types of risk into five different "Reasons for Concern" (RFC, O'Neill et al., 2017). This phrase is chosen because it is important to consider whether expected changes due to climate change actually create risks, i.e., do they actually matter, in terms of the expected consequences to human and natural systems. As O'Neill et al. (2017) explain, in considering this it is also important to take into account the magnitude, irreversibility, likelihood, and ability to adapt to the risk in question. The levels of risk are considered "moderate" once impacts are detectable and attributable to anthropogenic climate change, when the "ember" transitions from white to yellow; while levels of risk are considered "high" once risks become severe and widespread, when the "ember" transitions to "red"; and finally "very high" once impacts are severe and persistent, with limited ability to adapt, when the ember transitions to purple.

The assessed level of risk at a given level of warming has tended to increase with successive IPCC reports, owing to the gradual accumulation of literature over time (Zommers et al., 2020). Two of the RFCs, RFC1 and RFC4, are directly linked with the effects of climate change on biodiversity—while a third (RFC3, the distribution of impacts upon human societies across the world) is indirectly linked through the effects that biodiversity loss has on human systems due to losses in ecosystem services, which add to the ways in which climate change directly affects those human systems. RFC1 relates to "unique and threatened systems" which includes biodiversity hotspots and the Arctic (see Section 7). On the basis of the published peer reviewed literature on the subject, a small part of which is cited in this chapter, Hoegh-Guldberg et al. (2018) concluded that risks to terrestrial ecosystems are presently moderate, but will transition to high with 1.5°C of warming, and to very high between 1.5°C and 2°C, reflecting the limited capacity of ecosystems to adapt to climate change. RFC4 relates to global aggregate impacts, considering the economy as a whole, supplemented by an element that is difficult to accurately reflect in economic models—and that is the economic consequences of a global biodiversity decline (Fig. 3). Taking into account the effects of climate change on human and natural systems together, Hoegh-Guldberg et al. (2018) concluded that global aggregate risks would reach a moderate level with 1.5°C warming, and transition to high risks between 1.5°C and 2.5°C: it should be noted that the Paris Agreement aspires to restricting global warming to "well below 2°C" and to "pursue efforts" to limit to 1.5°C.

## 8.2 What does this mean for extinction risks overall?

Species extinction risks are already high due to a range of drivers, in particular changes in land use with approximately 1 in 8 million species currently at risk (IPBES, 2019). 322 vertebrate species have become extinct since 1500, and in the past four decades, vertebrate species have shown a 25% average decline in abundance; while 67% of monitored insect populations show a mean abundance decline of 45% (Dirzo et al., 2014). Climate change is projected to greatly increase extinction risks (Urban, 2015). Fischlin et al. (2007) concluded that as global temperature rise exceeds 2–3°C, 20%–30% of species assessed by that time were expected to be at increasingly high risk of extinction. This conclusion was based on much of the evidence provided in other sections of this chapter. Hoegh-Guldberg et al. (2018) explain how even small additional amounts of global warming matter, with increases in extinction risk being lower at 1.5°C warming than they are at 2°C warming.

# 9 Solutions

IPCC (2018) explore alternative climate change mitigation pathways for limiting warming to 1.5°C. Typically, climate mitigation involves humans making changes to their energy system to avoid emitting carbon dioxide into the atmosphere. It also involves stopping deforestation, which is still ongoing and contributes to biodiversity loss—overall, emissions from land use change (largely deforestation) are presently contributing 23% of the total greenhouse gas emissions emitted annually (IPCC, 2019). In terms of biodiversity, halting deforestation

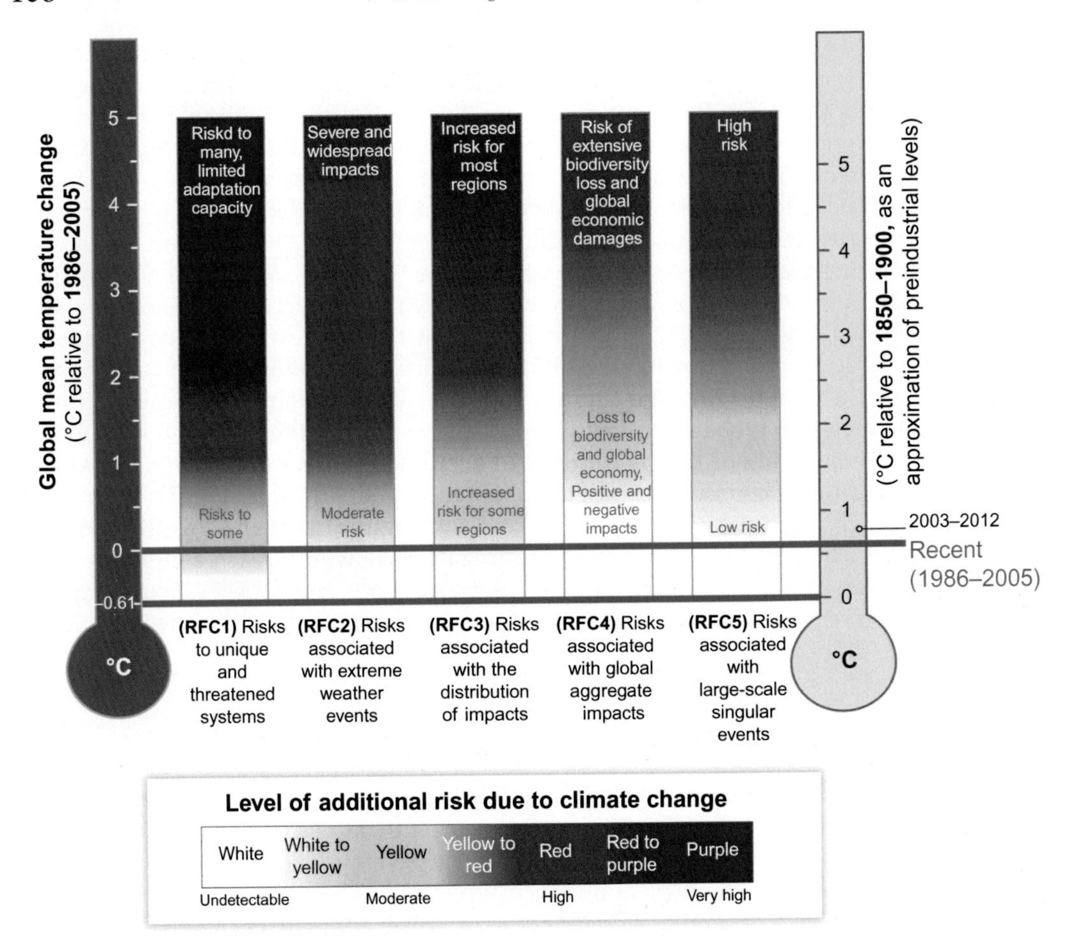


**FIG. 3** Five integrative Reasons for Concern (RFCs) provide a framework for summarizing key impacts and risks across sectors and regions, and were introduced in the IPCC Third Assessment Report. RFCs illustrate the implications of global warming for people, economies and ecosystems. Impacts and/or risks for each RFC are based on assessment of the new literature that has appeared. As in AR5, this literature was used to make expert judgments to assess the levels of global warming at which levels of impact and/or risk are undetectable, moderate, high or very high. The selection of impacts and risks to natural, managed and human systems in the lower panel is illustrative and is not intended to be fully comprehensive. RFC1 Unique and threatened systems: ecological and human systems that have restricted geographic ranges constrained by climate-related conditions and have high endemism or other distinctive properties. Examples include coral reefs, the Arctic and its indigenous people, mountain glaciers and biodiversity hotspots. RFC2 Extreme weather events: risks/impacts to human health, livelihoods, assets and ecosystems from extreme weather events such as heat waves, heavy rain, drought and associated wildfires, and coastal flooding. RFC3 Distribution of impacts: risks/impacts that disproportionately affect particular groups due to uneven distribution of physical climate change hazards, exposure or vulnerability. RFC4 Global aggregate impacts: global monetary damage, global-scale degradation and loss of ecosystems and biodiversity. RFC5 Large-scale singular events: are relatively large, abrupt and sometimes irreversible changes in systems that are caused by global warming. Examples include disintegration of the Greenland and Antarctic ice sheets. *Source: Reproduced with permission from IPCC, 2018. Summary for policymakers. In: Masson-Delmotte, V., Zhai, P., Pörtner, H-O., Roberts, D., Skea, J., Shukla, P.R., Pirani, A., Moufourma-Okia, W., Péan, C., Pidcock, R., Connors, S.,Matthews, J.B.R., Chen, Y., Zhou, X., Gomis, M.I., Lonnoy, E., Maycock, T., Tignor, M., Waterfield, T. World Meteorological Organization, Geneva, Switzerland.*

and other detrimental land use change is clearly a first step in working towards solutions: the UN program REDD+ (reducing emissions from deforestation and degradation) aims to do so. Indeed, without addressing deforestation, the Paris Agreement goals will likely not be attainable.

However, so much carbon dioxide has now accumulated into the atmosphere that all of the pathways to 1.5°C now need to ALSO use carbon dioxide removal (CDR), in which carbon is taken out of the atmosphere. Plants are effective at so doing. Hence, one way to do this is by using bioenergy crops and burning them to produce energy while capturing the released carbon dioxide and storing it underground: this is known as "Bioenergy with Carbon Capture and Storage" (or BECCS for short). Bioenergy crops, if grown on large scales however, would compete for land use with food production and biodiversity conservation (IPCC, 2019), thus raising significant concerns as to whether such mitigation pathways would appropriately deliver on Article 2 of the UNFCCC. Furthermore, carbon capture and storage (CCS) is not yet a mature widely deployed technology and there is doubt as to its large scale applicability on the required timescales, owing to lack of investment and issues related to public acceptability, although there has been some success in storing $CO_2$ in depleted oil and gas fields. Some of the pathways require much larger amount of BECCS than others. Fortunately, there are ways to reduce the amount of BECCS that is needed. Indeed, it is possible to achieve the Paris Agreement goal without BECCS entirely if energy demand is reduced (IPCC, 2018; Grubler et al., 2018). Another way to reduce the amount of BECCS required is to use ecosystem restoration to capture the carbon. There is a common perception that the solution is "planting trees"; however, this is not a panacea and a more holistic ecological approach is more desirable. "Afforestation" is often touted as a useful way forward as it can combine carbon sequestration with timber or fuelwood production. However, afforestation policies can have un-intended consequences, such as when the replacement of native vegetation with monocultures of bioenergy crops has highly detrimental outcomes for biodiversity and water use. Biodiversity benefits are generally not realized unless the "afforestation" uses native trees and occurs in locations where forest previously existed. Furthermore, forest is not the only ecosystem that stores carbon: large amounts of carbon are also stored in peatland and grassland, and thus "ecosystem restoration" is a more generally positive solution. This has the added benefit of real synergy with the goals of the United Nations Convention on Biological Diversity (CBD), which set the Aichi targets to protect biodiversity, which includes the aim of protecting 17% of terrestrial (including inland water) ecosystems by 2020: only a few countries have achieved this, and most have not; meanwhile, the UN has named the 2020s the "decade of ecosystem restoration".

The various approaches such as reforestation with native trees, restoration of degraded lands, agroforestry, and other techniques such as improved soil management in agricultural land are together known as "Nature-based solutions" and have been estimated to be able to provide 35%–40% of the climate change mitigation effort needed between now and 2030 to limit global warming to 2°C (IPBES, 2019). Large-scale restoration of forests are also an effective way to reduce the prevalence of fire, counteracting the tendency for climate change to increase fire frequency and increasing the permanency of carbon storage in forests (Liang et al., 2018). A synergy between the goals of the UNFCCC and the UNCBD can be achieved through using ecosystem restoration and protection in climatic refugia, the locations where biodiversity is most resilient to climate change.

## References


Abatzoglou, J.T., Williams, A.P., Barbero, R., 2019. Global emergence of anthropogenic climate change in fire weather indices. Geophys. Res. Lett. 46 (1), 326–336.

Ahmadi, M., Hemami, M.R., Kaboli, M., Malekian, M., Zimmermann, N.E., 2019. Extinction risks of a Mediterranean neo-endemism complex of mountain vipers triggered by climate change. Sci. Rep. 9 (1), 1–12.

Barbero, R., Abatzoglou, J.T., Pimont, F., Ruffault, J., Curt, T., 2020. Attributing increases in fire weather to anthropogenic climate change over France. Front. Earth Sci. 8, 104.

Bartomeus, I., Ascher, J.S., Wagner, D., Danforth, B.N., Colla, S., Kornbluth, S., Winfree, R., 2011. Climate-associated phenological advances in bee pollinators and bee-pollinated plants. Proc. Natl. Acad. Sci. 108 (51), 20645–20649.

Beard, K.H., Kelsey, K.C., Leffler, A.J., Welker, J.M., 2019. The missing angle: ecosystem consequences of phenological mismatch. Trends Ecol. Evol. 34 (10), 885–888.

Biesmeijer, J.C., Roberts, S.P., Reemer, M., Ohlemüller, R., Edwards, M., Peeters, T., Schaffers, A.P., Potts, S.G., Kleukers, R., Thomas, C.D., Settele, J., 2006. Parallel declines in pollinators and insect-pollinated plants in Britain and the Netherlands. Science 313 (5785), 351–354.

Bellard, C., et al., 2014. Vulnerability of biodiversity hotspots to global change. Glob. Ecol. Biogeogr. 23 (12), 1376–1386. https://doi.org/10.1111/geb.12228.

Boer, M.M., de Dios, Bradstock, R.A., 2020. Unprecedented burn area of Australian mega forest fires. Nat. Clim. Change 10 (3), 171–172.

Both, C., Bouwhuis, S., Lessells, C., Visser, M.E., 2006. Climate change and population declines in a long-distance migratory bird. Nature 441, 81–83.

Bowman, D.M., Balch, J.K., Artaxo, P., Bond, W.J., Carlson, J.M., Cochrane, M.A., D'Antonio, C.M., DeFries, R.S., Doyle, J.C., Harrison, S.P., Johnston, F.H., 2009. Fire in the earth system. Science 324 (5926), 481–484.

Brando, P.M., Soares-Filho, B., Rodrigues, L., Assunção, A., Morton, D., Tuchschneider, D., Fernandes, E.C.M., Macedo, M.N., Oliveira, U., Coe, M.T., 2020. The gathering firestorm in southern Amazonia. Sci. Adv. 6 (2), eaay1632.

Brown, S.C., Wigley, T.M., Otto-Bliesner, B.L., Rahbek, C., Fordham, D.A., 2020. Persistent quaternary climate refugia are hospices for biodiversity in the anthropocene. Nat. Clim. Chang. 10 (3), 244–248.

Buitenwerf, R., Bond, W.J., Stevens, N., Trollope, W.S.W., 2012. Increased tree densities in S outh African savannas: >50 years of data suggests $CO_2$ as a driver. Glob. Change Biol. 18 (2), 675–684.

Burkle, L.A., Marlin, J.C., Knight, T.M., 2013. Plant-pollinator interactions over 120 years: loss of species, co-occurrence, and function. Science 339 (6127), 1611–1615.

CAFF, 2013. Arctic Biodiversity Assessment. Status and Trends in Arctic Biodiversity. Conservation of Arctic Flora and Fauna, Akureyri.

Calinger, K.M., Queenborough, S., Curtis, P.S., 2013. Herbarium specimens reveal the footprint of climate change on flowering trends across north-Central North America. Ecol. Lett. 16 (8), 1037–1044.

Carozza, D.A., Bianchi, D., Galbraith, E.D., 2019. Metabolic impacts of climate change on marine ecosystems: implications for fish communities and fisheries. Glob. Ecol. Biogeogr. 28 (2), 158–169.

Caruso, N.M., Sears, M.W., Adams, D.C., Lips, K.R., 2014. Widespread rapid reductions in body size of adult salamanders in response to climate change. Glob. Chang. Biol. 20 (6), 1751–1759.

Casazza, G., Giordani, P., Benesperi, R., Foggi, B., Viciani, D., Filigheddu, R., Farris, E., Bagella, S., Pisanu, S., Mariotti, M.G., 2014. Climate change hastens the urgency of conservation for range-restricted plant species in the Central-Northern Mediterranean region. Biol. Conserv. 179, 129–138.

Cassell, B.A., Scheller, R.M., Lucash, M.S., Hurteau, M.D., Loudermilk, E.L., 2019. Widespread severe wildfires under climate change lead to increased forest homogeneity in dry mixed-conifer forests. Ecosphere. 10(11), e02934.

Chen, I.C., Hill, J.K., Ohlemüller, R., Roy, D.B., Thomas, C.D., 2011. Rapid range shifts of species associated with high levels of climate warming. Science 333 (6045), 1024–1026.

Chichorro, F., Juslén, A., Cardoso, P., 2019. A review of the relation between species traits and extinction risk. Biol. Conserv. 237, 220–229.

Clarke, H., Penman, T., Boer, M., Cary, G.J., Fontaine, J.B., Price, O., Bradstock, R., 2020. The proximal drivers of large fires: a pyrogeographic study. Front. Earth Sci. 8, 90.

Cohen, J.M., Lajeunesse, M.J., Rohr, J.R., 2018. A global synthesis of animal phenological responses to climate change. Nat. Clim. Chang. 8 (3), 224–228.

Costa, W.F., Ribeiro, M., Saraiva, A.M., Imperatriz-Fonseca, V.L., Giannini, T.C., 2017. Bat diversity in Carajás National Forest (Eastern Amazon) and potential impacts on ecosystem services under climate change. Biol. Conserv. 218, 200–210.

de Groot, W.J., Flannigan, M.D., Cantin, A.S., 2013. Climate change impacts on future boreal fire regimes. For. Ecol. Manag. 294, 35–44.

DeVictor, V., van Swaay, C., Brereton, T., Brotons, L., Chamberlain, D., Heliola, J., Herrando, S., Julliard, R., Kuussaari, M., Lindstrom, A., Reif, J., Roy, D.B., Schweiger, O., Settele, J., Stefanescu, C., Van Strien, A., Van Turnhout, C., Vermouzek, Z., Wallis, M., DeVries, I., Wynhoff, I., Jiguet, F., 2012. Differences in the climatic debts of birds and butterflies at a continental scale. Nat. Clim. Chang. 2, 121–124.

Dirnböck, T., Essl, F., Rabitsch, W., 2011. Disproportional risk for habitat loss of high-altitude endemic species under climate change. Glob. Change Biol. 17 (2), 990–996.

Dirzo, R., Young, H.S., Galetti, M., Ceballos, G., Isaac, N.J., Collen, B., 2014. Defaunation in the anthropocene. Science 345 (6195), 401–406.

Doi, H., Takahashi, M., 2008. Latitudinal patterns in the phenological responses of leaf colouring and leaf fall to climate change in Japan. Glob. Ecol. Biogeogr. 17 (4), 556–561.

Dowdy, A.J., Ye, H., Pepler, A., Thatcher, M., Osbrough, S.L., Evans, J.P., Di Virgilio, G., McCarthy, N., 2019. Future changes in extreme weather and pyroconvection risk factors for Australian wildfires. Sci. Rep. 9 (1), 1–11.

Dudgeon, D., et al., 2006. Freshwater biodiversity: importance, threats, status and conservation challenges. Biol. Rev. 81 (2), 163–182. https://doi.org/10.1017/s1464793105006950.

Dulle, H.I., Ferger, S.W., Cordeiro, N.J., Howell, K.M., Schleuning, M., Böhning-Gaese, K., Hof, C., 2016. Changes in abundances of forest understorey birds on Africa's highest mountain suggest subtle effects of climate change. Divers. Distrib. 22, 288–299.

Enquist, B.J., Feng, X., Boyle, B., Maitner, B., Newman, E.A., Jørgensen, P.M., Roehrdanz, P.R., Thiers, B.M., Burger, J.R., Corlett, R.T., Couvreur, T.L., 2019. The commonness of rarity: global and future distribution of rarity across land plants. Sci. Adv. 5 (11), eaaz0414.

Esquivel-Muelbert, A., Baker, T.R., Dexter, K.G., Lewis, S.L., Brienen, R.J., Feldpausch, T.R., Lloyd, J., Monteagudo-Mendoza, A., Arroyo, L., Álvarez-Dávila, E., Higuchi, N., 2019. Compositional response of Amazon forests to climate change. Glob. Chang. Biol. 25 (1), 39–56.

Estrella, N., Menzel, A., 2006. Responses of leaf colouring in four deciduous tree species to climate and weather in Germany. Clim. Res. 32 (3), 253–267.

Feurdean, A., Florescu, G., Tanţău, I., Vannière, B., Diaconu, A.C., Pfeiffer, M., Warren, D., Hutchinson, S.M., Gorina, N., Gałka, M., Kirpotin, S., 2020. Recent fire regime in the southern boreal forests of western Siberia is unprecedented in the last five millennia. Quat. Sci. Rev. 244, 106495.

Fischlin, A, Midgley, G.F., Price, J.T., et al., 2007. Ecosystems, their properties, goods, and services. In: Parry, M.L., Canziani, O.F., Palutikof, J.P., van der Linden, P.J., Hanson, C.E. (Eds.), IPCC Climate Change 2007: Impacts, Adaptation and Vulnerability. Cambridge Univ. Press, Cambridge, pp. 211–272.

Fitchett, J.M., Grab, S.W., Portwig, H., 2019. Progressive delays in the timing of sardine migration in the southwest Indian Ocean. S. Afr. J. Sci. 115 (7–8), 1–6.

Foden, W.B., et al., 2013. Identifying the world's most climate change vulnerable species: a systematic trait-based assessment of all birds, amphibians and corals. PLoS One. 8(6), e65427. https://doi.org/10.1371/journal.pone.0065427.

Freeman, B.G., Scholer, M.N., Ruiz-Gutierrez, V., Fitzpatrick, J.W., 2018. Climate change causes upslope shifts and mountaintop extirpations in a tropical bird community. Proc. Natl. Acad. Sci. U. S. A. 115, 11982–11987.

Gaston, K.J., Fuller, R.A., 2008. Commonness, population depletion and conservation biology. Trends Ecol. Evol. 23 (1), 14–19.

Gonzalez, P., Neilson, R.P., Lenihan, J.M., Drapek, R.J., 2010. Global patterns in the vulnerability of ecosystems to vegetation shifts due to climate change. Glob. Ecol. Biogeogr. 19 (6), 755–768.

Grab, S., Craparo, A., 2011. Advance of apple and pear tree full bloom dates in response to climate change in the southwestern cape, South Africa: 1973–2009. Agric. For. Meteorol. 151 (3), 406–413.

Grabowski, M.M., Doyle, F.I., Reid, D.G., Mossop, D., Talarico, D., 2013. Do Arctic-nesting birds respond to earlier snowmelt? A multi-species study in North Yukon, Canada. Polar Biol. 36 (8), 1097–1105.

Greenspan, S.E., Migliorini, G.H., Lyra, M.L., Pontes, M.R., Carvalho, T., Ribeiro, L.P., Moura-Campos, D., Haddad, C.F., Toledo, L.F., Romero, G.Q., Becker, C.G., 2020. Warming drives ecological community changes linked to host-associated microbiome dysbiosis. Nat. Clim. Change 10, 1–5.

Grubler, A., Wilson, C., Bento, N., Boza-Kiss, B., Krey, V., McCollum, D.L., Rao, N.D., Riahi, K., Rogelj, J., De Stercke, S., Cullen, J., 2018. A low energy demand scenario for meeting the 1.5 C target and sustainable development goals without negative emission technologies. Nat. Energy 3 (6), 515–527.

Habel, J.C., Rasche, L., Schneider, U.A., Engler, J.O., Schmid, E., Rödder, D., Meyer, S.T., Trapp, N., Sos del Diego, R., Eggermont, H., Lens, L., 2019. Final countdown for biodiversity hotspots. Conserv. Lett. 12(6), e12668.

Hannah, L., Roehrdanz, P.R., Marquet, P.A., Enquist, B.J., Midgley, G., Foden, W., et al., 2020. 30% land conservation and climate action reduces tropical extinction risk by more than 50%. Ecography. 43(7). https://doi.org/10.1111/ecog.05166.

Herrera, J., Solari, A., Lucifora, L.O., 2015. Unanticipated effect of climate change on an aquatic top predator of the Atlantic rainforest. Aquat. Conserv. Mar. Freshwat. Ecosyst. 25 (6), 817–828.

Hoegh-Guldberg, O., Jacob, D., Taylor, M., et al., 2018. Impacts of 1.5oC global warming on natural and human systems. In: Masson-Delmotte, V. et al., (Ed.), Global Warming of 1.5°C. An IPCC special report on the impacts of global warming of 1.5°C above pre-industrial levels and related global greenhouse gas emission pathways, in the context of strengthening the global response to the threat of climate change, sustainable development, and efforts to eradicate poverty. World Meteorological Organization, Geneva, Switzerland.

Hoffmann, A.A., Rymer, P.D., Byrne, M., Ruthrof, K.X., Whinam, J., McGeoch, M., Bergstrom, D.M., Guerin, G.R., Sparrow, B., Joseph, L., Hill, S.J., 2019. Impacts of recent climate change on terrestrial flora and fauna: some emerging Australian examples. Austral Ecol. 44 (1), 3–27.

Hoover, S.E., Ladley, J.J., Shchepetkina, A.A., Tisch, M., Gieseg, S.P., Tylianakis, J.M., 2012. Warming, CO2, and nitrogen deposition interactively affect a plant-pollinator mutualism. Ecol. Lett. 15 (3), 227–234.

Illan, J.G., Gutierrez, D., Diez, S.B., Wilson, R.J., 2012. Elevational trends in butterfly phenology: implications for species responses to climate change. Ecol. Entomol. 37 (2), 134–144.

IPBES, 2019. Díaz, S., Settele, J., Brondízio, E.S., Ngo, H.T., Guèze, M., Agard, J., … Zayas, C.N. (Eds.), Summary for Policymakers of the Global Assessment Report on Biodiversity and Ecosystem Services of the Intergovernmental Science-Policy Platform on Biodiversity and Ecosystem Services. IPBES Secretariat, Bonn, Germany 56 pp.

IPCC, 2018. Summary for policymakers. In: Masson-Delmotte, V., Zhai, P., Pörtner, H-O., Roberts, D., Skea, J., Shukla, P.R., Pirani, A., Moufourma-Okia, W., Péan, C., Pidcock, R., Connors, S., Matthews, J.B.R., Chen, Y., Zhou, X., Gomis, M.I., Lonnoy, E., Maycock, T., Tignor, M., Waterfield, T. (Eds.), Global Warming of 1.5°C. An IPCC special report on the impacts of global warming of 1.5°C above pre-industrial levels and related global greenhouse gas emission pathways, in the context of strengthening the global response to the threat of climate change, sustainable development, and efforts to eradicate poverty. World Meteorological Organization, Geneva, Switzerland.

IPCC, 2019. Summary for policymakers. In: Shukla, P.R., Skea, J., Calvo Buendia, E., Masson-Delmotte, V., Pörtner, H.-O., Roberts, D.C., Zhai, P., Slade, R., Connors, S., van Diemen, R., Ferrat, M., Haughey, E., Luz, S., Neogi, S., Pathak, M., Petzold, J., Portugal Pereira, J., Vyas, P., Huntley, E., Kissick, K., Belkacemi, M., Malley, J. (Eds.), Climate change and land: an IPCC special report on climate change, desertification, land degradation, sustainable land management, food security, and greenhouse gas fluxes in terrestrial ecosystems. Cambridge Univ. Press.

James, C.S., Reside, A.E., Van Der Wal, J., Pearson, R.G., Burrows, D., Capon, S.J., Harwood, T.D., Hodgson, L., Waltham, N.J., 2017. Sink or swim? Potential for high faunal turnover in Australian rivers under climate change. J. Biogeogr. 44 (3), 489–501.

Jia, G.E., Shevliakova, P., Artaxo, N., et al., 2019. Land–climate interactions. In: Shukla, P.R., Skea, J., Calvo Buendia, E., Masson-Delmotte, V., Pörtner, H.-O., Roberts, D.C., Zhai, P., Slade, R., Connors, S., van Diemen, R., Ferrat, M., Haughey, E., Luz, S., Neogi, S., Pathak, M., Petzold, J., Portugal Pereira, J., Vyas, P., Huntley, E., Kissick, K., Belkacemi, M., Malley, J. (Eds.), Climate change and land: an IPCC special report on climate change, desertification, land degradation, sustainable land management, food security, and greenhouse gas fluxes in terrestrial ecosystems. Cambridge Uni. Press.

Kallimanis, A.S., 2010. Temperature dependent sex determination and climate change. Oikos 119 (1), 197–200.

Kaplan, J.O., New, M., 2006. Arctic climate change with a 2 °C global warming: timing, climate patterns and vegetation change. Clim. Change 79, 213–241.

Khorshidi, M.S., Dennison, P.E., Nikoo, M.R., AghaKouchak, A., Luce, C.H., Sadegh, M., 2020. Increasing concurrence of wildfire drivers tripled megafire critical danger days in Southern California between 1982 and 2018. Environ. Res. Lett. 15 (10), 104002.

Kier, G., Kreft, H., Lee, T.M., Jetz, W., Ibisch, P.L., Nowicki, C., Mutke, J., Barthlott, W., 2009. A global assessment of endemism and species richness across island and mainland regions. Proc. Natl. Acad. Sci. 106 (23), 9322–9327.

Kirchmeier-Young, M.C., Zwiers, F.W., Gillett, N.P., Cannon, A.J., 2017. Attributing extreme fire risk in Western Canada to human emissions. Clim. Change 144 (2), 365–379.

Kishimoto-Yamada, K., Itioka, T., Sakai, S., Momose, K., Nagamitsu, T., Kaliang, H., Meleng, P., Chong, L., Karim, A.H., Yamane, S., Kato, M., 2009. Population fluctuations of light-attracted chrysomelid beetles in relation to supra-annual environmental changes in a Bornean rainforest. Bull. Entomol. Res. 99 (3), 217.

Krabbenhoft, T.J., Platania, S.P., Turner, T.F., 2014. Interannual variation in reproductive phenology in a riverine fish assemblage: implications for predicting the effects of climate change and altered flow regimes. Freshw. Biol. 59 (8), 1744–1754.

Krikken, F., Lehner, F., Haustein, K., Drobyshev, I., van Oldenborgh, G.J., 2019. Attribution of the role of climate change in the forest fires in Sweden 2018. Nat. Hazards Earth Syst. Sci. 1–24.

Kudo, G., Ida, T.Y., 2013. Early onset of spring increases the phenological mismatch between plants and pollinators. Ecology 94 (10), 2311–2320.

Laloë, J.O., Esteban, N., Berkel, J., Hays, G.C., 2016. Sand temperatures for nesting sea turtles in the Caribbean: implications for hatchling sex ratios in the face of climate change. J. Exp. Mar. Biol. Ecol. 474, 92–99.

Le Page, Y., Morton, D., Hartin, C., Bond-Lamberty, B., Pereira, J.M.C., Hurtt, G., Asrar, G., 2017. Synergy between land use and climate change increases future fire risk in Amazon forests. Earth Syst. Dynam. 8.

Lee, A.T., Barnard, P., 2016. Endemic birds of the fynbos biome: a conservation assessment and impacts of climate change. Bird Conserv. Int. 26 (1), 52–68.

Lewis, S.C., Perkins-Kirkpatrick, S.E., King, A.D., 2019. Approaches to attribution of extreme temperature and precipitation events using multi-model and single-member ensembles of general circulation models. Adv. Statist. Climatol. Meteorol. Oceanogr. 5 (2), 133–146.

Liang, S., Hurteau, M.D., Westerling, A.L., 2018. Large-scale restoration increases carbon stability under projected climate and wildfire regimes. Front. Ecol. Environ. 16 (4), 207–212.

Lozano, O.M., Salis, M., Ager, A.A., Arca, B., Alcasena, F.J., Monteiro, A.T., Finney, M.A., Del Giudice, L., Scoccimarro, E., Spano, D., 2017. Assessing climate change impacts on wildfire exposure in Mediterranean areas. Risk Anal. 37 (10), 1898–1916.

MacDonald, G.M., 2010. Some Holocene palaeoclimatic and palaeoenvironmental perspectives on Arctic/Subarctic climate warming and the IPCC 4th Assessment Report. J. Quat. Sci. 25, 39–47.

Malcolm, J.R., Liu, C., Neilson, R.P., Hansen, L., Hannah, L.E.E., 2006. Global warming and extinctions of endemic species from biodiversity hotspots. Conserv. Biol. 20 (2), 538–548.

Marcelino, J., Silva, J.P., Gameiro, J., Silva, A., Rego, F.C., Moreira, F., Catry, I., 2020. Extreme events are more likely to affect the breeding success of lesser kestrels than average climate change. Sci. Rep. 10 (1), 1–11.

Martens, C., Hickler, T., Davis-Reddy, C., Engelbrecht, F., Higgins, S.I., von Maltitz, G.P., Midgley, G.F., Pfeiffer, M., Scheiter, S., 2021. Large uncertainties in future biome changes in Africa call for flexible climate adaptation strategies. Glob. Change Biol. 27 (2), 340–358.

Maxwell, S.L., Butt, N., Maron, M., McAlpine, C.A., Chapman, S., Ullmann, A., Segan, D.B., Watson, J.E., 2019. Conservation implications of ecological responses to extreme weather and climate events. Divers. Distrib. 25 (4), 613–625.

McClelland, G.T., Altwegg, R., Van Aarde, R.J., Ferreira, S., Burger, A.E., Chown, S.L., 2018. Climate change leads to increasing population density and impacts of a key island invader. Ecol. Appl. 28 (1), 212–224.

McDermott-Long, O., Warren, R., Price, J., Brereton, T.M., Botham, M.S., Franco, A.M.A., 2016. Sensitivity of UK butterflies to local climatic extremes: which life stages are most at risk? J. Anim. Ecol. 86, 108–116.

Milne, R., Cunningham, S.J., Lee, A.T.K., Smit, B., 2015. The role of thermal physiology in recent declines of birds in a biodiversity hotspot. Conserv. Physiol. 3, 1–17.

Midgley, G.F., Hannah, L., Millar, D., Rutherford, M.C., Powrie, L.W., 2002. Assessing the vulnerability of species richness to anthropogenic climate change in a biodiversity hotspot. Glob. Ecol. Biogeogr. 11 (6), 445–451.

Midgley, G.F., Bond, W.J., 2015. Future of African terrestrial biodiversity and ecosystems under anthropogenic climate change. Nat. Clim. Chang. 5 (9), 823–829.

Mohandass, D., Zhao, J.-L., Xia, Y.-M., Campbell, M.J., Li, Q.-J., 2015. Increasing temperature causes flowering onset time changes of alpine ginger Roscoea in the Central Himalayas. J. Asia Pac. Biodivers. 8 (3), 191–198. https://doi.org/10.1016/j.japb.2015.08.003.

Moret, P., Aráuz, M.L.A., Gobbi, M., Barragán, A., 2016. Climate warming effects in the tropical Andes: first evidence for upslope shifts of Carabidae (Coleoptera) in Ecuador. Insect Conserv. Divers. 9 (4), 1–9.

Myers, B.J., Lynch, A.J., Bunnell, D.B., Chu, C., Falke, J.A., Kovach, R.P., Krabbenhoft, T.J., Kwak, T.J., Paukert, C.P., 2017. Global synthesis of the documented and projected effects of climate change on inland fishes. Rev. Fish Biol. Fish. 27 (2), 339–361.

Morton, E., Rafferty, N., 2017. Plant-pollinator interactions under climate change: the use of spatial and temporal transplants. Appl. Plant Sci. 5(6). https://doi.org/10.3732/apps.1600133.

Myers, N., et al., 2000. Biodiversity hotspots for conservation priorities. Nature 403 (6772), 853–858. https://doi.org/10.1038/35002501.

Niang, I., Ruppel, O.C., Abdrabo, M.A., Essel, A., Lennard, C., Padgham, J., Urquhart, P., 2014. Africa. In: Barros, V.R., Field, C.B., Dokken, D.J., Mastrandrea, M.D., Mach, K.J., Bilir, T.E., … White, L.L. (Eds.), Climate Change 2014: Impacts, Adaptation, and Vulnerability. Part B: Regional Aspects. Contribution of Working Group II to the Fifth Assessment Report of the Intergovernmental Panel on Climate Change. Cambridge University Press, Cambridge, United Kingdom and New York, NY, USA, pp. 1199–1265.

O'Neill, B.C., Oppenheimer, M., Warren, R., Hallegatte, S., Kopp, R.E., Pörtner, H.O., Scholes, R., Birkmann, J., Foden, W., Licker, R., Mach, K.J., 2017. IPCC reasons for concern regarding climate change risks. Nat. Clim. Chang. 7 (1), 28–37.

Olaka, L.A., Odada, E.O., Trauth, M.H., Olago, D.O., 2010. The sensitivity of East African rift lakes to climate fluctuations. J. Paleolimnol. 44 (2), 629–644.

Olson, D.M., Dinerstein, E., 2002. The global 200: priority ecoregions for global conservation. Ann. Mo. Bot. Gard. 89, 199–224.

Palmer, G., Platts, P.J., Brereton, T., Chapman, J.W., Dytham, C., Fox, R., Pearce-Higgins, J.W., Roy, D.B., Hill, J.K., Thomas, C.D., 2017. Climate change, climatic variation and extreme biological responses. Philos. Trans. R. Soc. B 372 (1723), 20160144.

Panchen, Z.A., Gorelick, R., 2017. Prediction of Arctic plant phenological sensitivity to climate change from historical records. Ecol. Evol. 7 (5), 1325–1338.

Parmesan, C., Ryrholm, N., Stefanescu, C., Hill, J.K., Thomas, C.D., Descimon, H., Huntley, B., Kaila, L., Kullberg, J., Tammaru, T., 1999. Poleward shifts in geographical ranges of butterfly species associated with regional warming. Nature 399 (6736), 579–583.

Park-Williams, A., Abatzoglou, J.T., Gershunov, A., Guzman-Morales, J., Bishop, D.A., Balch, J.K., Lettenmaier, D.P., 2019. Observed impacts of anthropogenic climate change on wildfire in California. Earth's Future 7 (8), 892–910.

Parmesan, C., 2006. Ecological and evolutionary responses to recent climate change. Annu. Rev. Ecol. Evol. Syst. 37, 637–669.

Parmesan, C., Yohe, G., 2003. A globally coherent fingerprint of climate change impacts across natural systems. Nature 421 (6918), 37–42.

Partain Jr., J.L., Alden, S., Strader, H., Bhatt, U.S., Bieniek, P.A., Brettschneider, B.R., Walsh, J.E., Lader, R.T., Olsson, P.Q., Rupp, T.S., Thoman Jr., R.L., 2016. An assessment of the role of anthropogenic climate change in the Alaska fire season of 2015. Bull. Am. Meteorol. Soc. 97 (12), S14–S18.

Pearce-Higgins, J.W., Dennis, P., Whittingham, M.J., Yalden, D.W., 2010. Impacts of climate on prey abundance account for fluctuations in a population of a northern wader at the southern edge of its range. Glob. Change Biol. 16 (1), 12–23.

Pearce-Higgins, J.W., Yalden, D.W., Whittingham, M.J., 2005. Warmer springs advance the breeding phenology of golden plovers Pluvialis apricaria and their prey (Tipulidae). Oecologia 143 (3), 470–476.

Peterson, A.T., Ortega-Huerta, M.A., Bartley, J., Sánchez-Cordero, V., Soberón, J., Buddemeier, R.H., Stockwell, D.R., 2002. Future projections for Mexican faunas under global climate change scenarios. Nature 416 (6881), 626–629.

Phillips, S.J., Dudík, M., 2008. Modeling of species distributions with Maxent: new extensions and a comprehensive evaluation. Ecography 31 (2), 161–175.

Ploquin, E.F., Herrera, J.M., Obeso, J.R., 2013. Bumblebee community homogenization after uphill shifts in montane areas of northern Spain. Oecologia 173 (4), 1649–1660.

Polato, N.R., Gill, B.A., Shah, A.A., Gray, M.M., Casner, K.L., Barthelet, A., Messer, P.W., Simmons, M.P., Guayasamin, J.M., Encalada, A.C., Kondratieff, B.C., 2018. Narrow thermal tolerance and low dispersal drive higher speciation in tropical mountains. Proc. Natl. Acad. Sci. 115 (49), 12471–12476.

Rafferty, N.E., Ives, A.R., 2011. Effects of experimental shifts in flowering phenology on plant–pollinator interactions. Ecol. Lett. 14 (1), 69–74.

Reed, T.E., Jenouvrier, S., Visser, M.E., 2013. Phenological mismatch strongly affects individual fitness but not population demography in a woodland passerine. J. Anim. Ecol. 82 (1), 131–144.

Renner, S.S., Zohner, C.M., 2018. Climate change and phenological mismatch in trophic interactions among plants, insects, and vertebrates. Annu. Rev. Ecol. Evol. Syst. 49, 165–182.

Richardson, A.D., Keenan, T.F., Migliavacca, M., Ryu, Y., Sonnentag, O., Toomey, M., 2013. Climate change, phenology, and phenological control of vegetation feedbacks to the climate system. Agric. For. Meteorol. 169, 156–173.

Roberts, J.J., Fausch, K.D., Hooten, M.B., Peterson, D.P., 2017. Non-native trout invasions combined with climate change threaten persistence of isolated cutthroat trout populations in the southern Rocky Mountains. N. Am. J. Fish Manag. 37 (2), 314–325.

Romps, D.M., Seeley, J.T., Vollaro, D., Molinari, J., 2014. Projected increase in lightning strikes in the United States due to global warming. Science 346 (6211), 851–854.

Root, T.L., Price, J.T., Hall, K.R., Schneider, S.H., Rosenzweig, C., Pounds, J.A., 2003. Fingerprints of global warming on wild animals and plants. Nature 42, 57–60.

Roy, D.B., Sparks, T.H., 2000. Phenology of British butterflies and climate change. Glob. Change Biol. 6, 407–416.

Ruffault, J., Thomas, C., Moron, V., Trigo, R., Mouillot, F., et al., 2020. Increased likelihood of heat-induced large wildfires in the Mediterranean Basin. Sci. Rep. 10(1). https://doi.org/10.1038/s41598-020-70069-z.

Rushing, C.S., Royle, J.A., Ziolkowski, D.J., Pardieck, K.L., 2020. Migratory behavior and winter geography drive differential range shifts of eastern birds in response to recent climate change. Proc. Natl. Acad. Sci. 117 (23), 12897–12903.

Sakai, S., Kitajima, K., 2019. Tropical phenology: recent advances and perspectives. Ecol. Res. 34 (1), 50–54.

Samplonius, J.M., Atkinson, A., Hassall, C., Keogan, K., Thackeray, S.J., Assmann, J.J., Burgess, M.D., Johansson, J., Macphie, K.H., Pearce-Higgins, J.W., Simmonds, E.G., 2020. Strengthening the evidence base for temperature-mediated phenological asynchrony and its impacts. Nat. Ecol. Evol. 1–10.

Saracco, J.F., Siegel, R.B., Helton, L., Stock, S.L., DeSante, D.F., 2019. Phenology and productivity in a montane bird assemblage: trends and responses to elevation and climate variation. Glob. Change Biol. 25 (3), 985–996.

Settele, J., Scholes, R., Betts, R., Bunn, S., Leadley, P., Nepstad, D., Overpeck, J.T., Taboada, M.A., 2014. Terrestrial and inland water systems. In: Field, C.B., Barros, V.R., Dokken, D.J., Mach, K.J., Mastrandrea, M.D., Bilir, T.E., … White, L.L. (Eds.), Climate Change 2014: Impacts, Adaptation, and Vulnerability. Part A: Global and Sectoral Aspects. Contribution of Working Group II to the Fifth Assessment Report of the Intergovernmental Panel on Climate Change. Cambridge University Press, Cambridge, United Kingdom and New York, NY, USA, pp. 271–359 (Chapter 6).

Sheridan, J.A., Bickford, D., 2011. Shrinking body size as an ecological response to climate change. Nat. Clim. Chang. 1 (8), 401–406.

Shipley, J.R., Twining, C.W., Taff, C.C., Vitousek, M.N., Flack, A., Winkler, D.W., 2020. Birds advancing lay dates with warming springs face greater risk of chick mortality. Proc. Natl. Acad. Sci. 117 (41), 25590–25594.

Singh, P., Negi, G.C.S., Pant, G.B., 2016. Impact of climate change on phenological responses of major forest tree of Kumaun Himalaya. ENVIS Bull. Himal. Ecol. 24, 112–116.

Stephenson, N.L., Das, A.J., Condit, R., Russo, S.E., Baker, P.J., Beckman, N.G., Coomes, D.A., Lines, E.R., Morris, W.K., Rüger, N., Alvarez, E., 2014. Rate of tree carbon accumulation increases continuously with tree size. Nature 507 (7490), 90–93.

Soroye, P., Newbold, T., Kerr, J., 2020. Climate change contributes to widespread declines among bumble bees across continents. Science 367 (6478), 685–688.

Talbot, C.J., Bennett, E.M., Cassell, K., Hanes, D.M., Minor, E.C., Paerl, H., Raymond, P.A., Vargas, R., Vidon, P.G., Wollheim, W., Xenopoulos, M.A., 2018. The impact of flooding on aquatic ecosystem services. Biogeochemistry 141 (3), 439–461.

Tanner, C.E., Marco, A., Martins, S., Abella-Perez, E., Hawkes, L.A., 2019. Highly feminised sex-ratio estimations for the world's third-largest nesting aggregation of loggerhead sea turtles. Mar. Ecol. Prog. Ser. 621, 209–219.

Tao, J., He, D., Kennard, M.J., Ding, C., Bunn, S.E., Liu, C., Jia, Y., Che, R., Chen, Y., 2018. Strong evidence for changing fish reproductive phenology under climate warming on the Tibetan plateau. Glob. Chang. Biol. 24 (5), 2093–2104.

Thackeray, S.J., Henrys, P.A., Hemming, D., Bell, J.R., Botham, M.S., Burthe, S., Helaouet, P., Johns, D.G., Jones, I.D., Leech, D.I., Mackay, E.B., Massimino, D., Atkinson, S., Bacon, P.J., Brereton, T.M., Carvalho, L., Clutton-Brock, T.H., Duck, C., Edwards, M., Elliott, J.M., Hall, S.J.G., Harrington, R., Pearce-Higgins, J.W., Hoye, T.T., Kruuk, L.E.B., Pemberton, J.M., Sparks, T.H., Thompson, P.M., White, I., Winfield, I.J., Wanless, S., 2016. Phenological sensitivity to climate across taxa and trophic levels. Nature 535, 241–U94.

Thackeray, S.J., Sparks, T.H., Frederiksen, M., Burthe, S., Bacon, P.J., Bell, J.R., Botham, M.S., Brereton, T.M., Bright, P.W., Carvalho, L., Clutton-Brock, T.I.M., 2010. Trophic level asynchrony in rates of phenological change for marine, freshwater and terrestrial environments. Glob. Change Biol. 16 (12), 3304–3313.

Thieme, M.L., Lehner, B., Abell, R., Matthews, J., 2010. Exposure of Africa's freshwater biodiversity to a changing climate. Conserv. Lett. 3 (5), 324–331.

Thomas, C.D., Cameron, A., Green, R.E., Bakkenes, M., Beaumont, L.J., Collingham, Y.C., Erasmus, B.F., De Siqueira, M.F., Grainger, A., Hannah, L., Hughes, L., 2004. Extinction risk from climate change. Nature 427 (6970), 145–148.

Thuiller, W., Lavorel, S., Araújo, M.B., Sykes, M.T., Prentice, I.C., 2005. Climate change threats to plant diversity in Europe. Proc. Natl. Acad. Sci. 102 (23), 8245–8250.

Thuiller, W., Broennimann, O., Hughes, G., Alkemade, J.R.M., Midgley, G.F., Corsi, F., 2006. Vulnerability of African mammals to anthropogenic climate change under conservative land transformation assumptions. Glob. Chang. Biol. 12 (3), 424–440.

Todd, C.D., Hughes, S.L., Marshall, C.T., Maclean, J.C., Lonergan, M.E., Biuw, E.M., 2008. Detrimental effects of recent ocean surface warming on growth condition of Atlantic salmon. Glob. Chang. Biol. 14 (5), 958–970.

Turco, M., Rosa-Cánovas, J.J., Bedia, J., Jerez, S., Montávez, J.P., Llasat, M.C., Provenzale, A., 2018. Exacerbated fires in Mediterranean Europe due to anthropogenic warming projected with non-stationary climate-fire models. Nat. Commun. 9 (1), 1–9.

Urban, M.C., 2015. Accelerating extinction risk from climate change. Science 348 (6234), 571–573.

Wan, J.Z., Wang, C.J., Yu, F.H., 2016. Risk hotspots for terrestrial plant invaders under climate change at the global scale. Environ. Earth Sci. 75 (12), 1012.

Warren, R., Van Der Wal, J., Price, J., Welbergen, J.A., Atkinson, I., Ramirez-Villegas, J., Osborn, T.J., Jarvis, A., Shoo, L.P., Williams, S.E., Lowe, J., 2013. Quantifying the benefit of early climate change mitigation in avoiding biodiversity loss. Nat. Clim. Chang. 3 (7), 678–682.

Warren, R., Price, J., Graham, E., Forstenhaeusler, N., Vanderwal, J., 2018a. The projected effect on insects, vertebrates and plants of limiting global warming to 1.5°C rather than 2°C. Science 360, 791–795.

Warren, R., Price, J., Van Der Wal, J., Cornelius, S., Sohl, H., 2018b. The implications of the United Nations Paris agreement on climate change for globally significant biodiversity areas. Clim. Chang. 147 (3–4), 395–409.

Wigley, B.J., Bond, W.J., Hoffman, M.T., 2010. Thicket expansion in a South African savanna under divergent land use: local vs. global drivers? Glob. Change Biol. 16 (3), 964–976.

Williams, S.E., Bolitho, E.E., Fox, S., 2003. Climate change in Australian tropical rainforests: an impending environmental catastrophe. Proc. R. Soc. Lond. B Biol. Sci. 270 (1527), 1887–1892.

Wingfield, J.C., Pérez, J.H., Krause, J.S., Word, K.R., González-Gómez, P.L., Lisovski, S., Chmura, H.E., 2017. How birds cope physiologically and behaviourally with extreme climatic events. Philos. Trans. R. Soc. B. 372(20160140).

Wu, J., 2016. Detection and attribution of the effects of climate change on bat distributions over the last 50 years. Clim. Chang. 134 (4), 681–696.

Yom-Tov, Y., Yom-Tov, S., 2004. Climatic change and body size in two species of Japanese rodents. Biol. J. Linn. Soc. 82 (2), 263–267.

Yom-Tov, Y., Yom-Tov, S., 2006. Decrease in body size of Danish goshawks during the twentieth century. J. Ornithol. 147 (4), 644–647.

Zhao, F., Liu, Y., Shu, L., 2020. Change in the fire season pattern from bimodal to unimodal under climate change: the case of Daxing'anling in Northeast China. Agric. For. Meteorol. 291, 108075.

Ziska, L.H., Pettis, J.S., Edwards, J., Hancock, J.E., Tomecek, M.B., Clark, A., Dukes, J.S., Loladze, I., Polley, H.W., 2016. Rising atmospheric $CO_2$ is reducing the protein concentration of a floral pollen source essential for north American bees. Proc. R. Soc. B Biol. Sci. 283(1828), 20160414.

Zommers, Z., Marbaix, P., Fischlin, A., Ibrahim, Z.Z., Grant, S., Magnan, A.K., Pörtner, H.O., Howden, M., Calvin, K., Warner, K., Thiery, W., 2020. Burning embers: towards more transparent and robust climate-change risk assessments. Nat. Rev. Earth Environ. 1, 1–14.

CHAPTER

# 5

# Effect of climate change on marine ecosystems

*Phillip Williamson*[a] *and Valeria A. Guinder*[b]

[a]University of East Anglia, Norwich, United Kingdom [b]Argentine Institute of Oceanography, National Scientific and Technical Research Council, Bahía Blanca, Argentina

OUTLINE




https://doi.org/10.1016/B978-0-12-822373-4.00024-0

# 1 How climate change affects marine ecosystems

## 1.1 Introduction: The ocean and climate change

Climate is usually considered to be a property of the atmosphere, described in terms of long-term records of air temperature, and the occurrence of rain, snow, sunshine, and wind. These atmospheric parameters directly affect our daily activities, our food supply, and the land-based natural ecosystems with which we are most familiar. Furthermore, changes in atmospheric chemistry have naturally driven climate change in the past and are currently doing so as a result of human activities. Nevertheless, the ocean is arguably the fundamental regulator of environmental conditions on Earth: without it, the planet would be uninhabitable. In particular, the ocean's role in energy redistribution and water cycling is responsible for global weather and climate patterns; ocean volume determines sea level, that has varied by more than 100 m since the evolution of *Homo sapiens*; the ocean uptake of fossil-fuel derived carbon dioxide has to date kept global surface temperatures below critical thresholds; and the natural variety of ocean conditions supports an impressively rich biodiversity, mostly unknown, whose holistic functioning provides many crucial ecosystem services and benefits for human society, mostly unappreciated.

In this chapter we initially consider how climate-linked alterations in ocean physics and chemistry—warming, ocean acidification, deoxygenation, sea level rise, loss of sea ice cover, stratification, and nutrient changes—affect marine life, acting together, and on their own (Sections 1.3–1.8). Climate change impacts on specific coastal (Section 2), shelf sea, and open ocean ecosystems (Section 3) are then considered in greater detail, covering both the impacts observed to date and those that are projected under different climate change scenarios.

Three complex cross-cutting issues are relevant to all aspects of climate change in the ocean. First, many climate change impacts occur together, not as single factors, and are superimposed on other human-induced stressors, such as fishing, pollution, and coastal development. Different combinations of different impacts occur in different parts of the ocean, and different species will show different sensitivities to those combinations; such issues are discussed further in Section 1.2.

The second cross-cutting issue is temporal variability, arising from the uncertain rate of future climate change (a function of political decisions, determining future anthropogenic emissions), and the occurrence of fast and slow ocean processes. The speed of climate change necessarily affects the severity of biological impacts. Slow climate change may allow either acclimatization within an individual organism's lifetime (intra-generational phenotypic change) or evolutionary adaptation (intergenerational genetic changes, arising from selection); however, such responses may become impossible under conditions of more rapid and potentially chaotic warming (Schoepf et al., 2019; Bennett et al., 2019). Actual impacts will depend on the location of individuals and populations within their distributional range, as well as species' acclimatization and adaptation capacities (Fox et al., 2019).

The third complication is spatial variability, both vertically and horizontally within and between ocean basins, closely linked to the occurrence of differential time scales (Schlunegger et al., 2020). Near-surface waters and shallow shelf seas are warming relatively rapidly, matching the rate of change in many other associated physico-chemical factors and human influences. In contrast, changes in the ocean interior (and the cryosphere; Sections 1.6

and 1.7) occur more slowly, yet have long-term momentum (Long et al., 2014; Collins et al., 2013). Continuing marine climate impacts, additional to those that have occurred already, are therefore inevitable. Yet, there is also institutional inertia in the policy response (Munck af Rosenschöld et al., 2014); as a result, many impacts of worst case and best case climate change scenarios are closely similar for the next 20–30 years (Schwalm et al., 2020). The benefits of rapid action in emission reduction are therefore mostly in the far future, avoiding potentially catastrophic feedbacks and tipping points (Steffen et al., 2018; Lenton et al., 2019) by keeping within the "safe" temperature limits of the Paris Agreement (Raftery et al., 2017; Streck, 2020).

Much of the information provided here is based on the comprehensive synthesis of marine climate change and its consequences provided by the Special Report on the Ocean and Cryosphere in a Changing Climate (SROCC) of the Intergovernmental Panel on Climate Change (IPCC, 2019; Bindoff et al., 2019). This chapter also takes account of results from more recent research, bringing together complementary information from direct observations, experimental studies, and modeling.

## 1.2 Multiple drivers and combined effects

The many different impacts of climate change on the marine environment are superimposed on natural variability. While most impacts involve a progressive change in the mean state of the ocean system (e.g., increasing heat content, ocean acidification), such changes increase the severity of extreme events arising from natural fluctuations (e.g., storms, heatwaves, and climate oscillations like El Niño) (Boyd et al., 2016). Climate change impacts are also additional to other human influences on the ocean, such as pollution or over-fishing, with all marine species—totaling at least a million (Bouchet, 2017)—potentially at risk from these multiple stressors. The interaction of their effects at individual and population levels can amplify or decrease the effects of single drivers, subsequently causing ecosystem changes at regional and global scales (Piggott et al., 2015; Turner et al., 2020). There is now increasing evidence that species groups that might actually benefit from climate change occurring on its own become disadvantaged under combined impacts (Coll et al., 2020; Ullah et al., 2020).

Different organisms and ecosystems are affected in different ways, not only due to their level of exposure but also their underlying processes and sensitivities (Zscheischler et al., 2018). For example, tolerance to environmental changes is wider for marine microbes characterized by phenotypic plasticity and more rapid genetic adaptation, due to shorter generation times. In contrast, multicellular plants and animals may be more niche-specific, with different life cycle stages showing variable sensitivity (Pörtner, 2012; Dahlke et al., 2020), and slower evolutionary change due to smaller populations (lower genetic diversity) and longer generation times (Foo and Byrne, 2016). Acclimatization and adaptation responses include modifications in the temporal and spatial distribution of species, together with changes in gene expression, physiology, reproduction, or behavior. These responses are complex, and their within-species variation makes it difficult to extrapolate from short-term, local studies to long-term, global effects (Munday, 2014; Bennett et al., 2019). Nevertheless, the most frequent outcome of the interactive effects of multiple impacts on marine ecosystems is likely to be a loss of biodiversity, since species that are unable to compete successfully under changed environmental conditions will be locally excluded (Boyd et al., 2016; Reddin et al., 2020). The consequent changes in assemblage composition will then affect ecosystem functionality, changing productivity, trophic structure, and nutrient cycling, with potential feedbacks to the climate system as well as reductions in ecosystem services (Kordas et al., 2011; Rocha et al., 2015).

Global analyzes have highlighted where the combined effects of climate change and direct human change are currently strongest (Halpern et al., 2015). The occurrence of these (mostly adverse) impacts arising from multiple stressors is expected to become much more pervasive as warming intensifies (Henson et al., 2017; Fig. 1). Food-web interactions accentuate such effects: climate change impacts on phytoplankton will result in changes in primary production that then become a driver for additional impacts, either positive or negative, for other trophic levels. The overall effect is the development of novel biogeochemical regimes (Reygondeau et al., 2020), not just latitudinal movements in response to temperature changes.

The potentially high economic costs of such major re-arrangements of ocean functioning, from pole to pole and surface to abyssal depths, are currently poorly quantified. That knowledge gap needs to be urgently addressed, so that these issues can be given appropriate attention in climate policy ( Jin et al., 2020; Narita et al., 2020). The maintenance and further development of long-term ocean observing programs are crucial in that context, since they provide the data to detect and attribute ecosystem changes to climatic drivers worldwide (Doo et al., 2020; Melet et al., 2020). Greater interdisciplinarity and research connectivity are also required. For example, bridging outputs from laboratory experiments on single and multiple drivers to the observed conditions and behaviors of natural populations in the ocean (Boyd et al., 2018; Turner et al., 2020); and more comprehensive modeling, combining future climate scenarios with other human-induced perturbations (such as eutrophication, and relevant changes in land use and river conditions), in order to develop more efficient mitigation and restoration plans (Palmer et al., 2019; Herrera-R et al., 2020; Gissi et al., 2021). While there may be some scope for "ocean-based solutions" (Gattuso et al., 2018; Hoegh-Guldberg et al., 2019), the overall need is to minimize the projected degradation of all natural marine ecosystems in a warming climate, by achieving global net zero emissions for greenhouse gases as rapidly as possible.

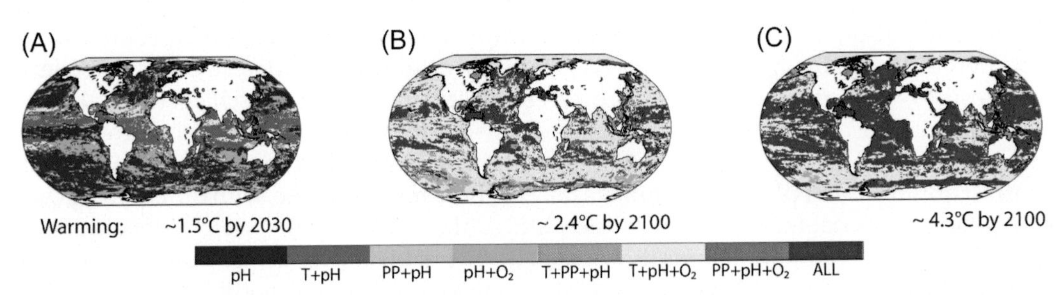


**FIG. 1** The development of multistressor conditions in the global ocean, showing areas that are expected to depart from historical seasonal variability in pH, temperature, and primary production (in upper ocean) and oxygen (ocean interior) in various combinations. (A) by 2030, under RCP 8.5 scenario with global surface warming of ~1.5°C compared to preindustrial; (B) by 2100, under RCP 4.5 with warming of ~2.4°C; and (C) by 2100 under RCP 8.5 with warming of ~4.3°C. *Based on Henson, S.A., Beaulieu, C., Ilyina, T., John, J.G., Long, M., Séférian, R., Tijiputra, J., Sarmiento, J.L., 2017. Rapid emergence of climate change in environmental drivers of marine ecosystems. Nat. Commun. 8, 14682.*

## 1.3 Warming

Directly observed temperature data unequivocally shows a global warming trend at the ocean surface since 1900 (Fig. 2A). Over that period, there has been a temperature increase of ~1.0°C, with evidence for an accelerating rate of increase since 1960, when data reliability has been highest. Although there was an apparent slowing or "hiatus" in ocean surface warming between 2000 and 2010 (Trenberth and Fasullo, 2013) that was less pronounced than in air surface temperature records, and does not appear in warming records that include deeper ocean waters.

For water depths below 2000 m, warming to date has generally been less than 0.1°C. Nevertheless, a consistent long-term trend of increasing temperature increase is apparent, even in abyssal depths (Meinen et al., 2020), and small increases in water temperature for large volumes represent large amounts of heat energy absorbed. The total heat content of the ocean has increased by more than $30 \times 10^{22}$ J since 1960, representing around 93% of all of the extra heat absorbed by the Earth as a result of human-driven increases in greenhouse gas concentrations (Fig. 2B). This dataset is mostly based on direct measurements throughout the water column, using research ships and autonomous sensors that provide high-quality temperature profiles (Argo floats; Riser et al., 2016). Its main features are consistent with modeling analyzes and have also been confirmed by independent estimates of ocean heat content obtained from satellite-based measurements of ocean heat fluxes and ocean volume (Meyssignac et al., 2019).

Not all ocean basins have warmed by the same amount, with circulation patterns and decadal scale variability affecting recent regional warming rates (Bindoff et al., 2019; Schlunegger et al., 2020). For example, the Southern Ocean warmed more slowly at the surface than the global rate for 1981–2019, but much more rapidly at water depths of 2000–5000 m. Longer time series obtained from marine sediments in the North Atlantic show that present-day sea surface temperatures are the highest for at least 2900 years (Lapointe et al., 2020), and that recent seafloor warming around Iceland, accentuated by basin-scale circulation changes, is unprecedented in the last 10,000 years (Spooner et al., 2020).

Warming impacts on marine organisms can initially be either direct, through temperature-dependent physiological processes, or indirect, through physical effects on ocean mixing (altering the supply of nutrients and oxygen, and affecting dispersal and population connectivity), sea ice cover, sea level and storm events. These physical effects are considered in later sections. All responses subsequently contribute to further indirect impacts through ecological interactions, including those driven by changes in seasonal timing (phenology).

The direct biological effects of ocean warming are closely similar to those on land: an increasing mismatch between the thermal conditions experienced by an organism (plant, animal or microbe) and the upper limits of its species-specific optimal range or "thermal window," determined by a range of molecular, cellular and physiological processes (Pörtner, 2012). At temperature extremes, these effects include protein denaturation and anaerobiosis; before then, metabolic impairment is manifested in reduced performance and fitness, affecting growth, reproduction, feeding, immune competence, and behaviors. The overall effect is a decrease in individual and population-level competitiveness. For fish and many other animal species, thermal sensitivity is greatest for eggs, larvae, and spawning adults (Pörtner and Farrell, 2008).

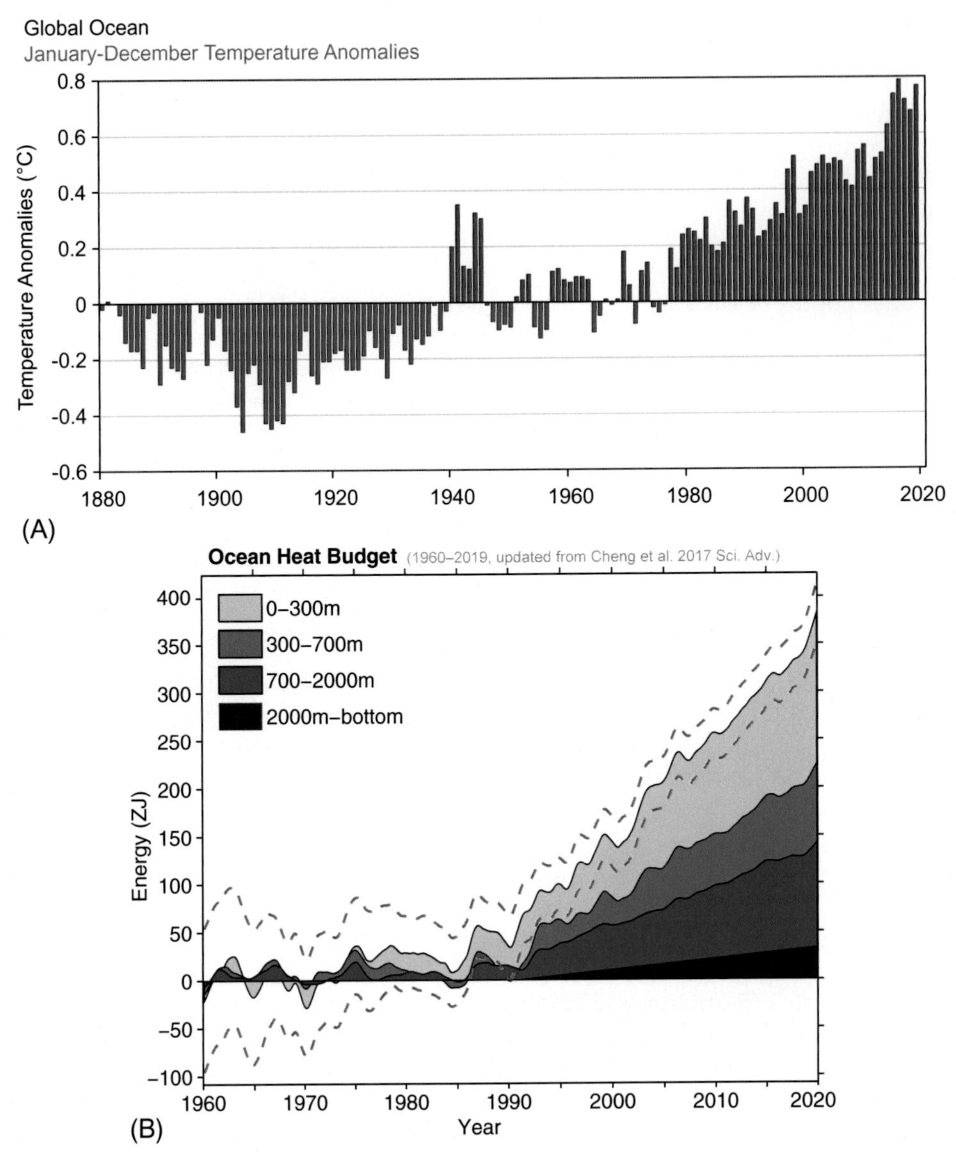


**FIG. 2** Observed warming of the ocean. (A) Ocean surface warming (°C) 1880–2020 in comparison to the 20th century average. (B) Changes in observed ocean heat content ($10^{22}$ J) 1960–2019, for four depth layers. *Dashed lines* indicate 95% confidence interval for total energy with that amount representing additional energy retained by the Earth system primarily as a result of anthropogenic greenhouse gas emissions. *(A) Courtesy NOAA, 2020. Climate at a Glance; Surface Ocean. https://www.ncdc.noaa.gov/cag/global/time-series/globe/ocean/ann/7/1880-2020. (B) Reproduced with permission from Cheng, L., Abraham, J., Zhu, J., Trenberth, K.E., Fasullo, J., Bover, T., Locarnini, R., Zhang, B., Yu, F., Wan, L., Chen, X., 2020. Record-setting warmth continued in 2019. Adv. Atmos. Sci. 37, 137–142.*

Such temperature-related factors (particularly the occurrence of extreme events, e.g., marine heatwaves, Laufkötter et al., 2020) are therefore expected to have adverse impacts at warmer (lower-latitude) distributional limits, while providing potential benefits at cooler (higher-latitude) range limits. The overall outcome is therefore poleward population movement, as already observed for many marine macrophytes, zooplankton, and fish (e.g., Duarte et al., 2018; Gregory et al., 2009; Perry et al., 2005). The rate of that movement will depend on species' mobility and reproductive strategies, and may also be constrained by temperature extremes, hydrodynamic conditions, and trophic interactions (Poloczanska et al., 2016). For marine ecosystems structured by foundation species (coastal vegetation, coral reefs), their mobility—or lack of it—will be critical for the community as a whole.

On land, many animal species have access to "thermal refuges," where extreme temperature conditions can be avoided; however, these are generally lacking in the ocean, and marine ectotherms (fish, benthic invertebrates, and zooplankton) may therefore be more vulnerable to warming than terrestrial species, even though the measured temperature change may be less for the former (Pinsky et al., 2019).

One further impact of warming involves phenology: changes to seasonally determined events that have may different effects on different ecosystem components, with particularly important implications for trophic links in pelagic systems (Sydeman and Bograd, 2009; Thackeray et al., 2016). Earlier seasonal occurrence of peak phytoplankton abundance, linked to warming, has been widely observed in satellite data and direct large-scale surveys, with dinoflagellates most affected (Chivers et al., 2020). However, the timings of the spring phytoplankton bloom, high zooplankton abundance (e.g., the end of overwintering for calanoid copepods) and the annual reproductive cycle of fish may be set by different factors, both exogenous and endogenous; as a result, the egg hatching and hence early larval development of taxa higher in the trophic chain can be misaligned with the times of greatest availability of their food supply (Rogers and Dougherty, 2019). Under high emission scenarios, the occurrence of serious mismatches in these events (>30 days) is projected to increase 10-fold for fish whose spawning grounds are determined geographically, resulting in recruitment failure (Asch et al., 2019). Such effects are expected to be greatest in mid- to high-latitude waters, with strongest seasonality; however, tropical systems can also be affected (Gittings et al., 2018).

## 1.4 Acidification

Ocean acidification is considered to be a component of climate change (e.g., by the IPCC) on the basis of its linkage to carbon dioxide ($CO_2$) emissions, and the crucial role of the ocean carbon cycle to the Earth's climate system. Atmospheric $CO_2$ levels have to date increased by nearly 50% as a result of human-driven emissions, from ~280 ppm (mean for 1720–1800; Hawkins et al., 2017) to the 2019 global average of 410 ppm (Lindsey, 2020). As a result, more $CO_2$ dissolves in the ocean, reacting with it to form carbonic acid and releasing hydrogen ions ($H^+$), thereby reducing seawater pH. The observed long-term trend is an average global decrease of around 0.002 pH units per year in the upper open ocean (Bates et al., 2014), superimposed on daily, seasonal, and decadal variability in ocean carbonate chemistry (Bates and Johnson, 2020). Natural variability is highest in coastal waters and shelf seas (Duarte et al., 2013; CBD, 2014), where short-term acidification rates can be as much as

45 times higher, driven by local and regional factors (Scanes et al., 2020). Projected end of the century pH decreases at the ocean surface range from 0.15–0.50 compared with 1900 values, depending on future $CO_2$ emission scenarios (Bopp et al., 2013; Gehlen et al., 2014; Kwiatkowski et al., 2020). Since pH measurements are made on an inverse, logarithmic scale, a decrease of one pH unit represents a 10-fold increase in $H^+$. Associated chemical changes include increased concentration of bicarbonate ions and dissolved inorganic carbon, and reduction in the concentration of carbonate ions.

The realization that global-scale anthropogenic ocean acidification is underway is relatively recent, with research on its potentially severe ecological impacts starting around 20 years ago (Gattuso and Hansson, 2011). A wide range of consequences have since been identified, mostly adverse, with differential vulnerability of key taxonomic groups (Kroeker et al., 2013; Wittmann and Pörtner, 2013). The production of shells and external structures based on calcium carbonate ($CaCO_3$) is particularly sensitive to ocean acidification: warm- and cold-water corals, together with many species of phytoplankton, crustaceans and mollusks, require greater energy to produce and maintain their exoskeletons under low pH conditions (CBD, 2014; Sunday et al., 2017). As a result, around half of benthic species have lower growth and survival under projected future acidification. However, these responses are variable. Some species and populations can tolerate acidification, e.g., mollusks living close to volcanic $CO_2$ seeps or in naturally-low pH waters, if they have sufficient food supply to support increased energy requirements (Tunnicliffe et al., 2009; Thomsen et al., 2013) and/or with the opportunity for long-term adaptation (Thomsen et al., 2017).

There is strong evidence that low pH conditions may affect animal behavior (Clements and Hunt, 2015), with many (but not all) studies indicating that coral reef fish are particularly sensitive. While a recent high-profile study reported no effects when earlier experiments were repeated (Clark et al., 2020), there were many differences between the original and "replicated" studies. More than 80 other papers have reported significant pH-related changes in sensory discrimination and other neuro-physiological impacts, for both marine and freshwater fish (Munday et al., 2020).

Like warming, ocean acidification necessarily interacts with other climatic and nonclimatic stressors, with direct impacts on organisms and populations subsequently affecting communities and ecosystems (Riebesell and Gattuso, 2015; Kroeker et al., 2017; Doney et al., 2020). Further research on such issues is ongoing, using mesocosms and Free Ocean $CO_2$ Enrichment (FOCE) for multistressor field manipulations (e.g., Alguero-Muniz et al., 2017; Stark et al., 2019), other multifactorial experiments (Boyd et al., 2018), integrated modeling (Ullah et al., 2020) and analyzes of geological extinction events that involved ocean acidification (Jurikova et al., 2020).

## 1.5 Deoxygenation

Ocean deoxygenation has both climatic and nonclimatic causes, with the former dominating at the global scale. Its importance is summarized by a recent review (Laffoley and Baxter, 2019) that quotes the medical motto "if you can't breathe, nothing else matters." The main climatic drivers are a combination of decreasing solubility of oxygen in warmer surface water, and increased stratification, reducing the mixing of the relatively well oxygenated ocean surface with deeper layers. Other climate-linked processes include basin-scale circulation changes and increased respiration by microbes and other marine biota (Oschlies et al., 2018; Robinson, 2019). The overall result has been a ~2% decrease in global ocean oxygen over the period 1960–2010, representing an estimated oxygen loss of 145 thousand million tonnes (Gt, $10^9$), with an associated quadrupling in the global volume of anoxic water (levels below instrumental detection limits) (Schmidtko et al., 2017; Stramma and Schmidtko, 2019). Particularly rapid changes have occurred in the tropical eastern Atlantic Ocean (Fig. 3A and B) and other mid-water oxygen minimum zones in the tropics. In the Pacific, decadal variability has been linked to El Niño-Southern Ocean Oscillation events and regional wind patterns; e.g., affecting coastal upwelling off South America (Bindoff et al., 2019; Garçon et al., 2019).

Future projections of climate-driven ocean deoxygenation show continuing, and strengthening, trends for the high emission scenario, RCP 8.5. Warming-driven deoxygenation cannot be easily reversed, because of the slow timescale of global circulation processes; the ocean oxygen inventory is therefore likely to take centuries to recover from the effects of major perturbations. However, if the very low emissions scenario RCP 2.6 is achieved, global ocean oxygen loss is projected to be much slower; indeed, it might change to net uptake in the upper ocean by 2100, although with high associated uncertainties (Fig. 3C; Bindoff et al., 2019).

Most nonphotosynthetic marine organisms require oxygen levels of at least 20 μmol $O_2$ $kg^{-1}$ to stay active, with large predatory fish, e.g., sharks and tuna, having much higher oxygen demands, >100 μmol $O_2$ $kg^{-1}$ (Sims, 2019). Several squid species can, however, tolerate lower oxygen levels, and waters that are severely oxygen-depleted may support rich microbial communities. Unfortunately, most of the latter are considered undesirable from a human perspective. They include: denitrifying species that reduce the availability of nitrogen, decreasing overall productivity while also producing the powerful greenhouse gas nitrous oxide ($N_2O$); methanogens that produce methane ($CH_4$), another greenhouse gas; and species producing toxic compounds, e.g., hydrogen sulfide, ($H_2S$) (Conley and Slomp, 2019).

The main nonclimatic cause of deoxygenation is anthropogenic nutrient supply by rivers and via the atmosphere causing eutrophication, particularly in temperate coastal waters. While the resulting higher primary production, mostly in spring and summer, releases oxygen in near-surface waters, oxygen is subsequently depleted as a result of higher decomposition, with strongest effects in near-bottom waters. In areas with large river inputs, the delivery of biologically available nitrogen and phosphorus increased by ~30% from 1970 to 2000 (Seitzinger et al., 2010). Unless regulated, such effects are likely to intensify (Rabalais, 2019) with adverse impacts on benthic fauna and coastal fisheries. The interaction with temperature is of particular concern (Deutsch et al., 2015; Breitburg et al., 2019), since warming decreases oxygen solubility, while simultaneously increasing the oxygen demand of metabolically active organisms; e.g., by as much as 70% for coral reef fish for a 3°C temperature rise (Nilsson et al., 2010).

Deoxygenation is inherently concurrent with increasing dissolved $CO_2$ and ocean acidification, with additive adverse impacts for many marine species (Steckbauer et al., 2020). Nevertheless, there is variability in response, and higher $CO_2$ levels may increase hypoxia tolerance for some fish (Montgomery et al., 2019). An interaction between deoxygenation and fishing arises from habitat compression: increasing areas of low oxygen water constrain

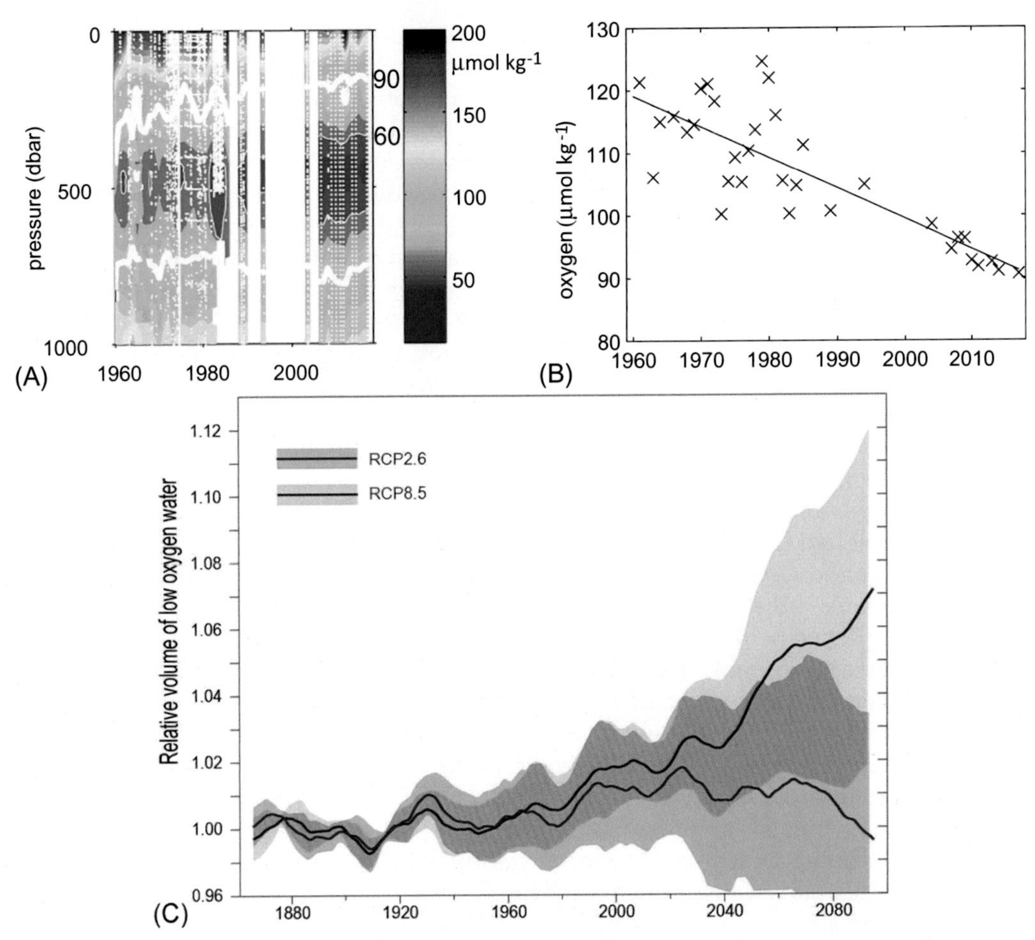


**FIG. 3** Observed and projected ocean deoxygenation. (A) changes in dissolved oxygen concentration as a function of depth and time (1960–2016) for the tropical eastern North Atlantic; (B) decrease in mean oxygen concentration for the water layer 50–300m for that region over that time period; (C) modeled global hindcasts and future projections for the volume of low oxygen (<80 $\mu$mol kg$^{-1}$) water in the 100–600m layer for high emission (RCP 8.5, *red line*) and low emission (RCP 2.6, *blue line*) scenarios, normalized to the volume in 1850–1900. *Shaded areas* indicate 90% confidence intervals for eight CMIP5 models, corrected for drift using their control simulations. *From (A, B) Stramma, L., Schmidtko, S., 2019. Global evidence of ocean deoxygenation. In: Laffoley, D., Baxter, J.M. (Eds.), Ocean Deoxygenation: Everyone's Problem—Causes, Impacts, Consequences and Solutions. IUCN, Gland, Switzerland, pp. 25–36. (C) Bindoff, N.L., Cheung, W.W.L., Kairo, J.G., Arístegui, J., Guinder, V.A., Hallberg, R., Hilmi, N., Jiao, N., Karim, M.S., Levin, L., O'Donoghue, S., Purca Cuicapusa, R., Rinkevich, B., Suga, T., Tagliabue, A., Williamson, P., 2019. Changing ocean, marine ecosystems, and dependent communities. In: Pörtner, H.-O., Roberts, D.C., Masson-Delmotte, V., Zhai, P., Tignor, M., Poloczanska, E., Mintenbeck, K., Alegría, A., Nicolai, M., Okem, A., Petzold, J., Rama, B., Weyer, N.M. (Eds.), IPCC Special Report on the Ocean and Cryosphere in a Changing Climate. Online and in press, pp. 447–587 (Chapter 5).*

fish and mobile crustacea to refuge locations, where they can be increasingly vulnerable to over-fishing and hence population collapse (Breitburg et al., 2019; Sims, 2019).

## 1.6 Sea level rise and storm events

Global mean sea level has risen by 16 cm since 1900; it is currently increasing at the rate of ~0.4 cm per year, and there are built-in commitments for further rises (Oppenheimer et al., 2019). End-of-the-century increases could be as much as 2 m under high emission scenarios (Bamber et al., 2019), but are uncertain due to incomplete understanding of nonlinearities in ice sheet behavior in Greenland and Antarctica. If global surface temperatures warm by 2°C, the collapse of the West Antarctic ice sheet seems inevitable on century-to-millennium time-scales, contributing 2.5 m to global sea level rise; 4°C warming would add 6.5 m, as a result of wider Antarctic ice loss (Garbe et al., 2020). The scale of these projected impacts is concerning, yet conservative when compared with the geological record. Thus global mean sea level was 6–9 m higher than now during the last interglacial (around 120,000 years ago) and up to 25 m higher in the mid-Pliocene (around 3 million years ago), when global temperatures were around 0.5–1.0°C and 2–4°C warmer, respectively (Oppenheimer et al., 2019). Additional details of cryospheric responses to global climate change are given in Chapter 3.

Melting ice is not the only driver of sea level change. A range of other factors have been responsible for around half of the global sea level increase observed to date and have caused ±30% local variability relative to the global mean. These factors include: ocean thermal expansion, with basin-scale differential warming of different water masses (Section 1.3); coastal sediment consolidation and anthropogenic ground-water extraction; changes in river flows and inland water storage; and regional scale uplift or sinking of the Earth's crust (e.g., in response to de-glaciation, volcanic activity and tectonic subduction). The overall outcome presents an existentialist threat to many major coastal cities and small island states (Brown et al., 2018; Kulp and Strauss, 2019; Oppenheimer et al., 2019), as well as having high potential impacts on coastal ecosystems. Nevertheless, if such ecosystems remain functionally intact and appropriately managed, they could play an important role in coastal protection (Bindoff et al., 2019).

Extreme sea level rise events, occurring episodically and superimposed on the long-term trends, are directly responsible for most adverse impacts, societally and ecologically. In particular, storm surges and extreme wave heights not only cause flooding and coastal erosion, but can severely damage coastal wetlands, coral reefs and kelp ecosystems (Cahoon, 2006; Little et al., 2015; Puotinen et al., 2016; Smale and Vance, 2016). The frequency of such extreme events is projected to increase dramatically: under the high emission scenario RCP 8.5, events that until recently had a 1-in-100 year frequency are expected to occur annually on a world-wide basis by the end of the century. For some locations, even the low emissions scenario RCP 2.6 is projected to result in the annual occurrence of historically rare events by mid-century (Oppenheimer et al., 2019). That effect is partly caused by the trend for increased storm intensity, with associated higher storm surge levels; that pattern is expected to continue, with 1%–10% more tropical cyclones reaching Category 4–5 level under 2°C warming. However, neither observations nor models indicate any changes in the frequency of tropical hurricanes or cyclones (Collins et al., 2019). Ecosystem-specific consequences of rising sea level are limited to coasts and shallow seas, as considered in Section 2.

## 1.7 Sea ice loss

The different distributions of land and ocean in the Arctic and the Antarctic have resulted in different responses to date to global climate change. Surface warming has been strongest in the Arctic, at rates twice as high as the global mean (Richter-Menge et al., 2018). Such warming has resulted in a ~50% reduction in the extent of multiyear sea ice since 1980, with a corresponding increase in the ice-free area during summer months. The amount of sea ice cover in September has shown the greatest change; it is currently decreasing at the rate of 83,000 $km^2$ $year^{-1}$ (~13% per decade) (Meredith et al., 2019), with a record delay in 2020 for the onset of re-freezing (Bamber, 2020). These changes produce positive feedbacks to the climate system, with the exposed ocean absorbing and retaining additional heat (Overland, 2020). The near-total loss of summer Arctic sea ice is expected within decades, with major implications for Arctic ecosystems (discussed in Section 3.6).

In contrast, the Antarctic region and Southern Ocean have shown less uniform temperature change (although based on fewer measurements), with greater regional and decadal variability. There is no overall trend with regard to sea ice cover: while there was a decrease between 1982 and 2017 in the Pacific sector, particularly in the summer, increases occurred over that period in the South Atlantic sector (Meredith et al., 2019). Note that sea ice is formed by the freezing of seawater, with strong seasonality, whereas the large ice shelves characteristic of Antarctica originated on land.

## 1.8 Stratification and nutrient supply

Human activities have greatly increased the supply of nutrients to coastal waters and shelf seas (Seitzinger et al., 2010). Yet in near-surface waters of the open ocean, climate change is causing the opposite effect, resulting in a projected net decline in global ocean productivity—although with strong latitudinal variability. The overall effect is a consequence of ocean warming increasing vertical density gradients, strengthening stratification, and thereby reducing mixing in the top 200 m by ~5% during the past 60 years (Bindoff et al., 2019; Li et al., 2020). Less mixing means slower recycling of nutrients from the ocean interior to the upper ocean, where they are needed for phytoplankton growth (the basis for marine productivity). Decreases in net primary production have amplified effects on higher trophic levels, especially in tropical and subtropical waters (Kwiatkowski et al., 2019; Lotze et al., 2019), with changes in near-surface community structure and productivity also affecting the quantity and quality of organic material exported by the "biological pump" to the ocean interior and seafloor communities (Boyd et al., 2019). A counteracting factor is additional atmospheric nitrogen input, from anthropogenic sources; however, this has a relatively small influence on productivity due to nitrogen cycle feedbacks and phosphorus limitation (Yang and Gruber, 2016).

A review of Earth system model projections indicated that mean end-of-the-century declines in the nitrate content of the upper 100 m would be in the range 1.5%–6% for low emissions (RCP 2.6) or 9%–14% for a high emission scenario (RCP 8.5), relative to recent values (Bindoff et al., 2019). More recent multimodel analyzes gave a higher mean projection for nitrate reduction, although projected declines in productivity were slightly less (Kwiatkowski et al., 2020). For both, the largest absolute declines are expected to occur in present-day upwelling zones.

# 2 Climate change impacts on shallow coastal ecosystems

## 2.1 Introduction

The definition of an ecosystem considered here is relatively broad, noting that interactions of marine organisms and their physical environment occur over many spatial scales. For the shallow coastal environment, ecosystems have not only been delineated on the basis of their foundation species (e.g., kelp ecosystems, mangrove forests and coral reefs) but also according to their geomorphological structure (e.g., estuaries). Both groupings are used here, to match Bindoff et al. (2019). While mostly in water depths <50 m, intertidal habitats are also included. Two of those, saltmarshes and mangrove forests, are frequently grouped with seagrass meadows to comprise coastal "blue carbon" ecosystems (Windham-Myers et al., 2019), characterized by high carbon storage in their sediments, high biodiversity, and the delivery of a wide range of other ecosystem benefits.

Other shallow coastal ecosystems given separate attention comprise warm-water coral reefs, kelp ecosystems, rocky and sandy intertidal, and estuaries. All seven of these display different levels of sensitivity, exposure, and vulnerability to climate hazards across their distributional range, with strong regional variability in their adaptive capacity. They are considered here in the sequence of their likely global vulnerability to future climate change, as assessed by Bindoff et al. (2019), with the most sensitive considered first.

The detection and attribution of changes in coastal ecosystems' biodiversity, structure, and functioning to climate-related impacts are challenging (Doo et al., 2020). On the one hand, biotic and abiotic components that define each ecosystem, such as sediments, benthic fauna, or vegetation, display nonlinearity in their responses to environmental changes. On the other hand, interactive effects between climatic-related impacts such as ocean warming and marine heatwaves, and other human driven impacts such as coastal eutrophication and land use, increase the complexity of the ecosystems' responses. Multiple lines of evidence reveal how different combinations of environmental hazards exert different degree of pressures on natural ecosystems, affecting both their resilience and vulnerability. The vulnerability to climate hazards is therefore linked to the distinct responses of the key ecosystem components, as well as to the degree of exposure to climate hazards, related to the geographical setting and the local anthropogenic disturbances.

Coastal ecosystems are particularly vulnerable to biological invasions. The introduction of nonindigenous species, either due to human transport, natural dispersal, or distributional range expansion enable by global warming, may lead to their establishment and colonization of new habitats in detriment of native and endemic species (Bailey et al., 2020). Conversely, hybridization (Gallego-Tévar et al., 2020) or contemporary evolution (Des Roches et al., 2020) may occur in some cases to facilitate the adaptation of species to changing climate and habitat.

## 2.2 Coral reefs

In this section, climate change impacts on reef-forming warm-water corals, occurring in tropical and subtropical shallow waters, are reviewed; cold water corals, occurring in cooler and deeper water, are considered in Section 3.5. Coral reefs are among the world's most biodiverse habitats and are recognized as particularly vulnerable to climate change. Such reefs are primarily built by coral polyps (sessile invertebrates containing intracellular algae) with calcified exoskeletons, together with other reef-building organisms like calcareous algae. Coral reef ecosystems support high numbers of other animal and plant species: very many mollusks, echinoderms, crustaceans, sponges, tunicates, anemones, fish, sea birds, marine mammals and turtles (including endangered species) depend on coral reefs, to some degree (Roberts et al., 2002). Phyto- and zooplankton communities contribute to the ecosystem productivity, providing food for benthic filter-feeding organisms (Smith et al., 2016; Russell et al., 2019). Coral reefs not only protect the coastline from storms and erosion, but also underpin large fisheries, tourism, and recreational activities (e.g., diving) (Moberg and Folke, 1999), with emergent interest on social-ecological aspects (Woodhead et al., 2019).

Healthy coral reefs need clean, circulating waters for the well-functioning of benthic-pelagic interactions and the maintenance of underwater light conditions for benthic photosynthetic organisms (Russell et al., 2019). Several persistent human threats on habitat integrity superimpose on the effects of warming, ocean acidification and other climate hazards. These human impacts include sediment and nutrient loads from land runoff and sewage (e.g., nitrate enrichment, Burkepile et al., 2020), pollution and overfishing, including the removal of keystone species in structuring the ecosystem, such as grazers on benthic algae, releasing competition stress on corals (Taylor et al., 2020).

Even if global warming is limited to 1.5°C above preindustrial temperatures, a very high risk of irreversible change to coral reefs is projected (IPCC, 2019). Heat stress causes mortality through coral bleaching: the loss of the endosymbiotic algae (intracellular zooxanthellae), responsible for corals' food supply, through photosynthesis, and their color. Marine heatwaves have already caused worldwide reef degradation since the 1980s at increasing frequency and severity, enhanced in many regions by concurrent nonclimatic stressors such as pollution, and overfishing (Hughes et al., 2018). Within a region, the severity and extent of coral bleaching is directly correlated with the duration of marine heatwave events (Fig. 4A; Smale et al., 2019).

Coral reef recovery after severe bleaching episodes is possible, but occurs slowly: from years to a decade for faster-growing species, with longer recovery for fully mature communities (Dalton et al., 2020; Hughes et al., 2019). At Lord Howe Island, Australia, moderate bleaching in the marine heatwave of 2010–11 was followed by recovery to predisturbance conditions within 3 years (Dalton et al., 2020). However, more severe bleaching in 2015–17 in the northern Great Barrier Reef and in the central Indian Ocean was followed by a shift in foundational species toward algal-dominated reefs and increased numbers of herbivorous parrotfish, whose abundance peaked 2 years after the heat disturbance (Taylor et al., 2020). Full recovery may now be impossible, due to the widespread and severe bleaching that occurred in 2019–20 (AIMS, 2020) and the increasing future likelihood of such events (Heron et al., 2018).

Coral reef recovery after bleaching-induced mortality is already slowed by ocean acidification, with calcification, and hence coral growth rates, under field conditions currently reduced by ~7% compared to preindustrial water chemistry (Albright et al., 2016). Recovery will be further constrained by increased future acidification (Albright et al., 2018), as well as by changes in sedimentation and ocean currents affecting the dispersal (planktonic) stages of corals, thereby impairing their capacity to settle and rebuild the destroyed reef (Espinel-Velasco et al., 2018; Hughes et al., 2019). Nevertheless, the large genetic, physiological and behavioral diversity that characterize coral reef assemblages results in a mosaic of ecological

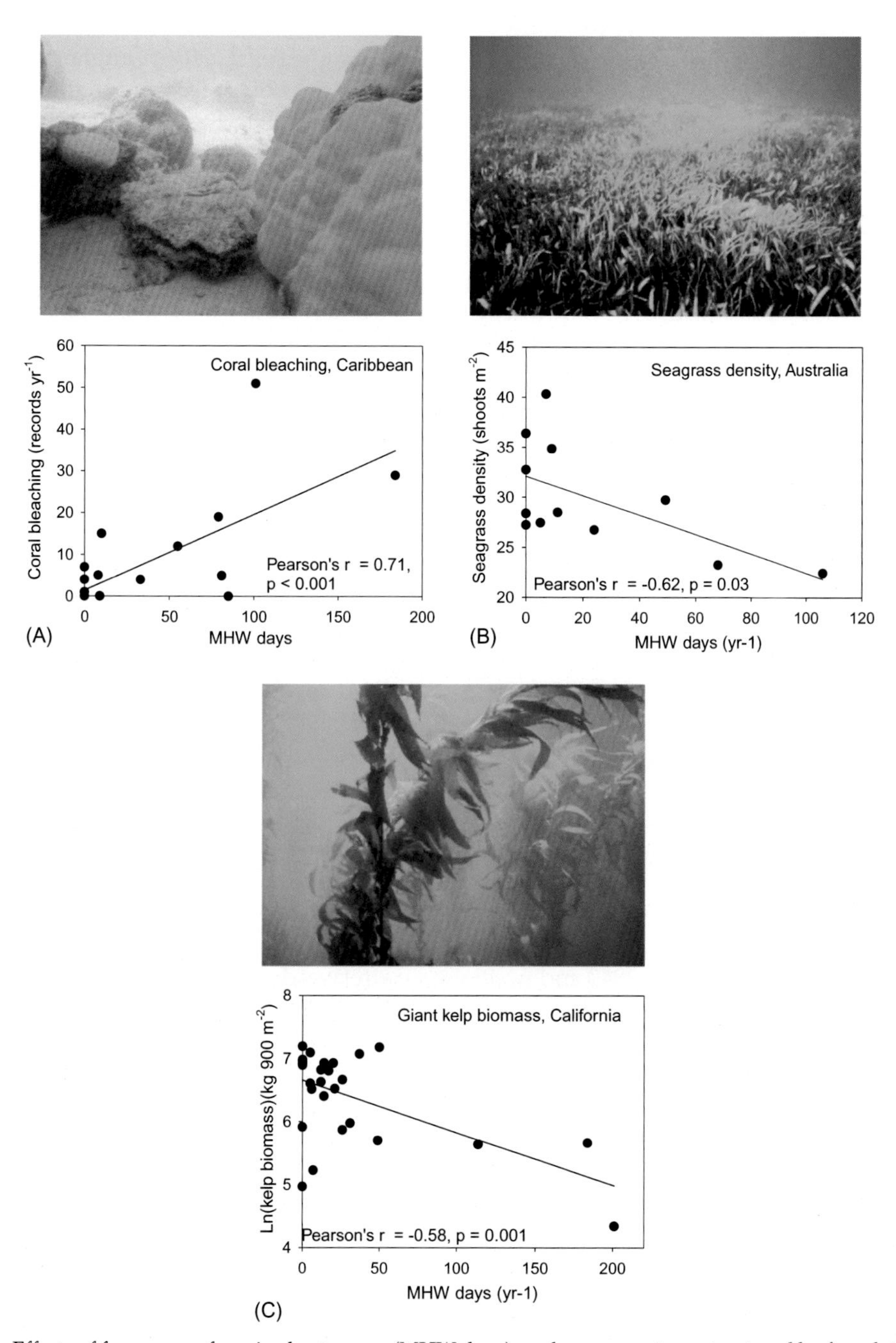


**FIG. 4** Effects of frequency of marine heat waves (MHW days) on three ecosystems structured by foundation species. (A) Coral bleaching in the Caribbean; (B) kelp biomass in Californian coastal waters; and (C) seagrass density in Australia. *Reproduced with permission from Smale, D.A., Wernberg, T., Oliver, E.C., Thomsen, M., Harvey, B.P., Straub, S.C., Burrows, M.T., Alexandre, L.V., Benthuysem, J.A., Donat, M.G., Feng, M., Hoibday, A.J., Holbrook, N.J., Perkins-Kirkpatrick, S.E., Scannell, H.A., Gupta, A.S., Payne, B.L., Moore, P.J., 2019. Marine heatwaves threaten global biodiversity and the provision of ecosystem services. Nat. Clim. Chang. 9(4), 306–312.*

outcomes within reef habitats, promoting species acclimation and ecosystem resilience to repeated global change disturbances (Camp et al., 2018; Hughes et al., 2019; Quigley et al., 2019).

Although high sensitivity to heat stress and acidification is overall common for coral reefs, some species seem resistant to these climate threats, likely related to their evolutionary and migration history, such as in northern Red Sea (Krueger et al., 2017). Different community assemblages, species-specific traits and environmental settings, e.g., acidification effects on coral reef zooplankton (Smith et al., 2016), coral proximity to seaweeds (Clements et al., 2020), and temporal variability in water chemistry (Enochs et al., 2018) may interact and drive variable ecosystem outputs and resilience under climate hazards (Bahr et al., 2020; Koester et al., 2020; Schoepf et al., 2019; Teixidó et al., 2020). Coral reefs in volcanic $CO_2$ seeps provide insights on species acclimatization and local adaptation to future ocean acidification (Teixidó et al., 2020). For example, corals are able to survive extreme pH levels as low as 6.5, with undersaturation of aragonite, at times of reduced water flow during slack tide at Mayreau Island in the Caribbean; however, they have poorly-developed, thin exoskeletons (Enochs et al., 2020).

The vulnerability of shallow coral reefs to sea level rise and changes in wave exposure depends on coastal geomorphology, local hydrodynamics, carbonate sediment supply, and coral species assemblages. Tropical cyclones physically destroy reef habitat (Puotinen et al., 2016), and associated enhanced sedimentation can also reduce live coral cover. Reduced coral fitness and calcification, due to compound climate and nonclimatic stressors, make it unlikely that most near-surface reefs will be able to keep pace with sea level rise projections (Cacciapaglia and van Woesik, 2020), while severe sea level rise (under RCP 8.5 scenario) could also displace coral communities in steep slopes beyond the euphotic depth, with impacts on benthic-pelagic interactions (Morgan et al., 2020). Nevertheless, vertical accommodation and reef accretion may still be possible in coral reef islands that have sufficient sediment supply (Masselink et al., 2020).

Future ocean warming, acidification, extreme climate events and sea level rise will continue to cause widespread coral bleaching, mortality, and turnover of the foundation species of coral reefs, with severe implications for ecosystem biodiversity, sustainability, and provision of services. Under a high emissions scenario (RCP 8.5), severe coral bleaching is projected to occur twice-per-decade by 2040 and up to yearly afterwards (Heron et al., 2018). Mitigating local impacts such as coastal eutrophication could broaden the areas providing ecological refugia for coral reef species and facilitate their adaptation and persistence (Guan et al., 2020). Re-introductions, relocation and assisted evolution through "coral gardening" initiatives could also assist in ensuring coral reef survival, although in modified form and limited in scale (Rinkevich, 2015, 2019; Bindoff et al., 2019). While the efficiency of different conservation and restoration practices is still under evaluation (Walsworth et al., 2019; Foo and Asner, 2020), habitat heterogeneity and connectivity, and evolutionary traits, arise as key aspects with regard to actions that enhance coral resilience (Tittensor et al., 2019).

## 2.3 Kelp ecosystems

Kelp ecosystems (also known as kelp forests) span from temperate to polar regions on rocky substrates worldwide, and have been estimated to cover a quarter of the global coastline. Dense populations of kelps (large canopy-forming brown macroalgae) are the engineers of structurally complex and highly productive submerged ecosystems in the shallow areas of continental shelves (< 50 m depth), supporting high marine biodiversity, from invertebrates to fish and marine mammals (Vásquez et al., 2014). Unlike plants in coastal wetlands (Sections 2.4, 2.6, and 2.8), seaweeds lack a root system and hence the biomass produced by kelps is not stored below ground at the source area but exported by currents to neighboring soft sediment habitats, deep canyons and the deep sea. The exported organic carbon (~80% of their production, Krumhansl and Scheibling, 2012), can serve as food in adjacent habitats for secondary production, or may reach the deep seafloor as allochthonous detritus and be buried in the sediments for long periods, contributing to $CO_2$ sequestration (Krause-Jensen et al., 2018; Filbee-Dexter and Wernberg, 2020).

The high vulnerability of kelp ecosystems to climate change has put them on the spotlight along with corals reefs, due to the growing risk imposed to their integrity and to the many services they provide. Periods of excessively high temperature (marine heatwaves) are the main climate hazards; with critical thresholds for many expected to occur at around 1.5°C of global warming, depending on population assemblages and regional environmental conditions (Fig. 4B; IPCC, 2019). Two main ecological traits define the low adaptive capacity of kelps to climatic stress: their high physiological sensitivity to rising temperatures and their low dispersal capabilities, which make them highly vulnerable to extreme heat events, with low recovery after mass mortalities (Straub et al., 2019; Smale et al., 2019). Across latitudinal patterns, genetic diversity among kelp populations and species physiological plasticity shape the ability of species to withstand heat stress and enhance the ecosystem resilience (Wernberg et al., 2018; Liesner et al., 2020). However, more intense extreme climatic events pose high risk of genetic diversity depletion especially in marginal populations already at higher risk of exclusion (Gurgel et al., 2020). Increased storm surges are also important climate threats because they can cause massive canopy damage and detachment of kelps from the hard substrates affecting the ecosystem resilience (Byrnes et al., 2011; Hanley et al., 2020). High wave energy environments may hinder the re-establishment of kelps, and ultimately lead to a shift in the community structure toward a younger stage with reduced canopy (Schoenrock et al., 2020).

At a global scale, loss of kelp forests has occurred at a rate of ~2% per year over the past half century, with large variability in biomass trends at local and regional scales (Poloczanska et al., 2016; Wernberg et al., 2019). Range reductions of kelps at the warm end of distributional margins and expansions at the poleward end are expected to occur at a faster rate under higher emission scenarios (Assis et al., 2018; Wilson et al., 2019). Poleward shifts lead to spread of invasive species with changes in habitat structure and functioning, including reduction in species richness and carbon stocks (Franco et al., 2018; Pessarrodona et al., 2019). The loss of kelp forests, e.g., due to marine heatwaves, is often followed by the invasion of turf-forming seaweeds, reducing habitat complexity (Filbee-Dexter and Wernberg, 2018). Biodiversity loss is associated with the simplification of the ecosystem at the same time that reduced productivity may have negative cascade effects on the food web.

The historical and continuing large-scale deterioration of kelp ecosystems due to warming is aggravated by other environmental stressors at regional scales (Franco et al., 2018; Smale et al., 2019). These include alterations in turbidity (hence light availability), nutrient loads and enhanced grazing. Low frequency climate signals (such as the North Pacific Gyre Oscillation) can lead to long-term trends in kelp biomass through changes in water temperature and nutrient levels, as observed along the coast of California (Bell et al., 2020). The introduction of warmer-water herbivorous fishes into temperate kelp forests or the occurrence of disease in grazers or their predators may impose strong control on kelp biomass (Vergés et al., 2016; Eisaguirre et al., 2020). Important human-driven impacts include eutrophication, overfishing, and pollution (Filbee-Dexter and Wernberg, 2018; Smale et al., 2019).

In polar regions, light is a key driver of kelp productivity. Summertime reduction in ice cover allows higher sunlight penetration in the water column and favors kelp growth (Deregibus et al., 2020). However, climate-driven loss of sea-ice will also enhance sediment loads and reduce salinity, leading to different ecosystem responses depending on species assemblages (Bartsch et al., 2016; Springer et al., 2017). Ocean acidification and warming will be most pronounced in the Arctic, where kelps show variable species-specific responses under end of the century climate change scenarios for $CO_2$ and temperature. Higher resilience of Arctic kelp communities to these climate hazards is expected than their cold-temperate counterparts (Gordillo et al., 2016).

## 2.4 Seagrass meadows

Seagrasses are flowering plants that reside in subtidal shallow zones with soft sediments. They are widely distributed from tropical to temperate regions, occurring in estuaries, lagoons, coastal waters, and enclosed marginal seas (Short et al., 2007). Seagrass meadows are highly productive, and their associated sediments are very effective in accumulating carbon; they also provide many other ecosystem services, including habitat for large populations of marine species, fisheries production, coastal protection, and nutrient absorption. As a coastal blue carbon ecosystem (Crooks et al., 2019), they contribute to climate regulation by the net uptake of atmospheric $CO_2$ and by countering ocean acidification (Camp et al., 2016). The protection and restoration of seagrass meadows is therefore considered as a natural climate solution (Oreska et al., 2020), although with many uncertainties (Johannessen and Macdonald, 2016).

Seagrass meadows are highly vulnerable to warming and marine heatwaves (Fig. 4C, Smale et al., 2019; Arias-Ortiz et al., 2018). Physical disturbance driven by increasing storminess and wave energy also threatens the structure and functioning of seagrass ecosystems and the services they supply (Hanley et al., 2020). Large areas of seagrasses have been lost worldwide over the past decades due to compound effects of warming, erosion, and eutrophication, with several massive mortalities following heat stress events (e.g., de los Santos et al., 2019; Strydom et al., 2020). The loss of coastal vegetation and soil erosion releases $CO_2$ back to the atmosphere, fueling climate warming (Salinas et al., 2020, Sections 2.6 and 2.8). Current climate-related risks to these sensitive ecosystems will continue to rise, with high likelihood of severe and widespread impacts (and reduced natural capacity to adapt) if global warming exceeds 2°C above preindustrial temperatures (IPCC, 2019). Although ecosystem responses vary across regions and between meadows composed of different seagrass species, the global trend is range contraction and reduced density at low latitudes and gain at higher latitudes. Some seagrass species are therefore expected to colonize new areas, while others are expected to retract (de los Santos et al., 2019), with an overall loss of habitat complexity in warm-temperate regions, such as Western Australia (Hyndes et al., 2016) and the Mediterranean (Chefaoui et al., 2018).

The vulnerability of seagrasses to climate threats is directly related to: their position within the distributional thermal range; the tight dependence on water temperature of their life cycle stages (e.g., flowering, seed production) and life-story strategies (e.g., opportunistic, colonizing, and perennial species), which defines the recovery capacity of populations after perturbations (Repolho et al., 2017; York et al., 2017; Ruiz et al., 2018); the local sediment type, light conditions, microbial activity and other ecological stressors such as grazers (Hyndes et al., 2016; Hernán et al., 2017; Lima et al., 2020); and the pervasive human-mediated perturbations that increase the sensitivity of seagrasses to warming (Orth et al., 2006). Complex ecosystem outputs can therefore arise from the interplay of these multiple-drivers at species and community levels (Pagès et al., 2018). Short-term benefits may arise from low rates of rising $CO_2$ and warming, e.g., enhanced photosynthesis or plant flowering (Ruiz et al., 2018) or relieved pathogen pressure (Olsen et al., 2015); however, in the long-term and at higher $CO_2$ emissions, climate-driven changes are expected to have detrimental impacts on the holistic structure and functioning of seagrass ecosystems.

The intensity of herbivory on seagrasses is globally expected to increase with warming, particularly in temperate seagrass meadows due to the inflow of tropical consumers (Pergent et al., 2014; Hyndes et al., 2016). Alterations in the structure of food webs may also elicit a reduction in the tolerance of plants to rising temperature and biological invasions (Scott et al., 2018). Tropical seagrass meadows are commonly in oligotrophic waters, and nutrient limitation, for example phosphorous and iron in the Red Sea (Anton et al., 2020) is therefore an important factor in the regulation of plant metabolism and distribution. Anthropogenically driven eutrophication aggravated by warming (as commonly observed in estuaries; Section 2.7), leads to hypoxic conditions that may reduce vegetation fitness. In addition, enhanced sediment run-off and organic matter accumulation decrease the underwater light availability, affecting the overall carrying capacity of seagrass meadows. Furthermore, rapid salinity changes may have additive long-term unfavorable impacts on the submerged vegetation and the carbon sediment stocks, reducing the ecosystem resilience to warming (Douglass et al., 2020; Lima et al., 2020).

Large-scale global losses of seagrass meadows are projected to continue with rising ocean temperatures, at higher rates under higher future $CO_2$ emission scenarios (Waycott et al., 2009). Most vulnerable ecosystems will be those with high endemism in enclosed coastal areas with limited expansion capacity and high risk of species invasions (Chefaoui et al., 2018), and those exposed to more frequent and intense extreme temperature events and storm surges (Smale et al., 2019). The Mediterranean Sea, the Western coast of Australia, and seagrass meadows exposed to tropical cyclones, represent global hotspots of high risk of habitat loss and severe declines in biodiversity and carbon storage.

## 2.5 Rocky and sandy intertidal

Two kinds of intertidal ecosystem are considered here, rocky, and sandy. Although their physical structures and biotas are different, they are grouped together because the climate change hazards they experience are closely similar. Both are vulnerable (to different degrees) to sea level rise, increased storminess, and changes in alongshore currents, warming and marine heatwaves, and ocean acidification in the case of calcifiers (Bindoff et al., 2019). Other impacts such as over-fishing, pollution, habitat deterioration and urbanization pose additional risk of reducing suitable habitat for intertidal organisms and the capacity of the ecosystem to adapt to rapid climate change.

Rocky intertidal ecosystems represent over 50% of the global coastline and are dominated by sessile (immobile) organisms, most frequently mussels, limpets, and barnacles in co-occurrence with seaweeds, coralline algae and lichens, and other mollusks, bryozoans, echinoderms, and sponges. These organisms display marked vertical zonation across the intertidal, from the shallow subtidal (submerged kelp-dominated rocky shores are reviewed in Section 2.3) to the supralittoral zone. Zonation depends on species tolerance to withstand submersion during high tide and exposure to air during low tide, as well as on ecological interactions such as competition and grazing. Rocky shores embrace cliffs, platforms and ponds, and bio-engineering organisms further build three dimensional structures, providing habitat complexity (Helmuth et al., 2006a; Hawkins et al., 2016).

Sandy intertidal ecosystems comprise over 30% of the ice-free shorelines of the world and are characterized by flatter topography with burrowing benthic fauna such as clams, polychaetes, isopoda, marine snails and crabs, and other crustaceans and mollusks, with zonation along the beach profile (Carcedo et al., 2019). Dunes with halophyte vegetation are common in the supralittoral zone. Sandy beaches are highly dynamic systems due to sediment movement and wave action varying with geomorphological aspects, i.e., profile inclination, sediment type (fine or coarse sand) and width of the surf zone (McLachlan and Defeo, 2017). Rocky and sandy shores expand globally—sometimes in co-occurrence—providing nesting, breeding, and feeding habitats for numerous species including seabirds, marine mammals, turtles, and fishes. They support regional fisheries, deliver coastal protection, and also provide space for recreation and tourism, esthetic appreciation of nature, and cultural identity (Miloslavich et al., 2016; Drius et al., 2019; Gaylard et al., 2020).

The resilience (or vulnerability) of sandy beaches to shoreline erosion depends on beach topography, sediment supply and available space for lateral and inland sediment relocation. Surprisingly, over the past 33 years, vegetation cover over coastal dunes has increased across different global regions, enhancing coastal protection and dune stability (Jackson et al., 2019). But ambient modifications such as changes in storminess patterns and pervasive human activities are increasing erosion risk, limiting the landward migration of the beach profile. A significant loss of global sandy coastlines (nearly 50%) is projected by the end of the century under different sea level rise scenarios (Vousdoukas et al., 2020). Low-lying densely populated areas are especially vulnerable, although with a high regional component in ecosystem responses that includes gains in beach area at accretive shorelines (Cooper et al., 2020; Vousdoukas et al., 2020).

Sea level rise and warming are expected to have most effect on organisms in the upper- and mid-shore levels of the rocky intertidal, which are more vulnerable to desiccation and overheating during low tide (Hawkins et al., 2016; Schaefer et al., 2020). For example, in vermetid reefs in the Mediterranean Sea, long aerial exposure periods mostly driven by winds can cause mass mortalities of fleshy and calcareous algae and mobile gastropods (Zamir et al., 2018). Local exclusions or seaward retraction may be expected, with consequent impacts on ecological interactions. In sandy beaches, coastal erosion due to more severe storms, wave action and offshore winds can remove large quantities of sediments causing severe impacts on the benthic communities and dune vegetation (Delgado-Fernandez et al., 2019), with variable recovery (Scapini et al., 2019).

Poleward distributional shifts of intertidal benthic species have occurred widely due to warming (Helmuth et al., 2006b; Sanford et al., 2019), and critical thresholds for rocky shores are expected to be reached at global temperature increases of >1.5°C (IPCC, 2019). More intense and prolonged synoptic extreme events, such as marine heatwaves (Smale et al., 2019) and storms (Hanley et al., 2020) have accelerated this distributional shift. Changes in benthic species assemblages can disrupt complex trophic linkages and competitive interactions, affecting ecosystem productivity. For instance, the decreased abundance of the blue mussel in coastal Maine, United States and its replacement by weedy algae (Sorte et al., 2017) has reduced habitat complexity and caused biodiversity loss. Species' range contraction and expansion have also occurred in rocky shores in northern California (Sanford et al., 2019), and there have been poleward shifts and mass mortality of the keystone yellow clam (Carcedo et al., 2019) along sandy beaches of Uruguay and northern Argentina (Ortega et al., 2016; Franco et al., 2020a). Larval mobility and transport are crucial in colonization success, and climate-driven shifts in alongshore currents have the potential to either promote or hinder the dispersal of intertidal species.

Episodes of mass occurrence of drifting mats of *Sargassum* brown alga have recently become more frequent and severe in the tropical Atlantic (Caribbean) and the West Africa upwelling (Wang et al., 2019; Johns et al., 2020). Accumulation of this macroalgae in coastal environments has undesirable impacts on the ecosystem and the local economy (Wang et al., 2019); changes in wind patterns and ocean circulation are apparently responsible.

Warming and beach erosion are expected to have variable impacts on habitat suitable for the nesting of marine turtles (Varela et al., 2019; Fuentes et al., 2020) and on their temperature-dependent sex determination (Patrício et al., 2019). Shifts in sex ratios and in the abundance and distribution of nesting sites are a function of turtle species; future scenarios of sea level rise and human development; and other site-specific physical settings (e.g., local weather) and ecological interactions. For instance, the current nesting areas of green and loggerhead turtles in the Mediterranean Sea are expected to be reduced by 59% and 67%, respectively, under a projected sea level rise of 1.2m (Varela et al., 2019), with larger losses (78%–81% by 2050) of currently-suitable nesting areas along the Atlantic coast of US for green turtle and leatherback turtles, depending on sea level rise scenarios and species (Fuentes et al., 2020). New potentially suitable areas may, however, become available.

Vulnerability to heat stress will be exacerbated in areas where ocean acidification poses additional risk to calcifying organisms (e.g., mussels), while enhancing the performance of fleshy algae in rocky intertidal (Milazzo et al., 2019). For some coralline algae, the combination of warming and acidification may increase their growth rates (Kim et al., 2020). Although different trophic pathways can lead to variable ecosystems outputs (Ullah et al., 2018), the overall result of any reduction of the calcified bio-engineers of rocky shores will be habitat simplification and biodiversity loss (Agostini et al., 2018). Calcareous benthic organisms in sandy beaches could also be affected, especially their dispersal and settlement stages (Espinel-Velasco et al., 2018). Such impacts on beach species are not well-studied (Scapini et al., 2019); however, it is possible that the large natural deposits of biogenic carbonate in sandy sediments may greatly reduce in situ ocean acidification effects (Schoeman et al., 2014).

The spatial heterogeneity of rocky intertidal ecosystems provide different microhabitats with variable temperature, salinity, pH and oxygen (Helmuth et al., 2006a; Duarte and Krause-Jensen, 2018; Rilov et al., 2020). For example, maximum temperature can vary as much as 25°C in closely located Californian mussel beds at the same tidal height (Denny et al., 2011). These local thermal extremes could greatly influence the impacts of heatwaves on sessile species. In a similar way, mean pH in Greenland tidal pools can be up to 0.55 higher than nearby coastal water, due to photosynthetic activity of seaweeds providing a natural buffer system against acidification (Duarte and Krause-Jensen, 2018). These examples highlight the importance of long-term, high frequency monitoring at contrasting sites in rocky intertidal ecosystems for global change studies, in order to cover the spatial and temporal variability in their physico-chemical properties and multiple-scale biological responses (Rilov et al., 2020; Kroeker et al., 2020).

## 2.6 Saltmarshes

Saltmarshes are productive intertidal ecosystems structured by salt-tolerant herbaceous plants, with characteristic zonation patterns in biodiversity according to shoreline height and tidal range. They support rich benthic communities of invertebrates (e.g., crabs and polychaetes) and provide foraging and nursery habitat for fish and sea birds. Saltmarshes are most abundant in mid-latitude, temperate regions, but also occur in tropical and polar regions, with latitudinal variation in plant species assemblages, biomass, and coverage density (Bortolus et al., 2015). Saltmarshes are commonly found in estuaries and bays with muddy sediments, where they play an active role in nutrient cycling between land and sea, and between seawater and sediment (Negrin et al., 2016). Moreover, tidal saline wetlands play important roles in coastal carbon budgets due to high belowground productivity (Najjar et al., 2018).

Saltmarshes are highly vulnerable to warming, sea level rise, storm events and increasing wave energy (Adams, 2020; Bacino et al., 2020), with their stability affected by sediment erodibility and vegetation type and cover. Sea level rise poses highest risk of saltmarsh loss where there is limited sediment supply and a lack of space for landward relocation (Bindoff et al., 2019; FitzGerald and Hughes, 2019), either constrained by topography or human land use, such as urbanization or flood barriers causing "coastal squeezing" (Schuerch et al., 2018; Valiela et al., 2018). As in estuaries (Section 2.7), direct human activities exacerbate climate vulnerabilities of saltmarshes, causing very high losses over the past 100 years, particularly in Europe and China (losses of 60%–90%) (Bindoff et al., 2019).

Habitat degradation includes transformation into mudflats or shifts in the vegetation structure due to species turnover, where less tolerant species are locally displaced and more tolerant, invasive species colonize the open niches (e.g., Bortolus et al., 2015; Raposa et al., 2017; Gu et al., 2018). Nevertheless, some genetic adaptation may be possible, with experimental hybridization between native and nonnative species resulting in greater tolerance to inundation and salinity (Gallego-Tévar et al., 2020). Under stress conditions, hybrids produced more plant biomass than native species, and might therefore be more competent to deal with future climate change.

The ability of saltmarshes to keep pace with sea level rise depends on their plant biomass and species composition, the rate of sediment deposition, and the rate of change. Saltmarshes can be initially resilient to low rates of sea level rise due to plant-mediated organic soil accretion (Gonneea et al., 2019). However, they are not expected to be able to withstand the high rates of future sea level rise projected under high $CO_2$ emission scenarios, particularly where they cannot move laterally and upward in elevation (Schuerch et al., 2018). Increased microbial respiration rates as a result of warming, prolonged submersion, and increased eutrophication, are likely to contribute to the reduction of vegetated area and carbon storage (Carey et al., 2017; Mueller et al., 2018). The release of the stored carbon back to the atmosphere—together with methane and possibly nitrous oxide, both strong greenhouse gases—would provide positive feedbacks to the climate system (Lovelock et al., 2017; Gonneea et al., 2019; Rosentreter et al., 2021).

Some local and regional expansion is however possible. In Baja California, Mexico, marshes have gained area as the result of the formation of new lagoons after dune overwash events (Watson and Hinojosa Corona, 2017). There has also been lateral expansion of saltmarshes in the northern United Kingdom over 150 years (Ladd et al., 2019), and increased range of the marsh species *Spartina alterniflora* over the 20th century along the Altantic coast of South America, toward higher latitudes, probably linked to increased temperatures (Bortolus et al., 2015). Conversely, the warming-driven poleward expansion of tropical mangrove forests is threating subtropical marshes. Mangrove encroachment into adjacent saltmarshes leads to a radical shift in the type of coastal ecosystem to the detriment of open areas with herbaceous plants (Saintilan et al., 2014; Armitage et al., 2015), although increasing carbon storage and coastal storm protection (Kelleway et al., 2017).

When saltmarshes are exposed to combined long-term climate shifts and local stressors such as nutrient loads, overfishing, and species invasions (Crotty et al., 2017; Legault et al., 2018; Mueller et al., 2018), the ecosystem risk of habitat deterioration is magnified together with the loss of associated biota (Johnson and Williams, 2017). Minimizing human local impacts on saltmarshes will therefore relieve (to some extent) climate change impacts; nevertheless, climate change will continue to impose high risk at the global level to saltmarshes, surpassing critical physiological tolerance thresholds and reducing their above- and below-ground biomass and carbon storage.

## 2.7 Estuaries

The hydrological and morphological complexity of estuaries along the river-sea transition offers a wide diversity of habitats which hold multiple freshwater, brackish, and marine species. Tidal flats and marshes are common features at the mouth and margins of these ecosystems, contributing to habitat heterogeneity. Estuaries act as buffer areas between environmental impacts on rivers, land, and coastal ocean. In consequence, estuaries require a landscape-widespread approach to ensure a thorough understanding of ecosystem functioning and vulnerability under climate change scenarios (Murray et al., 2019; Palmer et al., 2019). Since estuaries occur worldwide, they are experiencing diverse range of changes in climate hazards. The most common climate threats to global estuaries are sea level rise, ocean warming and deoxygenation, with ocean acidification becoming a more important hazard in eutrophic estuaries of mid and high latitudes (Bindoff et al., 2019; Cai et al., 2021).

Changes in the balance between river runoff and sea intrusion affect the distribution of salinity, turbidity, and nutrients, with mean water column depth and tidal amplitude also affecting ecosystem responses (Scanes et al., 2020). Salinization, increased inundation, and erosion of coastlines are the main impacts of joint effects of raising temperature and sea level, especially in shallow, microtidal estuaries. Deep estuaries with high tidal exchange and associated well-developed sediment are expected to be more resilient to these climate-related impacts. Salinization will intensify in drought regions, leading to upstream relocation of brackish and marine benthic and pelagic organisms, depending on their physiological tolerance range to salinity and substrate type (Hallett et al., 2018). More recent evidence reports shifts in estuarine species range distribution following global warming and regional scale-climate variability in multiple abiotic gradients (Lauchlan and Nagelkerken, 2020; Franco et al., 2020a). The observed interspecific phenotypic differentiation in the Threespine Stickleback *Gasterosteus aculeatus* along contrasting estuaries in the California coast (Des Roches et al., 2020) provides an example of how climate-driven habitat transformation can affect natural populations.

Increased precipitation and river discharges can also occur, due to more frequent anomalous extreme climate events or regional climate signals such as El Niño-Southern Oscillation (ENSO) (Dogliotti et al., 2016; Wang et al., 2018). Effects of increased river flushing can include habitat shifts, together with increased organic and inorganic nutrients, and sediment loads. For example, in the large La Plata River basin, the recent trend of rising precipitation and river flow has altered water mass exchange with the adjacent ocean and fish populations offshore (Franco et al., 2020a). Observed long-term shifts in plankton phenology and changes in trophic interactions have been linked to water warming, increasing turbidity and nutrient loads, coupled with shifts in wind patterns driven by El Niño-Southern Oscillation (ENSO) (Guinder et al., 2017; Lopez-Abbate et al., 2019).

Since estuaries are the main coastal areas for growing human settlements, local anthropogenic impacts have important interactions with climate change (Trolle et al., 2019; Herrera-R et al., 2020). For instance, in the large San Francisco Bay-Delta estuary, land fill for urban expansion or agriculture have caused the loss of 90% of tidal wetland in recent decades (Parker and Boyer, 2019). Similarly, as reviewed in Section 2.6, increasing seawater intrusions and wave energy fluxes are causing erosion of tidal marshes, leading to reduced soil accretion and plant growth, and significantly constraining the capacity of vegetation to retreat inland in response to sea level rise (Bacino et al., 2020).

Increasing estuarine eutrophication similarly combines the effects of local human impacts and climate change. Large depletions in dissolved oxygen are projected in eutrophic estuaries under future climate change scenarios (Ni et al., 2019). Warming and deoxygenation together enhance bacterial degradation of accumulated organic matter in eutrophic estuaries, leading to expansion of hypoxic and anoxic volumes with detrimental effects for water quality and estuarine biota. Organic matter accumulation results from massive microalgae proliferations, including harmful algal blooms (HABs) (Trainer et al., 2020; Griffith and Gobler, 2020), especially in estuaries that already experience high levels of seasonal hypoxia and acidification, variable river discharge and associated biogeochemical alterations in the estuarine zone (Lehman et al., 2020). Severe pathogenic outbreaks of *Vibro* spp. in eutrophic estuaries are also projected to intensify in the warmer future (Vezzulli et al., 2020). More frequent and intense microalgae and pathogenic blooms can result in trophic mismatch and changes in food webs, while together with oxygen-deficient conditions can induce mass mortality events of invertebrates and fishes (Hallett et al., 2018).

Acidification is another key climate-related hazard for estuaries, especially in those already experiencing oxygen loss driven by both natural and anthropogenic drivers

(Steckbauer et al., 2020). A 12-year monitoring program in Australia has revealed that estuaries are warming and acidifying at variable rates depending on the estuarine morphological features, with most rapid changes in shallow, enclose ecosystems (e.g., coastal lakes) with constrained interaction with the open ocean (Scanes et al., 2020). Eutrophic, stratified, large estuaries with long residence times are likely to be highly vulnerable to changes in the subsurface carbonate system, with intensifying risk for calcifiers (Cai et al., 2021).

## 2.8 Mangroves

Mangroves are long-living, large size woody trees and shrubs in the intertidal zone of tropical and subtropical coasts, in salty and brackish waters. Mangroves can tolerate wide ranges in salinity, temperature, and moisture. Their massive root systems are partly aerial, which confers the ability to withstand stress from salinization and anoxic conditions in the muddy sediments. Mangrove forests provide a variety of micro-environments for numerous species, including fish nursery areas. They reduce the impacts of storms, tsunamis, wave action, and coastline erosion (Alongi, 2008). They also regulate nutrients and trace metal contamination, as well as assisting in climate mitigation as "blue carbon" ecosystems, through accumulating organic carbon in their sediments (Crooks et al., 2019).

These coastal wetlands are under high pressure from a variety of pervasive anthropogenic disturbances, such as eutrophication and pollution, coastal development (causing squeezing) and aquaculture. Mangroves are commonly used for firewood, construction, and medicine products (de Lacerda et al., 2019). In addition, their deterioration is anticipated to continue with global warming, sea level rise and extreme climatic events, adding risk of habitat loss (Borchert et al., 2018; Mafi-Gholami et al., 2020). Local management action for sustainable use and protection may help conserve biodiversity, restore local hydrodynamics, and sediment delivery, and thereby contribute to climate change adaptation and mitigation (de Lacerda et al., 2019).

Relative sea level rise is the greatest climate hazard for mangrove ecosystems, threatening their capacity to build soil elevation and keep pace with sediment subsidence and salinization. As described for saltmarshes (Section 2.6), hydro-geomorphological settings (primarily sediment supply, soil accretion and space for relocation) control the responses of the intertidal vegetation to rising sea level. Mangroves are resilient to current rates of mean sea level rise (of $\sim$0.4 cm per year) and future rates projected to 2100 under a low emissions scenario (RCP2.6); their pioneer-phase traits also facilitate their ability to recover from storms and hurricanes (Alongi, 2015; Krauss and Osland, 2020). However, their soil accretion rates are unlikely to cope beyond mid-century under the high emission scenario (RCP8.5) (Bindoff et al., 2019). In particular, fringe mangroves under microtidal regimes, in steep topography lacking rivers or creeks, are more vulnerable to sea level rise because they are constrained in sediment supply, soil accretion and inland migration (Borchert et al., 2018).

There is more uncertainty over other climate hazards on mangrove forests, because severe human impacts may hinder the emergence of ecosystem responses to long-term changes (Gilman et al., 2008; Krauss and Osland, 2020). Over the past century, massive losses of mangrove forests have occurred worldwide, primarily due to deforestation and other direct anthropogenic impacts (land use change to agriculture, aquaculture, human settlement and port development) (Saintilan et al., 2014; Sippo et al., 2018). Warming and large-scale extreme temperature events (marine heatwaves) provide additional risks (Duke et al., 2017; Cavanaugh et al., 2019); however, warming also enables mangrove range expansion at their cooler latitudinal limits (Cohen et al., 2020; Whitt et al., 2020). This range expansion results in the loss of the invaded subtropical or temperate saltmarshes, and their associated biota. After establishment, young mangrove plants are greater carbon sinks in their biomass than older ones, although soil carbon accumulation rates are not age dependent (Walcker et al., 2018). The further poleward expansion of mangroves will be controlled by the local environmental conditions that define the suitable microclimate (e.g., winter temperatures) (Osland et al., 2019); the long-term ecological impacts of such changes warrant further attention (Cavanaugh et al., 2019).

Additional climate-related threats include more intense and frequent synoptic extreme climatic events, such as storm surges (Servino et al., 2018) or rainy/drought periods. Impacts will be greatest in localities already experiencing other stresses, e.g., increased eutrophication and deoxygenation (Manna et al., 2010), or nutrient limitation in oligotrophic mangrove ecosystems (Anton et al., 2020). Rainfall patterns can also strongly influence mangrove distributions (Eslami-Andargoli et al., 2009), together with temperature and cyclone frequency (Alongi, 2015; Simard et al., 2019). Positive impacts in mangrove area and biomass are expected in regions where precipitation is projected to increase, and negative impacts on more arid regimes related to hypersalinization and sea-level fluctuations (Adame et al., 2020).

# 3 Ecosystem impacts in shelf seas and the open ocean

## 3.1 Introduction

The five ecosystem groups considered here are shelf sea benthos; upper ocean plankton; fish and fisheries; cold water corals; ice-influenced ecosystems; and deep seafloor ecosystems. As in previous sections, the concept of ecosystem is used flexibly, recognizing that all ocean ecosystems are connected, and that climate change impacts on other categories of physically- or biologically-structured communities and habitats could be assessed. For example, the shelf edge, seamounts, deep sea vent systems and ocean midwater each have characteristic biotas and could be considered separately; however, climate change impacts on such ecosystems have generally not been well-studied.

As in near-coastal ecosystems, climate-related drivers in shelf seas and the open ocean have direct physiological effects on organisms, in different combinations; they also indirectly affect marine biogeochemistry and species interactions, with cascading effects on communities, food webs and ecosystem services (Boyd et al., 2018; Lotze et al., 2019). However, in open ocean ecosystems, attribution of shifts in biodiversity, species distribution and ecosystem functioning to climate change can be more straightforward, since a narrower set of climate-related changes are involved—primarily warming, ocean acidification, deoxygenation, and other consequences of intensified stratification. Human influences in the open ocean are less than at the land-sea transition, although the impacts from fishing (Christensen et al., 2014) and plastic pollution (Peng et al., 2020) are now globally ubiquitous.

Shelf sea ecosystems are intermediate: their local and regional heterogeneity can be high, arising from complex hydrodynamic processes both on the shelf (river plumes, tides and frontal systems) and at the shelf edge (boundary currents and upwelling), as well as regional-scale climate signals (Huthnance, 1995; Liu et al., 2003; Tinker et al., 2016; Franco et al., 2020b). They also experience multiple anthropogenic disturbances affecting their functioning (Levin et al., 2015; Martinetto et al., 2020). While large changes in many shelf sea ecosystems have been observed during the past 50–100 years, they are not necessarily caused by global-scale climate drivers (Kenchington et al., 2007; Yang et al., 2017).

## 3.2 Shelf sea benthos

Shelf seas or continental margins comprise around 7% of the global ocean, in polar, temperate, and tropical regions, with their boundaries conventionally defined as 200 m water depth. They are highly productive, supporting around 90% of global fisheries; they are also extensively used for other purposes, including hydrocarbon extraction, renewable energy resources and waste disposal (Kröger et al., 2018). The taxonomic composition of seafloor communities varies according to latitude and sediment type, typically comprising mollusks, crustacea, polychaetes and echinoderms, together with sediment-dwelling meso-fauna and microbes. Demersal finfish are also a component of the benthos but are here considered separately (Section 3.4).

Because of their limited water depth, many, but not all, shelf seas are warming more rapidly than the open ocean, with tidal mixing limiting stratification and the development of strong vertical temperature gradients. Belkin (2009) compared changes in sea surface temperature for 64 "Large Marine Ecosystems" between 1982 and 2006: the fastest warming occurred in the Baltic Sea and North Sea, where there were increases of ~1.3°C, a rate three times greater than the global mean surface ocean warming over that period. Relatively rapid warming has also occurred in East Asian Seas, around Newfoundland, in western Greenland and off north-west Alaska. These regional differences in warming rates have been ascribed to the North Atlantic Oscillation (NAO), nearby land warming, and changes in freshwater runoff (affecting stratification).

Hobday and Pecl (2014) used a longer data set (50 years) combined with global climate models to identify 24 global marine hotspots, with relatively rapid warming; 23 of these fully or partly included continental shelf seas or occurred around islands. More detailed analyzes were subsequently carried out on six of these (off the coasts of Australia, South Africa, Madagascar, India, and Brazil) by Popova et al. (2016), using a high resolution global model to project future conditions under a high emissions scenario, not only for sea surface temperature, but also for ocean acidification, deoxygenation, stratification, nutrient supply, primary production and ocean circulation. They found that recent rapid warming was not necessarily indicative of high values for other stressors; that regional patterns could be highly sensitive to circulation changes; and that other parts of the ocean could warm more rapidly in future, particularly Arctic shelf seas. Recent studies in specific shelf sea hotspots include observed changes in the Patagonia Large Marine Ecosystem, where benthic productivity has been strongly influenced by circulation changes and (increasing) primary production over a ~20 year period (Franco et al., 2020b; Marrari et al., 2017).

The above example emphasizes the need for caution with regard to the linkage between observed changes in the shelf sea benthos, warming and other environmental trends. Additional complexities arise from interactions with anthropogenic deoxygenation (Rabalais, 2019; Section 1.5), and widespread disturbance caused by demersal trawling in many areas (Amoroso et al., 2018). In European shelf seas, around 50% of the seafloor is estimated to be trawled at least once a year, with some areas re-fished more than 10 times a year (Eigaard et al., 2017). Under such circumstances, long-term benthic monitoring at undisturbed sites, e.g., Marine Protected Areas, is needed to distinguish climate change impacts from other causes of variability. Those datasets are currently lacking, or are short term only—noting that time-series of several decades may be necessary for reliable detection of climatic signals in "noisy" shelf sea environments (Duarte et al., 2013; Carter et al., 2019).

An alternative, spatially integrated approach used data from the Continuous Plankton Recorder survey to determine long-term changes in the abundance of planktonic phases of major benthic invertebrate groups for the central North Sea (Beare et al., 2013). Over the period 1958–2010, the abundance of bivalve mollusk larvae decreased, while the abundance of echinoderm larvae increased. However, there was high year-to-year variability; it was not possible to identify changes in individual species; and temporal correlations between larval abundance and temperature were not statistically significant, nor for changes in pH. There is stronger evidence for temperature-induced impacts in Australian shelf seas, with warming considered to be the main factor causing major declines in rock lobster recruitment (Johnson et al., 2011) and in the range expansion of an ecologically important sea urchin (Ling et al., 2009). Effects of warming on tropical corals have already been considered (Section 2.2).

Significant yet variable effects of ocean acidification on a wide range of benthic fauna have been demonstrated in the laboratory (e.g., Wittmann and Pörtner, 2013; CBD, 2014; Doney et al., 2020), including interactions with warming (Byrne and Hernández, 2020), metal pollution (Lewis et al., 2016), and differential effects on life-cycle stages (Gibson et al., 2011; Espinel-Velasco et al., 2018). Major changes occur in seafloor communities living under naturally high $CO_2$ levels, near vent systems, are also well-documented (Hall-Spencer et al., 2008; Kroeker et al., 2011). While direct field evidence for climate-related, long-term acidification effects on benthic ecosystems remains limited (Doo et al., 2020), a combination of ecosystem modeling and multispecies experimental studies provides testable hypotheses with regard to effects on food web structure (Nagelkerken et al., 2020) and identifying locations in shelf seas where acidification effects are expected to be greatest (e.g., Fay et al., 2017; Hodgson et al., 2018).

## 3.3 Upper ocean plankton

The upper ocean (also known as the epipelagic) has been physically defined as the top 200 m (Bindoff et al., 2019). It comprises the light-influenced euphotic zone and the relatively well-mixed surface layer; in stratified waters, the latter is separated from the deep ocean by the thermocline, a strong temperature gradient. In the upper ocean, biomass and production processes are dominated by photosynthetic phytoplankton. Their abundance, species composition, and size structures strongly influence biota at all other trophic levels, including zooplankton, fish (here considered separately in Section 3.4), turtles, sea mammals, and microbial heterotrophs (protists, bacteria, archaea, fungi, and viruses). The quantity and quality of phytoplankton also determine the export of organic carbon to deeper waters and the seafloor—hence the food supply to those ecosystems and their productivity—and provide critical global feedbacks to the global climate system, through the removal and return of $CO_2$ over a range of timescales. Climate change impacts summarized here relate to changes in planktonic species' ranges and their seasonality (phenology); changes in phytoplankton abundance and productivity, at global and regional levels; implications for food web structure; and effects on organic carbon export. Ocean acidification impacts are also briefly considered.

Temperature provides the first order control of taxonomic distributions in the upper ocean, with species' thermal tolerances set by their metabolic rates and other physiological processes. The two main effects of warming are an overall poleward shift in species' latitudinal ranges, with seasonal cycles occurring earlier at any particular location within those distributions; both these effects have been widely observed. Latitudinal range shifts of 40–50 km per year have been observed for phytoplankton, and 14–20 km per year for zooplankton (Beaugrand, 2009; Poloczanska et al., 2013). Differences occur between different phytoplankton groups, e.g., dinoflagellates have moved more rapidly than diatoms (Chivers et al., 2017). Decadal variability can be high, and its causes are not well-understood; as a result, temperature-based species distribution models are relatively poor in hind-casting distributional changes for representative plankton in the North Atlantic (Brun et al., 2016). Future projections for that region indicate that east-west shifts may be greater than northern movements (Barton et al., 2016), with community re-arrangement involving new combinations of co-existing species. The above rates of movement are up to an order of magnitude higher than the average of ~5 km per year for all marine species since the 1960s (Poloczanska et al., 2013, 2016).

Differential changes in biogeographical distributions have increased importance when combined with changes in the timing of seasonal events (phenology), increasing the likelihood of mis-matches between the annual productivity of phytoplankton and the reproduction cycles of their primary and secondary consumers (zooplankton and fish)—hence reducing food availability for the early life stages of the latter (Kharouba et al., 2018; Rogers and Dougherty, 2019; Asch et al., 2019). Observed timing of seasonal life cycle events has advanced more for zooplankton (mean of ~1.1 days per year, over 50 years) than it has for phytoplankton (~0.6 days per year); fish show more variable data, covering that range (Lindley and Kirby, 2010; Poloczanska et al., 2013). There is limited evidence for ongoing adaptation (Neuheimer et al., 2018), and outcomes of specific interactions are complex (Guinder et al., 2017). Overall, pelagic organisms show greater phenological change than other marine groups, with the average seasonal advancement for all taxa being 0.4 days per year (Poloczanska et al., 2013).

Spatial patterns in ocean mixing processes, linked to nutrient availability, are a major cause of inter- and intra-regional variability in the above effects, creating novel biogeochemical provinces (Reygondeau et al., 2020). Changes in wind stress, upwelling and stratification (Elsworth et al., 2020) determine the re-supply from deeper water of macro-nutrients (biologically-available nitrogen, phosphorus and silica) and micro-nutrients (e.g., iron; Tagliabue et al., 2020) essential for phytoplankton growth, while also affecting the levels of dissolved oxygen that can limit the depth distributions of animals and microbes. Such local-to-regional physico-chemical changes are considered responsible for the variability in phytoplankton abundance responses for the period 1998–2015, determined by satellite remote sensing (Hammond et al., 2020). Significant positive trends in chlorophyll occurred over that period in the Southern Ocean and most of the North Atlantic, with negative trends in the central Pacific and Indian Ocean. The overall global trend was barely discernible (increase of <0.1% per year) and not significant, confirming an earlier global ocean-color analysis (Gregg et al., 2017). Future projections indicate an intensification of regional differences, more strongly based on latitude (Bopp et al., 2013; Bindoff et al., 2019; Kwiatkowski et al., 2020). Thus a 7%–16% reduction in net primary production is expected in tropical regions by 2100 relative to 2006–15 under high emission scenarios, although that decrease is expected to be largely (but not wholly) offset by increases in polar and subpolar regions (Fig. 5A) (IPCC, 2019).

Even if climatic impacts on phytoplankton are small, any reductions—but not necessarily increases—in their biomass are near-certain to be amplified at higher levels in the trophic system (Lotze et al., 2019; Kwiatkowski et al., 2019). In addition to phenological disruption, three other factors apply: first, higher temperatures increase the metabolic rates of herbivores and carnivores, increasing their food requirements through grazing and predation (Nagelkerken and Connell, 2015). The same quantity of food therefore supports less biomass higher in the food chain. Second, warmer temperatures, increased stratification, and ocean acidification (see below) all favor smaller-celled phytoplankton, such as the cyanobacteria *Prochlococcus* and *Synechococcus* (Flombaum et al., 2013; Schmidt et al., 2020). Being smaller, more trophic steps are needed before their organic material is made available to larger organisms; they are also inherently less nutritious (lacking essential fatty acids and sterols) (Brander and Kiørboe, 2020). Both effects contribute to further amplification. Thirdly, warming interacts synergistically with ocean acidification for many marine invertebrates, with both stressors increase energy demands (Kroeker et al., 2013). For all these processes, species-specific differences in sensitivity are likely to apply, increasing the risk of breakdown of community structure and the unpredictability of the overall effect. Not all these amplifications and interactions are currently included in whole-ecosystem models; nevertheless, such models indicate overall reductions of 4%–15% in total marine animal biomass by 2100 relative to recent levels, with the range indicating estimates for low and high emission scenarios, RCP 2.6 and RCP 8.5 respectively (Fig. 5B) (Bindoff et al., 2019).

Ocean acidification impacts on upper ocean plankton communities have been investigated separately from warming using in situ mesocosm studies, in Arctic and European waters (Riebesell et al., 2013; Bach et al., 2016; Gazeau et al., 2017). A metaanalysis of 49 papers arising from these studies (Dutkiewicz et al., 2015) showed a wide range of phytoplankton responses, with two groups of nitrogen-fixing cyanobacteria (nitrogen-fixing diazotrophs and *Synechococcus* spp.) generally increasing their growth rates under high $CO_2$ conditions, by 18%–25%, out-competing other species groups. However, reduced pH may adversely affect heterotrophic nanoflagellates (Deppeler et al., 2020) and bacterial nitrifiers (Beman et al., 2011), with highly variable effects on other heterotrophic bacteria (Burrell et al., 2017; Harvey et al., 2020).

Coccolithophores are a group of calcifying phytoplankton that are frequently considered vulnerable to ocean acidification, based on single-species studies (Meyer and Riebesell, 2015). However, "anomalous" results have also been reported (Iglesias-Rodriguez et al., 2008), and the overall effect in multispecies studies has been a slight, nonsignificant increase in their growth rates (Dutkiewicz et al., 2015). Observational data from field surveys support that

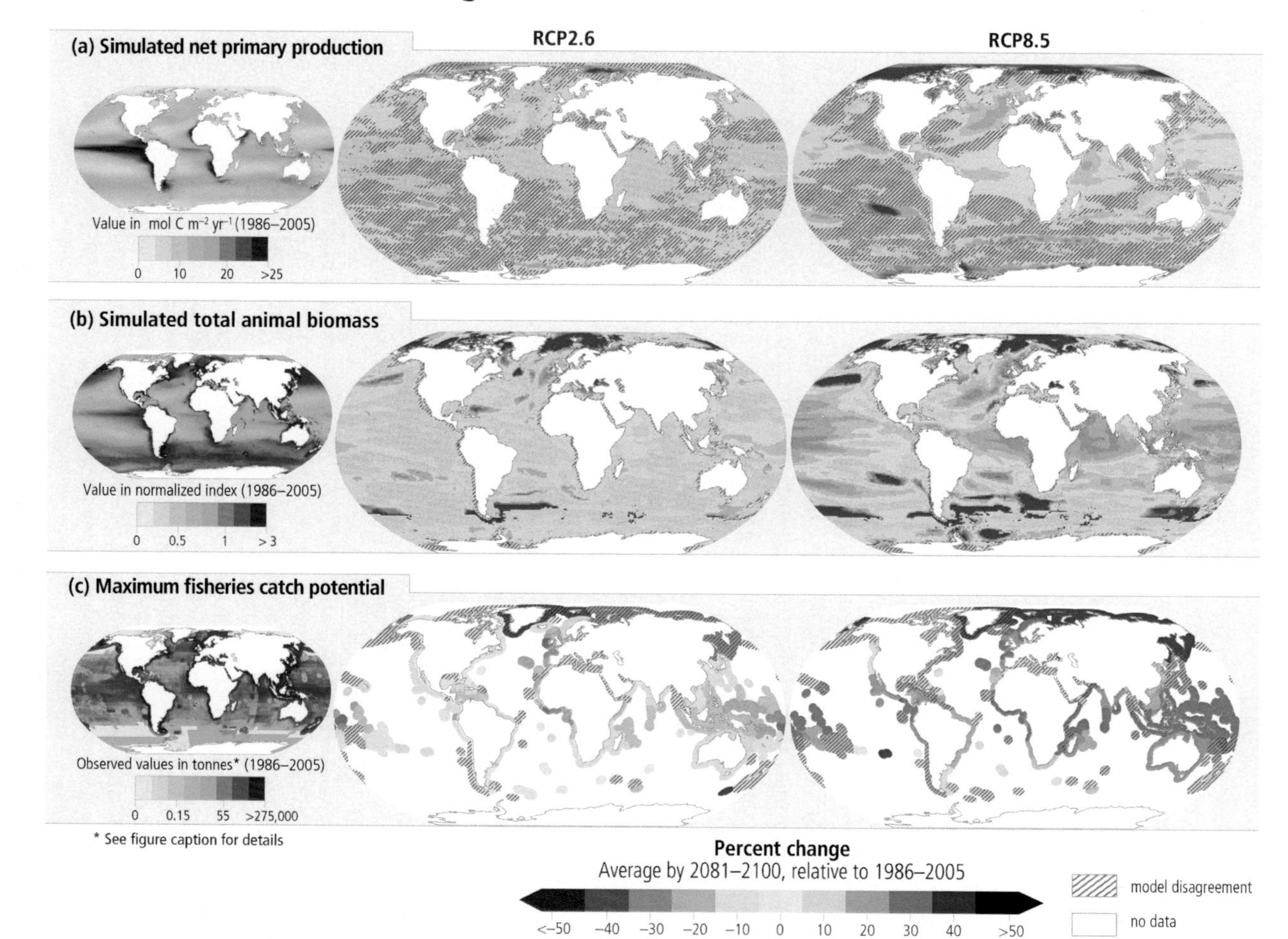


**FIG. 5** See figure legend in next page

counter-intuitive response, indicating an increased frequency of coccolithophorid blooms in the North Atlantic in recent decades (Beare et al., 2013; Rivero-Calle et al., 2015; Krumhardt et al., 2016).

Calcified pteropods (planktonic mollusks, also known as sea butterflies) are the zooplankton group considered most at risk from ocean acidification, particularly in cooler waters where low pH levels can cause erosion of their aragonite shells (Bednaršek et al., 2012a,b) and may also be already affecting the growth and survival of early life stages (Lischka et al., 2011; Gardner et al., 2018). A metaanalysis of experimental studies (Bednaršek et al., 2019) has identified a risk range of aragonite saturation state from 1.5 (early warning) to 0.9 (lethal impacts). Saturation state values in this range already occur in upwelling waters off the California coast and in the Arctic (Bednaršek et al., 2019; Robbins et al., 2013; Qi et al., 2017), and values $<1.0$ are projected to occur widely in the Southern Ocean, even under emission-stabilizing scenarios (Hauri et al., 2016). These effects are considered further in Section 3.6.

Changes in phytoplankton, zooplankton, and bacterial communities resulting from ocean acidification, warming and other climate-related impacts will also affect the quality and quantity of organic material exported to the ocean interior and seafloor (Bach et al., 2019), and hence the effectiveness of climate regulation by the biological carbon pump (Ducklow et al., 2001; Gehlen et al., 2011; Barange et al., 2017). In particular, increasing $CO_2$ availability could be expected to increase the carbon-nitrogen ratio in falling organic particle. In experimental mesocosm studies, such effects were significant (changing the C:N ratio by up to 20%) but variable, affected more by grazing zooplankton and microbial decomposition than by changes in phytoplankton taxonomic structure (Taucher et al., 2021). That finding reinforces observational evidence for the importance of fragmentation (Briggs et al., 2020) and microbial processes (Legendre et al., 2015) in controlling the flux of organic material through the water column, necessarily interacting with gradients, and climate-driven changes, in temperature and oxygenation (Laufkötter et al., 2017).

The overall effect of climate change on the downward flux of carbon at 1000 m in the North Atlantic has been estimated to be a reduction of 27%–43% by 2100, depending on emission

**FIG. 5** Projected changes, impacts, and risks for ocean regions and ecosystems. (A) Depth integrated net primary production (from CMIP5), (B) total animal biomass, depth integrated including fish and invertebrates (from FISHMIP), and (C) maximum fisheries catch potential. The three left panels represent the simulated (A, B) and observed (C) mean values for the recent past (1986–2005), the middle, and right panels represent projected changes (%) by 2081–2100 relative to recent past under low (RCP2.6) and high (RCP8.5) greenhouse gas emissions scenario, respectively. Total animal biomass in the recent past (B, left panel) represents the projected total animal biomass by each spatial pixel relative to the global average. (C) *Average observed fisheries catch in the recent past (based on data from the Sea Around Us global fisheries database); projected changes in maximum fisheries catch potential in shelf seas are based on the average outputs from two fisheries and marine ecosystem models. To indicate areas of model inconsistency, *shaded areas* represent regions where models disagree in the direction of change for more than: (A) and (B) 3 out of 10 model projections, and (C) one out of two models. Although *unshaded*, the projected change in the Arctic and Antarctic regions in (B) total animal biomass and (C) fisheries catch potential have low confidence due to uncertainties associated with modeling multiple interacting drivers and ecosystem responses. Projections presented in (B) and (C) are driven by changes in ocean physical and biogeochemical conditions, e.g., temperature, oxygen level, and net primary production projected from CMIP5 Earth system models. *Reproduced with permission from IPCC (Intergovernmental Panel on Climate Change), 2019. Summary for policy makers. In: Pörtner, H.-O., Roberts, D.C., Masson-Delmotte, V., Zhai, P., Tignor, M., Poloczanska, E., Mintenbeck, K., Alegría, A., Nicolai, M., Okem, A., Petzold, J., Rama, B., Weyer, N.M. (Eds.), IPCC Special Report on the Ocean and Cryosphere in a Changing Climate. IPCC/WMO, pp. 1–35. https://www.ipcc.ch/srocc (Accessed 7 October 2020).*

scenarios (Barange et al., 2017). However, there are many uncertainties regarding relevant mid-water processes (Martin et al., 2020), with additional complications arising from the different methodologies used to determine the current global value for that flux (Buesseler et al., 2020) and recent re-assessments of the relative importance of different flux components (Boyd et al., 2019).

## 3.4 Fish and fisheries

Fish are necessarily part of wider marine ecosystems, in coastal waters, shelf seas, and the open ocean, occurring in the full range of water depths. Their role in providing food, nutrition, income, and livelihoods for many millions of people, particularly in developing countries is considered in greater detail in Chapter 8 (Lotze et al., 2021). Here they are primarily considered in the context of their role in marine food webs and biogeochemical cycles, and the interactions of fishing with climate change. Landings of harvested species currently have an average trophic level of ~3.0 (representing links in the food chain, with primary producers at 1.0), reduced from ~3.4 in the 1950s due to the removal of larger species (Pauly et al., 1998). The impacts to date of fishing, not just on fish but on other components of marine ecosystems, are considered to be at least as high as climate change (Watson et al., 2013).

As discussed in Section 3.3 above, a poleward range shift in fish species' distributions can be expected as the main response to warming. Such effects have already been observed in both northern and southern hemispheres (Montero-Serra et al., 2015; Robinson et al., 2015; Poloczanska et al., 2016), and are projected to continue, with rates depending on emission scenarios (Jones and Cheung, 2015; Morley et al., 2018). As a result, species richness and productivity decreases in tropical oceans, but can increase in mid to high-latitude regions (Molinos et al., 2016). Fish species distribution models show greatest decreases in species richness in the Indo-Pacific region, and semienclosed seas such as the Red Sea and Persian Gulf (Burrows et al., 2014; Wabnitz et al., 2018), with projected losses in catch potential of 25%–50% by 2100 relative to the recent past in waters around Pacific islands, depending on future emission scenarios (Asch et al., 2018).

Mid-water deoxygenation can also influence fish distributions (Jones and Cheung, 2015; Gallo and Levin, 2016). More pervasive decreases in dissolved oxygen slow fish growth and reduce adult body size (Deutsch et al., 2015; Pauly and Cheung, 2017) and thereby increasing the dominance of smaller-bodied fishes in the upper ocean (Lefort et al., 2015). A wide range of impacts of ocean acidification on fish neuro-physiology have been reported in experimental studies, including potential impairments of olfaction, hearing, vision, homing, and predator avoidance (Kroeker et al., 2013; Clements and Hunt, 2015; Munday et al., 2020). These behavioral impacts are likely to exacerbate deleterious effects of other stressors, while noting that the experimental studies mostly relate to end of the century pH levels under high emissions, and some species seem resilient.

Despite the pervasive nature of climate change impacts, fishing intensity, and exploitation status are likely to continue to be the main driver of fishery yields in many regions (Cheung et al., 2018). Better fish management could therefore make it possible for fish catches to increase if future climate change can be limited (Barange, 2019). Changes in fish catches between 1998 and 2006 were significantly related to "bottom-up" forcing, by either temperature or phytoplankton biomass (estimated from levels of chlorophyll *a*) in 20 large marine ecosystems (Mcowen et al., 2015), with "top-down" forcing (fishing effort) the dominant factor in 16 regions and a further 11 being inconclusive. Other global analyzes also provide evidence for climate-driven declines in stock recruitment in recent decades (Britten et al., 2016; Free et al., 2019).

A summary of projected geographical patterns in climate-driven changes for maximum fish catch potential is shown in panel (C) of Fig. 5, for both high and low emissions scenarios (IPCC, 2019). The main regional differences (tropical decreases; polar increases, particularly in the Arctic Ocean) reflect the latitudinal influences on pelagic production processes, as already discussed. Overall, decreases in fish catch potential of 21%–24% are projected by 2100 relative to 1986–2005 under high emission scenarios (RCP 8.5), compared to decreases of <8% under low emissions (RCP 2.6) (IPCC, 2019). However, as noted in the figure legend, the simulated increase in maximum catch potential projected for the Arctic is uncertain due to the complex nature of the interacting processes occurring there. Other factors that are not yet well-simulated in the global model include variation in the relative importance of progressive warming and the occurrence of regional-scale marine heatwaves. For the northeast Pacific, the effect of the latter on fisheries has been estimated as four times greater than the former (Cheung and Frölicher, 2020).

## 3.5 Cold water corals

Cold water corals lack the algal symbionts found in warm-water coral species (Section 2.2). They occur over a very wide latitudinal range (polar to tropical), typically in water depths of 150–1500 m where they favor hard substrates subject to strong currents, e.g., at the shelf edge and on seamounts (Roberts and Cairns, 2014). Their calcified structures provide foundation habitat for many other marine organisms (Buhl-Mortensen et al., 2010). The lower depth limits of cold water corals generally match aragonite saturation values (that decrease with water depth) of ~1.0, i.e., the threshold for unprotected carbonate structures to remain stable, suggesting that these organisms might be particularly vulnerable to ocean acidification (Guinotte et al., 2006). That linkage is supported by geological evidence (Hebbeln et al., 2019) and limited observational records (Boolukos et al., 2019). It has also by been investigated by at least 24 experimental studies (reviewed by Maier et al., 2019), although several uncertainties regarding ecological implications remain.

In the laboratory, cold water corals regulate their internal pH, and are therefore able to grow and calcify under conditions of aragonite undersaturation (Georgian et al., 2016; Kurman et al., 2017; Maier et al., 2019). While additional energy is required, this may represent less than 3% of total metabolic requirements (Maier et al., 2016). Nevertheless, the strength of the coral exoskeletons can be substantially weaker under high $CO_2$ conditions (Hennige et al., 2015); bioerosion is likely to be enhanced (Schönberg et al., 2017); and direct effects of undersaturation on the structural integrity of dead coral may be as important, or more important, for coral-structured habitats than the indirect effects on living organisms (Jackson et al., 2014; Hennige et al., 2015, 2020). When the cold water coral *Lophelia pertusa* was collected from a low pH environment, at ~300 m water depth off southern California, its calcification rate was greatly enhanced in the laboratory under more favorable water chemistry conditions (Gomez et al., 2018).

Many populations of cold water corals may also be close to their species-specific critical thresholds for temperature (~14°C for *L. pertusum*) and dissolved oxygen (Lunden et al., 2014; Hanz et al., 2019), while also subject to localized deep-water pollution (Järnegren et al., 2020) and salinity changes, e.g., arising from the increasing salinization of the Mediterranean (Skliris, 2019). Physico-chemical evidence for increasing aragonite undersaturation at the seafloor is provided both by modeling (Hauri et al., 2013; Fransner et al., 2020) and direct observations (Perez et al., 2018; Fontela et al., 2020). The biological impacts of such changes in carbonate chemistry may be accentuated by warming, regional changes in ocean circulation patterns (Puerta et al., 2020), as well as other stressors. Projections of future suitable habitat in the North Atlantic for six species of cold water corals under a high emission scenario (RCP 8.5) indicated areal declines by 2100 of 79% for *L. pertusum* and 99% for the octocoral *Paragorgia arborea* (Morato et al., 2020); projections for low emission scenarios have not yet been made.

## 3.6 Ice-influenced ecosystems

Global climate models consistently project the greatest temperature changes in polar regions, with profound consequences for the cryosphere. Ice-influenced marine ecosystems, in both the Arctic Ocean and Southern Ocean, will necessarily be strongly affected by those changes, occurring on land (Chapter 3; Shahgenova, 2021) as well as at sea (Gutt et al., 2015; AMAP, 2019; Meredith et al., 2019). The focus here is on changes in sea-ice cover; however, the consequences of glacial retreat and the collapse of ice-shelves are also briefly summarized, since those changes affect salinity (that in turn affects ocean circulation, nutrient levels, acidification, and biological productivity), as well as other physico-chemical impacts of drifting icebergs.

As noted in Section 1.7, pronounced warming has already been observed in the Arctic, with summer sea-surface temperature anomalies of up to 5°C (Steele et al., 2008) and an associated late summer reduction in sea-ice cover from around 7.5 million $km^2$ in 1980 to 3.7 million $km^2$ in 2020 (NSIDC, 2020). This has already had major ecological impacts, that—from different perspectives—can be considered as either beneficial or damaging. The most direct effect of sea-ice loss is in providing greater habitat area, and a longer growing season, for open-water phytoplankton. As a result, net primary production in the Arctic Ocean is estimated to have increased by >30% since 1998 (Arrigo and van Dijken, 2015), with large regional variations, primarily determined by nutrient availability (Li et al., 2019; Ardyna and Arrigo, 2020). There are, however, uncertainties in the overall scale of that increase, since under ice phytoplankton productivity (that may have either increased or have been reduced) is not detected by satellite remote-sensing. Such productivity can be substantial (Arrigo et al., 2012; Zhang et al., 2015), yet until recently had been considered negligible (Ardyna and Arrigo, 2020). Micro-algae growing within sea-ice may also make a significant, and apparently increasing, contribution to total productivity (Song et al., 2016).

Increased advection in the main Arctic Ocean gateways (Fram Strait/European Arctic Corridor and the Bering Strait) also contributes to ice-related biogeographic changes, not only in providing nutrients from the North Atlantic and Pacific, but also in changing plankton community composition through the introduction of temperate species (Neukermans et al., 2018; Oziel et al., 2020). The larger-scale loss of sea-ice will enable trans-Arctic interchange for a wide range of species in the northern hemisphere, primarily from the Pacific to the Atlantic (Reid et al., 2007; Vermeij and Roopnarine, 2008). There is genetic evidence from closely related fish species that such interchange also occurred at the time of the previous interglacial warm period, ~100 kyr ago (Coulson et al., 2006).

Changes in phytoplankton species, abundance, and production necessarily affect wider biogeochemical processes and other ecosystem components. While increased vertical export of organic particles can be expected, the overall effect may be small (Randelhoff and Guthrie, 2016; Meredith et al., 2019). Zooplankton biomass has generally increased, but not at all localities (Sigler et al., 2017; Rutzen and Hopcroft, 2018). Future changes will depend on complex trophic interactions, including the outcome of interspecific competition (e.g., between the polar copepod *Calanus glacialis* and the boreal *C. finmarchicus*; Dalpadado et al., 2016) and interactions with other stressors. Impacts on fish and benthic invertebrates are expected to be similarly mixed, while likely to favor range expansion by sub-Arctic species at the expense of native, cold-adapted species (Kjesbu et al., 2014; Steiner et al., 2019). A range of life-history characteristics will determine outcomes (Hollowed et al., 2013), but there is currently low confidence in projecting the species-specific composition of future Arctic marine communities (Meredith et al., 2019).

Ocean acidification in the Arctic is relatively well-documented (Qi et al., 2017; AMAP, 2018). Aragonite undersaturation is projected to occur throughout the Arctic Ocean by 2100 under a high emission scenario (RCP 8.5), with basin-wide average saturation state ($\Omega$) values <0.8 in the top 1000 m and below 3000 m water depth (Terhaar et al., 2020). The ecological impacts of these changes remain uncertain, yet are likely to be increasingly important for pteropods and benthic calcifiers (AMAP, 2018).

Arctic marine mammals that use sea-ice for breeding and foraging will be strongly disadvantaged by its further reduction and future loss, since they will have little scope for further northward movement; e.g., walrus (Kovacs et al., 2016) and polar bears (Hamilton et al., 2017). However, not all populations of the latter are decreasing (Stapleton et al., 2016), and adaptive behavior may be possible for some species, e.g., ringed seals (Lydersen et al., 2017).

Many of the above considerations also apply to climate change impacts in the Southern Ocean, although without an equivalent scale of warming or sea-ice loss occurring to date (Meredith et al., 2019). There are also several unique features, including the importance of krill (*Euphausia superba*) as a phytoplankton grazer and food for higher trophic levels (fish, squid, marine mammals, and seabirds). Regional reductions in krill abundance have occurred, but may not be climate-related (Steinberg et al., 2015); future reductions are projected by food web models linking growth rates to warming scenarios (Klein et al., 2018), with biomass increases in gelatinous salps (Suprenand and Ainsworth, 2017). However, different climatic drivers apply in different zonal bands of the Southern Ocean (Leung et al., 2015), and there are also regional uncertainties regarding the wide range of other observed and projected impacts of climate change on Southern Ocean ecosystems (Gutt et al., 2015; Ingels et al., 2012; Meredith et al., 2019).

Iceberg formation, through the collapse of coastal ice shelves, is a more important physical driver in the Southern Ocean than in the Arctic (Constable et al., 2014). Ice shelves fringe 75% of the Antarctic coastline, and cover around 30% of Antarctica's continental shelf. When they break-up, new seafloor habitat is created (Ingels et al., 2018, 2020), and large amounts of freshwater are delivered elsewhere to the ocean, affecting primary production through stratification and the delivery of nutrients. Up to a fifth of the Southern Ocean's downward carbon flux is estimated to currently originate with giant iceberg fertilization, and this is expected to increase in future (Duprat et al., 2016; Ingels et al., 2020).

## 3.7 Deep seafloor ecosystems

The deep seafloor is here considered as >200 m, covering the full range of bathyal (200–3000 m), abyssal (3000–6000 m) and hadal (>6000 m) depth zones widely used in oceanographic literature. Their combined seabed area is 335 million km$^2$—more than twice the land area, and therefore representing the largest surface-based ecosystem on Earth. Less than 20% of the deep seafloor has been mapped in detail using modern sonar technology (Seabed2030, 2020). Although much of the global seafloor is topographically uniform, comprising the abyssal plains of the major ocean basins, its features include the slope and canyons of the continental shelf, seamounts, mid-ocean ridges, fracture zones, underwater vents and volcanoes, and hadal trenches to a depth of 10,900 m. Each of those could be considered as a separate ecosystem (biogeographic unit), with their biotas determined both by local physico-chemical conditions and biological factors in the overlying water column (Watling et al., 2013).

Considered together, all parts of the deep seafloor are potentially vulnerable to climate change, acting indirectly through organisms' food supply and changes in oxygen and pH, as well directly through temperature (Levin and Le Bris, 2015; Sweetman et al., 2017; Bindoff et al., 2019). Information on such impacts is, however, limited. In particular, the relative inaccessibility of the deep sea environment has constrained the mapping, monitoring and experimental studies that are needed to identify the factors determining species' distributions and abundance changes, as well as providing basic knowledge of their life-cycles and ecological interactions. A high proportion of deep seafloor organisms, particularly smaller species, have yet to be described (Costello et al., 2010; Bouchet, 2017).

Temperature changes due to ongoing global warming are generally much less at the deep seafloor than at the surface (Section 1.3), rapidly declining between 1000 and 2000 m. However, different patterns occur in different ocean basins, and in the Southern Ocean the highest rate of deep-sea (and seafloor) warming is at water depths of between 4000 and 5000 m, at ~0.03°C per decade (Bindoff et al., 2019). This rate of change would be insignificant in a coastal (or terrestrial) context, yet may be ecologically significant for the deep sea, where seasonal temperature fluctuations are lacking and organisms may be more sensitive to small thermal changes (Yasuhara and Danovaro, 2016). For example, in the deep Mediterranean, temperature shifts of 0.1°C or less are considered responsible for major changes in the biodiversity and community structure of sediment nematodes, a group of high functional importance (Danovaro et al., 2004).

Spatial and temporal variation in surface productivity, and hence the associated flux of organic material through the water column, are regarded as more important than temperature in determining the biomass, biodiversity, and productivity of the deep seabed (Smith et al., 2008, 2018; Hartman et al., 2015). Most climate change impacts are therefore likely to be indirect, including the effects of temperature, deoxygenation, and ocean acidification on midwater fish, zooplankton and microbial heterotrophs, affecting the downward flux of exported material (Belcher et al., 2019; Kelly et al., 2019; Boyd et al., 2019). These interactions and their outcomes are not well understood (Martin et al., 2020), with very limited observations on their natural variability and temporal trends. Nevertheless, a model-based best estimate under the high emission scenario (RCP 8.5) indicated a global decline of 11.4% in the food supply to the deep seafloor (bathyal, abyssal, and hadal) by 2100 compared to present values, with basin-specific values ranging from an increase of 9.6% in the Southern Ocean to decreases of up to 15.4% in the Atlantic, with high subbasin variability (between +62.3% and −61.0% in different parts of the North Atlantic) (Jones et al., 2014). Broadly similar values were found in a more recent study, with abyssal decreases in particulate organic flux of up to 27% in the Atlantic, and between 31% and 40% in the Pacific and Indian Oceans (Sweetman et al., 2017). However, the two analyzes disagreed as to whether organic carbon fluxes to the seafloor would increase or decrease in the Arctic Ocean.

The adverse effects of any decrease in food supply to deep seafloor biota will be accentuated by warming, since that will increase metabolic rates, requiring more food to support the same amount of biomass. The overall effect is estimated to be a reduction of 5%–18% of global benthic biomass by 2100 under a high emissions scenario, considered to be a "medium confidence" assessment by IPCC (Bindoff et al., 2019). Bathyal ecosystems, such as productive seamounts and shelf slopes could be at greater risk, since their current ecological zonation reflects vertical changes in physico-chemical conditions, that are known to be undergoing relatively rapid changes (Goericke et al., 2015; Sato et al., 2017; Levin, 2018) that are expected to continue (Morato et al., 2020). Nevertheless, there may be opposing responses by different meiofaunal groups to the interactions between hypoxia and low pH (Wit et al., 2016; van Dijk et al., 2017) and mid-depth seafloor ecosystems that have already adapted to low oxygen conditions, in oxygen minimum zones, may be less vulnerable to ocean acidification impacts (Taylor et al., 2014).

Deep-sea fishing effort has mostly focused on the shelf slope and seamounts, where seabed trawling has resulted in declines in faunal biodiversity, cover, and abundance (Clark et al., 2016). Other nonclimatic pressures include cable laying; mining for polymetallic nodules and at vent sites (Niner et al., 2018; Toro et al., 2020), and the accumulation of anthropogenic debris, including micro-plastics (Jamieson et al., 2019; Zhang et al., 2020). Because of the slow growth of many deep sea invertebrates and fish, recovery from such disturbance and pollution may take decades to centuries, potentially slowed by future climate change.

# 4 Conclusions

This review shows that there is now no doubt that all parts of the global ocean, and all the organisms living there, are already affected by some aspect of anthropogenic climate change. Such pressures will intensify in future, superimposed on and potentially interacting with a wide range of other human-driven stressors. While the exact consequences of further marine climate change impacts at global, regional, or local level cannot be known in advance, they are overwhelmingly disadvantageous from a human perspective. Nevertheless, the scale of the associated damage to ocean ecosystems, and the services they provide, can be limited—through the successful implementation of international commitments to rapidly achieve net zero for greenhouse gas emissions and subsequent stabilization of the global climate.

Table 1 summarizes the overall vulnerability to climate change for the 13 ocean components considered here as ecosystems: seven in shallow coastal waters, and six broader groupings for shelf seas and the open ocean. These vulnerability assessments are based on the IPCC Special Report on the Ocean and Cryosphere (IPCC, 2019; Bindoff et al., 2019), together with additional, more recent information. A noteworthy feature of this review is the amount of

**TABLE 1** Summary of the vulnerability of marine ecosystems in shallow coastal seas, shelf seas and the open ocean to climate change, and the relative importance of specific drivers in the context of other human-driven stressors.

| Ecosystems | Climate drivers and associated changes | | | | | | | Other important human-driven stressors |
|---|---|---|---|---|---|---|---|---|
| | Overall climatic vulnerability | Warming | Acidification | Deoxygenation | Sea level rise and storms | Sea ice loss | Stratification & nutrient supply | |
| **Shallow coastal ecosystems** | | | | | | | | |
| Coral reefs | | | | | | | | Over-fishing, pollution, eutrophication |
| Kelp forests | | | | | | | | Pollution, eutrophication, overfishing, kelp harvesting |
| Seagrass meadows | | | | | | | | Eutrophication, coastal development, over-fishing |
| Intertidal: rocky[a] | | | | | | | | Pollution |
| Intertidal: sandy[a] | | | | | | | | Coastal development, pollution |
| Saltmarshes | | | | | | | | Coastal development |
| Estuaries | | | | | | | | Coastal development, pollution, eutrophication |
| Mangroves | | | | | | | | Coastal development pollution; aquaculture |
| **Shelf seas and the open ocean** | | | | | | | | |
| Shelf sea benthos | | | | | | | | Over-fishing; pollution dredging |
| Upper ocean plankton | | | | | | | | Pollution |
| Fish and fisheries | | | | | | | | Over-fishing, pollution |
| Cold water corals | ? | | | | | | | Fishing; pollution; |
| Ice-influenced ecosystems | | | | | | | | Shipping |
| Deep seafloor ecosystems | ? | | | | | | | Mineral exploitation; deep sea fishing |

[a]*Considered together in main text; separated here to show difference in climate drivers.*

*Overall vulnerability:* red, *high (moderate-high impacts likely at global surface temperature increase of 1.5–2.5°C relative to preindustrial, with associated changes in other drivers)*; orange, *medium (moderate-high impacts at 2.5–3.5°C)*; yellow, *low (moderate-high impacts at >3.5°C). Specific drivers:* dark gray, *high contribution to overall effect;* mid-gray, *medium contribution;* light gray, *low contribution. "?," low-medium confidence for overall vulnerability assessments.*

*Based on Bindoff, N.L., Cheung, W.W.L., Kairo, J.G., Arístegui, J., Guinder, V.A., Hallberg, R., Hilmi, N., Jiao, N., Karim, M.S., Levin, L., O'Donoghue, S., Purca Cuicapusa, R., Rinkevich, B., Suga, T., Tagliabue, A. and Williamson, P., 2019. Changing ocean, marine ecosystems, and dependent communities. In: Pörtner, H.-O., Roberts, D.C., Masson-Delmotte, V., Zhai, P., Tignor, M., Poloczanska, E., Mintenbeck, K., Alegría, A., Nicolai, M., Okem, A., Petzold, J., Rama, B., Weyer, N.M. (Eds.), IPCC Special Report on the Ocean and Cryosphere in a Changing Climate. Online and in press, pp. 447–587 (Chapter 5) and more recent studies, as reviewed here.*

new material that has been taken into account, using information from more than 190 papers (42% of the total references) published since 2019. While some of these recent studies challenge the consensus, providing exceptions or new concepts, most strengthen existing evidence, increasing overall confidence in previous conclusions.

As shown in Table 1, four ecosystems (coral reefs, kelp ecosystems, seagrass meadows, and ice-influenced ecosystems) are regarded as highly vulnerable, being badly affected by the "optimistic" level of climate change represented by a global surface temperature increase of 1.5–2.5°C relative to preindustrial (i.e., a low emissions scenario, equivalent to RCP 2.6). Most of the others (rocky and sandy intertidal, saltmarshes, estuaries, mangroves, shelf sea sediment communities, upper ocean plankton, fish and fisheries, and cold water corals) are considered to be at medium risk, with major impacts not occurring until global temperatures increase by 2.5–3.5°C (medium emissions; RCP 4.5), while deep seafloor ecosystems may not be severely impacted until global temperatures rise by more than 3.5°C (high emissions; RCP 6.0 and RCP 8.5). These assessments are global-scale, with regional variability; effects may differ markedly across species' ranges and within ocean basins. There is also variability in response relating to the speed of warming and the occurrence of nonclimatic stressors (particularly fishing, pollution and coastal development), with more fundamental uncertainties relating to incomplete understanding of the processes involved. For example, the major knowledge gaps with regard to the biodiversity and functioning of deep sea ecosystems, affected by many complex interactions occurring throughout the water column.

Two further considerations are important with regard to the ecosystem categories provided here. First, the amount of scientific attention given to different ocean components is closely related to their accessibility to researchers, as well as their direct socio-economic value; second, that all ocean components/ecosystems are linked, physically and biologically, without clear boundaries between them, particularly with regard to the open ocean. Indeed, arguably the largest and most important marine ecosystem of all, the ocean interior (mesopelagic), has not been given explicit attention. Key issues have, however, been considered in two sections, on upper ocean plankton and deep seafloor ecosystems, with the biological carbon pump providing the main ecological connection between the two. How that connection might change in response to climate drivers is highly uncertain: Bindoff et al. (2019) identified 15 relevant processes that might increase or decrease in strength, yet projected changes for 12 of those were assessed as low confidence, two as medium and only one as high.

To address those knowledge gaps, and others, there is clearly need for more observational data, from monitoring studies on a worldwide basis, e.g., through the auspices of the Global Ocean Observing System (Bax et al., 2018; Weller et al., 2019). Such information needs to be brought together with experimental studies through an integrated modeling framework, considering both climatic and nonclimatic drivers, to inform policy action in the wider context of sustainable development and societal needs, as discussed in other chapters of this volume.

## References


Adame, M.F., Reef, R., Santini, N.S., Najera, E., Turschwell, M.P., Hayes, M.A., Masque, P., Lovelock, C.E., 2020. Mangroves in arid regions: ecology, threats, and opportunities. Estuar. Coast. Shelf Sci., 106796.

Adams, J., 2020. Salt marsh at the tip of Africa: patterns, processes and changes in response to climate change. Estuar. Coast. Shelf Sci. 237, 106650.

Agostini, S., Harvey, B.P., Wada, S., Kon, K., Milazzo, M., Inaba, K., Hall-Spencer, J.M., 2018. Ocean acidification drives community shifts towards simplified non-calcified habitats in a subtropical–temperate transition zone. Sci. Rep. 8 (1), 1–11.

AIMS (Australian Institute of Marine Science), 2020. Long-Term Reef Monitoring Program. Annual Summary Report on Coral Reef Condition for 2019/20. https://www.aims.gov.au/reef-monitoring/gbr-condition-summary-2019-2020. (Accessed 16 November 2020).

Albright, R., Caldeira, L., Hosfelt, J., Kwiatkowski, L., Maclaren, J.K., Mason, B.M., Nebuchina, Y., Ninokawa, A., Pongratz, J., Ricke, K.L., Rivlin, T., 2016. Reversal of ocean acidification enhances net coral reef calcification. Nature 531 (7594), 62–365.

Albright, R., Takeshita, Y., Koweek, D.A., Ninokawa, A., Wolfe, K., Rivlin, T., Nebuchina, Y., Young, J., Caldeira, K., 2018. Carbon dioxide addition to coral reef waters suppresses net community calcification. Nature 555, 516–519.

Alguero-Muniz, M., Alvarez-Fernandez, S., Thor, P., Bach, L.T., Esposito, M., Horn, H.G., Ecker, U., Langer, J.A., Taucher, J., Malzahn, A.M., Riebesell, U., 2017. Ocean acidification effects on mesozooplankton community development: results from a long-term mesocosm experiment. PLoS One 12 (4), e0175851.

Alongi, D.M., 2008. Mangrove forests: resilience, protection from tsunamis, and responses to global climate change. Estuar. Coast. Shelf Sci. 76 (1), 1–13.

Alongi, D.M., 2015. The impact of climate change on mangrove forests. Curr. Clim. Chang. Res. 1 (1), 30–39.

AMAP (Arctic Monitoring and Assessment Programme), 2018. AMAP Assessment 2018: Arctic Ocean Acidification. AMAP, Tromsø. 187 pp.

AMAP (Arctic Monitoring and Assessment Programme), 2019. Arctic Climate Change Update 2019. https://www.amap.no/documents/download/3295/inline. (Accessed 11 November 2020).

Amoroso, R.O., Pitcher, C.R., Rijnsdorp, A.D., McConnaughey, R.A., Parma, A.M., Suuronen, P., Eigaard, O.R., Bastardie, F., Hintzen, N.T., Althaus, F., et al., 2018. Bottom trawl fishing footprints on the world's continental shelves. Proc. Natl. Acad. Sci. 115 (43), E10275–E10282.

Anton, A., Baldry, K., Coker, D.J., Duarte, C.M., 2020. Drivers of the low metabolic rates of seagrass meadows in the Red Sea. Front. Mar. Sci. 7, 69.

Ardyna, M., Arrigo, K.R., 2020. Phytoplankton dynamics in a changing Arctic Ocean. Nat. Clim. Chang. 10, 892–903.

Arias-Ortiz, A., Serrano, O., Masqué, P., Lavery, P.S., Mueller, U., Kendrick, G.A., Rozaimi, M., Esteban, A., Fourqurean, J.W., Marbà, N.J.N.C.C., Mateo, M.A., 2018. A marine heatwave drives massive losses from the world's largest seagrass carbon stocks. Nat. Clim. Chang. 8 (4), 338–344.

Armitage, A.R., Highfield, W.E., Brody, S.D., Louchouarn, P., 2015. The contribution of mangrove expansion to salt marsh loss on the Texas Gulf Coast. PLoS One 10 (5), e0125404.

Arrigo, K.R., van Dijken, G.L., 2015. Continued increases in Arctic Ocean primary production. Prog. Oceanogr. 136, 60–70.

Arrigo, K.R., Perovich, D.K., Pickart, R.S., Brown, Z.W., Van Dijken, G.L., Lowry, K.E., Mills, M.M., Palmer, M.A., Balch, W.M., Bahr, F., Bates, N.R., 2012. Massive phytoplankton blloms under Arctic sea ice. Science 336 (6087), 1408.

Asch, R.G., Cheung, W.W.L., Reygondeau, G., 2018. Future marine ecosystem drivers, biodiversity, and fisheries maximum catch potential in Pacific Island countries and territories under climate change. Mar. Policy 88, 285–294.

Asch, R.G., Stock, C.A., Sarmiento, J.L., 2019. Climate change impacts on mismatches between phytoplankton blooms and fish spawning phenology. Glob. Chang. Biol. 25 (8), 2544–2559.

Assis, J., Araújo, M.B., Serrão, E.A., 2018. Projected climate changes threaten ancient refugia of kelp forests in the North Atlantic. Glob. Chang. Biol. 24 (1), e55–e66.

Bach, L.T., Taucher, J., Boxhammer, T., Ludwig, A., Kristineberg KOSMOS Consortium, Achterberg, E.P., Algueró-Muñiz, M., Anderson, L.G., Bellworthy, J., Büdenbender, J., Czerny, J., Ericson, Y., et al., 2016. Influence of ocean acidification on a natural winter-to-summer plankton succession: first insights from a long-term mesocosm study draw attention to periods of low nutrient concentrations. PLoS One 11 (8), e0159068.

Bach, L.T., Stange, P., Taucher, J., Achterberg, E.P., Algueró-Muñiz, M., Horn, H., Esposito, M., Riebesell, U., 2019. The influence of plankton community structure on sinking velocity and remineralization rate of marine aggregates. Glob. Biogeochem. Cycles 33 (8), 971–994.

Bacino, G.L., Dragani, W.C., Codignotto, J.O., Pescio, A.E., Farenga, M.O., 2020. Shoreline change rates along Samborombón Bay, Río de la Plata estuary, Argentina. Estuar. Coast. Shelf Sci. 237, 106659.

Bahr, K.D., Tran, T., Jury, C.P., Toonen, R.J., 2020. Abundance, size, and survival of recruits of the reef coral *Pocillopora acuta* under ocean warming and acidification. PLoS One 15 (2), e0228168.

Bailey, S.A., Brown, L., Campbell, M.L., Canning-Clode, J., Carlton, J.T., Castro, N., Chainho, P., Chan, F.T., Creed, J. C., Creed, A., Darling, J., et al., 2020. Trends in the detection of aquatic non-indigenous species across global marine, estuarine and freshwater ecosystems: a 50-year perspective. Divers. Distrib. 26 (12), 1780–1797.

Bamber, J., 2020. Arctic Ocean: why winter sea ice has stalled, and what it means for the rest of the world. In: The Conversation. https://theconversation.com/arctic-ocean-why-winter-sea-ice-has-stalled-and-what-it-means-for-the-rest-of-the-world-148753. (Accessed 29 October 2020).

Bamber, J.L., Oppenheimer, M., Kopp, R.E., Aspinall, W.P., Cooke, R.M., 2019. Ice sheet contributions to future sea-level rise from structured expert judgment. Proc. Natl. Acad. Sci. 116 (23), 11195–11200.

Barange, M., 2019. Avoiding misinterpretation of climate change projections of fish catches. ICES J. Mar. Sci. 76 (6), 1390–1392.

Barange, M., Butenschön, M., Yool, A., Beaumont, N., Fernandes, J.A., Martin, A.P., Allen, J., 2017. The cost of reducing the North Atlantic Ocean biological carbon pump. Front. Mar. Sci. 3, 290.

Barton, A.D., Irwin, A.J., Finkel, Z.V., Stock, C.A., 2016. Anthropogenic climate change drives shift and shuffle in North Atlantic phytoplankton communities. Proc. Natl. Acad. Sci. 113 (11), 2964–2969.

Bartsch, I., Paar, M., Fredriksen, S., Schwanitz, M., Daniel, C., Hop, H., Wiencke, C., 2016. Changes in kelp forest biomass and depth distribution in Kongsfjorden, Svalbard, between 1996-1998 and 2012-2014 reflect Arctic warming. Polar Biol. 39 (11), 2021–2036.

Bates, N.R., Johnson, R.J., 2020. Acceleration of ocean warming, salinification, de-oxygenation and acidification in the surface subtropical North Atlantic Ocean. Commun. Earth Environ. 1, 33.

Bates, N.R., Astor, Y.M., Church, M.J., Currie, K., Dore, J.E., González-Dávila, M., Lorenzoni, L., Muller-Karger, F., Olafsson, J., Santana-Casiano, J.M., 2014. A time-series view of changing surface ocean chemistry due to ocean uptake of anthropogenic $CO_2$ and ocean acidification. Oceanography 27 (1), 126–141.

Bax, N.J., Appeltans, W., Brainard, R., Duffy, J.E., Dunstan, P., Hanich, Q., Harden Davies, H., Hills, J., Miloslavich, P., Muller-Karger, F.E., Simmons, S., 2018. Linking capacity development to GOOS monitoring networks to achieve sustained ocean observation. Front. Mar. Sci. 5, 346.

Beare, D., McQuatters-Gollop, A., van der Hammen, T., Machiels, M., Teoh, S.J., Hall-Spencer, J.M., 2013. Long-term trends in calcifying plankton and pH in the North Sea. PLoS One 8 (5), e61175.

Beaugrand, G., 2009. Decadal changes in climate and ecosystems in the North Atlantic Ocean and adjacent seas. Deep-Sea Res. II Top. Stud. Oceanogr. 56, 656–673.

Bednaršek, N., Tarling, G.A., Bakker, D.C., Fielding, S., Cohen, A., Kuzirian, A., McCorkle, D., Lézé, B., Montagna, R., 2012a. Description and quantification of pteropod shell dissolution: a sensitive bioindicator of ocean acidification. Glob. Chang. Biol. 18 (7), 2378–2388.

Bednaršek, N., Tarling, G.A., Bakker, D.C.E., Fielding, S., Jones, E.M., Venables, H.J., Ward, P., Kuzirian, A., Lézé, B., Feely, R.A., Murphy, E.J., 2012b. Extensive dissolution of live pteropods in the Southern Ocean. Nat. Geosci. 5 (12), 881–885.

Bednaršek, N., Feely, R.A., Howes, E.L., Hunt, B.P., Kessouri, F., León, P., Lischka, S., Maas, A.E., McLaughlin, K., Nezlin, N.P., Sutula, M., 2019. Systematic review and meta-analysis toward synthesis of thresholds of ocean acidification impacts on calcifying pteropods and interactions with warming. Front. Mar. Sci. 6, 227.

Belcher, A., Saunders, R.A., Tarling, G.A., 2019. Respiration rates and active carbon flux of mesopelagic fishes (Family Myctophidae) in the Scotia Sea, Southern Ocean. Mar. Ecol. Prog. Ser. 610, 149–162.

Belkin, I.M., 2009. Rapid warming of large marine ecosystems. Prog. Oceanogr. 81, 207–213.

Bell, T.W., Allen, J.G., Cavanaugh, K.C., Siegel, D.A., 2020. Three decades of variability in California's giant kelp forests from the Landsat satellites. Remote Sens. Environ. 238, 110811.

Beman, J.M., Chow, C.E., King, A.L., Feng, Y., Fuhrman, J.A., Andersson, A., Bates, N.R., Popp, B.N., Hutchins, D.A., 2011. Global declines in oceanic nitrification rates as a consequence of ocean acidification. Proc. Natl. Acad. Sci. 108 (1), 208–213.

Bennett, S., Duarte, C.M., Marbà, N., Wernberg, T., 2019. Integrating within-species variation in thermal physiology into climate change ecology. Philos. Trans. R. Soc. B 374, 20180550.

Bindoff, N.L., Cheung, W.W.L., Kairo, J.G., Arístegui, J., Guinder, V.A., Hallberg, R., Hilmi, N., Jiao, N., Karim, M.S., Levin, L., O'Donoghue, S., Purca Cuicapusa, R., Rinkevich, B., Suga, T., Tagliabue, A., Williamson, P., 2019. Changing ocean, marine ecosystems, and dependent communities. In: Pörtner, H.-O., Roberts, D.C., Masson-Delmotte, V., Zhai, P., Tignor, M., Poloczanska, E., Weyer, N.M. (Eds.), IPCC Special Report on the Ocean and Cryosphere in a Changing Climate, pp. 447–587. Online and in press (Chapter 5).

Boolukos, C.M., Lim, A., O'Riordan, R.M., Wheeler, A.J., 2019. Cold-water corals in decline—a temporal (4 year) species abundance and biodiversity appraisal of complete photomosaiced cold-water coral reef on the Irish Margin. Deep-Sea Res. I Oceanogr. Res. Pap. 146, 44–54.

Bopp, L., Resplandy, L., Orr, J.C., Doney, S.C., Dunne, J.P., Gehlen, M., Halloran, P., Heinze, C., Ilyina, T., Seferian, R., Tjiputra, J., 2013. Multiple stressors of ocean ecosystems in the 21st century: projections with CMIP5 models. Biogeosciences 10, 6225–6245.

Borchert, S.M., Osland, M.J., Enwright, N.M., Griffith, K.T., 2018. Coastal wetland adaptation to sea level rise: quantifying potential for landward migration and coastal squeeze. J. Appl. Ecol. 55 (6), 2876–2887.

Bortolus, A., Carlton, J.T., Schwindt, E., 2015. Reimagining South American coasts: unveiling the hidden invasion history of an iconic ecological engineer. Divers. Distrib. 21 (11), 1267–1283.

Bouchet, P., 2017. Chapter 17, Marine biodiversity. What is there still to discover? In: Euzen, A., Gaill, F., Lacroix, D., Cury, P. (Eds.), The Ocean Revealed. CNRS Editions, Paris.

Boyd, P.W., Cornwall, C.E., Davison, A., Doney, S.C., Fourquez, M., Hurd, C.L., Lima, I.D., McMinn, A., 2016. Biological responses to environmental heterogeneity under future ocean conditions. Glob. Chang. Biol. 22, 2633–2650.

Boyd, P.W., Collins, S., Dupont, S., Fabricius, K., Gattuso, J.-P., Havenhand, J., Hutchins, D.A., Riebesell, U., Rintoul, M.S., Vichi, M., Biswas, H., Ciotti, A., Gao, K., Gehlen, M., Hurd, C.L., Kurihara, H., McGraw, C.M., Navarro, J.M., Nilsson, G.E., Passow, U., Pörtner, H.-O., 2018. Experimental strategies to assess the biological ramifications of multiple drivers of global ocean change—a review. Glob. Chang. Biol. 24, 2239–2261.

Boyd, P.W., Claustre, H., Levy, M., Siegel, D.A., Weber, T., 2019. Multi-faceted particle pumps drive carbon sequestration in the ocean. Nature 568 (7752), 327–335.

Brander, K., Kiørboe, T., 2020. Decreasing phytoplankton size adversely affects ocean food chains. Glob. Chang. Biol. 26, 5356–5357.

Breitburg, D.L., Baumann, H., Sokolova, I.M., Frieder, C.A., 2019. Multiple stressors—forces that combine to worsen deoxygenation and its effects. In: Laffoley, D., Baxter, J.M. (Eds.), Ocean Deoxygenation: Everyone's Problem—Causes, Impacts, Consequences and Solutions. IUCN, Gland, Switzerland, pp. 225–248.

Briggs, N., Dall'Olmo, G., Claustre, H., 2020. Major role of particle fragmentation in regulating biological sequestration of $CO_2$ by the oceans. Science 367 (6479), 791–793.

Britten, G.L., Dowd, M., Worm, B., 2016. Changing recruitment capacity in global fish stocks. Proc. Natl. Acad. Sci. 113 (1), 134–139.

Brown, S., Nicholls, R.J., Goodwin, P., Haigh, I.D., Lincke, D., Vafeidis, A.T., Hinkel, J., 2018. Quantifying land and people exposed to sea-level rise with no mitigation and 1.5 C and 2.0 C rise in global temperatures to year 2300. Earth's Future 6 (3), 583–600.

Brun, P., Kiørboe, T., Licandro, P., Payne, M.R., 2016. The predictive skill of species distribution models for plankton in a changing climate. Glob. Chang. Biol. 22 (9), 3170–3181.

Buesseler, K.O., Boyd, P.W., Black, E.E., Siegel, D.A., 2020. Metrics that matter for assessing the ocean biological carbon pump. Proc. Natl. Acad. Sci. 117 (18), 9679–9687.

Buhl-Mortensen, L., Vanreusel, A., Gooday, A.J., Levin, L.A., Priede, I.G., Buhl-Mortensen, P., Gheerardyn, H., King, N.J., Raes, M., 2010. Biological structures as a source of habitat heterogeneity and biodiversity on the deep ocean margins. Mar. Ecol. 31 (1), 21–50.

Burkepile, D.E., Shantz, A.A., Adam, T.C., Munsterman, K.S., Speare, K.E., Ladd, M.C., Rice, M.M., Ezzat, L., McIlroy, S., Wong, J.C.Y., Baker, D.M., Brooks, A.J., Schmidt, R.J., Holbrook, S.J., 2020. Nitrogen identity drives differential impacts of nutrients on coral bleaching and mortality. Ecosystems 23, 798–811.

Burrell, T.J., Maas, E.W., Hulston, D.A., Law, C.S., 2017. Variable response to warming and ocean acidification by bacterial processes in different plankton communities. Aquat. Microb. Ecol. 79 (1), 49–62.

Burrows, M.T., Schoeman, D.S., Richardson, A.J., Molinos, J.G., Hoffmann, A., Buckley, L.B., Moore, P.J., Brown, C.J., Bruno, J.F., Duarte, C.M., Halpern, B.S., Hoegh-Guldbrg, O., Kappel, C.V., Kiessling, W., O'Connor, M.I., Pandolfi, J.M., Parmesan, C., Sydeman, W.J., Ferrier, S., Williams, K.J., Poloczanska, E.S., 2014. Geographical limits to species-range shifts are suggested by climate velocity. Nature 507 (7493), 492–495.

Byrne, M., Hernández, J.C., 2020. Sea urchins in a high $CO_2$ world: impacts of climate warming and ocean acidification across life history stages. Dev. Aquac. Fish. Sci. 43, 281–297.

Byrnes, J.E., Reed, D.C., Cardinale, B.J., Cavanaugh, K.C., Holbrook, S.J., Schmitt, R.J., 2011. Climate-driven increases in storm frequency simplify kelp forest food webs. Glob. Chang. Biol. 17, 2513–2524.

Cacciapaglia, C.W., van Woesik, R., 2020. Reduced carbon emissions and fishing pressure are both necessary for equatorial coral reefs to keep up with rising seas. Ecography 43 (6), 789–800.

Cahoon, D.R., 2006. A review of major storm impacts on coastal wetland elevations. Estuaries Coast 29 (6), 889–898.

Cai, W.-J., Feely, R.A., Testa, J.M., Li, M., Evans, W., Alin, S.R., Xu, Y.-Y., Pelletier, G., Ahmed, A., Greeley, D.J., Newton, J.A., Bednaršek, N., 2021. Natural and anthropogenic drivers of acidification in large estuaries. Annu. Rev. Mar. Sci. 13, 19.1–19.33.

Camp, E.F., Suggett, D.J., Gendron, G., Jompa, J., Manfrino, C., Smith, D.J., 2016. Mangrove and seagrass beds provide different biogeochemical services for corals threatened by climate change. Front. Mar. Sci. 3, 52.

Camp, E.F., Schoepf, V., Mumby, P.J., Hardtke, L.A., Rodolfo-Metalpa, R., Smith, D.J., Suggett, D.J., 2018. The future of coral reefs subject to rapid climate change: lessons from natural extreme environments. Front. Mar. Sci. 5, 4.

Carcedo, M.C., Fiori, S.M., Scotti, M., Ito, M., Dutto, M.S., Carbone, M.E., 2019. Dominant bivalve in an exposed sandy beach regulates community structure through spatial competition. Estuaries Coast 42 (7), 1912–1923.

Carey, J.C., Moran, S.B., Kelly, R.P., Kolker, A.S., Fulweiler, R.W., 2017. The declining role of organic matter in New England saltmarshes. Estuaries Coast 40 (3), 626–639.

Carter, B.R., Williams, N.L., Evans, W., Fassbender, A.J., Barbero, L., Hauri, C., Feely, R.A., Sutton, A.J., 2019. Time of detection as a metric for prioritizing between climate observation quality, frequency, and duration. Geophys. Res. Lett. 46 (7), 3853–3861.

Cavanaugh, K.C., Dangremond, E.M., Doughty, C.L., Williams, A.P., Parker, J.D., Hayes, M.A., Rodriguez, W., Feller, I.C., 2019. Climate-driven regime shifts in a mangrove–salt marsh ecotone over the past 250 years. Proc. Natl. Acad. Sci. 116 (43), 21602–21608.

CBD (Secretariat of the Convention on Biological Diversity), 2014. In: Hennige, S., Roberts, J.M., Williamson, P. (Eds.), An Updated Synthesis of the Impacts of Ocean Acidification on Marine Biodiversity. CBD Technical Series No 75, CBD, Montreal, 99 pp.

Chefaoui, R.M., Duarte, C.M., Serrão, E.A., 2018. Dramatic loss of seagrass habitat under projected climate change in the Mediterranean Sea. Glob. Chang. Biol. 24 (10), 4919–4928.

Cheung, W.W.L., Frolicher, T.L. 2020. Marine heatwaves exacerbate climate change impacts for fisheries in the northeast Pacific. Sci. Rep. 10, 6678.

Cheung, W.W.L., Bruggeman, J., Butenschön, M., 2018. Projected changes in global and national potential marine fisheries catch under climate change scenarios in the twenty-first century. In: Barange, M., Bahri, T., Beveridge, M.C.M., Cochrane, K.L., Funge-Smith, S., Poulain, F. (Eds.), Impacts of Climate Change on Fisheries and Aquaculture. FAO, Rome, Italy, pp. 63–86. FAO Fisheries and Aquaculture Technical Paper.

Cheung, W.W.L., Frolicher, T.L. 2020. Marine heatwaves exacerbate climate change impacts for fisheries in the northeast Pacific. Sci. Rep. 10, 6678.

Chivers, W.J., Walne, A.W., Hays, G.C., 2017. Mismatch between marine plankton range movements and the velocity of climate change. Nat. Commun. 8, 14434.

Chivers, W.J., Edwards, M., Hays, G.C., 2020. Phenological shuffling of major marine phytoplankton groups over the last six decades. Divers. Distrib. 26 (5), 536–548.

Christensen, V., Coll, M., Piroddi, C., Steenbeek, J., Buszowski, J., Pauly, D., 2014. A century of fish biomass decline in the ocean. Mar. Ecol. Prog. Ser. 512, 155–166.

Clark, M.R., Althaus, F., Schlacher, T.A., Williams, A., Bowden, D.A., Rowden, A.A., 2016. The impacts of deep-sea fisheries on benthic communities: a review. ICES J. Mar. Sci. 73 (suppl 1), i51–i69.

Clark, T.D., Raby, G.D., Roche, D.G., Binning, S.A., Speers-Roesch, B., Jutfelt, F., Sundin, J., 2020. Ocean acidification does not impair the behaviour of coral reef fishes. Nature 577 (7790), 370–375.

Clements, J.C., Hunt, H.L., 2015. Marine animal behaviour in a high $CO_2$ ocean. Mar. Ecol. Prog. Ser. 536, 259–279.

Clements, C.S., Burns, A.S., Stewart, F.J., Hay, M.E., 2020. Seaweed-coral competition in the field: effects on coral growth, photosynthesis and microbiomes require direct contact. Proc. R. Soc. B 287 (1927), 20200366.

Cohen, M.C., Rodrigues, E., Rocha, D.O., Freitas, J., Fontes, N.A., Pessenda, L.C.R., Bendassolli, J.A. (Eds.), 2020. Southward migration of the austral limit of mangroves in South America. Catena *195*, 104775.

Coll, M., Steenbeck, J., Pennino, M.G., Buszowski, J., Kaschner, K., Lotze, H.K., Rousseau, Y., Tittensor, D.P., Walters, C., Watson, R.A., Christensen, 2020. Advancing global ecological modelling capabilities to simulate future trajectories of change in marine ecosystems. Front. Mar. Sci. 7, 567877.

Collins, M., Knutti, R., Arblaster, J., Dufresne, J.L., Fichefet, T., Friedlingstein, P., Gao, X., Gutowski, W.J., Johns, T., Krinner, G., Shongwe, M., Tebaldi, C., Weaver, A.J., Wehner, M., 2013. Long-term climate change: projections, commitments and irreversibility. In: Climate Change 2013—The Physical Science Basis: Contribution of Working Group I to the Fifth Assessment Report of the Intergovernmental Panel on Climate Change. Cambridge University Press, Cambridge, pp. 1029–1136 (Chapter 12).

Collins, M., Sutherland, M., Bouwer, L., Cheong, S.-M., Frölicher, T., Jacot Des Combes, H., Koll Roxy, M., Losada, I., McInnes, K., Ratter, B., Rivera-Arriaga, E., Susanto, R.D., Swingedouw, D., Tibig, L., 2019. Extremes, abrupt changes and managing risk. In: Portner, H.-O., Roberts, D.C., Masson-Delmotte, V., Zhai, P., Tignor, M., Poloczanska, E., Weyer, N.M. (Eds.), IPCC Special Report on the Ocean and Cryosphere in a Changing Climate, pp. 589–655. Online and in press.

Conley, D.J., Slomp, C.P., 2019. Ocean deoxygenation impacts on microbial processes, biogeochemistry and feedbacks. In: Laffoley, D., Baxter, J.M. (Eds.), Ocean Deoxygenation: Everyone's Problem—Causes, Impacts, Consequences and Solutions. IUCN, Gland, Switzerland, pp. 249–262.

Constable, A.J., Melbourne-Thomas, J., Corney, S.P., Arrigo, K.R., Barbraud, C., Barnes, D.K., Bindoff, N.L., Boyd, P. W., Brandt, A., Costa, D.P., Davidson, A.T., 2014. Climate change and Southern Ocean ecosystems I: how changes in physical habitats directly affect marine biota. Glob. Chang. Biol. 20 (10), 3004–3025.

Cooper, J.A.G., Masselink, G., Coco, G., Short, A.D., Castelle, B., Rogers, K., Anthony, E., Green, A.N., Kelley, J.T., Pilkey, O.H., Jackson, D.W.T., 2020. Sandy beaches can survive sea-level rise. Nat. Clim. Chang. 10, 993–995.

Costello, M.J., Coll, M., Danovaro, R., Halpin, P., Ojaveer, H., Miloslavich, P., 2010. A census of marine biodiversity knowledge, resources, and future challenges. PLoS One 5 (8), e12110.

Coulson, M.W., Marshall, H.D., Pepin, P., Carr, S.M., 2006. Mitochondrial genomics of gadine fishes: implications for taxonomy and biogeographic origins from whole-genome data sets. Genome 49, 1115–1130.

Crooks, S., Windham-Myers, L., Troxler, T.G., 2019. Defining blue carbon: the emergence of a climate context for coastal carbon dynamis. In: Windham-Myers, L., Crooks, S., Troxler, T.G. (Eds.), A Blue Carbon Primer: The State of Coastal Wetland Carbon Science, Practice and Policy. CRC Press/Taylor & Francis, Boca Raton, pp. 1–8. 481 pp.

Crotty, S.M., Angelini, C., Bertness, M.D., 2017. Multiple stressors and the potential for synergistic loss of New England saltmarshes. PLoS One 12 (8), e0183058.

Dahlke, F.T., Wohlrab, S., Butzin, M., Pörtner, H.O., 2020. Thermal bottlenecks in the life cycle define climate vulnerability of fish. Science 369, 65–70.

Dalpadado, P., Hop, H., Rønning, J., Pavlov, V., Sperfeld, E., Buchholz, F., Rey, A., Wold, A., 2016. Distribution and abundance of euphausiids and pelagic amphipods in Kongsfjorden, Isfjorden and Rijpfjorden (Svalbard) and changes in their relative importance as key prey in a warming marine ecosystem. Polar Biol. 39 (10), 1765–1784.

Dalton, S.J., Carroll, A.G., Sampayo, E., Roff, G., Harrison, P.L., Entwistle, K., Huang, Z., Salih, A., Diamond, S.L., 2020. Successive marine heatwaves cause disproportionate coral bleaching during a fast phase transition from El Niño to La Niña. Sci. Total Environ. 715, 136951.

Danovaro, R., Dell'Anno, A., Pusceddu, A., 2004. Biodiversity response to climate change in a warm deep sea. Ecol. Lett. 7, 821–828.

de Lacerda, L.D., Borges, R., Ferreira, A.C., 2019. Neotropical mangroves: conservation and sustainable use in a scenario of global climate change. Aquat. Conserv. Mar. Freshwat. Ecosyst. 29 (8), 1347–1364.

de los Santos, C.B., Krause-Jensen, D., Alcoverro, T., Marbà, N., Duarte, C.M., Van Katwijk, M.M., Pérez, M., Romero, J., Sánchez-Lizaso, J.L., Roca, G., Jankowska, E., Pérez-Lloréns, J.L., Fournier, J., Montefalcone, M., Pergent, G., Ruiz, J.M., Cabaço, S., Cook, K., Wilkes, R.J., Moy, F.E., Trayter, G.M.-R., Arañó, X.S., de Jong, D.J., Fernández-Torquemada, Y., Aubry, I., Vergara, J.J., Santos, R., 2019. Recent trend reversal for declining European seagrass meadows. Nat. Commun. 10 (1), 1–8.

Delgado-Fernandez, I., O'Keeffe, N., Davidson-Arnott, R.G., 2019. Natural and human controls on dune vegetation cover and disturbance. Sci. Total Environ. 672, 643–656.

Denny, M.W., Dowd, W.W., Bilir, L., Mach, K.J., 2011. Spreading the risk: small-scale body temperature variation among intertidal organisms and its implications for species persistence. J. Exp. Mar. Biol. Ecol. 400 (1–2), 175–190.

Deppeler, S., Schulz, K.G., Hancock, A., Pascoe, P., McKinlay, J., Davidson, A., 2020. Ocean acidification reduces growth and grazuing impact of Antarctic heterotrophic flagellates. Biogeosciences 17, 4153–4171.

Deregibus, D., Zacher, K., Bartsch, I., Campana, G.L., Momo, F.R., Wiencke, C., Gómez, I., Quartino, M.L., 2020. Carbon balance under a changing light environment. In: Gómez, I., Huovinen, P. (Eds.), Antarctic Seaweeds. Springer, Cham, pp. 173–191.

Des Roches, S., Bell, M.A., Palkovacs, E.P., 2020. Climate-driven habitat change causes evolution in Threespine Stickleback. Glob. Chang. Biol. 26, 597–606.

Deutsch, C., Ferrel, A., Seibel, B., Pörtner, H.O., Huey, R.B., 2015. Climate change tightens a metabolic constraint on marine habitats. Science 348 (6239), 1132–1135.

Dogliotti, A.I., Ruddick, K., Guerrero, R., 2016. Seasonal and inter-annual turbidity variability in the Río de la Plata from 15 years of MODIS: El Niño dilution effect. Estuar. Coast. Shelf Sci. 182, 27–39.

Doney, S.C., Busch, D.S., Cooley, S.R., Kroeker, K.J., 2020. The impacts of ocean acidification on marine ecosystems and reliant human communities. Annu. Rev. Environ. Resour. 45, 83–112.

Doo, S., Kealoha, A., Andersson, A., Cohen, A.L., Hicks, T.L., Johnson, Z.I., Long, M.H., McElhany, P., Mollica, N., Shamberger, K.E.F., Silbiger, N.J., Takeshita, Y., Busch, D.S., Browman, H., 2020. The challenges of detecting and attributing ocean acidification impacts on marine ecosystems. ICES J. Mar. Sci. 77 (7–8), 2411–2422.

Douglass, J.G., Chamberlain, R.H., Wan, Y., Doering, P.H. 2020. Submerged vegetation responses to climate variation and altered hydrology in a subtropical estuary: interpreting 33 years of change. Est. Coasts 43, 1406–1424.

Drius, M., Jones, L., Marzialetti, F., de Francesco, M.C., Stanisci, A., Carranza, M.L., 2019. Not just a sandy beach. The multi-service value of Mediterranean coastal dunes. Sci. Total Environ. 668, 1139–1155.

Duarte, C.M., Krause-Jensen, D., 2018. Greenland tidal pools as hot spots for ecosystem metabolism and calcification. Estuaries Coast 41 (5), 1314–1321.

Duarte, C.M., Hendriks, I.E., Moore, T.S., Olsen, Y.S., Steckbauer, A., Ramajo, L., Carstensen, J., Trotter, J.A., McCulloch, M., 2013. Is ocean acidification an open-ocean syndrome? Understanding anthropogenic impacts on seawater pH. Estuaries Coast 36 (2), 221–236.

Duarte, B., Martins, I., Rosa, R., Matos, A.R., Roleda, M.Y., Reusch, T.B., Engelen, A.H., Serrão, E.A., Pearson, G.A., Marques, J.C., Caçador, I., 2018. Climate change impacts on seagrass meadows and macroalgal forests: an integrative perspective on acclimation and adaptation potential. Front. Mar. Sci. 5, 190.

Ducklow, H.W., Steinberg, D.K., Buesseler, K.O., 2001. Upper ocean carbon export and the biological pump. Oceanography 14 (4), 50–58.

Duke, N.C., Kovacs, J.M., Griffiths, A.D., Preece, L., Hill, D.J., Van Oosterzee, P., Mackenzie, J., Morning, H.S., Burrows, D., 2017. Large-scale dieback of mangroves in Australia's Gulf of Carpentaria: a severe ecosystem response, coincidental with an unusually extreme weather event. Mar. Freshw. Res. 68 (10), 1816–1829.

Duprat, L.P., Bigg, G.R., Wilton, D.J., 2016. Enhanced Southern Ocean marine productivity due to fertilization by giant icebergs. Nat. Geosci. 9 (3), 219–221.

Dutkiewicz, S., Morris, J.J., Follows, M.J., Scott, J., Levitan, O., Dyhrman, S.T., Berman-Frank, I., 2015. Impact of ocean acidification on the structure of future phytoplankton communities. Nat. Clim. Change 5, 1002–1006.

Eigaard, O.R., Bastardie, F., Hintzen, N.T., Buhl-Mortensen, L., Buhl-Mortensen, P., Catarino, R., Dinesen, G.E., Egekvist, J., Fock, H.O., Geitner, K., et al., 2017. The footprint of bottom trawling in European waters: distribution, intensity, and seabed integrity. ICES J. Mar. Sci. 74 (3), 847–865.

Eisaguirre, J.H., Eisaguirre, J.M., Davis, K., Carlson, P.M., Gaines, S.D., Caselle, J.E., 2020. Trophic redundancy and predator size class structure drive differences in kelp forest ecosystem dynamics. Ecology 101 (5), e02993.

Elsworth, G.W., Lovenduski, N.S., McKinnon, K.A., Krumhardt, K.M., Brady, R.X., 2020. Finding the fingerprint of anthropogenic climate change in marine phytoplankton abundance. Curr. Clim. Chang. Rep. 6 (2), 37–46.

Enochs, I.C., Manzello, D.P., Jones, P.J., Aguilar, C., Cohen, K., Valentino, L., Schopmeyer, S., Kolodziej, G., Jankulak, M., Lirman, D., 2018. The influence of diel carbonate chemistry fluctuations on the calcification rate of *Acropora cervicornisunder* present day and future acidification conditions. J. Exp. Mar. Biol. Ecol. 506, 135–143.

Enochs, I.C., Formel, N., Manzello, D., Morris, J., Mayfield, A.B., Boyd, A., Kolodziej, G., Adams, G., Hendee, J., 2020. Coral persistence despite extreme periodic pH fluctuations at a volcanically acidified Caribbean reef. Coral Reefs 39, 523–528.

Eslami-Andargoli, L., Dale, P.E.R., Sipe, N., Chaseling, J., 2009. Mangrove expansion and rainfall patterns in Moreton Bay, southeast Queensland, Australia. Estuar. Coast. Shelf Sci. 85 (2), 292–298.

Espinel-Velasco, N., Hoffmann, L., Agüera, A., Byrne, M., Dupont, S., Uthicke, S., Webster, N.S., Lamare, M., 2018. Effects of ocean acidification on the settlement and metamorphosis of marine invertebrate and fish larvae: a review. Mar. Ecol. Prog. Ser. 606, 237–257.

Fay, G., Link, J.S., Hare, J.A., 2017. Assessing the effects of ocean acidification in the Northeast US using an end-to-end marine ecosystem model. Ecol. Model. 347, 1–10.

Filbee-Dexter, K., Wernberg, T., 2018. Rise of turfs: a new battlefront for globally declining kelp forests. Bioscience 68 (2), 64–76.

Filbee-Dexter, K., Wernberg, T., 2020. Substantial blue carbon in overlooked Australian kelp forests. Sci. Rep. 10 (1), 1–6.

FitzGerald, D.M., Hughes, Z., 2019. Marsh processes and their response to climate change and sea-level rise. Annu. Rev. Earth Planet. Sci. 47, 481–517.

Flombaum, P., Gallegos, J.L., Gordillo, R.A., Rincón, J., Zabala, L.L., Jiao, N., Karl, D.M., Li, W.K., Lomas, M.W., Veneziano, D., Vera, C.S., 2013. Present and future global distributions of the marine Cyanobacteria Prochlorococcus and Synechococcus. Proc. Natl. Acad. Sci. 110 (24), 9824–9829.

Fontela, M., Perez, F.F., Carracedo, L.I., Padin, X.A., Velo, A., Garcia-Ibanez, M.I., Lherminier, P., 2020. The northeast Atlantic is running out of excess carbonate in the horizon of cold-water coral communities. Sci. Rep. 10, 14714.

Foo, S.A., Asner, G.P., 2020. Sea surface temperature in coral reef restoration outcomes. Environ. Res. Lett. 15 (7), 074045.

Foo, S.A., Byrne, M., 2016. Acclimatization and adaptive capacity of marine species in a changing ocean. Adv. Mar. Biol. 74, 69–116.

Fox, R.J., Donelson, J.M., Schunter, C., Ravasi, T., Gaitán-Espitia, J.D., 2019. Beyond buying time: the role of plasticity in phenotypic adaptation to rapid environmental change. Philos. Trans. R. Soc. B 374, 20180174.

Franco, J.N., Tuya, F., Bertocci, I., Rodríguez, L., Martínez, B., Sousa-Pinto, I., Arenas, F., 2018. The 'golden kelp' *Laminaria ochroleuca* under global change: integrating multiple eco-physiological responses with species distribution models. J. Ecol. 106 (1), 47–58.

Franco, B.C., Defeo, O., Piola, A.R., Barreiro, M., Yang, H., Ortega, L., Gianelli, I., Castello, J.P., Vera, C., Buratti, C., Pájaro, M., Pezzi, L.p., Möller, O.O., 2020a. Climate change impacts on the atmospheric circulation, ocean, and fisheries in the southwest South Atlantic Ocean: a review. Clim. Chang. 162, 2359–2377.

Franco, B.C., Combes, V., González Carman, V., 2020b. Subsurface ocean warming hotspots and potential impacts on marine species: the southwest South Atlantic Ocean case study. Front. Mar. Sci. 7, 824.

Fransner, F., Fröb, F., Tjiputra, J., Chierici, M., Fransson, A., Jeansson, E., Johannessen, T., Jones, E., Lauvset, S.K., Ólafsdóttir, S.R., Omar, A., Skjelvan and Olsen, A., 2020. Nordic seas acidification. Biogeosci. Discuss. https://doi.org/10.5194/bg-2020-339.

Free, C.M., Thorson, J.T., Pinsky, M.L., Oken, K.L., Wiedenmann, J., Jensen, O.P., 2019. Impacts of historical warming on marine fisheries production. Science 363 (6430), 979–983.

Fuentes, M.M., Allstadt, A.J., Ceriani, S.A., Godfrey, M.H., Gredzens, C., Helmers, D., Ingram, D., Pate, M., Radeloff, V.C., Shaver, D.J., Wildermann, N., Taylor, L., Bateman, B.L., 2020. Potential adaptability of marine turtles to climate change may be hindered by coastal development in the USA. Reg. Environ. Chang. 20 (3), 1–14.

Gallego-Tévar, B., Grewell, B.J., Futrell, C.J., Drenovsky, R.E., Castillo, J.M., 2020. Interactive effects of salinity and inundation on native *Spartina foliosa*, invasive *S. densiflora* and their hybrid from San Francisco Estuary, California. Ann. Bot. 125, 377–389.

Gallo, N.D., Levin, L.A., 2016. Fish ecology and evolution in the world's oxygen minimum zones and implications of ocean deoxygenation. Adv. Mar. Biol. 74, 117–198.

Garbe, J., Albrecht, T., Levermann, A., Donges, J.F., Winkelmann, R., 2020. The hysteresis of the Antarctic Ice Sheet. Nature 585 (7826), 538–544.

Garçon, V., Dewitte, B., Montes, I., Goubanova, K., 2019. Land-sea-atmosphere interactions exacerbating ocean deoxygenation in Eastern Boundary Upwelling Systems (EBUS). In: Laffoley, D., Baxter, J.M. (Eds.), Ocean Deoxygenation: Everyone's Problem—Causes, Impacts, Consequences and Solutions. IUCN, Gland, Switzerland, pp. 155–170.

Gardner, J., Manno, C., Bakker, D.C., Peck, V.L., Tarling, G.A., 2018. Southern Ocean pteropods at risk from ocean warming and acidification. Mar. Biol. 165 (1), 8.

Gattuso, J.-P., Hansson, L. (Eds.), 2011. Ocean Acidification. Oxford University Press, Oxford. 326 pp.

Gattuso, J.-P., Magnan, A.K., Bopp, L., Cheung, W.W., Duarte, C.M., Hinkel, J., Mcleod, E., Micheli, F., Oschlies, A., Williamson, P., Billé, R., Chalastani, V.I., Gates, R.D., Irisson, J.-O., Middelburg, J.J., Pörtner, H.-O., Rau, G.H., 2018. Ocean solutions to address climate change and its effects on marine ecosystems. Front. Mar. Sci. 5, 337.

Gaylard, S., Waycott, M., Lavery, P., 2020. Review of coast and marine ecosystems in temperate Australia demonstrates a wealth of ecosystem services. Front. Mar. Sci. 7, 453.

Gazeau, F., Sallon, A., Maugendre, L., Louis, J., Dellisanti, W., Gaubert, M., Lejeune, P., Gobert, S., Borges, A.V., Harlay, J., Champenois, W., 2017. First mesocosm experiments to study the impacts of ocean acidification on plankton communities in the NW Mediterranean Sea (MedSeA project). Estuar. Coast. Shelf Sci. 186, 11–29.

Gehlen, M., Gruber, N., Gangstø, R., Bopp, L., Oschlies, A., 2011. Biogeochemical consequences of ocean acidification and feedbacks to the earth system. In: Gattuso, J.-P., Hansson, L. (Eds.), Ocean Acidification. Oxford University Press, Oxford, pp. 230–248.

Gehlen, M., Séférian, R., Jones, D.O., Roy, T., Roth, R., Barry, J., Bopp, L., Doney, S.C., Dunne, J.P., Heinze, C., Joos, F., 2014. Projected pH reductions by 2100 might put deep North Atlantic biodiversity at risk. Biogeosciences 11 (23), 6955–6967.

Georgian, S.E., Dupont, S., Kurman, M., Butler, A., Strömberg, S.M., Larsson, A.I., Cordes, E.E., 2016. Biogeographic variability in the physiological response of the cold-water coral *Lophelia pertusa* to ocean acidification. Mar. Ecol. 37 (6), 1345–1359.

Gibson, R., Atkinson, R., Gordon, J., Smith, I., Hughes, D., 2011. Impact of ocean warming and ocean acidification on marine invertebrate life history stages: vulnerabilities and potential for persistence in a changing ocean. Oceanogr. Mar. Biol. Annu. Rev. 49, 1–42.

Gilman, E.L., Ellison, J., Duke, N.C., Field, C., 2008. Threats to mangroves from climate change and adaptation options: a review. Aquat. Bot. 89 (2), 237–250.
Gissi, E., Manea, E., Mazaris, A.D., Fraschetti, S., Almpanidou, V., Bevilacqua, S., Coll, M., Guarnieri, G., Lloret-Loret, E., Pascual, M., Petza, D., Rilov, G., Schonwald, M., Stelzenmüller, V., Katsanevakis, S., 2021. A review of the combined effects of climate change and other local human stressors on the marine environment. Sci. Total Environ. 755, 142564.
Gittings, J.A., Raitsos, D.E., Krokos, G., Hoteit, I., 2018. Impacts of warming on phytoplankton abundance and phenology in a typical tropical marine ecosystem. Sci. Rep. 8 (1), 1–12.
Goericke, R., Bograd, S.J., Grundle, D.S., 2015. Denitrification and flushing of the Santa Barbara Basin bottom waters. Deep-Sea Res. II Top. Stud. Oceanogr. 112, 53–60.
Gomez, C.E., Wickes, L., Deegan, D., Etnoyer, P.J., Cordes, E.E., 2018. Growth and feeding of deep-sea coral *Lophelia pertusa* from the California margin under simulated ocean acidification conditions. PeerJ 6, e5671.
Gonneea, M.E., Maio, C.V., Kroeger, K.D., Hawkes, A.D., Mora, J., Sullivan, R., Madsen, S., Buzard, R.M., Cahill, N., Donnelly, J.P., 2019. Salt marsh ecosystem restructuring enhances elevation resilience and carbon storage during accelerating relative sea-level rise. Estuar. Coast. Shelf Sci. 217, 56–68.
Gordillo, F.J.L., Carmona, R., Vinegla, B., Wiencke, C., Jiménez, C., 2016. Effects of simultaneous increase in temperature and ocean acidification on biochemical composition and photosynthetic performance of common macroalgae from Kongsfjorden (Svalbard). Polar Biol. 39, 1993–2007.
Gregg, W.W., Rousseaux, C.S., Franz, B.A., 2017. Global trends in ocean phytoplankton: a new assessment using revised ocean colour data. Remote. Sens. Lett. 8 (12), 1102–1111.
Gregory, B., Christophe, L., Martin, E., 2009. Rapid biogeographical plankton shifts in the North Atlantic Ocean. Glob. Chang. Biol. 15 (7), 1790–1803.
Griffith, A.W., Gobler, C.J., 2020. Harmful algal blooms: a climate change co-stressor in marine and freshwater ecosystems. Harmful Algae 91, 101590.
Gu, J., Luo, M., Zhang, X., Christakos, G., Agusti, S., Duarte, C.M., Wu, J., 2018. Losses of salt marsh in China: trends, threats and management. Estuar. Coast. Shelf Sci. 214, 98–109.
Guan, Y., Hohn, S., Wild, C., Merico, A., 2020. Vulnerability of global coral reef habitat suitability to ocean warming, acidification and eutrophication. Glob. Chang. Biol. 26 (10), 5646–5660.
Guinder, V.A., Molinero, J.C., Abbate, C.M.L., Berasategui, A.A., Popovich, C.A., Spetter, C.V., Marcovecchio, J.E., Freije, R.H., 2017. Phenological changes of blooming diatoms promoted by compound bottom-up and top-down controls. Estuaries Coast 40 (1), 95–104.
Guinotte, J.M., Orr, J., Cairns, S., Freiwald, A., Morgan, L., George, R., 2006. Will human-induced changes in seawater chemistry alter the distribution of deep-sea scleractinian corals? Front. Ecol. Environ. 4 (3), 141–146.
Gurgel, C.F.D., Camacho, O., Minne, A.J.P., Wernberg, T., Coleman, M.A., 2020. Marine heatwave drives cryptic loss of genetic diversity in underwater forests. Curr. Biol. 30 (7), 1199–1206.
Gutt, J., Bertler, N., Bracegirdle, T.J., Buschmann, A., Comiso, J., Hosie, G., Isla, E., Schloss, I.R., Smith, C.R., Tournadre, J., Xavier, J.C., 2015. The Southern Ocean ecosystem under multiple climate change stresses—an integrated circumpolar assessment. Glob. Chang. Biol. 21 (4), 1434–1453.
Hallett, C.S., Hobday, A.J., Tweedley, J.R., Thompson, P.A., McMahon, K., Valesini, F.J., 2018. Observed and predicted impacts of climate change on the estuaries of south-western Australia, a Mediterranean climate region. Reg. Environ. Chang. 18 (5), 1357–1373.
Hall-Spencer, J.M., Rodolfo-Metalpa, R., Martin, S., Ransome, E., Fine, M., Turner, S.M., Rowley, S.J., Tedesco, D., Buia, M.C., 2008. Volcanic carbon dioxide vents show ecosystem effects of ocean acidification. Nature 454 (7200), 96–99.
Halpern, B.S., Frazier, M., Potapenko, J., Casey, K.S., Koenig, K., Longo, C., Lowndes, J.S., Rockwood, R.C., Selig, E.R., Selkoe, K.A., Walbridge, S., 2015. Spatial and temporal changes in cumulative human impacts on the world's ocean. Nat. Commun. 6, 1–7.
Hamilton, C.D., Kovacs, K.M., Ims, R.A., Aars, J., Lydersen, C., 2017. An Arctic predator–prey system in flux: climate change impacts on coastal space use by polar bears and ringed seals. J. Anim. Ecol. 86 (5), 1054–1064.
Hammond, M.L., Beaulieu, C., Henson, S.A., Sahu, S.K., 2020. Regional surface chlorophyll trends and uncertainties in the global ocean. Sci. Rep. 10 (1), 1–9.
Hanley, M.E., Bouma, T.J., Mossman, H.L., 2020. The gathering storm: optimizing management of coastal ecosystems in the face of a climate-driven threat. Ann. Bot. 125 (2), 197–212.

Hanz, U., Wienberg, C., Hebbeln, D., Duineveld, G., Lavaleye, M., Juva, K., Dullo, W.C., Freiwald, A., Tamborrino, L., Reichart, G.J., Flögel, S., 2019. Environmental factors influencing benthic communities in the oxygen minimum zones on the Angolan and Namibian margins. Biogeosciences 16 (22), 4337–4356.

Hartman, S.E., Jiang, Z.P., Turk, D., Lampitt, R.S., Frigstad, H., Ostle, C., Schuster, U., 2015. Biogeochemical variations at the Porcupine Abyssal Plain sustained Observatory in the northeast Atlantic Ocean, from weekly to inter-annual timescales. Biogeosciences 12 (3), 845–853.

Harvey, B.P., Kerfahi, D., Jung, Y., Shin, J.H., Adams, J.M., Hall-Spencer, J.M., 2020. Ocean acidification alters bacterial communities on marine plastic debris. Mar. Pollut. Bull. 161, 111749.

Hauri, C., Gruber, N., McDonnell, A.M.P., Vogt, M., 2013. The intensity, duration, and severity of low aragonite saturation state events on the California continental shelf. Geophys. Res. Lett. 40 (13), 3424–3428.

Hauri, C., Friedrich, T., Timmermann, A., 2016. Abrupt onset and prolongation of aragonite undersaturation events in the Southern Ocean. Nat. Clim. Chang. 6 (2), 172–176.

Hawkins, S., Evans, A.J., Firth, L.B., Genner, M.J., Herbert, R.J., Adams, L.C., Moore, P.J., Mieszkowska, N., Thompson, R.C., Burrows, M.T., Fenburg, P.B., 2016. Impacts and effects of ocean warming on intertidal rocky habitats. In: Laffoley, D., Baxter, J.M. (Eds.), Explaining Ocean Warming: Cause, Scale, Effects and Consequences. IUCN, Gland, Switzerland, pp. 147–176.

Hawkins, E., Ortega, P., Suckling, E., Schurer, A., Hegerl, G., Jones, P., Joshi, M., Osborn, T.J., Masson-Delmotte, V., Mignot, J., Thorne, P., van Oldenborgh, G.J., 2017. Estimating changes in global temperature since the preindustrial period. Bull. Am. Meteorol. Soc. 98, 1841–1856.

Hebbeln, D., Portilho-Ramos, R.D.C., Wienberg, C., Titschack, J., 2019. The fate of cold-water corals in a changing world: a geological perspective. Front. Mar. Sci. 6, 119.

Helmuth, B., Broitman, B.R., Blanchette, C.A., Gilman, S., Halpin, P., Harley, C.D.G., O'Donnell, M.J., Hofmann, G.E., Menge, B., Strickland, D., 2006a. Mosaic patterns of thermal stress in the rocky intertidal zone: implications for climate change. Ecol. Monogr. 76 (4), 461–479.

Helmuth, B., Mieszkowska, N., Moore, P., Hawkins, S.J., 2006b. Living on the edge of two changing worlds: forecasting the responses of rocky intertidal ecosystems to climate change. Annu. Rev. Ecol. Evol. Syst. 37, 373–404.

Hennige, S.J., Wicks, L.C., Kamenos, N.A., Perna, G., Findlay, H.S., Roberts, J.M., 2015. Hidden impacts of ocean acidification to live and dead coral framework. Proc. R. Soc. B Biol. Sci. 282 (1813), 20150990.

Hennige, S.J., Wolfram, U., Wickes, L., Murray, F., Roberts, J.M., Kamenos, N.A., Schofield, S., Groetsch, A., Spiesz, E. M., Aubin-Tam, M.E., Etnoyer, P.J., 2020. Crumbling reefs and cold-water coral habitat loss in a future ocean: evidence of "coralporosis" as an indicator of habitat integrity. Front. Mar. Sci. 7, 668.

Henson, S.A., Beaulieu, C., Ilyina, T., John, J.G., Long, M., Séférian, R., Tijiputra, J., Sarmiento, J.L., 2017. Rapid emergence of climate change in environmental drivers of marine ecosystems. Nat. Commun. 8, 14682.

Hernán, G., Ortega, M.J., Gándara, A.M., Castejón, I., Terrados, J., Tomás, F., 2017. Future warmer seas: increased stress and susceptibility to grazing in seedlings of a marine habitat-forming species. Glob. Chang. Biol. 23, 4530–4543.

Heron, S.F., van Hooidonk, R., Maynard, J., Anderson, K., Day, J.C., Geiger, E., Hoegh-Guldberg, O., Hughes, Y., Marshall, P., Obura, D., Eakin, C.M., 2018. Impacts of Climate Change on World Heritage Coral Reefs: Update to the First Global Scientific Assessment. UNESCO World Heritage Centre, Paris. 16 pp.

Herrera-R, G.A., Oberdorff, T., Anderson, E.P., Brosse, S., Carvajal-Vallejos, F.M., Frederico, R.G., Hidalgo, M., Jézéquel, C., Maldonado, M., Maldonaldo-Ocampo, J., Ortega, H., Radinger, J., Torrente-Vilara, G., Zuanon, J., Tedesco, P.A., 2020. The combined effects of climate change and river fragmentation on the distribution of Andean Amazon fishes. Glob. Chang. Biol. 26, 5509–5523.

Hobday, A.J., Pecl, G.T., 2014. Identification of global marine hotspots: sentinels for change and vanguards for adaptation action. Rev. Fish Biol. Fish. 24 (2), 415–425.

Hodgson, E.E., Kaplan, I.C., Marshall, K.N., Leonard, J., Essington, T.E., Busch, D.S., Fulton, E.A., Harvey, C.J., Hermann, A.J., McElhany, P., 2018. Consequences of spatially variable ocean acidification in the California current: lower pH drives strongest declines in benthic species in southern regions while greatest economic impacts occur in northern regions. Ecol. Model. 383, 106–117.

Hoegh-Guldberg, O., Caldeira, K., Chopin, T., Gaines, S., Haugan, P., Hemer, M., Howard, J., Konar, M., Krause-Jensen, D., Lindstad, E., Lovelock, C.E., Michelin, M., et al., 2019. The Ocean as a Solution to Climate Change: Five Opportunities for Action. World Resources Institute. 116 pp. http://www.oceanpanel.org/climate.

Hollowed, A.B., Planque, B., Loeng, H., 2013. Potential movement of fish and shellfish stocks from the sub-Arctic to the Arctic Ocean. Fish. Oceanogr. 22 (5), 355–370.

Hughes, T.P., Anderson, K.D., Connolly, S.R., Heron, S.F., Kerry, J.T., Lough, J.M., Baird, A.H., Baum, J.K., Berumen, M.L., Bridge, T.C., Claar, D.C., Eakin, C.M., et al., 2018. Spatial and temporal patterns of mass bleaching of corals in the Anthropocene. Science 359 (6371), 80–83.

Hughes, T.P., Kerry, J.T., Baird, A.H., Connolly, S.R., Chase, T.J., Dietzel, A., Hill, T., Hoey, A.S., Hoogenboom, M.O., Jacobson, M., Kerswell, A., Madin, J.S., et al., 2019. Global warming impairs stock–recruitment dynamics of corals. Nature 568 (7752), 387–390.

Huthnance, J.M., 1995. Circulation, exchange and water masses at the ocean margin: the role of physical processes at the shelf edge. Prog. Oceanogr. 35 (4), 353–431.

Hyndes, G.A., Heck Jr., K.L., Vergés, A., Harvey, E.S., Kendrick, G.A., Lavery, P.S., McMahon, K., Orth, R.J., Pearce, A., Vanderklift, M., Wernberg, T., Whiting, S., Wilson, S., 2016. Accelerating tropicalization and the transformation of temperate seagrass meadows. Bioscience 66, 938–948.

Iglesias-Rodriguez, M.D., Halloran, P.R., Rickaby, R.E., Hall, I.R., Colmenero-Hidalgo, E., Gittins, J.R., Green, D.R., Tyrrell, T., Gibbs, S.J., von Dassow, P., Rehm, E., 2008. Phytoplankton calcification in a high-$CO_2$ world. Science 320 (5874), 336–340.

Ingels, J., Vanreusel, A., Brandt, A., Catarino, A.I., David, B., De Ridder, C., Dubois, P., Gooday, A.J., Martin, P., Pasotti, F., Robert, H., 2012. Possible effects of global environmental changes on Antarctic benthos: a synthesis across five major taxa. Ecol. Evol. 2 (2), 453–485.

Ingels, J., Aronson, R.B., Smith, C.R., 2018. The scientific response to Antarctic ice-shelf loss. Nat. Clim. Chang. 8 (10), 848–851.

Ingels, J., Aronson, R.B., Smith, C.R., Baco, A., Bik, H.M., Blake, J.A., Brandt, A., Cape, M., Demaster, D., Dolan, E., Domack, E., Fire, S., et al., 2020. Antarctic ecosystem responses following ice-shelf collapse and iceberg calving: science review and future research. WIREs Clim. Change, 12, e682.

IPCC (Intergovernmental Panel on Climate Change), 2019. Summary for policy makers. In: P örtner, H.-O., Roberts, D. C., Masson-Delmotte, V., Zhai, P., Tignor, M., Poloczanska, E., Weyer, N.M. (Eds.), IPCC Special Report on the Ocean and Cryosphere in a Changing Climate. IPCC/WMO, pp. 1–35. https://www.ipcc.ch/srocc. (Accessed 7 October 2020).

Jackson, E.L., Davies, A.J., Howell, K.L., Kershaw, P.J., Hall-Spencer, J.M., 2014. Future-proofing marine protected area networks for cold water coral reefs. ICES J. Mar. Sci. 71 (9), 2621–2629.

Jackson, D.W., Costas, S., González-Villanueva, R., Cooper, A., 2019. A global 'greening' of coastal dunes: an integrated consequence of climate change? Glob. Planet. Chang. 182, 103026.

Jamieson, A.J., Brooks, L.S.R., Reid, W.D., Piertney, S.B., Narayanaswamy, B.E., Linley, T.D., 2019. Microplastics and synthetic particles ingested by deep-sea amphipods in six of the deepest marine ecosystems on Earth. R. Soc. Open Sci. 6 (2), 180667.

Järnegren, J., Brooke, S., Jensen, H., 2020. Effects and recovery of larvae of the cold-water coral *Lophelia pertusa* (*Desmophyllum pertusum*) exposed to suspended bentonite, barite and drill cuttings. Mar. Environ. Res. 158, 104996.

Jin, D., Hoagland, P., Buesseler, K.O., 2020. The value of scientific research on the ocean's biological carbon pump. Sci. Total Environ. 749, 141357.

Johannessen, S.C., Macdonald, R.W., 2016. Geoengineering with seagrasses: is credit due where credit is given? Environ. Res. Lett. 11 (11), 113001.

Johns, E.M., Lumpkin, R., Putman, N.F., Smith, R.H., Muller-Karger, F.E., Rueda-Roa, D.T., Hu, C., Wang, M., Brooks, M.T., Gramer, L.J., Werner, F.E., 2020. The establishment of a pelagic *Sargassum* population in the tropical Atlantic: biological consequences of a basin-scale long distance dispersal event. Prog. Oceanogr. 182, 102269.

Johnson, D.S., Williams, B.L., 2017. Sea level rise may increase extinction risk of a saltmarsh ontogenetic habitat specialist. Ecol. Evol. 7 (19), 7786–7795.

Johnson, C.R., Banks, S.C., Barrett, N.S., Cazassus, F., Dunstan, P.K., Edgar, G.J., Frusher, S.D., Gardner, C., Haddon, M., Helidoniotis, F., Hill, K.L., Holbrook, N.L., et al., 2011. Climate change cascades: shifts in oceanography, species' ranges and subtidal marine community dynamics in eastern Tasmania. J. Exp. Mar. Biol. Ecol. 400, 17–32.

Jones, D.O.B., Yool, A., Wei, C.-L., Henson, S.A., Ruhl, H., Watson, R.A., Gehlen, M., 2014. Global reductions in seafloor biomass in response to climate change. Glob. Change Biol. 20 (6), 1861–1872.

Jones, M.C., Cheung, W.W.L., 2015. Multi-model ensemble projections of climate change effects on global marine biodiversity. ICES J. Mar. Sci. 72 (3), 741–752.

Jurikova, H., Gutjahr, M., Wallmann, K., Flögel, S., Liebetrau, V., Posenato, R., Angiolini, L., Garbelli, C., Brand, U., Wiedenbeck, M., Eisenhauer, A., 2020. Permian–Triassic mass extinction pulses driven by major marine carbon cycle perturbations. Nat. Geosci. 13, 745–750.

Kelleway, J.J., Cavanaugh, K., Rogers, K., Feller, I.C., Ens, E., Doughty, C., Saintilan, N., 2017. Review of the ecosystem service implications of mangrove encroachment into saltmarshes. Glob. Chang. Biol. 23 (10), 3967–3983.

Kelly, T.B., Davison, P.C., Goericke, R., Landry, M.R., Ohman, M., Stukel, M.R., 2019. The importance of mesozooplankton diel vertical migration for sustaining a mesopelagic food web. Front. Mar. Sci. 6, 508.

Kenchington, E.L., Kenchington, T.J., Henry, L.A., Fuller, S., Gonzalez, P., 2007. Multi-decadal changes in the megabenthos of the Bay of Fundy: the effects of fishing. J. Sea Res. 58 (3), 220–240.

Kharouba, H.M., Ehrlén, J., Gelman, A., Bolmgren, K., Allen, J.M., Travers, S.E., Wolkovich, E.M., 2018. Global shifts in the phenological synchrony of species interactions over recent decades. Proc. Natl. Acad. Sci. USA 115 (20), 521–5216.

Kim, J.-H., Kim, N., Moon, H., Lee, S., Jeong, S.Y., Diaz-Pulido, G., Edwards, M.S., Kang, J.-H., Kang, E.J., Oh, H.-J., Hwamg, J.D., Kim, I.-N., 2020. Global warming offsets the ecophysiological stress of ocean acidification on temperate crustose coralline algae. Mar. Pollut. Bull. 157, 111324.

Kjesbu, O.S., Bogstad, B., Devine, J.A., Gjøsæter, H., Howell, D., Ingvaldsen, R.B., Nash, R.D., Skjæraasen, J.E., 2014. Synergies between climate and management for Atlantic cod fisheries at high latitudes. Proc. Natl. Acad. Sci. USA 111 (9), 3478–3483.

Klein, E.S., Hill, S.L., Hinke, J.T., Phillips, T., Watters, G.M., 2018. Impacts of rising sea temperature on krill increase risks for predators in the Scotia Sea. PLoS One 13 (1), e0191011.

Koester, A., Migani, V., Bunbury, N., Ford, A., Sanchez, C., Wild, C., 2020. Early trajectories of benthic coral reef communities following the 2015/16 coral bleaching event at remote Aldabra Atoll, Seychelles. Sci. Rep. 10 (1), 1–14.

Kordas, R.L., Harley, C.D.G., O'Connor, M.I., 2011. Community ecology in a warming world: the influence of temperature on interspecific interactions in marine systems. J. Exp. Mar. Biol. Ecol. 400, 218–226.

Kovacs, K.M., Lemons, P., MacCracken, J.G., Lydersen, C., 2016. Walruses in a Time of Climate Change. Arctic Report Card: Update for 2015. www.arctic.noaa.gov/Report-Card.

Krause-Jensen, D., Lavery, P., Serrano, O., Marba, N., Masque, P., Duarte, C.M., 2018. Sequestration of macroalgal carbon: the elephant in the Blue Carbon room. Biol. Lett. 14, 20180236.

Krauss, K.W., Osland, M.J., 2020. Tropical cyclones and the organization of mangrove forests: a review. Ann. Bot. 125 (2), 213–234.

Kroeker, K.J., Micheli, F., Gambi, M.C., Martz, T.R., 2011. Divergent ecosystem responses within a benthic marine community to ocean acidification. Proc. Natl. Acad. Sci. USA 108 (35), 14515–14520.

Kroeker, K.J., Kordas, R.L., Crim, R., Hendriks, I.E., Ramajo, L., Singh, G.S., Duarte, C.M., Gattuso, J.-P., 2013. Impacts of ocean acidification on marine organisms: quantifying sensitivities and interaction with warming. Glob. Chang. Biol. 19, 1884–1896.

Kroeker, K.J., Kordas, R.L., Harley, C.D., 2017. Embracing interactions in ocean acidification research: confronting multiple stressor scenarios and context dependence. Biol. Lett. 13 (3), 20160802.

Kroeker, K.J., Bell, L.E., Donham, E.M., Hoshijima, U., Lummis, S., Toy, J.A., Willis-Norton, E., 2020. Ecological change in dynamic environments: accounting for temporal environmental variability in studies of ocean change biology. Glob. Change Biol. 26, 54–67.

Kröger, S., Parker, R., Cripps, G., Williamson, P. (Eds.), 2018. Shelf Seas: The Engine of Productivity. Cefas Lowestoft. Policy Report. 24 pp. https://www.uk-ssb.org/shelf_seas_report.pdf.

Krueger, T., Horwitz, N., Bodin, J., Giovani, M.E., Escrig, S., Meibom, A., Fine, M., 2017. Common reef-building coral in the Northern Red Sea resistant to elevated temperature and acidification. R. Soc. Open Sci. 4 (5), 170038.

Krumhansl, K., Scheibling, R., 2012. Production and fate of kelp detritus. Mar. Ecol. Prog. Ser. 467, 281–302.

Krumhardt, K.M., Lovenduski, N.S., Freeman, N.M., Bates, N.R., 2016. Apparent increase in coccolithophore abundance in the subtropical North Atlantic from 1990 to 2014. Biogeosciences 13, 1163–1177.

Kulp, S.A., Strauss, B.H., 2019. New elevation data triple estimates of global vulnerability to sea-level rise and coastal flooding. Nat. Commun. 10 (1), 1–12.

Kurman, M.D., Gómez, C.E., Georgian, S.E., Lunden, J.J., Cordes, E.E., 2017. Intra-specific variation reveals potential for adaptation to ocean acidification in a cold-water coral from the Gulf of Mexico. Front. Mar. Sci. 4, 111.

Kwiatkowski, L., Aumont, O., Bopp, L., 2019. Consistent trophic amplification of marine biomass declines under climate change. Glob. Chang. Biol. 25 (1), 218–229.

Kwiatkowski, L., Torres, O., Bopp, L., Aumont, O., Chamberlain, M., Christian, J.R., Dunne, J.P., Gehlen, M., Ilyina, T., John, J.G., Lenton, A., Li, H., et al., 2020. Twenty-first century ocean warming, acidification, deoxygenation, and upper-ocean nutrient and primary production decline from CMIP6 model projections. Biogeosciences 17, 3439–3470.

Ladd, C.J., Duggan-Edwards, M.F., Bouma, T.J., Pagès, J.F., Skov, M.W., 2019. Sediment supply explains long-term and large-scale patterns in salt marsh lateral expansion and erosion. Geophys. Res. Lett. 46 (20), 11178–11187.

Laffoley, D., Baxter, J.M. (Eds.), 2019. Ocean Deoxygenation: Everyone's Problem—Causes, Impacts, Consequences and Solutions. IUCN, Gland, Switzerland. 580 pp.

Lapointe, F., Bradley, R.S., Francus, P., Balascio, N.L., Abbott, M.B., Stoner, J.S., St-Onge, G., De Coninch, A., Labarre, T., 2020. Annually resolved Atlantic sea surface temperature variability over the past 2,900 y. Proc. Natl. Acad. Sci. 117 (44), 27171–27178.

Lauchlan, S.S., Nagelkerken, I., 2020. Species range shifts along multistressor mosaics in estuarine environments under future climate. Fish Fish. 21 (1), 32–46.

Laufkötter, C., John, J.G., Stock, C.A., Dunne, J.P., 2017. Temperature and oxygen dependence of the remineralization of organic matter. Glob. Biogeochem. Cycles 31 (7), 1038–1050.

Laufkötter, C., Zscheischler, J., Frölicher, T.L., 2020. High-impact marine heatwaves attributable to human-induced global warming. Science 369, 1621–1625.

Lefort, S., Aumont, O., Bopp, L., Arsouze, T., Gehlen, M., Maury, O., 2015. Spatial and body-size dependent response of marine pelagic communities to projected global climate change. Glob. Chang. Biol. 21 (1), 154–164.

Legault, R., Zogg, G.P., Travis, S.E., 2018. Competitive interactions between native *Spartina alterniflora* and non-native *Phragmites australis* depend on nutrient loading and temperature. PLoS One 13 (2), e0192234.

Legendre, L., Rivkin, R.B., Weinbauer, M.G., Guidi, L., Uitz, J., 2015. The microbial carbon pump concept: potential biogeochemical significance in the globally changing ocean. Prog. Oceanogr. 134, 432–450.

Lehman, P.W., Kurobe, T., Teh, S.J., 2020. Impact of extreme wet and dry years on the persistence of *Microcystis* harmful algal blooms in San Francisco Estuary. Quat. Int. https://doi.org/10.1016/j.quaint.2019.12.003.

Lenton, T.M., Rockström, J., Gaffney, O., Rahmstorf, S., Richardson, K., Steffen, W., Schellnhuber, H.J., 2019. Climate tipping points—too risky to bet against. Nature 575, 592–595.

Leung, S., Cabré, A., Marinov, I., 2015. A latitudinally banded phytoplankton response to 21st century climate change in the Southern Ocean across the CMIP5 model suite. Biogeosciences 12 (19), 5715–5734.

Levin, L.A., 2018. Manifestation, drivers, and emergence of open ocean deoxygenation. Annu. Rev. Mar. Sci. 10 (1), 229–260.

Levin, L.A., Le Bris, N., 2015. The deep ocean under climate change. Science 350, 766–768.

Levin, L.A., Liu, K.K., Emeis, K.C., Breitburg, D.L., Cloern, J., Deutsch, C., Giani, M., Goffart, A., Hofmann, E.E., Lachkar, Z., Limburg, K., Liu, S.-M., Montes, E., Naqvi, W., Ragueneau, O., Rabouille, C., Sarkar, S.K., Swaney, D.P., Wassman, P., Wishner, K.F., 2015. Comparative biogeochemistry–ecosystem–human interactions on dynamic continental margins. J. Mar. Syst. 141, 3–17.

Lewis, C., Ellis, R.P., Vernon, E., Elliot, K., Newbatt, S., Wilson, R.W., 2016. Ocean acidification increases copper toxicity differentially in two key marine invertebrates with distinct acid-base responses. Sci. Rep. 6, 21554.

Li, H., Ke, C., Zhu, Q., Shu, S., 2019. Spatial-temporal variations in net primary productivity in the Arctic from 2003 to 2016. Acta Oceanol. Sin. 38 (8), 111–121.

Li, G., Cheng, L., Zhu, J., Trenberth, K.E., Mann, M.E., Abraham, J.P., 2020. Increasing ocean stratification over the past half-century. Nat. Clim. Chang. 10, 1116–1123.

Liesner, D., Fouqueau, L., Valero, M., Roleda, M.Y., Pearson, G.A., Bischof, K., Valentin, K., Bartsch, I., 2020. Heat stress responses and population genetics of the kelp *Laminaria digitata* (Phaeophyceae) across latitudes reveal differentiation among North Atlantic populations. Ecol. Evol. 10 (17), 9144–9177.

Lima, M.D.A.C., Ward, R.D., Joyce, C.B., 2020. Environmental drivers of sediment carbon storage in temperate seagrass meadows. Hydrobiologia 847 (7), 1773–1792.

Lindley, J.A., Kirby, R.R., 2010. Climate-induced changes in the North Sea Decapoda over the last 60 years. Clim. Res. 42 (3), 257–264.

Lindsey, R., 2020. Climate Change: Atmospheric Carbon Dioxide. www.climate.gov/print/8431.

Ling, S.D., Johnson, C.R., Ridgway, K., Hobday, A.J., Haddon, M., 2009. Climate-driven range extension of a sea urchin: inferring future trends by analysis of recent population dynamics. Glob. Chang. Biol. 15, 719–731.

Lischka, S., Büdenbender, J., Boxhammer, T., Riebesell, U., 2011. Impact of ocean acidification and elevated temperatures on early juveniles of the polar shelled pteropod *Limacina helicina*: mortality, shell degradation, and shell growth. Biogeosciences 8, 919–932.

Little, C.M., Horton, R.M., Kopp, R.E., Oppenheimer, M., Vecchi, G.A., Villarini, G., 2015. Joint projections of US East Coast sea level and storm surge. Nat. Clim. Chang. 5 (12), 114–1120.
Liu, K.K., Peng, T.H., Shaw, P.T., Shiah, F.K., 2003. Circulation and biogeochemical processes in the East China Sea and the vicinity of Taiwan: an overview and a brief synthesis. Deep-Sea Res. II Top. Stud. Oceanogr. 50, 1055–1064.
Lopez-Abbate, M.C., Molinero, J.C., Perillo, G.M.E., Barria de Cao, M.S., Pettigrosso, R.E., Guinder, V.A., Uibrig, R., Berasategui, A.A., Vitale, A., Marcovecchio, J.E., Hoffmeyer, M.S., 2019. Long-term changes on estuarine ciliates linked with modifications on wind patterns and water turbidity. Mar. Env. Res. 144, 46–55.
Long, S.M., Xie, S.P., Zheng, X.T., Liu, Q., 2014. Fast and slow responses to global warming: sea surface temperature and precipitation patterns. J. Clim. 27, 285–299.
Lotze, H.K., Tittensor, D.P., Bryndum-Buchholz, A., Eddy, T.D., Cheung, W.W., Galbraith, E.D., Galbraith, E.D., Barange, M., Barrier, N., Bianchi, D., Blanchard, J.L., Bopp, L., et al., 2019. Global ensemble projections reveal trophic amplification of ocean biomass declines with climate change. Proc. Natl. Acad. Sci. USA 116, 12907–12912.
Lotze, H.K., Bryndum-Buchholz, A., Boyce, D., 2021. Effect of climate change on food production (fishing) (Chapter 8). In: Letcher, T. (Ed.), The Impacts of Climate Change: A Comprehensive Study of Physical, Biophysical, Social and Political Issues. Elsevier Amsterdam.
Lovelock, C.E., Fourqurean, J.W., Morris, J.T., 2017. Modeled $CO_2$ emissions from coastal wetland transitions to other land uses: tidal marshes, mangrove forests, and seagrass beds. Front. Mar. Sci. 4, 143.
Lunden, J.J., McNicholl, C.G., Sears, C.R., Morrison, C.L., Cordes, E.E., 2014. Acute survivorship of the deep-sea coral *Lophelia pertusa* from the Gulf of Mexico under acidification, warming, and deoxygenation. Front. Mar. Sci. 1, 78.
Lydersen, C., Vaquie-Garcia, J., Lydersen, E., Christensen, G.N., Kovacs, K.M., 2017. Novel terrestrial haul-out behaviour by ringed seals (*Pusa hispida*) in Svalbard, in association with harbour seals (*Phoca vitulina*). Polar Res. 36 (1), 1374124.
Mafi-Gholami, D., Zenner, E.K., Jaafari, A., Bui, D.T., 2020. Spatially explicit predictions of changes in the extent of mangroves of Iran at the end of the 21st century. Estuar. Coast. Shelf Sci. 237, 106644.
Maier, C., Popp, P., Sollfrank, N., Weinbauer, M.G., Wild, C., Gattuso, J.-P., 2016. Effects of elevated $pCO_2$ and feeding on net calcification and energy budget of the Mediterranean cold-water coral *Madrepora oculata*. J. Exp. Biol. 219, 3208–3217.
Maier, C., Weinbauer, M.G., Gattuso, J.P., 2019. Fate of Mediterranean scleractinian cold-water corals as a result of global climate change. A synthesis. In: Orejas, C., Jiménez, C. (Eds.), Mediterranean Cold-Water Corals: Past, Present and Future. Springer, Cham, pp. 517–529.
Manna, S., Chaudhuri, K., Bhattacharyya, S., Bhattacharyya, M., 2010. Dynamics of Sundarban estuarine ecosystem: eutrophication induced threat to mangroves. Saline Syst. 6 (1), 8.
Marrari, M., Piola, A.R., Valla, D., 2017. Variability and 20-year trends in satellite-derived surface chlorophyll concentrations in Large Marine Ecosystems around South and Western Central America. Front. Mar. Sci. 4, 372.
Martin, A., Boyd, P., Buesseler, K., Cetinic, I., Claustre, H., Giering, S., Henson, S., Irigoien, X., Kriest, I., Memery, L., Robinson, C., 2020. The oceans' twilight zone must be studied now, before it is too late. Nature 580, 26–28.
Martinetto, P., Alemany, D., Botto, F., Mastrángelo, M., Falabella, V., Acha, E.M., Antón, G., Bianci, A., Campagna, C., Cañete, G., Filippo, P., Iribarne, O., Laterra, P., Martínez, P., Negri, R., Piola, A.R., Romero, S.I., Santos, D., Saraceno, M., 2020. Linking the scientific knowledge on marine frontal systems with ecosystem services. Ambio 49, 541–556.
Masselink, G., Beetham, E., Kench, P., 2020. Coral reef islands can accrete vertically in response to sea level rise. Sci. Adv. 6 (24), eaay3656.
McLachlan, A., Defeo, O., 2017. The Ecology of Sandy Shores. Elsevier/Academic Press, London. 560 pp.
Mcowen, C.J., Cheung, W.W.L., Rykaczewski, R.R., Watson, R.A., Wood, L.J., 2015. Is fisheries production within large marine ecosystems determined by bottom-up or top-down forcing? Fish Fish. 16 (4), 623–632.
Meinen, C.S., Perez, R.C., Dong, S., Piola, A.R., Campos, E., 2020. Observed ocean bottom temperature variability at four sites in the northwestern Argentine basin: evidence of decadal deep/abyssal warming amidst hourly to interannual variability during 2009–2019. Geophys. Res. Lett. 47. e2020GL089093.
Melet, A., Teatini, P., Le Cozannet, G., Jamet, C., Conversi, A., Benveniste, J., Almar, R., 2020. Earth observations for monitoring marine coastal hazards and their drivers. Surv. Geophys. 41, 1489–1534.
Meredith, M., Sommerkorn, M., Cassotta, S., Derksen, C., Ekaykin, A., Hollowed, A., Kofinas, G., Mackintosh, A., Melbourne-Thomas, J., Muelbert, M.M.C., Ottersen, G., Pritchard, H., Schuur, E.A.G., 2019. Polar regions. In: Pörtner, H.-O., Roberts, D.C., Masson-Delmotte, V., Zhai, P., Tignor, M., Poloczanska, E., Weyer, N.M. (Eds.), IPCC Special Report on the Ocean and Cryosphere in a Changing Climate, pp. 203–320. Online and in press.

Meyer, J., Riebesell, U., 2015. Reviews and syntheses. Responses of coccolithophores to ocean acidification: a meta-analysis. Biogeosciences 12 (6), 1671–1682.

Meyssignac, B., Boyer, T., Zhao, Z., Hakuba, M.Z., Landerer, F.W., Stammer, D., Köhl, A., Kato, S., L'Ecuyer, T., Ablain, M., Abraham, J.P., 2019. Measuring global ocean heat content to estimate the Earth energy imbalance. Front. Mar. Sci. 6, 432.

Milazzo, M., Alessi, C., Quattrocchi, F., Chemello, R., D'Agostaro, R., Gil, J., Vaccaro, A.M., Mirto, S., Gristina, M., Badalamenti, F., 2019. Biogenic habitat shifts under long-term ocean acidification show nonlinear community responses and unbalanced functions of associated invertebrates. Sci. Total Environ. 667, 41–48.

Miloslavich, P., Cruz-Motta, J.J., Hernández, A., Herrera, C., Klein, E., Barros, F., Bigatti, G., Cárdenas, M., Carranza, A., Flores, A., Gil-Kodaka, P., Gobin, J., et al., 2016. Benthic assemblages in South American intertidal rocky shores: biodiversity, services, and threats. In: Riosmena-Rodriguez, R. (Ed.), Marine Benthos: Biology, Ecosystem Functions and Environmental Impact. Nova Science Publishing, New York, pp. 83–137.

Moberg, F., Folke, C., 1999. Ecological goods and services of coral reef ecosystems. Ecol. Econ. 29 (2), 215–233.

Molinos, J.G., Halpern, B.S., Schoeman, D.S., Brown, C.J., Kiessling, W., Moore, P.J., Pandolfi, J.M., Poloczanska, E.S., Richardson, A.J., Burrows, M.T., 2016. Climate velocity and the future global redistribution of marine biodiversity. Nat. Clim. Chang. 6 (1), 83–88.

Montero-Serra, I., Edwards, M., Genner, M.J., 2015. Warming shelf seas drive the subtropicalization of European pelagic fish communities. Glob. Chang. Biol. 21 (1), 144–153.

Montgomery, D.W., Simpson, S.D., Engelhard, G.H., Birchenough, S.N.R., Wilson, R.W., 2019. Rising CO2 enhances hypoxia tolerance in a marine fish. Sci. Rep. 9, 15152.

Morato, T., González-Irusta, J.M., Dominguez-Carrió, C., Wei, C.L., Davies, A., Sweetman, A.K., Taranto, G.H., Beazley, L., García-Alegre, A., Grehan, A., Laffargue, P., 2020. Climate-induced changes in the suitable habitat of cold-water corals and commercially important deep-sea fishes in the North Atlantic. Glob. Chang. Biol. 26 (4), 2181–2202.

Morgan, K.M., Perry, C.T., Arthur, R., Williams, H.T., Smithers, S.G., 2020. Projections of coral cover and habitat change on turbid reefs under future sea-level rise. Proc. R. Soc. B 287 (1929), 20200541.

Morley, J.W., et al., 2018. Projecting shifts in thermal habitat for 686 species on the North American continental shelf. PLoS One 13 (5), e0196127.

Mueller, P., Schile-Beers, L.M., Mozdzer, T.J., Chmura, G.L., Dinter, T., Kuzyakov, Y., de Groot, A.V., Esselink, P., Smit, C., D'Alpaos, A., Ibanez, C., Lazarus, M., et al., 2018. Global-change effects on earlsy-stage decomposition processes in tidal wetlands–implications from a global survey using standardized litter. Biogeosciences 15 (10), 3189–3202.

Munck af Rosenschöld, J., Rozema, J.G., Frye-Levine, L.A., 2014. Institutional inertia and climate change: a review of the new institutionalist literature. Wiley Interdiscip. Rev. Clim. Chang. 5, 639–648.

Munday, P.L., 2014. Transgenerational acclimation of fishes to climate change and ocean acidification. F1000Prime Rep. 6, 99.

Munday, P.L., Dixson, D.L., Welch, M.J., Chivers, D.P., Domenici, P., Grosell, M., Heuer, R.M., Jones, G.P., McCormick, M.I., Meekan, M., Nilsson, G.E., Ravasi, T., Watson, S.-A., 2020. Methods matter in repeating ocean acidification studies. Nature 586, E20–E24.

Murray, N.J., Phinn, S.R., DeWitt, M., Ferrari, R., Johnston, R., Lyons, M.B., Clinton, N., Thau, D., Fuller, R.A., 2019. The global distribution and trajectory of tidal flats. Nature 565 (7738), 222–225.

Nagelkerken, I., Connell, S.D., 2015. Global alteration of ocean ecosystem functioning due to increasing human $CO_2$ emissions. Proc. Natl. Acad. Sci. USA 112 (43), 13272–13277.

Nagelkerken, I., Goldenberg, S.U., Ferreira, C.M., Ullah, H., Connell, S.D., 2020. Trophic pyramids reorganize when food web architecture fails to adjust to ocean change. Science 369 (6505), 829–832.

Najjar, R.G., Herrmann, M., Alexander, R., Boyer, E.W., Burdige, D.J., Butman, D., Cai, W.-J., Canuel, E.A., Chen, R.F., Friedrichs, M.A.M., Feagin, R.A., Griffith, P.C., et al., 2018. Carbon budget of tidal wetlands, estuaries, and shelf waters of Eastern North America. Glob. Biogeochem. Cycles 32 (3), 389–416.

Narita, D., Pörtner, H.O., Rehdanz, K., 2020. Accounting for risk transitions of ocean ecosystems under climate change: an economic justification for more ambitious policy responses. Clim. Chang., 162, 1–11.

Negrin, V.L., Botté, S.E., Pratolongo, P.D., Trilla, G.G., Marcovecchio, J.E., 2016. Ecological processes and biogeochemical cycling in saltmarshes: synthesis of studies in the Bahía Blanca estuary (Argentina). Hydrobiologia 774 (1), 217–235.

Neuheimer, A.B., MacKenzie, B.R., Payne, M.R., 2018. Temperature dependent adaptation allows fish to meet their food across their species' range. Sci. Adv. 4 (7), eaar4349.

Neukermans, G., Oziel, L., Babin, M., 2018. Increased intrusion of warming Atlantic water leads to rapid expansion of temperate phytoplankton in the Arctic. Glob. Chang. Biol. 24 (6), 2545–2553.

Ni, W., Li, M., Ross, A.C., Najjar, R.G., 2019. Large projected decline in dissolved oxygen in a eutrophic estuary due to climate change. J. Geophys. Res. Oceans 124 (11), 8271–8289.

Nilsson, G.E., Östlund-Nilsson, S., Munday, P.L., 2010. Effects of elevated temperature on coral reef fishes: loss of hypoxia tolerance and inability to acclimate. Comp. Biochem. Physiol. A Mol. Integr. Physiol. 156, 389–393.

Niner, H.J., Ardron, J.A., Escobar, E.G., Gianni, M., Jaeckel, A., Jones, D.O., Levin, L.A., Smith, C.R., Thiele, T., Turner, P.J., Van Dover, C.L., 2018. Deep-sea mining with no net loss of biodiversity—an impossible aim. Front. Mar. Sci. 5, 53.

NSIDC (National Snow and Ice Data Center), 2020. Arctic Sea Ice Decline Stalls Out at Second Lowest Minimum. http://nsidc.org/arcticseaicenews/2020/09/. (Accessed 11 November 2020).

Olsen, Y.S., Potouroglou, M., Garcias-Bonet, N., Duarte, C.M., 2015. Warming reduces pathogen pressure on a climate-vulnerable seagrass species. Estuaries Coast 38, 659–667.

Oppenheimer, M., Glalovovic, B.C., Hinkel, J., van dee Waal, R., Magnan, A.K., Abd-Elgawad, A., Cai, R., Cifuentes-Jara, M., DeConto, R.M., Ghosh, T., Hay, J., Isla, F., Marzeion, B., Meyssignac, B., Sebesvari, Z., 2019. Sea level rise and implications for low-lying islands, coasta and communinities. In: P örtner, H.-O., Roberts, D.C., Masson-Delmotte, V., Zhai, P., Tignor, M., Poloczanska, E., Weyer, N.M. (Eds.), IPCC Special Report on the Ocean and Cryosphere in a Changing Climate, pp. 321–455. Online and in press.

Oreska, M.P., McGlathery, K.J., Aoki, L.R., Berger, A.C., Berg, P., Mullins, L., 2020. The greenhouse gas offset potential from seagrass restoration. Sci. Rep. 10 (1), 1–15.

Ortega, L., Celentano, E., Delgado, E., Defeo, O., 2016. Climate change influences on abundance, individual size and body abnormalities in a sandy beach clam. Mar. Ecol. Prog. Ser. 545, 203–213.

Orth, R.J., Carruthers, T.J., Dennison, W.C., Duarte, C.M., Fourqurean, J.W., Heck, K.L., Hughes, A.R., Kendrick, G.A., Kenworthy, W.J., Olyarnik, S., Short, F.T., Waycott, M., Williams, S.L., 2006. A global crisis for seagrass ecosystems. Bioscience 56 (12), 987–996.

Oschlies, A., Brandt, P., Stramma, L., Schmidtko, S., 2018. Drivers and mechanisms of ocean deoxygenation. Nat. Geosci. 11 (7), 467–473.

Osland, M.J., Hartmann, A.M., Day, R.H., Ross, M.S., Hall, C.T., Feher, L.C., Vervaeke, W.C., 2019. Microclimate influences mangrove freeze damage: implications for range expansion in response to changing macroclimate. Estuaries Coast 42 (4), 1084–1096.

Overland, J.E., 2020. Less climatic resilience in the Arctic. Weather Clim. Extrem. 30, 100275.

Oziel, L., Baudena, A., Ardyna, M., Massicotte, P., Randelhoff, A., Sallée, J.B., Ingvaldsen, R.B., Devred, E., Babin, M., 2020. Faster Atlantic currents drive poleward expansion of temperate phytoplankton in the Arctic Ocean. Nat. Commun. 11 (1), 1–8.

Pagès, J.F., Smith, T.M., Tomas, F., Sanmarti, N., Boada, J., De Bari, H., Pérez, M., Romero, J., Arthur, R., Alcoverro, T., 2018. Contrasting effects of ocean warming on different components of plant-herbivore interactions. Mar. Pollut. Bull. 134, 55–65.

Palmer, K., Watson, C., Fischer, A., 2019. Non-linear interactions between sea-level rise, tides, and geomorphic change in the Tamar Estuary, Australia. Estuar. Coast. Shelf Sci. 225, 106247.

Parker, V.T., Boyer, K.E., 2019. Sea-level rise and climate change impacts on an urbanized Pacific Coast estuary. Wetlands 39 (6), 1219–1232.

Patrício, A.R., Varela, M.R., Barbosa, C., Broderick, A.C., Catry, P., Hawkes, L.A., Regalia, A., Godley, B.J., 2019. Climate change resilience of a globally important sea turtle nesting population. Glob. Chang. Biol. 25 (2), 522–535.

Pauly, D., Cheung, W.W.L., 2017. Sound physiological knowledge and principles in modeling shrinking of fishes under climate change. Glob. Chang. Biol. 24, e15–e26.

Pauly, D., Christensen, V., Dalsgaard, J., Froese, R., Torres, F., 1998. Fishing down marine food webs. Science 279 (5352), 860–863.

Peng, G., Bellerby, R., Zhang, F., Sun, X., Li, D., 2020. The ocean's ultimate trashcan: hadal trenches as major depositories for plastic pollution. Water Res. 168, 115121.

Perez, F.F., Fontela, M., García-Ibáñez, M.I., Mercier, H., Velo, A., Lherminier, P., Zunino, P., de la Paz, M., Alonso-Pérez, F., Guallart, E.F., Padin, X.A., 2018. Meridional overturning circulation conveys fast acidification to the deep Atlantic Ocean. Nature 554, 515–518.

Pergent, G., Bazairi, H., Bianchi, C., Boudouresque, C., Buia, M., Calvo, S., Clabaut, P., Harmelin-Vivien, M., Mateo, M., Montefalcone, M., Morri, C., Orfandis, S., Pergent-Martini, C., semroud, R., Serrano, O., Thibault, T., Tomasello, A., Verlaque, M., 2014. Climate change and Mediterranean seagrass meadows: a synopsis for environmental managers. Mediterr. Mar. Sci. 15 (2), 462–473.

Perry, A.L., Low, P.J., Ellis, J.R., Reymolds, J.D., 2005. Climate change and distribution shifts in marine fishes. Science 308, 1912–1915.

Pessarrodona, A., Foggo, A., Smale, D.A., 2019. Can ecosystem functioning be maintained despite climate-driven shifts in species composition? Insights from novel marine forests. J. Ecol. 107 (1), 91–104.

Piggott, J.J., Townsend, C.R., Matthaei, C.D., 2015. Reconceptualizing synergism and antagonism among multiple stressors. Ecol. Evol. 5, 1538–1547.

Pinsky, M.L., Eikeset, A.M., McCauley, D.J., 2019. Greater vulnerability to warming of marine versus terrestrial ectotherms. Nature 569, 108–111.

Poloczanska, E.S., Brown, C.J., Sydeman, W.J., Kiessling, W., Schoeman, D.S., Moore, P.J., Brander, K., Bruno, J.F., Buckley, L.B., Burrows, M.T., Duarte, C.M., Halpern, B.S., Holding, J., Kappel, C.V., O'Connor, M.I., Pandolfi, J.M., Parmesan, C., Schwing, F., Thompson, S.A., Richardson, A.J., 2013. Global imprint of climate change on marine life. Nat. Clim. Chang. 3 (10), 919–925.

Poloczanska, E.S., Burrows, M.T., Brown, C.J., García Molinos, J., Halpern, B.S., Hoegh-Guldberg, O., Kappel, C.V., Moore, P.J., Richardson, A.J., Schoeman, D.S., Sydeman, W.J., 2016. Responses of marine organisms to climate change across oceans. Front. Mar. Sci. 3, 62.

Popova, E., Yool, A., Byfield, V., Cochrane, K., Coward, A.C., Salim, S.S., Gasalla, M.A., Henson, S.A., Hobday, A.J., Pecl, G.T., Sauer, W.H., 2016. From global to regional and back again: common climate stressors of marine ecosystems relevant for adaptation across five ocean warming hotspots. Glob. Chang. Biol. 22 (6), 2038–2053.

Pörtner, H.O., 2012. Integrating climate-related stressor effects on marine organisms: unifying principles linking molecule to ecosystem-level changes. Mar. Ecol. Prog. Ser. 470, 273–290.

Pörtner, H.O., Farrell, A.P., 2008. Physiology and climate change. Science 322, 690–692.

Puerta, P., Johnson, C., Carreiro-Silva, M., Henry, L.A., Kenchington, E., Morato, T., Kazanidis, G., Rueda, J.L., Urra, J., Ross, S., Wei, C.L., 2020. Influence of water masses on the biodiversity and biogeography of deep-sea benthic ecosystems in the North Atlantic. Front. Mar. Sci. 7, 239.

Puotinen, M., Maynard, J.A., Beeden, R., Radford, B., Williams, G.J., 2016. A robust operational model for predicting where tropical cyclone waves damage coral reefs. Sci. Rep. 6, 26009.

Qi, D., Chen, L., Chen, B., Gao, Z., Zhong, W., Feely, R.A., Anderson, L.G., Sun, H., Chen, J., Chen, M., Zhan, L., Zhang, Y., Cai, W.-J., 2017. Increase in acidifying water in the western Arctic Ocean. Nat. Clim. Chang. 7 (3), 195–199.

Quigley, K.M., Willis, B.L., Kenkel, C.D., 2019. Transgenerational inheritance of shuffled symbiont communities in the coral *Montipora digitata*. Sci. Rep. 9 (1), 1–11.

Rabalais, N.N., 2019. Ocean deoxygenation from eutrophication (human nutrient inputs). In: Laffoley, D., Baxter, J.M. (Eds), Ocean Deoxygenation: Everyone's Problem—Causes, Impacts, Consequences and Solutions, IUCN, Gland, Switzerland, pp. 117–135.

Raftery, A.E., Zimmer, A., Frierson, D.M.W., Startz, R., Liu, P., 2017. Less than 2°C warming by 2100 unlikely. Nat. Clim. Chang. 7, 637–641.

Randelhoff, A., Guthrie, J.D., 2016. Regional patterns in current and future export production in the central Arctic Ocean quantified from nitrate fluxes. Geophys. Res. Lett. 43 (16), 8600–8608.

Raposa, K.B., Weber, R.L., Ekberg, M.C., Ferguson, W., 2017. Vegetation dynamics in Rhode Island saltmarshes during a period of accelerating sea level rise and extreme sea level events. Estuaries Coast 40 (3), 640–650.

Reddin, C.J., Nätscher, P.S., Kocsis, Á.T., Pörtner, H.O., Kiessling, W., 2020. Marine clade sensitivities to climate change conform across timescales. Nat. Clim. Chang. 10, 249–253.

Reid, P.C., Johns, D.G., Edwards, M., Starr, M., Poulin, M., Snoeijs, P., 2007. A biological consequence of reducing Arctic ice cover: arrival of the Pacific diatom Neodenticula seminae in the North Atlantic for the first time in 800 000 years. Glob. Chang. Biol. 13 (9), 1910–1921.

Repolho, T., Duarte, B., Dionisio, G., Paula, J.R., Lopes, A.R., Rosa, I.C., Grilo, T.F., Cacador, I., Calado, R., Rosa, R., 2017. Seagrass ecophysiological performance under ocean warming and acidification. Sci. Rep. 7, 41443.

Reygondeau, G., Cheung, W.W.L., Wabnitz, C.C.C., Lam, V.W.Y., Frölicher, T., Maury, O., 2020. Climate-change induced emergence of novel biogeochemical provinces. Front. Mar. Sci. 7, 657.

Richter-Menge, J., Jeffries, M.O., Osborne, E., 2018. The arctic. Bull. Am. Meteorol. Soc. 99 (8), S143.

Riebesell, U., Gattuso, J.P., 2015. Lessons learned from ocean acidification research. Nat. Clim. Chang. 5 (1), 12–14.

Riebesell, U., Gattuso, J.P., Thingstad, T.F., Middelburg, J.J., 2013. Preface "Arctic ocean acidification: pelagic ecosystem and biogeochemical responses during a mesocosm study". Biogeosciences 10 (8), 5619–5626.

Rilov, G., Peleg, O., Guy-Haim, T., Yeruham, E., 2020. Community dynamics and ecological shifts on Mediterranean vermetid reefs. Mar. Environ. Res. 160, 105045.

Rinkevich, B., 2015. Climate change and active reef restoration—ways of constructing the "reefs of tomorrow". J. Mar. Sci. Eng. 3 (1), 111–127.

Rinkevich, B., 2019. Coral chimerism as an evolutionary rescue mechanism to mitigate global climate change impacts. Glob. Chang. Biol. 25 (4), 1198–1206.

Riser, S.C., Freeland, H.J., Roemmich, D., Wijffels, S., Troisi, A., Belbéoch, M., Gilbert, D., Xu, J., Pouliquen, S., Thresher, A., Le Traon, P.Y., 2016. Fifteen years of ocean observations with the global Argo array. Nat. Clim. Chang. 6 (2), 145–153.

Rivero-Calle, S., Gnanadesikan, A., Del Castillo, C.E., Balch, W.M., Guikema, S.D., 2015. Multidecadal increase in North Atlantic coccolithophores and the potential role of rising CO2. Science. 350 (6267), 1533–1537.

Robbins, L.L., Wynn, J.G., Lisle, J.T., Yates, K.K., Knorr, P.O., Byrne, R.H., Liu, X., Patsavas, M.C., Azetsu-Scott, K., Takahashi, T., 2013. Baseline monitoring of the Western Arctic Ocean estimates 20% of Canadian Basin surface waters are undersaturated with respect to aragonite. PLoS One 8 (9), e73796.

Roberts, J.M., Cairns, S.D., 2014. Cold-water corals in a changing ocean. Curr. Opin. Environ. Sustain. 7, 118–126.

Roberts, C.M., McClean, C.J., Veron, J.E., Hawkins, J.P., Allen, G.R., McAllister, D.E., Mittermeier, C.G., Schueler, F. W., Spalding, M., Wells, F., Vynne, C., Werner, T.B., 2002. Marine biodiversity hotspots and conservation priorities for tropical reefs. Science 295 (5558), 1280–1284.

Robinson, C., 2019. Microbial respiration, the engine of ocean deoxygenation. Front. Mar. Sci. 5, 533.

Robinson, L.M., Gledhill, D.C., Moltschaniwskyj, N.A., Hobday, A.J., Frusher, S., Barrett, N., Stuart-Smith, J., Pecl, G. T., 2015. Rapid assessment of an ocean warming hotspot reveals "high" confidence in potential species' range extensions. Glob. Environ. Chang. 31, 28–37.

Rocha, J., Yletyinen, J., Biggs, R., Blenckner, T., Peterson, G., 2015. Marine regime shifts: drivers and impacts on ecosystems services. Philos. Trans. R. Soc. B 370, 20130273.

Rogers, L.A., Dougherty, A.B., 2019. Effects of climate and demography on reproductive phenology of a harvested marine fish population. Glob. Chang. Biol. 25 (2), 708–720.

Rosentreter, J.A., Al-Haj, A.N., Fulweiler, R.W., Williamson, P., 2021. Methane and nitrous oxide emissions complicate coastal blue carbon assessments. Global Biogeochem. Cycles 35 (2), e2020GB006858.

Ruiz, J.M., Marín-Guirao, L., García-Muñoz, R., Ramos-Segura, A., Bernardeau-Esteller, J., Pérez, M., Sanmarti, N., Ontoria, Y., Romero, J., Arthur, R., Alcoverro, T., Procaccini, G., 2018. Experimental evidence of warming-induced flowering in the Mediterranean seagrass *Posidonia oceanica*. Mar. Pollut. Bull. 134, 49–54.

Russell, B.J., Dierssen, H.M., Hochberg, E.J., 2019. Water column optical properties of Pacific coral reefs across geomorphic zones and in comparison to offshore waters. Remote Sens. 11 (15), 1757.

Rutzen, I., Hopcroft, R.R., 2018. Abundance, biomass and community structure of epipelagic zooplankton in the Canada Basin. J. Plankton Res. 40 (4), 486–499.

Saintilan, N., Wilson, N.C., Rogers, K., Rajkaran, A., Krauss, K.W., 2014. Mangrove expansion and salt marsh decline at mangrove poleward limits. Glob. Chang. Biol. 20 (1), 147–157.

Salinas, C., Duarte, C.M., Lavery, P.S., Masque, P., Arias-Ortiz, A., Leon, J.X., Callaghan, D., Kendrick, G.A., Serrano, O., 2020. Seagrass losses since mid-20th century fuelled $CO_2$ emissions from soil carbon stocks. Glob. Chang. Biol. 26 (9), 4772–4784.

Sanford, E., Sones, J.L., García-Reyes, M., Goddard, J.H., Largier, J.L., 2019. Widespread shifts in the coastal biota of northern California during the 2014–2016 marine heatwaves. Sci. Rep. 9 (1), 1–14.

Sato, K.N., Levin, L.A., Schiff, K., 2017. Habitat compression and expansion of sea urchins in response to changing climate conditions on the California continental shelf and slope (1994–2013). Deep-Sea Res. II Top. Stud. Oceanogr. 137, 377–389.

Scanes, E., Scanes, P.R., Ross, P.M., 2020. Climate change rapidly warms and acidifies Australian estuaries. Nat. Commun. 11 (1), 1–11.

Scapini, F., Degli, E.I., Defeo, O., 2019. Behavioral adaptations of sandy beach macrofauna in face of climate change impacts: a conceptual framework. Estuar. Coast. Shelf Sci. 225, 106236.

Schaefer, N., Mayer-Pinto, M., Griffin, K.J., Johnston, E.L., Glamore, W., Dafforn, K.A., 2020. Predicting the impact of sea-level rise on intertidal rocky shores with remote sensing. J. Environ. Manag. 261, 110203.

Schlunegger, S., Rodgers, K.B., Sarmiento, J.L., Ilyina, T., Dunne, J.P., Takano, Y., Christian, J.R., Long, M.C., Frölicher, T.L., Slater, R., Lehner, F., 2020. Time of emergence and large ensemble intercomparison for ocean biogeochemical trends. Glob. Biogeochem. Cycles 34 (8). e2019GB006453.

Schmidt, K., Birchill, A.J., Atkinson, A., Brewin, R.J., Clark, J.R., Hickman, A.E., Johns, D.G., Lohan, M.C., Milne, A., Pardo, S., Polimene, L., 2020. Increasing picocyanobacteria success in shelf waters contributes to long-term food web degradation. Glob. Chang. Biol. 26, 5574–5587.

Schmidtko, S., Stramma, L., Visbeck, M., 2017. Decline in global oceanic oxygen content during the past five decades. Nature 542, 335–339.

Schoeman, D.S., Schlacher, T.A., Defeo, O., 2014. Climate-change impacts on sandy-beach biota: crossing a line in the sand. Glob. Chang. Biol. 20 (8), 2383–2392.

Schoenrock, K.M., Chan, K.M., O'Callaghan, T., O'Callaghan, R., Golden, A., Krueger Hadfield, S.A., Power, A.M., 2020. A review of subtidal kelp forests in Ireland: from first descriptions to new habitat monitoring techniques. Ecol. Evol. 10 (13), 6819–6832.

Schoepf, V., Carrion, S.A., Pfeifer, S.M., Naugle, M., Dugal, L., Bruyn, J., McCulloch, M.T., 2019. Stress-resistant corals may not acclimatize to ocean warming but maintain heat tolerance under cooler temperatures. Nat. Commun. 10 (1), 1–10.

Schönberg, C.H., Fang, J.K., Carreiro-Silva, M., Tribollet, A., Wisshak, M., 2017. Bioerosion: the other ocean acidification problem. ICES J. Mar. Sci. 74 (4), 895–925.

Schuerch, M., Spencer, T., Temmerman, S., Kirwan, M.L., Wolff, C., Lincke, D., McOwen, C.J., Pickering, M.D., Reef, R., Vafeidis, A.T., Hinkel, J., Nicholls, R.J., Brown, S., 2018. Future response of global coastal wetlands to sea-level rise. Nature 561 (7722), 231–234.

Schwalm, C.R., Glendon, S., Duffy, P.B., 2020. RCP 8.5 tracks cumulative $CO_2$ emissions. Proc. Natl. Acad. Sci. USA 117, 19656–19657.

Seabed2030 (Nippon Foundation-GEBCO Seabed 2030 Project), 2020. Nearly A Fifth of the World's Ocean Floor Now Mapped. https://seabed2030.org/news/nearly-fifth-worlds-ocean-floor-now-mapped. (Accessed 13 November 2020).

Scott, A.L., York, P.H., Duncan, C., Macreadie, P.J., Connolly, R.M., Ellis, M.T., Jarvis, J.C., Jinks, K.I., Marsh, H., Rasheed, M.A., 2018. The role of herbivory in structuring tropical seagrass ecosystem service delivery. Front. Plant Sci. 9, 127.

Seitzinger, S.P., Mayorga, E., Bouwman, A.F., Kroeze, C., Beusen, A.H.W., Billen, G., Van Drecht, G., Dumont, E., Fekete, B.M., Garnier, J., Harrison, J.A., 2010. Global river nutrient export: a scenario analysis of past and future trends. Glob. Biogeochem. Cycles 24, GB0A08.

Servino, R.N., de Oliveira Gomes, L.E., Bernardino, A.F., 2018. Extreme weather impacts on tropical mangrove forests in the Eastern Brazil Marine Ecoregion. Sci. Total Environ. 628, 233–240.

Shahgenova, M., 2021. Climate change and melting glaciers. Chapter 3, In: Letcher, T. (Ed.), The Impacts of Climate Change: A Comprehensive Study of Physical, Biophysical, Social and Political Issues. Elsevier.

Short, F., Carruthers, T., Dennison, W., Waycott, M., 2007. Global seagrass distribution and diversity: a bioregional model. J. Exp. Mar. Biol. Ecol. 350 (1–2), 3–20.

Sigler, M.F., Mueter, F.J., Bluhm, B.A., Busby, M.S., Cokelet, E.D., Danielson, S.L., De Robertis, A., Eisner, L.B., Farley, E.V., Iken, K., Kuletz, K.J., Lauth, R.R., Logerwell, E.A., Pinchuk, A.I., 2017. Late summer zoogeography of the northern Bering and Chukchi seas. Deep-Sea Res. II Top. Stud. Oceanogr. 135, 168–189.

Simard, M., Fatoyinbo, L., Smetanka, C., Rivera-Monroy, V.H., Castañeda-Moya, E., Thomas, N., Van der Stocken, T., 2019. Mangrove canopy height globally related to precipitation, temperature and cyclone frequency. Nat. Geosci. 12 (1), 40–45.

Sims, D.W., 2019. The significance of ocean deoxygenation for Elasmobranchs. In: Laffoley, D., Baxter, J.M. (Eds.), Ocean Deoxygenation: Everyone's Problem—Causes, Impacts, Consequences and Solutions. IUCN, Gland, Switzerland, pp. 431–448.

Sippo, J.Z., Lovelock, C.E., Santos, I.R., Sanders, C.J., Maher, D.T., 2018. Mangrove mortality in a changing climate: an overview. Estuar. Coast. Shelf Sci. 215, 241–249.

Skliris, N., 2019. The Mediterranean is getting saltier: from the past to the future. In: Orejas, C., Jiménez, C. (Eds.), Mediterranean Cold-Water Corals: Past, Present and Future. Springer, Cham, pp. 507–512.

Smale, D.A., Vance, T., 2016. Climate-driven shifts in species' distributions may exacerbate the impacts of storm disturbances on North-east Atlantic kelp forests. Mar. Freshw. Res. 67 (1), 65–74.

Smale, D.A., Wernberg, T., Oliver, E.C., Thomsen, M., Harvey, B.P., Straub, S.C., Burrows, M.T., Alexandre, L.V., Benthuysem, J.A., Donat, M.G., Feng, M., Hoibday, A.J., Holbrook, N.J., Perkins-Kirkpatrick, S.E., Scannell, H. A., Gupta, A.S., Payne, B.L., Moore, P.J., 2019. Marine heatwaves threaten global biodiversity and the provision of ecosystem services. Nat. Clim. Chang. 9 (4), 306–312.

Smith, C.R., De Leo, F.C., Bernardino, A.F., Sweetman, A.K., Arbizu, P.M., 2008. Abyssal food limitation, ecosystem structure and climate change. Trends Ecol. Evol. 23 (9), 518–528.

Smith, J.N., De'ath, G., Richter, C., Cornils, A., Hall-Spencer, J.M., Fabricius, K.E., 2016. Ocean acidification reduces demersal zooplankton that reside in tropical coral reefs. Nat. Clim. Chang. 6 (12), 1124–1129.

Smith, K.L., Ruhl, H.A., Huffard, C.L., Messié, M., Kahru, M., 2018. Episodic organic carbon fluxes from surface ocean to abyssal depths during long-term monitoring in NE Pacific. Proc. Natl. Acad. Sci. USA 115 (48), 12235–12240.

Song, H.J., Lee, J.H., Kim, G.W., Ahn, S.H., Joo, H.M., Jeong, J.Y., Yang, E.J., Kang, S.H., Lee, S.H., 2016. *In-situ* measured primary productivity of ice algae in Arctic sea ice floes using a new incubation method. Ocean Sci. J. 51 (3), 387–396.

Sorte, C.J.B., Davidson, V.E., Franklin, M.C., Benes, K.M., Doellman, M.M., Etter, R.J., Hannigan, R.E., Lubchenco, J., Menge, B.A., 2017. Long-term declines in an intertidal foundation species parallel shifts in community composition. Glob. Chang. Biol. 23 (1), 341–352.

Spooner, P.T., Thornally, D.J.R., Oppo, D.W., Fox, A.D., Radionovskaya, S., Rose, N.L., Mallett, R., Cooper, E., Roberts, J.M., 2020. Exceptional 20th century ocean circulation in the Northeast Atlantic. Geophys. Res. Lett. 47, e2020GL087577.

Springer, K., Lütz, C., Lütz-Meindl, U., Wendt, A., Bischof, K., 2017. Hyposaline conditions affect UV susceptibility in the Arctic kelp *Alaria esculenta* (Phaseophyceae). Phycologia 56 (6), 675–685.

Stapleton, S., Peacock, E., Garshelis, D., 2016. Aerial surveys suggest long-term stability in the seasonally ice-free Foxe Basin (Nunavut) polar bear population. Mar. Mamm. Sci. 32 (1), 181–201.

Stark, J.S., Peltzer, E.T., Kline, D.I., Queirós, A.M., Cox, T.E., Headley, K., Barry, J., Gazeau, F., Runcie, J.W., Widdicombe, S., Milnes, M., 2019. Free Ocean $CO_2$ Enrichment (FOCE) experiments: scientific and technical recommendations for future in situ ocean acidification projects. Prog. Oceanogr. 172, 89–107.

Steckbauer, A., Klein, S.G., Duarte, C.M., 2020. Additive impacts of deoxygenation and acidification threaten marine biota. Glob. Chang. Biol. 26 (10), 5602–5612.

Steele, M., Ermold, W., Zhang, J., 2008. Arctic Ocean surface warming trends over the past 100 years. Geophys. Res. Lett. 35 (2), L02614.

Steffen, W., Rockström, J., Richardson, K., Lenton, T.M., Folke, C., Liverman, D., Summerhayes, C.P., Barnosky, A.D., Cornell, S.E., Crucifix, M., Donges, J.F., 2018. Trajectories of the earth system in the anthropocene. Proc. Natl. Acad. Sci. USA 115, 8252–8259.

Steinberg, D.K., Ruck, K.E., Gleiber, M.R., Garzio, L.M., Cope, J.S., Bernard, K.S., Stammerjohn, S.E., Schofield, O.M., Quetin, L.B., Ross, R.M., 2015. Long-term (1993–2013) changes in macrozooplankton off the Western Antarctic Peninsula. Deep-Sea Res. I Oceanogr. Res. Pap. 101, 54–70.

Steiner, N.S., Cheung, W.W.L., Cisneros-Montemayor, A.M., Drost, H., Hayashida, H., Hoover, C., Lam, J., Sou, T., Sumaila, U.R., Suprenad, P., Tai, T.C., VanderZwaag, D.L., 2019. Impacts of the changing ocean-ice system on the key forage fish Arctic cod (*Boreogadus saisa*) and subsistence fisheries in the western Canadian Arctic—evaluating linked climatre, ecosystem and economic (CEE) models. Front. Mar. Sci. 6, 179.

Stramma, L., Schmidtko, S., 2019. Global evidence of ocean deoxygenation. In: Laffoley, D., Baxter, J.M. (Eds.), Ocean Deoxygenation: Everyone's Problem—Causes, Impacts, Consequences and Solutions. IUCN, Gland, Switzerland, pp. 25–36.

Straub, S.C., Wernberg, T., Thomsen, M.S., Moore, P.J., Burrows, M., Harvey, B.P., Smale, D.A., 2019. Resistance to obliteration; responses of seaweeds to marine heatwaves. Front. Mar. Sci. 6, 763.

Streck, C., 2020. The mirage of Madrid: elusive ambition on the horizon. Clim. Pol. 20 (2), 143–148.

Strydom, S., Murray, K., Wilson, S., Huntley, B., Rule, M., Heithaus, M., Bessey, C., Kendrick, G.A., Burkholder, D., Fraser, M.W., Zdunic, K., 2020. Too hot to handle: unprecedented seagrass death driven by marine heatwave in a World Heritage Area. Glob. Chang. Biol. 26 (6), 3525–3538.

Sunday, J.M., Fabricius, K.E., Kroeker, K.J., Anderson, K.M., Brown, N.E., Barry, J.P., Connell, S.D., Dupont, S., Gaylord, B., Hall-Spencer, J.M., Klinger, T., 2017. Ocean acidification can mediate biodiversity shifts by changing biogenic habitat. Nat. Clim. Chang. 7 (1), 81–85.

Suprenand, P.M., Ainsworth, C.H., 2017. Trophodynamic effects of climate change-induced alterations to primary production along the western Antarctic Peninsula. Mar. Ecol. Prog. Ser. 569, 37–54.

Sweetman, A.K., Thurber, A.R., Smith, C.R., Levin, L.A., Mora, C., Wei, C.-L., Gooday, A.J., Jones, D.O.B., Rex, M., Yasuhara, M., Ingels, J., Ruhl, H.A., et al., 2017. Major impacts of climate change on deep-sea benthic ecosystems. Elementa (Wash. D.C.) 5, 4.

Sydeman, W.J., Bograd, S.J., 2009. Marine ecosystems, climate and phenology: introduction. Mar. Ecol. Prog. Ser. 393, 185–188.

Tagliabue, A., Barrier, N., Du Pontavice, H., Kwiatkowski, L., Aumont, O., Bopp, L., Cheung, W.W., Gascuel, D., Maury, O., 2020. An iron cycle cascade governs the response of equatorial Pacific ecosystems to climate change. Glob. Chang. Biol. 26 (11), 6168–6179.

Taucher, J., Boxhammer, T., Bach, L.T., Paul, A.J., Schartau, M., Stange, P., Riebesell, U., 2021. Changing carbon-to-nitrogen ratios of organic-matter export under ocean acidification. Nat. Clim. Chang 11, 52–57.

Taylor, J.R., Lovera, C., Whaling, P.J., Buck, K.R., Pane, E.F., Barry, J.P., 2014. Physiological effects of environmental acidification in the deep-sea urchin *Strongylocentrotus fragilis*. Biogeosciences 11 (5), 1413.

Taylor, B.M., Benkwitt, C.E., Choat, H., Clements, K.D., Graham, N.A., Meekan, M.G., 2020. Synchronous biological feedbacks in parrotfishes associated with pantropical coral bleaching. Glob. Chang. Biol. 26 (3), 1285–1294.

Teixidó, N., Caroselli, E., Alliouane, S., Ceccarelli, C., Comeau, S., Gattuso, J.P., Fici, P., Micheli, F., Mirasole, A., Monismith, S.G., Munari, M., Palumbi, S.R., et al., 2020. Ocean acidification causes variable trait shifts in a coral species. Glob. Chang. Biol. 26 (12), 6813–6830.

Terhaar, J., Kwiatkowski, L., Bopp, L., 2020. Emergent constraint on Arctic Ocean acidification in the twenty-first century. Nature 582 (7812), 379–383.

Thackeray, S.J., Henrys, P.A., Hemming, D., Bell, J.R., Botham, M.S., Burthe, S., Helaouet, P., Johns, D.G., Jones, I.D., Leech, D.I., Mackay, E.B., Massimino, D., et al., 2016. Phenological sensitivity to climate across taxa and trophic levels. Nature 535 (7611), 241–245.

Thomsen, J., Casties, I., Pansch, C., Körtzinger, A., Melzner, F., 2013. Food availability outweighs ocean acidification effects in juvenile *Mytilus edulis*: laboratory and field experiments. Glob. Chang. Biol. 19 (4), 1017–1027.

Thomsen, J., Stapp, L.S., Haynert, K., Schade, H., Danelli, M., Lannig, G., Wegner, K.M., Melzner, F., 2017. Naturally acidified habitat selects for ocean acidification–tolerant mussels. Sci. Adv. 3 (4), e1602411.

Tinker, J., Lowe, J., Pardaens, A., Holt, J., Barciela, R., 2016. Uncertainty in climate projections for the 21st century northwest European shelf seas. Prog. Oceanogr. 148, 56–73.

Tittensor, D.P., Beger, M., Boerder, K., Boyce, D.G., Cavanagh, R.D., Cosandey-Godin, A., Crespo, G.O., Dunn, D.C., Ghiffary, W., Grant, S.M., Hannah, L., Halpin, P.N., et al., 2019. Integrating climate adaptation and biodiversity conservation in the global ocean. Sci. Adv. 5 (11), eaay9969.

Toro, N., Robles, P., Jeldres, R.I., 2020. Seabed mineral resources, an alternative for the future of renewable energy: a critical review. Ore Geol. Rev. 126, 103699.

Trainer, V.L., Moore, S.K., Hallegraeff, G., Kudela, R.M., Clement, A., Mardones, J.I., Cochlan, W.P., 2020. Pelagic harmful algal blooms and climate change: lessons from nature's experiments with extremes. Harmful Algae 91, 101591.

Trenberth, K.E., Fasullo, J.T., 2013. An apparent hiatus in global warming? Earth's Future 1 (1), 19–32.

Trolle, D., Nielsen, A., Andersen, H.E., Thodsen, H., Olesen, J.E., Børgesen, C.D., Refsgaard, J.C., Sonnenborg, T.O., Karlsson, I.B., Christensen, J.P., Markager, S., Jeppesen, E., 2019. Effects of changes in land use and climate on aquatic ecosystems: coupling of models and decomposition of uncertainties. Sci. Total Environ. 657, 627–633.

Tunnicliffe, V., Davies, K.T., Butterfield, D.A., Embley, R.W., Rose, J.M., Chadwick Jr., W.W., 2009. Survival of mussels in extremely acidic waters on a submarine volcano. Nat. Geosci. 2 (5), 344–348.

Turner, M.G., Calder, W.J., Cumming, G.S., Hughes, T.P., Jentsch, A., LaDeau, S.L., Lenton, T.M., Shuman, B.N., Turetsky, M.R., Ratajczak, Z., Williams, J.W., Williams, A.P., Carpenter, S.R., 2020. Climate change, ecosystems and abrupt change: science priorities. Philos. Trans. R. Soc. B 375, 20190105.

Ullah, H., Nagelkerken, I., Goldenberg, S.U., Fordham, D.A., 2018. Climate change could drive marine food web collapse through altered trophic flows and cyanobacterial proliferation. PLoS Biol. 16 (1), e2003446.

Ullah, H., Nagelkerken, I., Goldenberg, S.U., Fordham, D., 2020. Combining mesocosms with models to unravel the effects of global warming and ocean acidification on temperate marine ecosystems. EcoEvoRxiv. https://doi.org/10.32942/osf.io/zs78v. Preprints (online).

Valiela, I., Lloret, J., Bowyer, T., Miner, S., Remsen, D., Elmstrom, E., Cogswell, C., Thieler, E.R., 2018. Transient coastal landscapes: rising sea level threatens saltmarshes. Sci. Total Environ. 640, 1148–1156.

van Dijk, I., Bernhard, J.M., de Nooijer, L.J., Nehrke, G., Wit, J.C., Reichart, G.J., 2017. Combined impacts of ocean acidification and dysoxia on survival and growth of four agglutinating foraminifera. J. Foraminifer. Res. 47 (3), 294–303.

Varela, M.R., Patrício, A.R., Anderson, K., Broderick, A.C., DeBell, L., Hawkes, L.A., Tilley, D., Snape, R.T.E., Westoby, M.J., Godley, B.J., 2019. Assessing climate change associated sea-level rise impacts on sea turtle nesting beaches using drones, photogrammetry and a novel GPS system. Glob. Chang. Biol. 25 (2), 753–762.

Vásquez, J.A., Zuñiga, S., Tala, F., Piaget, N., Rodríguez, D.C., Vega, J.A., 2014. Economic valuation of kelp forests in northern Chile: values of goods and services of the ecosystem. J. Appl. Phycol. 26 (2), 1081–1088.

Vergés, A., Doropoulos, C., Malcolm, H.A., Skye, M., Garcia-Pizá, M., Marzinelli, E.M., Campbell, A.H., Ballesteros, E., Hoey, A.S., Vila-Concejo, A., Bozec, Y.M., Steinberg, P.D., 2016. Long-term empirical evidence of ocean warming leading to tropicalization of fish communities, increased herbivory, and loss of kelp. Proc. Natl. Acad. Sci. USA 113 (48), 13791–13796.

Vermeij, G.J., Roopnarine, P.D., 2008. The coming Arctic invasion. Science 321 (5890), 780–781.

Vezzulli, L., Baker-Austin, C., Kirschner, A., Pruzzo, C., Martinez-Urtaza, J., 2020. Global emergence of environmental non-O1/O139 *Vibrio cholerae* infections linked with climate change: a neglected research field? Environ. Microbiol. 22 (10), 4342–4355.

Vousdoukas, M.I., Ranasinghe, R., Mentaschi, L., Plomaritis, T.A., Athanasiou, P., Luijendijk, A., Feyen, L., 2020. Sandy coastlines under threat of erosion. Nat. Clim. Chang. 10 (3), 260–263.

Wabnitz, C.C.C., Lam, V.W.Y., Reygondeau, G., Teh, L.C.L., Al-Abdulrazzak, D., Khalfallah, M., Pauly, D., Palomares, M.L.D., Zeller, D., Cheung, W.W.L., 2018. Climate change impacts on marine biodiversity, fisheries and society in the Arabian Gulf. PLoS One 13 (5), e0194537.

Walcker, R., Gandois, L., Proisy, C., Corenblit, D., Mougin, E., Laplanche, C., Ray, R., Fromard, F., 2018. Control of "blue carbon" storage by mangrove ageing: evidence from a 66-year chronosequence in French Guiana. Glob. Chang. Biol. 24 (6), 2325–2338.

Walsworth, T.E., Schindler, D.E., Colton, M.A., Webster, M.S., Palumbi, S.R., Mumby, P.J., Essington, T., Pinsky, M.L., 2019. Management for network diversity speeds evolutionary adaptation to climate change. Nat. Clim. Chang. 9 (8), 632–636.

Wang, X.Y., Li, X., Zhu, J., Tanajura, C.A., 2018. The strengthening of Amazonian precipitation during the wet season driven by tropical sea surface temperature forcing. Environ. Res. Lett. 13 (9), 094015.

Wang, M., Hu, C., Barnes, B.B., Mitchum, G., Lapointe, B., Montoya, J.P., 2019. The great Atlantic *Sargassum* belt. Science 365 (6448), 83–87.

Watling, L., Guinotte, J., Clark, M.R., Smith, C.R., 2013. A proposed biogeography of the deep ocean floor. Prog. Oceanogr. 111, 91–112.

Watson, E.B., Hinojosa Corona, A., 2017. Assessment of blue carbon storage by Baja California (Mexico) tidal wetlands and evidence for wetland stability in the face of anthropogenic and climatic impacts. Sensors 18 (1), 32.

Watson, R.A., Cheung, W.W., Anticamara, J.A., Sumaila, R.U., Zeller, D., Pauly, D., 2013. Global marine yield halved as fishing intensity redoubles. Fish Fish. 14 (4), 493–503.

Waycott, M., Duarte, C.M., Carruthers, T.J., Orth, R.J., Dennison, W.C., Olyarnik, S., Callardine, A., Fourquean, J.W., Heck, K.L., Hughes, A.R., Kendrick, G.A., Kenworthy, W.J., Short, F.T., Williams, S.A., 2009. Accelerating loss of seagrasses across the globe threatens coastal ecosystems. Proc. Natl. Acad. Sci. USA 106 (30), 12377–12381.

Weller, R.A., Baker, D.J., Glackin, M.M., Roberts, S.J., Schmitt, R.W., Twigg, E.S., Vimont, D.J., 2019. The challenge of sustaining ocean observations. Front. Mar. Sci. 6, 105.

Wernberg, T., Coleman, M.A., Bennett, S., Thomsen, M.S., Tuya, F., Kelaher, B.P., 2018. Genetic diversity and kelp forest vulnerability to climatic stress. Sci. Rep. 8 (1), 1–8.

Wernberg, T., Krumhansl, K., Filbee-Dexter, K., Pedersen, M.F., 2019. Status and trends for the world's kelp forests. In: Sheppard, C. (Ed.), World Seas: An Environmental Evaluation. Elsevier, New York, pp. 57–78.

Whitt, A.A., Coleman, R., Lovelock, C.E., Gillies, C., Ierodiaconou, D., Liyanapathirana, M., Macreadie, P.I., 2020. March of the mangroves: drivers of encroachment into sothern temperate saltmarsh. Est. Coast. Shelf. Sci. 240, 106776.

Wilson, K.L., Skinner, M.A., Lotze, H.K., 2019. Projected 21st-century distribution of canopy-forming seaweeds in the Northwest Atlantic with climate change. Divers. Distrib. 25 (4), 582–602.

Windham-Myers, L., Crooks, S., Troxler, T.G. (Eds.), 2019. A Blue Carbon Primer: The State of Coastal Wetland Carbon Science, Practice and Policy. CRC Press/Taylor & Francis Group, Boca Raton. 481 pp.

Wit, J.C., Davis, M.M., Mccorkle, D.C., Bernhard, J.M., 2016. A short-term survival experiment assessing impacts of ocean acidification and hypoxia on the benthic foraminifer *Globobulimina turgida*. J. Foraminifer. Res. 46 (1), 25–33.

Wittmann, A.C., Pörtner, H.O., 2013. Sensitivities of extant animal taxa to ocean acidification. Nat. Clim. Chang. 3, 995–1001.

Woodhead, A.J., Hicks, C.C., Norström, A.V., Williams, G.J., Graham, N.A., 2019. Coral reef ecosystem services in the Anthropocene. Funct. Ecol. 33 (6), 1023–1034.

Yang, S., Gruber, N., 2016. The anthropogenic perturbation of the marine nitrogen cycle by atmospheric deposition: nitrogen cycle feedbacks and the $^{15}N$ Haber-Bosch effect. Glob. Biogeochem. Cycles 30, 1418–1440.

Yang, S., Yang, Q., Qu, K., Sun, Y., 2017. Regional differences in decadal changes of diatom primary productivity in the eastern Chinese shelf sea over the past 100 years. Quat. Int. 441, 140–146.

Yasuhara, M., Danovaro, R., 2016. Temperature impacts on deep-sea biodiversity. Biol. Rev. 91, 275–287.

York, P.H., Smith, T.M., Coles, R.G., McKenna, S.A., Connolly, R.M., Irving, A.D., Jackson, E.L., McMahon, K., Runcie, J.W., Sherman, C.D.H., Sullivan, B.K., Trevathan-Trackett, S.M., et al., 2017. Identifying knowledge gaps in seagrass research and management: an Australian perspective. Mar. Environ. Res. 127, 163–172.

Zamir, R., Alpert, P., Rilov, G., 2018. Increase in weather patterns generating extreme desiccation events: implications for Mediterranean rocky shore ecosystems. Estuaries Coast 41 (7), 1868–1884.

Zhang, J., Ashjian, C., Campbell, R., Spitz, Y.H., Steele, M., Hill, V., 2015. The influence of sea ice and snow cover and nutrient availability on the formation of massive under-ice phytoplankton blooms in the Chukchi Sea. Deep-Sea Res. II Top. Stud. Oceanogr. 118, 122–135.

Zhang, D., Liu, X., Huang, W., Li, J., Wang, C., Zhang, D., Zhang, C., 2020. Microplastic pollution in deep-sea sediments and organisms of the Western Pacific Ocean. Environ. Pollut. 259, 113948.

Zscheischler, J., Westra, S., Van Den Hurk, B.J., Seneviratne, S.I., Ward, P.J., Pitman, A., Agha Kouchak, A., Bresch, D. N., Leonard, M., Wahl, T., Zhang, X., 2018. Future climate risk from compound events. Nat. Clim. Chang. 8, 469–477.

CHAPTER

# 6

# Natural disasters linked to climate change

*Raktima Dey*[a] *and Sophie C. Lewis*[b]

[a]Fenner School of Environment and Society, Australian National University, Canberra, ACT, Australia [b]School of Science, University of New South Wales, Canberra, ACT, Australia

OUTLINE



## 1 Introduction

The World Health Organization (WHO) defines a disaster as "an act of nature of such magnitude as to create a catastrophic situation in which the day-to-day patterns of life are suddenly disrupted, and people are plunged into helplessness and suffering, and, as a result, food, clothing, shelter, medical and nursing care and other necessities of life, and protection against unfavorable environmental factors and conditions" (WHO, 1971). Natural disasters are often confused with natural hazards, although subtle differences are crucial. Leroy (2020) defines natural hazards as unexpected and uncontrollable, relatively rare events that



*The Impacts of Climate Change*
**https://doi.org/10.1016/B978-0-12-822373-4.00004-5**

have substantial impacts on the environment at a global- or regional-scale, whereas, natural hazards may not have any direct impact on human lives. Thus, all-natural disasters are considered natural hazards; however, the opposite is not valid.

Natural hazards predate human existence. The Earth's atmosphere has undergone rapid changes through geological timescales, temperatures have fluctuated substantially, and the land surface has changed from periods of molten lava at the surface to being covered with a thick layer of ice. Natural hazards have also accompanied human evolution. Archeological and paleoclimate evidence shows that many disasters are linked to the collapse of civilizations, such as the collapse of Norse settlement in Greenland in CE 1540 because of the onset of Little Ice Age, the collapse of ancient Maya Indian civilization in Central America because of extended drought and degradation of the environment which existed between CE 250 and CE 900 (Leroy, 2020).

The Earth's climate has varied significantly through time, mainly because of the natural variability of climate. According to the Intergovernmental Panel on Climate Change (IPCC, 2013), the mechanisms behind natural climate variations include radiation from the Sun and global energy balance, Earth's natural cycle and variability because of its position and distance from the Sun, climate feedbacks, and interactions between components within the climate. Many disasters occur naturally within the climate, and its effects can interact with the climatic components at a large/small scale and lead to a change in long-term climate over a period followed by the disaster. For example, volcanic eruptions can cause a temporary drop in global temperature because of volcanic aerosols injected into the atmosphere. These aerosols reduce sunlight reaching the Earth's surface, hence it causes abrupt cooling. However, to cause a significant change in global temperature, the magnitude of the eruption has been to massive. An enormous eruption of Mount Pinatubo in 1991 in the Philippines led to a drop in global surface temperature by 0.5°C (Parker et al., 1996).

Natural disasters can be triggered by climate extreme events, however, even moderate climate events (e.g., those that are not defined as extreme meteorologically) can lead to large-scale disasters because of a range of reasons including policy failure, social vulnerabilities, and poor disaster management (Roth et al., 2017). The aim of this chapter is to assess historical and future changes in the key driving physical mechanisms of natural disasters from a climate perspective. Understanding changes in these fundamental meteorological variables allows quantification of the future risks of natural disasters because of the occurrence of one or more climate extremes.

Although natural variability plays a key role in modulating natural disasters, the recent increase in greenhouse gas concentration in the atmosphere and the associated changes in climate cannot be explained by the natural variability. The unusual nature of the recent changes will be discussed in-depth in this chapter. In addition, this chapter will examine observed trends in various natural disasters post-1900 (also known as the postinstrumental period where reliable observational record of meteorological parameters started). Various climate extremes associated with natural disasters are discussed in Sections 2–6. Lastly, major conclusions and key research gaps are summarized in Section 7.

# 2 Temperature related disasters

## 2.1 Heatwaves

Humans need to maintain a narrow range of core body temperature between 37°C and 37.2°C to function properly. Our bodies maintain this temperature by releasing and exhausting excessive high or low temperatures. Thus, abnormally high or low temperatures pose a great threat to human health. Heatwaves are one of the deadliest and most frequently occurring natural disasters in the world. Extreme heatwaves lead to high rates of mortality, causes stress to animals and plants, restricts outdoor activities, aggravates bushfire risk, and puts extreme pressure on electricity grids and water management sectors. Due to its highly damaging impacts on society, it is also one of the most studied natural disasters by the scientific community. However, there is no singular definition of heatwaves. The World Meteorological Organization (WMO) define heatwave broadly as a period of scorching temperature that causes physical discomfort, increases mortality and affects other sectors such as electricity, agriculture, and ecology. Although this definition is general and broad, it varies regionally to incorporate human adaptability and local climate.

Most generally, heatwaves are calculated using consecutive 3 days of high maximum, minimum, or mean daily temperature, or a combination of them (Nairn and Fawcett, 2014). There are various ways meteorologists define the "high" temperatures, such as taking fixed thresholds, thresholds relative to the climatological mean, or based on exceedance of percentile values calculated over a period. However, recent studies show that more complex metrics are required to understand heatwave impacts on human lives. For example, consecutive days of high nighttime temperature followed by high daily temperatures create a greater threat to vulnerable people than consecutive days of high maximum temperatures. As in the former case, the heat does not discharge overnight, creating a period of higher discomfort (Nairn and Fawcett, 2014). Similarly, when hot days coincide with high humidity, it restricts our ability to cool off by sweating and reducing internal temperature. Thus, studying just the air temperature (also known as the dry-bulb temperature) does not include the full impacts of a heatwave. Scientists proposed a metric, known as the heat stress, which is more commonly used to study the impacts of heatwaves on human health by taking into account humidity and dry bulb temperature. There are various other metrics available to study heatwaves and impacts, which are more tailored to best suit the local climate and human adaptability (Perkins-Kirkpatrick and Lewis, 2020; Tong et al., 2014).

### *2.1.1 Heatwaves and climate change*

Earth's temperature has varied strongly with periods of warm and cool phases. However, these periods of changes in temperatures can be explained by natural factors such as variations in the global energy balance because of changes in incoming solar radiation because of variations in the orbit (also known as the Milankovitch cycles). Globally, an increase in average temperatures of approximately 1°C has been observed since 1880. The rate of increase is much steeper in recent decades, with two-thirds of the warming being occurring since 1975.[a] The 14 years between 2000 and 2013 are within the 18 warmest years on record which

[a]https://earthobservatory.nasa.gov/world-of-change/global-temperatures.

indicates that the recent increase in temperature is unusual compared to what has been seen in the observational record as well as in paleoclimate context (Hansen and Sato, 2012). Temperature records in many land locations are being broken at an unprecedented rate (Power and Delage, 2019). Irrespective of which specific metric is used, a robust increase in frequency and intensity of heatwaves has been observed (Perkins et al., 2012). Temperature and pressure are key drivers of the global circulation; thus, a slight change in these parameters can cause a shift in atmospheric circulation and water cycle. For example, increased temperature causes an increase in evaporation, resulting in higher moisture in the atmosphere—thus making extreme rainfall more likely. However, these relationships are complex, as an increase in temperature also leads to higher atmospheric stability by increasing the moisture-holding capacity of the atmosphere.

Similar extremes are also observed in ocean temperature measures, which are known as marine heatwaves. They are characterized by abnormally high temperatures that are associated with severe loss of marine life. From 1925 to 2016, globally averaged marine heatwaves frequency and duration have increased by 34% and 17%, respectively (Oliver et al., 2018). Notorious tropical storms are fueled by high atmospheric moisture content and latent heat release from evaporation over the warm ocean surface. Thus, marine heatwaves play a key role in increasing intensity and frequency of tropical storms along with an increase in seasonal extent in which these storms occur. Overall, there is robust evidence of an increase in the numerous characteristics of heatwaves over land and ocean.

### 2.1.2 *Future projections of heatwaves*

Heatwaves characteristics such as intensity, duration, and frequency are also projected to increase globally, with increases likely to be even larger at regional scales. While the human body may be able to adapt to the increase in dry bulb temperature, the threshold for wet bulb temperature, which is a combination of dry-bulb temperature and humidity, remains below 35°C. Climate models show that in a "business as usual emissions scenario" (the representative concentration pathway 8.5, RCP8.5), many regions will cross the critical wet-bulb temperature threshold by 2100, making some regions inhabitable for humans (Pal and Eltahir, 2016). The Paris agreement is a global joint effort to keep the global temperatures rise by the end of this century well below 2°C compared to preindustrial temperatures. An increase in temperature by 1.5°C will lead to 13.8% increase in the world population exposure to severe heatwaves, whereas an increase of 2°C will lead to 36.9% of the population to be exposed to such heatwaves (Dosio et al., 2018). Climate change related damages are assessed using integrated modeling that combines damages in a large range of societal and economic sectors. However, a recent study by Schewe et al. (2019) shows that the majority of the impact models underestimate the severity of the impacts of extreme events in most sectors. This indicates that more research is required to understand and model the role of changes in extreme climate events due to climate change in natural disasters in a future warmer climate.

## 2.2 Cold waves and climate change

Cold waves are defined as a rapid decline in temperature causing damage to crop production, industry, and social activities. There are two major parameters that are important for measuring a cold wave; firstly, the rate at which temperature falls and secondly, the minimum temperature to which it drops, which is dependent on the time of the year and geographic location (American Meteorological Society, 2020). Similar to heatwaves, there are many ways to characterize and describe cold waves. A recent study by Chen et al. (2020a) showed that both heatwave and cold wave increase the cardiovascular mortality rate, while on average heatwave increased mortality rate in 31 cities in China by 19.03% (95% confidence interval: 11.92%–26.59%), cold waves increased cardiovascular mortality rate by 54.72% (95% confidence interval: 21.20%–97.51%). Similar results were found in Spain, where higher mortality rates were associated with cold waves compared to heat waves (Carmona et al., 2016). However, there are more awareness and mitigation plans available for heatwaves, and relatively fewer studies that explore the impacts of cold waves. Thus, a specific cold wave prevention plan at a regional scale is crucial to reduce mortality attributed to cold waves.

Have cold extremes increased when considered in the context of a global, and ubiquitous, increase in mean and maximum temperatures? When looking at the global scale, cold extremes have become milder (van Oldenborgh et al., 2019). Analysis using global station dataset over the last 50 years show that extreme cold temperatures have become milder at a faster rate than the trend in maximum temperatures (Johnson et al., 2018). However, there is still regional and decadal variability observed in the changes in cold extremes. The frequency, intensity and duration of cold waves are projected to decrease; however, there is still considerable uncertainty about the future changes in mortality due to cold waves. Downscaled fifth phase of the coupled model intercomparison project (CMIP5) models show a small decrease in the mortality associated with cold waves in future; however, it is too small to offset increased human mortality because of heatwaves (Carmona et al., 2016).

## 3 Bushfires

Wildfire smoke, also known as bushfire smoke, is known to cause significant respiratory damage. On average, 5%–8% of the total premature deaths from poor quality of air is attributed to wildfire smokes (Lelieveld et al., 2015). An extensive review by Milton and White (2020) summarized the impacts of bushfire on brain health. Particulate matter (PM) from bushfire is more harmful than ambient PM. Bushfire smoke is particularly toxic for neuronal and glial cells, which are essential for the proper functioning of the human body. Besides, prolonged exposure to bushfire smoke could accelerate cognitive decline.

### 3.1 Bushfire and climate change

There are many ways climate change increases the fire risks both in direct and indirect ways. For example, climate change causes warmer than average temperature, increases evaporation, creating a moisture deficit at the surface. Low soil moisture increases the amount of available dry bushfire fuel (defined as dead or living vegetation that modulates fire intensity and speed). Extreme hot and dry conditions are one of the major triggers for the widespread bushfire. Anthropogenic climate change has made bushfires at least 30% more likely than the preindustrial period, mainly driven by extreme heat (van Oldenborgh et al., 2020). Climate change also impacts bushfire season indirectly through impacting large-scale and regional scale drivers. For example, the El Nino Southern Oscillation (ENSO) is the biggest driver of variability in temperature and rainfall. The positive phase of ENSO, El Niño (anomalous high sea surface temperature in the central and east Pacific) and negative phase of ENSO, La Niña (anomalous high sea surface temperature in the west Pacific) leads increase/decrease in rainfall and temperature depending on the region across the globe. Extreme El Niño and La Niña conditions are the dominant underlying drivers of severe weather conditions in many parts of the world. With climate change, the frequency and intensity of extreme ENSO events have increased, and the extreme ENSO events are projected to intensify in future (Cai et al., 2015). Therefore, changes in large-scale driver such as ENSO will increase the future risk of extreme events and the likelihood of natural disasters. A similar driver in the Indian ocean is known as the Indian Ocean Dipole (IOD), where the positive phase of IOD is associated with above average rainfall in east Africa and parts of India and below average rainfall in Australia. Ummenhofer et al. (2009) showed that IOD plays the biggest role in driving droughts and bushfires in Australia. Historically, all major droughts in southeast Australia have been associated with a positive phase of IOD. In recent years, the frequency of positive IOD events has increased while the frequency of negative IOD events has decreased (Cai et al., 2009). In future more intensified positive IOD events with an increase in frequency is expected, which could imply more dry, hot, and severe bushfire conditions for Australia (Abram et al., 2008; Cai et al., 2013). Climate change impacts bushfire severity, both directly and indirectly.

Climate change extends fire weather season in large areas globally by extending the summer type hot weather. An 18.7% increase in fire weather seasonal extent has been observed between 1979 and 2013. The global burnable area has doubled, and the frequency of long fire weather season has increased by 53% of global vegetated land area ( Jolly et al., 2015). One of the bushfire mitigation technologies is prescribed burning; this is a routine procedure in which controlled burning is conducted to reduce the available bushfire fuel in advance of a fire season to reduce bushfire spread. The prescribed burning is conducted generally in cold and relatively wet weather, however, climate change shifts and reduces the window of the time period when it is safe to perform prescribed burning (Di Virgilio et al., 2020). Thus, in addition to making sever fire weather more frequent, climate change affects the implication of mitigation techniques to reduce bushfire risk.

## 3.2 Bushfire projection

As global warming continues, bushfires are expected to intensify, become frequent, and extend in seasonality. Existing research shows that the mortality rate because of respiratory and cardiovascular disease associated with bushfire smoke will increase in future (Reid et al., 2016). Recent research shows that bushfire has a significant impact on brain health, however, to date, there have been very few studies investigating this (Milton and White, 2020). Studying the impacts of bushfire on human health poses a great interdisciplinary challenge. Research efforts are limited because of the poor quality of air quality observation and sparse spatial and temporal coverage (Black et al., 2017). However, understating the long-term health impacts of bushfire smoke in the general population with a focus on the vulnerable community remains a major and urgent research gap that needs to be addressed.

# 4 Sea-level increase

More than two-thirds area of our planet is covered by ocean surface. Almost 40% of the world's population lives within 100 km of the coastline, and the majority of their livelihoods are dependent on coastal and marine resources (Thomson, 2017). Thus, studying sea-level changes and other marine characteristics has huge implications on coastal and marine ecosystems, and the livelihood of coastal communities. In 1848, technologies were first introduced to map ocean currents and wind directions (Verbeek et al., 2020), while first consistent sea-level measurement using tide gauge was recorded in 1831 (Bradshaw et al., 2015). However, observations over the ocean were sparse because of the challenges associated with covering vast ocean areas with observational instruments. The introduction of satellite dataset opened new horizons for oceanographic research. Since the 1970s, researchers have been able to study changes in global sea level, sea surface temperature, ocean ecosystem and many other crucial parameters using high-resolution satellite products.

## 4.1 Sea-level increase and climate change

Climate change has increased the rate of global mean sea-level rise since 1900. Sea level rise has major impacts in coastal and low-lying areas and poses a great threat to the population in these regions by causing major floods, erosion, and coastal inundation. The global sea level has increased by 1.7 mm per year over the period 1900–2010 (Church et al., 2013). Two major reasons of global sea-level increase are: the thermal expansion of water because of the increase in ocean temperatures, and ice melting (such as glaciers, ice sheets, and continental ice shelves)—both are exacerbated due to global warming The sea level rise post-1993 using satellite and ground observation is approximately double (3.1 mm per year) than the long-term increase (Dangendorf et al., 2017). We have experienced an increase of 11–16 cm in global mean sea level over the 20th century, largely driven by climate change. There is much regional variation in sea-level rise; for example, some ocean basins have already experienced 15–20 cm of increase since 1990, much higher than the global average (Church et al., 2013). Currently, 250 million people live below the annual flood line areas; this number is projected to reach 630 million by 2100 (Kulp and Strauss, 2019). Many major coastal cities in the world are experiencing a threat to infrastructure and business because of the increase in the frequency of coastal inundation. Between 1970 and 2015, the frequency of coastal inundation days in Sydney, Australia, has increased from 1.6 to 7.8 days per year. Coastal inundation can be because of daily astronomical tides, tsunami, storm surge, which cause an increase in sea level. By 2050, Sydney is projected to experience weekly coastal inundation, and by 2100, daily (Hague et al., 2020). An increase of 8–16 mm per year is projected by 2081–2100 in climate models (Church et al., 2013). Northwest Europe and Asia are identified as the global "hotspots" where the most significant increase in coastal flooding is expected in future because of storm surge, sea-level rise and diurnal tides (Kirezci et al., 2020). The recent study also showed large increases in global land area, population and global assets at risk because of coastal flooding by 2100 in the representative concentration pathways 8.5 (RCP8.5) scenario. Immediate adaptation measures and coastal defense strategies are required to avoid such large damages in the near future.

# 5 Extreme rainfall events and disasters

## 5.1 Wet extremes

The water cycle is defined as the continuous movement of water on Earth, from vapor state in the atmosphere to precipitation (in the form of rain and snow) on land and water surface, passing through the ground or discharged into water bodies and finally getting back into the atmosphere in a gaseous state. These steps occur through a complex system of different processes and phase change of water. It is difficult to study the changes in components of the water cycle, mainly because observations are less complete, compared to the temperature observational network. However, climate change has likely sped up the increase in some components of the water cycle (Durack et al., 2012). For example, hotter temperature leads to an increase in evaporation, resulting in higher moisture level in the atmosphere. Higher moisture availability increases the chance of frequent intense rainfall events and increases the amount of rain that falls from a system (Trenberth et al., 2003). However, rainfall is one of the most challenging parameters to study because of its highly variable nature. Although both satellites and models show a 7% increase in total atmospheric water with per degree warming of the atmosphere, global mean precipitation increases at a smaller rate, 1%–3% (Held and Soden, 2006). The lower rate of increase in precipitation compared to atmospheric water vapor is a result of the constraint of energy balance in the atmosphere (Pendergrass and Hertmann, 2014; Held and Soden, 2006).

The changes in the hydrological cycle are not spatially homogeneous. A "wet getting wetter" and "dry getting drier" pattern has been already observed globally, where dry regions have experienced a decline in rainfall, and the opposite is true for wet regions. The increase in global mean rainfall is small, and the observed magnitude of the increase is debated (Trenberth, 2011). However, an increase in extreme rainfall has been observed with climate change. "When it rains, it rains harder" type of pattern is now emerging across the globe, even in water limited regions (Contractor et al., 2018; Donat et al., 2019).

Large knowledge gaps still exist in understanding the variability in extreme rainfall within the scientific community. The mechanisms behind the large variability in extreme rainfall include large discrepancy in defining extreme rainfall, data inhomogeneities and regional variations (Pendergrass, 2018, science). Irrespective of the challenges, an abundance of research shows an overall increase in frequency and intensity of extreme rainfall events (Contractor et al., 2018; Donat et al., 2019; Westra et al., 2014). However, most studies simply focus on the intensity and frequency of extreme rainfall events. Recent studies show that focusing on just the intensity and frequency may not show the entire picture (Dey et al., 2020; Pendergrass and Knutti, 2018). For example, an analysis using global station dataset show that more than 50% of stations receive half of the total annual rainfall in just the 12 wettest days of the year. With climate change, the temporal distribution of rainfall is expected to become more skewed, that is, the majority of the rainfall is expected to fall in fewer days. This skewed temporal nature of rainfall can lead to major flooding when it rains and can stretch the period with no rain, risking the crop production. The time of the year when extreme rainfall occurs has shifted to later in the year at a global scale (Marelle et al., 2018), which implies, extreme rainfall is now occurring later in the year in most regions. Consistent with the timing of extreme rainfall, the timing of floods has shifted to later in the year in some regions.

However, in Australia, the timing of extreme rainfall does not show any significant shift as it is largely dependent on the Pacific drivers (ENSO and IPO). These types of information are crucial for the agriculture sector, as this can potentially affect the crop production cycle. More research is required in other regions of the world to understand the shift in the timing of mean and extreme rainfall.

A recent study by Dey et al. (2020) shows that rainfall has become sporadic in Australia, where rainfall events that last for 1–2 days have intensified and have become more frequent. While rainfall events that last >6 days have significantly decreased in Australia. Prolonged long duration events can lead to widespread flooding as well these events are potentially crucial for increasing soil moisture and breaking existing long-term droughts.

The global water cycle is projected to intensify with global warming, which means an increase in both evaporation and precipitation (Allen and Ingram, 2002). However, regional fine-scale projection of rainfall characteristics remains difficult to study as climate models often don't simulate rainfall well compared to temperature. In addition, due to large natural variability in rainfall and short observational record, it is challenging to identify changes in rainfall characteristics in response to climate change. However, intensification of extreme rainfall is becoming more and more evident lately (Hawkins et al., 2020). More in-depth analysis is required using additional rainfall characteristics other than intensity and frequency to understand impacts of extreme rainfall on the society.

Climate change has caused changes in various aspects of extreme events, such as frequency, intensity, duration, spatial extent, and timing. In this section, changes in three major rainfall systems (monsoon, cyclone, and mid-latitude storms) are described in brief, which provides an overview of how rainfall is changing overall. Monsoon is one of the most complex large-scale systems studied in the field of meteorology. This gigantic system is the "lifeblood" for more than half of the world's population (Chang, 2011). Both CMIP5 (Lee and Wang, 2014) and CMIP6 (Chen et al., 2020b) models show an intensification in monsoon rainfall in future, however, no changes in the extent of the monsoonal area. Climate change has caused changes in global circulation patterns. The Hadley cell has expanded because of an increase in stratospheric stability, which pushes the subtropical ridge southward. This poleward expansion of Hadley circulation pushes rain-bearing polar systems further southward, causing widespread droughts in mid-latitude regions (Seneviratne et al., 2012). Cyclones are destructive low-pressure systems formed over tropical oceans, however, named differently depending on the basin they form. Although the consensus is that these systems have reduced in numbers, robust trends in frequency are not available. The possible reasons are poor understanding of cyclones, because of the low quality of data before the satellite era, and large natural variability in cyclone frequency (Lavender and Walsh, 2011). The intensity of cyclones is expected to increase with climate change, although, substantial uncertainty exists (Chen et al., 2020b). Although the frequency of cyclone is projected to reduce in future, the risk associated with these systems will likely exacerbate, mainly because of an increase in population exposure and cyclone intensity (Peduzzi et al., 2012).

## 5.2 Floods and climate change

Between 1980 and 2003, direct economic loss because of floods was greater than $1 trillion, and more than 220,000 people lost their lives because of floods (Winsemius et al., 2016). There are three major types of floods, Fluvial or river flood, Pluvial flood (flash floods and surface water floods), and coastal flood (such as storm surge).[b] Fluvial or river floods are caused by an overflowing river, lake or large water bodies due to excessive rainfall over an extended period or snowmelt. Pluvial floods occur because of extreme rainfall events and result in flooding with or without overflowing a water body. Thus, pluvial floods can occur anywhere, even without a water body. Flash flood is one such example of a pluvial flood that causes massive damage. These types of floods occur because of a large amount of rain falling over a very short time over a relatively small area, causing a localized high level of damage (Wasko et al., 2019).

Although extreme rainfall has intensified and has become more frequent, the relationship between rainfall and floods are not linear. Globally the frequency of long-duration floods (duration >21 days) has increased over the period 1985–2015, the increase is maximum in the northern hemisphere mid-latitudes (Najibi and Devineni, 2018). In the tropical region, flood frequency has increased by four folds since the 2000s. However, Najibi and Devineni (2018) do not find any trend in short duration floods. With a robust increase in short-duration extreme rainfall event intensity and frequency, an increase in flood intensity and frequency is often expected (Yin et al., 2018). However, there is a vast literature with minimum consensus on the changes in various type of floods globally (Wasko et al., 2019; Yin et al., 2018). Global data on catchments shows an increase in extreme rainfall is not followed by a similar increase in extreme streamflow (Wasko and Sharma, 2017). This is mostly because extreme streamflow or flash flooding is dependent on many other characteristics, such as the size of the catchment, rarity of the rainfall event, and existing moisture (Wasko and Sharma, 2017; Wasko et al., 2019). Thus, changes in flash floods with climate change are not straight forward. Although there is still debate about the future changes in flood characteristics depending on its temporal and spatial scale, the flood risk has increased steeply in recent decades (Hirabayashi et al., 2013). Flood risk is defined as a product of the likelihood of flooding and the population and assets at risk to flooding (Merz et al., 2007). Global warming will strongly increase flood risk in future climate due to higher exposure of assets, building infrastructure in flood-prone areas (Hirabayashi et al., 2013). Thus, future mitigation and adaption plans are extremely crucial to avoid large damages from flooding.

## 5.3 Dry extreme events

Drought is one of the most challenging meteorological phenomena to investigate, as it is caused by complex interactions between atmosphere, land, and ocean (Henley et al., 2019). Two most commonly studied droughts are meteorological and hydrological.[c] Meteorological drought is defined as a period of below average rainfall condition, whereas hydrological drought is explained in terms of low streamflow and below average groundwater availability/supply. This chapter will focus on meteorological droughts.

Meteorological droughts can persist from a few weeks, months, years to decades. Because of its slow nature, in some cases, it can be challenging to precisely mark the beginning and end of droughts. Some extreme droughts experienced since the 1900s are the infamous "Dust

[b]https://www.zurich.com/en/knowledge/topics/flood-and-water-damage/three-common-types-of-flood.

[c]https://drought.unl.edu/Education/DroughtIn-depth/TypesofDrought.aspx.

bowl" drought in northern America (1931–39) that started a series of natural disasters (Cook et al., 2009), the California drought during 2011–15, which is considered the worst in the last 1200 years (Griffin and Anchukaitis, 2014), the Millennium drought in Australia, which is unprecedented in the last thousand years (1996–2009) (Vance et al., 2015). Although these types of events are rare in a paleoclimate context, paleoclimate reconstructions show that widespread long-duration droughts were not unusual in the past (Cook et al., 2009; Vance et al., 2015). This indicates that droughts can be much worse than what we have seen in the instrumental period, and our current understanding of droughts using datasets post-1900 may not cover the full range of climate variability.

Globally 42% of the total land area is covered by dryland, home to almost 3 billion population. Dryland is collectively defined as arid, semiarid, and dry subhumid areas. The drylands have increased in area by 9.2% between 1980 and 2000 (Mbow et al., 2017). Increase in drylands has manifolds impacts on human lives and the environment, such as a reduction in agricultural land, mortality due to cardiopulmonary diseases from dust storms, and loss of biodiversity. We have already experienced the ramifications of wet getting wetter and dry getting drier pattern, where an increase in both dry and wet extremes have been observed. An increase in drought frequency, intensity and duration have been seen in the south of Australia, east of Asia, Africa, and the Mediterranean region while a decrease in the three characteristics is observed in Americas and Russia over the period 1951–2010 (Spinoni et al., 2014). There are still large-degree of variability in identifying changes in droughts due to discrepancies in indices used to analyze droughts. Many studies argue that climate change is expected to aggravate droughts, mostly because of the lack of rainfall and an increase in evaporation with increasing temperature in dry regions. These changes are expected to continue in future with global warming.

Historical datasets show that climate change has aggravated the risk of desertification (Peters et al., 2012). The increase in desertification is mainly attributed to the lack of precipitation, strong increase in temperature, and increased evapotranspiration. Climate models suggest that it is highly likely that the increase in desertification will increase in future, given that the prerequisites are projected to continue their observed trends (Collins, 2020).

Multiyear droughts are often a result of a shift in the circulation that causes a decline in mean rainfall because of a shift in rainfall intensity and/or frequency. Recently, South Africa (Cape town drought) experienced the worst drought since the 20th century in 2015, which led to the "Day Zero" water crisis. Cape Town became the first major world city to run out of water, leaving its near 4 million population with extreme water restrictions of 90 s shower and half a gallon of drinking water to each person. The drought started in 2015 and persisted until 2017; the persistent dry conditions were unusual at a centennial scale (Wolski, 2018). Detailed climate scale analysis showed the impact of long-term change in circulation because of climate change and increased population density as the major drivers of the "Day Zero" water crisis. Similar persistent drought has been experienced in southern latitudes in Australia. The south of Australia has experienced a persistent decline in rainfall since 1970 (Pepler et al., 2019). Water usage restrictions have been enforced in Perth, Western Australian, since 2010. The dominant mechanism for rainfall in both south of South Africa and south of Australia are similar, winter cold fronts bringing the majority of the total annual rainfall. Cold fronts are low pressure systems embedded in westerlies from Poles toward mid-latitude, bringing snow and rainfall to the landmass. Recent research shows that because of climate change, Hadley circulation has expanded pushing the polar fronts southward (discussed in Section 2). As a result, fewer fronts reach the land area, reducing rainfall in many areas that receive the majority of rainfall from these fronts. Thus, future projections of drought partly depend on how well the climate models simulate the observed large-scale circulation and their shifts because of climate change. Overall climate models project a further poleward expansion of the Hadley cell, pushing these rain-bearing systems further south—posing a greater risk of droughts of the extended period (Catto et al., 2014).

A newly emerging field of research is flash drought, which has severe impacts on seasonal crop production. Flash droughts are identified as rapid onset of extreme dry conditions because of high temperature, strong winds and high incoming solar radiation, extremely low soil moisture which impedes crop production (Pendergrass et al., 2020). In contrary to droughts, flash droughts can build over a period of 2 weeks and can disappear rapidly. In 2012, Midwest United States, one of the most agriculturally productive regions in the world experienced a drought onset in May and rapidly intensified by July, covering 76% of the crop production region (Jin et al., 2019). This extreme loss of crop production led to an increase in maize prices by 53% compared to the previous 5-year average, and an increase of 146% compared to the average price over 2000–09 (Boyer et al., 2013). Similar droughts have also been observed in China and Australia (Yuan et al., 2015). Given the rapid nature of flash drought, the current seasonal forecast system and drought monitoring systems are unable to predict such droughts; hence extensive focus is required urgently in this research field.

## 6 Compound events

The mathematical definition of a compound event is the probability of occurring more than one event at the same time. In climate science, it translates to the same scenario where more than one extreme event occurs simultaneously, where the risks associated with natural disasters are many fold. Seneviratne et al. (2012) define three scenarios of compound events: (1) two or more extreme events occur successively or at once; (2) concurrent extreme events with underlying conditions that can have high impacts; or (3). two or more events that may not be extreme individually but can result in severe damages. Some historical examples of compound events are: in 2012, Hurricane Sandy made landfall in New Jersey and New York. The hurricane coincided with a high spring tide, resulted in the worst storm surge seen in 300 years, causing total damage of US$50 billion and 223 fatalities (Zscheischler et al., 2018). A recent example is the three consecutive winter rainfall deficit years in east Australia over the period 2017–19, mainly because of a record strong positive Indian Ocean Dipole. Coincident with the severe and protracted drought, Australia experienced its hottest year on record in 2019 with temperature 1.52°C above climatological average (base period 1961–90). Moreover, 2019 was the driest year on record, with the nation receiving 40% below average rainfall (The 2019–20 Bushfires: A CSIRO Explainer, 2020). All these extreme conditions combined together led to a record-breaking bushfire season in Australia, burning 30–40 million hectares of land (the estimation changes depending on the data and method used) (Bowman et al., 2020). The 2019/2020 bushfire smoke related health cost exceeded AU$1.95 billion in Australia, with 429 premature deaths because of smoke, 1523 emergency hospitalization because of Asthma, and 3230 hospital admission because of cardiovascular and respiratory issues (Johnston et al., 2020). Thus, combining the multiyear drought, extreme fires, and health hazards from smoke and heatwaves compounded the socioeconomic impacts. Other examples of compound events are destructions from strong winds, and flooding due to cyclones, storm surges and coastal inundation. Few contrasting compound events are heavy rainfall events after a period of drought, or future projection of an increase in both droughts and heavy precipitation. Compound events can be related to events that are not necessarily related to climate events—for example, drinking water shortage and onset of a pandemic after heavy floods. Although compound events have been identified within climate science for some time, the impacts of these events are often studied individually, leading to underestimation of the risks. Improved projection of compound events and planning in advance can reduce high impact events and associated natural disasters, although multidisciplinary research, implementation and decision making are essential.

## 7 Summary and concluding remarks

The planet is experiencing an increase in heatwave intensity, frequency, and duration. The prediction is that heatstroke, heat stress, hyperthermia, heat exhaustion will increase as a result. Cold waves have become milder and less frequent, however, extreme cold waves still occur in many regions and the future health risk from cold waves remains uncertain. Global mean rainfall is projected to increase by a small amount; however, increases in extreme rainfall characteristics have become more evident. "Wet getting wetter" and "dry getting drier" pattern will continue in future. Changes in extreme rainfall and floods do not always go hand in hand. While long duration widespread flood is projected to increase in future, there is little consensus about the future changes in short duration flash floods. Coastal flooding is projected to become more frequent as a result of global sea-level rise. Drought frequency, intensity, and duration are projected to increase in future, with an increase in the rate of desertification. The rapidly of flash droughts has been investigated by the climate science community over the past couple of years; historical and future changes is an active area of research. Recent research efforts have identified new ways to tackle natural disaster risk by studying the likelihood of compound events; this is again an active area of research. There is a clear signal of climate change in all climate extremes discussed in this chapter. The argument that natural disasters have always existed does not hold anymore, with climate change signals becoming stronger every year. Regional disaster risks as a result of climate change can be many folds higher compared to the global average. The poorest and population-dense countries, especially tropical islands will experience the greatest climate change impacts. United international efforts are essential to prevent poverty increase, mitigate climate change and taking adaptation measures.

## References

Abram, N.J., Gagan, M.K., Cole, J.E., Hantoro, W.S., Mudelsee, M., 2008. Recent intensification of tropical climate variability in the Indian Ocean. Nat. Geosci. 1 (12), 849. https://doi.org/10.1038/ngeo357.

Allen, M.R., Ingram, W.J., 2002. Constraints on future changes in climate and the hydrologic cycle. Nature 419 (6903), 228–232. https://doi.org/10.1038/nature01092.

American Meteorological Society, 2020. "Cold wave". Glossary of Meteorology. https://glossary.ametsoc.org/wiki/Cold_wave. Accessed on 18 Jan. 2021.

Black, C., Tesfaigzi, Y., Bassein, J.A., Miller, L.A., 2017. Wildfire smoke exposure and human health: significant gaps in research for a growing public health issue. Environ. Toxicol. Pharmacol. 55, 186–195. Retrieved from: http://www.sciencedirect.com/science/article/pii/S1382668917302478. https://doi.org/10.1016/j.etap.2017.08.022.

Bowman, D., Williamson, G., Yebra, M., Lizundia-Loiola, J., Pettinari, M.L., Shah, S., Chuvieco, E., 2020. Wildfires: Australia needs national monitoring agency. Nature 584, 188–191.

Boyer, J., Byrne, P., Cassman, K., Cooper, M., Delmer, D., Greene, T., Kenny, N., 2013. The US drought of 2012 in perspective: a call to action. Glob. Food Secur. 2 (3), 139–143.

Bradshaw, E., Rickards, L., Aarup, T., 2015. Sea level data archaeology and the Global Sea Level Observing System (GLOSS). GeoResJ 6, 9–16. https://doi.org/10.1016/j.grj.2015.02.005.

Cai, W., Cowan, T., Sullivan, A., 2009. Recent unprecedented skewness towards positive Indian Ocean Dipole occurrences and its impact on Australian rainfall. Geophys. Res. Lett. 36 (11). https://doi.org/10.1029/2009GL037604.

Cai, W., Zheng, X.-T., Weller, E., Collins, M., Cowan, T., Lengaigne, M., Yamagata, T., 2013. Projected response of the Indian Ocean Dipole to greenhouse warming. Nat. Geosci. 6 (12), 999. https://doi.org/10.1038/ngeo2009.

Cai, W., Wang, G., Santoso, A., McPhaden, M.J., Wu, L., Jin, F.-F., Guilyardi, E., 2015. Increased frequency of extreme La Niña events under greenhouse warming. Nat. Clim. Chang. 5 (2), 132–137. https://doi.org/10.1038/nclimate2492.

Carmona, R., Díaz, J., Mirón, I.J., Ortíz, C., León, I., Linares, C., 2016. Geographical variation in relative risks associated with cold waves in Spain: the need for a cold wave prevention plan. Environ. Int. 88, 103–111. Retrieved from: http://www.sciencedirect.com/science/article/pii/S0160412015301318. https://doi.org/10.1016/j.envint.2015.12.027.

Catto, J., Nicholls, N., Jakob, C., Shelton, K., 2014. Atmospheric fronts in current and future climates. Geophys. Res. Lett. 41 (21), 7642–7650. https://doi.org/10.1002/2014GL061943.

Chang, C.-P., 2011. The Global Monsoon System: Research and Forecast. vol. 5. World Scientific, Singapore.

Chen, J., Zhou, M., Yang, J., Yin, P., Wang, B., Ou, C.-Q., Liu, Q., 2020. The modifying effects of heat and cold wave characteristics on cardiovascular mortality in 31 major Chinese cities. Environ. Res. Lett. 15 (10), 105009.

Chen, Z., Zhou, T., Zhang, L., Chen, X., Zhang, W., Jiang, J., 2020b. Global land monsoon precipitation changes in CMIP6 projections. Geophys. Res. Lett. 47 (14). e2019GL086902.

Church, J.A., Clark, P., Cazenave, A., Gregory, J.M., Jevrejeva, S., Levermann, A., Nunn, P., 2013. Sea level change. Climate change 2013: the physical science basis. In: Contribution of Working Group I to the Fifth Assessment Report of the Intergovernmental Panel on Climate Change. Cambridge University Press, Cambridge, United Kingdom and New York, NY, USA, pp. 1137–1216.

Collins, M., 2020. Extremes, abrupt changes and managing risks. In: Paper Presented at the Ocean Sciences Meeting 2020.

Contractor, S., Donat, M.G., Alexander, L.V., 2018. Intensification of the daily wet day rainfall distribution across Australia. Geophys. Res. Lett. 45, 8568–8576.

Cook, B.I., Miller, R.L., Seager, R., 2009. Amplification of the North American "dust bowl" drought through human-induced land degradation. Proc. Natl. Acad. Sci. 106 (13), 4997–5001.

Dangendorf, S., Marcos, M., Wöppelmann, G., Conrad, C.P., Frederikse, T., Riva, R., 2017. Reassessment of 20th century global mean sea level rise. Proc. Natl. Acad. Sci. 114 (23), 5946–5951.

Dey, R., Gallant, A.J.E., Lewis, S.C., 2020. Evidence of a continent-wide shift of episodic rainfall in Australia. Weather Clim. Extremes 29, 100274. Retrieved from: http://www.sciencedirect.com/science/article/pii/S2212094719302403. https://doi.org/10.1016/j.wace.2020.100274.

Di Virgilio, G., Evans, J.P., Clarke, H., Sharples, J., Hirsch, A.L., Hart, M.A., 2020. Climate change significantly alters future wildfire mitigation opportunities in southeastern Australia. Geophys. Res. Lett. 47 (15). https://doi.org/10.1029/2020gl088893. e2020GL088893.

Donat, M.G., Angélil, O., Ukkola, A.M., 2019. Intensification of precipitation extremes in the world's humid and water-limited regions. Environ. Res. Lett. 14 (6), 065003.

Dosio, A., Mentaschi, L., Fischer, E.M., Wyser, K., 2018. Extreme heat waves under 1.5 C and 2 C global warming. Environ. Res. Lett. 13 (5), 054006.

Durack, P.J., Wijffels, S.E., Matear, R.J., 2012. Ocean salinities reveal strong global water cycle intensification during 1950 to 2000. Science 336 (6080), 455–458.

Griffin, D., Anchukaitis, K.J., 2014. How unusual is the 2012–2014 California drought? Geophys. Res. Lett. 41 (24), 9017–9023.

Hague, B.S., McGregor, S., Murphy, B.F., Reef, R., Jones, D.A., 2020. Sea-level rise driving increasingly predictable coastal inundation in Sydney, Australia. Earth's Future 8. e2020EF001607.

Hansen, J.E., Sato, M., 2012. Paleoclimate implications for human-made climate change. In: Climate Change. Springer, Vienna, pp. 21–47.

Hawkins, E., Frame, D., Harrington, L., Joshi, M., King, A., Rojas, M., Sutton, R., 2020. Observed emergence of the climate change signal: from the familiar to the unknown. Geophys. Res. Lett. 47 (6). e2019GL086259.

Held, I.M., Soden, B.J., 2006. Robust responses of the hydrological cycle to global warming. J. Clim. 19 (21), 5686–5699.

Henley, B., KIng, A., Ukkola, A., Peel, M., Wang, Q.J., Nathan, R., 2019. The Science of Drought Is Complex But the Message on Climate Change Is Clear. Retrieved from: https://theconversation.com/the-science-of-drought-is-complex-but-the-message-on-climate-change-is-clear-125941. (Accessed 10 January 2020).

Hirabayashi, Y., Mahendran, R., Koirala, S., Konoshima, L., Yamazaki, D., Watanabe, S., Kanae, S., 2013. Global flood risk under climate change. Nat. Clim. Chang. 3 (9), 816–821. https://doi.org/10.1038/nclimate1911.

IPCC, 2013. Climate change 2013: the physical science basis. In: Stocker, T.F., Qin, D., Plattner, G.-K., Tignor, M., Allen, S.K., Boschung, J., et al. (Eds.), Contribution of working group I to the fifth assessment report of the intergovernmental panel on climate change. Cambridge University Press, Cambridge.

Jin, C., Luo, X., Xiao, X., Dong, J., Li, X., Yang, J., Zhao, D., 2019. The 2012 flash drought threatened US midwest agroecosystems. Chin. Geogr. Sci. 29 (5), 768–783.

Johnson, N.C., Xie, S.-P., Kosaka, Y., Li, X., 2018. Increasing occurrence of cold and warm extremes during the recent global warming slowdown. Nat. Commun. 9 (1), 1724. https://doi.org/10.1038/s41467-018-04040-y.

Johnston, F.H., Borchers-Arriagada, N., Morgan, G.G., Jalaludin, B., Palmer, A.J., Williamson, G.J., Bowman, D.M.J.S., 2020. Unprecedented health costs of smoke-related PM2.5 from the 2019–20 Australian megafires. Nat. Sustain. https://doi.org/10.1038/s41893-020-00610-5.

Jolly, W.M., Cochrane, M.A., Freeborn, P.H., Holden, Z.A., Brown, T.J., Williamson, G.J., Bowman, D.M.J.S., 2015. Climate-induced variations in global wildfire danger from 1979 to 2013. Nat. Commun. 6 (1), 7537. https://doi.org/10.1038/ncomms8537.

Kirezci, E., Young, I.R., Ranasinghe, R., Muis, S., Nicholls, R.J., Lincke, D., Hinkel, J., 2020. Projections of global-scale extreme sea levels and resulting episodic coastal flooding over the 21st century. Sci. Rep. 10 (1), 11629. https://doi.org/10.1038/s41598-020-67736-6.

Kulp, S.A., Strauss, B.H., 2019. New elevation data triple estimates of global vulnerability to sea-level rise and coastal flooding. Nat. Commun. 10 (1), 1–12.

Lavender, S., Walsh, K., 2011. Dynamically downscaled simulations of Australian region tropical cyclones in current and future climates. Geophys. Res. Lett. 38 (10), L10705.

Lee, J.-Y., Wang, B., 2014. Future change of global monsoon in the CMIP5. Clim. Dyn. 42 (1–2), 101–119.

Lelieveld, J., Evans, J.S., Fnais, M., Giannadaki, D., Pozzer, A., 2015. The contribution of outdoor air pollution sources to premature mortality on a global scale. Nature 525 (7569), 367–371. https://doi.org/10.1038/nature15371.

Leroy, S.A.G., 2020. Natural hazards, landscapes and civilisations. In: Reference Module in Earth Systems and Environmental Sciences., https://doi.org/10.1016/B978-0-12-818234-5.00003-1.

Marelle, L., Myhre, G., Hodnebrog, Ø., Sillmann, J., Samset, B.H., 2018. The changing seasonality of extreme daily precipitation. Geophys. Res. Lett. 45 (20), 11,352–311,360.

Mbow, H.-O.P., Reisinger, A., Canadell, J., O'Brien, P., 2017. Special Report on Climate Change, Desertification, Land Degradation, Sustainable Land Management, Food Security, and Greenhouse Gas Fluxes in Terrestrial Ecosystems (SR2). IPCC, Ginevra.

Merz, B., Thieken, A., Gocht, M., 2007. Flood risk mapping at the local scale: concepts and challenges. In: Flood Risk Management in Europe. Springer, Dordrecht, pp. 231–251.

Milton, L.A., White, A.R., 2020. The potential impact of bushfire smoke on brain health. Neurochem. Int. 139, 104796. Retrieved from: http://www.sciencedirect.com/science/article/pii/S019701862030187X. https://doi.org/10.1016/j.neuint.2020.104796.

Nairn, J.R., Fawcett, R.J., 2014. The excess heat factor: a metric for heatwave intensity and its use in classifying heatwave severity. Int. J. Environ. Res. Public Health 12 (1), 227–253. https://doi.org/10.3390/ijerph120100227.

Najibi, N., Devineni, N., 2018. Recent trends in the frequency and duration of global floods. Earth Syst. Dyn. 9 (2), 757–783. https://doi.org/10.5194/esd-9-757-2018.

Oliver, E.C., Donat, M.G., Burrows, M.T., Moore, P.J., Smale, D.A., Alexander, L.V., Hobday, A.J., 2018. Longer and more frequent marine heatwaves over the past century. Nat. Commun. 9 (1), 1–12.

Pal, J.S., Eltahir, E.A., 2016. Future temperature in Southwest Asia projected to exceed a threshold for human adaptability. Nat. Clim. Chang. 6 (2), 197.

Parker, D.E., Wilson, H., Jones, P.D., Christy, J.R., Folland, C.K., 1996. The impact of Mount Pinatubo on world-wide temperatures. Int. J. Climatol 16, 487–497. https://doi.org/10.1002/(SICI)1097-0088(199605)16:5<487::AID-JOC39>3.0.CO;2-J.

Peduzzi, P., Chatenoux, B., Dao, H., De Bono, A., Herold, C., Kossin, J., Nordbeck, O., 2012. Global trends in tropical cyclone risk. Nat. Clim. Chang. 2 (4), 289–294.

Pendergrass, A.G., 2018. What precipitation is extreme? Science 360 (6393), 1072–1073

Pendergrass, A.G., Hartmann, D.L., 2014. Changes in the distribution of rain frequency and intensity in response to global warming. J. Climate 27 (22), 8372–8383.

Pendergrass, A.G., Knutti, R., 2018. The uneven nature of daily precipitation and its change. Geophys. Res. Lett. 45 (21), 11,980–911,988.

Pendergrass, A.G., Meehl, G.A., Pulwarty, R., Hobbins, M., Hoell, A., AghaKouchak, A., Hoffmann, D., 2020. Flash droughts present a new challenge for subseasonal-to-seasonal prediction. Nat. Clim. Chang. 10 (3), 191–199.

Pepler, A., Hope, P., Dowdy, A., 2019. Long-term changes in southern Australian anticyclones and their impacts. Clim. Dyn. 53 (7–8), 4701–4714.

Perkins, S.E., Alexander, L.V., Nairn, J.R., 2012. Increasing frequency, intensity and duration of observed global heatwaves and warm spells. Geophys. Res. Lett. 39 (20). https://doi.org/10.1029/2012gl053361.

Perkins-Kirkpatrick, S.E., Lewis, S.C., 2020. Increasing trends in regional heatwaves. Nat. Commun. 11 (1), 3357. https://doi.org/10.1038/s41467-020-16970-7.

Peters, D.P., Yao, J., Sala, O.E., Anderson, J.P., 2012. Directional climate change and potential reversal of desertification in arid and semiarid ecosystems. Glob. Chang. Biol. 18 (1), 151–163.

Power, S.B., Delage, F.P.D., 2019. Setting and smashing extreme temperature records over the coming century. Nat. Clim. Chang. 9 (7), 529–534. https://doi.org/10.1038/s41558-019-0498-5.

Reid, C.E., Brauer, M., Johnston, F.H., Jerrett, M., Balmes, J.R., Elliott, C.T., 2016. Critical review of health impacts of wildfire smoke exposure. Environ. Health Perspect. 124 (9), 1334–1343. https://doi.org/10.1289/ehp.1409277.

Roth, F., Eriksen, C., Prior, T., 2017. Understanding the root causes of natural disasters. In: Faculty of Social Sciences-Papers. 3180 https://ro.uow.edu.au/sspapers/3180.

Schewe, J., Gosling, S.N., Reyer, C., Zhao, F., Ciais, P., Elliott, J., Seneviratne, S.I., 2019. State-of-the-art global models underestimate impacts from climate extremes. Nat. Commun. 10 (1), 1–14.

Seneviratne, S., Nicholls, N., Reichstein, M., Sorteberg, A., Vera, C., Zhang, X., 2012. Changes in climate extremes and their impacts on the 1 natural physical environment 2. In: Managing the Risks of Extreme Events and Disasters to Advance Climate Change Adaptation. Cambridge University Press, Cambridge, UK/New York, NY, USA, pp. 109–230.

Spinoni, J., Naumann, G., Carrao, H., Barbosa, P., Vogt, J., 2014. World drought frequency, duration, and severity for 1951–2010. Int. J. Climatol. 34 (8), 2792–2804.

Anon., 2020. The 2019–20 Bushfires: A CSIRO Explainer. CSIRO. Retrieved from: https://www.csiro.au/en/Research/Environment/Extreme-Events/Bushfire/preparing-for-climate-change/2019-20-bushfires-explainer.

Thomson, P., 2017. Ocean Conference: Our Best and Last Chance to Get Things Right Interview—Peter Thomson, President of the UN General Assembly., https://doi.org/10.18356/4081bc1a-en.

Tong, S., Wang, X.Y., FitzGerald, G., McRae, D., Neville, G., Tippett, V., Verrall, K., 2014. Development of health risk-based metrics for defining a heatwave: a time series study in Brisbane, Australia. BMC Public Health 14 (1), 435. https://doi.org/10.1186/1471-2458-14-435.

Trenberth, K.E., 2011. Changes in precipitation with climate change. Clim. Res. 47 (1–2), 123–138.

Trenberth, K.E., Dai, A., Rasmussen, R.M., Parsons, D.B., 2003. The changing character of precipitation. Bull. Am. Meteorol. Soc. 84 (9), 1205–1217.

Ummenhofer, C.C., England, M.H., McIntosh, P.C., Meyers, G.A., Pook, M.J., Risbey, J.S., Taschetto, A.S., 2009. What causes Southeast Australia's worst droughts? Geophys. Res. Lett. 36 (4), L04706.

van Oldenborgh, G.J., Mitchell-Larson, E., Vecchi, G.A., De Vries, H., Vautard, R., Otto, F., 2019. Cold waves are getting milder in the northern midlatitudes. Environ. Res. Lett. 14 (11), 114004.

van Oldenborgh, G.J., Krikken, F., Lewis, S., Leach, N.J., Lehner, F., Saunders, K.R., Otto, F.E.L., 2020. Attribution of the Australian bushfire risk to anthropogenic climate change. Nat. Hazards Earth Syst. Sci. Discuss. 2020, 1–46. https://doi.org/10.5194/nhess-2020-69.

Vance, T., Roberts, J., Plummer, C., Kiem, A., Van Ommen, T., 2015. Interdecadal Pacific variability and eastern Australian megadroughts over the last millennium. Geophys. Res. Lett. 42 (1), 129–137.

Verbeek, L., Fritz, S., Keeble, S., Keeble, K., Wang, D., 2020. Ocean Observation Our Challenge to Learn Everything We Can About the Ocean Environment. Retrieved from: https://www.atlantos-h2020.eu/ocean-observation/#:~:text=Ocean%20observation%20%E2%80%93%20A%20Brief%20History&text=In%201848%20Matthew%20Fontaine%20Maury,America%20quicker%2C%20cheaper%20and%20safer.

Wasko, C., Sharma, A., 2017. Global assessment of flood and storm extremes with increased temperatures. Sci. Rep. 7 (1), 7945. https://doi.org/10.1038/s41598-017-08481-1.

Wasko, C., Sharma, A., Lettenmaier, D.P., 2019. Increases in temperature do not translate to increased flooding. Nat. Commun. 10 (1), 1–3.

Westra, S., Fowler, H., Evans, J., Alexander, L., Berg, P., Johnson, F., Roberts, N., 2014. Future changes to the intensity and frequency of short-duration extreme rainfall. Rev. Geophys. 52 (3), 522–555.

WHO, 1971. Guide to Sanitation in Natural Disasters. Retrieved from: https://www.who.int/environmental_health_emergencies/natural_events/en/.

Winsemius, H.C., Aerts, J.C., Van Beek, L.P., Bierkens, M.F., Bouwman, A., Jongman, B., Van Vuuren, D.P., 2016. Global drivers of future river flood risk. Nat. Clim. Chang. 6 (4), 381–385.

Wolski, P., 2018. How severe is Cape Town's "day zero" drought? Significance 15 (2), 24–27. https://doi.org/10.1111/j.1740-9713.2018.01127.x.

Yin, J., Gentine, P., Zhou, S., Sullivan, S.C., Wang, R., Zhang, Y., Guo, S., 2018. Large increase in global storm runoff extremes driven by climate and anthropogenic changes. Nat. Commun. 9 (1), 1–10.

Yuan, X., Ma, Z., Pan, M., Shi, C., 2015. Microwave remote sensing of short-term droughts during crop growing seasons. Geophys. Res. Lett. 42 (11), 4394–4401.

Zscheischler, J., Westra, S., van den Hurk, B.J.J.M., Seneviratne, S.I., Ward, P.J., Pitman, A., Zhang, X., 2018. Future climate risk from compound events. Nat. Clim. Chang. 8 (6), 469–477. https://doi.org/10.1038/s41558-018-0156-3.

CHAPTER

# 7

# Climate change and microbes

*Stanley Maloy*

San Diego State University, San Diego, CA, United States

OUTLINE



## 1 Introduction

Microbes are critical for life on earth. Microbes catalyze the key biogeochemical cycles that make our planet hospitable—providing forms of carbon and nitrogen that are essential for life. Different microbes can both trap greenhouse gases like carbon dioxide ($CO_2$), methane ($CH_4$), and nitrous oxide ($N_2O$), and produce these greenhouse gases. The balance of these microbial processes plays an important role in the abundance of these chemicals on earth and in our atmosphere.

Microbes have been both changing the climate and responding to climate change since life first developed on earth. As the climate changes, microbes can adapt to the changes or evolve new traits that enhance their survival, playing contradictory roles that can either speed-up climate change or mitigate the impacts of climate change.


https://doi.org/10.1016/B978-0-12-822373-4.00022-7

## 2 Does 1–2° matter?

The mean global temperature increase because of global warming, so far, is only about 0.6–1.2°C (1–2°F). Nevertheless, the global warming that we have already experienced has had major impacts on our planet. Many of these impacts involve microbes. Some examples include: thawing of the Arctic permafrost; altered geographical distribution of insects that act as vectors for microbial diseases in plants and animals; emergence of pathogens in new environments; invasive species that reduce biodiversity and decrease resilience of the ecosystem; nutrient run-off into freshwater lakes where the nutrients promote "blooms" of microbes that are often accompanied by production of toxins that can kill fish, animals, and humans; nutrient run-off into the ocean that promotes microbial growth, reducing oxygen in marine environments and killing fish; and many other examples (Maloy et al., 2016). In addition, these seemingly small temperature changes result in increasing incidence of extreme weather events that promote microbial diseases.

## 3 Generation of greenhouse gases

Microbes both incorporate carbon into large molecules that enter the food web, and break down large organic molecules, recycling waste and dead matter to be reused. Both halves of this process are critical for life on earth with a balanced equilibrium between the incorporation of carbon into organic material and the breakdown of organic materials with the release of $CO_2$ (Maloy et al., 2016).

However, as climate change increases the global temperature, increased metabolism of organic compounds by microbes can stimulate the release of $CO_2$ and $CH_4$ and thereby stimulate the rate of climate change. This is observed in many places, including the arctic, the tropics, and the ocean. For example, there is a tremendous amount of organic material and an equally large number of microbes in the arctic tundra. When covered by layers of ice the metabolism of these microbes is greatly slowed. The permafrost maintained the limited microbial activity. As climate change causes the overlying ice to melt, metabolism of the microbes increases, breaking down the organic material in the tundra and releasing large amounts of greenhouse gases (Jansson and Taş, 2014). Likewise, tropical forests also store a large amount of carbon in the trees, but deforestation and global warming can reverse the uptake of $CO_2$ during photosynthesis, releasing $CO_2$ into the atmosphere (Sullivan et al., 2020).

## 4 Altered geographical distribution of insect vectors

Insects and ticks are often adapted to particular ranges of temperature, humidity, and rainfall, which are influenced by climate change. When the environment changes, they often shift their range—as the global temperature increases many of these vectors have begun to shift to higher elevations or to cooler regions closer to the poles (see Fig. 1; Levy and Patz, 2015).

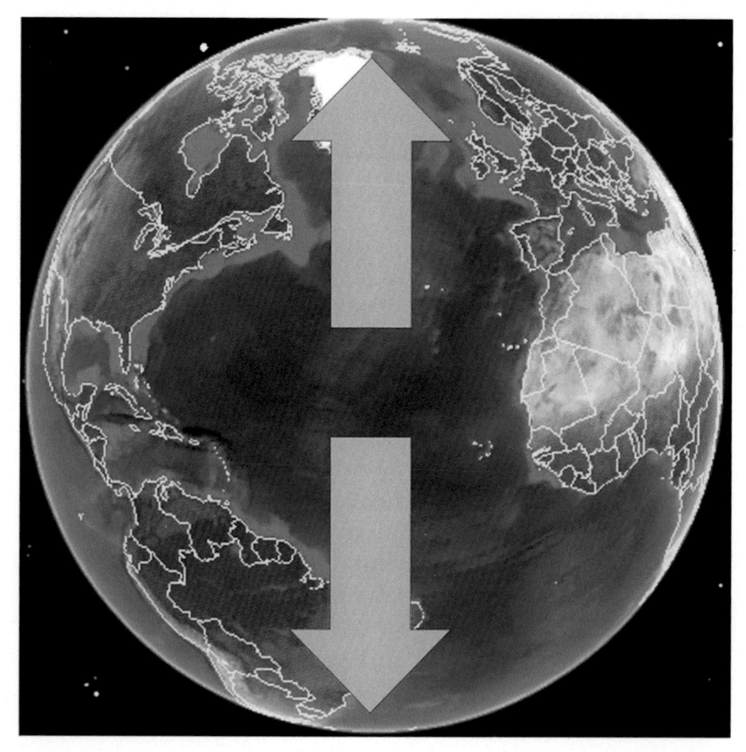

**FIG. 1** As the global temperature increases, many disease vectors are migrating from tropical regions toward the poles.

Mosquitos spread a wide variety of serious microbial diseases, including Malaria, Dengue, Chikungunya, Yellow Fever, Zika, and West Nile, as well as numerous other diseases. These diseases are transmitted to humans through mosquito bites (or from the mosquito's perspective, a blood meal). The growth of mosquitos and the frequency that they take a blood meal is influenced by the temperature (Paaijmans et al., 2010). In addition, mosquitos require standing water to reproduce. For this reason, small changes in temperature and rainfall can influence the range of mosquitos. Global warming has expanded the range of mosquitos to higher altitudes and latitudes. As the temperature has increased, Malaria has become common in high mountain villages in Africa and South America where it was previously rare (Siraj et al., 2014), and other diseases transmitted by mosquitos that were previously limited to the tropics are beginning to appear in temperate climates (Ryan et al., 2019).

Ticks also spread many human diseases, including Lyme Disease, Rocky Mountain Fever, Tularemia, and many others. Lyme disease is caused by an infection with the bacterium Borrelia. The bacteria are transferred to humans and animals by the bite of certain types of ticks. To reproduce, these ticks must take blood meals from multiple hosts, including birds, squirrels, and mice for early development, and deer for later development. If a tick takes a blood meal from a host infected with Borrelia, they can transmit the bacteria to a new host when they take their next blood meal. When the ticks bite humans or dogs, the bacteria can cause an infection that has short-term symptoms, but if untreated, the person can later develop serious chronic diseases including arthritis, heart disease, and neurological symptoms. Like mosquitoes, the tick vectors are very sensitive to temperature and humidity. In regions with winter snow, the ticks do not begin to seek out new hosts until the temperature rises to about 7°C (45°F). As global warming has shortened the winter, the ticks have been emerging earlier—an observation seen in North America, in Northern Europe, and at the opposite pole in South America (Bouchard et al., 2019). The early emergence allows the reproduction of the ticks to go through more reproductive cycles, increasing the transmission to animals and humans and expanding the infection range closer to the two poles. In addition, global warming increases the humidity which enhances the survival of the immature ticks. The rate of reproduction of the ticks that carry Lyme Disease in Old Lyme, Connecticut—the location this disease was named after—nearly doubled between 1980 and 2014 (Ogden et al., 2014).

The shift in insect vectors is not limited to animals. Insect vectors that transmit plant diseases are also increasing in response to climate change (Garrett et al., 2016). Since 1980, the average growing season in the United States has increased by approximately 12 days because of an earlier Spring thaw and later Fall frost. Freezing winter climates at higher latitudes diminished the abundance of agricultural pests by reducing their survival over the winter season, and the shorter growing season limited their reproductive cycles. With earlier warming, both the survival and number of reproductive cycles of insect pests has increased. This both has direct impacts on agriculture and also increases the transmission of plant diseases via insect vectors. In addition, changes in rainfall patterns and increased humidity associated with climate change promote the growth of fungal diseases ("rust") that harm mature plants.

## 5 Altered precipitation

Climate change has also accelerated the hydrological cycle—the process where water evaporates from the ocean, accumulates in clouds that move to a landmass, and release rain that subsequently flows back to the ocean. As the temperature increases, this cycle is disrupted resulting in periodic drought followed by flooding. Both drought and flooding increase the transmission of microbial diseases (Cann et al., 2013).

Flooding can contaminate fresh water supplies with sewage, which increases pathogens like *Salmonella* and *Vibrio cholerae* in drinking water and agricultural water. Although we know how to decontaminate food and water to prevent these diseases, *Salmonella*, pathogenic *E. coli*, *Vibrio*, and many other microbes continue to cause a tremendous amount of illness and death. People in regions of the world with reliable clean water supplies and sewage systems may view these diseases as a rare annoyance, but in many parts of the world where effective sanitary systems are not available, diseases transmitted by food and water are responsible for a large proportion of childhood deaths. The multiple problems that accompany extreme weather events further limit the access to potable water, often for extended periods. Following such natural disasters, water-borne diseases like Cholera kill many people around the world (Relman et al., 2008).

It seems counterintuitive, but drought also increases the incidence of pathogens in water. During a drought, the supplies of potable water are limited, enticing people to store containers of fresh water in containers for their personal use. However, frequent tapping into these water supplies increases the likelihood that an infected person will contaminate the entire container. In addition, animals drawn to dwindling water sources are more likely to remain closer to the water, potentially contaminating the upstream water supply.

Flooding also leaves areas of still water that provides a breeding ground for mosquitos, which promotes the transmission of diseases such as Malaria, Dengue, and West Nile Disease.

Periodic drought followed by periods of heavy rain often increases the population of rodents that can transmit disease. An excellent example of this is the outbreak of a new disease in a remote region where the states of Utah, Colorado, New Mexico, and Arizona meet, called the "Four Corners" (Engelthaler et al., 1999). This rural, arid region had an extended drought for several years in the early 1990s, followed by a heavy snowfall and spring rains that stimulated the rapid growth of plants. The increased vegetation produced abundant seeds that were consumed by deer mice, resulting in a deer mouse population explosion. The increase in mice led to more encounters with humans and transmission of a disease carried by the mice—infection with the Sin Nombre virus was often lethal (Centers for Disease Control and Prevention, National Center for Emerging and Zoonotic Infectious Diseases (NCEZID), Division of High-Consequence Pathogens and Pathology (DHCPP), 2020, https://www.cdc.gov/hantavirus/outbreaks/history.html). As climate change causes more of these drought-flood cycles, we have seen many other examples of enhanced transmission of diseases from animals to humans (see Fig. 2; Atlas and Maloy, 2014).

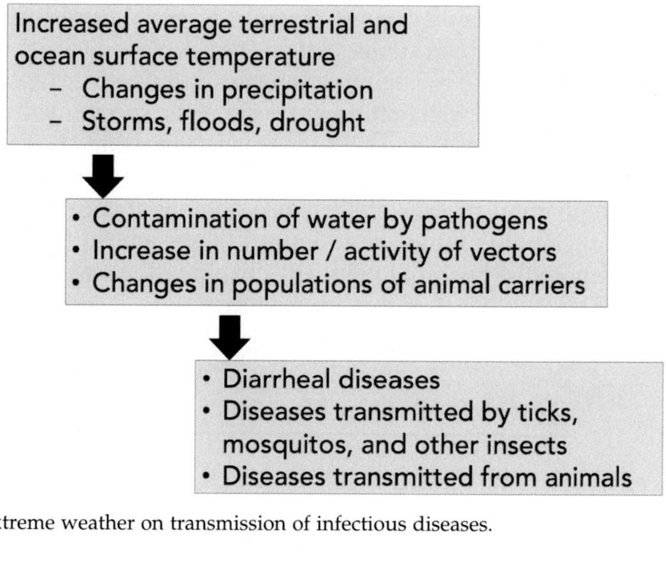

Increased average terrestrial and ocean surface temperature
- Changes in precipitation
- Storms, floods, drought

- Contamination of water by pathogens
- Increase in number / activity of vectors
- Changes in populations of animal carriers

- Diarrheal diseases
- Diseases transmitted by ticks, mosquitos, and other insects
- Diseases transmitted from animals

FIG. 2 Effect of extreme weather on transmission of infectious diseases.

## 6 Elevated ocean temperature

Like the terrestrial temperature, the temperature of the ocean has increased by about 0.6–1.2°C (1–2°F). Does this small increase in ocean temperature matter?

Certain microbes found in seawater can cause serious human diseases. One example is a bacterium called *Vibrio parahaemolyticus*. This pathogen can cause a very serious gastrointestional disease. It accumulates in shellfish and is most often acquired via consumption of infected raw oysters. However, *Vibrio parahaemolyticus* does not grow below 15°C (59°F) so oysters grown in the cold waters off of Alaska had always been safe—that is, until 2002 when the water temperature exceeded 15°C. The water has continued to warm since then, and so these once pristine oyster beds are now a source of occasional outbreaks (McLaughlin et al., 2005).

Another serious pathogen that lives in water is *Vibrio cholerae*, which causes the disease Cholera. Careful analysis of changes in ocean temperature have shown that Cholera outbreaks closely lag increases in water temperature, and the incidence of Cholera is strongly correlated with climate change (de Magny and Colwell, 2009).

## 7 Changes in biodiversity

Changes in temperature and water availability influence the growth and survival of different plants and animals in specific environments. As climate change has altered the physical environment, certain species move to more favorable environments while other species take over their previous habitat. Sometimes people think of this change in biodiversity as a natural occurrence that doesn't have major human consequences. However, changes in biodiversity can impact both agriculture and health. As certain species move out of a habitat that is no longer hospitable, new species can move in to take their place, including agricultural pests and weeds that reduce the yield of important crops.

Biodiversity can also impact animal and human health. A nice example of this is the observed changes in animals occupying different parts of Yosemite National Park (Santos et al., 2017). As the temperature increased at lower elevations, certain species moved to higher elevations. This change in species diversity can have a major impact on the transmission of Lyme disease, because the bacterium cannot reproduce as well in different hosts. If the species diversity shifts to more animals that are sensitive to infection, there will be a greater chance that a tick will bite an infected host and transmit disease. Hence, species diversity can reduce the spread of disease.

The impact of climate change on biodiversity is also a potential problem for microbes themselves. It is possible that climate change could impact the beneficial microbial communities (microbiomes) that promote the health of plants by changing the local environment in ways that restricts growth of key members of the community (Reese, 2016).

## 8 One Health

A unifying concept underlying the impact of climate change on increases in infectious disease is called "One Health." This concept is based on the observation that disruption of the environment has negative impacts on plants, wild and domesticated animals, and humans. The negative impacts may be manifested in many ways, from reducing agricultural productivity to an increase in infectious diseases. The role of the environment in emerging infectious diseases is shown in Fig. 3.

The One Health concept not only describes the relationship between environmental health, animal health, and human health, but also makes predictions that may be important for adaptation to climate change. Instead of identifying and treating disease, the One Health approach focuses on recognizing environmental changes and predicting downstream outcomes (Fig. 4). In some cases, this may allow us to make upstream environmental changes that avoid the negative impacts on agriculture, animals, and humans (Atlas et al., 2010).

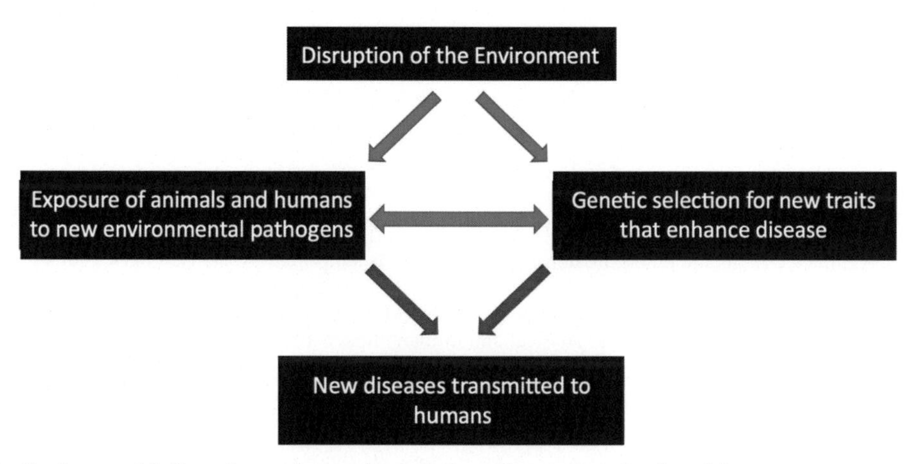


FIG. 3 Environmental disruption and emerging infectious diseases in animals and humans.

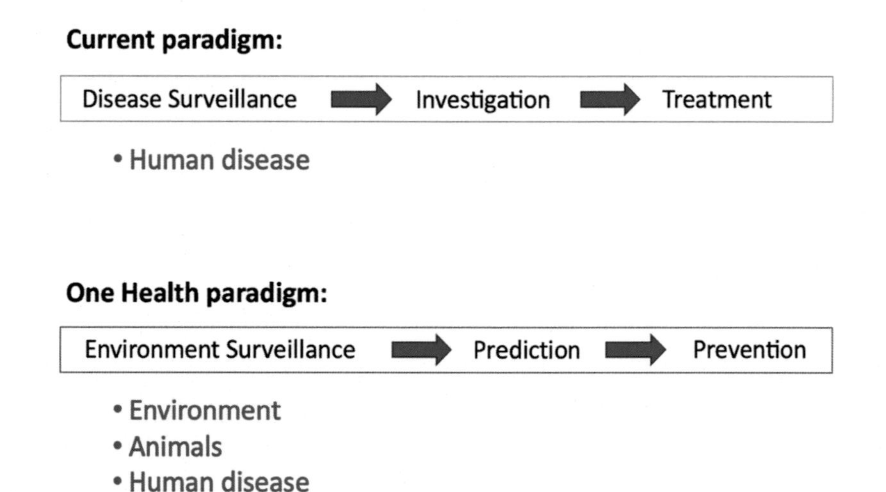


FIG. 4 How the One Health concept can facilitate to changes in infectious disease prevalence in response to climate change.

## 9 Take-home points

Microbes mediate many of the health impacts of climate change. The spread of infectious diseases is amplified and accelerated by the increase in terrestrial and ocean temperatures, extreme weather, flooding and drought, and altered humidity caused by climate change. These impacts may either be driven by directly influencing the microbe or indirectly by influencing a vector that transmits the microbe, the local biodiversity, or even the plant, animal, or human host. Microbes also play beneficial roles in plant, animal, and human health, so the health impacts of climate change could occur by disrupting these beneficial microbiomes.

In addition to their roles in health and disease, microbes play crucial roles in the biogeochemical cycles that are essential for life on earth. This includes consuming greenhouse gases and generating greenhouse gases, depending on the microbe and environment. The ability of microbes to consume greenhouse gases provides promise that microbes can be engineered to efficiently scrub greenhouse gases from the atmosphere (Mangodo et al., 2020). So, in addition to being part of the problem, microbes are likely to be part of the solution.

## Acknowledgments

Our work on climate change and emerging infectious diseases was supported by the WM Keck Foundation (Grant #G00009442). I also want to thank my colleagues Al Sweedler PhD, who is an expert in the physics of climate change, and Nancy Marlin PhD, who is an expert on human behavior. Our interdisciplinary discussions have expanded my understanding of the many nuances of the physical and human impacts of climate change.

## References

Atlas, R., Maloy, S., 2014. One Health: People, Animals, and Environment. ASM Press, Washington, DC.

Atlas, R., Rubin, C., Maloy, S., Daszak, P., Colwell, R., Hyde, B., 2010. One health—attaining optimal health for people, animals, and the environment. Microbe 5, 383–389.

Bouchard, C., Dibernardo, A., Koffi, J., Wood, H., Leighton, P., Lindsay, L., 2019. Increased risk of tick-borne diseases with climate and environmental changes. Can. Commun. Dis. Rep. 45, 81089.

Cann, K., Thomas, D., Salmon, R., Wyn-Jones, A., Kay, D., 2013. Extreme water-related weather events and waterborne disease. Epidemiol. Infect. 141, 671–686.

de Magny, G., Colwell, R., 2009. Cholera and climate: a demonstrated relationship. Trans. Am. Clin. Climatol. Assoc. 120, 119–128.

Engelthaler, D.M., Mosley, D.G., Cheek, J.E., Levy, C.E., Komatsu, K.K., Ettestad, P., Davis, T., Tanda, D., Miller, L., Frampton, J., Porter, R., Bryan, R., 1999. Climatic and environmental patterns associated with hantavirus pulmonary syndrome, four corners region, United States. Emerg. Infect. Dis. 5, 87–94.

Garrett, K., Nita, M., De Wolf, E., Esker, P., Gomez-Montano, L., Sparks, A., 2016. Plant pathogens as indicators of climate change. In: Climate Change. second ed. Elsevier, Oxford, UK, pp. 325–338.

Jansson, J., Taş, N., 2014. The microbial ecology of permafrost. Nat. Rev. Microbiol. 12, 414–425.

Levy, B., Patz, J. (Eds.), 2015. Climate Change and Public Health. Oxford University Press, Oxford, UK.

Maloy, S., Moran, M., Mulholland, M., Sosik, H. (Eds.), 2016. Microbes and Climate Change: Report on an American Academy of Microbiology and American Geophysical Union Colloquium Held in Washington, DC..

Mangodo, C., Adeyemi, T., Bakpolor, V., Adegboyega, D., 2020. Impact of microorganisms on climate change: a review. World News Nat. Sci. 31, 36–47.

McLaughlin, J., DePaola, A., Bopp, C., Martinek, K., Napolili, N., Allison, C., Murray, S., Thompson, E., Bird, M., Middaugh, J., 2005. Outbreak of *Vibrio parahaemolyticus* gastroenteritis associated with Alaskan oysters. N. Engl. J. Med. 353, 1463–1470.

Ogden, N., Radojević, M., Wu, X., Venkata, R., Duvvuri, V., Leighton, P., Wu, J., 2014. Estimated effects of projected climate change on the basic reproductive number of the Lyme disease vector *Ixodes scapularis*. Environ. Health Perspect. 122, 631–638.

Paaijmans, K., Blanford, S., Bell, A., Blanford, J., Read, A., Thomas, M., 2010. Influence of climate on malaria transmission depends on daily temperature variation. Proc. Natl Acad. Sci. USA 107, 15135–15139.

Relman, D., Hamburg, M., Choffnes, E., Mack, A. (Eds.), 2008. Global Climate Change and Extreme Weather Events. National Academies Press, Washington, DC.

Ryan, S.J., Carlson, C.J., Mordecai, E.A., Johnson, L.R., 2019. Global expansion and redistribution of *Aedes*-borne virus transmission risk with climate change. PLoS Negl. Trop. Dis. 13 (3), e0007213.

Santos, M., Smith, A., Thorne, J., Moritz, C., 2017. The relative influence of change in habitat and climate on elevation range limits in small mammals in Yosemite National Park, California, USA. Clim. Change Responses 4, 7–19.

Siraj, A., Santos-Vega, M., Bouma, M., Yadeta, D., Carrascal, D., Pascual, M., 2014. Altitudinal changes in malaria incidence in highlands of Ethiopia and Colombia. Science 43, 1154–1158.

Sullivan, M., Lewis, S., Affum-Baffoe, K., Castilho, C., Costa, F., Sanchez, A., Ewango, C., Hubau, W., Beatriz, M., et al., 2020. Long-term thermal sensitivity of earth's tropical forests. Science 368, 869–874.

# Web references

Centers for Disease Control and Prevention, National Center for Emerging and Zoonotic Infectious Diseases (NCEZID), Division of High-Consequence Pathogens and Pathology (DHCPP), 2020. Tracking a Mystery Disease: The Detailed Story of Hantavirus Pulmonary Syndrome (HPS). https://www.cdc.gov/hantavirus/outbreaks/history.html (Last modified March 11, 2020).

Reese, A., 2016. How climate change endangers microbes—and why that's not a good thing. Sci. Am.. https://blogs.scientificamerican.com/guest-blog/how-climate-change-endangers-microbes-and-why-that-s-not-a-good-thing/ (Last reviewed September 25, 2020).

CHAPTER

# 8

# Effects of climate change on food production (fishing)

*Heike K. Lotze*[a], *Andrea Bryndum-Buchholz*[a], *and Daniel G. Boyce*[b]

[a]Department of Biology, Dalhousie University, Halifax, NS, Canada [b]Ocean Frontier Institute, Dalhousie University, Halifax, NS, Canada

OUTLINE



## 1 Introduction

The warming of the Earth's climate has a range of effects on the biogeochemical properties of the ocean that support marine fisheries, seafood production, and other benefits to human societies (IPBES, 2019; IPCC, 2019a; Letcher, 2016, 2021). There are thousands of marine species that are of commercial importance in fisheries around the world. Together, they supply millions of people with seafood, an important source of nutrition and protein, as well as essential income and livelihoods (FAO, 2020; Hicks et al., 2019). A better understanding of the

*The Impacts of Climate Change*
**https://doi.org/10.1016/B978-0-12-822373-4.00017-3**

effects of climate change on marine species and ocean ecosystems will help project the consequences for fisheries and seafood production into the future and provide insight into necessary changes in fisheries management, ocean governance, and marine conservation (Blanchard et al., 2017; Boyce et al., 2020; Lotze et al., 2019; Tittensor et al., 2019).

The most prominent effects of climate change on the physical properties of the ocean include the increase in water temperatures from the surface to the deep sea, and changes in ocean currents, circulation patterns, stratification, sea level, and the extent of sea ice (Bopp et al., 2013; Kwiatkowski et al., 2020; Worm and Lotze, 2016, 2021). Thereby, increasing climatic variability can lead to more frequent extreme events, including storms and marine heatwaves (Oliver et al., 2018). The most prominent climate-change effects on the ocean's chemical properties include the increase in carbon dioxide resulting in ocean acidification, and the decrease in subsurface oxygen concentrations because of warming and altered circulation patterns (Bopp et al., 2013; Kwiatkowski et al., 2020; Worm and Lotze, 2016, 2021). These physical and chemical changes influence all levels of biology in the ocean (Fig. 1). This includes the basic survival, growth, and reproduction of individual organisms from microscopic plankton to giant whales, as well as the regional and global distribution of populations and species, the composition and dynamics of marine food webs, and the functioning and

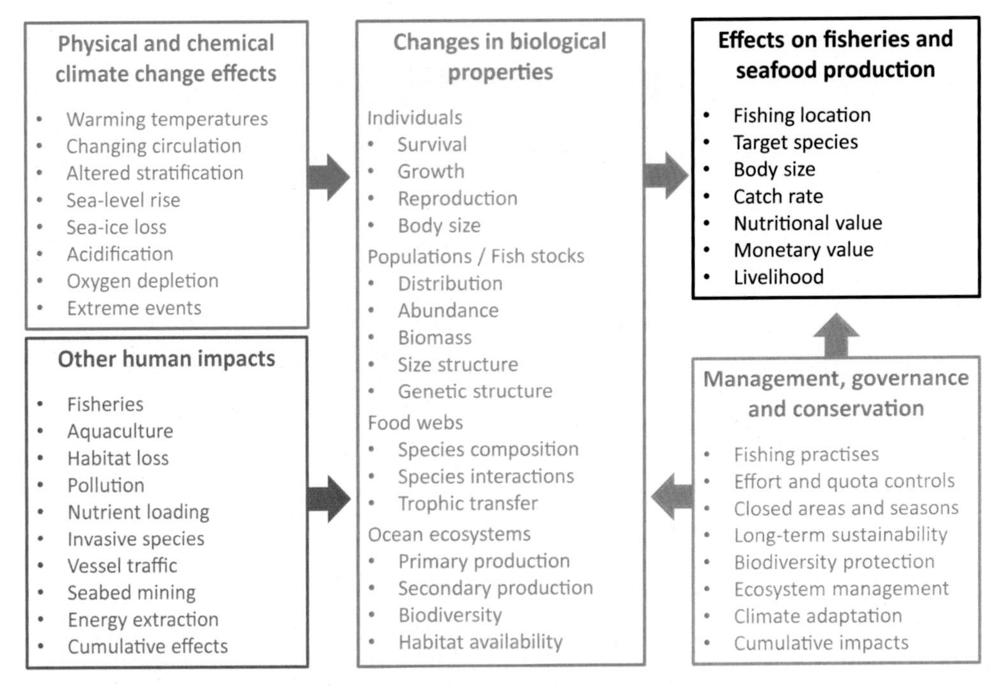


**FIG. 1** Overview of climate change effects and other human impacts on different biological levels of organization in the ocean, and the consequences of those changes for fisheries and seafood production, as well as potential management and governance strategies to mitigate these changes. *Adapted from Lotze, H.K., Froese, R., Pauly, D., 2018. The future of marine fisheries. In: Pauly, D., Ruiz-Leotaud, V. (Eds.), Marine and Freshwater Miscellanea. Fisheries Centre Research Reports. vol. 26, 5–13.*

production of ocean ecosystems (IPCC, 2019a; IPBES, 2019; Worm and Lotze, 2016, 2021) (Chapter 5 this volume). All these changes could affect fisheries and the provision of other ocean ecosystem goods and services.

Over evolutionary time scales, changes in climate regimes are not new. There have been enormous variations over the past millions of years that have shaped life in the ocean, including several mass extinctions (Finnegan et al., 2015; Harnik et al., 2012). Most often, these changes were caused by changing ocean temperatures, either warming or cooling, but also by changes in circulation, chemistry, productivity, and sea-level rise or fall. For example, during the warmer "greenhouse" period (34–66 million years ago), ocean temperatures reached a maximum of 34°C, and marine ecosystems were supported by smaller phytoplankton cells, which constrained the size of large predators (Norris et al., 2013). Currently, we are in another period of rapid climate change, but this time it is accompanied by a suite of other human impacts and, therefore, climate change is not acting in isolation (Fig. 1) (Harnik et al., 2012). Fisheries exert significant pressure on ocean ecosystems around the globe (Pauly and Zeller, 2016; Worm and Branch, 2012), and of all fish stocks that were assessed by the Food and Agriculture Organization in 2017, 34.2% were fished at biologically unsustainable levels (FAO, 2020). In addition, pollution, nutrient loading, habitat alteration, aquaculture, seabed mining, energy production, invasive species, and other human stressors are increasing and expanding in many parts of the ocean (Halpern et al., 2015; McCauley et al., 2015). These multifold human influences are superimposed on climate-change impacts and may lead to complex, unexpected, and unpredictable outcomes—an important difference compared to previous climate-change episodes (Boyce et al., 2020; Crain et al., 2008; Harley et al., 2006; Harnik et al., 2012). Such cumulative stressors may reduce the adaptive capacity of marine species, biodiversity, and fisheries to climate change (Sumaila and Tai, 2020). On the other hand, there have been many improvements in marine management and conservation over past decades that have helped sustain biodiversity and fisheries production in some regions (Hilborn et al., 2020; Worm et al., 2009; Worm and Branch, 2012). All these efforts, however, will need to adapt to climate change to remain successful into the future (Bocye, 2020; Bryndum-Buchholz, 2020; Tittensor et al., 2019; Wilson et al., 2020).

In this chapter, we will first describe observed changes in the biology, distribution, and abundance of marine species, followed by changes in marine food webs and ecosystem production that are relevant to fisheries and seafood production. Next, we describe the consequences of climate change for fisheries and fishing nations around the world. We then evaluate the management of fisheries and ocean governance and how they can adapt to climate change. Lastly, we briefly look into potential marine conservation strategies that can help mitigate some climate-change consequences on ocean ecosystems.

## 2 Biological changes in marine organisms

Global climate change is causing a range of physical and chemical changes in the ocean that influence marine organisms directly or indirectly through changes in their survival, growth, or reproduction (Fig. 1) (Bocye, 2020; Cheung et al., 2013a; Worm and Lotze, 2016, 2021). These changes can act on various temporal and spatial scales and differ among species, life-history stages, or age classes. Most organisms have a physiological tolerance for a range of environmental conditions but also a preferred range or optimum where performance is highest. Therefore, the effects of climate change depend on where organisms are relative to their optimum. For example, ocean warming will often lead to decreases in individual growth and body size in warmer ocean regions due to temperature-dependent metabolism but can enhance growth in colder regions where temperature has limiting effects (Cheung et al., 2013a; Drinkwater, 2005; Sheridan and Bickford, 2011). When increasing water temperatures exceed a species' physiological tolerance, individuals may die or move away, and if warming becomes more common in a specific region, the entire population's distribution, abundance, biomass, size structure, or genetic makeup may change (Fig. 2) (Cheung et al., 2009, 2013a; Pinsky et al., 2013, 2019; Scheffers et al., 2016; Trisos et al., 2020). Moreover, if conditions reach extreme levels, populations may collapse or become extinct in affected regions (Caputi et al., 2016; Cavole et al., 2016; Pershing et al., 2015; Oliver et al., 2017). Thereby, certain species may be sensitive to individual climate-change effects or the cumulative effects of multiple climate drivers (Fig. 1). Often, younger age groups or life-history stages are more sensitive to changes in their environment than adults (Britten et al., 2016; Dahlke et al., 2020). Furthermore, some species may not be directly affected by climate change but indirectly

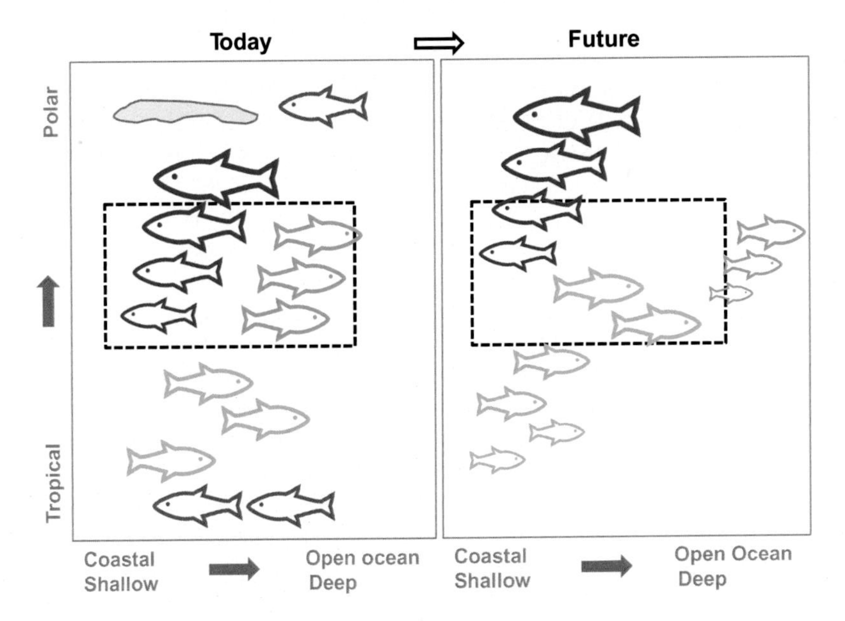


**FIG. 2** Conceptual overview of expected changes in fish communities with climate change. Some specialist species in polar and tropical regions may be faced with extinction *(red)*; many cold-adapted species will shift their distribution poleward *(dark blue)* or into deeper or more offshore waters *(light blue)*; many warm-adapted species will expand their distribution *(green)*, and some species may shift to smaller body sizes *(green, light blue)*. Consequently, the species composition and size structure in a distinct region *(dashed black box)* may change with impacts on local fisheries. *Adapted from Cheung, W.W.L., Sarmiento, J.L., Dunne, J., Frölicher, T.L., Lam, V.W.Y., Palomares, M.L.D., Watson, R., Pauly, D., 2013a. Shrinking of fishes exacerbates impacts of global ocean changes on marine ecosystems. Nat. Clim. Chang. 3, 254–258; Lotze, H.K., Froese, R., Pauly, D., 2018. The future of marine fisheries. In: Pauly, D., Ruiz-Leotaud, V. (Eds.), Marine and Freshwater Miscellanea. Fisheries Centre Research Reports. vol. 26, 5–13.*

through climate-induced changes to their habitat, food availability, predator abundance, or disease prevalence. For example, many open-ocean animals are influenced by changes in the location or strength of thermal fronts or upwelling areas that strongly influence food availability, which affects growth and survival (Worm et al., 2005; Worm and Lotze, 2016). Climate warming can also increase the prevalence of infectious aquatic diseases, which can cause rapid population declines or extinctions (Harvell et al., 2002). Many climate-change effects are superimposed on other human impacts in the ocean (Fig. 1), which can result in additive or synergistic effects that are often difficult to predict (Boyce et al., 2020; Crain et al., 2008; Halpern et al., 2015; McCauley et al., 2015).

All these changes can affect individual organisms as well as entire populations, including those important to commercial fisheries. Over the past decade, some commonly observed changes in fish and fish stocks around the world include reductions in growth and body size (Cheung et al., 2013a; Shackell et al., 2010; Sheridan and Bickford, 2011), decreases in reproduction (Britten et al., 2016), changes in phenology (Asch, 2015; Platt et al., 2003; Poloczanska et al., 2013, 2016), changes in abundance, biomass, and productivity (Britten et al., 2017; Cheung et al., 2010, 2013b; Free et al., 2019; Tai et al., 2019), as well as shifts in spatial distribution including local disappearances (Cheung et al., 2010; MacKenzie et al., 2014; Nye et al., 2009, 2011; Pershing et al., 2015; Pinsky et al., 2013). These changes have important implications for fisheries operations (see Section 6 below), as fishers need to adapt to changes in preferred fishing locations, available target species and body sizes, species-specific and overall catch rates, and the nutritional and monetary value of caught seafood species, all of which have impacts on their livelihoods (Figs. 1 and 2) (Cheung et al., 2010, 2013b; Lotze et al., 2018).

Faced with significant climate change, some species may be able to adapt or show evolutionary responses that may alter their tolerance for warming waters or other climate-change effects. Increasing evidence suggests both short-term acclimatory and long-term adaptive acquisition of climate resistance in some marine species (Scheffers et al., 2016; Worm and Lotze, 2016, 2021). Unfortunately, larger species with slower life histories, such as many commercial fish species, likely adapt more slowly to changing environmental conditions than fast-growing species and may thus be more vulnerable to rapid climate change (Perry et al., 2005). This has important implications for fisheries because many species with slower life histories are already more strongly affected by overexploitation and potentially less able to compensate for added climate-change effects (Dulvy et al., 2003; Hutchings and Reynolds, 2004; Pershing et al., 2015).

## 3 Changes in species distribution and abundance

As a result of climate-induced changes in marine organisms, many populations and species will shift their local, regional, or global distribution and increase or decrease in abundance within certain areas. Range shifts have been observed for many fish and invertebrates that are important for commercial fisheries (Cheung et al., 2009; Jones and Cheung 2015; Shackell et al., 2014), but also in their prey, predators, and habitat-providing species (Worm and Lotze, 2016, 2021) which can alter food-web dynamics (see Section 4 below). In many temperate ocean regions, warming is leading to an influx and increased abundance of warm-adapted species (tropicalization) and to a decline or disappearance of cold-adapted species (Fig. 2). In the North Sea, for example, bottom trawl surveys documented an almost 50% increase in the number of fish species with increasing temperatures from 1985 to 2006, mostly driven by the influx of small-bodied southern species (Hiddink and Ter Hofstede, 2008). Similar trends have been observed in the Gulf of Maine (Friedland et al., 2020) and the Bristol Channel in the United Kingdom (Henderson, 2007). On a regional scale, ocean warming has been associated with poleward range shifts in 17 out of 36 commercial fish species on the Northeast coast of the United States from 1968 to 2007 (Nye et al., 2009) and altered distribution and abundance in 35 out of 50 fish species in the Northeast Atlantic (Simpson et al., 2011). Not all species move poleward and range shifts can be altered by local and regional conditions, with some species moving into deeper or further offshore waters to remain at their preferred temperature (Fig. 2) (Cheung et al., 2013a; Pinsky et al., 2013). In addition, not all species move at the same pace, as some species are more sensitive than others, and species thriving under new abiotic or biotic conditions will increase in abundance or expand their distribution, while those suffering will decrease or shrink their distribution. Such varied species-specific responses have been shown in projections of thermal habitat indices for a range of commercial fish and invertebrate species in Atlantic Canada (Fig. 3) and the Northeast United States with climate change (Shackell et al., 2014). Results suggest that 55% of common commercial species in Atlantic Canada would lose habitat by 2060, compared to 21% gaining habitat and 24% remaining constant (Fig. 3), whereas in the United States, 65% of species would lose habitat, compared to 20% gaining and 15% remaining constant. These changes were much lower by 2030, highlighting the long-term changes that are determined by the current trajectories of climate change (Shackell et al., 2014). Overall, these varied responses among species will lead to a restructuring of local and regional communities and food webs (Worm and Lotze, 2016, 2021).

On a global scale, observed and predicted range shifts generally result in a poleward shift of many marine species (Bocye, 2020; Worm and Lotze, 2016, 2021). Within polar regions, warming leads to the influx of cold-adapted species, whereas sea-ice dependent species may suffer from melting ice sheets (Michel et al., 2012). In tropical regions, warming can result in the disappearance or extinction of species if maximum temperature tolerances are exceeded (Worm and Lotze, 2016, 2021). A recent study estimated that temperatures were projected to exceed the upper thermal tolerances for more than 30,000 marine and terrestrial species beginning in the 2030s, with the most rapid responses in the tropics but also in parts of the Northwest Atlantic (Trisos et al., 2020). Results from species distribution models suggest hotspots of species invasion in polar regions compared to hotspots of species extinctions in the tropics, and areas of high species turnover in temperate to subtropical regions in between (Cheung et al., 2009; Jones and Cheung, 2015). Such shifts in distribution also affect habitat-building species, such as corals, seagrass, kelp, and other macroalgae (Hughes et al., 2017; Wernberg et al., 2011, 2016; Wilson and Lotze, 2019; Wilson et al., 2019), which are essential breeding, spawning, nursery, and foraging grounds for many organisms and provide shelter from predation. Together, all these small to large changes in species distribution and abundance alter biodiversity patterns and habitat availability on local, regional and global scales with consequences for the structure, functions and services of ocean ecosystem (Worm et al., 2006; Worm and Lotze, 2016, 2021).

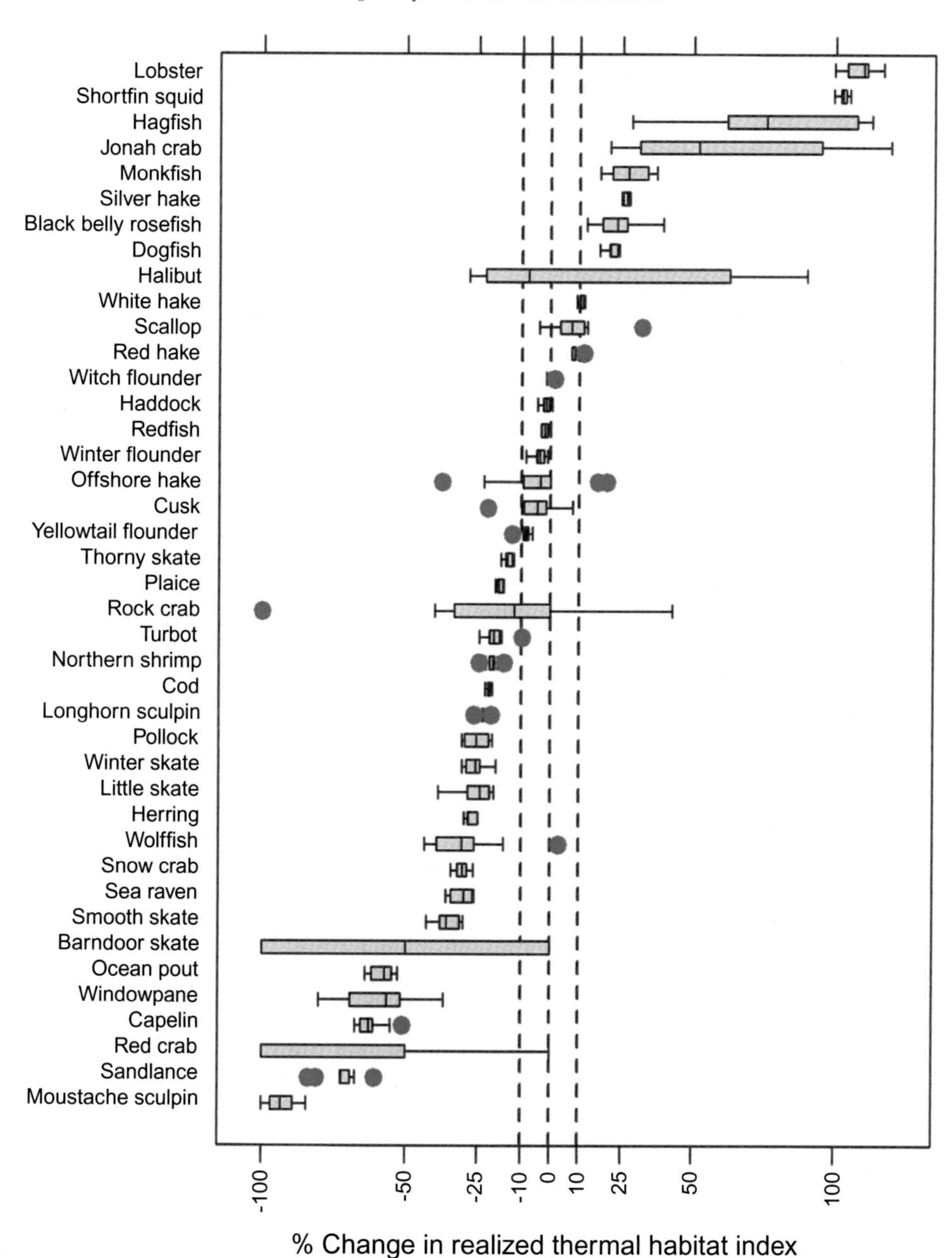


**FIG. 3** Projected regional range shifts in commercial fish and invertebrate species in Atlantic Canada with climate change by the year 2060, expressed as the % change in realized thermal habitat index. Box plots indicate the median *(black line)*, interquartile range *(gray box)*, minimum and maximum *(whiskers)*, and outliers *(blue dots)*; stippled lines indicate ±10% change. *Redrawn with permission from Shackell, N.L., Ricard, D., Stortini, C., 2014. Thermal habitat index of many Northwest Atlantic temperate species stays neutral under warming projected for 2030 but changes radically by 2060. PLoS ONE 9(3), e90662.*

## 4 Changes in marine food webs

The climate-induced changes across a wide suite of species, including commercially important fish and invertebrates but also their prey, predators and competitors will alter food-web dynamics with consequences for fisheries and seafood production (Coll et al., 2020; Lotze et al., 2019). At the bottom of marine food webs, climate change is causing a range of alterations to plankton and microbial communities, most of which are not well understood, but are likely to have repercussions for higher trophic levels (Cavicchioli et al., 2019). For example, phytoplankton are critical to supporting fisheries yet their levels have declined over the past century in association with warming oceans, and climate modeling suggest such declines may continue into the future unless emissions are abated (Boyce et al., 2010, 2014; Boyce and Worm, 2015). Climate change is also altering the distribution of bacteria and viruses leading to new host-parasite interactions and disease outbreaks with consequences for fisheries (Burge et al., 2014). The frequency and extent of harmful algal blooms that can lead to fishery closures are also projected to increase with climate change (Howard et al., 2013).

Across all trophic levels, shifts in the composition of species affect the fundamental structure of regional food webs and, together with changes in the abundance, size, and age of organisms, also their functioning. This includes the temporal and spatial overlap and interactions of species, such as predator-prey relationships, competitive or facilitative relationships, as well as the transfer of energy from lower to higher trophic levels. For example, a replacement of cold- with warm-adapted species has been observed at several food-web levels in the North Atlantic, including zooplankton (Beaugrand et al., 2002; Edwards and Richardson, 2004) and fish communities (Cheung et al., 2013b; Friedland et al., 2020; Perry et al., 2005). In the North Sea, these changes have not always been synchronized and have resulted in a growing mismatch between zooplankton availability and the emergence of Atlantic cod larvae and juveniles, compromising their growth and survival (Beaugrand et al., 2003). Consequently, cold populations are affected both directly by changes in water temperature as well as indirectly by changes in the planktonic prey essential for larval growth (Beaugrand et al., 2003). Similar mismatches in the seasonal variation or phenology of abiotic and biotic environmental conditions that can include several food-web levels have been observed in other regions around the world. On the Scotian Shelf in Atlantic Canada, the survival of haddock larvae is strongly influenced by the timing of the spring phytoplankton bloom, which influences zooplankton availability, and haddock survival can be substantially reduced when the spring bloom is delayed (Platt et al., 2003). In the Northwest Pacific Ocean, species-specific shifts in the seasonal timing of spawning have been observed across the entire fish community driven by changes in water temperatures, ocean mixing, and nutrient availability (Asch, 2015).

Such differential changes in the timing of certain life-history events and the spatial overlap of populations will influence species interactions and food-web dynamics in multiple and often complex ways. The more changes and species are involved, the more difficult it is to identify, understand or even project changes for a whole food web. Generally, however, there is growing evidence that water temperature and climate warming strongly influence predator-prey interactions and trophic dynamics. This has been shown in the North Atlantic and Arctic Oceans (Frank et al., 2006, 2007; Petrie et al., 2009) as well as globally based on observations and modeling (Boyce et al., 2015; Coll et al., 2020; Kwiatkowski et al., 2018; Lefort et al., 2015).

Water temperatures affect the metabolic rates of species and can thus influence predator-prey dynamics. In Atlantic Canada, decreases in the average body size of predatory fish with ocean warming have reduced their predation efficiency leading to strong increases in the biomass of their prey (Shackell et al., 2010). Warming waters have also been shown to increase the metabolic rates of secondary producers, such as zooplankton and other consumers, more rapidly than those of primary producers (e.g., phytoplankton), which would shift trophic relationships (O'Connor et al., 2009; Lewandowska et al., 2014). Studies have also shown that the capture efficiencies, per-capita prey encounter rates, and maximum capture rates of cold-blooded species, such as most finfish and invertebrates, would change with warming, while those of warm-blooded species, such as some tunas, sharks, billfish, and mammals would remain constant (Grady et al., 2019). Thus, cold-blooded species may benefit from warming and potentially consume a larger share of available prey than warm-blooded species, thereby changing food-web composition and dynamics.

The above examples highlight that resolving the varied climate-change effects on individual and multiple interacting species is highly challenging and, therefore, one of the key uncertainties in projecting climate-change effects on marine food webs and ocean ecosystems (Coll et al., 2020; Lotze et al., 2019; Tittensor et al., 2018). Recent advances in global ocean ecosystem modeling and multimodel ensemble approaches, however, have allowed some comparison of general climate-change effects. On a global scale, climate-change projections across multiple food web levels suggest that the effects of climate change amplify at higher trophic levels, meaning that higher food web levels will show stronger declines in response to warming than lower trophic levels (Fig. 4). This has been shown for plankton communities (Kwiatkowski et al., 2018), fish communities (Lefort et al., 2015) and entire marine food webs (Coll et al., 2020; Lotze et al., 2019) with consequences for overall ocean ecosystem production (see Section 5 below). Observational studies also suggest that basic environmental changes are amplified at higher food-web levels (Taylor et al., 2002). Climate-change projections with an individual ecosystem model (EcoOcean) that resolves species- and functional group responses showed that, generally, most groups of marine fish and invertebrates but also marine mammals, seabirds, and sea turtles declined in biomass with climate change, although the magnitude of responses varied widely (Coll et al., 2020). In addition, animals with larger body size (>30 cm length) declined more strongly than smaller animals (Coll et al., 2020). These results highlight important consequences of climate change on the structure and functioning of marine food webs.

## 5 Changes in ocean ecosystem production

At the ocean ecosystem level, the physical and chemical effects of climate change will alter the basic amount and quality of primary production, with consequences for the flux of organic matter and energy to higher trophic levels (Chassot et al., 2010). Warming has been shown to reduce phytoplankton biomass, cell size, and primary production and alter species composition (Boyce and Worm, 2015; Lewandowska et al., 2014; Morán et al., 2010), all of which can have critical consequences for higher trophic-level production. Changes at the foundation of marine food webs will be accompanied by changes in individuals, populations,

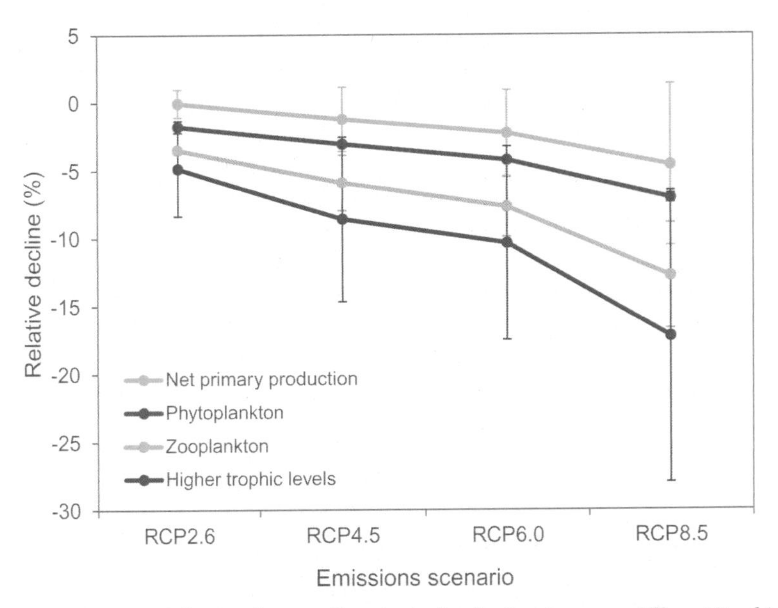


**FIG. 4** Trophic amplification of climate-change effects in marine food webs across different trophic levels, including net primary production (NPP, *light green*) and phytoplankton biomass *(dark green)*, zooplankton biomass *(in orange)* and higher trophic level animal biomass *(blue)*. Shown are projected changes (% relative decline, ±1 standard deviation) under four different emissions scenarios (RCPs) in the 2090s relative to 1990s derived from an ensemble of multiple global marine ecosystem models. *Reprinted with permission from Lotze, H.K., Tittensor, D.P., Bryndum-Buchholz, A., Eddy, T.D., Cheung, W.W.L., Galbraith, E.D., Barange, M., Barrier, N., Bianchi, D., Blanchard, J.L., Bopp, L., Büchner, M., Bulman, C.M., Carozza, D.A., Christensen, V., Coll, M., Dunne, J.P., Fulton, E.A., Jennings, S., Jones, M.C., Mackinson, S., Maury, O., Niiranen, S., Oliveros-Ramos, R., Roy, T., Fernandes, J.A., Schewe, J., Shin, Y.-J., Silva, T.A.M., Steenbeek, J., Stock, C.A., Verley, P., Volkholz, J., Walker, N.D., Worm, B., 2019. Global ensemble projections reveal trophic amplification of ocean biomass declines with climate change. Proc. Natl. Acad. Sci. 116, 12907–12912.*

and species across all trophic levels leading to overall changes in secondary production, including potential fisheries catch and seafood production. Globally, most climate or Earth system models project a long-term decline in net primary production (NPP) over the 21st century, ranging from an average of −2% (±4% standard deviation, SD) under a low emissions scenario (RCP2.6) to −9% (±8% SD) under a high emissions scenario (RCP8.5) by 2100 (Bopp et al., 2013). Yet there is considerable regional variation, with general increases in polar and decreases in temperate and tropical areas (Figs. 5 and 6A). These changes in primary production together with the effects of changing temperatures, ocean acidification, oxygen reduction, sea ice loss, and other climate-change drivers, are projected to reduce global-scale secondary production, but again with considerable regional variation (Figs. 5 and 6B).

Recent state-of-the-art ensemble modeling approaches have combined multiple Earth system with marine ecosystem models to project the climate-change effects on marine animal biomass (all vertebrates and invertebrates except zooplankton) under different emissions scenarios (Fish-MIP, https://www.isimip.org/about/marine-ecosystems-fisheries/, Tittensor et al., 2018). Results indicate that globally, marine animal biomass is projected to decrease by 5% (±4% SD) under a low emissions or strong mitigation scenario (RCP2.6) and by

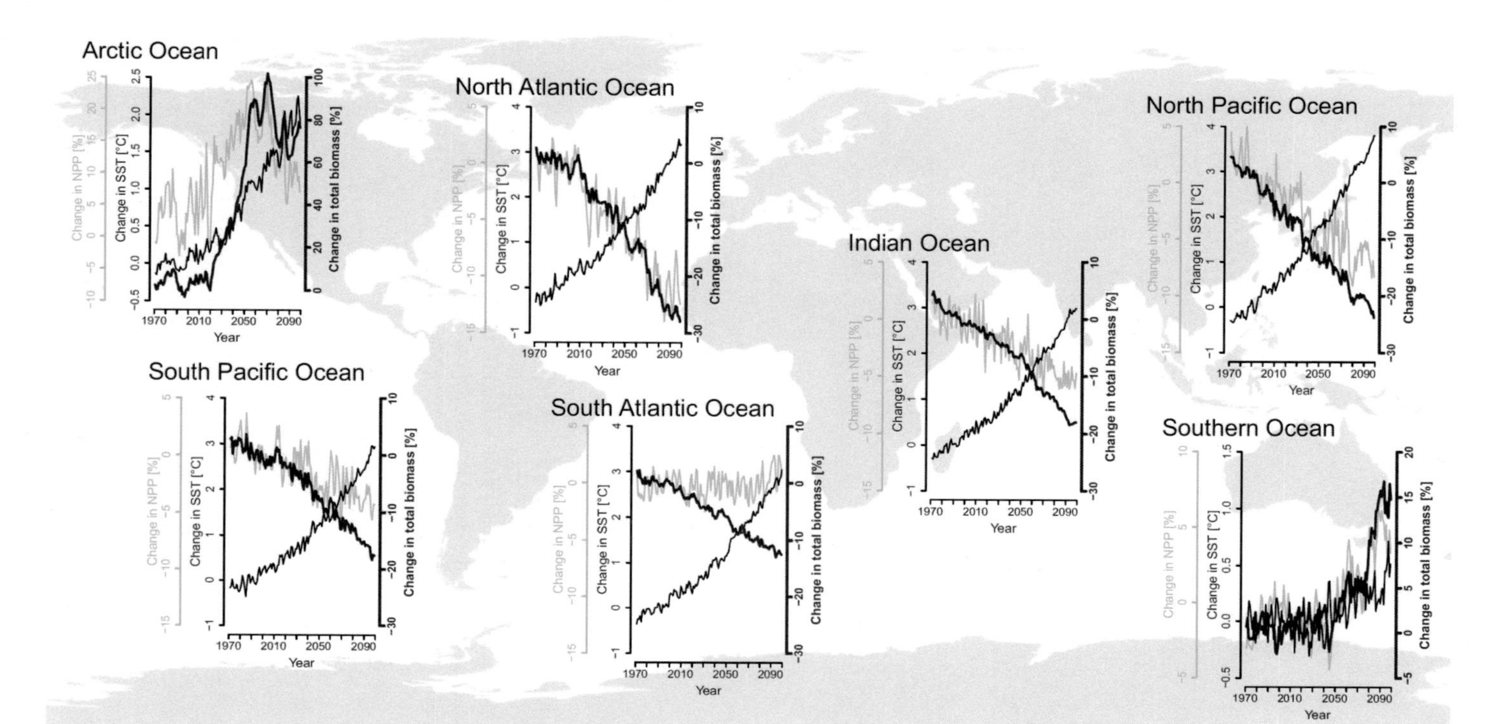


**FIG. 5** Multimodel ensemble mean changes of projected historical and future total marine animal biomass *(red)* from 1970 to 2100 and associated climate-change drivers including sea surface temperature (SST, *black*) and net primary production (NPP, *green*) across different ocean basins under the worst-case emissions scenario (RCP8.5). Total animal biomass and NPP trends are expressed in percent change and SST trends in degree °C relative to the mean of the 1990s. *Reprinted with permission from Bryndum-Buchholz, A., Tittensor, D., Blanchard, J.L., Cheung, W.W.L., Coll, M., Galbraith, E.D., Jennings, S., Maury, O., Lotze, H.K., 2019. Twenty-first-century climate change impacts on marine animal biomass and ecosystem structure across ocean basins. Glob. Chang. Biol. 25, 459–472.*

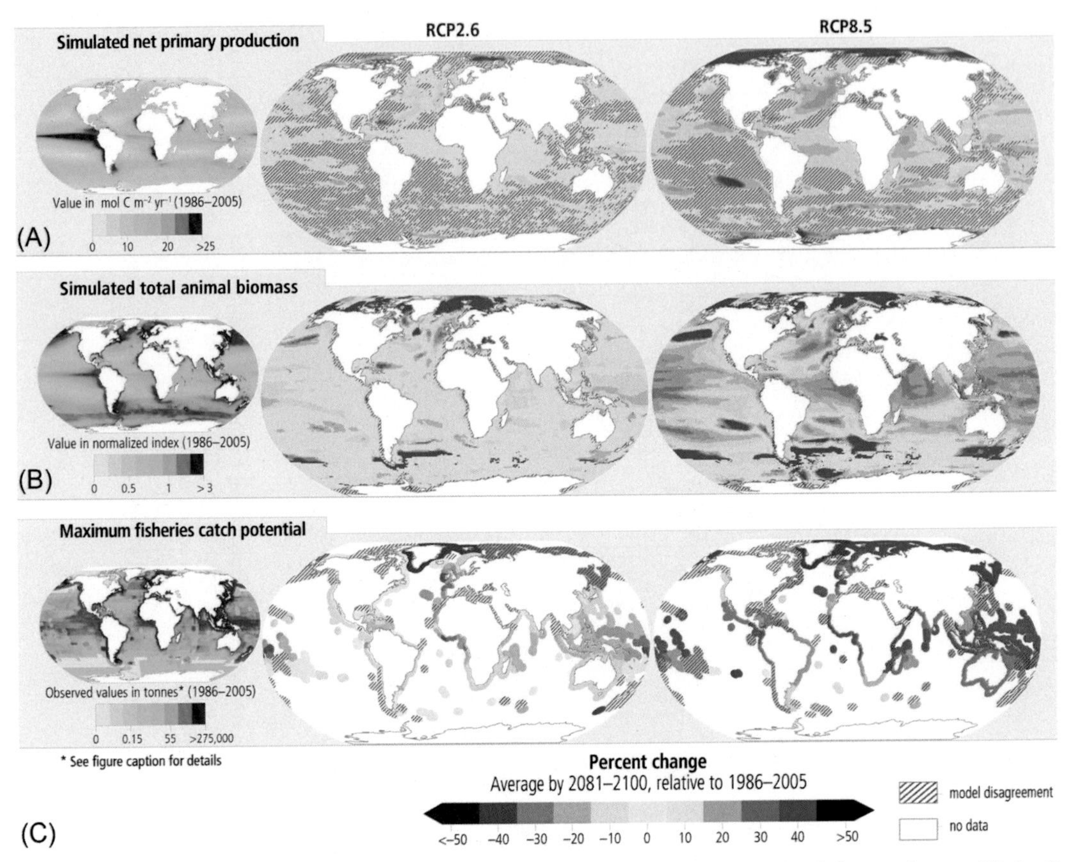


**FIG. 6** Projected changes and impacts for ocean regions and ecosystems as a result of climate change: (A) depth integrated net primary production (NPP from CMIP5), (B) total animal biomass (depth integrated, including fishes and invertebrates from Fish-MIP), (C) maximum fisheries catch potential. The three left panels represent the simulated (A, B) and observed (C) mean values for the recent past (1986–2005), the middle and right panels represent projected changes (%) by 2081–2100 relative to recent past under low (RCP2.6) and high (RCP8.5) greenhouse gas emissions scenarios. Total animal biomass in the recent past (B, *left panel*) represents the projected total animal biomass by each spatial pixel relative to the global average. (C) *Average observed fisheries catch in the recent past (based on data from the Sea Around Us global fisheries database); projected changes in maximum fisheries catch potential in shelf seas are based on the average outputs from two fisheries and marine ecosystem models. To indicate areas of model inconsistency, shaded areas represent regions where models disagree in the direction of change for more than: (A) and (B) 3 out of 10 model projections, and (C) one out of two models. Although unshaded, the projected change in the Arctic and Antarctic regions in (B) total animal biomass and (C) fisheries catch potential have low confidence due to uncertainties associated with modeling multiple interacting drivers and ecosystem responses. Projections presented in (B) and (C) are driven by changes in ocean physical and biogeochemical conditions, e.g., temperature, oxygen level, and net primary production projected from CMIP5 Earth system models. For more details see (IPCC, 2019a). *From Figure SPM.3(abc) of IPCC, 2019. Summary for policymakers. In: Pörtner, H.-O., Roberts, D.C., Masson-Delmotte, V., Zhai, P., Tignor, M., Poloczanska, E., Mintenbeck, K., Alegría, A., Nicolai, M., Okem, A., Petzold, J., Rama, B., Weyer, N.M. (Eds.), IPCC Special Report on the Ocean and Cryosphere in a Changing Climate. In press. https://www.ipcc.ch/srocc/chapter/summary-for-policymakers/.*

17% (±11% SD) under a worst-case emissions scenario (RCP8.5) by the 2090s relative to the 1990s (Lotze et al., 2019). These changes result in an average 5% decline in global marine animal biomass for every 1°C of warming (Lotze et al., 2019). Comparing these changes across different ocean basins revealed that the North Atlantic Ocean basin showed the strongest decline in marine animal biomass (−32% ± 14% SD) under RCP8.5 (Fig. 5), followed by the North Pacific Ocean basin (−26% ± 17% SD), whereas declines in the South Atlantic, South Pacific, and Indian Ocean basins were between −10% and −20% (Bryndum-Buchholz et al., 2019). In contrast, marine animal biomass was projected to increase by 19% (±36% SD) in the Southern Ocean basin and by 82% (±201% SD) in the Arctic Ocean basins, but the large standard deviations indicate high uncertainty of projections in polar regions (Bryndum-Buchholz et al., 2019). Nevertheless, these patterns of increased production in polar regions and decreased production in temperate to tropical regions have also been shown for phytoplankton and zooplankton communities (Kwiatkowski et al., 2018), different components of pelagic ocean ecosystems, including epipelagic, mesopelagic, and migratory communities (Lefort et al., 2015), and the total biomass of marine animal communities (Fig. 6B) (IPCC, 2019b; Lotze et al., 2019). These changes will affect both regional and global fisheries and seafood production.

## 6 Consequences for fisheries and seafood production

All the above-described changes will affect fish and invertebrate stocks that are important resources for commercial fisheries and seafood supply. Generally, climate variability has affected fish stocks and fisheries for most of history; however, fishing pressure has far outweighed the climate effect over past decades (Worm et al., 2009). Growing human populations and seafood demand, as well as increasing fishing effort and efficiency, have led to a rapid rise in global catches from 1950 to 1990, followed by stabilizing or even slightly decreasing catches while fishing effort continued to increase (Pauly and Zeller, 2016; Worm and Branch, 2012). The high fishing pressure has caused strong declines in fished stocks in many parts of the world's ocean, and today one- to two-thirds of stocks are considered overfished (Christensen et al., 2014; Costello et al., 2012; FAO, 2020; Froese et al., 2012; Worm et al., 2009). Although increased monitoring, assessment and more effective fisheries management and conservation have helped to rebuild some stocks in some regions (Hilborn et al., 2020; Worm et al., 2009), the added impacts of climate change and other human activities will alter the future of fisheries and seafood production (Bocye, 2020; Boyce et al., 2020; Cheung et al., 2010; Galbraith et al., 2017; Lotze et al., 2018).

Growing evidence indicates that climate change is already altering the distribution and abundance of commercial species as well as their body size, reproduction, and potential yield. As described above, many stocks are shifting further north, deeper, or offshore (Pinsky et al., 2013; Shackell et al., 2014), and some stocks show regional collapses or disappearances (Pershing et al., 2015). Warming waters have been associated with declines in average body size (Shackell et al., 2010), and continued climate change is expected to further decrease maximum body size of commercially important fish and invertebrates by 14%–24% globally from

2000 to 2050, with stronger declines in tropical than polar regions (Cheung et al., 2013a). Also, warming waters and decreasing plankton concentrations have been associated with a 3% reduction of new recruitment per decade across 262 fish stocks in 39 large marine ecosystems and high sea areas (Britten et al., 2016), and rising temperatures have been related to a 4% reduction in the maximum sustainable yield of 235 major fish stocks across 38 ocean ecoregions from 1930 to 2010 (Free et al., 2019). Of these, five ecoregions experienced reductions between 15% and 35% in maximum sustainable yield; however, not all fish stocks showed decreases in maximum sustainable yield, and some significantly increased with warming (Free et al., 2019).

All these changes will affect the temporal and spatial availability of fish stocks, the amount of potential catch and seafood, and its nutritional and monetary value. Therefore, fisheries increasingly need to adapt to where and when they can fish, what target species are available, what gear they can use, how much they can catch, and how large and valuable those species are (Figs. 1 and 2). Unfortunately, the effects of climate change do not act in isolation and may interact with and exacerbate the effects of fisheries exploitation. Most fisheries truncate the size and age structure of target species by preferentially removing larger, older individuals from the population. Then, the fishery becomes increasingly dependent on new recruitment, which is more strongly affected by climate variability than adults (Britten et al., 2016; Stenseth et al., 2002). Removing larger and older age classes, therefore, increases the susceptibility of both the target stock and the fishery to climate variability and change (Brander, 2007; Sumaila and Tai, 2020). Harvesting larger individuals can also affect the depth inhabited by fish populations and could create the illusion of climate-driven deepening (Frank et al., 2018). In addition, the increasing frequency and intensity of climate extremes, such as marine heatwaves (Oliver et al., 2018), have been associated with widespread ecological and socioeconomic effects, including mass mortality events, altered community structure, and fisheries disruption (Caputi et al., 2016; Cavole et al., 2016; Oliver et al., 2017).

With continued climate change, there will likely be winners and losers globally and regionally in terms of individual fisheries as well as fishing nations (Blanchard et al., 2017; Boyce et al., 2020). To get a better idea of how these may play out, there have been increasing efforts to project future changes in fish abundance and potential catches using various fisheries and ecosystem modeling approaches. Available results highly depend on the choice of model, its spatial, temporal, and taxonomic resolution, and the simulated climate change or fisheries scenarios (Lotze et al., 2018; Tittensor et al., 2018). However, most studies project some decreases in fish stock production and potential catch at lower tropical to temperate latitudes and increases at higher polar latitudes (Barange et al., 2014; Blanchard et al., 2012; Cheung et al., 2010; Tai et al., 2019). Recent ensemble approaches that combine different climate and marine ecosystem models project an overall 20%–24% global reduction of the maximum fisheries catch potential under a worst-case emissions scenario (RCP8.5) by the end of the 21st century relative to 1986–2005 (Fig. 6C), which is three to four times larger than that projected under a low emissions scenario (RCP2.6) (IPCC, 2019b). The regional distribution of projected changes in catch potential is related to trends projected for primary production (Fig. 6A) and animal biomass (Fig. 6B). We caution, however, that climate change will continue to restructure marine food webs and ecosystems in ways that are not yet fully understood, and there is

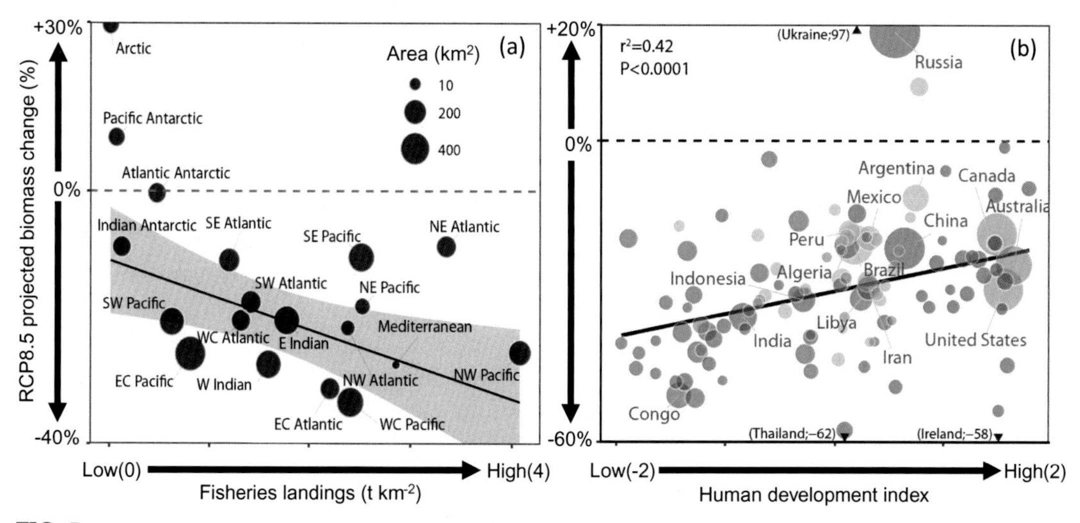


FIG. 7 Consequences of projected changes in marine animal biomass with climate change for (A) fisheries regions and (B) maritime nations. Shown are the relationships of projected biomass changes under RCP8.5 over the 21st century in relation to (A) total fisheries landings from 1990 to 2010 in each of 18 FAO statistical regions, and (B) the human development index of 107 maritime nations. *Circle size* indicates they area of the FAO region or exclusive economic zone, respectively, and colors in (B) refer to *red*=Africa, *blue*=Asia, *green*=Oceania, *orange*=North America, *yellow*=South America, *purple*=Europe. *The data were derived and redrawn from Boyce, D.G., Lotze, H.K., Tittensor, D. P., Carozza, D.A., Worm, B., 2020. Future Ocean biomass losses may widen socioeconomic equity gaps. Nat. Commun. 11, 2235.*

significant uncertainty in projecting climate-change effects on fish and fisheries, particularly in polar regions (Fig. 6B and C) (Brander, 2007; Bryndum-Buchholz et al., 2019; Cheung et al., 2016; Niemi et al., 2019).

The consequences of such changes for fisheries operations can be severe. Across major FAO fishing areas, almost all areas will experience a reduction of marine animal biomass over the 21st century, except the Arctic and the Pacific and Atlantic Antarctic (Fig. 7A) (Boyce et al., 2020). Moreover, regions that have had historically high fisheries landings, such as the Northwest Pacific and Mediterranean (Fig. 7A), will likely experience greater reductions in marine animal biomass than regions with historically lower fisheries landings (Boyce et al., 2020). Similar patterns have been shown on a regional scale across management divisions within the Northwest Atlantic Fisheries Organization (Bryndum-Buchholz et al., 2020). In terms of monetary consequences, available estimates suggest that climate change could reduce gross revenues in global fisheries between 17 and 41 billion US dollars per year (Sumaila and Cheung, 2010). Unfortunately, the geography of climate-change impacts on projected trends in animal biomass, potential catch and seafood production, as well as fisheries revenue show disproportionate effects on nations with lower human development index (Fig. 7B), which may widen existing socioeconomic equity gaps and requires attention in terms of fisheries management and global ocean governance (Blanchard et al., 2017; Boyce et al., 2020; Sumaila and Cheung, 2010).

## 7 Challenges for fisheries management and ocean governance

The effects of climate change will add another layer on already existing challenges for fisheries management and ocean governance. Over past decades, fishing pressure has had much stronger effects on fish biomass and catches than climate variability; thus, a lot will depend on future developments in fisheries and other ocean uses (Costello et al., 2016; Galbraith et al., 2017). A continuing rise in fishing effort and efficiency will likely result in further declines in fish stocks, whereas more sustainable uses of ocean resources, rebuilding of depleted stocks, and marine conservation strategies may help to halt or even reverse some declines (Costello et al., 2016; Hilborn et al., 2020; Quaas et al., 2015; Worm and Branch, 2012; Worm et al., 2009). Unfortunately, there are considerable regional differences worldwide in the level of management effectiveness, capacity for stock assessments, and implementation of conservation initiatives that are similar to the regional variation in climate-change effects (Fig. 7). This will hit some nations much harder than others and need to be addressed by the international community (Blanchard et al., 2017; Boyce et al., 2020; Worm and Branch, 2012).

Several studies have tested potential future interactions between different management and climate-change scenarios. For example, a study evaluating four contrasting economic scenarios based on fishing fleets and industries in the Northeast Atlantic found that the level of fisheries regulation was the most important factor for long-term fisheries development (Mullon et al., 2016). The effects of governance and trade decisions were also important, although less important than fisheries regulation, and stronger than the effects of climate change (Mullon et al., 2016). A multimodel approach exploring the interactions among multiple human activities in the Baltic Sea found that regional management of fishing intensity and nutrient loading was more important in determining the future of the marine ecosystem than climate change (Niiranen et al., 2013). On a global scale, simulations with a coupled human-Earth model revealed that further technological progress in fisheries operations, although explaining most of the historical increase in catch, would lead to long-term declines in future harvests because of overfishing (Galbraith et al., 2017). Climate change would also lead to gradual declines in global fish production over the 21st century, yet this was less important than social and economic factors (Galbraith et al., 2017). Lastly, a global species distribution model was used to assess the impacts of fishing and climate change on the extinction risk of 825 exploited fish species (Cheung et al., 2018). Results suggest that under a business-as-usual fishing and worst-case climate scenario, 60% of species would experience a high risk of extinction, but this amount could be substantially reduced under a sustainable fishing and low emissions scenario. Species at risk from climate change were concentrated in tropical and subtropical regions, while those at risk from overfishing were distributed more broadly (Cheung et al., 2018).

Although fishing pressure and fisheries management currently have much stronger effects on fish stocks than climate change, their interactions still need to be considered, particularly at low population levels. Both the risk of stock collapse and the rebuilding of overfished stocks are strongly impacted by climate change, whereas a reduction of fishing mortality and rebuilding of spawning stocks can help mitigate some climate-change effects (Britten et al., 2017; Free et al., 2020; Le Bris et al., 2018; Pershing et al., 2015). Another important issue to consider are time lags in the social-ecological responses to climate change. For example, many species or populations are shifting their ranges faster than their associated fisheries, which may allow stocks to increase at their leading-edge because of low exploitation rates while placing too much pressure on already declining stocks at their trailing edge (Pinsky and Fogarty, 2012). In these situations, fisheries management would need to proactively adjust fishing mortality for species known to be on the move. Furthermore, shifts in stock distribution and production levels will increase potential transboundary conflicts over fishery resources, which need to be negotiated among nations (Mendenhall et al., 2020).

Currently, there is an increasing recognition for the need to incorporate climate-change considerations into fisheries management, although formal mandates for climate-adaptation objectives are still largely missing in most countries (Bryndum-Buchholz, 2020). Several global initiatives exist that are influencing the development of fisheries management and ocean governance on national and international scales (Barange et al., 2018; Bryndum-Buchholz, 2020). This includes the Paris Agreement of the Conference of the Parties of the United Nations Framework Convention on Climate Change, the 2030 Agenda for Sustainable Development, the Sustainable Development Goals (SDGs), and FAO's Blue Growth Initiative (FAO, 2016). The latter promotes the Code of Conduct for Responsible Fisheries and the Ecosystem Approach to Fisheries and Aquaculture and aims to help countries implement the new global agenda (FAO, 2016). Responding to these developments, several nations have amended their national Fisheries Acts over the past decade and have produced climate-change adaptation reports (Barange et al., 2018; Bryndum-Buchholz, 2020), such as in the United States (Gregg et al., 2016), the United Kingdom (Defra, 2013), and Ireland (Kopke and O'Mahoney, 2011). In Australia, fisheries governance already includes several key climate-adaptive management components, which have been applied to varying degrees and have the potential to enhance the resilience of fisheries to climate change (Bryndum-Buchholz, 2020; Ogier et al., 2016). In the United States, the National Marine Fisheries Service has developed a climate science strategy to ensure that fisheries management is robust to climate-change uncertainties (Busch et al., 2016). Moreover, the European Green Deal and its Farm to Fork Strategy propose an increased shift toward sustainable fish and seafood production as well as an assessment of how climate-change adaptation is being addressed within the European Union's Common Fisheries Policy (European Commission, 2020).

Generally, many agencies aim to develop best practices for incorporating climate change into fish stock assessments and fisheries management (Bryndum-Buchholz, 2020). Ideally, fisheries management should be prepared to quickly respond to climate-induced changes in fish stocks and marine ecosystems, such as through dynamic management decisions, climate-informed catch allocations, or responsive time-area fisheries closures (Dunn et al., 2016; Karp et al., 2019). However, this requires explicit climate-change adaptation objectives in fisheries management policies and legislations, which depend on political awareness and willingness to prepare for the long-term consequences of climate change (Bryndum-Buchholz, 2020; Lindegren and Brander, 2018). Currently, adaptive management strategies are increasingly being implemented in terms of harvest control rules, which can act as predefined thresholds based on biological reference points (biomass of a given target stock) that, if approached, reduce or cease fishing pressure, regardless of the cause of change in biomass levels (Chavez et al., 2017; Kelly et al., 2015). This approach is naturally adaptive and can facilitate decision-making despite of scientific uncertainty and high environmental variability (Burden and Fujita, 2019). Several fisheries in the United States (e.g., Alaska pollock, Pacific sardine), South Africa (e.g., deep-water hake), and all domestic Australian fisheries are already managed by harvest control rules (Burden and Fujita, 2019; Bryndum-Buchholz, 2020).

An urgent need also exists for climate-informed reference points for fish stock and environmental assessments (Link et al., 2015; Busch et al., 2016), as well as climate-informed ecosystem and multispecies reference points to adapt ecosystem-based fisheries management (Bryndum-Buchholz, 2020; Holsman et al., 2020; Marshall et al., 2018; Tanaka, 2019). To achieve this would require enhanced temporal and spatial monitoring of the marine environment, harvested stocks, and ecosystem properties to provide the necessary data, as well as new quantitative tools and approaches to incorporate them into assessments and decision making. There is also a general paradigm shift needed to move current fisheries management objectives away from short-term economic profits toward climate-robust sustainability, including long-term ecological and economic sustainability with climate change (Marshall et al., 2018). Decision-making will need to adapt to higher levels of uncertainty because of the variability in climate-change effects and the possibility of extreme events (Lawler et al., 2010; Skern-Mauritzen et al., 2018). As fish stocks increasingly shift in their distribution across borders, there is also a need for strengthened or new multilateral fisheries agreements (Gaines et al., 2018), as well as international agreements for some compensation among nations winning and losing from climate change (Fig. 7.) (Blanchard et al., 2017; Boyce et al., 2020). Lastly, fish stocks and fisheries depend on healthy, productive, and diverse marine ecosystems that require strong national and international ocean governance aiming at the protection of marine biodiversity, the conservation of ocean habitats, the reduction of cumulative human impacts, and the rebuilding of fish stock and ecosystem resilience, all of which also need to adapt to climate change (see Section 8 below) (Tittensor et al., 2019; Wilson et al., 2020).

## 8 Marine conservation strategies for climate change mitigation

Intact levels of marine biodiversity and healthy ocean ecosystems are essential to ensure the continued provision of the goods and services humans receive from the ocean (Duarte et al., 2020; Worm et al., 2006), yet can also serve as an insurance against climate change or other environmental disturbances (Worm and Lotze, 2016, 2021). A growing number of field, laboratory, and theoretical studies have demonstrated that higher levels of genetic, population, species, or habitat diversity can enhance ecosystem productivity, stability, and resilience, as well as the adaptive capacity in the face of climate change (Duffy et al., 2016; Ehlers et al., 2008; Elmqvist et al., 2003; Schindler et al., 2010; Reusch et al., 2005; Yachi and Loreau, 1999). For example, higher diversity and biocomplexity of salmon stocks in the Northwest Pacific have been linked to higher productivity and stability, and thus greater long-term sustainability in associated fisheries (Hilborn et al., 2003; Schindler et al., 2010). Due to a range of local life-history adaptations, the overall stock complex shows greater resilience to climatic variability than any individual stock; therefore, overall productivity can be maintained by different substocks as environmental conditions change (Hilborn et al., 2003; Schindler et al., 2010). Globally, ocean regions with higher species diversity have shown faster recovery of depleted fish stocks, as there are more players and potential life-history options available to thrive under changing conditions (Worm et al., 2006). Also, higher levels of genetic diversity have led to faster recovery in seagrass habitats after severe heatwave damage, as different genotypes varied in their level of survival and re-growth (Ehlers et al., 2008; Reusch et al., 2005). These portfolio effects of diversity have been recognized as important insurance against climate-driven fluctuation in the marine realm analogous to the effects of asset diversity on the stability of financial portfolios (Schindler et al., 2010; Worm et al., 2006). Therefore, maintaining high levels of marine biodiversity at different levels of organization is essential to ensure the productivity, stability, and resilience of ocean ecosystems with continued climate and global change (Loreau, 2010; Pecl et al., 2017; Worm et al., 2006).

Another important asset of marine ecosystems is their role in regulating the climate by acting as a major carbon sink. From 1800 to 1994, almost half (48%) of all anthropogenic carbon emissions were absorbed by the ocean (Sabine et al., 2004) through the primary production of marine phytoplankton and macrophytes, such as salt marshes, mangroves, and seagrass meadows, and the burial of organic matter in shallow or deep-sea sediments (Donato et al., 2011; Field et al., 1998; Fourqurean et al., 2012). Moreover, large-bodied and long-lived marine organisms can serve as significant carbon reservoirs throughout their lifespans; thus, preserving large fish or other predators can also contribute to climate-change mitigation (Atwood et al., 2015).

All these issues have important implications for marine conservation planning. Over past decades and centuries, multiple human activities have strongly impacted marine biodiversity and ocean ecosystems and have caused or contributed to a number of local, regional, and global extinctions (Dulvy et al., 2003; Harnik et al., 2012; Lotze et al., 2006). These losses in marine biodiversity have resulted in reduced productivity and resilience (Britten et al., 2014; Lotze et al., 2011a; Worm et al., 2006), which could magnify the impacts of climate change. In turn, efforts to protect and enhance marine biodiversity through recovery, restoration, and rebuilding efforts (Duarte et al., 2020; Lotze et al., 2011b; Worm et al., 2009) should have the opposite effect. Therefore, conservation strategies aiming at reversing biodiversity loss and enhancing climate resilience are highly interconnected and need to be considered together (Tittensor et al., 2019). For example, marine protected areas have been identified as important management tools for protecting biodiversity as well as for enhancing climate resilience (Davies et al., 2017; Roberts et al., 2017). Nevertheless, the future effectiveness of marine protected areas to maintain biodiversity will depend on the trajectories of climate change. More adaptive and dynamic management approaches may be needed to future-proof marine conservation efforts and safeguard marine biodiversity and ecosystems in the face of rapid climate change (Dunn et al., 2016; Tittensor et al., 2019; Wilson et al., 2020).

From a global perspective, tropical and polar regions may require the most immediate conservation strategies to mitigate climate-change effects, since these regions are anticipated to experience the highest biodiversity losses and most rapid ecosystem changes. Tropical regions are already hotspots of multiple and increasing human pressures (Halpern et al., 2015), highly depend on seafood, and will suffer most climate-driven species extinctions and reductions in ocean productivity, yet often lack the necessary resources and management capacity for mitigation and adaptation (Blanchard et al., 2017; Boyce et al., 2020; Worm and Branch, 2012). On the other hand, polar ecosystems experience the most rapid climate-change effects with the loss of ice sheets, extinction of ice-dependent species, and high levels of species invasions, yet also face growing human development and associated cumulative impacts (Michel et al., 2012; Tai et al., 2019). Polar and tropical regions, therefore, require comprehensive conservation strategies to maintain the integrity and functions of their ocean ecosystems. Nevertheless, complex, unexpected, and surprising effects of climate change on marine biodiversity and ocean ecosystems may occur throughout the global ocean that will require dynamic responses, adaptive planning, and progressive rethinking of marine conservation approaches (Tittensor et al., 2019; Wilson et al., 2020).

## Acknowledgments

The authors wish to thank all contributors to the Fisheries and Marine Ecosystem Model Intercomparison Project (Fish-MIP) for their efforts in improving climate-change projections in ocean ecosystems. Thanks also to D. Tittensor and B. Worm for valuable discussions and insight. Financial support for this work was provided by Canada's Natural Sciences and Engineering Research Council (NSERC), the Ocean Frontier Institute (Module G) and a MEOPAR Postdoctoral Fellowship Award 2020–21.

## References

Asch, R.G., 2015. Climate change and decadal shifts in the phenology of larval fishes in the California Current ecosystem. Proc. Natl. Acad. Sci. 112 (30), E4065–E4074.

Atwood, T.B., Connolly, R.M., Ritchie, E.G., Lovelock, C.E., Heithaus, M.R., Hays, G.C., Fourqurean, J.W., Macreadie, P.I., 2015. Predators help protect carbon stocks in blue carbon ecosystems. Nat. Clim. Chang. 5, 1038–1045.

Barange, M., Merino, G., Blanchard, J.L., Scholtens, J., Harle, J., Allison, E.H., Allen, J.I., Holt, J., Jennings, S., 2014. Impacts of climate change on marine ecosystem production in societies dependent on fisheries. Nat. Clim. Chang. 4, 211–216.

Barange, M., Bahiri, T., Beveridge, M.C.M., Cochrane, K.L., Funge-Smith, S., Poulain, F. (Eds.), 2018. Impacts of climate change on fisheries and aquaculture: synthesis of current knowledge, adaptation and mitigation options. In: FAO Fisheries and Aquaculture Technical Paper 627, Rome, Italy, p. 628.

Beaugrand, G., Reid, P.C., Ibañez, F., Lindley, J.A., Edwards, M., 2002. Reorganization of North Atlantic marine copepod biodiversity and climate. Science 296, 1692–1694.

Beaugrand, G., Brander, K.M., Lindley, J.A., Souissi, S., Reid, P.C., 2003. Plankton effect on cod recruitment in the North Sea. Nature 426, 661–664.

Blanchard, J.L., Jennings, S., Holmes, R., Harle, J., Merino, G., Allen, J.I., Holt, J., Dulvy, N.K., Barange, M., 2012. Potential consequences of climate change for primary production and fish production in large marine ecosystems. Philos. Trans. R. Soc. B 367, 2979–2989.

Blanchard, J.L., Watson, R.A., Fulton, E.A., Cottrell, R.S., Nash, K.L., Bryndum-Buchholz, A., Büchner, M., Carozza, D., Cheung, W., Elliott, J., Davidson, L., Dulvy, N.K., Dunne, J.P., Eddy, T.D., Galbraith, E., Lotze, H.K., Maury, O., Müller, C., Tittensor, D., Jennings, S., 2017. Linked sustainability challenges and trade-offs among fisheries, aquaculture and agriculture. Nat. Ecol. Evol. 1, 1240–1249.

Bocye, D.G., 2020. Incorporating Climate Change into Fisheries Management in Atlantic Canada and the Eastern Arctic. Oceans North Conservation Society, Halifax, NS, Canada.

Bopp, L., Resplandy, L., Orr, J.C., Doney, S.C., Dunne, J.P., Gehlen, M., Halloran, P., Heinze, C., Ilyina, T., Seferian, R., Tjiputra, J., Vichi, M., 2013. Multiple stressors of ocean ecosystems in the 21st century: projections with CMIP5 models. Biogeosciences 10, 6225–6245.

Boyce, D.G., Worm, B., 2015. Patterns and ecological implications of historical marine phytoplankton change. Mar. Ecol. Prog. Ser. 534, 251–272.

Boyce, D.G., Lewis, M.R., Worm, B., 2010. Global phytoplankton decline over the past century. Nature 466, 591–596.

Boyce, D.G., Dowd, M., Lewis, M.R., Worm, B., 2014. Estimating global chlorophyll changes over the past century. Prog. Oceanogr. 122, 163–173.

Boyce, D.G., Frank, K.T., Worm, B., Leggett, W.C., 2015. Spatial patterns and predictors of trophic control across marine ecosystems. Ecol. Lett. 18, 1001–1011.

Boyce, D.G., Lotze, H.K., Tittensor, D.P., Carozza, D.A., Worm, B., 2020. Future Ocean biomass losses may widen socioeconomic equity gaps. Nat. Commun. 11, 2235.

Brander, K.M., 2007. Global fish production and climate change. Proc. Natl. Acad. Sci. 104, 19709–19714.

Britten, G.L., Dowd, M., Minto, C., Ferretti, F., Boero, F., Lotze, H.K., 2014. Predator decline leads to decreased stability in a coastal fish community. Ecol. Lett. 17, 1518–1525.

Britten, G.L., Dowd, M., Worm, B., 2016. Changing recruitment capacity in global fish stocks. Proc. Natl. Acad. Sci. 113, 134–139.

Britten, G.L., Dowd, M., Kanary, L., Worm, B., 2017. Extended fisheries recovery timelines in a changing environment. Nat. Commun. 8, 15325.

Bryndum-Buchholz, A., 2020. Marine Ecosystem Impacts and Management Responses Under 21st Century Climate Change (PhD thesis). Dalhousie University, Halifax, NS, Canada.

Bryndum-Buchholz, A., Tittensor, D., Blanchard, J.L., Cheung, W.W.L., Coll, M., Galbraith, E.D., Jennings, S., Maury, O., Lotze, H.K., 2019. Twenty-first-century climate change impacts on marine animal biomass and ecosystem structure across ocean basins. Glob. Chang. Biol. 25, 459–472.

Bryndum-Buchholz, A., Boyce, D.G., Tittensor, D.P., Christensen, V., Bianchi, D., Lotze, H.K., 2020. Climate change impacts and fisheries management challenges in the North Atlantic Ocean. Mar. Ecol. Prog. Ser. 648, 1–17.

Burden, M., Fujita, R., 2019. Better fisheries management can help reduce conflict, improve food security, and increase economic productivity in the face of climate change. Mar. Policy 108, 103610.

Burge, C.A., Mark Eakin, C., Friedman, C.S., Froelich, B., Hershberger, P.K., Hofmann, E.E., Petes, L.E., Prager, K.C., Weil, E., Willis, B.L., Ford, S.E., Harvell, C.D., 2014. Climate change influences on marine infectious diseases: implications for management and society. Annu. Rev. Mar. Sci. 6, 249–277.

Busch, D.S., Griffis, R., Link, J., Abrams, K., Baker, J., Brainard, R.E., Ford, M., Hare, J.A., Himes, A., Hollowed, A., Mantuah, N.J., McClatchie, S., McClure, M., Nelson, M.W., Osgood, K., Peterson, J.O., Rust, M., Sabal, V., Merrick, R., 2016. Climate science strategy of the US National Marine Fisheries Service. Mar. Policy 74, 58–67.

Caputi, N., Kangas, M., Denham, A., Feng, M., Pearce, A., Hetzel, Y., Chandrapavan, A., 2016. Management adaptation of invertebrate fisheries to an extreme marine heat wave event at a global warming hot spot. Ecol. Evol. 6, 3583–3593.

Cavicchioli, R., Ripple, W.J., Timmis, K.N., Azam, F., Bakken, L.R., Baylis, M., Behrenfeld, M.J., Boetius, A., Boyd, P. W., Classen, A.T., Crowther, T.W., Danovaro, R., Foreman, C.M., Huisman, J., Hutchins, D.A., Jansson, J.K., Karl, D.M., Koskella, B., Welch, D.B.M., Martiny, J.B.H., Moran, M.A., Orphan, V.J., Reay, D.S., Remais, J.V., Rich, V.I., Singh, B.K., Stein, L.Y., Stewart, F.J., Sullivan, M.B., van Oppen, M.J.H., Weaver, S.C., Webb, E.A., Webster, N.S., 2019. Scientists' warning to humanity: microorganisms and climate change. Nat. Rev. Microbiol. 17 (9), 569–586.

Cavole, L.-C.M., Demko, A.M., Diner, R.E., Giddings, A., Koester, I., Pagniello, C.M.L.S., Paulsen, M.-L., Ramirez-Valdez, A., Schwenck, S.M., Yen, N.K., Zill, M.E., Franks, P.J.S., 2016. Biological impacts of the 2013-2015 warm-water anomaly in the Northeast Pacific. Oceanography 29, 273–285.

Chassot, E., Bonhommeau, S., Dulvy, N.K., Mélin, F., Watson, R., Gascuel, D., Le Pape, O., 2010. Global marine primary production constrains fisheries catches. Ecol. Lett. 13, 495–505.

Chavez, F.P., Costello, C., Aseltine-Neilson, D., Doremus, D., Field, J.C., Gaines, S.D., Hall-Arber, M., Manuta, N.J., McCovey, B., Pomeroy, C., Sievanen, L., Sydeman, W., Wheeler, S.A., 2017. Readying California Fisheries for Climate Change. California Ocean Science Trust, Oakland, CA, USA. https://escholarship.org/uc/item/2kr7839k.

Cheung, W.W.L., Lam, V.W.Y., Sarmiento, J.L., Kearney, K., Watson, R., Pauly, D., 2009. Projecting global marine biodiversity impacts under climate change scenarios. Fish Fish. 10, 235–251.

Cheung, W.W.L., Lam, V.W.Y., Sarmiento, J.L., Kearney, K., Watson, R., Zeller, D., Pauly, D., 2010. Large-scale redistribution of maximum fisheries catch potential in the global ocean under climate change. Glob. Chang. Biol. 16, 24–35.

Cheung, W.W.L., Sarmiento, J.L., Dunne, J., Frölicher, T.L., Lam, V.W.Y., Palomares, M.L.D., Watson, R., Pauly, D., 2013a. Shrinking of fishes exacerbates impacts of global ocean changes on marine ecosystems. Nat. Clim. Chang. 3, 254–258.

Cheung, W.W.L., Watson, R., Pauly, D., 2013b. Signature of ocean warming in global fisheries catch. Nature 497, 365–368.

Cheung, W.W.L., Jones, M.C., Reygondeau, G., Stock, C.A., Lam, V.W.Y., Frölicher, T.L., 2016. Structural uncertainty in projecting global fisheries catches under climate change. Ecol. Model. 325, 57–66.

Cheung, W.W.L., Jones, M.C., Reygondeau, G., Frölicher, T.L., 2018. Opportunities for climate-risk reduction through effective fisheries management. Glob. Chang. Biol. 24, 5149–5163.

Christensen, V., Coll, M., Piroddi, C., Steenbeek, J., Buszowski, J., Pauly, D., 2014. A century of fish biomass decline in the ocean. Mar. Ecol. Prog. Ser. 512, 155–166.

Coll, M., Steenbeek, J., Pennino, M.G., Buszowski, J., Garilao, C., Kaschner, K., Lotze, H.K., Rousseau, Y., Tittensor, D. P., Walters, C., Watson, R., Christensen, V., 2020. Advancing global ecological modelling capabilities to simulate future trajectories of change in marine ecosystems. Front. Mar. Sci. 7, 567877.

Costello, C., Ovando, D., Ray Hilborn, R., Gaines, S.D., Deschenes, O., Lester, S.E., 2012. Status and solutions for the world's unassessed fisheries. Science 338, 517–520.

Costello, C., Ovando, D., Clavelle, T., Strauss, C.K., Hilborn, R., Melnychuk, M.C., Branch, T.A., Gaines, S.D., Szuwalskia, C.S., Cabral, R.B., Rader, D.N., Leland, A., 2016. Global fishery prospects under contrasting management regimes. Proc. Natl. Acad. Sci. 113, 5125–5129.

Crain, C.M., Kroeker, K., Halpern, B.S., 2008. Interactive and cumulative effects of multiple human stressors in marine systems. Ecol. Lett. 11, 1304–1315.

Dahlke, F.T., Wohlrab, S., Butzin, M., Poertner, H.-O., 2020. Thermal bottlenecks in the life cycle define climate vulnerability of fish. Science 369, 65–70.

Davies, T., Maxwell, S., Kaschner, K., Garilao, C., Ban, N.C., 2017. Large marine protected areas represent biodiversity now and under climate change. Sci. Rep. 7, 9569.

Defra, 2013. Economics of Climate Resilience: Natural Environment Theme: Sea Fish. Department for Environment, Food & Rural Affairs London, United Kingdom, p. 99.

Donato, D.C., Kauffman, J.B., Murdiyarso, D., Kurnianto, S., Stidham, M., Kanninen, M., 2011. Mangroves among the most carbon-rich forests in the tropics. Nat. Geosci. 4, 293–297.

Drinkwater, K.F., 2005. The response of Atlantic cod (*Gadus morhua*) to future climate change. ICES J. Mar. Sci. 62, 1327–1337.

Duarte, C.M., Agusti, S., Barbier, E., Britten, G.L., Castilla, J.C., Gattuso, J.-P., Fulweiler, R.W., Hughes, T.P., Knowlton, N., Lovelock, C.E., Lotze, H.K., Predragovic, M., Poloczanska, E., Roberts, C., Worm, B., 2020. Rebuilding marine life. Nature 580, 39–51.

Duffy, J.E., Lefcheck, J.S., Stuart-Smith, R.D., Navarrete, S.A., Edgar, G.J., 2016. Biodiversity enhances reef fish biomass and resistance to climate change. Proc. Natl. Acad. Sci. 113, 6230–6235.

Dulvy, N.K., Sadovy, Y., Reynolds, J.D., 2003. Extinction vulnerability in marine populations. Fish Fish. 4, 25–64.

Dunn, D.C., Maxwell, S.M., Boustany, A.M., Halpin, P.N., 2016. Dynamic ocean management increases the efficiency and efficacy of fisheries management. Proc. Natl. Acad. Sci. 113, 668–673.

Edwards, M., Richardson, A.J., 2004. Impact of climate change on marine pelagic phenology and trophic mismatch. Nature 430, 881–884.

Ehlers, A., Worm, B., Reusch, T.B.H., 2008. Importance of genetic diversity in eelgrass *Zostera marina* for its resilience to global warming. Mar. Ecol. Prog. Ser. 355, 1–7.

Elmqvist, T., Folke, C., Nyström, M., Peterson, G., Bengtsson, J., Walker, B., Norberg, J., 2003. Response diversity, ecosystem change, and resilience. Front. Ecol. Environ. 1, 488–494.

European Commission, 2020. Farm to Fork Strategy. For a Fair, Healthy and Environmentally-Friendly Food System. https://ec.europa.eu/food/sites/food/files/safety/docs/f2f_action-plan_2020_strategy-info_en.pdf.

FAO, 2016. The State of World Fisheries and Aquaculture 2016. Food and Agriculture Organization, Rome.

FAO, 2020. The State of World Fisheries and Aquaculture 2020. Food and Agriculture Organization, Rome.

Field, C.B., Behrenfeld, M.J., Randerson, J.T., Falkowski, P., 1998. Primary production of the biosphere: integrating terrestrial and oceanic components. Science 281, 237–240.

Finnegan, S., Anderson, S.C., Harnik, P.G., Simpson, C., Tittensor, D.P., Byrnes, J.E., Finkel, Z.V., Lindberg, D.R., Liow, L.H., Lockwood, R., Lotze, H.K., McClain, C.R., McGuire, J.L., O'Dea, A., Pandolfi, J.M., 2015. Paleontological baselines for evaluating extinction risk in the modern oceans. Science 348, 567–570.

Fourqurean, J.W., Duarte, C.M., Kennedy, H., Marba, N., Holmer, M., Angel Mateo, M., Apostolaki, E.T., Kendrick, G. A., Krause-Jensen, D., McGlathery, K.J., Serrano, O., 2012. Seagrass ecosystems as a globally significant carbon stock. Nat. Geosci. 5, 505–509.

Frank, K.T., Petrie, B., Shackell, N.L., Choi, J.S., 2006. Reconciling differences in trophic control in mid-latitude marine ecosystems. Ecol. Lett. 9, 1096–1105.

Frank, K.T., Petrie, B., Shackell, N.L., 2007. The ups and downs of trophic control in continental shelf ecosystems. Trends Ecol. Evol. 22, 236–242.

Frank, K.T., Petrie, B., Leggett, W.C., Boyce, D.G., 2018. Exploitation drives an ontogenetic-like deepening in marine fish. Proc. Natl. Acad. Sci. 115 (25), 6–11.

Free, C.M., Thorson, J.T., Pinsky, M.L., Oken, K.L., Wiedenmann, J., Jensen, O.P., 2019. Impacts of historical warming on marine fisheries production. Science 363, 979–983.

Free, C.M., Mangin, T., Molinos, J.G., Ojea, E., Burden, M., Costello, C., Gaines, S.D., 2020. Realistic fisheries management reforms could mitigate the impacts of climate change in most countries. PLoS ONE 15 (3), e0224347.

Friedland, K.D., Langan, J.A., Large, S.I., Selden, R.L., Link, J.S., Watson, R.A., Collie, J.S., 2020. Changes in higher trophic level productivity, diversity and niche space in a rapidly warming continental shelf ecosystem. Sci. Total Environ. 704, 135270.

Froese, R., Zeller, D., Kleisner, K., Pauly, D., 2012. What catch data can tell us about the status of global fisheries. Mar. Biol. 159, 1283–1292.

Gaines, S.D., Costello, C., Owashi, B., Mangin, T., Bone, J., Molinos, J.G., Burden, M., Dennis, H., Halpern, B.S., Kappel, C.V., Kleisner, K.M., Ovando, D., 2018. Improved fisheries management could offset many negative effects of climate change. Sci. Adv. 4 (8), eaao1378.

Galbraith, E.D., Carozza, D.A., Bianchi, D., 2017. A coupled human-Earth model perspective on long-term trends in the global marine fishery. Nat. Commun. 8, 1–7.

Grady, J.M., Maitner, B.S., Winter, A.S., Kaschner, K., Tittensor, D.P., Record, S., Smith, F.A., Wilson, A.M., Dell, A.I., Zarnetske, P.L., Wearing, H.J., Alfaro, B., Brown, J.H., 2019. Metabolic asymmetry and the global diversity of marine predators. Science 363 (6425), eaat4220.

Gregg, R.M., Score, A., Pietri, D., Hansen, L., 2016. The State of Climate Adaptation in US Marine Fisheries Management. EcoAdapt, Bainbridge Island, WA, USA, p. 110.

Halpern, B.S., Frazier, M., Potapenko, J., Casey, K.S., Koenig, K., Longo, C., Lowndes, J.S., Rockwood, R.C., Selig, E.R., Selkoe, K.A., Walbridge, S., 2015. Spatial and temporal changes in cumulative human impacts on the world's ocean. Nat. Commun. 6, 7615.

Harley, C.D.G., Hughes, A.R., Hultgren, K.M., Miner, B.G., Sorte, C.J.B., Thornber, C.S., Rodriguez, L.F., Tomanek, L., Williams, S.L., 2006. The impacts of climate change in coastal marine systems. Ecol. Lett. 9, 228–241.

Harnik, P.G., Lotze, H.K., Anderson, S.C., Finkel, Z.V., Finnegan, S., Lindberg, D.R., Liow, L.H., Lockwood, R., McClain, C.R., McGuire, J.L., O'Dea, A., Pandolfi, J.M., Simpson, C., Tittensor, D.P., 2012. Extinctions in ancient and modern seas. Trends Ecol. Evol. 27, 608–617.

Harvell, C.D., Mitchell, C.E., Ward, J.R., Altizer, S., Dobson, A.P., Ostfeld, R.S., Samuel, M.D., 2002. Climate warming and disease risks for terrestrial and marine biota. Science 296, 2158–2162.

Henderson, P.A., 2007. Discrete and continuous change in the fish community of the Bristol Channel in response to climate change. J. Mar. Biol. Assoc. U. K. 87, 589–598.

Hicks, C.C., Cohen, P.J., Graham, N.A.G., Nash, K.L., Allison, E.H., D'Lima, C., Mills, D.J., Roscher, M., Thilsted, S.H., Thorne-Lyman, A.L., MacNeil, M.A., 2019. Harnessing global fisheries to tackle micronutrient deficiencies. Nature 574, 95–98.

Hiddink, J.G., Ter Hofstede, R., 2008. Climate induced increases in species richness of marine fishes. Glob. Chang. Biol. 14, 453–460.

Hilborn, R., Quinn, T.P., Schindler, D.E., Rogers, D.E., 2003. Biocomplexity and fisheries sustainability. Proc. Natl. Acad. Sci. 100, 6564–6568.

Hilborn, R., Amoroso, R.O., Anderson, C.M., Baum, J.K., Branch, T.A., Costello, C., De Moor, C.L., Faraj, A., Hively, D., Jensen, O.P., Kurota, H., Little, L.R., Mace, P., McClanahan, T., Melnychuk, M.C., Minto, C., Osio, G.C., Parma, A.M., Pons, M., Segurado, S., Szuwalski, C.S., Wilson, J.R., Ye, Y., 2020. Effective fisheries management instrumental in improving fish stock status. Proc. Natl. Acad. Sci. 117, 2218–2224.

Holsman, K.K., Haynie, A.C., Hollowed, A.B., Reum, J.C.P., Aydin, K., Hermann, A.J., Cheng, W., Faig, A., Ianelli, J. N., Kearney, K.A., Punt, A.E., 2020. Ecosystem-based fisheries management forestalls climate-driven collapse. Nat. Commun. 11, 4579.

Howard, J., Babij, E., Griffis, R., Helmuth, B., Himes-Cornell, A., Niemier, P., Orbach, M., Petes, L., Allen, S., Auad, G., Auer, C., Beard, R., Boatman, M., Bond, N., Boyer, T., Brown, D., Clay, P., Crane, K., Cross, S., Dalton, M., Diamond, J., Diaz, R., Dortch, Q., Duffy, E., Fauquier, D., Fisher, W., Graham, M., Halpern, B., Hansen, L., Hayum, B., Herrick, S., Hollowed, A., Hutchins, D., Jewett, E., Jin, D., Kowlton, N., Kotowicz, D., Kristiansen, T., Little, P., Lopez, C., Loring, P., Lumpkin, R., Mace, A., Menerink, K., Morrison, J.R., Murray, J., Norman, K., O'Donnell, J., Overland, J., Parsons, R., Pettigrew, N., Pfeiffer, L., Pidgeon, E.Y., Plummer, M., Polovina, J., Quintrell, J., Rowles, T., Runge, J., Rowles, T., Rust, M., Sanford, E., Send, U., Singer, M., Speir, C., Stanitski, D., Thornber,

C., Wilson, C., Zue, Y., 2013. Oceans and marine resources in a changing climate. Oceanogr. Mar. Biol. Annu. Rev. 51, 71–192.

Hughes, T.P., Kerry, J.T., Álvarez-Noriega, M., Álvarez-Romero, J.G., Anderson, K.D., Baird, A.H., Babcock, R.C., Beger, M., Bellwood, D.R., Berkelmans, R., 2017. Global warming and recurrent mass bleaching of corals. Nature 543, 373–377.

Hutchings, J.A., Reynolds, J.D., 2004. Marine fish population collapses: consequences for recovery and extinction risk. Bioscience 54, 297–309.

IPBES, 2019. In: Brondizio, E.S., Settele, J., Díaz, S., Ngo, H.T. (Eds.), Global Assessment Report on Biodiversity and Ecosystem Services of the Intergovernmental Science-Policy Platform on Biodiversity and Ecosystem Services. IPBES Secretariat, Bonn, Germany.

IPCC, 2019a. In: Pörtner, H.-O., Roberts, D.C., Masson-Delmotte, V., Zhai, P., Tignor, M., Poloczanska, E., Weyer, N. M. (Eds.), IPCC Special Report on the Ocean and Cryosphere in a Changing Climate. Intergovernmental Panel on Climate Change. In press.

IPCC, 2019b. Summary for policymakers. In: Pörtner, H.-O., Roberts, D.C., Masson-Delmotte, V., Zhai, P., Tignor, M., Poloczanska, E., Mintenbeck, K., Alegría, A., Nicolai, M., Okem, A., Petzold, J., Rama, B., Weyer, N.M. (Eds.), IPCC Special Report on the Ocean and Cryosphere in a Changing Climate. In press. https://www.ipcc.ch/srocc/chapter/summary-for-policymakers/.

Jones, M.C., Cheung, W.W.L., 2015. Multi-model ensemble projections of climate change effects on global marine biodiversity. ICES J. Mar. Sci. 72, 741–752.

Karp, M.A., Peterson, J.O., Lynch, P.D., Griffis, R.B., Adams, C.F., Arnold, W.S., Barnett, L.A., deReynier, Y., DiCosimo, J., Fenske, K.H., Gaichas, S.K., Hollowed, A., Holsman, K., Karnauskas, M., Kobayashi, D., Leising, A., Manderson, J.P., McClure, M., Morrison, W.E., Schnettler, E., Thompson, A., Thorson, J.T., Walter, J.F., Yau, A.J., Methot, R.D., Link, J.S., 2019. Accounting for shifting distributions and changing productivity in the development of scientific advice for fishery management. ICES J. Mar. Sci. 76, 1305–1315.

Kelly, R.P., Erickson, A.L., Mease, L.A., Battista, W., Kittinger, J.N., Fujita, R., 2015. Embracing thresholds for better environmental management. Philos. Trans. R. Soc. B 370, 20130276.

Kopke, K., O'Mahoney, C., 2011. Preparedness of key coastal and marine sectors in Ireland to adapt to climate change. Mar. Policy 35, 800–809.

Kwiatkowski, L., Aumont, O., Bopp, L., 2018. Consistent trophic amplification of marine biomass declines under climate change. Glob. Chang. Biol. 25, 218–229.

Kwiatkowski, L., Torres, O., Bopp, L., Aumont, O., Chamberlain, M., Christian, J., Dunne, J.P., Gehlen, M., Ilyina, T., John, J.G., Lenton, A., Li, H., Lovenduski, N.S., Orr, J.C., Palmieri, J., Schwinger, J., Séférian, R., Stock, C.A., Tagliabue, A., Takano, Y., Tjiputra, J., Toyama, K., Tsujino, H., Watanabe, M., Yamamoto, A., Yool, A., Ziehn, T., 2020. Twenty-first century ocean warming, acidification, deoxygenation, and upper ocean nutrient decline from CMIP6 model projections. Biogesciences. https://doi.org/10.5194/bg-2020-16.

Lawler, J.J., Tear, T.H., Pyke, C., Shaw, M.R., Gonzalez, P., Kareiva, P., Hansen, L., Hannah, L., Klausmeyer, K., Aldous, A., Bienz, C., Pearsall, S., 2010. Resource management in a changing and uncertain climate. Front. Ecol. Environ. 8, 35–43.

Le Bris, A., Mills, K.E., Wahle, R.A., Chen, Y., Alexander, M.A., Allyn, A.J., Schuetz, J.G., Scott, S.G., Pershing, A.J., 2018. Climate vulnerability and resilience in the most valuable North American fishery. Proc. Natl. Acad. Sci. 115, 1831–1836.

Lefort, S., Aumont, O., Bopp, L., Arsouze, T., Gehlen, M., Maury, O., 2015. Spatial and body-size dependent response of marine pelagic communities to projected global climate change. Glob. Chang. Biol. 21, 154–164.

Letcher, T., 2016. Climate and Global Change: Observed Impacts on Planet Earth, second ed. Elsevier.

Letcher, T., 2021. Climate and Global Change: Observed Impacts on Planet Earth, third ed. Elsevier.

Lewandowska, A.M., Boyce, D.G., Hofmann, M., Matthiessen, B., Sommer, U., Worm, B., 2014. Effects of sea surface warming on marine plankton. Ecol. Lett. 17, 614–623.

Lindegren, M., Brander, K., 2018. Adapting fisheries and their management to climate change: a review of concepts, tools, frameworks, and current progress toward implementation. Rev. Fish. Sci. Aquac. 26, 400–415.

Link, J.S., Griffis, R., Busch, S., 2015. NOAA fisheries climate science strategy. In: NOAA Technical Memorandum NMFS-F/SPO-155, p. 70.

Loreau, M., 2010. Linking biodiversity and ecosystems: towards a unifying ecological theory. Philos. Trans. R. Soc. B 365, 49–60.

Lotze, H.K., Lenihan, H.S., Bourque, B.J., Bradbury, R.H., Cooke, R.G., Kay, M.C., Kidwell, S.M., Kirby, M.X., Peterson, C.H., Jackson, J.B.C., 2006. Depletion, degradation, and recovery potential of estuaries and coastal seas. Science 312, 1806–1809.

Lotze, H.K., Coll, M., Dunne, J., 2011a. Historical changes in marine resources, food-web structure and ecosystem functioning in the Adriatic Sea. Ecosystems 14, 198–222.

Lotze, H.K., Coll, M., Magera, A.M., Ward-Paige, C.A., Airoldi, L., 2011b. Recovery of marine animal populations and ecosystems. Trends Ecol. Evol. 26, 595–605.

Lotze, H.K., Froese, R., Pauly, D., 2018. The future of marine fisheries. In: Pauly, D., Ruiz-Leotaud, V. (Eds.), Marine and Freshwater Miscellanea. 26. Fisheries Centre Research Reports, pp. 5–13.

Lotze, H.K., Tittensor, D.P., Bryndum-Buchholz, A., Eddy, T.D., Cheung, W.W.L., Galbraith, E.D., Barange, M., Barrier, N., Bianchi, D., Blanchard, J.L., Bopp, L., Büchner, M., Bulman, C.M., Carozza, D.A., Christensen, V., Coll, M., Dunne, J.P., Fulton, E.A., Jennings, S., Jones, M.C., Mackinson, S., Maury, O., Niiranen, S., Oliveros-Ramos, R., Roy, T., Fernandes, J.A., Schewe, J., Shin, Y.-J., Silva, T.A.M., Steenbeek, J., Stock, C.A., Verley, P., Volkholz, J., Walker, N.D., Worm, B., 2019. Global ensemble projections reveal trophic amplification of ocean biomass declines with climate change. Proc. Natl. Acad. Sci. 116, 12907–12912.

MacKenzie, B.R., Payne, M.R., Boje, J., Hoyer, J.L., Siegstad, H., Hoyer, J.L., Siegstad, H., 2014. A cascade of warming impacts brings bluefin tuna to Greenland waters. Glob. Chang. Biol. 20, 2484–2491.

Marshall, K.N., Koehn, L.E., Levin, P.S., Essington, T.E., Jensen, O.P., 2018. Inclusion of ecosystem information in US fish stock assessments suggests progress toward ecosystem-based fisheries management. ICES J. Mar. Sci. 76, 1–9.

McCauley, D.J., Pinsky, M.L., Palumbi, S.R., Estes, J.A., Joyce, F.H., Warner, R.R., 2015. Marine defaunation: animal loss in the global ocean. Science 347, 1255641.

Mendenhall, E., Hendrix, C., Nyman, E., Roberts, P.M., Hoopes, J.R., Watson, J.R., Lam, V.W.Y., Sumaila, U.R., 2020. Climate change increases the risk of fisheries conflict. Mar. Policy 117, 103954.

Michel, C., Bluhm, B., Gallucci, V., Gaston, A.J., Gordillo, F.J.L., Gradinger, R., Hopcroft, R., Jensen, N., Mustonen, T., Niemi, A., Nielsen, T.G., 2012. Biodiversity of Arctic marine ecosystems and responses to climate change. Biodiversity 13, 200–214.

Morán, X.A.G., López-Urrutia, A., Calvo-Díaz, A., Li, W.K.W., 2010. Increasing importance of small phytoplankton in a warmer ocean. Glob. Chang. Biol. 16, 1137–1144.

Mullon, C., Steinmetz, F., Merino, G., Frenandes, J.A., Cheung, W.W.L., Butenschoen, M., Barange, M., 2016. Quantitative pathways for Northeast Atlantic fisheries based on climate, ecological-economic and governance modelling scenarios. Ecol. Model. 320, 273–291.

Niemi, A., Ferguson, S., Hedges, K., Melling, H., Michel, C., Ayles, B., Azetsu-Scott, K., Coupel, P., Deslauriers, D., Devred, E., Doniol-Valcroze, T., Dunmall, K., Eert, J., Galbraith, P., Geoffroy, M., Gilchrist, G., Hennin, H., Howland, K., Kendall, M., Kohlbach, D., Lea, E., Loseto, L., Majewski, A., Marcoux, M., Matthews, C., McNicholl, D., Mosnier, A., Mundy, C.J., Ogloff, W., Perrie, W., Richards, C., Richardson, E., Reist, J., Roy, V., Sawatzky, C., Scharffenberg, K., Tallman, R., Tremblay, J.-E., Tufts, T., Watt, C., Williams, W., Worden, E., Yurkowski, D., Zimmerman, S., 2019. State of Canada's Arctic seas. Can. Tech. Rep. Fish. Aquat. Sci. 3344, 189.

Niiranen, S., Yletyinen, J., Tomczak, M.T., Blenckner, T., Hjerne, O., MacKenzie, B.R., Mueller-Karulis, B., Neumann, T., Meier, H.E.M., 2013. Combined effects of global climate change and regional ecosystem drivers on an exploited marine food web. Glob. Chang. Biol. 19, 3327–3342.

Norris, R.D., Kirtland Turner, S., Hull, P.M., Ridgwell, A., 2013. Marine ecosystem responses to cenozoic global change. Science 341, 492–498.

Nye, J.A., Link, J.S., Hare, J.A., Overholtz, W.J., 2009. Changing spatial distribution of fish stocks in relation to climate and population size on the Northeast United States continental shelf. Mar. Ecol. Prog. Ser. 393, 111–129.

Nye, J.A., Joyce, T.M., Kwon, Y.-O., Link, J.S., 2011. Silver hake tracks changes in Northwest Atlantic circulation. Nat. Commun. 2, 412.

O'Connor, M.I., Piehler, M.F., Leech, D.M., Anton, A., Bruno, J.F., 2009. Warming and resource availability shift food web structure and metabolism. PLoS Biol. 7, e1000178.

Ogier, E.M., Davidson, J., Fidelman, P., Haward, M., Hobday, A.J., Holbrook, N.J., Pecl, G.T., 2016. Fisheries management approaches as platforms for climate change adaptation: comparing theory and practice in Australian fisheries. Mar. Policy 71, 82–93.

Oliver, E.C.J., Benthuysen, J.A., Bindoff, N.L., Hobday, A.J., Holbrook, N.J., Mundy, C.N., Perkins-Kirkpatrick, S.E., 2017. The unprecedented 2015/16 Tasman Sea marine heatwave. Nat. Commun. 8, 16101.

Oliver, E.C.J., Donat, M.G., Burrows, M.T., Moore, P.J., Smale, D.A., Alexander, L.V., Benthuysen, J.A., Feng, M., Gupta, A.S., Hobday, A.J., Holbrook, N.J., Perkins-Kirkpatrick, S.E., Scannell, H.A., Straub, S.C., Wernberg, T., 2018. Longer and more frequent marine heatwaves over the past century. Nat. Commun. 9, 1324.

Pauly, D., Zeller, D., 2016. Catch reconstructions reveal that global marine fisheries catches are higher than reported and declining. Nat. Commun. 7, 10244.

Pecl, G.T., Araújo, M.B., Bell, J.D., Blanchard, J., Bonebrake, T.C., Chen, I.-C., Clark, T.D., Colwell, R.K., Danielsen, F., Evengård, B., Falconi, L., Ferrier, S., Frusher, S., Garcia, R.A., Griffis, R.B., Hobday, A.J., Janion-Scheepers, C., Jarzyna, M.A., Jennings, S., Lenoir, J., Linnetved, H.I., Martin, V.Y., McCormack, P.C., McDonald, J., Mitchell, N.J., Mustonen, T., Pandolfi, J.M., Pettorelli, N., Popova, E., Robinson, S.A., Scheffers, B.R., Shaw, J.D., Sorte, C.J.B., Strugnell, J.M., Sunday, J.M., Tuanmu, M.-N., Vergés, A., Villanueva, C., Wernberg, T., Wapstra, E., Williams, S.E., 2017. Biodiversity redistribution under climate change: impacts on ecosystems and human well-being. Science 355, eaai9214.

Perry, A.L., Low, P.J., Ellis, J.R., Reynolds, J.D., 2005. Climate change and distribution shifts in marine fishes. Science 308, 1912–1915.

Pershing, A.J., Alexander, M.A., Hernandez, C.M., Kerr, L.A., Le Bris, A., Mills, K.E., Nye, J.A., Record, N.R., Scannell, H.A., Scott, J.D., Sherwood, G.D., Thomas, A.C., 2015. Slow adaptation in the face of rapid warming leads to collapse of the Gulf of Maine cod fishery. Science 350, 809–812.

Petrie, B., Frank, K.T., Shackell, N.L., Leggett, W.C., 2009. Structure and stability in exploited marine fish communities: quantifying critical transitions. Fish. Oceanogr. 18, 83–101.

Pinsky, M.L., Fogarty, M.J., 2012. Lagged social-ecological responses to climate and range shifts in fisheries. Clim. Chang. 115, 883–891.

Pinsky, M.L., Worm, B., Fogarty, M.J., Sarmiento, J.L., Levin, S.A., 2013. Marine taxa track local climate velocities. Science 341, 1239–1242.

Pinsky, M.L., Eikeset, A.M., McCauley, D.J., Payne, J.L., Sunday, J.M., 2019. Greater vulnerability to warming of marine versus terrestrial ectotherms. Nature 569, 108–111.

Platt, T., Fuentes-Yaco, C., Frank, K.T., 2003. Spring algal bloom and larval fish survival. Nature 423, 398–399.

Poloczanska, E.S., Brown, C.J., Sydeman, W.J., Kiessling, W., Schoeman, D.S., Moore, P.J., Brander, K., Bruno, J.F., Buckley, L.B., Burrows, M.T., Duarte, C.M., Halpern, B.S., Holding, J., Kappel, C.V., O'Connor, M.I., Pandolfi, J.M., Parmesan, C., Schwing, F., Thompson, S.A., Richardson, A.J., 2013. Global imprint of climate change on marine life. Nat. Clim. Chang. 3, 919–925.

Poloczanska, E.S., Burrows, M.T., Brown, C.J., Garcia Molinos, J., Halpern, B.S., Hoegh-Guldberg, O., Kappel, C.V., Moore, P.J., Richardson, A.J., Schoeman, D.S., Sydeman, W.J., 2016. Responses of marine organisms to climate change across oceans. Front. Mar. Sci. 3, 62.

Quaas, M.F., Reusch, T.B.H., Schmidt, J.O., Tahvonen, O., Voss, R., 2015. It is the economy, stupid! Projecting the fate of fish populations using ecological–economic modeling. Glob. Chang. Biol. 22, 264–270.

Reusch, T.B.H., Ehlers, A., Hämmerli, A., Worm, B., 2005. Ecosystem recovery after climatic extremes enhanced by genotypic diversity. Proc. Natl. Acad. Sci. 102, 2826–2831.

Roberts, C.M., O'Leary, B.C., McCauley, D.J., Cury, P.M., Duarte, C.M., Lubchenco, J., Pauly, D., Sáenz-Arroyo, A., Sumaila, U.R., Wilson, R.W., Worm, B., Castilla, J.C., 2017. Marine reserves can mitigate and promote adaptation to climate change. Proc. Natl. Acad. Sci. 114, 6167–6175.

Sabine, C.L., Feely, R.A., Gruber, N., Key, R.M., Lee, K., Bullister, J.L., Wanninkhof, R., Wong, C.S., Wallace, D.W.R., Tilbrook, B., Millero, F.J., Peng, T.-H., Kozyr, A., Ono, T., Rios, A.F., 2004. The oceanic sink for anthropogenic $CO_2$. Science 305, 367–371.

Scheffers, B.R., De Meester, L., Bridge, T.C.L., Hoffmann, A.A., Pandolfi, J.M., Corlett, R.T., Butchart, S.H.M., Pearce-Kelly, P., Kovacs, K.M., Dudgeon, D., Pacifici, M., Rondinini, C., Foden, W.B., Martin, T.G., Mora, C., Bickford, D., Watson, J.E.M., 2016. The broad footprint of climate change from genes to biomes to people. Science 354, aaf7671.

Schindler, D.E., Hilborn, R., Chasco, B., Boatright, C.P., Quinn, T.P., Rogers, L.A., Webster, M.S., 2010. Population diversity and the portfolio effect in an exploited species. Nature 465, 609–613.

Shackell, N.L., Frank, K.T., Fisher, J.A.D., Petrie, B., Leggett, W.C., 2010. Decline in top predator body size and changing climate alter trophic structure in an oceanic ecosystem. Proc. R. Soc. B Biol. Sci. 277, 1353–1360.

Shackell, N.L., Ricard, D., Stortini, C., 2014. Thermal habitat index of many Northwest Atlantic temperate species stays neutral under warming projected for 2030 but changes radically by 2060. PLoS ONE 9 (3), e90662.

Sheridan, J.A., Bickford, D., 2011. Shrinking body size as an ecological response to climate change. Nat. Clim. Chang. 1, 401–406.

Simpson, S.D., Jennings, S., Johnson, M.P., Blanchard, J.L., Schoen, P.-J., Sims, D.W., Genner, M.J., 2011. Continental shelf-wide response of a fish assemblage to rapid warming of the sea. Curr. Biol. 21, 1565–1570.

Skern-Mauritzen, M., Olsen, E., Huse, G., 2018. Opportunities for advancing ecosystem-based management in a rapidly changing, high latitude ecosystem. ICES J. Mar. Sci. 75 (7), 2425–2433.

Stenseth, N.C., Mysterud, A., Ottersen, G., Hurrell, J.W., Chan, K.-S., Lima, M., 2002. Ecological effects of climate fluctuations. Science 297, 1292–1296.

Sumaila, U.R., Cheung, W.W.L., 2010. Cost of adapting fisheries to climate change. In: Development and Climate Change, Discussion Paper No. 5. World Bank.

Sumaila, U.R., Tai, T.C., 2020. End overfishing and increase the resilience of the ocean to climate change. Front. Mar. Sci. 7 (523), 1–8.

Tai, T.C., Steiner, N.S., Hoover, C., Cheung, W.W.L., Sumaila, U.R., 2019. Evaluating present and future potential of arctic fisheries in Canada. Mar. Policy 108, 103637.

Tanaka, K.R., 2019. Integrating environmental information into stock assessment models for fisheries management. In: Cisneros-Montemayor, A.M., Cheung, W.W.L., Ota, Y. (Eds.), Predicting Future Oceans. Elsevier, pp. 193–206.

Taylor, A.H., Allen, J.I., Clark, P.A., 2002. Extraction of a weak climatic signal by an ecosystem. Nature 416, 629–632.

Tittensor, D.P., Eddy, T.D., Lotze, H.K., Galbraith, E.D., Cheung, W.W.L., Barange, M., Walker, N. (Eds.), 2018. A protocol for the intercomparison of marine fishery and ecosystem models: Fish-MIP v1.0. Geosci. Model Dev. 11, 1421–1442.

Tittensor, D.P., Beger, M., Börder, K., Boyce, D., Cavanagh, R., Cosandey-Godin, A., Crespo, G.O., Dunn, D., Ghiffary, W., Grant, S.M., Hannah, L., Halpin, P., Harfoot, M., Heaslip, S.G., Jeffery, N.W., Kingston, N., Lotze, H.K., McLeod, E., McGowan, J., McOwen, C.J., O'Leary, B.C., Schiller, L., Stanley, R.R.E., Westhead, M., Wilson, K. L., Worm, B., 2019. Integrating climate adaptation and biodiversity conservation in the global ocean. Sci. Adv. 5 (11), eaay9969.

Trisos, C.H., Merow, C., Pigot, A.L., 2020. The projected timing of abrupt ecological disruption from climate change. Nature 580, 496–501.

Wernberg, T., Russell, B.D., Thomsen, M.S., Gurgel, C.F.D., Bradshaw, C.J.A., Poloczanska, E.S., Connell, S.D., 2011. Seaweed communities in retreat from ocean warming. Curr. Biol. 21, 1828–1832.

Wernberg, T., Bennett, S., Babcock, R.C., Bettignies, T.D., Cure, K., Depczynski, Dufois, F., Fromont, J., Fulton, C.J., Hovey, R.K., Harvey, E.S., Holmes, T.H., Kendrick, G.A., Radford, B., Santana-Garcon, J., Saunders, B.J., Smale, D. A., Thomsen, M.S., Tuckett, C.A., Tuya, F., Vanderklift, M.A., Wilson, S., 2016. Climate-driven regime shift of a temperate marine ecosystem. Science 353, 169–172.

Wilson, K.L., Lotze, H.K., 2019. Projected range shift of eelgrass (*Zostera marina*) in the Northwest Atlantic with climate change. Mar. Ecol. Prog. Ser. 620, 47–62.

Wilson, K.L., Skinner, M.A., Lotze, H.K., 2019. Projected 21st century distribution of canopy-forming seaweeds in the Northwest Atlantic with climate change. Divers. Distrib. 25, 582–602.

Wilson, K.L., Tittensor, D.P., Worm, B., Lotze, H.K., 2020. Incorporating climate change adaption into marine protected area planning. Glob. Chang. Biol. 26, 3251–3267.

Worm, B., Branch, T.A., 2012. The future of fish. Trends Ecol. Evol. 27, 594–599.

Worm, B., Lotze, H.K., 2016. Marine biodiversity and climate change. In: Letcher, T. (Ed.), Climate and Global Change: Observed Impacts on Planet Earth, second ed. Elsevier, pp. 195–212.

Worm, B., Lotze, H.K., 2021. Marine biodiversity and climate change. In: Letcher, T. (Ed.), Climate and Global Change: Observed Impacts on Planet Earth, third ed. Elsevier. In press.

Worm, B., Sandow, M., Oschlies, A., Lotze, H.K., Myers, R.A., 2005. Global patterns of predator diversity in the open ocean. Science 309, 1365–1369.

Worm, B., Barbier, E.B., Beaumont, N., Duffy, J.E., Folke, C., Halpern, B.S., Jackson, J.B.C., Lotze, H.K., Micheli, F., Palumbi, S.R., Sala, E., Selkoe, K.A., Stachowicz, J.J., Watson, R., 2006. Impacts of biodiversity loss on ocean ecosystem services. Science 314, 787–790.

Worm, B., Hilborn, R., Baum, J.K., Branch, T.A., Collie, J.S., Costello, C., Fogarty, M.J., Fulton, E.A., Hutchings, J.A., Jennings, S., Jensen, O.P., Lotze, H.K., Mace, P.M., McClanahan, T.R., Minto, C., Palumbi, S.R., Parma, A.M., Ricard, D., Rosenberg, A.A., Watson, R., Zeller, D., 2009. Rebuilding global fisheries. Science 325, 578–585.

Yachi, S., Loreau, M., 1999. Biodiversity and ecosystem productivity in a fluctuating environment: the insurance hypothesis. Proc. Natl. Acad. Sci. 96, 1463–1468.

CHAPTER

# 9

# Effect of climate change on food production (animal products)

*Haorui Wu*[a] *and Florence Etienne*[b]

[a]School of Social Work, Dalhousie University, Halifax, NS, Canada [b]Independent Researcher, Vancouver, BC, Canada

OUTLINE



## 1 Introduction

Around the globe, climate change is a major threat to human and nonhuman inhabitants (e.g., livestock, wildlife, and plants). Combating climate change remains a long-standing debate and an ongoing action item for the United Nations (UN) and its broad supporting organizations and networks. The UN Framework Convention on Climate Change

*The Impacts of Climate Change*
https://doi.org/10.1016/B978-0-12-822373-4.00019-7

(United Nations, 1992), mobilized all countries to reach a common understanding of the climate emergency, its potential solutions through binding instruments such as the 1995 Kyoto Protocol (United Nations Framework Convention on Climate Change (UNFCCC), 2008), the 2015 Paris Agreement (UN, 2016). These initiatives cumulated to the objectives set under the 2030 Sustainable Development Goals (SDGs) (n.d.). Among the17 SDGs, the Goal 13 is the only one that intrinsically consolidates all other 16 goals, serving, thus, as the cornerstone for parties to build upon to achieve compliance by 2030 (Sustainable Development Goals, n.d.).

Greenhouse gases (GHGs) are globally deemed the major factor exacerbating the climate crisis. Human activity, particularly fossil fuel combustion, agriculture, urbanization, and deforestation are the key GHGs sources (Government of Canada, 2020). The United Nations Environment Program (UNEP) (2009) states that "we have less than 10 years to halt the global rise in GHGs if we are to avoid catastrophic consequences" (p. 5). In view of the ascending curve of GHGs since 2010 and the alarming annual records, the UN called for a Climate Action Summit in 2019 urging that all parties reinforce their commitments regarding global warming (UN, 2019a). Worldwide cases have illustrated how current mitigation protocols always got implemented, too late, too little, or not at all. The events unfolding have brought upon catastrophic, unalterable results on the cultural, ecological, economic, social, and political realities of many in international communities (Jakob et al., 2012). "The window is closing"; humans must act now (para. 1) (Chemnick, 2020).

In 2019, the United Nations launched ActNow, a global appeal for each and every one to take due action for climate change (UN, 2019b). The campaign is a critical part of the UN's coordinated vision and efforts to implement the Paris Agreement. Every citizen is then called upon to alleviate climate change in earnest to opt for "Meat-Free Meals" (para. 8) (UN, 2019b) because the livestock sector has been confirmed as "one of the top two or three most significant contributors to the most serious of environmental problems, from local to global" (p. xx) (FAO, 2020). Communities worldwide are, thus, becoming more informed of the undeniable impact of animal products, with their significant contribution toward the climate emergency. Some nations proactively consider measures that will reduce meat consumption in order to combat climate change and protect the environment. For example, Germany planned to introduce a "meat tax" (Weston, 2019) and China projected to reduce by 50% meat consumption throughout the country (Milman and Leavenworth, 2016). Such measures shed light once more on the intense albeit unresolved debate around climate change and its impact on animal consumption and production.

Internationally, during the 50-year period (1961–2014), meat consumption per capita has nearly tripled (from 23 kg in 1961 to 64 kg in 2014) (Graca, 2016) while the world's population has actually doubled (UN, 2019c). This hastened "appetite for meat" worsens: soil conditions, freshwater supply, greenhouse gas (GHG) emissions, and methane emanations. The stages of producing animal-based ingredients to answer human needs (e.g., dairy, meat, leather, etc.) require abundant natural resources. The higher the demand on outputs in answer to market demand the more serious the correlation between the climate crisis and animal products. Such interaction makes for what can be considered a rather "vicious cycle" of negative reflexivity. This seemingly unending wheel of cause-effect places tremendous burdens on natural and built environments (Devlin, 2018). As humans intensify their interventions on nature, especially by razing down forests to make for pasturelands and spraying new chemical formulas for greater yields on crops already impoverished by pesticides, their effects both cause and compound environmental degradation, consequently and negatively influencing agriculture, animal product notably, as well as jeopardizing further human and nonhuman residents' health and well-being (Carter and Woodworth, 2018).

Demand for animal-derived goods is both a catalyst and a constraint for climate change. According to the United States Department of Agriculture (n.d.), animal products (as derived from animal husbandry and livestock production) signify all animals farmed, wild, and domestic (excluding fishing) exploited for meat, dairy, wool, and leather. The United Nations Food and Agriculture Organization (FAO) (2019) states that "livestock contributes to nearly 40 percent of total agricultural output in developed countries and 20 percent in developing countries" (para. 6). Cattle herds represent over 17% of the livelihood and 34% globally of the key resource for food protein of citizens globally (United Nations Food and Agriculture Organization, 2019).

Freshwater, land, and forest, all-natural resources essential to our very existence are dwindling. We must imperatively re-evaluate our concepts and actions if we are to ensure food security and proper nutrition while implementing long-term solutions to solve climate emergencies. We must review our current modes of feeding, transporting, educating, and earning our existence. The findings and our decisions will determine whether we survive or thrive and if we will set the grounds for a livable Earth for generations forth.

## 2 Climate change and animal food productions

Agriculture contributes to and is affected by climate change (Legg and Huang, 2020). As we tackle and resolve the interconnectivity part of the equation, we will succeed in alleviating its critical effects on ourselves, animals, and the planet. Humans have depended on animals for nutrition. Animals, in turn, depend on water, soil, and the conditions set by humans to graze and develop—both seemingly linked by a common denominator of "food and feed." Climate change is quantified through its diverse and numerous exteriorizations. Extreme weather patterns and long-term environmental degradation seem the most widely known aspects of climate change: unusual floods, earthquakes, tornados, typhoons, sea-level rise, the disappearance of water bodies, and dead sea zones. Far too many, around the globe grapple daily with a more dire and direct experiencing climate change: food and water storages, forced agriculture-related migration and animal-borne and water-borne diseases (Intergovernmental Panel on Climate Change (IPCC), 2014).

The Oxford dictionary announced "climate emergency" as the Oxford Word of 2019 (Pytel, 2019). In light of the climate events of this decade alone, urgent actions are indeed needed to reverse the environmental damage. This section analyzes how the world fared when it comes to plant- and animal-based proteins from human agriculture interventions (National Geographic, 2014). We then reflect on the incidence of climate change on animal products based on extreme weather events, extreme temperatures, and livestock diseases.

### 2.1 Climate-related disasters

With climate change, we observe a considerable surge in the intensity, scope, and frequency of extreme weather events (e.g., hurricanes, tornados, storms, and floods). These events are compounded by secondary disasters (e.g., floods can happen due to hurricanes, and infections can ensue due to floods). These events continue to greatly threaten the lives of humans and livestock, to devastate crops, and to cause heavy damage to man-made infrastructures.

When environmental disasters or extreme weather events occur, both domestic and wild animals find themselves in immediate great danger, with little or no recourse to save their lives contrarily to the human population. Flying debris, the actual structure of a farmhouse, and other facilities may prevent the animals from escaping during these extreme events and thus lead to higher incidences of injury or even death (Texas A&M Agrilife Extension, 2016). As an illustration, hundreds of livestock were drowned or stranded during floods throughout the US Midwest in 2019 (Hanna and Baldacci, 2019). When Hurricane Dorian plummeted the Eastern littoral of North America, wild horses and cattle operations established in Cedar Island, North Carolina, were swirled away and only a handful survived (Katz, 2019). Again in 2019, unprecedented bush fires in Australia caused the disappearance of 480 million wild animals, including mammals, birds, and reptiles (BBC News, 2019).

Extreme weather events leave in their wake a toll of heavy losses when it comes to animals (domestic and wild) and the products we derive from them (meat, dairy, leather, etc.). Disasters always place the surviving animals in precarious situations. The first concern is their feed as damaged fields signify lower yields—if a harvest remains possible or maligned crops unfit for fodder (Babinszky et al., 2011). Correlatively, a loss of harvest or lower agricultural yields indirectly affects the quantity and quality of food available to humans as well as to animals. As humans face food insecurity, so do their domestic animals.

Extreme weather patterns may cause wild animals to be displaced from their habitat, relocating close to animal farms. Domestic animals could find themselves in proximity to their natural enemies, and a farmer unable to affect predator control. This poses additional threats to already weakened disaster survivors of animals. As an example, floods forced snakes to seek higher and dryer grounds causing them to move into farms and bite domestics animals (Frank, 2017). The US Federal Emergency Management Agency (2007) published an emergency tool kit to help farmers formulate emergency preparedness plans for farm animals.

## 2.2 Extreme temperatures

In addition to weather events, animals are sensitive to extreme temperatures (Cho, 2018). Other temperature-relevant factors, such as humidity, radiation, air movements are to be taken into account (Food and Agriculture Organization, 1986). According to the Food and Agriculture Organization (1986) "all domestic livestock are homoeothermic" (para. 4). Their comfort zones vary from 10°C to 20°C as temperatures play an important role in their metabolic heat production, their capacity to grow, develop, and produce (Food and Agriculture Organization, 1986). Nearly all animals can innately react adequately to the given fluctuation of temperatures; however, extreme highs (heat waves) or lows (freezing conditions) seriously affect their welfare and even cause death.

It is understood that heat stress will have costly effects on decreasing the animals' feed intake and increasing their water intake (Rotter and Van de Geijn, 1999). Meanwhile, heat stress will change the animals' respiration rate by lowering their reproductive efficiency (Nardone et al., 2010). Hot weather will downgrade cows' feed intake by 10%–20%, reflected as a 15%–40% reduced milk yield (Manitoba, n.d.). Heat stress is also known to impact the animals' behaviors and even raise livestock mortality (Rojas-Downing et al., 2017). According to Morignat et al. (2014) in France, in 2003, after a heatwave lasting 2-weeks, the mortality rate hit a historical record of 24%. Responding interventions such as providing abundant shade and shelter, cool water, and daily monitoring will reduce animal loss (PIRSA, 2015).

What can be said of temperature tables in today's context of the climate crisis? Most animals residing in the temperate and polar zones are able to sustain low temperatures when qualitative and quantitative feed is provided (Food and Agriculture Organization, 1986). Animals in these zones may even develop "extreme cold hardiness in Ectotherms" (p. 3) (Costanzo, 2011). A cold environment jeopardizes the survival of young animals (Babinszky et al., 2011), while in a tropical zone even brief exposure to extreme ambient temperature sets off a "risk of irreparable injury or death" (p. 3) (Costanzo, 2011). As global temperature turns unpredictable, it becomes increasingly difficult and costly for farmers to shield their animals at a reasonable level to ensure survival.

## 2.3 Livestock diseases

Climate change impacts microbial communities by accelerating the growth of pathogens and/or parasites, inducing vector-borne and food-borne diseases in domestic animals and at times in humans (see the next section) (Rojas-Downing et al., 2017; Slenning, 2010). In areas affected by floods, rising temperatures could activate pathogens contaminating the water and feed available. This phenomenon known as "moldy feed" can easily infect the animals, particularly those weakened from surviving a disaster (Schattenberg et al., 2017). Common vector-borne insects, such as mosquitoes and ticks, seem evermore present with global warming (Thornton et al., 2009). These pests could serve as super spreaders in transmitting fatal pathogens (e.g., anaplasmosis) to livestock (Texas AgriLife Extension Service, n.d.). Moreover, when climate change events cause mass animal deaths, dead bodies and warmer conditions can lead to pathogens to develop and disease to be transmitted among animal hosts (Thornton et al., 2009) and even between animals and humans (University of Queensland, 2019). Consequently, the World Health Organization (WHO) (2004) stresses the importance of disposing corpses effectively in order to reduce instances of disease and loss of livestock.

Climate change also affects animal products through other dimensions such as water supply (Ritchie and Roser, 2018), soil condition (Scheelbeeka et al., 2018), biological diversity and biological evolution (FAO, 2007), as well as crop yields and food security (US Climate Change Science Program, 2008). In fact, farmers all over the globe rely on rainwater to irrigate 80% of crops cultivated for humans and/or animals (Smith et al., 2013). Climate change has dangerously reduced not only such vital water supply (lower or unpredictable rainfall, drying of water bodies, droughts) but also the very soil (loss of important nutrients, of arable land mass), all cumulating in degraded food products. Globally, the yield in agricultural products (vegetal and legume) could decline by 35% by 2100 if GHGs emissions remain at current levels (Scheelbeeka et al., 2018). The Intergovernmental Panel on Climate Change (IPCC) establishes that most species of plants and animals must evolve 10,000 times faster than habitually to avoid extinction (Intergovernmental Panel on Climate Change (IPCC), 2014). The IPCC estimates that 20% to 30% of species assessed face extinction should global temperatures attain the levels predicted for 2100 (Cho, 2018). Animal agriculture depletes natural resources massively accelerating environmental degradation and compounding climate change (Grossi et al., 2019). The next section transposes the equation by analyzing the effect of animal production on climate change.

# 3 Animal production impacts climate change

According to Poore and Nemecek (2018), food and agriculture account for 26% of global GHGs emissions and 78% of global freshwater and ocean pollution. The animal sector alone tallies to 61% of the total GHGs generated for food production. This statistic rises to 81% if deforestation is factored in Poore and Nemecek (2018). Farming production (plant and animal) fetched 70% of all freshwater available globally (FAO, 2011), covering 50% of habitable land on the globe (Owen, 2005). Ritchie and Roser (2020) further confirmed that "food, therefore, remains central to discussions centered on solving climate change by reducing water stress, pollution, restoring lands to forests or grasslands, and protecting the world's wildlife" (para. 2). This section will examine the animal productions' catastrophic imprint on climate through four aspects: GHGs, land degradation, water shortage and water pollution, and loss of biodiversity.

## 3.1 A comprehensive calculation of GHGs from the animal sector

It is well documented that GHGs are a fundamental catalyst of climate change; animal production, especially livestock, remains a major supplier of GHGs on a global scale (Bailey et al., 2020). In 2006, a widely-cited report from the FAO determined that livestock produces 18% of annual global GHGs (FAO, 2016). Goodland and Anhang (2019) argue this number did not encompass the total GHGs truly associated with animal products when certain factors are excluded, such as "respiration by livestock," "land use," and "uncountable methane" (p. 11). All determinants once tallied up, livestock (and by-products) would thus contribute to more than half of global GHGs (Goodland and Anhang, 2019). To illustrate, Table 1

**TABLE 1** Environmental impact of cow milk and plant-based milk.

| Name (1 L) | GHG emissions (kg $CO_2$ eq) | Land use ($M^2$) | Fresh water (L) |
|---|---|---|---|
| **Cow's milk** | **3.2** | **9** | **628** |
| Rice milk | 1.2 | 0.3 | 270 |
| Soy milk | 1 | 0.7 | 28 |
| Oat milk | 0.9 | 0.8 | 49 |
| Almond milk | 0.7 | 0.5 | 371 |

*Data from Poore, J., Nemecek, T., 2018. Reducing food's environmental impacts through producers and consumers. Science 360(6392), 987–992.*

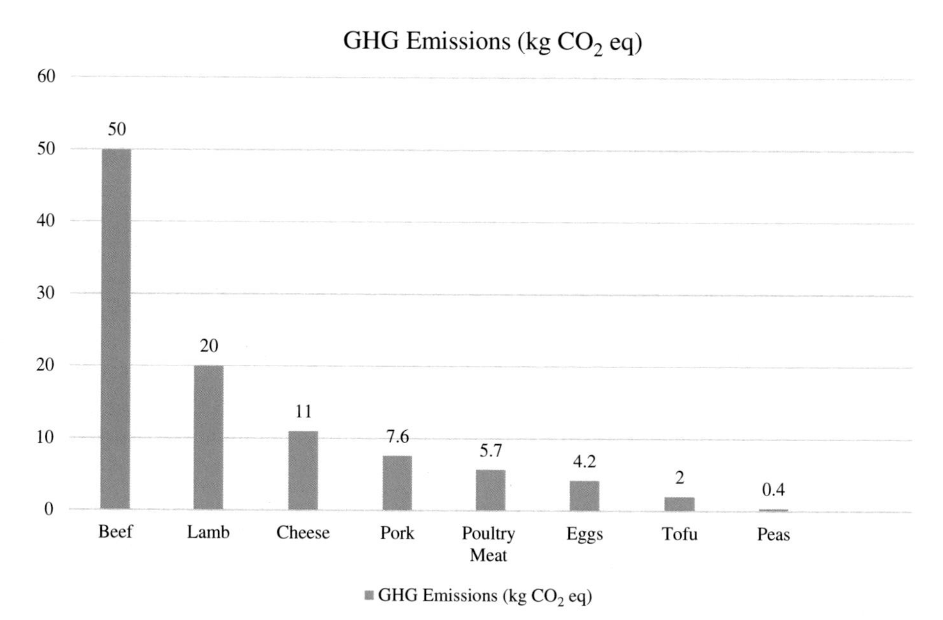


**CHART 1** GHG emissions per 100-g protein. *Data from Poore, J., Nemecek, T., 2018. Reducing food's environmental impacts through producers and consumers. Science 360(6392), 987–992.*

compares GHGs, land use, and water use in producing 1 L of cow milk with producing the same volume using four types of plant-based milk. It can be clearly observed that producing 1 L of cow-based milk releases at least 3 times more carbon dioxide than plant-based milk. Similarly, as shown in Chart 1, producing 100 g of protein from cattle (beef) generates 25 times more GHGs than the same volume of tofu. Producing cheese from dairy generates 27.5 times more GHGs than cheese made from legumes (pea).

Tackling climate variation by reducing GHGs is everyone's responsibility. Goodland and Anhang (2019) further pinpoint that "livestock (like automobiles) are a human invention meant for convenience; livestock was not inscribed in man's prehistory; moreover, a molecule of carbon dioxide exhaled by livestock is no more natural than one from an auto tailpipe" (p. 12). Livestock generates more direct emissions than the transportation sector (Bailey et al., 2020). Humans are by nature resourceful; they hold the creativity and capacity necessary to deploy solutions to create a sustainable earth for all.

## 3.2 Land degradation and deforestation

Forests and natural habitats have dominated global landscapes for most of human history, stabilizing the climate and sequester GHGs (Ritchie and Roser, 2020). Advances in agriculture have caused significant alterations to arable land use, land management, and land cover. Converting forests and natural habitats into agricultural forms (e.g., pastureland, plantations, and cropland) evolved with man's needs and civilizations, spanning thousands of years (Goldewijk and Battjes, 1997). Since the mid-19th century, the demand for animal products continues its upward spiral (Pimm, 2020).

Animal production drives deforestation all over the world, including Brazil, Indonesia, Madagascar, and in the Congo Basin (Africa) (Tollefson, 2013). Extreme deforestation is reported ongoing in the world's largest unfragmented forest, the Amazon. The Amazon Basin plays a vital role in regulating the climate globally as it absorbs nearly 100 billion metric tons of carbon, "more than ten times the annual global emissions from fossil fuels" (para. 3) (Greenpeace, n.d.). However, in just 50 years (until 2018), 20% of the Amazon forest has vanished; with most of the cleared area converted into farmland to cultivate grains and graze livestock (World Wide Fund for Nature (WWF), 2018). Cattle ranching, therefore, must contribute 80% of the responsibility of deforestation throughout Amazon countries (Yale University, 2020). Specifically, as a nation dwelling on deforested land, Brazil, with about 200 million heads of cattle, holds a quarter of the global beef market (Yale University, 2020). Brazilian soy (used mostly for livestock feed) is cultivated principally on lands that were before forests and savannahs, thus, exacerbating this converting process as well (University of Bonn, 2020).

Processing and transporting animal products also demand additional landmass. Extreme deforestation and the subsequent reshaping of former forest space continues to swiftly and steadily release plant-stored carbon into the air, accelerating, and intensifying global warming (Union of Concerned Scientists, 2020). Cederberg et al. (2011) noted the quest of humans for animal protein simply propels deforestation further. At present, it is widely documented that any kilogram of meat brought to market leaves an unfathomable carbon footprint in its trail.

Would growing crops neutralize the effects of deforestation? In modern times, the advancement of agricultural technologies means that 30% less farmland than in 1961 is required to generate an identical yield (Ritchie and Roser, 2020). As shown in Chart 2, producing 100 g of protein from beef requires approximately 164 $m^2$/year, which is more than 74 times the landmass needed to produce tofu. Likewise, producing 100 g of protein from lamb and mutton requires 185 $m^2$/year, which proves to bear more than 40 times the carbon footprint of grains cultivated on the very same surface. In essence, existing croplands worldwide are

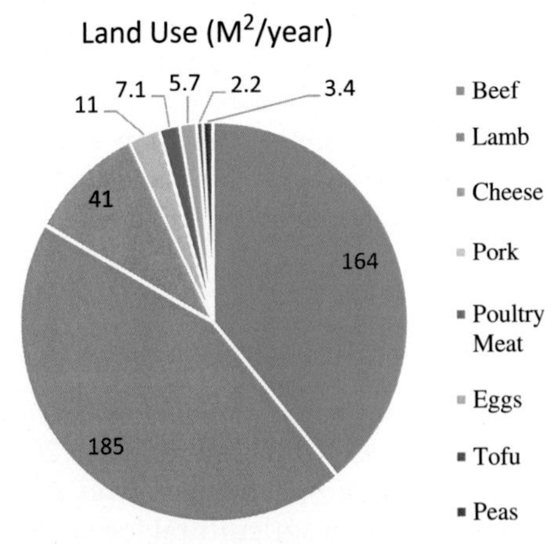


**CHART 2** Land use for producing 100-g protein. *Data from Poore, J., Nemecek, T., 2018. Reducing food's environmental impacts through producers and consumers. Science 360(6392), 987–992.*

estimated sufficient to produce enough plant-based protein to sustain humans. Deforestation and its effects, such as soil erosion and the loss of species (both plant and animals), releasing rather than capturing $CO_2$ weighs too heavily for a balanced equation. Creating landmass through deforesting gravely disrupts an equilibrium set by nature itself while costing tremendously to humans in many dimensions (health, social, economic, and political). Therefore, ending animal protein production saves forests and species, and protects climate because only intact forests and natural habitats will guarantee their climate-related benefits (Walker, 2016).

## 3.3 Water shortage and water pollution

Four billion people are increasingly dealing with water scarcity worldwide (Climate Central, 2019). Water scarcity, at both global and regional levels albeit driven by population growth and the steep demand for animal products, is worsened by the irregular global rainfall patterns that are the norm with climate change (Doreau et al., 2012). Meat-driven deforestation accelerates climatological modification, which in turn depletes water bodies.

Globally, agriculture draws 92% of the freshwater supply available (Gerbens-Leenes et al., 2013), representing 70% of the global surface and groundwater (Doreau et al., 2012). Of this water usage, approximately one third goes to generate animal-based food, mostly beef, pork, and poultry (Gerbens-Leenes et al., 2013). In the United States, cattle-feed farms (mainly fulfilling beef and dairy markets) use up to 23% of the nation's water, resulting in downsizing, receding, and disappearing rivers and lakes (Farah, 2020). In modern times, to produce animal-based foodstuff and related items, require a masses of energy at various stages such as "extracting, conveying, treating, distributing, and cleaning" (para. 4) (Buxton, 2014). The energy demands per process leave a long trail of GHGs deepening global warming. The only satisfactory response to the climate crisis includes foregoing animal-based protein in favor for more eco-friendly, less energy-hungry plant-based protein.

All over the globe freshwater usage for the same amount of animal products averages much higher than required for a similar yield of plant-based products. Illustratively, producing 1 kg of bovine meat absorbs 15,415 L of fresh water, which is approximately 9.4 times as much as what is needed to grow the same volume of cereals (Armstrong, 2017). Transferring to plant-based agriculture can successfully lower freshwater consummation, contributing to "strengthening our planet's ecosystem and reducing the risk of extreme weather events that make water supply sources more unpredictable, more polluted, and scarcer" (para. 3.) (Henkel, 2020). Another long-term climatological benefit of ceasing animal production is the return to regular rainfall pattern, a vital solution to water scarcity.

Owing to organic matter deposits, excess nutrients, and pathogen contamination, the animal industry plays an essential role in the pollution of streams, lakes, and groundwater (FAO, 2017; Lake Conservation Notes, 2002). The US Environmental Protection Agency (EPA) 2000 National Water Quality Inventory Report to Congress illustrated that approximately 40% of national rivers and streams were polluted at different levels due to agriculture, especially animal farms (US Environmental Protection Agency (EPA), 2000). Animal farms generate tremendous amount of animal wastes (e.g., urine and poop) in a very limited space, overflowing contaminating groundwater and streams (Centers for Disease Control and

Prevention, 2016). Other animal-related industries, such as slaughterhouses, tanneries, and wool mills, perpetrate this contaminating process. According to the EPA survey of 98 US slaughterhouses from 2016 to 2018, 60 of them dumped their wastewater, including enormous bacteria and pathogens, thus polluting rivers, streams, and other waterways (Environmental Integrity Project, 2018).

## 3.4 Biodiversity

All life on this shared planet informs biological diversity (biodiversity for short) (Carrington, 2018). Biodiversity services the ecosystem and contributes essentially to the well-being of humans (Millennium Ecosystem Assessment, 2005). Human interventions (such as urbanization, industrialization, and deforestation) and climate change (directly and indirectly resulting from human actions) remain the two key reasons for the destruction or loss of biodiversity (Haenggi, 2012; Hoffman, 2020). Human interventions and climate change are intertwined; consequently, animal production, the largest vector of climate change remains a major threat to biodiversity directly and indirectly.

Livestock requires much land, water, and energy in order to produce food for humans; all in a noncircular pattern. As discussed above, animal farming, therefore, causes very significant surges of GHGs that in turn intensify global warming. Rosane (2020) alerts that 15% of ecosystems would lose 20% of their biodiversity should global temperatures increases by4°C. Deforestation is increasingly linked to intensive meat and dairy productions. According to the World Wide Fund for Nature (WWF) (2018) Living Planet Report, between 1970 and 2014, 60% of wildlife had declined as a direct result of human activities as forests and natural habitats became decimated. Interestingly, as wild species were going extinct during this very period, the number of animals raised for food was tripling (Greenpeace International, 2018). This trend is to be reversed for humans to thrive and not merely survive; we should remove animal proteins from our plates and, thus, save millions of lives per year (Greenpeace International, 2018). The United Nations Environment Programme and International Livestock Research Institute (2020) both predict that habitat loss will render humans more vulnerable to animal-to-human disease transmission. As all components of the ecosystem are closely connected, dysfunction in one node triggers dire ripple effects as climate change demonstrates.

This section presents some key arguments around the major role of animal production on climate change. Animal farming and its subelements, including the elaboration, transportation, use and distribution of chemical-based (fertilizers), and natural means used to increase yield in animal production are considered indirect factors that aggravate global warming and endanger humans' health due to the high content of nitrous oxide ($N_2O$), which is a long-lived GHG, third in importance as a trigger vector of global warming, and also water pollution, and is the top deplete of stratospheric ozone. Animal production and climate change are, indeed, correlated for both affect human health and well-being.

# 4 Effect of animal production and climate change on human health and well-being

Animal production aims to fulfill human health and well-being principally through food and nutrition (e.g., protein and vitamin B12). Scientific evidence and the points made earlier raise the question as to whether humans derive in the end benefits sought from animal products-despite all the natural resources put into cause to generate them (Moekti, 2020). Human and natural systems are interdependent, and answering our human needs sustainably is conducive to optimal health and well-being for all on the planet Earth (Pinillos, 2018).

## 4.1 Food security, nutrition, and physical health

The FAO's mission of "no hunger" encompasses food availability, accessibility, sufficiency, safety, and nutrition for all of the Earth's inhabitants in order to support their health and well-being (FAO, 2020). Animal production has worsened food insecurity worldwide notwithstanding climate change (World Animal Foundation, 2020). Smith et al. (2013) point that livestock used more than 50% of global grains to supply only 13% of energy for the world's nutrition. Diets rich in animal-based protein (in red meat intake notably) largely augment the risk of cardiovascular diseases, obesity, diabetes, and kidney failure (Wang et al., 2019).

Historically, several civilizations gave the spotlight to animal products (meats) on the table, while relegating plant-based dishes to the status of far-removed cuisines. In traditional Chinese culture, blood and energy from animals are touted as the best way to fill what the human body does not have (Lian, 2020). Although scriptures of Western religions, such as Judaism, Christianity, and Islam, seem to sanction meat consumption, most Eastern religions (Hinduism, Zoroastrian, Jainism, and others) advocate meatless nutrition (Srivastava, 2007).

Debates abound to establish whether ethnic minority groups have derived key benefits (in social, economic, and health aspects), from their traditional skills of hunting for and eating meat. For the sake of comparison: Bang et al. (1971) believed that Greenland Eskimos' "health diet" (associated with a high intake of seal and whale fats) lowers the risk of coronary artery disease, which made for a lucrative fish oil market in modern times. Fodor et al. (2014) on the other end are of the opinion that the "Eskimos diet" does not necessarily correlate to coronary heart disease, it contributes rather to a higher risk of cerebral infarction and kidney cancer. Ironically, research highlights plant-based omega-3 fatty acids as robust shields against cardiovascular illnesses (Abdelhamid et al., 2020). Must we concur with English philosopher, Francis Bacon (1617–21), who argues that "man prefers to believe what he prefers to be true!"

Animal-based proteins harm inherently the environment and pose health risks and even death to humans. In regards to human health and safety, for estimating the cost of industrial farming (including animal production as a key component), the Union of Concerned Scientists took into account the toxicity of herbicides and insecticides customarily found to cause poisoning and chronic illnesses, their runoff, which contaminates potable water supplies (Smit and Heederik, 2017). In fact, the use of antibiotics in industrial animal agriculture has triggered antimicrobial resistance, declared as a "global health emergency" by the WHO (2020a). In some countries, more than 80% of the nation's medical antibiotics were used in the animal sector, a critical source of food-borne pathogens affecting human beings (WHO, 2020a). This might trigger ecosystem-wide antimicrobial resistance as all these systems are interconnected.

## 4.2 Zoonotic disease

The brisk expansion of animal production is more than ever keenly associated with environmental degradation and the upward tendency of zoonotic diseases to spread. Most if not all pandemics and endemics seem to stem from the contamination of humans through animal-based proteins (Neuman, 2020). To enumerate: the human immunodeficiency virus (HIV, cited to have been transmitted from the consumption of chimpanzee flesh) (Sharp and Hahn, 2011), to the severe acute respiratory syndrome (SARS, China, 2003, likely transmitted from eating Himalayan palm civets and raccoon dogs) (Shi and Hu, 2008), to the Middle East respiratory syndrome (MERS, Saudi Arabia, 2012, derived from consuming camel meat) (Hemida et al., 2017), and to the current COVID-19 (China, 2019, likely traced to wet markets selling the meat of minks, bat, and pangolins) (Mallapaty, 2020). The World Health Organization (WHO) (2005) confirms that in the past decade more than 60% of all human pathogens are zoonotic in origin; among them, over 75% of pathogens induced zoonotic disease that continues to threaten human physical health, mental wellness, and overall wellbeing. Ms. Inger Andersen, the executive director of the United Nations Environment Program (UNEP) indicated that zoonotic diseases have caused over $100 billion USD economic damage during the past two decades (Earth.org, 2020).

What cause-and-effect lessons can we come to when analyzing such patterns? What is to be distinguished as we inspect facts around animal production: wild and domestic, the loss of habitat, and the proximity of human and animal? All are indeed to be factored in; this dramatic prevalence of zoonotic diseases in our times in definitive ways can be traced to the "high demand for animal protein, unsustainable agricultural practices, and climate change" (para. 2) (BBC News, 2020). During the past decade, global meat production increased by 26% from 271.51 million tons in 2007 to 341.16 million tons in 2018 (Ritchie and Roser, 2017). This rapid market growth in industrial animal agriculture along with fossil fuels caused GHGs to soar resulting in a record-breaking rise in temperatures on land and ocean (National Oceanic and Atmospheric Administration (NOAA), 2020).

Animal production-driven human interventions strengthen the animal-to-human virus transmission determinant. While Johnson et al. (2020) indicate that domestic animals harbor 83 times more zoonotic viruses than wildlife. In 2007, 12 years ago, Cheng et al. (2007) had already warned that "the presence of a large reservoir of SARS-CoV-like viruses in horseshoe bats, together with the culture of eating exotic mammals in southern China, makes for a ticking time bomb" (p. 683). Another example stems from the largest Ebola outbreak in West Africa from 2014 to 2016. Ebola, yet another zoonotic disease, was transmitted to people from fruit bats, a species, known to evolve in dense forests far away from human settlements (WHO, 2020b). Deforestation (to increase animal production by raising more livestock) and its related climate change consequences forced the bats to relocate in the vicinity of humans. This suite of events triggered a direct transmission of the virus to local residents through contaminated water, food, and other ways (WHO, 2020b). In parallel to West Africa, we see a recrudescence of parallel cause-and-effect events in developed countries, notably the United States, where Lyme disease, a tick-borne disease in small mammal reservoirs, had more than doubled from 2004 to 2016, especially in northeast States due to heavy deforestation (Bai, 2018). As more than 80% of animal, plant, and insect species live in forests (McFall-Johnsen, 2019), humans disregard a most precious ally in carbon offsetting of curative and nutritional plants by clear-cutting forested areas to the pace of a football-field-sized per minute in order to generate the second-hand, energy-guzzling, disease-prone protein found in animal products. Climate change and its unfavorable results will accelerate this process until we opt to consume more eco-friendly and health-inducing plant-based proteins (Huang et al., 2020).

As if to corroborate the data and conclusions advanced above, as we write these lines, the world is grappling with a COVID-19, which is a declared global public health emergency that brings into full focus on the link between animal products consumption and the market humans have created for animal-based proteins (VOA News, 2020). COVID-19 was traced to the consumption of pangolin and/or bat meat purchased in a wet market in the city of Wuhan, China. The Chinese government diligently implemented health measures and the banning of wildlife trade; however, loopholes are still existing (Gorman, 2020). The traditional and folkloric belief in the medicinal properties derived from these animal products still drives illegal wild markets not only in China but also internationally (such as Malaysia, the Philippines, and India) (Yu, 2020).

In Spain, several workers tested positive for COVID-19 in a mink farm; their breeding is geared to the high-end, luxury goods market. As over 87% of minks in that farm were carriers, local health authorities ordered the cull of nearly 100,000 minks to avoid animal-to-human transmission (Briggs, 2020). Bubonic plague was reported in Mongolia during the COVID-19 outbreak when a teenager died from eating marmot meat (Cruse, 2020). That virus has already spread to its southern neighbors in northern China, Inner Mongolia Autonomous Region (Munroe et al., 2020). Research confirms that zoonotic diseases are equally presented in marine mammals, birds, fish, and other seafood (Bogomolni et al., 2008; Smith et al., 1998). An increase in these ocean system "products" would mean increased human infections, destroying eco-system balance, and accelerating climate change.

## 4.3 Ethical and sustainable human feed—The key solution

The vision enshrined in the Global Health Security Agenda (GHSA) 2024 Framework (2018) advocates for "a world safe and secure from global health threats posed by infectious diseases, whether natural, deliberate, or accidental" (p. 4). Zoonotic diseases respect no geographies and spread death, food insecurity, social unrest, economic decline in their wake. Research is clear, proven, and equivocal in underlining the quest for animal-based proteins, the production and consumption of animal foodstuff as a source of harm to the climate and humans. The impact of animal-based proteins is not conducive to sound health and well-being, livable nature (as witnessed in the current climate emergency), and healthy and safe animal species (the emergence of one to many pandemic of zoonotic source). All this causes an unbearable deficit to mankind as individuals and entire nations around the globe must cope with and resolve climate change and the emergencies unfolding from it.

A reflection on social and legal ethics prompts that international communities at the lower end of the economic spectrum are disproportionately impacted by global warming while being the ones that actually contribute the least to the climate emergency (UN, 2019d). Case in point, the top two world economic giants, United States and China, also ranked as the top two countries emitting GHGs (International Energy Agency, 2019) and with the greatest number of meat consumers (Ritchie and Roser, 2017). The notion of environmental justice arises as emerging, underdeveloped countries with lesser animal products consumption also bear the brunt of disasters triggered by global warming (UN, 2019d). An ethical and sustainable switch to plant-base proteins as well as organic, green lifestyle choices is imperative to restore harmony with nature and rebuild societies and economies all the while ensuring a better life for all (The Supreme Master Ching Hai, 2020).

Mass killing is unethical. Animal production and climate crisis seem chained by cause and effect. Tremendous science-backed, notable expert-driven research all demonstrate the enormous suffering and losses animal production caused on the material, physical, economic, societal fronts (The Supreme Master Ching Hai, 2020). Perhaps lesser-examined, but of pervading impact are the mental and spiritual burden of ethics we have placed upon ourselves as humans, by culling the innocent, the infected animals that are not party to the conditions that led them to become disease transmitters. Discarding them through senseless culls brings the question: As part of the human species, are we not to reflect, to act, to react "humanely"? The mineral, the vegetable, the animal, and the human all equally share "Spaceship Earth." A better life for humans equates to a better life for all cohabitants.

Could our violent actions toward the environment and living beings such as animals and plants have resulted in the violence we observe within our societies, worldwide? Not as perpetrators but rather as victims of violence, the animals we raise to generate food for humans face dire living conditions. Could it be said that violence-as-a-disease could subliminally transmit through a transitivity relationship? The following incident may fuel that argument: it was reported that a worker known otherwise as a very nice and respectful person at a slaughterhouse in Fresno, California, United States shot four of his coworkers (McWilliams, 2012). Research illustrates that humans working in slaughterhouses and animal farmworkers have been suffering chronic physical and mental health issues related to the violent treatment of animals they perpetrate—or are exposed to (Dillard, 2007).

In mapping the spread of COVID-19 globally, countries in North America, South America, Europe, and Oceanic emerge as the most hit. They are the ones with the biggest livestock processing plants and meat and poultry production. Many of these plants have become epicenters of COVID-19 and most workers there claimed that COVID-19 worsened their already traumatizing work circumstances (Foote, 2020; Molteni, 2020). Evidence and business cases have demonstrated that closing livestock and related industries in favor of greener, healthier commodities or crops (particularly if organic and chemical-free) would enhance the staff's health and well-being, and provide green jobs incentives (Ewing-Chow, 2020). Plant-based sustainable agriculture helps the economy return to viability and improves environmental conditions in cities and communities (Nittle, 2020).

Millennials (young adults born between 1983 and 2000) speak the strongest for animal ethics, for changing the climate trajectory by solving global warming. The concept of health and well-being leads them in great numbers to favor a totally plant-based diet and all alternatives possible to animal products (Olayanju, 2019). This trend is particularly strong in European countries and present in the United States, where according to *the Economist*, over a quarter of young Americans (25–35-year-old) claim following a vegan or vegetarian diet (Parker, 2019). Markets and Markets (2019) predicts the global meat substitutes market will reach $3.5 billion by 2026 with 12% annual increases. The upward tendency also confirms that more employment opportunities are emerging, guaranteeing health and prosperity to those with the savvy to implement plant-based food production business models. Organic, plant-based agriculture establishes an effective approach to mitigate climate change (Cleveland and Gee, 2017). *The Economist* further confirmed that medical doctors and nutritionists highlighted the impressive benefits of choosing plant-based proteins for weight loss and lowering the risks of heart disease, cancer, and diabetes (Harvard Health Publishing, 2020). Hence, humans need to "reverse the destruction of the earth and save its sentient habitants, humans, and animals alike" (p. 1912) (Khoury, 2020).

## 5 Conclusion: A green future awaits everyone!

"Climate change requires urgent, coordinated, and consistent action" (para. 2) (UN, 2019a). Similar to the sides of the same coin showing the same value, animal production and climate change elicit the need to bring animal-based protein to a full stop to exit their vicious cycle. Impact analysis highlights that animal production is greatly affected by climate change while posing tremendous risks to natural and built environments. The global pandemic of COVID-19 sets the stage for all humans, in this era, to rethink our true footprint on one another, on animals, plants, and other inhabitants sharing the planet Earth. To advance their future, humans must re-evaluate their diet and implement measures bringing renewal.

Settling for environmentally, animal-friendly, and sustainable means of nutrition is the only solution to feed the fast-growing world population (World Economic Forum, 2017). The United Nations alerts us the world population will increase by 2 billion in the next 30 years, from 7.7 billion in 2020 to 9.7 billion in 2050 (UN, 2019c). Humans must curb in earnest and at the earliest their urge for animal proteins—a need made more critical to also avoid further pandemics?

Scientific evidence confirms that animal products contribute massively to GHGs, the root cause of climate change. The European Environment Agency (2015) advocated to stop eating meat and switch from animal-based protein to plant-based protein. Animal production contributes the most to an unviable carbon footprint, requires far too much energy, and notably fresh water, a natural resource essential to us. As extolled by the former UN Secretary-General, Ban Ki Moon in his address on World Environment Day (June 5, 2008): "Mitigating climate change, eradicating poverty and promoting economic and political stability all demand the same solution: we must kick the carbon habit" (para. 5) (UN News, 2008). We have the capacity to solve climate change and put an end to zoonotic diseases as well by deciding on plant-based protein. So that organic, low-carbon lifestyle proves effective, ethical, and prosperous.

## References

Abdelhamid, A.S., Brown, T.J., Brainard, J.S., Biswas, P., Thorpe, G.C., Moore, H.J., Hooper, L., 2020. Omega-3 fatty acids for the primary and secondary prevention of cardiovascular disease. Cochrane Database Syst. Rev. 3. https://doi.org/10.1002/14651858.CD003177.pub5.

Armstrong, M., 2017. How Thirsty Is Our Food? Statista. https://www.statista.com/chart/9483/how-thirsty-is-our-food/. (Accessed 27 June 2020).

Babinszky, L., Halas, V., Verstegen, M., 2011. Impacts of climate change on animal production and quality of animal food products. In: Blanco, J., Kheradmand, H. (Eds.), Climate Change—Socioeconomic Effects. IntechOpen, London, pp. 165–190.

Bai, N., 2018. Lyme Disease Is on the Rise—An Expert Explains Why. https://www.ucsf.edu/news/2018/05/410401/lyme-disease-rise-expert-explains-why. (Accessed 20 June 2020).

Bailey, R., Froggatt, A., Wellesley, L., 2020. Livestock—Climate Change's Forgotten Sector: Global Public Opinion on Meat and Dairy Consumption. Chatham House. https://www.chathamhouse.org/sites/default/files/field/field_document/20141203LivestockClimateChangeForgottenSectorBaileyFroggattWellesleyFinal.pdf. (Accessed 20 June 2020).

Bang, H.O., Dyerberg, J., Nielson, A.B., 1971. Plasma lipid and lipoprotein pattern in greenlandic west-coast eskimos. Lancet 1 (7710), 1143. https://doi.org/10.1016/s0140-6736(71)91658-8.

BBC News, 2019. Australia Fires: How Do We Know How Many Animals Have Died? https://www.bbc.com/news/50986293. (Accessed 25 June 2020).

BBC News, 2020. Coronavirus: Fear Over Rise in Animal-to-Human Diseases. https://www.bbc.com/news/health-53314432. (Accessed 27 July 2020).

Bogomolni, A.L., Gast, R.J., Ellis, J.C., Dennett, M., Pugliares, K.R., Lentell, B.J., Moore, M.J., 2008. Victims or vectors: a survey of marine vertebrate zoonoses from coastal waters of the Northwest Atlantic. Dis. Aquat. Org. 81 (1), 13–38. https://doi.org/10.3354/dao01936.

Briggs, H., 2020. Coronavirus: Spain Orders Culling of Almost 100,000 Mink. https://www.bbc.com/news/world-europe-53439263. (Accessed 20 June 2020).

Buxton, N., 2014, February 11. Why Conserving Water Also Helps Tackle Climate Change. https://www.cooldavis.org/2014/02/11/why-conserving-water-also-helps-tackle-climate-change/. (Accessed 01 July 2020).

Carrington, D., 2018. What Is Biodiversity and Why Does It Matter to Us? https://www.theguardian.com/news/2018/mar/12/what-is-biodiversity-and-why-does-it-matter-to-us. (Accessed 20 June 2020).

Carter, P.D., Woodworth, E., 2018. Unprecedented Crime: Climate Change Denial and Game Changers for Survival. Clarity Press, Atlanta.

Cederberg, C., Persson, U.M., Neovius, K., Molander, S., Clift, R., 2011. Including carbon emissions from deforestation in the carbon footprint of Brazilian beef. Environ. Sci. Technol. 45 (5), 1773–1779. https://doi.org/10.1021/es103240z.

Centers for Disease Control and Prevention, 2016. Animal Feeding Operations. https://www.cdc.gov/healthywater/other/agricultural/afo.html. (Accessed 03 July 2020).

Chemnick, J., 2020. The Window Is Closing to Avoid Dangerous Global Warming. https://www.scientificamerican.com/article/the-window-is-closing-to-avoid-dangerous-global-warming/. (Accessed 20 July 2017).

Cheng, V.C.C., Lau, S.K.P., Woo, P.C.Y., Yuen, K.Y., 2007. Severe acute respiratory syndrome coronavirus as an agent of emerging and reemerging infection. Clin. Microbiol. Rev. 20 (4), 660. https://doi.org/10.1128/cmr.00023-07.

Cho, R., 2018. How Climate Change Will Alter Our Food. https://blogs.ei.columbia.edu/2018/07/25/climate-change-food-agriculture/. (Accessed 25 June 2020).

Cleveland, D.A., Gee, Q., 2017. Plant-based diets for mitigating climate change. In: Mariotti, F. (Ed.), Vegetarian and Plant-Based Diets in Health and Disease Prevention. Elsevier, London, pp. 135–156.

Climate Central, 2019. Climate Change & Water Use. https://www.climatecentral.org/gallery/graphics/climate-change-water-use. (Accessed 22 June 2020).

Costanzo, J.P., 2011. Extreme cold hardiness in ectotherms. Nat. Educ. Knowl. 3 (10), 3.

Cruse, E., 2020. Teenager dies from bubonic plague in Mongolia after eating marmot meat. Evening Standard. https://www.standard.co.uk/news/world/teenager-dies-bubonic-plague-mongolia-a4497731.html. (Accessed 20 June 2020).

Devlin, H., 2018. Rising Global Meat Consumption 'Will Devastate Environment'. https://www.theguardian.com/environment/2018/jul/19/rising-global-meat-consumption-will-devastate-environment. (Accessed 20 July 2018).

Dillard, J., 2007. A slaughterhouse nightmare: psychological harm suffered by slaughterhouse employees and the possibility of redress through legal reform. Georgetown J. Poverty Law Policy. https://ssrn.com/abstract=1016401. (Accessed 20 June 2020).

Doreau, M., Corson, M.S., Wiedemann, S.G., 2012. Water use by livestock: a global perspective for a regional issue? Anim. Front. 2 (2), 9–16. https://doi.org/10.2527/af.2012-0036.

Earth.org, 2020. Zoonotic Diseases Will Continue to Rise Without Wildlife, Environment Conservation. UN. https://earth.org/zoonotic-diseases-will-continue-to-rise-without-conservation-un/. (Accessed 20 July 2020).

Environmental Integrity Project, 2018. Water Pollution From Slaughterhouses. Available from: https://environmentalintegrity.org/wp-content/uploads/2018/10/Slaughterhouse-report-2.14.2019.pdf.

European Environment Agency, 2015. Agriculture and Climate Change. https://www.eea.europa.eu/signals/signals-2015/articles/agriculture-and-climate-change. (Accessed 20 June 2020).

Ewing-Chow, D., 2020. A Shift to Plant-Based Diets Would Create 19 Million Jobs in Latin America & the Caribbean. https://www.forbes.com/sites/daphneewingchow/2020/07/31/a-shift-to-plant-based-diets-would-create-19-million-jobs-in-latin-america- -the-caribbean/#97c78e96dec9. (Accessed 20 June 2020).

FAO, 2007. The State of the World's Animal Genetic Resources for Food and Agriculture: In Brief. Available from: http://www.fao.org/3/a-a1260e.pdf.

FAO, 2011. The State of the World's Land and Water Resources for Food and Agriculture (SOLAW)—Managing Systems at Risk. Food and Agriculture Organization of the United Nations, Rome and Earthscan, London.

FAO, 2016. Livestock's Long Shadow: Environmental Issues and Options. http://www.fao.org/3/a0701e/a0701e.pdf. (Accessed 20 June 2020).

FAO, 2017. Water Pollution From Agriculture: A Global Review. http://www.fao.org/3/a-i7754e.pdf. (Accessed 20 June 2020).

FAO, 2020. World Food Summit: Five Years Later Reaffirms Pledge to Reduce Hunger. http://www.fao.org/worldfoodsummit/english/newsroom/news/8580-en.html. (Accessed 20 July 2017).

Farah, T., 2020, July 2. US rivers and lakes are shrinking for a surprising reason: cows. The Guardian. https://www.theguardian.com/environment/2020/jul/02/agriculture-cattle-us-water-shortages-colorado-river. (Accessed 20 June 2020).

Fodor, J.G., Helis, E., Yazdekhasti, N., Vohnout, B., 2014. "Fishing" for the origins of the "Eskimos and heart disease" story: facts or wishful thinking? Can. J. Cardiol. 30 (8), 864–868. https://doi.org/10.1016/j.cjca.2014.04.007.

Food and Agriculture Organization, 1986. Farm Structures in Tropical Climates: Animal Environmental Requirements. http://www.fao.org/docrep/s1250e/s1250e10.htm. (Accessed 20 June 2020).

Foote, N., 2020. Working Conditions in Meat Processing Plants Make Them Hotbed for COVID-19. EURACTIV. https://www.euractiv.com/section/agriculture-food/news/working-conditions-in-meat-processing-plants-make-them-hotbed-for-covid-19/. (Accessed 20 July 2017).

Frank, M., 2017. Snakes and Flooding. https://texashelp.tamu.edu/wp-content/uploads/2016/02/snakes-and-flooding.pdf. (Accessed 20 June 2020).

Gerbens-Leenes, P.W., Mekonnen, M.M., Hoekstra, A.Y., 2013. The water footprint of poultry, pork and beef: a comparative study in different countries and production systems. Water Resour. Ind. 1–2, 25–36. https://doi.org/10.1016/j.wri.2013.03.001.

Global Health Security Agenda (GHSA), 2024 Framework, 2018. Available from: https://ghsa2024.files.wordpress.com/2020/06/ghsa2024-framework.pdf.

Goldewijk, K., Battjes, J.J., 1997. A Hundred Year (1980–1990) Database for Integrated Environmental Assessments. National Institute of Public Health and the Environment, Bilthoven, The Netherlands.

Goodland, R., Anhang, J., 2019. Livestock and Climate Change: What If the Key Actors in Climate Change Are... Cows, Pigs and Chickens? WorldWatch. https://templatelab.com/livestock-and-climate-change/. (Accessed 20 June 2020).

Gorman, J., 2020. China's Ban on Wildlife Trade a Big Step, But Has Loopholes, Conservationists Say. https://www.nytimes.com/2020/02/27/science/coronavirus-pangolin-wildlife-ban-china.html. (Accessed 20 June 2020).

Government of Canada, 2020. Greenhouse Gas Emissions. https://www.canada.ca/en/environment-climate-change/services/environmental-indicators/greenhouse-gas-emissions.html. (Accessed 20 July 2017).

Graca, J., 2016. Towards an integrated approach to food behaviour: meat consumption and substitution, from context to consumers psychology. Community Dent. Health 5 (2). https://doi.org/10.5964/pch.v5i2.169.

Greenpeace, n.d. Brazil and the Amazon Forest. Greenpeace. https://www.greenpeace.org/usa/issues/brazil-and-the-amazon-forest/. (Accessed 20 June 2020).

Greenpeace International, 2018. Less Is More: Scientific Background for the Greenpeace Vision of the Meat and Dairy System Towards 2050. Available from: https://storage.googleapis.com/planet4-international-stateless/2018/03/6942c0e6-longer-scientific-background.pdf.

Grossi, G., Goglio, P., Vitali, A., Williams, A.G., 2019. Livestock and climate change: impact of livestock on climate and mitigation strategies. Anim. Front. 9 (1), 69–76. https://doi.org/10.1093/af/vfy034.

Haenggi, M., 2012. Diversity loss due to interference correlation. IEEE Commun. Lett. 16 (10), 1600–1603. https://doi.org/10.1109/lcomm.2012.082012.120863.

Hanna, J., Baldacci, M., 2019. The Midwest Flooding Has Killed Livestock, Ruined Harvests and Has Farmers Worried for Their Future. https://www.cnn.com/2019/03/21/us/floods-nebraska-iowa-agriculture-farm-loss/index.html. (Accessed 20 June 2020).

Harvard Health Publishing, 2020. The Right Plant-Based Diet for You. https://www.health.harvard.edu/staying-healthy/the-right-plant-based-diet-for-you. (Accessed 20 June 2020).

Hemida, M.G., Elmoslemany, A., Al-Hizab, F., Alnaeem, A., Almathen, F., Faye, B., Chu, D.K.W., Perera, R.A.P.M., Peiris, M., 2017. Dromedary camels and the transmission of Middle East Respiratory Syndrome Coronavirus (MERS-CoV). Transbound. Emerg. Dis. 64 (2), 344–353.

Henkel, 2020. Saving Water and Tackling Climate Change. https://www.henkel.com/spotlight/2020-03-20-saving-water-and-tackling-climate-change-1046204. (Accessed 20 June 2020).

Hoffman, A., 2020. Climate Change and Biodiversity. Australian Academy of Science. https://www.science.org.au/curious/earth-environment/climate-change-and-biodiversity. (Accessed 20 June 2020).

Huang, J.L., Liao, S.J., Weinstein, R., Sinha, B.I., Graubard, D., 2020. Association between plant and animal protein intake and overall and cause-specific mortality. JAMA Intern. Med. 180, 1173–1184.

Intergovernmental Panel on Climate Change (IPCC), 2014. Climate change 2014: synthesis report. In: Core Writing Team, Pachauri, R.K., Meyer, L.A. (Eds.), Contribution of Working Groups I, II and III to the Fifth Assessment Report of the Intergovernmental Panel on Climate Change. IPCC, Geneva, Switzerland.

International Energy Agency, 2019. CO2 Emissions From Fuel Combustion 2019 Overview. https://webstore.iea.org/co2-emissions-from-fuel-combustion-2019-overview. (Accessed 20 June 2020).

Jakob, M., Luderer, G., Steckel, J., Tavoni, M., Monjon, S., 2012. Time to act now? Assessing the costs of delaying climate measures and benefits of early action. Clim. Chang. 114, 79–99.

Johnson, C.K., Hitchens, P.L., Pandit, P.S., Rushmore, J., Evans, T.S., Young, C.C., Doyle, M.M., 2020. Global shifts in mammalian population trends reveal key predictors of virus spillover risk. Proc. Biol. Sci. 287 (1924), 20192736. https://doi.org/10.1098/rspb.2019.2736.

Katz, B., 2019. Three cows swept away by hurricane Dorian have been found alive. Smithsonian. https://www.smithsonianmag.com/smart-news/three-cows-swept-away-hurricane-dorian-have-been-found-alive-180973557/. (Accessed 20 June 2020).

Khoury, B., 2020. The root causes of COVID-19 screech for compassion. Mindfulness 3 (1–4). https://doi.org/10.1007/s12671-020-01412-8.

Lake Conservation Notes, 2002. Surface Water Pollution From Livestock Production. Available from: http://lshs.tamu.edu/docs/lshs/end-notes/surface%20water%20pollution%20from%20livestock%20production-2500205058/surface%20water%20pollution%20from%20livestock%20production.pdf.

Legg, W., Huang, H., 2020. Climate Change and Agriculture. OECD Trade and Agriculture Directorate, Paris.

Lian, Y.Z., 2020. Why Did the Coronavirus Outbreak Start in China? https://www.nytimes.com/2020/02/20/opinion/sunday/coronavirus-china-cause.html. (Accessed 20 June 2020).

Mallapaty, S., 2020. Animal source of the coronavirus continues to elude scientists. Nature. https://www.nature.com/articles/d41586-020-01449-8. (Accessed 20 June 2020).

Manitoba, n.d. Feeding Heat Stressed Dairy Cows. https://www.manitoba.ca/agriculture/livestock/dairy/feeding-heat-stressed-dairy-cows.html. (Accessed 20 June 2020).

Markets and Markets, 2019. Meat Substitutes Market by Source (Soy Protein, Wheat Protein, Pea Protein), Type (Concentrates, Isolates, and Textured), Product (Tofu, Tempeh, Seitan, and Quorn), Form (Solid and Liquid), and Region—Global Forecast to 2026. https://www.marketsandmarkets.com/Market-Reports/meat-substitutes-market-979.html. (Accessed 20 June 2020).

McFall-Johnsen, M., 2019. We've Killed Off More Than 50% of Forest Animals on Earth, a New Report Found—Even More Evidence of a 6th Mass Extinction. Business Insider. https://www.businessinsider.com/people-killed-half-of-forest-animals-on-earth-since-1970-2019-8. (Accessed 20 June 2020).

McWilliams, J., 2012. Shooting of Four Workers at Slaughterhouse and the Connection Between Violence to Animals and Humans. https://freefromharm.org/animal-products-and-psychology/shooting-of-four-workers-at-slaughterhouse-and-the-connection-between-violence-to-animals-and-humans/. (Accessed 20 June 2020).

Millennium Ecosystem Assessment, 2005. Ecosystems and Human Well-Being: Biodiversity Synthesis. World Resources Institute, Washington, DC.

Milman, O., Leavenworth, S., 2016. China's Plan to Cut Meat Consumption by 50% Cheered by Climate Campaigners. https://www.theguardian.com/world/2016/jun/20/chinas-meat-consumption-climate-change. (Accessed 20 July 2018).

Moekti, G.R., 2020. Industrial livestock production: a review on advantages and disadvantages. IOP Conf. Ser. Earth Environ. Sci. 492, 012094.

Molteni, M., 2020. Why meatpacking plants have become covid-19 hot spots. Wired. https://www.wired.com/story/why-meatpacking-plants-have-become-covid-19-hot-spots/. (Accessed 20 June 2020).

Morignat, E., Perrin, J.B., Gay, E., Vinard, J.L., Calavas, D., Hénaux, V., 2014. Assessment of the impact of the 2003 and 2006 heat waves on cattle mortality in France. PLoS One 9 (3), e93176. https://doi.org/10.1371/journal.pone.0093176.

Munroe, T., Liu, R., Zhang, M., 2020. City in China's Inner Mongolia Warns After Suspected Bubonic Plague Case and Tells People to Report Sick Marmots. https://www.independent.co.uk/news/world/asia/bubonic-plague-inner-mongolia-warning-marmots-a9602611.html. (Accessed 20 June 2020).

Nardone, A., Ronchi, B., Lacetera, N., Ranieri, M.S., Bernabucci, U., 2010. Effects of climate changes on animal production and sustainability of livestock systems. Livest. Sci. 130 (1–3), 57–69. https://doi.org/10.1016/j.livsci.2010.02.011.

National Geographic, 2014. Agriculture. https://www.nationalgeographic.org/encyclopedia/agriculture/. (Accessed 20 June 2020).

National Oceanic and Atmospheric Administration (NOAA), 2020. National Centers for Environmental Information. https://www.ncdc.noaa.gov/sotc/global/202001. (Accessed 20 June 2020).

Neuman, S.U.N., 2020. Predicts Rise in Diseases That Jump From Animals to Humans Due to Habitat Loss. https://www.npr.org/sections/coronavirus-live-updates/2020/07/06/888077232/u-n-predicts-rise-in-diseases-that-jump-from-animals-to-humans?utm_campaign=storyshare&fbclid=IwAR19VnkWTYC3JeYC68YUePVMqlsfPkLwZzsj5dHKuJPY5ivhR2IkOhHRTRs. (Accessed 20 June 2020).

Nittle, N., 2020. The Plant-Based Movement to Transition Farmers Away From Meat and Dairy Production. Civil Eats. https://civileats.com/2020/01/13/the-plant-based-movement-to-transition-farmers-away-from-meat-and-dairy-production/. (Accessed 20 July 2020).

Olayanju, J., 2019. Plant-Based Meat Alternatives: Perspectives on Consumer Demands and Future Directions. https://www.forbes.com/sites/juliabolayanju/2019/07/30/plant-based-meat-alternatives-perspectives-on-consumer-demands-and-future-directions/#797851966daa. (Accessed 20 June 2020).

Owen, J., 2005. Farming Claims Almost Half Earth's Land, New Maps Show. National Geographic. https://www.nationalgeographic.com/news/2005/12/agriculture-food-crops-land/. (Accessed 20 June 2020).

Parker, J., 2019. The Year of the Vegan. https://worldin2019.economist.com/theyearofthevegan. (Accessed 20 June 2020).

Pimm, S.L., 2020. Deforestation. Encyclopaedia Britannica. https://www.britannica.com/science/deforestation. (Accessed 25 July 2020).

Pinillos, R.G., 2018. One Welfare: A Framework to Improve Animal Welfare and Human Well-Being. CABI, Boston, MA.

PIRSA, 2015. Caring for Livestock During a Heatwave. https://www.pir.sa.gov.au/emergency_management/caring_for_livestock_during_a_heatwave. (Accessed 20 June 2020).

Poore, J., Nemecek, T., 2018. Reducing food's environmental impacts through producers and consumers. Science 360 (6392), 987–992.

Pytel, B., 2019. "Climate Emergency" Is Oxford Dictionaries' 2019 Word of the Year. https://www.earthday.org/climate-emergency-is-2019-oxford-word-of-the-year/. (Accessed 20 June 2020).

Ritchie, H., Roser, M., 2017. Meat and Dairy Production. https://ourworldindata.org/meat-production. (Accessed 20 June 2020).

Ritchie, H., Roser, M., 2018. Water Use and Stress. https://ourworldindata.org/water-use-stress. (Accessed 20 June 2020).

Ritchie, H., Roser, M., 2020. Environmental Impacts of Food Production. https://ourworldindata.org/environmental-impacts-of-food. (Accessed 20 July 2020).

Rojas-Downing, M.M., Nejadhashemi, A.P., Harrigan, T., Woznicki, S.A., 2017. Climate change and livestock: impacts, adaptation, and mitigation. Clim. Risk Manag. 16, 145–163. https://doi.org/10.1016/j.crm.2017.02.001.

Rosane, O., 2020. Climate-Driven Biodiversity Loss Will Be Sudden, Study Warns. EcoWatch. https://www.ecowatch.com/biodiversity-loss-climate-crisis-2645676582.html?rebelltitem=1#rebelltitem1. (Accessed 20 June 2020).

Rotter, R., Van de Geijn, S.C., 1999. Climate change effects on plant growth, crop yield and livestock. Clim. Chang. 43 (4), 651–681. https://doi.org/10.1023/a:1005541132734.

Schattenberg, P., Cleere, J., Paschal, J., Banta, J., 2017. AgriLife Extension Experts Offer Advice on Livestock Safety, Care After Harvey. https://agrilifetoday.tamu.edu/2017/09/01/agrilife-extension-experts-offer-advice-livestock-safety-care-harvey/. (Accessed 20 June 2020).

Scheelbeeka, P.F.D., Birda, F.A., Tuomistob, H.L., Greena, R., Harrisa, F.B., Joya, E.J.M., et al., Dangoura, A.D., 2018. Effect of environmental changes on vegetable and legume yields and nutritional quality. PNAS 116, 6804–6809.

Sharp, P.M., Hahn, B.H., 2011. Origins of HIV and the AIDS pandemic. Cold Spring Harb. Perspect. Med. 1 (1), a006841.

Shi, Z., Hu, Z., 2008. A review of studies on animal reservoirs of the SARS coronavirus. Virus Res. 133 (1), 74–87. https://doi.org/10.1016/j.virusres.2007.03.012.

Slenning, B.D., 2010. Global climate change and implications for disease emergence. Vet. Pathol. 47 (1), 28–33. https://doi.org/10.1177/0300985809354465.

Smit, L.A.M., Heederik, D., 2017. Impacts of intensive livestock production on human health in densely populated regions. GeoHealth 1, 272–277.

Smith, A.W., Skilling, D.E., Cherry, N., Mead, J.H., Matson, D.O., 1998. Calicivirus emergence from ocean reservoirs: zoonotic and interspecies movements. Emerg. Infect. Dis. 4 (1), 13–20. https://doi.org/10.3201/eid0401.980103.

Smith, J., Sones, K., Grace, D., MacMillan, S., Tarawali, S., Herrero, M., 2013. Beyond milk, meat, and eggs: role of livestock in food and nutrition security. Anim. Front. 3 (1), 6–13. https://doi.org/10.2527/af.2013-0002.

Srivastava, J., 2007. Vegetarianism and Meat-Eating in 8 Religions. https://www.hinduismtoday.com/modules/smartsection/item.php?itemid=1541. (Accessed 20 June 2020).

Sustainable Development Goals, n.d. Climate Action. https://www.un.org/sustainabledevelopment/climate-action/ (Accessed 20 July 2017).

Texas A&M Agrilife Extension, 2016. Care and Treatment of Livestock After a Hurricane. https://texashelp.tamu.edu/wp-content/uploads/2016/02/care-and-treatment-of-livestock-after-a-hurricane-1.pdf. (Accessed 20 June 2020).

Texas AgriLife Extension Service, n.d. Mosquito. https://livestockvetento.tamu.edu/insectspests/mosquito/. (Accessed 21 July 2020).

The Supreme Master Ching Hai, 2020. From Crises to Peace: The Organic Vegan Way Is the Answer. Love Ocean, Taipei.

The US Federal Emergency Management Agency, 2007. Animals in Emergencies. FEMA K-673 https://www.fema.gov/media-library/assets/documents/12960. (Accessed 23 June 2020).

Thornton, P.K., van de Steeg, J., Notenbaert, A., Herrero, M., 2009. The impacts of climate change on livestock and livestock systems in developing countries: a review of what we know and what we need to know. Agric. Syst. 101 (3), 113–127. https://doi.org/10.1016/j.agsy.2009.05.002.

Tollefson, J., 2013. A Light in the Forest: Brazil's Fight to Save the Amazon and Climate-Change Diplomacy. https://www.foreignaffairs.com/articles/brazil/2013-02-11/light-forest. (Accessed 23 June 2020).

UN, 2016. Report of the Conference of the Parties on Its Twenty-First Session, Held in Paris From 30 November to 13 December 2015. Available from: https://unfccc.int/resource/docs/2015/cop21/eng/10a01.pdf. (Accessed 20 July 2017).

UN, 2019a. UN Climate Action Summit 2019. https://www.un.org/en/climatechange/un-climate-summit-2019.shtml. (Accessed 20 July 2017).

UN, 2019b. Act Now. https://www.un.org/en/actnow/. (Accessed 20 July 2017).

UN, 2019c. Growing at a Slower Pace, World Population is Expected to Reach 9.7 Billion in 2050 and Could Peak at Nearly 11 Billion Around 2100. https://www.un.org/development/desa/en/news/population/world-population-prospects-2019.html. (Accessed 20 July 2018).

UN, 2019d. Unprecedented Impacts of Climate Change Disproportionately Burdening Developing Countries, Delegate Stresses, as Second Committee Concludes General Debate. https://www.un.org/press/en/2019/gaef3516.doc.htm. (Accessed 20 June 2020).

UN News, 2008. On World Environment Day, UN Officials Call for End to Carbon Addiction. https://news.un.org/en/story/2008/06/261792-world-environment-day-un-officials-call-end-carbon-addiction. (Accessed 20 June 2020).

Union of Concerned Scientists, 2020. Tropical Deforestation and Global Warming. Union of Concerned Scientists. https://www.ucsusa.org/resources/tropical-deforestation-and-global-warming. (Accessed 20 June 2020).

United Nations, 1992. What Is the United Nations Framework Convention on Climate Change? https://unfccc.int/files/essential_background/background_publications_htmlpdf/application/pdf/conveng.pdf. (Accessed 20 July 2017).

United Nations Environment Program (UNEP), 2009. Moving Towards a Climate Neutral UN The UN System's Footprint and Efforts to Reduce It. https://www.icao.int/environmental-protection/Documents/CNUN_report_09.pdf. (Accessed 20 July 2017).

United Nations Environment Programme and International Livestock Research Institute, 2020. Preventing the Next Pandemic: Zoonotic Diseases and How to Break the Chain of Transmission. Nairobi, Kenya. Available from: https://wedocs.unep.org/bitstream/handle/20.500.11822/32316/ZP.pdf?sequence=1&isAllowed=y.

United Nations Food and Agriculture Organization, 2019. Animal Production. http://www.fao.org/animal-production/en/. (Accessed 20 June 2020).

United Nations Framework Convention on Climate Change (UNFCCC), 2008. Kyoto Protocol Reference Manual. https://unfccc.int/resource/docs/publications/08_unfccc_kp_ref_manual.pdf. (Accessed 20 July 2017).

United States Department of Agriculture, n.d. Animal Production. https://nifa.usda.gov/topic/animal-production (Accessed 20 July 2018).

University of Bonn, 2020. Global Trade in Soy Has Major Implications for Climate. https://www.sciencedaily.com/releases/2020/05/200507104446.htm. (Accessed 01 July 2020).

University of Queensland, 2019. Changing Climate May Affect Animal-to-Human Disease Transfer. www.sciencedaily.com/releases/2019/05/190501114619.htm. (Accessed 20 June 2020).

US Climate Change Science Program, 2008. The Effects of Climate Change on Agriculture, Land Resources, Water Resources, and Biodiversity in the United States. Available from: https://www.fs.fed.us/rm/pubs_other/rmrs_2008_backlund_p003.pdf.

US Environmental Protection Agency (EPA), 2000. National Water Quality Inventory Report to Congress. https://www.epa.gov/waterdata/2000-national-water-quality-inventory-report-congress. (Accessed 20 June 2020).

VOA News, 2020. WHO Expert Believes Wuhan Wet Market Played Role in COVID Outbreak. https://www.voanews.com/covid-19-pandemic/who-expert-believes-wuhan-wet-market-played-role-covid-outbreak. (Accessed 20 June 2020).

Walker, C., 2016. Making the Case for Forest Restoration: A Guide to Engaging Companies. IUCN, Global Forest and Climate Change Program, Gland, Switzerland.

Wang, Z., Bergeron, N., Levison, B.S., Li, X.S., Chiu, S., Jia, X., Hazen, S.L., 2019. Impact of chronic dietary red meat, white meat, or non-meat protein on trimethylamine N-oxide metabolism and renal excretion in healthy men and women. Eur. Heart J. 40 (7). https://doi.org/10.1093/eurheartj/ehy79.

Weston, P., 2019. Germany may introduce 'meat tax' to protect the environment. Independent. https://www.independent.co.uk/environment/german-meat-tax-environment-animal-welfare-a9045271.html. (Accessed 20 July 2018).

WHO, 2004. Management of Dead Bodies in Disaster Situations. Disaster Manuals and Guidelines Series, No. 5. Available from: https://www.who.int/hac/techguidance/management_of_dead_bodies.pdf.

WHO, 2020a. Stop Using Antibiotics in Healthy Animals to Prevent the Spread of Antibiotic Resistance. https://www.who.int/news-room/detail/07-11-2017-stop-using-antibiotics-in-healthy-animals-to-prevent-the-spread-of-antibiotic-resistance. (Accessed 20 June 2020).

WHO, 2020b. Ebola Virus Disease. https://www.who.int/news-room/fact-sheets/detail/ebola-virus-disease. (Accessed 20 June 2020).

World Animal Foundation, 2020. Factory Farms Cause Hunger. https://www.worldanimalfoundation.com/advocate/farm-animals/params/post/1280003/factory-farms-cause-hunger. (Accessed 01 July 2020).

World Economic Forum, 2017. Here's How We're Ensuring There'll Be Enough Sustainable and Nutritious Food for 9.8 Billion People by 2050. https://www.weforum.org/our-impact/feeding-the-world-nutritiously-and-sustainably. (Accessed 20 June 2020).

World Health Organization (WHO), 2005. A Route to Poverty Alleviation. https://apps.who.int/iris/bitstream/handle/10665/43485/9789241594301_eng.pdf;jsessionid=3B8974B0715D6614C2BF02C97785F35F?sequence=1. (Accessed 20 June 2020).

World Wide Fund for Nature (WWF), 2018. Living Planet Report—2018: Aiming Higher. Gland, Switzerland.

Yale University, 2020. Cattle Ranching in the Amazon Region. https://globalforestatlas.yale.edu/amazon/land-use/cattle-ranching. (Accessed 20 June 2020).

Yu, W., 2020. Coronavirus: Revenge of the Pangolins? https://www.nytimes.com/2020/03/05/opinion/coronavirus-china-pangolins.html. (Accessed 20 June 2020).

CHAPTER

# 10

# Emerging typology and framing of climate-resilient agriculture in South Asia☆

*Rajesh S. Kumar*[a], *Shilpi Kundu*[b,c], *Bishwajit Kundu*[d], *N.K. Binu*[e], and *M. Shaji*[e]

[a]Indian Forest Service (IFS), New Delhi, India [b]Cities Research Institute & School of Environment and Science, Griffith University, Brisbane, QLD, Australia [c]Sher-e-Bangla Agricultural University, Dhaka, Bangladesh [d]Bangladesh Jute Research Institute, Dhaka, Bangladesh [e]College of Forestry, Kerala Agricultural University, Thrissur, India

OUTLINE



☆The views/analysis/observations presented in this chapter are personal and based on the domain experience of the authors and may not be construed/reflected as the official positions of the organizations under which the authors are employed presently or in the past.


https://doi.org/10.1016/B978-0-12-822373-4.00021-5

# 1 Introduction

Climate change is threatening the food and livelihood security of millions across the world (IPCC, 2014). The agriculture systems and such similar ecosystem-based food production systems have been facing a crisis due to ongoing climate change (Raza et al., 2019; FAO et al., 2018; Pound et al., 2018; Lobell et al., 2012; Mall et al., 2006; Brida et al., 2013; Gourdji et al., 2013). South Asian (SA) geography—the home for 1.75 billion people in just 5 million $km^2$ with nearly 4% of global Gross Domestic Product (GDP)—occupies a significant place in the global engagement to respond to climate change (Pound et al., 2018). The regional landscape presents a challenging scenario on account of rapid urbanization, growth of emerging economies, wide prevalence of poverty, high economic inequality, extreme geophysical vulnerability to climate change impacts, etc. (ADB, 2020). The varied climatic zones in the regional landscapes have been experiencing multiple impacts of climate change evidenced by sea levels rising, unprecedented changes in temperature and rainfall regimes, frequent recurrence of extreme climatic events such as drought, floods, cyclones, etc. (ADB, 2020; Prakash, 2018). The region has been witnessing decadal rise in the average temperature regime since 1960 (Prakash, 2018). According to the International Monetary Fund (IMF), unmitigated climate change may shave off 11% of the regional GDP. The Global Climate Risk Index 2019 ranks Sri Lanka, Nepal, Bangladesh, and India in the second, fourth, ninth, and fourteenth positions, respectively, for climate vulnerability (German Watch, 2019). It is reported that the region has already suffered economic damage to the tune of $USD 147.27 billion due to natural hazards in the period from 2000 to 2019. In addition, nearly 700 million people were affected by one or more climate-related disasters in the same period. The ongoing changes in average weather patterns and frequent recurrence of extreme weather events are slated to impact the lives and livelihoods of more than 800 million people across the region by 2030 (World Bank, 2019). Such a scenario assumes a serious development concern since the region houses more than 75% of the poor and vulnerable groups relying on rainfed agriculture, animal husbandry, and forestry resources for their food, nutrition, and livelihood security (IFPRI, 2018). The emerging context becomes increasingly challenging as the regional population is also expected to grow by 40% by the mid-century. This will squarely demand for increased provisioning of food and other resources to meet the development requirements (Jaffery, 2019). The scenario therefore demands augmenting cereal food production itself by 1.6% per year. And much of this food production target needs to be met by increasing productivity of agriculture systems rather than by expanding the area under cultivation (ADB, 2013).

The Intergovernmental Panel on Climate Change (IPCC) predicts that the air temperature in the region will rise by 0.5–1.2°C by 2020, 0.88–3.16°C by 2050, and 1.56–5.44°C by 2080, with concomitant increase in the frequency and intensity of precipitation events (IPCC, 2007). Therefore, climate change will increasingly undermine the agricultural production systems due to the impacts of drought spells, floods, and desertification (Brida et al., 2013; Lobell et al., 2012). As per the reports, the historical and future impacts of climate change on cereal production will lead to yield declines to the extent of 35%, 20%, 50%, 13%, and 60% for rice, wheat, sorghum, barley, and maize crops, respectively (Porter et al., 2014). Such drastic productivity declines will be immense for the small and medium-scale agriculturists inhabiting the rainfed agricultural and pasture landscapes in South Asia

(Campbell et al., 2016). It is also highlighted in the current literature that climate change and its impacts may spill over to the next century as well. Such a cross-century extension of climate change and its impacts will leave South Asia compromised to realize development outcomes regarding food, nutrition, health, and urbanization (ADB, 2020). The above mentioned challenges and complexities demand adapting of agricultural systems in the region to deal with the risks and challenges posed by climate change through effective coping arrangements on an urgent basis.

## 1.1 Climate-resilient agriculture (CRA)

As mentioned already, adaptation of agricultural and other ecosystem-based food production systems to climate change is integral to climate change response strategies. The Third Assessment Report of IPCC defines climate change adaptation as an "adjustment in ecological, social, or economic systems in response to actual or expected climatic stimuli, and their effects or impacts. This term refers to changes in processes, practices or structures to moderate or offset potential damages or to take advantage of opportunities associated with changes in climate" (McCarthy et al., 2001). However, adapting agricultural systems to climate change will demand transformative interventions to enhance system resilience, efficiency, and sustainability to cope with the impacts of climate change. Climate-Smart Agriculture (CSA), therefore, refers to a set of composite adaptation and mitigation strategies for agricultural systems to cope with the challenges posed by climate change impacts (Dinesh et al., 2015). As a strategy, CSA endeavors to (a) sustainably increase agricultural productivity to maintain and enhance farm income and food security, (b) build landscape-level resilience to climate change to achieve development goals, and (c) mitigate greenhouse gas (GHG) emissions from agricultural, livestock, and fishery activities. CSA addresses climate risks to the agricultural systems both locally and regionally through policies, institutional arrangements, resources, and community ownership. While CSA encompasses both adaptation and mitigation actions, in climate change response measures, the Climate-Resilient Agriculture (CRA) strategy focuses on autonomous and planned coping strategies and interventions at ecosystem and social system levels from an adaptation perspective.

In this chapter, we have considered CRA as a subset of the CSA framework, focusing on improving the climate resilience of agricultural and social systems (Pound et al., 2018). CRA therefore encompasses a range of both on-farm and off-farm interventions targeting improvement of resilience at individual, community, and landscape levels. However, such interventions are designed and realized through/under climate adaptation policies which provide ways and means for the implementation of such interventions (Dessai and Hulme, 2004). Over the decades, CRA initiatives have been gaining much traction across the world, and the Global Alliance for CRA was formed in 2010 (FAO, 2013). Since then, several regional alliances have emerged in the global landscape to promote CRA. However, studies indicate that the benefits of CRA vary with crop(s), cultivation practices, changes in temperature, and rainfall regimes (Easterling et al., 2007; FAO, 2013). Therefore, the adoption of a CSA approach can improve crop yield, increase input use efficiency, enhance net farmer income, and reduce GHG emissions as compared to conventional agricultural practices. These measures are also found to offer an enhanced cost benefit ratio (CBR) of 1.81 as compared to a CBR of 1.02 under conventional farming systems (Rai et al., 2018; Khatri-Chhetri et al., 2016; Sapkota et al., 2014; Gathala et al., 2011). It is noteworthy that such adaptive transformational changes can be effected and scaled up with technology adoption to enhance crop yield and farmers' income (Parihar et al., 2016; Byjesh et al., 2010; Kumar et al., 2014; Kumar and Aggarwal, 2013; Smith et al., 2007).

The implementation of the abovementioned strategies and initiatives will, however, require promotion of synergy and trade-offs among the key input-side factors for the systems under consideration. Ironically, it is reported that the current level of adoption of CRA is much lower than expected, despite the benefits and prospects that it offered (Palanisami et al., 2015). CRA adoption is influenced by a host of factors including the socioeconomic status of farming communities, location-specific biophysical challenges, and access to and affordability of adaptation technologies (Below et al., 2012; Deresa et al., 2011). The availability of credible evidence on current and future climate adaptation options as well as awareness of such prospects of CRA across the relevant actor and stakeholder groups will foster the implementation of CRA (Pound et al., 2018). As such, the identification, prioritization, and promotion of existing CRA technologies are reported to be among the major challenges for up-scaling CRA across the diverse agro-ecological landscapes in the region. In addition, availability of such validated evidence will help planners in developing pragmatic investment portfolios for designing CRA strategies at landscape and sublandscape levels by evaluating and prioritizing options in the context of climate risks faced by the region. However, studies reported that evidence on farmers' prioritization, current debates, discourses on CRA, institutional arrangements and their governance quality, etc. are critical to the development of pragmatic adaptation roadmaps for successful long-term climate change adaptation in the agriculture sector. Driven by those objectives, we explored the emerging discourses and initiatives on CRA in South Asia, to address the following questions: (a) what is the emerging typology of CRA in the region?; (b) what barriers and challenges are affecting wide-scale adoption of CRA?; and (c) what are the strategic areas to focus on? In order to address these questions, we followed a systematic literature analysis based on secondary literature available in the open domain to address the abovementioned questions and present a typological framework of CRA interventions in the region. The study outputs are expected to contribute toward strengthening strategic areas for policy intervention and further resource investments for promotion of CRA in the region and beyond.

## 2 Data, method, and analysis

In this study, we followed a systematic literature analysis approach, as it is an established methodology to explore novel approaches, and appreciate critical aspects and gaps in the current knowledge in the domain of exploration. The methodology involves appraising the literature with predefined questions and established criteria (Nielsen and D'haen, 2014). We also took the support of the approach followed by Ford et al. (2011) and Berrang-Ford et al. (2015) for a climate change-related literature review. A review of literature was conducted between March and September 2020, and this covered research reports, technical documents, and policy documents providing evidence on CRA in the selected South Asian countries, viz., Afghanistan, Bangladesh, Bhutan, India, Nepal, and Sri Lanka. Study documents were identified by mining online databases including Science Direct, Web of Science, and Google scholar.

In the first phase, we reviewed published scientific and technical literature (journal articles, book chapters, technical reports, working papers, and technical articles) on CRA to develop a framework of keywords as well as to define the exclusion and inclusion criteria. The inclusion criteria included documents published or available in the public domain reflecting CRA in South Asia, evidence on typology elements, CRA interventions, barriers to CRA, and institutional and governance arrangements. The set of keywords used included Afghanistan, Bangladesh, Bhutan, India, Nepal, Sri Lanka, climate change, Climate Change Adaptation (CCA), Climate Resilient Agriculture (CRA), weather-smart activities, nutrient-smart activities, water-smart activities, energy-smart activities, knowledge-smart activities, stress-tolerant varieties, crop management interventions, social resilience, institutional resilience, and barriers to CRA. We considered the time frame from 2015 to 2020 for the literature for review. This approach not only helped us to maintain objectivity in the selection of literature, but also enabled us to capture an overview of the emerging typology of CRA in the region (Petticrew, 2001).

The exclusion criteria included publications reported before 2015, newspaper and social media articles and documents not published in English in the overall context of CRA. The scope of publications and time frame of published documents were necessary to identify the emerging narratives on CSA. The initial search retrieved 509 documents and these were subject to screening based on title, keywords, and abstract. The introduction part of each publication was read to identify the geographical location pertaining to the research, as we have already identified the countries considered in this study. In the final selection, 62 documents were identified. The selected documents were analyzed using NVIVO version 12. This software platform helped us to organize the evidence on CRA as per the thematic areas, adaptation domains, and barriers. The thematic areas were further divided into four subcategories, viz., weather-smart activities, cropping and cropland management, socioeconomic resilience, and institution and governance. The framework used to analyze the data is presented in Fig. 1. The text accordingly coded was retrieved and evaluated to identify and describe the typology, narratives, and framing of CRA in the region (Ford et al., 2011). This review does have some limitations, as the identification of the review documents was limited by title specifics, abstract, and keywords in the background of the study considerations already mentioned. However, the final selection of documents was restricted by the search criteria fixed for the current study. In addition, articles referring to CSA and climate change mitigation approaches in the agriculture sector may have been overlooked.

## 3 Results and discussion

In order to address the research questions, we examined 62 documents identified across the four strategic domains of CRA considered in the study, viz., weather-smart activities, cropping and crop land management, socioeconomic resilience, and institution and governance. The current analysis followed the major contours of the dominant narratives related to CRA on both planned adaptation and autonomous adaptation streams.

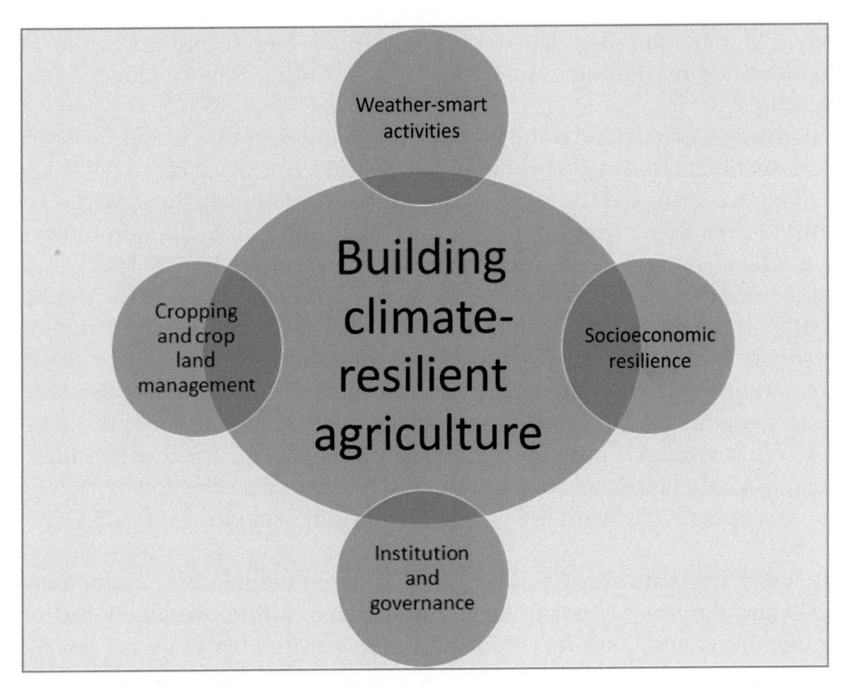


**FIG. 1** The thematic framework used in the study.

## 3.1 Weather-smart activities

Climate change and the associated unprecedented weather variations are increasingly felt in the agricultural landscapes across the countries in the region. They not only place a heavy economic burden on farmers but also undermine the resilience of farming communities to pursue tactical and strategic decisions on agricultural activities (Nesheim et al., 2017). Agro-meteorological services are disseminated over free or paid service platforms, and are reported to have enabled the agricultural communities to minimize damage to crops due to adverse weather conditions through effective planning of farming operations (Nesheim et al., 2017; Tall et al., 2014). The agro-met services in the region are, however, observed to have been functioning with a large variety of service features, delivery models, delivery platforms, and service providers. Some of the models in operation are briefly discussed below.

### *3.1.1 Integrated agro-meteorological advisory service*

This model, functional in India, is a mega-scale agro-meteorological service and assists farmers to improve both crop production as well as climate resilience. It is characterized by (a) coproduction of advisories by agricultural and meteorological experts, (b) institutional partnership at the local level, and (c) dissemination of advisories over Short Message Services (SMS), internet, voice messaging, bulletins, and training, and also through capacity building events (CGIAR, 2020). The model is reported to have assisted farmers in undertaking CRA practices such as climate-efficient modern agricultural technologies and practices, weather-based irrigation management, enhanced crop protection measures, and adoption of improved postharvest technologies (Maini and Rathore, 2011).

### 3.1.2 *Automatic weather station network*

The model functional in selected districts in India provides real-time information on rainfall, temperature, relative humidity, wind speed and direction, solar radiation, evapotranspiration, etc. based on observations collated from a network of mini weather observatories located in villages. The system provides customized agro advisories to farming communities to promote weather literacy-based farming practices and crop husbandry, leading to effective use of various agricultural inputs such as irrigation and fertilizers, in addition to undertaking appropriate crop protection measures. Thus, the network provides a great deal of support toward the development of resilience in agricultural systems (Nesheim et al., 2017).

### 3.1.3 *Drought forecasting and communication model*

The model operational in Sri Lanka aims to support farmers with appropriate seasonal forecasting and advisories to adapt with the stress conditions. In addition, the model supports community-based landslide risk identification and its management through a set of well-laid-out protocols supported by a network of telemetry rain gauges in sync with warning sirens in high landslide risk-prone areas. The model also provides clients with a menu of prescribed crops and cropping alternatives to cope with the impacts of climate-induced extreme rainfall events (Amalanathan et al., 2017).

### 3.1.4 *Famine early warning systems network*

This network, operational in Afghanistan and set up by the United States Agency for International Development (USAID) in collaboration with international, regional, and national partners, provides timely early warning on climatic events. It analyzes the probability of emerging and/or evolving food security concerns from the perspective of crop protection from adverse climatic events, for decision-making for guided responses. The network in particular provides early warnings on extreme weather events, such as heavy rains and flash floods in collaboration with the local meteorological department (MAIL, 2020).

### 3.1.5 *Partnership in climate services for resilient agriculture*

This collaborative model implemented by the USAID and Skynet Weather Services in India encompasses digitization of traditional methods practiced by the farming communities. It aims to develop new risk mitigation products for climate-resilient agriculture. The project started in 2015 and is reported to have benefited more than 8000 farmers across nine states in India. Under this model, hyperlocal weather information is provided by automated weather stations. The system provides customized crop advisory services and follows the concept of a community practice approach. The service delivery is done through Voice Broadcast Services (VBS) in regional languages, weather display boards, and toll-free numbers (Bano, 2019).

### 3.1.6 *Weather research and forecasting (WRF) service model*

Under this model, currently being experimented with in Bangladesh, local weather advisory services based on real-time weather data backed with simulations are provided to farmers through information and communications technology (ICT). The model is benefited by a multidisciplinary team comprised of agronomists, soil scientists, plant protection specialists, and data management experts. It is reported that the model had been assisting farmers in reducing cultivation costs by 3%–5% while gaining on the productivity front in rice cultivation by 9%–12% (Chowdhury et al., 2019).

#### *3.1.7 Local-level agro-met extension service model*

This model, prevalent in Bangladesh, offers weather advisory services by trained agro-met extension personnel to farmers, on aspects such as irrigation, disease pest management, fertilizer application, and other crop husbandry services backed by ICT. Local weather data are analyzed by experts in collaboration with meteorologists, and locality-specific advisories are issued to stakeholders. The service has been found to support farmers to adapt crop husbandry to the changing weather and climate parameters (NAEP, 2012).

#### *3.1.8 Community-based early warning systems*

This early warning system, operational in Nepal, assists communities in effective management of flash floods and mitigation of flood risks at community level. Communities are central to the functioning of the model, which is based on local-level information dissemination and rapid response. Early warnings are given by the system to vulnerable groups and communities over megaphones, sirens, and SMS via cell phones based on evolving meteorological information (Adhikari and Sitoula, 2018; Smith et al., 2017).

In addition to the abovementioned models, several generic systems have also been found providing agro-met services the study countries. These include the Bangladesh Agro-Meteorological Information Portal service, and a 24hour weather forecast service over national television, print media, etc. in Bhutan (Sihvola et al., 2014; Shrestha et al., 2016). As such, it may be observed that the weather forecasting systems prevalent in the region have been immensely assisting farming communities to assimilate climate and weather information to develop pragmatic resilience strategies to adapt to climate change and disaster risks, and thus to foster climate-resilient agricultural systems.

### 3.2 Cropping and crop land management

Cropping and crop land management are at the center of adaptation of agricultural systems to climate change, as they combine the synergy of various production inputs and management protocols. These autonomous and planned models of climate change adaptation aim to improve the resilience of agricultural systems and thus support sustainable crop production. However, the engagement follows an ecosystem approach, which conserves and enhances natural resources to increase both natural capital and ecosystem services while reducing the negative impacts on the environment (Azzu et al., 2013). Driven by those considerations, we describe in the following subsections various crop and crop land management strategies followed in the context of CRA in South Asia.

#### *3.2.1 Adoption of stress-tolerant crop varieties*

We observed that farmers and communities across the study countries in the region had been adapting to climate change under both autonomous and planned domains by modifying crop composition and cropping patterns. Most common and widely prevalent among them has been the development of a variety of stress-tolerant varieties of crops that offer a range of cobenefits and features. The current literature is rich with a wide spectrum of such crop varieties that have been developed to address specific adaptation requirements as well as food diversity. We have captured a snapshot of such varieties from different crops cultivated across the countries in the region, and this information is presented in Table 1.

It can be observed from Table 1 that a spectrum of crop varieties have been developed across the countries in the region to adapt to the various stresses on food production systems such as droughts, increasing temperature regimes, diseases, pest infestations, salt ingression, floods, and submergence. These varieties have been reported to be providing better yields than the traditionally used varieties as well (Sultana et al., 2019).

### 3.2.2 *Cropping system interventions*

Several cropping system interventions have been fast emerging in the agricultural adaptation landscapes in the region. These interventions are based on resilient cropping plans incorporating adaptive cropping patterns and appropriate crop rotation schedules. Against the backdrop of climate change risks, such decisions on crop choice, spatial distribution, and temporal succession are destined to enhance resilience and resource use efficiency in the farm landscapes (Dury et al., 2012). In the current review, we have come across a variety of such models, and these are summarized in Table 2. The sheer variety of cropping patterns

**TABLE 1** Different stress-tolerant crop varieties cultivated in the region.

| Stress | Stress-tolerant varieties and country | References |
|---|---|---|
| Drought | Tomato var. Arka Meghali (India); Aerobic Rice var. Anjali (India); Suwandal, Kirimurunga (Sri Lanka); Sukkhadhan-1 (Nepal); Bhur Kambja 2 (Bhutan); BRRI dhan 83 (Bangladesh); BARI Alu-15 (Bangladesh) | Venkateswarlu et al. (2012), Maheswari et al. (2019), Ghimiray and Katwal (2013), BRKB (2020), BARI (2020), and CDD (2015) |
| Diseases and pests | Groundnut var. GPBD-4 (India); Aerobic Rice var. Anjali (India); Khangma Maap (Bhutan); BRRI dhan 69 (Bangladesh); BARI Gom-33 (Bangladesh); and BARI Mung-8 (Bangladesh) | Ghimiray and Katwal (2013), BARI (2020), and BRKB (2020) |
| Temperature | Mustard var. Bonex Gold (India); BARI Hybrid Maize-15 (Bangladesh); BARI Alu-72 (Bangladesh); BARI Hybrid Tomato-3 (Bangladesh) | BARI (2020) |
| Salt | Rice var. Ac 810 (Sri Lanka); Rice var. Ac 5557 (Sri Lanka); BRRI dhan 78 (Bangladesh); BARI Gom-25 (Bangladesh); BJRI Deshi Pat-8 (Bangladesh); BARI Hybrid Maize-16 (Bangladesh); CSR 43 (India) | IRRI (2016), BRKB (2020), BJRI (2020), BARI (2020), and BINA (2020) |
| Flood/ submergence | Binadhan 23 (Bangladesh); BRRI dhan 91 (Bangladesh); Anadi (Nepal); Mansara (Nepal); Gorakhnath Gold (Nepal); BARI Sarisa-7 (Bangladesh) | Manandhar et al. (2011), BRKB (2020), BARI (2020), and BINA (2020) |

**TABLE 2** Cropping pattern/models adopted for climate-resilient agriculture in the agricultural landscapes in South Asia.

| Cropping pattern | Special features | Country | References |
|---|---|---|---|
| Ragi + Red gram | Ragi yield went up to 16.63 (q/ha) under Ragi Pigeon pea mixed cropping (4:2) compared to 10.97 (q/ha) under monocropping. | India | Jasna et al. (2017) |
| Maize + Red gram | Maize and Pigeon pea intercropping at 4:2 row ratio with 50% Pigeon pea population provided a maximum maize equivalent yield of 80.76 $q/ha^{-1}$ and net return (INR 30,492/ha). The combination offered a cost-benefit ratio of 2.75 compared to other intercropping systems and sole crops. | India | Marer et al. (2007) |
| Maize – Legume crop rotation | Maize legume rotation in dryland in subtropical humid agro-ecozones reported several add-on benefits and enhanced cash streams. The model further improves the soil nitrogen status and reduces soil erosion. | Bhutan | Katwal (2010) |
| Potato – Mustard crop rotation | This system has low labor requirement and mustard crop uses the residual nutrients applied for potato crop. However, mustard yield is estimated to range from 160 to 560 kg/acre. | Bhutan | Katwal (2010) |
| Potato + Maize intercropping system | Standing crop of maize reduces the impact of falling rain drops and the consequent runoff effects. It thus conserves soil, while the maize stubble improves soil's physical and biological conditions. | Bhutan | Katwal (2010) |
| Maize – Beans relay cropping system | This is a traditional cropping system and it is widely practiced across the country in drylands where maize is cultivated. The key feature of this system is that tall and robust maize plants are used as physical support for the beans to complete their life cycle. | Bhutan | Katwal (2010) |
| Rice – Mustard – Mungbean intercropping system | Legume helps to improve soil physico-chemical properties, and thus enhances soil productivity and crop production. This pattern can tolerate 4–8 DS $m^{-1}$ salinity. | Bangladesh | Hussain (2017) |
| Mustard – Potato – Green gram – Rice system | Increases crop productivity and additional income for the rural poor by utilizing fallow land. | Bangladesh | Hossain et al. (2014) |
| Mustard – Onion/ Maize – Rice (T. *Aman*) system | 37.63% higher production compared to single cropping pattern. Crop production as well as land use efficiency increased by 9.33% and 19.18%, respectively. | Bangladesh | Islam et al. (2018) |
| Maize + Groundnut system | Net return from 1:2 row of maize + groundnut yielded $USD 464 $ha^{-1}$ whereas the net return from single maize crop was $USD 315 $ha^{-1}$. | Nepal | Upreti (2002) and Mishra (2002) |
| Maize + Cowpea system | Net return from 1:1 row of Maize + Cowpea was $USD 478 whereas the net return from single maize was $USD 318 $ha^{-1}$. | Nepal | Upreti (2002) and Mishra (2002) |

and cropping models has been driving both autonomous as well as planned adaptation actions in the agricultural landscapes in the countries studied.

Apart from the abovementioned cropping models, the landscape is diverse with a large number of cultivation models adopted by the farmers to build crop system resilience. For example, in Nepal, farmers have been switching over from water-intensive crops, such as paddy, to high-value and low water-intensive horticultural and vegetable crops as an adaptation measure (Hussain et al., 2016). However, it has also been reported that the probability of farmers growing beans as a major crop to adapt to climate change is 94% higher than it is for those who do not grow such leguminous crops (Menike and Arachchi, 2016). The above discussion gives a sense of the field-level dynamics driving cropping model selection to cope with climate change risks.

## 3.3 Water-smart activities

As water is a basic input in the agricultural systems, water-smart activities stay at the core of the strategies toward climate-resilient agriculture (Teklewold et al., 2019). It needs no emphasis that water shortage is bound to escalate with rising global average temperatures. Agriculture is the single largest water-consuming bioproduction sector responsible for the utilization of nearly 70% of the water extracted from above-ground and below-ground sources (Batchelor and Schnetzer, 2018). The philosophy of *doing more with less* is particularly relevant to sustainable water use in agriculture. Water-smart agriculture as a paradigm thus encompasses a host of practices such as micro-irrigation, smart irrigation, direct-seeded rice cultivation, alternate wetting and drying, in-situ soil moisture conservation, etc. (AQUA4D, 2019). Driven by these aspects, we discuss in the following paragraphs the water-smart autonomous and planned adaptation measures undertaken across the study countries in the region.

### 3.3.1 *Micro-irrigation systems*

Micro-irrigation systems such as drip and sprinkler irrigation systems are increasingly being harnessed to improve irrigation efficiency in cropping scenarios. These systems are known to enhance crop yield, save water and energy, and are particularly suitable for marginal lands. The current literature evidences that around 50%–90%, 31%, and 29% savings in water, energy, and fertilizer, respectively, can be gained by resorting to micro-irrigation methods with a consequent 42% enhancement in farmer income levels. Such practices are reported from all study countries and have supported farmers to become resilient to extreme weather events, as they can cultivate larger areas during the delayed monsoon times or heatwave spells (Kishore et al., 2014; Zotarelli et al., 2015). Studies from India report that about 114% yield gain was obtained for cotton crop under drip irrigation compared to flood irrigation regimes. Similar results are also reported for crops such as banana, coconut, grape, turmeric, and apple, with productivity gains of 4%, 15%, 16%, 22%, and 35%, respectively, compared to traditional cultivation methods (Randev, 2015; Viswanathan et al., 2016). As such, micro-irrigation is reported to achieve a water-saving of 25%–80% across the cultivation practices and crops (Kumar et al., 2008) and a saving in electricity consumption of 25%–77%. Thus, it may be held that micro-irrigation measures enable farmers to contribute toward both climate change adaptation and mitigation measures.

### 3.3.2 *Direct seeded rice (DSR) cultivation method*

This method is found widely practiced in several agro-ecological landscapes across the countries in the region owing to the flexibility in sowing protocols and seasonal specifics. The technique works with less water and commercial energy, and thus reduces the water and carbon footprint of the crops produced. Several evaluation reports on DSR mention that average crop yield increased in Basmati rice systems by 51 quintal/ha compared to 46.3 q/ha under other sowing regimes. Similarly, DSR provided a cost saving of $USD 40–55 per hectare in the plains of Punjab in India (Prasad et al., 2014). The current literature, however, reports its scaling up across suitable agro-ecological landscapes in the region as well.

### 3.3.3 *Alternate wetting and drying technique*

Alternate wetting and drying (AWD) is an irrigation scheduling technique that reduces irrigation demand by about 14%–18% coupled with a productivity enhancement of 0.81–0.83 kg/m$^3$ compared to 0.69–0.73 kg/m$^3$ under traditional irrigation regimes. AWD practice requires 22% less irrigation water than conventional techniques. It is reported that AWD practice increased the yield of rice (6.0%) compared to traditional practices (Hasan et al., 2016). Moreover, it reduced carbon emission by 160 kg $CO_2$/ha and methane flux emission by 23%–36% from the rice fields compared to flooding irrigation arrangements (Maniruzzaman et al., 2019). In summary, AWD is found to be water-saving, irrigation cost-saving, and crop yield-enhancing, and thus supports CRA.

### 3.3.4 *In-situ soil moisture conservation practices*

In-situ soil moisture conservation is a highly resilience building practice for farmlands. It includes creation of raised beds and furrows with several local modifications. For instance, the *sorjan technique* practiced in Bangladesh involves raised beds for cultivation of crops and furrows to harvest water for irrigation. Some farmers also use it for multiple cropping with pisciculture. In addition, application of material has been found much widely practiced to harness efficiency in both soil moisture conservation and weed management. Similar practices involving changing planting methods and adjusting planting time are also reported to reduce irrigation requirements by 31%–37% during kharif and rabi seasons (Kumar et al., 2013). The literature is rich in several such local models, and these models are reported to support improved productivity, enhanced water availability, diversified enterprises, enhanced cropping intensity, and system resilience (Sikka et al., 2018).

### 3.3.5 *Zero tillage cultivation*

Zero tillage (ZT) is an integral feature of conservation agriculture. It is widely practiced in India, Bangladesh, and Nepal, where rice-wheat cultivation constitutes the predominant cereal production system. The technique has received much acceptance for its reduced irrigation demand, prospects for improved soil fertility, and higher profit margins (Erenstein, 2009). Studies from India report improvement in soil organic content by 21% under the ZT regime with the cobenefit of residue recycling. In the Basmati rice-wheat system, it increased productivity by 36% and net returns by 43% compared to conventional practices (Jat et al., 2019). However, maize systems yielded a 5% bonus and potato crops provided a 70% bonus under ZT regimes as per studies from Bangladesh (Alam et al., 2015; Gathala et al., 2016).

As such, it is reported that ZT reduced irrigation needs by 27% and 22% compared to traditional methods for rice and wheat/cotton crops, respectively, while water productivity was enhanced 12-fold (Alam et al., 2015; Choudhary et al., 2016).

## 3.4 Nutrient-smart practices

Integrated crop and nutrient management practices have tremendous potential to augment the productivity of agricultural systems. These systems offer positive gains in nutrient consumption, GHG mitigation, and farmer income. It is reported that integrated crop nutrition management improves crop yields by 20%, 40%, and 35% in cereals, legumes, and oilseeds, respectively, with a resultant increase in farmer income level by 35% (Wani et al., 2017). We briefly discuss in the following paragraphs the models in vogue in the study countries.

### 3.4.1 *Soil Quality Index (SQI)-based crop nutrient management*

The Soil Quality Index (SQI) helps to determine appropriate land management options for enhancing crop yield. For instance, in Bangladesh, water logging and salinity are very common challenges to agriculture productivity. Therefore, fertilizer management is key to maximize crop yield as well to safeguard soil health on a sustainable basis (Biswas et al., 2019). The Soil Health Card System, which is prevalent in India, provides information on farm level soil fertility status and other important parameters affecting crop productivity. This service covers macronutrients (N, P, and K), secondary nutrients (S), and micronutrients (Zn, Fe, Cu, Mn, and Bo), and other critical parameters such as pH, electrical conductivity (EC), and organic carbon (OC). These cards are issued biannually and enable farmers to take evidence-based decisions on fertilizer application on an effective and sustainable scale with a low carbon footprint (Islam et al., 2017a,b). It has been estimated that almost 7% of chemical fertilizers can be potentially saved, and farmer income can be increased to the tune of 5% by scaling up soil testing-based integrated nutrient management approaches. This will help farmers to operate farming systems with reduced carbon footprints. Apart from those measures, other methods such as leaf color charts and soil pH-based extension services are also prevalent in these countries, aiding farmers to move toward CRA.

### 3.4.2 *Organic farming*

Organic farming has been found to be practiced in most of the study countries to varying degrees. In Bhutan, it is practiced as a pillar of the green economy and is integral to carbon-neutral development policies. Farming in Bhutan has been largely a subsistence system, utilizing limited or minimal quantities of inorganic inputs. Currently, major crops under organic production are sweet buck wheat, herbs, and vegetables. Fruit crops are grown free of chemicals or with minimal use of inorganic inputs under organic farming regimes. The pest management strategies practiced include pheromone traps and bait traps for fruit flies and similar crop pests. However, nutrient management strategies followed include legume cropping, cultivation of fodder crops during fallow time, application of organic matter/farmyard manure, and residue management (Tenzin et al., 2019; Khanal, 2018; Sikka et al., 2018; Karki et al., 2020).

## 3.5 Knowledge-smart activities

A comprehensive understanding of climate change adaptation knowledge is key to secure an effective community response to climate change risks. Stakeholders with sufficient background information on CRA is highly catalytic in effective translation of CRA strategies to practice. Moreover, a knowledge-smart response is key to avoid the risk of maladaptation and the consequent socioeconomic and ecological disruption. Against this background, we discuss in the following subsections some of the major knowledge-smart activities observed in the region.

### 3.5.1 *Climate field schools (CFS) in Bangladesh*

Climate field schools (CFS) are a capacity-building initiative at grassroots level, currently operational in 52 subdistricts in Bangladesh. These schools aim to improve the socioeconomic conditions of farming communities by enabling them to acquire skills and expertise to respond pragmatically to climate change challenges especially in drought-prone, flood-prone, flash flood-prone, and salinity-affected areas. These gender-inclusive agricultural extension services have skill-building sessions on baseline survey, input supply, technology demonstration, etc. CFS are reported to have been empowering communities to address climate change and develop capability to comprehend disaster warnings and climate change impacts by following a *learning by doing and learning by experiencing approach* (Mahashin and Roy, 2017).

### 3.5.2 *Community leadership development*

Extension facilitator training and development is a platform for encouraging climate-resilient agriculture-related knowledge sharing among the cultivator communities through development of community leadership. Under this initiative, promising leaders are identified and are given scientific training and know-how, to implement locality-specific climate-resilient agriculture technologies and interventions. This engagement relies on the strength of local leadership resources and community learning. Such trained leaders are expected to act as a dynamic link between the communities and agricultural extension experts in furthering CRA-linked strategies and interventions in their command areas (NAEP, 2012).

## 3.6 Energy-smart technologies and practices

The relationships between food production systems and energy consumption have become much stronger and complex with the increasing use of energy intensive technologies in agriculture. However, such intrinsic relations often run the risk of setting on vicious circles in the backdrop of climate change as well. Since the agriculture sector is a significant contributor to greenhouse gases, a near-complete dependency of food production systems on fossil fuels needs to be transformed to tap cleaner energy sources (FAO, 2013). In addition, efficient harnessing and management of energy sources both for and from agrifood systems can potentially facilitate climate-resilient agriculture while providing food, energy, and climate security. However, such transformations will call for scaling up energy-smart technologies and practices. Guided by those interests, we have captured here a set of such smart models operating in the region.

### 3.6.1 *Adaptive crop residue management*

Under this model currently practiced in India and Bhutan, crop residues are used either in situ or locally as raw mulch to promote local nutrient cycling. In some cases, crop residue materials are supplemented by biochars as well. This model combines the twin benefits of soil productivity enhancement and carbon conservation in the agricultural fields. It is reported that the combined application of mulch and biochars led to a production enhancement of 36%–64% and carbon mitigation of 12% (Deya et al., 2020; Woolf et al., 2010; Lehmann and Joseph, 2015). The crop residue retention on soil surface helps to improve soil health as well as reduce greenhouse gas emission to an extent of nearly 13 t $ha^{-1}$ (Sharma et al., 2008).

### 3.6.2 *Solar power in remunerative crop production models*

This model, promoted in India under the Kisan *Urja Suraksha Evam Utthaan Mahaabhiyan* initiative, endeavors to combine the economic and climatic change mitigation dividends of application of solar power in farm irrigation. The model works on a cooperative arrangement in which groups of smallholder farmers are organized and provided with 7.5–10.8 k Wp solar-powered irrigation pumps. These units are interconnected to develop a local grid to distribute and manage surplus production. The model adds economic and ecological resilience to both the beneficiaries and the government while mitigating climate change by avoiding the use of fossil fuel-based energy for food production and development of associated value chains (Nexus, 2018).

### 3.6.3 *Energy-efficient agriculture pumps program*

This program, implemented in the agricultural sector in India, promotes energy efficiency by reducing the energy consumption in irrigation. This arrangement is delivered by efficiency upscaling of nearly 25–30 million irrigation pump sets operational in the country. The strategy is set to achieve an energy saving of nearly 30%–40% in agricultural energy consumption through wide adoption of energy-efficient pump sets. These smart pumps are operated remotely by remote-control devices at the comfort and convenience of farmers (BoEF, 2020).

### 3.6.4 *Laser land leveling (LLL) technique*

This water-saving technology aims to conserve groundwater pumped for irrigation by ensuring that water is applied evenly in the crop land. Thus, it aims to optimize the use of available irrigation water by minimizing runoff and water logging, and thus promotes irrigation efficiency (Aryal and Jat, 2015). Currently, this technology is used in more than 1.5 million ha in South Asia and accounts for 25%–30% water savings coupled with 5%–15% improvement in yield ( Jat et al., 2009; Chakhansuri, 2018). Studies from India report that application of laser leveling in rice fields reduced irrigation duration by 47–69 h/ha/season and improved yield by approximately 7% compared to fields leveled by traditional techniques. However, in laser leveled fields for wheat cultivation systems, the irrigation time was reduced by 10–12 h/ha/season, while the yield increased by 7%–9%. As such, it is reported that LLL is a scale-neutral technology and suitable for all farm holdings. The technology stands promising in the CRA practitioners with potential scope for earning an additional $USD 143.5/ha/year through increased crop yield and reduced electricity consumption (Aryal et al., 2015).

It is evident from the above discussion that several smart irrigation measures not only conserve precious irrigation water but also offer a set of financial, ecological, and climate change abatement benefits. As such, the wide-scale practice of such technologies is vital to CRA, as it is predicted that a 10% increase in irrigation water demand will occur with every 1°C rise in temperature in arid and semiarid regions of the subcontinent (Aryal and Jat, 2015).

## 3.7 Socioeconomic resilience

It is clear that the food systems are complex entities and signify integration of biophysical systems, economic systems, and social systems at different scales and intricacies (FAO, 2012). The existence of such complexity demands that climate change adaptation action operates beyond the biophysical systems concerned. It warrants engagements on strengthening income streams and livelihood sources at landscape level and beyond. Against this background, we profile some of the prominent models advancing socioeconomic resilience development operational in the study countries.

Evidence from the current literature indicates that the probability of a farmer resorting to CRA is significantly influenced by multiple socioeconomic determinants. These determinants are majorly linked to household size, disposable income, opportunity cost of investment, education level of farmers, access to cost-effective advisory services, gender dimensions, and social and cultural barriers. Studies report that in Sri Lanka, the probability of adapting to climate change decreases with decreasing household size, as the large-sized households are advantaged by technical and manpower skills and resources (Choden et al., 2020; Gupta et al., 2020; Sam et al., 2020; Alam, 2017). Similarly, education level and access to financial and social capital are also found to influence the selection of the adaptation methods (Menike and Arachchi, 2016). The availability of nonfarm income therefore potentially influences the development and strengthening of resilience to climate change among the agricultural communities (Sam et al., 2020). Similarly, codeterminants such as formal education and skill level of farmers are found to influence positively the choice of crops considered under the climate adaptation strategies (Williams and Carrico, 2017). In the following sections, we briefly discuss various models operating across the study countries toward building resilience at the community level.

### *3.7.1 Community-based response systems*

A study of minor irrigation systems in Sri Lanka shows that the adaptation decisions taken at the community level are more effective in implementation of adaptation measures such as conserving water, crop diversification, pest and disease management, selection of seed varieties, planting schedules, etc. Such community-based arrangements also provide local platforms for convergence action between government programs and community initiatives to develop and implement pragmatic adaptation action plans with community ownership (Government of Sri Lanka, 2010).

The Climate Smart Village is yet another model of community engagement in CRA promoted in India and Nepal. This model, actively facilitated by the Consultative Group on International Agricultural Research (CGIAR), involves a convergence of researchers, farmer cooperatives, local governments, private sector, and policy actors in the identification of appropriate CRA interventions. These interventions are, however, subsequently institutionalized under local village development plans to secure sustainability to ensure both food and livelihood security of the local communities (Pound et al., 2018).

### 3.7.2 *Promotion of agritourism*

Agritourism is one of the emerging areas of entrepreneurial engagements for the agriculturalists to cope with the economic risks posed by climate change. It essentially involves diversifying income streams from agricultural activities and the associated value chains. In a broader sense, agritourism envisages visits to a working agricultural setting for leisure, recreation, or education purposes (Arroyo et al., 2013). These engagements range from farm-based ventures to well-organized agro technology parks (Mahaliyanaarachchi et al., 2019). Agritourism in its various formats has been reported to be strengthening resilient building engagements from locations across the countries. However, the spread of agritourism is fraught with gaps in awareness, level identification, and networking of viable agritourism spots, etc. (Mahaliyanaarachchi et al., 2019).

### 3.7.3 *Participation in labor markets*

Major changes in labor participation in the production sectors are bound to create macroeconomic impacts. Climate change has been reported to reduce labor productivity in the agricultural sector, diminish livelihood options, and as such force farmers to attempt to create income diversification streams from other sectors of the economy (Day et al., 2018). It is widely reported that often small-scale and marginal-scale farmers and land-less farmers venture into other productive sectors such as side jobs, trade, construction, wage labor, salaried jobs, tool manufacturing units, small-scale industries, brick factories, etc. to compensate for the lost income from agricultural activities. In addition, such labor shifts have also been influenced by the increasing dependency of the agriculture communities on the public distribution system for food security. Such shifts in agricultural labor engagement will create vicious circles for local food production systems due to the development of shortages of agricultural labor. It will also constrain farming operations and thus will lead to a reduction in the net area sown in the localities. Overall, it may exacerbate the impact of climate change on the food and nutrition security of the marginalized communities (George and McKay, 2019).

### 3.7.4 *Erosive coping models*

It is evident from the foregone discussions that farming communities follow several strategies and measures to cope with climate change risks impacting their food production systems and livelihoods. However, the current study reveals that erosive coping measures are also undertaken in certain localities in the study countries. Under erosive coping models, farmers reduce their expenditure on food consumption, sell off farmlands, increase borrowing, and discontinue their children's education. Such erosive coping evidence is reported from the farming communities in Eastern India (Aryal et al., 2019). As such, unmitigated continuation of erosive coping measures can undermine the economic and social development gains already achieved by those communities. In addition, they can trigger vicious circles that would undermine the realization of the milestones fixed under the Sustainable Development Goals (SDGs) (Rao et al., 2018).

#### *3.7.5 Migration*

Migration is reported to be a critical livelihood diversification strategy cum coping measure resorted to by the members of communities deriving their livelihood from ecosystem-based food production systems (Maharjan et al., 2020). The current literature indicates that both internal and international migration are undertaken by farming communities as a strategy to secure as well as diversify their income streams. It is reported that internal migration has been as high as 37%, 14%, and 10% of the total populations in India, Nepal, and Bangladesh, respectively (Maharjan et al., 2020). In the region, internal migration has become a dominant coping recourse for agricultural communities. Internal migrations do happen seasonally, or across the year as influenced by a set of determinants such as household capability, increased climatic variability, livelihood opportunities, and proximity to towns and cities. Studies from India report that households with single-source cash income are very likely to resort to migration as an adaptation measure to the impacts of climate change on their agriculture and family income (Ojha et al., 2014; Bhatta et al., 2017; Brown et al., 2018; Thornton et al., 2018). Meanwhile, the literature also evidences that families with multiple income sources are very likely to undertake on-farm resilient measures as compared to migrating out.

### 3.8 Institution and governance

Support of enabling and robust institutional arrangements and high-quality governance are fundamental to upscale CRA for the millions practicing agriculture in the subcontinent. This proposition on CRA is often not very evidently carried by the agricultural and food policies (Thornton, 2020; Forch et al., 2014). It is reported that the success of CRA efforts is greatly shaped by the formal and informal institutions existing in the region or in the landscape (Mubaya and Mafongoya, 2017). Against the above backdrop, we have captured a quick overview of some of the schemes and programs operating in the study countries promoting CRA.

#### *3.8.1 National Mission on Sustainable Agriculture (NMSA)*

The NMSA, under the National Action Plan on Climate Change of India, is based on three pillars, viz., water use efficiency, nutrient management, and livelihood diversification. The NMSA promotes sustainable development through environmental conservation, energy efficiency, and natural resource conservation (NMSA, 2019). In addition, it aims to promote location-specific improved agronomic practices through soil health management, enhanced water use efficiency, judicious use of chemicals, crop diversification, animal husbandry, sericulture, agroforestry, fish farming, etc. (NMSA, 2019). However, the National Initiative on Climate Resilient Agriculture (NICRA) in India aims to enhance resilience in agriculture sector through strategic research and technology demonstration coupled with participatory evaluation of location-specific interventions in vulnerable districts of India (ICAR, 2020).

#### *3.8.2 Crop contingency plans*

In order to address climate-related risks to agriculture, crop contingency plans have been developed and administered. These plans are known to carry information packages for choosing alternate crops or cultivars in tune with the prevailing agro geo-climatic scenario. They also provide prescriptions on technological, institutional, and policy aspects concerning the implementation of contingency plans. In India, such plans have already been developed for more than 400 rural districts. Operationalization of these plans during aberrant monsoon years facilitated by the district/block level agricultural extension staff helps farmers to cope with climate variability (Reddy et al., 2014).

### 3.8.3 *Crop insurance schemes*

Agricultural insurance arrangements can potentially promote wide-scale adoption of CRA measures to cope with climate change impacts by poling the risks due to climatic change (Dolan et al., 2001; Mcleman and Smit, 2006; Garrido and Zilberman, 2008; Skees et al., 2008; Schwank et al., 2010; Garrido et al., 2011; Foudi and Erdlenbruch, 2012). Various insurance models are found in operation across the countries in the region. In India, such insurances are provided based on two models: (a) crop yield-based insurance schemes and (b) weather-based insurance schemes. Although the scheme is in operation, the adoption rate has been found to be only average, particularly for weather-based insurance schemes. However, in Sri Lanka, crop insurance is provided by the Agricultural and Agrarian Insurance Board (AAIB) and the National Insurance Trust Fund (NITF). The crop insurance schemes in Sri Lanka are indemnity based. AAIB in Sri Lanka, however, gives compulsory free-of-charge crop insurance cover for crops such as paddy, maize, and onion. The NITF is responsible for the indemnity-based Natural Agricultural Insurance Scheme, which covers damage caused to paddy crops by drought and flood events (Wijenayake et al., 2019). In Bangladesh, crop insurance was introduced through the State-owned insurance company Sadharan Bima Corporation (SBC) in 1977; however, this was discontinued in 1996 (CCC, 2009). The Asian Development Bank (ADB) has also recently piloted a weather index-based crop insurance for rice farmers in Bangladesh (Chatterjee, 2018).

### 3.8.4 *Community-based resource governance models*

In the water-scarce pockets of Sri Lanka, crop fields are fed by small-scale irrigation systems. The communities in the region foster equity and inclusion in their arrangements to ensure that each farming family gets a parcel of land in upstream, middle stream, and downstream stretches of the canals leaving the community irrigation tanks. This measure is reported to promote access, inclusion, and equity in the utilization of community water resources. In addition, several governance features such as local autonomy, effective monitoring, community-based management of perceived risks promoting crop diversification, fostering of community cohesion, and farmer experience explained much of the variation in the adoption of CRA measures across the communities. In Sri Lanka, increasing institutional support for cultivation of crops other than paddy has been found to reduce water use in dry regions while diversifying the portfolio of options available to farmers to manage drought (Burchfield and Gilligan, 2016, Wassmann et al., 2009; Babel et al., 2011; Varela-Ortega et al., 2014).

### 3.8.5 *Multistakeholder action in CRA*

Cross-cutting multistakeholder action is key to the success of planning and implementing strategies and programs to deal with complex problems in climate change governance, since the set of risks and benefits is shared by several actors and stakeholders. Multistakeholder action essentially brings in actor coherence, awareness, ownership, leadership, momentum, and alignment to the communities to respond to policy prescriptions targeting complex environmental challenges such as climate change (Winter et al., 2017). All across the study countries, we could observe collaboration of different agencies in the pursuit of CRA. In Nepal, the District Agriculture Development Office (DADO), Prayatnasil Community Development Society (PRAYAS Nepal), Community Self Reliance Center (CSRC), Resource Identification and Management Society Nepal (RIMS Nepal), and Christian Action Research and Education, Nepal (CARE Nepal) have been found to carry out interventions on soil, water, fertilizer management, rainwater harvesting, agricultural supporting infrastructure, promotion of drip irrigation, mulching, and integrated pest management (Karki et al., 2020). Capacity building initiatives such as training, strengthening the local community, and setting up of farmer business schools are also supported by the District Coordination Committee (DCC), Local Initiatives for Biodiversity, Research, and Development (LI-BIRD), and Rural Reconstruction Nepal (RRN) in Nepal (Karki et al., 2020). In Bhutan, several international organizations, such as the World Bank and the Food and Agriculture Organization, as well as local organizations like the Tarayana Foundation and Samdrupjongkhar, have been found actively working on promotion of climate-smart approaches by implementing programs and projects. They have also been funding initiatives targeting adaptation and mitigation efforts in the agricultural sector (CIAT, 2017). Similar collaborative multistakeholder actions for promotion of CRA are found in vogue across the study countries in the region.

## 3.9 Maladaptation

In the context of climate change adaptation, maladaptation can be understood as "an action taken ostensibly to avoid or reduce vulnerability to climate change, that impacts adversely or increases the vulnerability of other systems, sectors or social groups" (Barnett and O'Neill, 2010). The current review identified such maladaptation events occurring in some locations in study countries. For instance, in Bangladesh, prawn cultivation in coastal localities is reported to be more harmful to the environment than the immediate economic resilience it can create for the agricultural communities. Unscientific prawn cultivation is reported to be causing environmental degradation, environmental pollution, and loss of farmlands for marginal farmers (Afroz and Alam, 2013; Johnson et al., 2016; Shameem et al., 2015). It is also observed that the current literature has seldom reported about the negative impacts of adaptive measures, and these need to be considered for further investigations.

## 3.10 Emerging typology of CRA

Typological information provides an empirical reflection of the evidence from the field and sheds new insights on the emerging CRA landscape in the region. It profiles what adaptation actions are in vogue and, thus, provides an evidence base for further policy and program development on CRA (Biagini et al., 2014). Based on the current study, we have identified the following typology for the prevailing CRA landscape in the region. In this framework, four overarching dimensions of CRA actions are considered, and these are summarized in Table 3.

TABLE 3 Emerging typology of climate-resilient agriculture in selected countries of South Asia.

| Dimension | Models | Practiced country | References |
|---|---|---|---|
| **Weather-smart activities** | | | |
| | Integrated agro-meteorological advisory service | India | CGIAR (2020) and Maini and Rathore (2011) |
| | Automatic weather station model | India | Nesheim et al. (2017) |
| | Drought forecasting and communication model | Sri Lanka | Amalanathan et al. (2017) |
| | Famine early warning systems network | Afghanistan | MAIL (2020) |
| | Partnership in climate services for resilient agriculture | India | Bano (2019) |
| | Weather research and forecasting service model | Bangladesh | Chowdhury et al. (2019) |
| | Local-level agro-met extension service | Bangladesh | BAMIS (2016) |
| | Community-based early warning systems | Nepal | Adhikari and Sitoula (2018) and Smith et al. (2017) |
| **Cropping and cropland management** | | | |
| Water-smart activities | Micro-irrigation systems | Across the region | Kishore et al. (2014), Zotarelli et al. (2015), Randev (2015), Viswanathan et al. (2016), and Kumar et al. (2008) |
| | Direct seeded rice cultivation method | Across the region | Prasad et al. (2014) |
| | Alternate wetting and drying technique | Across the region | Hasan et al. (2016), Maniruzzaman et al. (2019), and Alauddin et al. (2020) |
| | In-situ soil moisture conservation practices | Across the region | Sikka et al. (2018) |
| | Zero tillage cultivation | India, Bangladesh, and Nepal | Erenstein (2009), Jat et al. (2019), Alam et al. (2015), Choudhary et al. (2016), and Gathala et al. (2016) |
| Nutrient-smart practices | Soil Quality Index-based crop nutrient management | Bangladesh and India | Biswas et al. (2019) and Islam et al. (2017a,b) |
| | Organic farming approaches | Bhutan | Tenzin et al. (2019), Khanal (2018), Sikka et al. (2018), and Karki et al. (2020) |

*Continued*

TABLE 3 Emerging typology of climate-resilient agriculture in selected countries of South Asia—Cont'd

| Dimension | Models | Practiced country | References |
|---|---|---|---|
| Knowledge-smart activities | Climate field schools | Bangladesh | Mahashin and Roy (2017) |
| | Community leadership development | Bangladesh | NAEP (2012) |
| Energy-smart technologies and practices | Adaptive crop residue management | India and Bhutan | Deya et al. (2020), Woolf et al. (2010), Lehmann and Joseph (2015), and Sharma et al. (2008) |
| | Solar power in remunerative crop production models | India | Nexus (2018) |
| | Energy-efficient agriculture pumps program | India | BoEF (2020) |
| | Laser land leveling technique | Afghanistan and India | Aryal and Jat (2015), Jat et al. (2009), Chakhansuri (2018), and Aryal et al. (2015) |
| **Socioeconomic resilience** | | | |
| | Community-based response systems | Sri Lanka | Government of Sri Lanka (2010) |
| | Promotion of agritourism | Sri Lanka | Mahaliyanaarachchi et al. (2019) |
| | Participation in labor market | Across the region | Day et al. (2018) and George and McKay (2019) |
| | Erosive coping models | India | Aryal et al. (2019) and Rao et al. (2018) |
| | Migration | India, Nepal, and Bangladesh | Maharjan et al. (2020), Ojha et al. (2014), Bhatta et al. (2017), Brown et al. (2018), and Thornton et al. (2018) |
| **Institution and governance** | | | |
| | National mission on sustainable agriculture | India | NMSA (2019), Karki et al. (2020), and CIAT (2017) |
| | Crop contingency plans | India | Reddy et al. (2014) |
| | Crop insurance schemes | India, Bangladesh, and Sri Lanka | Dolan et al. (2001), Mcleman and Smit (2006), Garrido and Zilberman (2008), Skees et al. (2008), Schwank et al. (2010), Garrido et al. (2011), Foudi and Erdlenbruch (2012), Chatterjee (2018), and Wijenayake et al. (2019) |
| | Community-based governance models | Sri Lanka | Burchfield and Gilligan (2016), Wassmann et al. (2009), Babel et al. (2011), and Varela-Ortega et al. (2014) |
| | Multiple stakeholder action in CRA | Across the region | Winter et al. (2017), Karki et al. (2020), and CIAT (2017) |

## 3.11 Analysis of the gaps observed

The current study evidences that the principles and practices of CRA are operational under both planned and autonomous domains of CRA. However, the adoption rate of CRA has been reported to be below average for several reasons. The access to and adoption of CRA technology and practices have been fraught with multiple constraints, and we discuss these in the following paragraphs.

Although there are several models found operational across the countries in the region providing meteorological advisory to the farmers, the service landscape still suffers from robust local arrangements to provide accurate real-time meteorological data to the farming communities. Gaps in the availability of such services will leave farmers underprepared to take quality decisions on the requirements of irrigation and other crop protection measures. However, such gaps need to be addressed in tandem with capacitation of farmers to comprehend effectively the risk management advisories and to implement them efficiently in the field. Expansion of a local meteorological service network backed with cost-efficient service delivery models will provide real-time meteorological information service to the farmers and thus will pave the way to augment efficiency on various input-side factors to achieve enhanced agricultural productivity.

Poor penetration of crop insurance and other such risk mitigation measures has also been constraining wide-scale adoption of CRA by the agricultural communities. Major gaps in insurance penetration are lack of awareness about various insurance products, low insurance product diversity, cumbersome administrative procedures, and indemnity-based validation procedures. In addition, the prevailing insurance schemes cover largely extreme weather events such as floods or droughts. These schemes by and large overlook other system impacts of climate change such as salinity intrusion, soil degradation, biodiversity, and ecosystem services erosion. However, it is frequently argued in the literature that linking local weather networks with crop insurance schemes will lead to higher inclusion and structuring of the planned adaptation measures with cost-effectiveness and scale.

It is also worth appreciating that significant advancements have been made toward the adaptation of crop production systems and crop management practices in order to cope with the impacts of climate change. The regional landscape showcases umpteen pieces of evidence in terms of the availability of crop varieties to adapt to climate change impacts. However, the current CRA landscape demands that the gaps be plugged across agricultural extension service, multistakeholder action for capacity building on CRA, and proactive research to develop new varieties of crops to keep pace with climate change. Similarly, measures toward capacity building of farmers to absorb and develop CRA methods, development of information systems for knowledge management, monitoring adaptation process to avoid maladaptation, and identification of transformative actions for planned adaptation at the landscape level are also integral to plug the gaps in CRA adoption by the farming communities.

As we have already mentioned, CRA does not limit its interventions to the agricultural systems. It is rather cross-cutting in approach and hence needs multistakeholder action. Therefore, it is imperative that robust institutional arrangements for effective governance of actors, stakeholders, and the process involved are highly integral to scale up the already evidenced highly effective CRA practices as well as to foster an ecosystem for wide-scale adoption of CRA. Although several governance arrangements are available across the countries, it is essential to develop monitoring, verification, and reporting (MRV) protocols to carry out objective analysis of CRA interventions and their impacts. Furthermore, it is also equally relevant to foster development of landscape and regional-level alliances of CRA initiatives to share best practices to promote collective learning and regional development (Chowdhury et al., 2019).

As regards water-smart activities, major gaps have been observed with regard to the level of adoption of micro-irrigation measures although several interventions on water conservation are undertaken in the region. Major reasons behind the low penetration of micro-irrigation may be the low availability of disposable income, poor access to cheap agricultural credit, low awareness level, and other access and scale-related constraints. Similar concerns are also found on the front of water conservation measures.

Apart from the abovementioned gaps, several other input and market-side factors have also been found inhibiting the rapid expansion of CRA. These are largely related to shortages in irrigation water, agricultural labor short supply, nonremunerative prices, access constraints to improved seeds, low market access, poor agricultural infrastructure development, and institutional inadequacy (Barnett and Webber, 2010). Thus, it may be noted that a cross-cutting approach is crucial to address the various constraints that are hindering the wide-scale expansion of CRA.

## 4 Conclusions and suggestions

The consequences of climate change on agriculture and food production systems are among major development concerns in South Asia. Nearly 800 million people living in the region are at the risk of declining income and living standards due to the increasing impacts of climate change (WBG, 2020). It is clear that CRA is a pragmatic way forward to adapt food production systems to the impact of climate change. This chapter provided a reflection of the major autonomous and planned adaptation measures undertaken across the varying agro-geographic landscapes in the region and the cross-cutting lessons learned. The sections in the chapter dwelled on the major pillars of CRA, such as weather-smart activities, smart crop and cropland management activities, measures undertaken for socioeconomic resilience, and institution and governance aspects. The current study emphasizes the *triple win* of increased agricultural productivity, enhanced climate resilience of agricultural systems, and the requirement for carbon conservation. Different interventional models profiled in the chapter have already demonstrated the advantage of CRA-based food production models as compared to the traditional cropping systems and practices.

However, the CRA landscape is fraught with several constraints and gaps. These concerns include the urgent need for mainstreaming CRA in the development sector policies and programs, accelerated inclusion of socioeconomically marginalized farmers and migratory communities practicing agriculture, strengthening agricultural extension service with cutting-edge information dissemination technology backed by local weather grids, promotion of supply-side market interventions to push the demand for technologies and equipment linked to CRA, promotion of agribusiness-based value chains, and addressing the research gaps in CRA for the development of pragmatic farm integrated business models.

The above-mentioned challenges however need to be addressed on priority for food and livelihood securities of the millions affected by the impacts of climate change. In addition, closing those gaps will also help to avoid the downturns in the socio-economic developmental milestones achieved by the countries in the region. As CRA is not a business-as-usual proposition and practice, it calls for a landscape-level integrated action from all the actors and stakeholders concerned with CRA under good governance arrangements. Thus, multistakeholder action is a characteristic feature and driver of CRA action across the landscapes. While summarizing, we suggest certain measures to strengthen and consolidate the foundations of CRA that have already been developed in the region. These include (i) scale up the proven CRA practices at landscape level; (ii) mainstream CRA in development projects and policies for convergence action; (iii) promote research with an active laboratory to land technology transfer arrangement to develop, validate evidence on CRA; (iv) promote cost-effective insurance models; (v) aggressive public campaign for wide-scale adoption of CRA; (vi) efficient, reliable, and cost-effective financial services and instruments; (vii) promotion of circular economies supported by ecosystem-based food production systems; (viii) catchment conservation; (ix) disaster proofing of agricultural storage infrastructure, positioning disaster response plans and teams; (x) promotion of social capital development, addressing gender gaps; (xi) conservation of local knowledge and traditional cultivars; (xii) controlling maladaptation and coercive adaptation; (xiii) harnessing the potential of the Internet and big data in management of complex agricultural systems; (xiv) promotion of investment-friendly environments for CRA; (xv) development of monitoring, reporting, and verification (MRV) protocols with a pragmatic set of criteria and indicators; and (xvi) transboundary collaboration and joint action. It should be noted that closing the gaps already reported in the domain, however, will also demand urgent intervention of actors and stakeholders of CRA at all levels. We conclude that CRA in South Asia has been providing benefits to the farming communities to develop climate resilience and prospects for sustainable development. As such, CRA needs to be further scaled up across the varying agro-climatic regions by following a systematic approach to secure the food and livelihood security of the millions living in the South Asia region and to realize the Sustainable Development Goals.

# References


ADB (Asian Development Bank), 2013. Food Security in Asia and the Pacific. Asian Development Bank, Philippines.

ADB (Asian Development Bank), 2020. Climate Change in South Asia-Strong Responses for Building a Sustainable Future. Asian Development Bank, Manila.

Adhikari, B.R., Sitoula, N.R., 2018. Community based flash flood early warning system: a low cost technology for Nepalese mountains. Bull. Dep. Geol. 20–21, 87–92.

Afroz, T., Alam, S., 2013. Sustainable shrimp farming in Bangladesh: a quest for an integrated coastal zone management. Ocean Coast. Manag. 2013 (71), 275–283. https://doi.org/10.1016/j.ocecoaman.2012.10.006.

Alam, G.M.M., 2017. Livelihood cycle and vulnerability of rural households to climate change and hazards in Bangladesh. Environ. Manag. 59 (5), 777–791. https://doi.org/10.1007/s00267-017-0826-3.

Alam, M.M., Ladha, J.K., Faisal, M.W., Sharma, S., Saha, A., Noor, S., Rahman, M.A., 2015. Improvement of cereal-based cropping systems following the principles of conservation agriculture under changing agricultural scenarios in Bangladesh. Field Crop Res. 175, 1–15. https://doi.org/10.1016/j.fcr.2014.12.015.

Alauddin, M., Sarker, M.A.R., Islam, Z., Tisdell, C., 2020. Adoption of alternate wetting and drying (AWD) irrigation as a water-saving technology in Bangladesh: economic and environmental considerations. Land Use Policy. 91, 104430, https://doi.org/10.1016/j.landusepol.2019.104430.

Amalanathan, S., Kadupitya, H.K., Induwage, L., Junaid, M., 2017. Drought Monitoring and Management in Sri Lanka. pp. 1–15. Retrieved from: https://www.gwp.org/globalassets/global/gwp-sas_files/news-and-activities/sadms/jan-2017/sri-lanka–sadms-presentation-india.pdf.

AQUA4D, 2019, October 30. Water-Smart Agriculture: The Efficiency Drive the World Needs. Retrieved from: https://aqua4d.com/water-smart-agriculture/.

Arroyo, C.G., Barbieri, C., Rich, S.R., 2013. Defining agritourism: a comparative study of stakeholders' perceptions in Missouri and North Carolina. Tour. Manag. 37, 39–47. https://doi.org/10.1016/j.tourman.2012.12.007.

Aryal, J., Jat, M.L., 2015, May 26. Laser Land Levelling: How it Strikes all the Right Climate-Smart Chords. International Maize and Wheat Improvement Center (CIMMYT). Retrieved from: https://ccafs.cgiar.org/research-highlight/laser-land-levelling-how-it-strikes-all-right-climate-smart-chords#.X4V4B9AzZPZ.

Aryal, J.P., Mehrotra, M.B., Jat, M.L., Sidhu, H.S., 2015. Impacts of laser land leveling in rice–wheat systems of the north–western indo-gangetic plains of India. Food Secur. 7 (3), 725–738. https://doi.org/10.1007/s12571-015-0460-y.

Aryal, J.P., Sapkota, T.B., Khurana, R., Khatri-Chhetri, A., Rahut, D.B., Jat, M.L., 2019. Climate change and agriculture in South Asia: adaptation options in smallholder production systems. Environ. Dev. Sustain. https://doi.org/10.1007/s10668-019-00414-4.

Azzu, N., Redfern, S., Friedrich, T., Gbehounou, G., 2013. Module 7: climate-smart crop production system. In: Climate-Smart Agriculture Sourcebook. pp. 191–205. Retrieved from: https://www.researchgate.net/publication/313928175_Module_7_Climate-smart_crop_production_system.

Babel, M., Agarwal, A., Swain, D., Herath, S., 2011. Evaluation of climate change impacts and adaptation measures for rice cultivation in Northeast Thailand. Clim. Res. 46, 137. https://doi.org/10.3354/cr00978.

BAMIS (Bangladesh Agro-Meteorological Information System), 2016. Agro-Meteorological Information Systems Development Project (Component -C of Bangladesh Weather and Climate Services Regional Project (BWCSRP). Department of Agriculture Extension, Ministry of Agriculture.

Bano, S., 2019. Partnership in Climate Services for Resilient Agriculture in India. pp. 57–60.

BARI (Bangladesh Agricultural Research Institute), 2020. E-Agriculture, Crop Variety. Retrieved from: http://www.bari.gov.bd/.

Barnett, J., O'Neill, S., 2010. Maladaptation. Glob. Environ. Chang. 20, 211–213. https://doi.org/10.1016/j.gloenvcha.2009.11.004.

Barnett, J., Webber, M., 2010. Accommodating Migration to Promote Adaptation to Climate Change. Policy Research Working Paper, 5270. Retrieved from: https://www.researchgate.net/publication/46443903_Accommodating_Migration_to_Promote_Adaptation_to_Climate_Change.

Batchelor, C., Schnetzer, J., 2018. Compendium on Climate-Smart Irrigation Concepts, Evidence and Options for a Climate Smart Approach to Improving the Performance of Irrigated Cropping Systems. Global Alliance for Climate-Smart Agriculture Rome.

Below, T.B., Mutabazi, K.D., Kirschke, D., Franke, C., Sieber, S., Siebert, R., Tscherning, K., 2012. Can Farmers' adaptation to climate change be explained by socio-economic household-level variables? Glob. Environ. Chang. 22 (1), 223–235.

Berrang-Ford, L., Pearce, T., Ford, J.D., 2015. Systematic review approaches for climate change adaptation research. Reg. Environ. Chang. 15 (5), 755–769. https://doi.org/10.1007/s10113-014-0708-7.

Bhatta, G.D., Ojha, H.R., Aggarwal, P.K., Sulaiman, V.R., Sultana, P., Thapa, D., 2017. Agricultural innovation and adaptation to climate change: empirical evidence from diverse agro-ecologies in South Asia. Environ. Dev. Sustain. 19, 497–525.

Biagini, B., Bierbaum, R., Stults, M., Dobardzic, S., McNeeley, S.M., 2014. A typology of adaptation actions: a global look at climate adaptation actions financed through the Global Environment Facility. Glob. Environ. Chang. 25, 97–108. https://doi.org/10.1016/j.gloenvcha.2014.01.003.

BINA (Bangladesh Institute of Nuclear Agriculture), 2020. Released Varieties. Retrieved from: http://www.bina.gov.bd/.

Biswas, J.C., Maniruzzaman, M., Naher, U., Zahan, A., Haque, T.M.M., Ali, M.H., Kabir, W., Kalra, N., Rahnamayan, S., 2019. Prospect of developing soil health index in Bangladesh. Curr. Investig. Agric. Curr. Res. https://doi.org/10.32474/CIACR.2019.06.000234.

BJRI (Bangladesh jute Research Institute), 2020. Agriculture, Technology and JTPDC Research. Agriculture, Breeding division. Retrieved from: http://www.bjri.gov.bd/.

BoEF (Bureau of Energy Efficient), 2020. Agricultural Demand Side Management. Retrieved from: https://beeindia.gov.in/content/agriculture-dsm-0.

Brida, A.B., Owiyo, T., Sokona, Y., 2013. Loss and damage from the double blow of flood and drought in Mozambique. Int. J. Glob. Warm. 5 (4), 514. https://doi.org/10.1504/ijgw.2013.057291.

BRKB (Bangladesh Rice knowledge bank), 2020. BRRI (Bangladesh Rice Research Institute) Rice Varieties. Retrieved from:http://knowledgebank-brri.org/.

Brown, P.R., Afroz, S., Chialue, L., Chiranjeevi, T., El, S., Grunbuhel, C.M., 2018. Constraints to the capacity of smallholder farming households to adapt to climate change in south and Southeast Asia. Clim. Dev., 1–18.

Burchfield, E., Gilligan, J.M., 2016. Dynamics of Individual and Collective Agricultural Adaptation to Water Scarcity. In: Winter Simulation Conference 2016 (July 20, 2016). SSRN Available from: https://ssrn.com/abstract=2807452.

Byjesh, K., Naresh Kumar, S., Aggarwal, P.K., 2010. Simulating impacts, potential adaptation and vulnerability of maize to climate change in India. Mitig. Adapt. Strateg. Glob. Chang. 15, 413–431. https://doi.org/10.1007/s11027-010-9224-3.

Campbell, B.M., Vermeulen, S.J., Aggarwal, P.K., Corner-Dolloff, C., Girvetz, E., Loboguerrero, A.M., Wollenberg, E., 2016. Reducing risks to food security from climate change. Glob. Food Secur. 11, 34–43.

CCC (Climate Change Cell), 2009. Climate Change and Health Impacts in Bangladesh. Climate Change Cell, DoE, MoEF; Component 4b, CDMP, MoFDM. June 2009, Dhaka. Retrieved from: http://ngof.org/wdb_new/sites/default/files/Climate_Change_and_Health%202009.pdf.

CDD (Crop Development Directorate), Department of Agriculture, 2015. Rice Varietal Mapping in Nepal: Implication for Development and Adoption. www.cddnepal.gov.np.

CGIAR (Climate Change, Agriculture and Food Security), 2020. Weather-Based Agricultural Advice Boosts Agricultural Production in India. Evidence of Success. Retrieved from: https://ccafs.cgiar.org/BIGFACTS. Accessed 11 September 2020.

Chakhansuri, S., 2018. Innovation Agricultural technologies in Afghanistan. In: Sultana, N., Jahan, F.N., Bokhtiar, S.M. (Eds.), Innovation Agricultural Technologies in South Asia. SAARC Agriculture Centre, Dhaka, Bangladesh, p. 17.

Chatterjee, A.K., 2018. Crop Insurance Lessons From My Field Trip to Rajshahi. Retrieved from: https://blogs.adb.org/blog/crop-insurance-lessons-my-field-trip-rajshahi#:~:text=ADB%20has%20piloted%20weather%20index,by%20the%20Bangladesh%20Meteorological%20Department.

Choden, K., Keenan, R.J., Nitschke, C.R., 2020. An approach for assessing adaptive capacity to climate change in resource dependent communities in the Nikachu watershed, Bhutan. Ecol. Indic. 114, 106293. https://doi.org/10.1016/j.ecolind.2020.106293.

Choudhary, R., Singh, P., Sidhu, H.S., Nandal, D.P., Jat, H.S., Yadvinder-Singh, Jat, M.L., 2016. Evaluation of tillage and crop establishment methods integrated with relay seeding of wheat mungbean for sustainable intensification of cotton-wheat system in South Asia. Field Crop Res. 199, 31–41. https://doi.org/10.1016/j.fcr.2016.08.011.

Chowdhury, A., Mamun, A.A., Rahman, N.M., 2019. Valuation of weather manifested rice cultivation in Bangladesh: a way forward. Am. Int. J. Agric. Stud.. 2641-41552 (1) E-ISSN 2641-418X Published by American Center of Science and Education, USA.

CIAT (International Center for Tropical Agriculture), 2017. Climate-Smart Agriculture in Bhutan. CSA Country Profiles for Asia SeriesThe World Bank, Washington, DC, pp. 1–26.

Day, E., Fankhauser, S., Kingsmill, N., Costa, H., Mavrogianni, A., 2018. Upholding labour productivity under climate change: an assessment of adaptation options. Clim. Pol. 1–19. https://doi.org/10.1080/14693062.2018.1517640.

Deresa, T.T., Hassan, R.M., Ringler, C., 2011. Perception of and adaptation to climate change by farmers in the Nile Basin of Ethiopia. J. Agric. Sci. 149, 23–31.

Dessai, S., Hulme, M., 2004. Does climate adaptation policy need probabilities? Clim. Pol. 4 (2), 107–128. https://doi.org/10.1080/14693062.2004.9685515.

Deya, D., Gyeltshen, T., Aich, A., Naskar, M., Roy, A., 2020. Climate adaptive crop-residue management for soil-function improvement; recommendations from field interventions at two agro-ecological zones in South Asia. Environ. Res. 183, 109164. https://doi.org/10.1016/j.envres.2020.109164.

Dinesh, D., Frid-Nielsen, S., Norman, J., Mutamba, M., Loboguerrero Rodriguez, A.M., Campbell, B., 2015. Is Climate-Smart Agriculture Effective? A Review of Selected Cases. CCAFS Working Paper no. 129CGIAR Research Program on Climate Change, Agriculture and Food Security (CCAFS), Copenhagen, Denmark. Available from: www.ccafs.cgiar.org.

Dolan, A.H., Smit, B., Skinner, M.W., Bradshaw, B., Bryant, C.R., 2001. Adaptation to Climate Change in Agriculture: Evaluation of Options, Occasional Paper No. 26. Department of Geography, University of Guelph, Ontario, Canada.

Dury, J., Schaller, N., Garcia, F., Reynaud, A., Bergez, J.E., 2012. Models to support cropping plan and crop rotation decisions. A review. Agron. Sustain. Dev. 32, 567–580. https://doi.org/10.1007/s13593-011-0037-x.

Easterling, D.R., Wallis, T.W.R., Lawrimore, J.H., Heim, R.R., 2007. Effects of temperature and precipitation trends on U.S. drought. Geophys. Res. Lett.. 34(20).

Erenstein, O., 2009. Zero Tillage in the Rice-Wheat Systems of the Indo-Gangetic Plains: A Review of Impacts and Sustainability Implications, 916. International Food Policy Research Institute.

FAO, 2012. Building resilience for adaptation to climate change in the agriculture sector. In: Proceedings of a Joint FAO/OECD Workshop. FAO Retrieved from: http://www.fao.org/3/i3084e/i3084e.pdf.

FAO, 2013. Module 5: sound management of energy for climate-smart agriculture. In: Climate-Smart Agriculture Sourcebook. FAO, pp. 139–169. Retrieved from: http://www.fao.org/3/a-i3325e.pdf.

FAO, IFAD, UNICEF, WEP, WHO, 2018. The state of food security and nutrition in the world. In: Building Climate Resilience for Food Security and Nutrition. FAO, Rome, Italy.

Forch, W., Thornton, P., Vasileiou, L., 2014. Governance & Institutions Across Scales in Climate Resilient Food Systems. Retrieved from: https://www.slideshare.net/cgiarclimate/gov-ws-brussels.

Ford, J.D., Berrang-Ford, L., Paterson, J., 2011. A systematic review of observed climate change adaptation in developed nations. Clim. Chang. 106 (2), 327–336. https://doi.org/10.1007/s10584-011-0045-5.

Foudi, S., Erdlenbruch, K., 2012. The role of irrigation in farmers' risk management strategies in France. Eur. Rev. Agric. Econ. 39, 439–457.

Garrido, A., Zilberman, D., 2008. Revisiting the demand of agricultural insurance: the case of Spain. Agric. Financ. Rev. 68, 43–66.

Garrido, A., Bielza, M., Rey, D., Mınguez, M.I., Ruiz-Ramos, M., 2011. Insurance as an adaptation to climate variability in agriculture. In: Hazell, P., Mendelsohn, R., Dinar, A. (Eds.), Handbook on Climate Change and Agriculture. Edward Elgar Publishing, Cheltenham, UK, pp. 420–445.

Gathala, M.K., Ladha, J.K., Kumar, V., Saharawat, Y.S., Kumar, V., Sharma, P.K., Sharma, S., Pathak, H., 2011. Tillage and crop establishment affects sustainability of South Asia Rice—Wheat system. Agron. J. 103 (4), 961–971.

Gathala, M.K., Timsina, J., Islam, M.S., Krupnik, T.J., Bose, T.R., Islam, N., McDonald, A., 2016. Productivity, profitability, and energetics: a multi-criteria assessment of farmers' tillage and crop establishment options for maize in intensively cultivated environments of South Asia. Field Crop Res. 186, 32–46. https://doi.org/10.1016/j.fcr.2015.11.008.

George, N.A., McKay, F.H., 2019. The public distribution system and food security in India. Int. J. Environ. Res. Public Health 16 (17), 3221. https://doi.org/10.3390/ijerph16173221.

German Watch, 2019. Global Climate Risk Index 2019. Retrieved August 3rd, 2020, from: https://germanwatch.org/en/16046.

Ghimiray, M., Katwal, T.B., 2013. Crop genetic resources for food security and adaptation to climate change: a review and way forward. J. Renew. Nat. Resour. Bhutan 9 (1), 1–14.

Gourdji, S.M., Mathews, K.L., Reynolds, M., Crossa, J., Lobell, D.B., 2013. An assessment of wheat yield sensitivity and breeding gains in hot environments. Proc. R. Soc. B 280, 20122190. https://doi.org/10.1098/rspb.2012.2190.

Government of Sri Lanka, 2010. Sector Vulnerability Profile: Agriculture and Fisheries. Supplement Document to the National Climate Change Adaptation Strategy for Sri Lanka 2011 to 2016. Government of Sri Lanka, Colombo, Sri Lanka.

Gupta, A.K., Negi, M., Nandy, S., Kumar, M., Singh, V., Valente, D., Petrosille, I., Pandey, R., 2020. Mapping socio-environmental vulnerability to climate change in different altitude zones in the Indian Himalayas. Ecol. Indic. 109, 105787. https://doi.org/10.1016/j.ecolind.2019.105787.

Hasan, K., Abdullah, A.H.M., Bhattacharjee, D., Afrad, S.I., 2016. Impact of alternate wetting and drying technique on rice production in the drought prone areas of Bangladesh. Indian Res. J. Exten. Educ. 16 (1), 39–48.

Hossain, I., Mondal, M.R.I., Islam, M.J., Aziz, M.A., Khan, A.S.M.M.R., Begum, F., 2014. Four crops-based cropping pattern studies for increasing cropping intensity and productivity in Rajshahi Region of Bangladesh. Bangladesh Agron. J. 17 (2), 55–60.

Hussain, S.G., 2017. Identification and Modeling of Suitable Cropping Systems and Patterns for Saline, Drought and Flood Prone Areas of Bangladesh. Christian Commission for Development in Bangladesh (CCDB), Dhaka, Bangladesh, pp. 1–166.

Hussain, A., Rasul, G., Mahapatra, B., Tuladhar, S., 2016. Household food security in the face of climate change in the Hindu-Kush Himalayan region. Food Secur 8 (5), 921–937.

ICAR (Indian Council for Agriculture Research), 2020. Towards climate resilient agriculture through adaptation and mitigation strategies. In: Enabling Farmers to Cope With Climate Variability Through Land, Water, Crop and Livestock Management in Vulnerable Districts of India.pp. 1–4. Retrieved from: http://www.nicra-icar.in/nicrarevised/images/Books/NICRA%20Climate%20Resilient%20Agriculture%20Brochure.pdf.

IFPRI (International Food Policy Research Institute), 2018, March. 2018 Global Food Policy Report: Anti-Globalism Threatens Progress in Hunger, Poverty Reduction; Requires Strong Policies and Global Leadership. International Food Policy Research Institute (IFPRI). Retrieved July 25, 2020, from: https://www.ifpri.org/news-release/2018-global-food-policy-report-anti-globalism-threatens-progress-hunger-poverty.

IPCC (Intergovernmental Panel on Climate Change), 2007. Summary for policymakers. In: Parry, M.L., Canziani, O.F., Palutikof, J.P., van der Linden, P.J., Hanson, C.E. (Eds.), Climate Change (2007): Impacts, Adaptation and Vulnerability. Contribution of Working Group II to the Fourth Assessment Report of the Intergovernmental Panel on Climate Change. Cambridge University Press, Cambridge, UK, pp. 7–22.

IPCC (Intergovernmental Panel on Climate Change), 2014. Climate Change 2014: Synthesis Report. IPCC Fifth Assessment Synthesis Report, Tech. rep.

IRRI (International Rice Research Institute), 2016. Climate Smart Rice. Available from: http://books.irri.org/Smart_rice_brochure.pdf.

Islam, S.M.D., Bhuiyan, M.A.H., Mohinuzzaman, M., Ali, M.H., Moon, S.R., 2017a. A Soil Health Card (SHC) for soil quality monitoring of agricultural lands in the south eastern coastal region of Bangladesh. Environ. Syst. Res. 6, 15. https://doi.org/10.1186/s40068-017-0092-7.

Islam, M.T., Ullah, M.M., Amin, M.G.M., Hossain, S., 2017b. Rainwater harvesting potential for farming system development in a hilly watershed of Bangladesh. Appl Water Sci 7, 2523–2532. https://doi.org/10.1007/s13201-016-0444-x.

Islam, M.A., Islam, M.J., Ali, M.A., Khan, A.S.M.M., Hossain, M.F., Moniruzzaman, M., 2018. Transforming triple cropping system to four crops pattern: an approach of enhancing system productivity through intensifying land use system in Bangladesh. Int. J. Agron.. 6, https://doi.org/10.1155/2018/7149835.

Jaffery, R., 2019, November 22. Climate Change and South Asia's Pending Food Crisis. The Diplomat. Retrieved July 25, 2020, from:https://thediplomat.com/2019/11/climate-change-and-south-asias-pending-food-crisis/.

Jasna, V.K., Burman, R.R., Padaria, R.N., Sharma, J.P., Varghese, E., Chakrabatri, B., Dixit, S., 2017. Impact of climate resilient technologies in rainfed agro ecosystems. Indian J. Agric. Sci. 87 (6), 816–824.

Jat, M.L., Gathala, M.K., Ladha, J.K., Saharawat, Y.S., Jat, A.S., Kumar, V., Sharma, S.K., Kumar, Y., Gupta, R., 2009. Evaluation of precision land leveling and double zero-till systems in the rice–wheat rotation: water use, productivity, profitability and soil physical properties. Soil Tillage Res. 105 (1), 112–121. https://doi.org/10.1016/j.still.2009.06.003.

Jat, H.S., Kumar, P., Sutaliya, J.M., Kumar, S., Choudhary, M., Singh, Y., Jat, M.L., 2019. Conservation agriculture based sustainable intensification of basmati rice-wheat system in North-West India. Arch. Agron. Soil Sci. 65 (10), 1370–1386. https://doi.org/10.1080/03650340.2019.1566708.

Johnson, F.A., Hutton, C.W., Hornby, D., Lazar, A.N., Mukhopadhyay, A., 2016. Is shrimp farming a successful adaptation to salinity intrusion? A geospatial associative analysis of poverty in the populous Ganges–Brahmaputra–Meghna Delta of Bangladesh. Sustain. Sci. 11, 423–439. https://doi.org/10.1007/s11625-016-0356-6.

Karki, S., Burton, P., Mackey, B., 2020. Climate change adaptation by subsistence and smallholder farmers: insights from three agro-ecological regions of Nepal. Cogent Soc. Sci. 6 (1), 1720555. https://doi.org/10.1080/23311886.2020.1720555.

Katwal, T., 2010. Multiple cropping in Bhutanese agriculture–present status and opportunities. In: Regional Consultative Meeting on Popularizing Multiple Cropping Innovations as a Means to raise Productivity and Farm Income in SAARC Countriespp. 1–36.

Khanal, U., 2018. Farmers' Perspectives on Autonomous and Planned Climate Change Adaptations: A Nepalese Case Study. (Thesis)Tribhuvan University, Nepal, pp. 1–283.

Khatri-Chhetri, A., Aryal, J.P., Sapkota, T.B., Khurana, R., 2016. Economic benefits of climate-smart agricultural practices to smallholders' farmers in the Indo-Gangetic Plains of India. Curr. Sci. 110 (7), 1251–1256.
Kishore, A., Shah, T., Tewari, N.P., 2014. Solar powered irrigation pumps: farmers' experience and state policy in Rajasthan. Econ. Polit. Wkly. 49 (10), 55–62.
Kumar, S.N., Aggarwal, P.K., 2013. Climate change and coconut plantations in India: impacts and potential adaptation gains. Agric. Syst. 117, 45–54.
Kumar, S.N., Aggarwal, P.K., Swaroopa Rani, D.N., Saxena, R., Chauhan, N., Jain, S., 2014. Vulnerability of wheat production to climate change in India. Clim. Res. 59, 173–187. https://doi.org/10.3354/cr01212.
Kumar, R.M., Rao, R.P., Somasekhar, N., Surekha, K., Padmavathi, C.H., et al., 2013. SRI—a method for sustainable intensification of rice production with enhanced water productivity. Agrotechnology S11 (009), 1–6. https://doi.org/10.4172/2168-9881.S11-009.
Kumar, M.D., Turral, H., Sharma, B., Amarasingh, U., Singh, O.P., 2008. Water Saving and Yield Enhancing Micro Irrigation Technologies in India: When and Where Can they Become Best Bet Technologies? (September 2015).
Lehmann, J., Joseph, S., 2015. Biochar for environmental management: an introduction. In: Lehmann, J., Joseph, S. (Eds.), Biochar for Environmental Management: Science, Technology and Implementation. Taylor and Francis, London, pp. 1–13.
Lobell, D.B., Sibley, A., Ivan Ortiz-Monasterio, J., 2012. Extreme heat effects on wheat senescence in India. Nat. Clim. Chang. 2 (3), 186–189. https://doi.org/10.1038/nclimate1356.
Mahaliyanaarachchi, R.P., Elapata, M.S., Esham, M., Madhuwanthi, B.C.H., 2019. Agritourism as a sustainable adaptation option for climate change. Open Agric. 4 (1), 737–742. https://doi.org/10.1515/opag-2019-0074.
Maharjan, A., de Campos, R.S., Singh, C., Das, S., Srinivas, A., Bhuiyan, M.R.A., Ishaq, S., Umar, M.A., Dilshad, T., Shrestha, K., Bhadwal, S., Ghosh, T., Suckall, N., Vincent, K., 2020. Migration and household adaptation in climate-sensitive hotspots in South Asia. Curr. Clim. Change Rep. 6, 1–16. https://doi.org/10.1007/s40641-020-00153-z.
Mahashin, M., Roy, R., 2017. Mapping practices and technologies of climate-smart agriculture in Bangladesh. J. Environ. Sci. Nat. Resour. 10 (2), 29–37.
Maheswari, M., Sarkar, B., Vanaja, M., Srinivasa Rao, M., Prasad, J.V.N.S., Prabhakar, M., Ravindra Chary, G., Venkateswarlu, B., Ray Choudhury, P., Yadava, D.K., Bhaskar, S., Alagusundaram, K., 2019. Climate Resilient Crop Varieties for Sustainable Food Production under Aberrant Weather Conditions. ICAR-Central Research Institute for Dryland Agriculture, Hyderabad, .p. 64. Retrieved from: http://www.nicra-icar.in/nicrarevised/images/publications/Climate%20Resilent%20Crop_All%20Pages_12-03-19_low.pdf.
MAIL (Ministry of Agriculture, Irrigation and Livestock), 2020. Retrieved from: https://www.mail.gov.af/en/node/595.
Maini, P., Rathore, L., 2011. Economic impact assessment of the Agrometeorological Advisory Service of India. Curr. Sci. 101 (10), 1296–1310. http://www.jstor.org/stable/24079638.
Mall, R.K., Singh, R., Gupta, A., Srinivasan, G., Rathore, L.S., 2006. Impact of climate change on Indian agriculture: a review. Clim. Chang. 78 (2–4), 445–478. https://doi.org/10.1007/s10584-005-9042-x.
Manandhar, S., Vogt, D.S., Perret, S.R., Kazama, F., 2011. Adapting cropping systems to climate change in Nepal: a cross-regional study of farmers' perception and practices. Reg. Environ. Chang. 11, 335–348. https://doi.org/10.1007/s10113-010-0137-1.
Maniruzzaman, M., Alam, M.M., Mainuddin, M., Islam, M.T., Kabir, M.J., Scobie, M., Schmidt, E., 2019. Technological Interventions for Improving Water Use Efficiency in the Northwest Region of Bangladesh. Report on the Project of Improving Dry Season Irrigation for Marginal and Tenant Farmers in the Eastern Gangetic Plains (LWR/2012/079)BRRI, Gazipur, Bangladesh.
Marer, S.B., Lingaraju, B.S., Shashidhara, G.B., 2007. Productivity and economics of maize and pigeonpea intercropping under rainfed conditions in the northern transitional zone of Karnataka. Karnataka J. Agric. Sci. 20 (1), 1–3.
McCarthy, J.J., Canziani, O.F., Leary, N.A., Dokken, D.J., White, K.S. (Eds.), 2001. Climate change 2001: impacts, adaptation and vulnerability. Contribution of working group II to the third assessment report of the intergovernmental panel on climate change. Cambridge University Press, pp. 1–1023.
Mcleman, R., Smit, B., 2006. Vulnerability to climate change hazards and risks: crop and flood insurance. Can. Geogr. 50, 217–226.
Menike, L.M.C.S., Arachchi, K.A.G.P.K., 2016. Adaptation to climate change by smallholder farmers in rural communities: evidence from Sri Lanka. Procedia Food Sci. 6, 288–292. International Conference of Sabaragamuwa University of Sri Lanka 2015. https://doi.org/10.1016/j.profoo.2016.02.057.

Mishra, M., 2002. Maize inter cropped with legumes in bariland under rainfed condition in the eastern hill of Nepal. In: Rajbhandari, N.P., Ransom, J.K., Adhikari, K., Palmer, A.F.E. (Eds.), Sustainable Maize Production Systems for Nepal: Proceedings of a Maize Symposium Held in December 3–5, 2001. NARC and CIMMYT, Kathmandu, Nepal.

Mubaya, C.P., Mafongoya, P., 2017. The role of institutions in managing local level climate change adaptation in semi-arid Zimbabwe. Clim. Risk Manag. 16, 93–105. https://doi.org/10.1016/j.crm.2017.03.003.

NAEP (National Agricultural Extension Policy), 2012. Draft National Agricultural Extension Policy. Ministry of Agriculture, Government of the People's Republic of Bangladesh.

Nesheim, I., Barkved, L., Bharti, N., 2017. What is the role of agro-met information services in farmer decision-making? Uptake and decision-making context among farmers within three case study villages in Maharashtra, India. Agriculture 7, 70. https://doi.org/10.3390/agriculture7080070.

Nexus, 2018. Nationwide Implementation of Solar Farming Planned // Climate-Smart Agriculture Made in India. Retrieved from: https://www.water-energy-food.org/news/nationwide-implementation-of-solar-farming-planned-climate-smart-agriculture-made-in-india/.

Nielsen, J.O., D'haen, S.A.L., 2014. Asking about climate change: reflections on methodology in qualitative climate change research published in Global Environmental Change since 2000. Glob. Environ. Chang. 24, 402–409. https://doi.org/10.1016/j.gloenvcha.2013.10.006.

NMSA (National Mission for Sustainable Agriculture), 2019, June 25. National Mission for Sustainable Agriculture (NMSA). Retrieved from:https://nmsa.dac.gov.in.

Ojha, H.R., Sulaiman, V.R., Sultana, P., Dahal, K., Thapa, D., Mittal, N., 2014. Is South Asian agriculture adapting to climate change? Evidence from the indo-gangetic plains. Agroecol. Sustain. Food Syst. 38, 505–531.

Palanisami, K., Kakumanu, K.R., Ranganathan, C.R., Udaya Sekhar, N., 2015. Farm-level cost of adaptation and expected cost of uncertainty associated with climate change impacts in major river basins in India. Int. J. Clim. Change Strategies Manage. 7 (1), 76–96. https://doi.org/10.1108/ijccsm-04-2013-0059.

Parihar, C.M., Jat, S.L., Singh, A.K., Kumar, B., Pradhan, S., Pooniya, V., Dhauja, A., Chaudhary, V., Jat, M.L., Jat, R.K., Yadav, O.P., 2016. Conservation agriculture in irrigated intensive maize-based systems of North-Western India: effects on crop yield, water productivity and economic profitability. Field Crop Res. 193, 104–116.

Petticrew, M., 2001. Systematic reviews from astronomy to zoology: myths and misconceptions. BMJ 322 (7278), 98–101. https://doi.org/10.1136/bmj.322.7278.98.

Porter, J.R., Xie, L., Challinor, A.J., Cochrane, K., Howden, S.M., Iqbal, M.M., Lobell, D.B., Travasso, M.I., 2014. Food security and food production systems. In: Climate Change 2014: Impacts, Adaptation, and Vulnerability. Part A: Global and Sectoral Aspects. Contribution of Working Group II to the Fifth Assessment Report of the Intergovernmental Panel on Climate Change. Cambridge University Press, pp. 485–533.

Pound, B., Lamboll, R., Croxton, S., Gupta, N., Bahadur, A.V., 2018. Climate-Resilient Agriculture in South Asia: An Analytical Framework and Insights From Practice. pp. 1–40.

Prakash, A., 2018, Sep. Finance and Development-Boiling Point. International Monitory Fund. Retrieved July 25, 2020, from: https://www.imf.org/external/pubs/ft/fandd/2018/09/pdf/southeast-asia-climate-change-and-greenhouse-gas-emissions-prakash.pdf.

Prasad, Y.G., Maheswari, M., Dixit, S., Srinivasarao, C.H., Sikka, A.K., Venkateswarlu, B., Sudhakar, N., Kumar, P.S., Singh, A.K., Gogoi, A.K., Singh, A.K., Singh, Y.V., Mishra, A., 2014. Smart Practices and Technologies for Climate Resilient Agriculture. Central Research Institute for Dryland Agriculture (ICAR), Hyderabad 76 p.

Rai, R.K., Bhatta, L.D., Acharya, U., Bhatta, A.P., 2018. Assessing climate-resilient agriculture for smallholders. Environ. Dev. 27, 26–33.

Randev, A.K., 2015. Analysis of crops' productivity potential and drip irrigation system in India–policy implications. In: 26th Euro Mediterranean Regional Conference and Workshops, October 12–15, pp. 1–5.

Rao, C.A.R., Raju, B.M.K., Rao, A.V.M.S., Rao, K.V., Ramachandran, K., Nagasree, K., Samuel, J., Ravi Shankar, K., Srinivasa Rao, M., Maheswari, M., Nagarjuna Kumar, R., Sudhakara Reddy, P., Yella Reddy, D., Rajeshwar, M., Hegde, S., Swapna, N., Prabhakar, M., Sammi Reddy, K., 2018. Climate Change Impacts, Adaptation and Policy Preferences: A Snapshot of Farmers' Perceptions in India. Policy Paper 01/2018ICAR-Central Research Institute for Dryland Agriculture, Hyderabad 34 p.

Raza, A., Razzaq, A., Mehmood, S., Zou, X., Zhang, X., Lv, Y., Xu, J., 2019. Impact of climate change on crops adaptation and strategies to tackle its outcome: a review. Plan. Theory 8 (2), 34. https://doi.org/10.3390/plants8020034.

Reddy, G.R., Sudhakar, N., Prasad, Y.G., Srinivasa Rao, C., Dattatri, K., Appaji, C., Prasad, J.V., Reddy, A.R., 2014. Manual on Contingency Agricultural Plan (NICRA Project). ICAR-Zonal Project Directorate, Hyderabad. 113

p. Retrieved from: http://zpd5hyd.nic.in/publication/Manual%20on%20Contingency%20Agricultural%20Plan.pdf.

Sam, A.S., Padmaja, S.S., Kachele, H., Kumar, R., Muller, K., 2020. Climate change, drought and rural communities: understanding people's perceptions and adaptations in rural eastern India. Int. J. Disaster Risk Reduct. 44, 101436. https://doi.org/10.1016/j.ijdrr.2019.101436.

Sapkota, T.B., Majumdar, K., Jat, M.L., Kumar, A., Bishnoi, D.K., Mcdonald, A.J., Pampolino, M., 2014. Precision nutrient management in conservation agriculture based wheat production of Northwest India: profitability, nutrient use efficiency and environmental footprint. Field Crop Res. 155, 233–244.

Schwank, O., Steinemann, M., Bhojwani, H., Holthaus, E., Norton, M., Osgood, D., Sharoff, J., Bresch, D., Spiegel, A., 2010. Insurance as an Adaptation Option Under UNFCCC, Background Paper. Final Version. INFRAS/IRI/SWISS RE, Zurich.

Shameem, M.I.M., Momtaz, S., Kiem, A.S., 2015. Local perceptions of and adaptation to climate variability and change: the case of shrimp farming communities in the coastal region of Bangladesh. Clim. Chang. 133, 253–266. https://doi.org/10.1007/s10584-015-1470-7.

Sharma, R.K., Chhokar, R.S., Jat, M.L., Singh, S., Mishra, B., Gupta, R.K., 2008. Direct drilling of wheat into rice residues: experiences in Haryana and western Uttar Pradesh. In: Humphreys, E., Roth, C.H. (Eds.), Permanent Beds and Rice-Residue Management for Rice- Wheat System of the Indo-Gangetic Plain.Australian Centre Int. Agric. Res. (ACIAR) ProcIn: vol. 127. pp. 147–158.

Shrestha, M.S., Goodrich, C.G., Udas, P., Rai, D.M., Gurung, M.B., Khadgi, V., 2016. Flood Early Warning Systems in Bhutan: a Gendered Perspective. ICIMOD Working Paper 2016/13ICIMOD, Kathmandu. https://doi.org/10.13140/RG.2.2.11116.97929.

Sihvola, K.P., Namgyal, P., Dorji, C., 2014. Socio-Economic Study on Improved Hydro-Meteorological Services in the Kingdom of Bhutan. Technical Report. Available from: https://www.researchgate.net/publication/301286763.

Sikka, A.K., Islam, A., Rao, K.V., 2018. Climate-smart land and water management for sustainable agriculture, irrigation and drainage. Irrig. Drain. 67, 72–81. https://doi.org/10.1002/ird.2162.

Skees, J.R., Barnett, B.J., Collier, B., 2008. Agricultural Insurance: Background and Context for Climate Adaptation Discussions. Global AgRisk, Inc., Lexington

Smith, P., Martino, D., Cai, Z., Gwary, D., Janzen, H., Kumar, P., McCarl, B., Ogle, S., O'Mara, F., Rice, C., Scholes, B., 2007. Policy and technological constraints to implementation of greenhouse gas mitigation options in agriculture. Agric. Ecosyst. Environ. 118 (1), 6–28.

Smith, P.J., Brown, S., Dugar, S., 2017. Community-based early warning systems for flood risk mitigation in Nepal. Nat. Hazards Earth Syst. Sci. 17, 423–437. https://doi.org/10.5194/nhess-17-423-2017.

Sultana, R., Rahman, M., Haque, M., Sarkar, M., Islam, S., 2019. Yield gap of stress tolerant rice varieties Binadhan-10 & Binadhan-11 in some selected areas of Bangladesh. Agric. Sci. 10, 1438–1452. https://doi.org/10.4236/as.2019.1011105.

Tall, A., Hansen, J., Jay, A., Campbell, B., Kinyangi, J., Aggarwal, P.K., Zougmoré, R., 2014. Scaling Up Climate Services for Farmers: Mission Possible. Learning From Good Practice in Africa and South Asia. CCAFS Report No. 13CGIAR Research Program on Climate Change, Agriculture and Food Security (CCAFS), Copenhagen, Denmark. Available from: https://cgspace.cgiar.org/handle/10568/42445.

Teklewold, H., Mekonnen, A., Kohlin, G., 2019. Climate change adaptation: a study of multiple climate-smart practices in the Nile Basin of Ethiopia. Clim. Dev. 11 (2), 180–192. https://doi.org/10.1080/17565529.2018.1442801.

Tenzin, J., Phuntsho, L., Lakey, L., 2019. Climate smart agriculture: adaptation and mitigation strategies to climate change in Bhutan. In: Climate Smart Agriculture: Strategies to Respond to Climate Change in South Asia. SAARC Agriculture Centre (SAC), Dhaka, Bangladesh, pp. 37–61.

Thornton, P., 2020. Governance and Institutions for Climate-Resilient Food Systems. Retrieved from: https://ccafs.cgiar.org/governance-and-institutions-climate-resilient-food-systems#.X4WQB9AzZPa.

Thornton, P.K., Kristjanson, P., Forch, W., Barahona, C., Cramer, L., Pradhan, S., 2018. Is agricultural adaptation to global change in lower-income countries on track to meet the future food production challenge? Glob. Environ. Chang. 52, 37–48.

Upreti, R.P., 2002. Inter cropping study on maize with different legumes under hill condition. In: Rajbhandari, N.P., Ransom, J.K., Adhikari, K., Palmer, A.F.E. (Eds.), Sustainable Maize Production Systems for Nepal: Proceedings of a Maize Symposium Held, December 3–5, 2001. NARC and CIMMYT, Kathmandu, Nepal.

Varela-Ortega, C., Blanco-Gutiérrez, I., Esteve, P., Bharwani, S., Fronzek, S., Downing, T., 2014. How can irrigated agriculture adapt to climate change? Insights from the Guadiana Basin in Spain. Reg. Environ. Chang. 16. https://doi.org/10.1007/s10113-014-0720-y.

Venkateswarlu, B., Kumar, S., Dixit, S., Srinivasa Rao, C., Kokate, K.D., Singh, A.K., 2012. Demonstration of Climate Resilient Technologies on Farmers' Fields Action Plan for 100 Vulnerable Districts. Central Research Institute for Dryland Agriculture, Hyderabad. 163 p. Retrieved from: http://www.nicra-icar.in/nicrarevised/images/Books/Action%20Plan%20100%20Vulnerable%20Districts%20.pdf.

Viswanathan, P.K., Kumar, M.D., Narayanamoorthy, A., 2016. Micro Irrigation Systems in India: Emergence, Status and Impacts. India Studies in Business and EconomicsSpringer Science Business Media, Singapore.

Wani, S.P., Chander, G., Anantha, K.H., 2017. Enhancing resource use efficiency through soil management for improving livelihoods. In: Adaptive Soil Management: From Theory to Practices. Springer, Singapore, pp. 413–451.

Wassmann, R., Jagadish, K., Heuer, S., Ismail, A.M., Redona, E., Serraj, R., Singh, R., Howell, G., Pathak, D.S., Sumfleth, K., 2009. Climate change affecting rice production: the physiological and agronomic basis for possible adaptation strategies. Adv. Agron.. 101.

WBG (World Bank Group), 2020. Open Learning Campus. Building a Climate-Resilient South Asia. Retrieved from: https://olc.worldbank.org/content/building-climate-resilient-south-asia.

Wijenayake, V., Wickramasinghe, B., Mombauer, D., Halkewela, M., 2019. Policy Brief: Climate Change and Agricultural Insurance in Sri Lanka. pp. 1–15.

Williams, N.E., Carrico, A., 2017. Examining adaptations to water stress among farming households in Sri Lanka's dry zone. Ambio 46, 532–542. https://doi.org/10.1007/s13280-017-0904-z.

Winter, S., Bijker, M., Carson, M., 2017. The Role of Multi-Stakeholder Initiatives in Promoting the Resilience of Smallholder Agriculture to Climate Change in Africa. pp. 1–30. Retrieved from: https://www.technoserve.org/wp-content/uploads/2017/02/the-role-of-multi-stakeholder-initiatives-in-promoting-the-resilience-of-smallholder-resilience-to-climate-change-report.pdf.

Woolf, D., Amonette, J.E., Street-Perrott, F.A., Lehmann, J., Joseph, S., 2010. Sustainable biochar to mitigate global climate change. Nat. Commun. 1, 56.

World Bank, 2019, November 11. South Asia Needs to Act as One to Fight Climate Change. World Bank Blogs. Retrieved August 04, 2020, from: https://blogs.worldbank.org/endpovertyinsouthasia/south-asia-needs-act-one-fight-climate-change.

Zotarelli, L., Fraisse, C., Dourte, D., 2015. Agricultural Management Options for Climate Variability and Change: Micro Irrigation. HS120300. University of Florida IFAS Extension. http://edis.ifas.ufl.edu/hs1203.

SECTION C

# Social impacts

CHAPTER

# 11

# Social issues related to climate change and food production (crops)

*Thandi F. Khumalo*

Department of Sociology and Social Work, University of Eswatini, Kwaluseni Campus, Kwaluseni, Eswatini

OUTLINE



## 1 Introduction

Food is life, it provides not only nourishment but sustains energy and promotes growth for all living organisms. Food production began around 13,000 years ago by our ancestors who were adapting to their new environment from a background of 200,000 years of hunting-gathering, horticulture lifestyle. As societies progressed through industrialization from the


https://doi.org/10.1016/B978-0-12-822373-4.00012-4

18th century, agricultural/food production methods also developed and modernized. Agriculture in Africa is dominated by subsistence food production. Mechanical agriculture is a preserve of large commercial farmers.

Climate change is the determinant of agricultural productivity. Climate change is influencing food and livestock production in the 21st century to greater proportions. The impact of climate change on food production manifests in crop failure emanating from drought, floods, harsh winds, severe hailstorms, and recently biting cold due to frost in Southern Africa. Food and Agriculture Organisation (FAO) (2009) had predicted the status of food production in Africa will decline because of the increasingly unpredictable and erratic nature of weather systems placing an extra burden on food security and rural livelihoods. Crop production is paying a significant cost from the ravages caused by climate change. Socio-cultural attitudes in food production have not corresponded with climate change and persistent and dynamic environmental factors impacting crop production. Traditional staple food crop varieties still dominate food production decisions against the reality of climate change.

Agriculture, particularly food production has transformed societies through new technologies to increased crop yields using pest- and climate-resistant crops/seeds to withstand climate change. However, some of these innovations have been viewed with suspicion by subsistence farmers who make up a large proportion of farmers in rural Africa. My perception is that subsistence farmers are undoubtedly generational farmers who have inherited farming skills through experience from their parents. Attitudinal change is very uncomfortable for traditional farmers, and to them, the unknown becomes unchartered waters which they dread to explore. As such, maize in many Southern African countries is a trusted staple grain, more than sorghum and wheat. On the contrary climate change favors drought-tolerant crops like sorghum and wheat leading to failing crop harvest in Southern Africa.

While other very crucial traditional farming skills have slowly been substituted by subsistence farmers, such as the intercropping method which ensured food security has increasingly been challenged by more space needed for cash crops. Subsistence farmers planted maize and other vegetable crops in-between while ensuring proper spacing of crops for good yields. Examples would include pumpkin patches, sorghum, beet, sugar cane, and so forth. At harvest time not only will the family be harvesting maize but also pumpkins and other crops for family consumption. Food crops would also be planted in other hectares including sweet potatoes, sugar beans, and vegetables, producing a surplus sold to cater for family needs.

As much as farmers acknowledge the impact of climate change and crop failure, much of their hopes are still on divine intervention and science to deal with crop failure. Basically, the solution is viewed as an external intervention and less as an adaptation to climate change. The future of crop production is partly in the adaptation techniques and in the attitudinal change of farmers, national governments, and the international community in responding to calls on climate change.

## 2 The impact of climate change on agriculture in Southern Africa

Southern Africa is defined here as the total geographical area occupied by the 16 member states of the Southern African Development Community (SADC): Angola, Botswana, Comoros, Democratic Republic of Congo, Lesotho, Madagascar, Malawi, Mauritius, Mozambique,

Namibia, Seychelles, South Africa, Eswatini (former Swaziland), Tanzania, Zambia, and Zimbabwe. Established in 1992, SADC is committed to regional integration and poverty eradication within Southern Africa through economic development and ensuring peace and security (SADC, 2020). This chapter will focus on eight countries: Botswana, Eswatini, Lesotho, Malawi, Mozambique, South Africa, Zambia, and Zimbabwe. Intergovernmental Panel on Climate Change (IPCC), in 2007, cited in (FAO, 2009) predicted that Africa will be the most vulnerable to climate change globally due to poor infrastructure, poverty, and governance. They predict that temperatures are likely to increase 1.5–4 degrees in the 21st century reducing yields by up to 50% and projected to fall as low as 90% by 2100. Indeed, the predicted droughts, floods and El Nino climate events have all occurred in Southern African countries. Agricultural losses of up to 1.3% have been felt in Southern Africa including crop pests and diseases (FAO, 2009).

It is worth mentioning that not all climate changes are necessarily bad for Africa. FAO (2009) predicted that changes in seasons and production cycles in Southern Africa are likely to extend the food growing season because of increased temperature and rainfall changes. According to Kuivanen et al. (2015), the climatic conditions of Southern Africa follow a broad gradient, with more arid conditions in the west and increasingly humid conditions toward the east. However, closer to the equator, the climate is largely humid. Precipitation patterns reveal lower annual rainfall in the south versus higher annual rainfall in the north. Thus, the climate ranges from the winter rainfall Mediterranean conditions around the tip of South Africa and semiarid summer rainfall savannah regions of the Kalahari in Namibia and Botswana to the subhumid rainfall regimes typical of Malawi (Kuivanen et al., 2015).

The Southern African agricultural sector can be roughly divided into two broad subsectors: commercial and subsistence. Commercial farmers occupy relatively large land areas and tend to be more integrated with the market (Kuivanen et al., 2015). Commercial farming is dominated by South Africa while other countries in southern Africa have a majority of smallholder farmers utilizing 2ha or less for subsistence. Most production is rainfed, except in South Africa, which is the largest maize producer in the region due to the contribution of irrigated farmlands (Kuivanen et al., 2015).

### 2.1 Temperatures, precipitation, extreme weather patterns

Southern Africa is getting warmer. Since the mid-20th century, most of the region has experienced an increase in annual average, maximum, and minimum temperatures. A rise between 0°C and 2°C of the average annual temperature (Kuivanen et al., 2015). Seasonal rainfall patterns, such as the onset or duration of rains, frequency of dry spells, and intensity of rainfall, as well as delays in the onset of rainfall have changed. Some countries have had late fewer summer rains and a modest decrease in rainfall (Kuivanen et al., 2015). Draught, floods, extreme temperatures, harsh winds, severe hailstorms, and frost, to mention a few, have characterized weather patterns making food production very risky.

## 3 Botswana

Botswana is a landlocked country in the middle of Southern Africa, bordered by Namibia on the west and Zambia and Zimbabwe on the northeast, and South Africa to the south.

The population of the country is about 1.8 million and over 64% urbanization has occurred due to increasing poverty and drought affecting rural livelihoods (DHFS, 2018/2019). Poverty levels have declined as gross domestic product (GDP) improved with diamond income although evidence of the widening income gap between rich and poor as well as rural and urban is increasing. Governance is relatively good and the country is stable (DHFS, 2018/2019).

Botswana has a semiarid climate characterized by warm winters, hot summers, and low rainfall, which on its own presents limitations in agriculture particularly food production contributing under 20% of GDP. In the center is the Kalahari Desert, where the climate is suitable for wildlife and livestock farming particularly cattle, which the country prides itself on, producing more than 80% of GDP (Zhou et al., 2013).

Rainfall is erratic and rainfall pattern varies each year with periods of very severe drought. In 2018/2019, the distribution of rainfall was poor both in space and time during the first half of the rainfall season (DHFS, 2018/2019). Consequently, crop production is a challenge and becomes an expensive exercise as water availability has to be improved by drilling boreholes and rehabilitating old ones (Zhou et al., 2013). Botswana's climate severely limits food production with only 5% of the country's vast land area of 581,730 $km^2$ is suitable for farming (DHFS, 2018/2019).

Botswana has reported droughts in 2001–2003; 2005–2006, 2006–2007 (Ministry of Agriculture, 2010). The dry season is between April and October but can sometimes be stretched to November. Some isolated areas receive heavy downpours while others receive little or no rain at all, with high temperatures of above 40°C in the west of the country. Only the Central part of the country received normal rainfall of about 100 mm from December 2018 to February 2019 (DHFS, 2018/2019). Temperatures were above normal in 2019/2019 with severe heat waves experienced and drought conditions particularly in the Southern and Western parts of the country. As such 2018/2019 can be termed to be a severe meteorological drought year (DHFS, 2018/2019). These harsh conditions affect the already fragile agricultural economy impacting a large proportion of rural dwellers whose livelihoods depend on agriculture.

The 2018/2019 planting season was, therefore, met with limiting climate change conditions, and an estimated 5356 MT (2%) of cereal compared with 66,093 MT in 2017/2018 was expected, far below the required national cereal harvest of 300,000 MT (DHFS, 2018/2019). Maize dominancy as the major planted and consumed cereal accounts for 45% (38,705 ha) of the planted area, followed by cowpeas 18% (16,699 ha) and sorghum 18% (16,445 ha) (DHFS, 2018/2019). As a result of the climatic conditions, poor distribution of rainfall, and scorching heat, most crops failed some at the geminating stage while others wilted at the vegetative stage (DHFS, 2018/2019). The commercial arable sector was not spared from climatic change conditions, the crops also wilted at vegetative and flowering stages due to heatwave conditions (DHFS, 2018/2019)

The Livestock sector was also impacted, thus, conditions dropping from good to fair in 2018/2019 (DHFS, 2018/2019) with severe heat and poor rains challenging water and grazing stocks. Botswana is relatively focusing on livestock farming rather than crop production as climate change has impacted them in different ways. Droughts have increased their water shortage vulnerability for animal production and slightly for crop production. The future for Botswana is in wildlife and livestock farming as well as in mining (DHFS, 2018/2019).

Food relief programs are now a permanent feature in the livelihood of many citizens, school feeding, and household supply of food. The centralized food items include pure sunflower oil, full cream milk, stewed steak with gravy, while the decentralized food includes sorghum grains, sorghum meal, white maize, samp, and beans (DHFS, 2018/2019). The dependence on relief food is increasing from 50.7% in 2016 to 63.24% in 2017 and 84% in 2018 (DHFS, 2018/2019). The food supplies to primary schools and health facilities have shown a decrease due to the unavailability of products and procurement processes (DHFS, 2018/2019).

Other social impacts of climate change and decreased food availability is the undernourishment of children below the age of 5, reduced school attendance due to hunger for primary school children, reduced food items for most households due to the poor performance of agriculture, and declining relief food procurement and distribution (DHFS, 2018/2019).

The government has put in place a social protection system to cushion Botswana from the negative impacts of various vulnerabilities, risks, and shocks. Vulnerable individuals and households are provided with food and cash transfers (DHFS, 2018/2019). Given the increasing crop failure and losses in agricultural jobs, many more are likely to register for social protection increasing the selection criteria of beneficiaries to the detriment of the missing middle that easily falls in the cracks.

A number of projects for the vulnerable citizen are disbursed countrywide to assist and boost individuals' and households' earning capacity. The coronavirus pandemic has compounded food shortage problems due to disruption of markets where the government procured relief food stocks to feed vulnerable populations.

# 4 Eswatini

The kingdom of Eswatini (formerly known as Swaziland) has a land area of 17,364 $km^2$ bordered on the north, west, and south by the Republic of South Africa and on the east by Mozambique. The country is landlocked with no access to the sea and is one of the smallest in the world. The population in 2017 was 1,124,805 and projected to be 1,160,164 in 2020 (World Population Review, 2020). Eswatini is predominantly a rural country with over 70% of the population living in rural areas, and the majority are poor and vulnerable to food insecurity (European Development Fund, 2019a,b). In 2020, between June and September which is the lean season, 292,794 (32%) people in rural areas and 37,424 (17%) in urban areas faced acute food insecurity (Integrated Food Security Phase Classification (IPC), 2020). In 2021, the situation will deteriorate to 379,000 people in acute food insecurity (IPC, 2020). The drivers of food insecurity include dry spells in November and December 2019, resulting in the late onset of summer rains delaying the start of the planting season, low food stocks from harvest, and the negative impact of COVID 19 on employment and household income deficit reducing expenditure on food supplies and farming inputs (IPC, 2020). In Manzini, the busy commercial city, the outbreak of COVID-19 has led to reduced income for about 70% of households due to temporary closures of businesses and/or reduced business activity. About 20% of households have suffered permanent loss of employment both in Manzini and Hhohho regions (IPC, 2020). Although food is available in the market, access is constrained by an increase in food prices (maize meal, rice, and beans) and poor purchasing power. There is considerable humanitarian assistance in the area, including some cash relief for urban households (IPC, 2020).

Eswatini rural areas are vulnerable and affected by climate change. The overall food insecurity situation is partly attributed to the poor performance of the agricultural sector owing to the high variability in weather patterns across the country despite normal to above-normal rainfall during the planting season countrywide. In addition to the late onset of the rainfall season (late October), the central and eastern regions of Lubombo, South-Western Shiselweni, and North-Western Hhohho received below normal rainfall. The combination of irregular rains and dry spells affected more than half (54%) of farming households in the country (IPC, 2020).

Under the current climate change trend, droughts are expected to become more frequent (Government of Eswatini, 2019). The climate change and variability in Eswatini have brought about erratic rainfall and severe drought and storms in some regions impacting food production, particularly maize which is the staple crop and the main starch consumed by the population and often used as an index of the availability of food in the country.

Maize is cultivated mainly in the Highveld and Middleveld in large proportions and in the Lowveld in smaller quantities by small scale subsistence farmers in the rural areas although the country has never been self-sufficient in maize production (Government of Eswatini, 2019). The country has recorded a shortfall in maize production of 71.93 metric tons to meet the 162.32 metric tons domestic requirement for the staple crop. Planned import requirements amount to 61.7 metric tons, with an uncovered food gap of 10.22 metric tons (IPC, 2020). The country is a net importer of maize, wheat, dairy products, and other food crops. Even though some regions especially the Lowveld experience drought, each planting season farmers still insist on the maize crop which is intolerant to drought (Government of Eswatini, 2019). According to Manyatsi et al. (2013), crop models show an opportunity for maize to adapt to climate change in other ecological zones of the country and, therefore, the need to shift maize production to areas that are becoming more productive as a result of climate change.

Compounding food insecurity is COVID-19 restrictions that continue to disrupt food supply chains in the country, negatively impacting food availability. Humanitarian assistance programs have been initiated to provide cash and food relief to ease the COVID 19-induced food challenges. The greater impact of the pandemic weighs heavily on the rural poor due to the disruption of the informal sector—the mainstay of rural livelihood (IPC, 2020). Access to food may be more challenging due to reduced purchasing power and complete loss of livelihoods for some households (IPC, 2020).

A number of interventions by the Ministry of Agriculture encourage the cultivation of drought-tolerant crops, such as sorghum and cotton on a commercial basis to acquire money to purchase maize. These attempts did not bear much fruit as farmers still try their luck with maize. Irrigation is mainly used for sugar cane production, a commercial crop grown by large-scale farmers and other small growers supplying the sugar production companies.

## 5 Lesotho

Lesotho is a landlocked mountainous country completely surrounded by the Republic of South Africa, with a population of above 1.96 million people. It is located on the plateau of Southern Africa, with altitudes ranging from 1400 to 3480 m above sea level exposing it to both the influences of the Indian and the Atlantic oceans, with differences in temperatures (Gwimbi et al., 2013). In Lesotho, temperatures are highly variable from very cold to very warm, and it faces high winds and summer thunderstorms. The high altitude means that Lesotho experiences some of the lowest temperatures in southern Africa with snow falling on the high ground annually. Only 13% of Lesotho's land area of 30,355 $km^2$ is suitable for crop production, the rest is rocky mountains and foothills (Gwimbi et al., 2013). Lesotho faces climate change challenges, and the key stressors are drought, land degradation, and loss of biodiversity. The major grains produced by Basotho are maize, sorghum, and wheat.

Agriculture is the lifeline of many Basothos, they rely on rainfed agriculture which makes crop production vulnerable to climate change. As such, erratic weather patterns (temperatures rising and rainfall decreasing) due to climate change are threatening Lesotho's food production and livestock production. With continuing climate change, more fields will lie fallow and yields will continue to decline discouraging more subsistence farmers to continue with food production. Like Eswatini, adapting to drought-resistant crop varieties face resistance from subsistence farmers.

Rainfall patterns have been good and above average since January 2020, leading to an improved harvest than 2018, but still below the 5-year average levels (FAO, 2020). The only risk was the early onset of frost becoming a major threat to a good harvest, forcing farmers to harvest their crop before full maturity or earlier than anticipated in 2020 (FAO, 2020). However, owing to delays in green harvests caused by 2019 late onset of rains to start the planting season in 2019, the 2020 lean season has been extended by one month from April to May where households comprising 433,000 (FAO, 2020) people faced poor to moderate food insecurity (Famine Early Warning Systems Network (FEWSN), 2020a,b). The dry harvest began by May 2020 creating a food access gap that needs to be filled by government food relief. May to August food is available again until September (FEWSN, 2020a,b).

Indirectly, climate change affects water quality, air, and food availability which have a direct impact on the nutrition and health of both children and adults. Winters are extremely cold and dry, and mountains are covered with snow in June, July, and August. The rainy season runs from October to March, but this pattern has been disturbed by climate change, in 2020 showers of rain were sufficient from January replenishing soil reserves moisture and helping raising yields to near average normal (FAO, 2020).

The majority of the population is poor and depends on subsistence agriculture for livelihood. The male population migrates to South Africa for employment and remit money back home. Recently, women have joined in the migration to South Africa to work as domestic workers. The reduced food production affects the nutrition of both children and adults, and this has become the major health challenge for Basothos (Ministry of Health and Social Welfare, 2009).

Owing to climate change, lower levels of precipitation result leading to reduced availability of freshwater for general domestic consumption and, thus, affecting the one major income earner in the Lesotho Water Project. Also, worth mentioning are the real threats of the coronavirus pandemic impacting job losses of migrant Basotho nationals in South Africa, disrupting cash remittances, and compounding food shortage during the lean period. As a response, the government has launched an economic mitigation package of (USD 58 million) as an expansion to existing social protection programs such as the Child Grant, assisting vulnerable households (FAO, 2020).

## 6 Malawi

Malawi is located in the eastern part of Southern Africa with a land area of 118,483 km$^2$, 94,275 km$^2$ is land and 24,208 km$^2$ is water (Saka et al., 2013). The topography of the country is highly varied; the Great Rift Valley runs from north to south containing Lake Malawi which is Africa's third-largest lake and the world's eleventh (Saka et al., 2013). The lake covers almost the entire country. It has a tropical climate, but due to its high elevation, the climate is relatively cool. Malawi is a landlocked country bordering Tanzania, Mozambique, and Zambia with no access to the Indian Ocean. Malawi has a population of over 19 million (National Statistical Office, 2015) projected to be 40 million by 2050, with fast-growing urbanization (18% in 2018 census) due to poverty in rural areas (Saka et al., 2013).

The country's economy has sluggish growth making the country highly dependent on aid. Over 70% of the population lives below the poverty line, relying on rainfed agriculture which is vulnerable to shocks and the informal sector for income (Famine Early Warning Systems Network (FEWS), 2020a,b). Households dependent on agriculture, trade, casual and farm employment, and tobacco income have been hardest hit by low levels of harvest in 2019 and COVID-19 impacts in 2020 (FEWSN, 2020a,b).

There are two distinct seasons, the rainy season (October–April) and the dry season (May–August), and September and October are hot and dry. Sharing borders with Mozambique has also made Malawi susceptible to cyclones. In March 2019, Malawi was struck by Cyclone Idai which first devastated Mozambique before making landfall in the southern region of Malawi impacting at least 13 districts (FEWSN, 2020a,b). More than 29,000 households were affected by the floods and displaced. Agriculture which is the main source of livelihood of rural populations was impacted as fields were inundated and recently planted crops were destroyed (FEWSN, 2020a,b).

Harvest typically takes place from April to June but it also depends on the region, for example, harvesting was completed in May in the southern region and the northern and western regions by June 2020. It is reported that grain stocks were sufficient for production till August 2020 although the national food stocks have a shortfall and the country will have to purchase about 200,000 metric tons to stock enough for its humanitarian distribution (FEWSN, 2020a,b).

Agriculture is the backbone of Malawi's economy providing over 50% of GDP. Malawi produced above-average maize in 2019 which was over 3.7 million tons (11%) above 2018 and (28%) above 5-year average (FEWSN, 2020a,b). Malawi has assisted most of SADC in maize exports to the region. Production of most other key food crops such as pulses is expected to be above average except for tobacco, which is expected to decline by 11% (FEWSN, 2020a,b). Despite the above-average production, some areas experienced reduced levels of production due to weather-related hazards. In northern Malawi, some parts of Rumphi and Karonga districts experienced some flooding and waterlogging that damaged crops. In central Malawi, Salima district experienced an early stop of rainfall in February, impacting crop maturation. In central Malawi, Nsanje and Chikwawa districts including others experienced localized dry spells and erratic rainfall, which resulted in below-average production (FEWSN, 2020a,b).

Even though Malawi's economy is dependent on agricultural productivity, only 5% of the land is irrigated and much food production depends on rainfall. Owing to climate change, the past three decades have seen unpredictability in rainfall patterns which has impacted food production in Malawi. Agricultural activity is concentrated in rural areas and is rainfed, making it vulnerable to climatic hazards such as drought (Saka et al., 2013).

Culturally, Malawians use a wide range of grain food crops and have not wholly depended on maize, making them less vulnerable than people in other countries in Southern Africa in experiencing food insecurity when there is drought (Saka et al., 2013). Maize is the staple food crop and is allocated a large land area, yet it is not all cultivated due to changing consumption patterns. Root and tuber crops such as cassava and potatoes have a high allocation, while the legumes and pulses have a slightly high allocation. Cassava and potatoes have a high household consumption value and supplements maize as a starch. It even surpasses other commercial crops such as tobacco and sugarcane. Cassava is now characterized as a food security crop, and this could be due to its high resistance to drought even when maize fails (Saka et al., 2013).

COVID-19 has compounded food security issues with over 1.9 million expected to face food hunger when current stocks get depleted from September to December 2020. Making the situation worse is the increased prices for maize that rose by 60%–100% in 2019 making food access difficult for the vulnerable populations (FEWSN, 2020a,b). The urban poor households will be most impacted by food insecurity during COVID-19 trade restrictions and low savings to purchase food. The situation is compounded by the falling market prices of tobacco which is a cash crop most families depend on for cash to buy food (FEWSN, 2020a,b). Also, to note is that no humanitarian food assistance program has been launched but there are plans to do so (FEWSN, 2020a,b).

## 7 Mozambique

Mozambique is in Southern Africa with the Indian Ocean to the east, Malawi and Zambia are to the northwest, Zimbabwe to the west, and Eswatini and South Africa to the north. The population is over 31 million people (2018). Geographically, the country is the world's 35th largest at 799,380 $km^2$. The Zambezi River divides the country into two topographic regions. To the river's north, the coast moves to inland hills and plateaus. Further west, it becomes rugged highlands. To the Zambezi River's south, there are broader lowlands and the Lebombo Mountains in the south (Briannica, 2020).

The country's climate is tropical with a wet season from October to March and dry from April to September. The climate changes based on the altitude, along the coast, rainfall is heavy. During the wet season, cyclones are common (Briannica, 2020).

Like other countries in Southern Africa, agriculture is an important sector in Mozambique employing a large proportion (about 80%) of the population (Maure et al., 2013). Economically, the country is ranked among the poor countries of the world. Civil wars and political conflict have worsened the economic situation of the country.

Unlike its neighbors, Mozambique has experienced increased yield in maize production and this is expected to continue till the year 2025 (Maure et al., 2013). The potential for export is even greater for maize. Cassava also has increased yields following maize. Cassava is one of the main crops consumed in Mozambique and climate-resistant varieties are needed (Maure et al., 2013). There is hope that rice production will be one of the drivers of Mozambique's agricultural economy.

The climate scenarios generated for Southern Africa show varying results for Mozambique with some predicting the rise in temperatures and drought while the other show a substantial increase in rainfall (Maure et al., 2013). Since its independence in 1975, Mozambique has been affected by numerous natural disasters including floods, in 2007 the Zambezia River broke its banks after heavy rains, and again Cyclone Idai made landfall in central Mozambique on 14–15 March 2019 bringing strong winds of 180–220 km/h and heavy rains. This disaster was followed immediately by Cyclone Kenneth 6 weeks later (April 25, 2019) in northern Mozambique with wind gusts of 220 km/h considered the worst cyclone in Africa, adding pressure to the already strained humanitarian agencies and the government of Mozambique response (Inter-Agency Humanitarian Evaluation (IAHE), 2020). Tropical Cyclone Kenneth made a worse impact as it fell at the end of the rainy season when rivers were already full.

Cyclone Kenneth left devastating consequences of crop losses (more than 500,000 ha) during the harvest season, hitting families in central and northern regions with food insecurity (IAHE, 2020). Cyclone Idai's wreckage came on top of an already serious food insecurity situation in Mozambique. From September to December 2018, an estimated 1.78 million people (IPC phase 3 and above) were severely food insecure in the country, according to the Integrated Phase Classification (IPC) analysis and the food security and nutrition assessment conducted by the Technical Secretariat for Food Security and Nutrition (SETSAN) in October 2018 (ROSEA, 2018–2019).

Previously, in 2017–2018 some parts (five provinces—Cabo Delgada, Gaza, Inhambane, Sofala, and Tete) of the country were affected by drought leaving 814,700 food insecure and causing acute malnutrition in children under 5 years due to low quantity and quality of feeding (ROSEA, 2018–2019). The 2017–2018 rainy season was characterized by a late start and extended midseason dry spell (December–January) and heavy rains. (ROSEA, 2018–2019). The dry spell resulted in moisture stress and wilting of early planted crops in many areas (ROSEA, 2018–2019). Drought has caused below-average yields, particularly in southern and some central parts of Mozambique in 2019 (ROSEA, 2018–2019).

Drought has had devastating impacts on the well-being of children's education, causing absenteeism and poor concentration in class and ultimately leading to a decrease in learning outcomes (ROSEA, 2018–2019). School feeding schemes have been introduced as a response measure. These disasters affect the most vulnerable populations dependent on agriculture for livelihoods and employment. The agricultural sector accounts for 25% of GDP and employs 71% of the labor force, of whom 94% are primarily engaged in agricultural production (IAHE, 2020). Food and cash transfers also provide relief to affected families. Distribution of farming inputs, vegetable seed are response measures for the agricultural recovery program (ROSEA, 2018–2019).

For northern Mozambique there is potential for productive agricultural conditions, however, the poor road network hampers such development. Although, rainfed agriculture faces risk from changing climatic conditions. It looks positive that food production will continue to prosper in Mozambique with irrigated agriculture in place of rainfed agriculture. Holding climate conditions constant, predictions show no effects on yields for cassava and maize up to 2050 with maize yields expected to double (Maure et al., 2013).

## 8 Republic of South Africa

South Africa is the southernmost country on the African continent, with varied topography inhabited by over 58 million people, with an area square of 471,359. South Africa unlike its neighbors has shown increased rates of urbanization. It has an urban population of 66.4% and a rural population of 33.6% (Britannica, 2020).

South Africa is unique in Southern Africa, it is more like the center to the peripheral neighboring countries. Even in terms of agricultural trends, rural subsistence agriculture is declining (with only 13% of land area) with trends moving toward urban agriculture (Johnston et al., 2013). As well as a predominance of large-scale commercial farming (with land area of 87% of agricultural land) which is supported by irrigation. There is great potential for rural subsistence agriculture to shift from labor-intensive to more mechanized production (Johnston et al., 2013). Historically, the Boers sustained the economy through commercial agriculture, mining, and manufacturing industries.

South Africa's land area has been utilized most for agriculture (food crops and livestock breeding), forests and orchards, and most vegetation used for grazing. Agricultural activities range from intensive export production of (fruit, vegetables, cereal crops—maize, sorghum, wheat) and grapes for wine exports. About 50% of available water is used in agriculture (Johnston et al., 2013). Maize is the main grain crop produced by South Africa and is the staple food of a majority of the population. The maize industry is also important for the country's export earnings. Maize is also important as a feed for the beef and poultry industry. It is estimated that over 9000 commercial farmers produce maize employing more than 130,000 workers (Johnston et al., 2013). The planting season is October to December but due to climate change, it varies from October, November, and December. Sustained production of maize depends on good rainfall and warm temperatures throughout the planting season.

Climate change has brought vulnerabilities of very high temperatures sucking all moisture from the soil and damaging the maize crop, as well as variable rainfall patterns. South Africa will need to increase its irrigation demand due to the high temperatures and evaporation impacting on crop yields. Given that agriculture already accounts for more than 60% of water use in South Africa, this increase is likely to result in added pressure on existing water resources, which may, in turn, lead to less certain supply and yields from irrigated agriculture (Wider, 2016). The prevailing energy crisis will also negatively affect irrigation power.

The country has the potential to be self-sufficient in all agricultural products, however, the rate of exports has been slower than that of imports. Major imports include wheat, rice, vegetable oils, and poultry meat. The largest exports are wines, citrus, sugar, grapes, and other fruit varieties (Johnston et al., 2013).

The impacts of global changes in climate are rapidly escalating in South Africa. Unless concerted action is taken to reduce greenhouse gas emissions, temperatures may rise by more than 4°C over the Southern African interior by 2100, and by more than 6°C over the western, central, and northern parts of South Africa (Chersich and Wright, 2019). Extreme weather events are the most noticeable effects to date, especially the drought in the Western Cape and wildfires (Chersich and Wright, 2019). Owing to its geographical location some parts of South Africa are prone to drought. At present, the country faces multiple stressors including variable rainfall, frosty cold winters damaging crops, hailstorms, floods, cyclones, and veld fires ( Johnston et al., 2013).

Food security is under threat with crop yields declining in some provinces with low precipitation like the North West province. The South African agricultural sector is, therefore, vulnerable to climate change conditions affecting agriculture, although unlike its neighbors varied strategies supported by financial backing have been initiated toward adaptation. Research is also very much advanced and supported financially to inform the climate change adaptation strategies for farmers ( Johnston et al., 2013). Efficient technologies such as irrigation is another advantage that South Africa has over its neighbors. However, climate change has affected the dam levels due to erratic rainfall impacting food production.

## 9 Zambia

Zambia is a landlocked country situated on a high plateau in south-central Africa, bordered by Angola on the west, Namibian Caprivi strip to the southwest, and three neighbors Namibia, Botswana, and Zimbabwe on the eastern end. Zambia has a population of over 17 million people. It has a total land area of 752,614 km$^2$. The rural population was 43.5% and urban 56.5% in 2018 (Williams, 2018). The man-made Lake Kariba forms part of the river border with Zimbabwe.

Much parts of the country are thinly populated, most of the population is concentrated in the most developed area—the Line Rail served by the railway linking the Copperbelt with Lusaka. The country is divided into three main agro-ecological regions based primarily upon annual precipitation: soil types, temperature, and elevation. The southwestern region is the driest and most prone to drought, and the soil contains low levels of organic matter, low nutrients, and high acidity. The central part with two subregions, the degraded plateau of the southeast, south-center, and southwest, and Kalahari Sands and the Zambezi floodplain in the west. Soils of the Kalahari Sands have little agricultural potential and are mainly under woodland. The third region is in the northern part; its soils tend to be highly weathered and leached with a low pH (Williams, 2018).

Zambia has a tropical climate and generally favorable to human settlement. The warm and wet season lasts from November until April with September and October hot and dry. December and January are the wettest with sun and rain at their peak and by April precipitation lowers with June and July weather dry and cold. Precipitation varies according to agroecological region but generally comes in storms with heavy raindrops that lead to hard soil surface and erosion.

Agriculture in Zambia is mostly small-scale and rainfed and it suffers from climate change. The economy is heavily dependent on manufacturing and mining copper although reserves are getting depleted. Agriculture is relatively poorly developed only about one-sixth of arable land is under cultivation (DMMU, 2019–2020). Subsistence farming dominates where traditional manual methods of cultivation still persist. Maize is the staple grain and accounts for much of the cultivated land area. Cassava, sorghum, and millet are also grown. Maize is highly vulnerable to climate change and this makes rural populations more vulnerable as they rely on maize as a staple grain. Adaptation is very difficult for rural populations as they have limited access to capital, technology, and extension services (Williams, 2018).

In 2019, most of Central, Eastern Lusaka, Southern, and Western Provinces were affected by prolonged dry spells and experienced a 50%–60% drop in maize production (Zambia Disaster Management and Mitigation Unit (DMMU), 2019–2020). According to the Zambian Meteorological Department (ZMD), 2018–2019 rainfall season was one of the poorest the southern half of Zambia has faced since 1981, negatively impacting crop production and consequently food access and availability. Leaving 39% of households with food to last for only 6 months, having to depend on the market for food access and limited by high commodity prices (DMMU, 2019–2020).

Pest infestation including the fall of armyworm compounded the situation. A total of 58 districts reported a decline in the maize crop, which is the staple food in Zambia, from 2,394,907 metric tons in 2017–2018 to 2,004,389 metric tons in 2018–2019 (DMMU, 2019–2020). The vulnerability assessment reported acute malnutrition of nearly 6% across nine provinces in Zambia (DMMU, 2019–2020). In response, government and food security partners like the United Nations International Children's Emergency Fund (UNICEF) and World Food Programme (WFP) provide pulses and emergency cash transfers to more than 66,000 households in affected regions, including the launch by WFP of the early recovery and resilience program targeting 104,000 smallholder farmers in five of the 58 districts affected by drought in the 2018–2019 season (DMMU, 2019–2020). The intervention is aimed to help the farmers to recover from the worst drought experienced by the country since the 1980s and provide school feeding for 33,000 children (WFP, 2020).

## 10 Zimbabwe

Zimbabwe is a landlocked country with a total area of 390,580 km$^2$ (Mugabe et al., 2013) with a total population of over 16 million, and the majority about 70% live in rural areas dependent on agriculture for a livelihood (USAID, 2019). The country shares borders with Mozambique, South Africa, Botswana, Zambia, and Namibia. Zimbabwe was once dubbed as the basket of Africa as its vast agricultural output fed many African countries in grain production. Political and economic stressors manifested in reduced agricultural production and subsequent climate change conditions affected food production (Mugabe et al., 2013).

Zimbabwe faces variable climatic risks challenging agricultural production, compounded by years of contentious land reform policies (USAID, 2019). Agriculture is largely rainfed and highly susceptible to variable climatic conditions. Rainfall is erratic resulting in crop failures every 3–5 years, making smallholder subsistence farmers in marginal regions face food security issues (USAID, 2019). Drought is the most prevalent climatic condition affecting the southern, western, and eastern regions of Zimbabwe (USAID, 2019).

The Masvingo district/zone's main livelihood source is the intensive farming of maize and groundnuts, which used to have 600–800 mm of rainfall annually, and is now attacked by droughts, dry spells, fires, crop pests such as armyworms. Manicaland is characterized by semiintensive farming rainfed agriculture, producing maize and small grains was normally self-sufficient. It also produces cash crops, which are an important source of income for the households. Climate change has impacted them with drought, fires, dry spells, and crop diseases and pests (USAID, 2019). Also, in the eastern highlands of Manicaland where there is commercial farming specializing in timber, fruits, vegetables, tea and coffee, and flowers with an annual rainfall of 650–1000 mm, cyclones and fires have impacted production (USAID, 2019). The above are just a few examples, in general, climate change in Zimbabwe has brought drought, dry spells, cyclones, heavy rainfall-induced flooding, occurring almost yearly in some parts of the country.

From the year 2000–2001 when significant land reforms took place to a period where agriculture became small-scale and less mechanized in Zimbabwe coupled with climate change conditions food production fell dramatically. The white commercial farmers engaged in large scale farming with all the equipment and technologies to raise yields. There is a recent move by the government in 2020 to return the land to the white farmers with the hope of reviving agricultural production in Zimbabwe which was once feeding most African countries and exporting abroad.

The significant crops to the poor which are most vulnerable to climate change are maize, sorghum, millet, and groundnuts (Mugabe et al., 2013). Maize is the staple food of Zimbabwe and is grown by smallholder farmers for subsistence. Noteworthy is that even commercial farmers have shifted their focus from maize to high-value cash crops such as tobacco or shifted to horticulture. This has exacerbated the shortage of maize production which has been left to smallholder farmers already struggling with declined production due to climate change, manifested in erratic rainfall. Although sorghum and millet are drought-tolerant, they are not as popular crops with the population, as is maize (Mugabe et al., 2013).

# 11 Discussion

## 11.1 Climate change and food security

Southern Africa is heavily affected by climate change and variability, and predictions suggest the impact will be more severe in the next decade. The most pronounced climate change will be an increase in temperature, leading to heat stress and reduced crop yield and the most prone crop being maize, the staple food; changes in rainfall patterns, increasingly erratic rainfall of high intensity, leading to floods and more frequent drought and dry spells; a delayed onset of rainfall season and an early tailing off, thus reducing the growing period of crops (SADC, 2019a,b,c). As temperatures increase and precipitation is reduced in the long term, a drastic fall in yields will be experienced in Southern Africa, impacting food security in the whole region.

Since much of Southern Africa relies on agriculture for food, income, and employment, the region is considered highly vulnerable to climate change. Agriculture contributes between 4% and 27% of GDP and approximately 13% of overall export earnings for SADC Member States (SADC, 2019a,b,c). Like many developing countries, the countries in Southern Africa are dependent on rainfed agriculture for food production that feeds the majority of their population in the rural areas engaged in subsistence production, as such accurate predictions of weather patterns are instrumental to livelihoods.

There have been repeated periods of droughts and floods influenced by the El Niño/La Niña Southern Oscillation (ENSO) from the nineteenth century onwards (FAO, 2011). The correlation between ElNino events and droughts is high, resulting in lower than average rainfall for the whole of Southern Africa. This event happened in 2015 resulting in the worst drought in 35 years in Southern Africa, affecting agriculture and livestock deaths including wildlife (USAID, 2019). Evidence from the countries shows that the rainy season is now starting progressively later than before, demanding a change in the planting season to move forward from August to October, and sometimes to November and December in certain areas. The changes in the timing and duration of the rainy season have had devastating effects on crop yields, the dry spells have also compounded the already stressful farming processes as crops, particularly maize wilt and dry out before maturation.

The impacts of climate are somewhat unique to each country and even within different parts of a country, lessons drawn from these Southern Africa case studies may provide useful insights for other countries in the region. Also, the magnitude of the climate change impact is compounded by the social, economic, and political vulnerability of the different countries and communities. The increasing frequency, magnitude, and duration of climate change has challenged available measures to cope leaving little or no time to recover from the last incident such as the floods in Mozambique and droughts in Zimbabwe. The predicted impact of climate change on precipitation, temperature, and the increased frequency and intensity of droughts and floods are likely to negatively affect water resources and the agricultural sector. The impacts are likely to be significant for subsistence farmers (Republic of South Africa, 2014).

From the case studies presented here, except for South Africa most food producers are the smallholder farmers mainly producing for subsistence and where there is a surplus it is sold in the local markets and the cash used to supplements household needs. It has also been noted that the majority of small-scale farmers depend on rainfed agriculture for their grain food, making food production vulnerable to changing weather patterns. Consequently, cereal production has not kept up with the population increase and more mouths to feed. From being self-sufficient in the 1960s many countries in Africa are now net importers of cereal grain food (NEPAD, 2013).

According to the International Water Management Institute (IWMI, 2020) rainfed agriculture produces much of the food consumed globally and by poor communities in developing countries. Rainfed agriculture accounts for more than 95% of farmed land in sub-Saharan Africa. It is apparent that food production losses are high due to the evaporation of moisture in heatwaves and extended drought periods, resulting in food insecurity and poverty for rural communities (IWMI, 2020). It is pertinent to understand the risks and trade-offs that farmers face in rainfed agriculture, as well as why they are not adopting agricultural practices to mitigate climate change. Sufficient adaptation measures need to be adopted and made available to communities.

Further, and perhaps more promising some analysts have noted that through climate change the planting season has been extended in some parts due to warmer conditions conducive for agriculture, while in some parts extreme heat, drought and cold have impacted agriculture negatively. Over a long period of time focus has been on building dams to assist both in agriculture and livestock production. The dam infrastructure has been very hastily implemented in many rural communities for political grandstanding during elections without the necessary infrastructure for small-scale irrigation introduced to the households.

The water management policy is also restrictive to small-scale farmers and benefiting more the commercial farmers. Water policies focus on river, groundwater, and lakes water resources with no focus on rainwater and how it is managed (IWMI, 2020). This has cost food production as in some seasons there is more than enough rainwater to quadruple yields even in water-constrained regions (IWMI, 2020). There is an untapped potential missed in small-scale irrigation which could increase yields and open possibilities of growing fruits and vegetables in all households. Supplemental irrigation would enhance the increasing vegetable garden farming in many rural communities that have water sources nearby with small-scale irrigation potential. Investment in water management can unlock the potential of rainfed agriculture (IWMI, 2020).

Food insecurity was largely a rural household phenomenon, but as the populations migrate to the urban centers congregating in informal settlements, food insecurity is now generally widespread. In the rural households, the longer away from the harvest season the more food poverty there is, yet in the urban informal settlements where food is purchased food insecurity is the order of the day as families live on low or no income for extended periods. Moreover, any negative impacts of climate change on the economies will have major implications for people's access to food, which is largely contingent on affordability. Food access is already tenuous given the existing levels of poverty and ownership of arable land is highly inequitable, reflecting the particular history of the country (Chersich and Wright, 2019).

COVID-19 pandemic has put more pressure on food security, and it is anticipated to trigger an increase in the prevalence of food insecurity from the third quarter of 2020 to early 2021. Regarding winter crops, mainly wheat and pulses, reports from the country indicate that planting operations, which normally take place in May, were delayed due to the COVID-19 pandemic-induced movement restrictions and stricter sanitary measures that impeded normal access to imported seeds and fertilizers (FAO, 2020). The effects of the pandemic on urban and rural household's livelihoods, who are heavily dependent on causal labor, remittances, and petty trade, are foreseen to be primarily channeled through a reduction in economic activities and associated income losses (FAO, 2020).

## 11.2 Adaptation measures and strategies (policies, institutions, and development partners)

In November 2019, SADC with development partners European Union (EU) launched a regional climate adaptation strategy to strengthen the capacity of SADC Member States to undertake Climate Change Adaptation and Mitigation Action. This is part of the SADC Regional Development Agenda (SADC, 2019a,b,c). This project is funded to the tune of 8 million euros and is aimed to run for 4 years with implementation to end in June 2023. The program aims to support SADC governments, regional organizations, private and public sector to deliver on the following areas:

- Strengthen the capacity of SADC Member States to undertake regional and national adaptation and mitigation actions in response to the challenges caused by the effects of global climate change and climate variability;

- Facilitate implementation of the provisions of the Paris Agreement on Climate Change in SADC;
- Facilitate sharing of knowledge and experience with other African, Caribbean, and Pacific (ACP) regional organizations, including South-South Cooperation;
- Assist to design pilot projects on adaptation in the several Member States; and
- Support Universities and Research Centers from the SADC Region in the development of innovative solutions to climate change challenges (SADC, 2019a,b,c).

While this effort is commendable, adaptation will be effective when the Member States deliberately plan and shape their policy space on climate change and make it a multisectoral response with an effective coordination mechanism, combining efforts from agriculture, energy, water, commercial/trade, as well as the financial sectors of society. Coherent policy approaches can lead to greater effectiveness and efficiency, and reduce competing for limited budgets and resources (England et al., 2018). The major challenge is the lack of overarching climate change policies in some Member States resulting in the lack of an effective coordination mechanism. Policy coordination remains weak across Southern Africa and needs to be strengthened to allow greater support to cross-sectoral planning (England et al., 2018).

Climate change has more negative effects on poorer households as they have the lowest capacity to adapt to changes in climatic conditions. Adaptation to climate change involves changes in agricultural management practices in response to climate change conditions (Nhemachema and Hassan, 2018). Farmers have had to adjust their farming practices to maintain healthy yields in their crops. While adaptation is classically defined as the ability to deal with change, it also encompasses the capacity to learn from it. Doing so requires investments in research and analytical systems, especially among public health practitioners, a collaboration across several countries, including South Africa (which has better capacity in research technologies) (Chersich and Wright, 2019).

Adaptation in agriculture includes producing crops that are drought tolerant and in Southern Africa, there is an overreliance on maize as the staple crop which is drought intolerant and vulnerable to pests like the fall armyworm. Sorghum is drought tolerant and the ideal crop in climate change conditions, but it is not a preferred grain crop by the majority of households. The ideal adaptation should include scientific technologies like gene-editing technology to come up with stress-tolerant seeds, particularly maize which has proven to be fragile in climate change. When these climate tolerant seeds are available, they need to be made accessible to subsistence farmers through farmer support initiatives. Postharvest processing capacity building and promotion of value chains through market linkages (DMMU, 2019–2020) need to be emphasized to enhance benefits and avoid postharvest losses.

Crop diversification or mixed cropping is another adaptation measure ensuring food security throughout the planting season. This requires farm labor which is not as available as it used to before migration. Adaptation also occurs in changing the planting calendar/season to coincide with the onset of rains, coupled with irrigation where possible to extend the planting season. Resource limitations and poor infrastructure limit the ability of smallholder farmers to take up adaptation measures (Nhemachema and Hassan, 2018).

## 11.3 Response to climate change

A number of strategies have been implemented to adapt and mitigate climate change in agriculture. The use of conservation agriculture techniques such as minimal or no-till methods, intercropping (mixing crop types in one field), and cover cropping (introducing alternative crops in successive years on the same field). These techniques conserve soil moisture, encourage soil health, and reduce dependence on fertilizers and herbicides. Integrated approaches range from seed, farming techniques, water management, storage and processing to markets. Focus on reducing postharvest losses through heat as well as excessive or unseasonal rainfall. Education information on climate change by the Southern Africa Regional Universities Association (SARUA), crop diversification, and climate adaptations has been fairly disseminated to farmers in Southern Africa, liaising with other farmers, academics, and agricultural organizations to keep abreast of the latest developments (Climate and Development Knowledge Network (C&DKN), 2014).

The Comprehensive Africa Agriculture Development Programme (CAADP) defining national priorities, as well as for the process of Africans' regaining control CAADP has established itself as the expression of reclaimed ownership of the agricultural policy by the African States and citizens of the continent (NEPAD, 2013). Nevertheless, public commitment to boosting agriculture has been limited and has failed to match the targets set. In 2010, out of the 44 countries for which data is available, only 9 have reached or exceeded the target of allocating 10% of public expenditure to agriculture (NEPAD, 2013). Economic institutions are lacking in Africa compared with other parts of the world, especially in the financial and insurance sectors. This hampers farmers' ability to take more risks and to increase investment (NEPAD, 2013).

Southern Africa like other countries globally has signed the United Nations Framework Convention on Climate Change (UNFCCC), the Kyoto Protocol, NAPA which includes adaptation projects, enhancements of agricultural productivity including irrigation systems, an early warning system for droughts, and water resource development, and research and development. Grants have been received from the international community to deal with climate change disasters (NEPAD, 2013).

However, challenges still exist with a public policy specifically designed to address climate change and subsequent budgeting for it. Public awareness is an on-going process, yet cultural influences present themselves as barriers for subsistence farmers to adapt. Another important element is the lack of technical expertise to deal with climate change in all sectors including government; academic research in the area to inform public policy is lacking, as well as there is lack of government financial capacity to resource climate change adaptation programs.

# 12 Conclusion

Climate models and scenarios indicate that the impacts of regional warming across sub-Saharan Africa will be most severe in the Western Sahel and Southern Africa (Bauer, 2010). Evidence points to the increased risks of climate change in southern Africa. There is meteorological evidence of increased warming (2–5 degrees), erratic rainfall, dry spells, and flooding from cyclones negatively impacting agricultural production. Worsening the situation is the farmer's overreliance on the maize crop which is drought intolerant. Maize is a staple food for the majority in Southern Africa and a rainfed crop mainly cultivated by subsistence small-holder farmers in the rural areas.

Consequently, Southern Africa should respond by reducing the overdependency on weather patterns for food production and enable irrigation farming as widely as possible reaching the majority of subsistence farmers in rural communities. This would require upscaling measures of water harvesting and water management to store water over longer periods. Diversification of crops including research on maize crop with a high degree of tolerance to conditions of dry spells, high temperatures, and drought.

Ultimately, SADC and the Member States need to prioritize timely planning and implementation of comprehensive adaptation measures with effective coordination mechanisms, for the efficient use of resources and equitable distribution to affected populations particularly in rural areas.

## References


Bauer, S., 2010. Adaptation to climate change in Southern Africa. Clim. Dev. 2 (2), 83–93 (Accessed 08.03.2020) https://doi.org/10.3763/cdev.2010.0040. https://www.researchgate.net/publication/250274791_Adaptation_to_climate_change_in_Southern_Africa_New_boundaries_for_sustainable_development.

Botswana, Ministry of Agriculture, 2010. Guidelines for Integrated Support Programme for Arable Agricultural Development (ISPAAD). Government Publishers, Gaborone.

Britannica, 2020. Culture, History & People (Accessed 08.03.2020) https://www.britannica.com.

Chersich, M.F., Wright, C.Y., 2019. Global Health., p. 10 (Accessed 08.03.2020) https://doi.org/10.1186/s12992-019-0466-x.

Climate and Development Knowledge Network (C&DKN), June 2014. Climate Change Counts: Strengthening SADC Universities' Contribution to Climate Compatible Development (Accessed 08.20.2020) https://cdkn.org/project/climate-change-counts-strengthening-university-contributions-to-climate-compatible-development-in-the-sadc-region/?loclang=en_gb.

DHFS, 2018/2019. Summary Report for the Drought and Household Food Security Vulnerability Assessment. Botswana Vulnerability Assessment Committee, Gaborone.

England, M.I., Dougill, A.J., Stringer, L.C., et al., 2018. Climate change adaptation and cross-sectoral policy coherence in southern Africa. Reg. Environ. Change 18, 2059–2071.

European Development Fund (EDF), 2019a. European Development Fund (Accessed 08.03.2020) https://gtai.de.

European Development Fund (EDF), 2019b. Summary Annual Action Programme 2019 in Favour of the Kingdom of Eswatini to be Financed From the 11th European Development Fund (EDF) (Accessed 03.08.2020) https://gtai.de.

Famine Early Warning Systems Network (FEWSN), 2020a. Southern Africa Lesotho, Key Updates on Delayed Green Consumption and Dry Harvest to Prolong Lean Season (Accessed 03.31.2020) http://reliefweb.int/sites/reliefweb.int/files/resources/Lesotho.pdf.

Famine Early Warning Systems Network (FEWSN), 2020b. Malawi Emergency Appeal May-October 2020. http://reliefweb.int/report/Malawi-emergency-appeal-may-october-2020.

FAO, 2020. Lesotho Country Profiles (Accessed 08.28.2020). fa0.org/countryprofiles/index/en/?iso3=LSO.

Food Agricultural Organisation (FAO), 2011. Strengthening Capacity for Climate Change Adaptation in Agriculture: Experience and Lessons from Lesotho (Accessed 08.28.2020) http://www.fao.org.

Food and Agricultural Organisation (FAO), 2009. GIEWS Brief on Lesotho (Accessed 07.10.2020) http://reliefweb.int/sites/reliefweb.int/files/resources/LSO_11.pdf.

Government of Eswatini, 2019. Annual Action Programme 2019 in Favour of the Kingdom of Eswatini to be Financed From the 22th European Development Fund (EDF). Government of Eswatini, Mbabane.

Gwimbi, P., Thomas, T.S., Hachigonta, S., Sibanda, L.M., 2013. Lesotho. In: Thomas, T.S., Sibanda, L.S., Hachingota, S., Nelson, G.C. (Eds.), Southern African Agriculture and Climate Change: A Comprehensive Analysis. International

Food Policy Research Institute (Accessed 08.02.2020) (Chapter 4) ifpri.org/publication/southern-african-agriculture-and-climate-change-comprehensive-analysis.

Integrated Food Security Phase Classification (IPC), August 2020. Eswatini IPC Acute Food Insecurity Analysis. June 2020–March 2021 (Accessed 08.03.2020) https://reliefweb.int/sites/reliefweb.int/files/resources/IPC%20Eswatini%20AcuteFoodInsecurity2020JuneMarch2021%20Report.pdf.

Inter-Agency Humanitarian Evaluation (IAHE), 2020. 2020 Response to Cyclone Idai in Mozambique (Accessed 07.10.2020) http://reliefweb.int/sites/reliefweb.int/files/resources/IAHE%20Mozambique%20Report%20%28English%29.pdf.

International Water Management Institute (IWMI), 2020. Summary (Accessed 08.20.2020) http://iwmi.cgiar.org/issues/rainfed-agriculture/summary.

Johnston, P., Thomas, T.S., Hachigonta, S., Sibanda, L.M., 2013. South Africa. In: Thomas, T.S., Sibanda, L.S., Hachingota, S., Nelson, G.C. (Eds.), Southern African Agriculture and Climate Change: A Comprehensive Analysis. International Food Policy Research Institute (Chapter 7) (Accessed 08.02.2020) http://ifpri.org/publication/southern-african-agriculture-and-climate-change-comprehensive-analysis.

Kuivanen, K., Alvarez, S., Langeveld, C., 2015. Climate Change in Southern Africa: Farmer's Perceptions and Responses. Farming Systems Ecology, Wageningen University, Wageningen, Netherlands.

Manyatsi, A.M., Thomas, T.S., Masarirambe, M.T., Sibanda, L.M., 2013. Swaziland. In: Thomas, T.S., Sibanda, L.S., Hachingota, S., Nelson, G.C. (Eds.), Southern African Agriculture and Climate Change: A Comprehensive Analysis. International Food Policy Research Institute (Chapter 8) (Accessed 08.02.2020) http://ifpri.org/publication/southern-african-agriculture-and-climate-change-comprehensive-analysis.

Maure, G.A., Thomas, T.S., Hachigonta, S., Sibanda, L.M., 2013. Mozambique. In: Thomas, T.S., Sibanda, L.S., Hachingota, S., Nelson, G.C. (Eds.), Southern African Agriculture and Climate Change: A Comprehensive Analysis. International Food Policy Research Institute (Chapter 6) (Accessed 08.02.2020) http://ifpri.org/publication/southern-african-agriculture-and-climate-change-comprehensive-analysis.

Ministry of Health and Social Welfare, 2009. Lesotho, Demographic and Health Survey, Maseru, Lesotho (Accessed 08.28.2020). https://nam03.safelinks.protection.outlook.com/?url=http%3A%2F%2Fdhsprogram.com%2Fpubs%2Fpdf%2FFR241.pdf&data=04%7C01%7CS.Viswam%40elsevier.com%7C3b3edb2498ee4c94ee9108d8bd4280b3%7C9274ee3f94254109a27f9fb15c10675d%7C0%7C0%7C637467441034542975%7CUnknown%7CTWFpbGZsb3d8eyJWIjoiMC4wLjAwMDAiLCJQIjoiV2luMzIiLCJBTiI6Ik1haWwiLCJXVCI6Mn0%3D%7C1000&sdata=7u8LQCm%2FpF%2B3aT1aVoDHpFGWl4iHR%2FxgLzJSkfp5bIQ%3D&reserved=0.

Mugabe, F.T., Thomas, T.S., Hachigonta, S., Sibanda, L.M., 2013. Zimbabwe. In: Thomas, T.S., Sibanda, L.S., Hachingota, S., Nelson, G.C. (Eds.), Southern African Agriculture and Climate Change: A Comprehensive Analysis. International Food Policy Research Institute (Chapter 10) (Accessed 08.02.2020) http://ifpri.org/publication/southern-african-agriculture-and-climate-change-comprehensive-analysis.

National Statistical Office, 2015. Malawi Demographic and Health Survey 2015-16, Zomba, Malawi (Accessed 08.28.2020). https://nam03.safelinks.protection.outlook.com/?url=http%3A%2F%2Fdhsprogram.com%2Fpubs%2Fpdf%2FFR319.pdf&data=04%7C01%7CS.Viswam%40elsevier.com%7C3b3edb2498ee4c94ee9108d8bd4280b3%7C9274ee3f94254109a27f9fb15c10675d%7C0%7C0%7C637467441034542975%7CUnknown%7CTWFpbGZsb3d8eyJWIjoiMC4wLjAwMDAiLCJQIjoiV2luMzIiLCJBTiI6Ik1haWwiLCJXVCI6Mn0%3D%7C1000&sdata=SL9ouE80WSSalIj5jDUVe1Pjn7jFv1FbTCndSxSx14M%3D&reserved=0.

New Partnership for African Development (NEPAD), 2013. Agriculture in Africa: Transformation and Outlook. Midrand, Johannesburg.

Nhemachema, R., Hassan, C., 2018. Micro-Level Analysis of Farmers' Adaptation to Climate Change in Southern Africa. International Food Policy Institute (IFPI) Sustainable Solutions for Ending Food Hunger (Discussion Paper 00714 August 2007) www.ifpri.org.

Republic of South Africa, 2014. South Africa's Second National Climate Change Report. Environment Forestry & Fisheries, Environment. gov.za/otherdocuments/reports/southafricas_secondnational_climatechange.

ROSEA, 2018–2019. Humanitarian Response Plan 2018-2019 (Accessed 08.02.2020) http://reliefweb.int/sites/reliefweb.int/files/resources/ROSEA_20190325_MozambiqueFlashAppeal.pdf.

SADC, 2019a. News 14 November 2019 (Accessed 09.20.2020) https://www.sadc.int/news-events/news/sadc-and-eu-launch-programme-strengthen-capacity-sadc-member-states-undertake-climate-change-adaptation-and-mitigation-actions/.

SADC, 2019b. Agriculture and Food Security in SADC (Accessed 08.28.2020) https://www.sadc.int/themes/agriculture-food-security/.
SADC, 2019c. Synthesis Report on the State of Food and Nutrition Security and Vulnerability in Southern Africa. SADC, Windoek, Namibia.
SADC, 2020. About SADC (Accessed 02.08.2020) https://sadc.int/about-sadc.
Saka, J.D.K., Sibale, P., Thomas, T.S., Hachigonta, S., Sibanda, L.M., 2013. Malawi. In: Thomas, T.S., Sibanda, L.S., Hachingota, S., Nelson, G.C. (Eds.), Southern African Agriculture and Climate Change: A Comprehensive Analysis. International Food Policy Research Institute (Chapter 5) (Accessed 08.02.2020) http://ifpri.org/publication/southern-african-agriculture-and-climate-change-comprehensive-analysis.
USAID, 2019. Climate Risks in Food for Peace Geographies Zimbabwe (Accessed 08.03.2020) https://www.usaid.gov/sites/default/files/documents/1866/Zimbabwe_FFP_CRP_WITHOUT_Adaptive_Measures.pdf.
Wider, U.N.U., 2016. Climate Change Effects on Irrigation Demand and Crop Yields in South Africa. Helsinki, UNU-WIDER.
Williams, G.J., 2018. Britannica (Accessed 08.03.2020) https://www.britannica.com.
World Food Programme (WFP), 2020. Zambia Country Brief, July 2020 (Accessed 08.03.2020) https://reliefweb.int/sites/reliefweb.int/files/resources/Zambia%20Country%20Brief%20-%20July%202020.pdf.
World Population Review, 2020. Swaziland Population (Accessed 08.03.2020) https://worldpopulationreview.com.
Zambia Disaster Management and Mitigation Unit (DMMU), 2019–2020. Humanitarian Appeal 2019-2020 Zambia (Accessed 08.03.2020) http://reliefweb.int/sites/reliefweb.int/files/resources/ROSEA_20191024_Zambia_Response__Plan.pdf.
Zhou, P.P., Simbini, T., Ramakgotlwane, G., Hachigonta, S., Sibanda, L.M., 2013. Botswana. In: Thomas, T.S., Sibanda, L.S., Hachingota, S., Nelson, G.C. (Eds.), Southern African Agriculture and Climate Change: A Comprehensive Analysis. International Food Policy Research Institute (Chapter 3) (Accessed 08.02.2020) http://ifpri.org/publication/southern-african-agriculture-and-climate-change-comprehensive-analysis.

# Further reading

Cobbing, J.R.D., 2020. Republic of South Africa, Union of South Africa. Britannica (Accessed 08.02.2020) https://www.britannica.com.
Department of Environmental Affairs. Republic of South, n.d. Government Website. (Accessed 08.28.2020). https://www.environment.gov.za/sites/default/files/reports/ltasfactssheet_perspectiveforSADC.pdf.
Integrated Food Security Phase Classification (IPC), 2020–2021. Eswatini IPC Acute Food Insecurity Analysis. IPC (June 2020–March 2021).
Kanyanga, J., Thomas, T.S., Hachigonta, S., Sibanda, L.M., 2013. Zambia. In: Thomas, T.S., Sibanda, L.S., Hachingota, S., Nelson, G.C. (Eds.), Southern African Agriculture and Climate Change: A Comprehensive Analysis. International Food Policy Research Institute (Chapter 9) (Accessed 08.02.2020) http://ifpri.org/publication/southern-african-agriculture-and-climate-change-comprehensive-analysis.
Lesotho. Ministry of Health and Social Welfare, 2005. Lesotho Demographic and Health Survey (DHS) 2004. Maseru.
Republic of South Africa. Department of Environmental Affairs, n.d. Climate Change Adaptation SADC. (Accessed 8.20.2020). https://www.environment.gov.za/sites/default/files/reports/ltasfactsheet_perspectiveforSADC.pdf.
USAID, 2012. Climate Change Adaptation in Southern Africa (Accessed 08.03.2020) https://www.climatelinks.org/sites/default/files/asset/document/southern_africa_adaptation_fact_sheet_jan2012.pdf.

CHAPTER

# 12

# Climate change and world population

*Jane O'Sullivan*

School of Agriculture and Food Sciences, University of Queensland, St Lucia Campus, Brisbane, QLD, Australia

OUTLINE



## 1 Introduction

Anthropogenic climate change, as a cumulative impact of human activity, is intimately entwined with human demography. The more people there are, the greater are our collective impacts on the environment. Yet none of the associations are simple, and many evoke sensitivities that cause population to be given little attention in climate change discourse.

There are three relevant dimensions to this relationship. The most obvious to our theme is how climate change might affect population change, through impacts on births, deaths, and movements of people. The second is the extent to which demographic factors influence the vulnerability of communities to the effects of climate change. The third is the effect of


https://doi.org/10.1016/B978-0-12-822373-4.00008-2

population change on the climate, greenhouse gas emissions, and our prospects for abating them. To orient ourselves, we will first review the current trends in global demography.

## 2 The human population in the 21st century

It took over 100,000 years for the population of humans to reach 1 billion. It took only about 220 years for it to double three times over to 8 billion. The global population will likely reach between 9.5 and 14 billion unless the burden on planetary systems becomes so great as to cause an unthinkable escalation in deaths.

For the past half-century, 1 billion people have been added roughly every 12 years. During 2021, somewhere between 80 million and 90 million people will be added to planet Earth's population. It is difficult to imagine where an extra country the size of Germany could be accommodated, this year and again the next year and the year after. In reality, these people are mainly accommodated in the burgeoning slums of cities in poor countries. They draw increasingly on their local resources, particularly freshwater, soil, forests, fisheries, and wildlife but also increasingly depend on global trade in food, influencing land-use decisions throughout the world.

It is commonly reported that global population growth is already decelerating toward a population peak, and will then decline in the not-too-distant future. This is misleading. While it is true that the growth rate has fallen, from its peak of 2.1% in 1968 to around 1.05% in 2020, this is entirely because the denominator—global population—has doubled, not because the increment added each year has declined. So, it is premature to say that population growth is tapering off. There was the beginning of a downward trend in the 1990s but since around the turn of the century, the annual increment has grown again. It might now be turning the corner, but we do not yet know.

It is impossible to know exactly how much the world population changes each year because the data are incomplete. In many countries, birth and death records are far from comprehensive. Censuses are infrequent and do not reach all households. However, thanks to "demographic and health surveys" (DHS) conducted in most developing countries every few years and overseen by the United Nations (UN), we have a fairly good idea of who is living where and about their fertility, mortality, living arrangements, and family size preferences.

The past 75 years have seen dramatic changes in world demography. A post-WWII baby boom in the West coincided with the rapid dissemination of antibiotics, immunization, and sanitation programs in developing countries. These two trends together generated an enormous increase in population growth. By the 1960s, there was mounting concern that population growth in developing countries would lead to major famines. This threat was averted by the rapid deployment of new crop varieties and fertilizers in an agricultural transformation known as the "green revolution." Norman Borlaug, the plant breeder most prominent in the development of green revolution crops, won the Nobel Prize in 1970 for "eradicating hunger." Many people still claim that the green revolution proved the "Malthusians" wrong. However, in his Nobel lecture, Borlaug insisted that the new technologies had only bought us a breathing space of perhaps three decades, as "the frightening power of human reproduction must also be curbed; otherwise the success of the green revolution will be ephemeral only" (Borlaug, 1970).

Five decades later, fertility rates have fallen dramatically in most countries but they remain persistently high in some regions. Although on average women are having half as many children, with more than twice as many women, the population is increasing by even greater numbers than in 1970. This fertility decline bought us some extra breathing space, but we could be approaching its limit: the incidence of hunger and undernutrition, after falling for decades, has been rising again since 2014 (FAO, IFAD, UNICEF, WFP, and WHO, 2020).

Despite the persistent specter of food insecurity and increasingly disturbing symptoms of environmental strain (UNEP, 2019), political attention to population growth has diminished. In less developed countries, fertility fell by 1.4 children per woman in the 1970s, but it fell by only 0.16 in the decade since 2010. Is complacency justified? Will the unfinished business of population stabilization solve itself?

## 2.1 The role of voluntary family planning programs

The contraceptive pill was a welcome innovation in the 1960s, and its uptake rapidly reduced fertility levels in developed countries, many of which achieved "below replacement" fertility by the mid-1970s. (The "replacement" level is that at which the children's generation equals in number the parents' generation—around 2.08 children per woman where child mortality rates are low.) New contraceptive technologies also enabled family planning to emerge as a major focus of development efforts, recognizing not only the threat of famine but also the economic challenge of population growth (Coale and Hoover, 1958). The United Nations hosted international conferences on population and development to garner donor support and disseminate best practice, and the United Nations Fund for Population Activities (UNFPA) was born in 1967.

Under well-promoted voluntary family planning and reproductive health programs, fertility fell rapidly in many countries (Fig. 1). In addition to ensuring contraception access to all parts of the country, programs generally included a range of measures aimed at changing

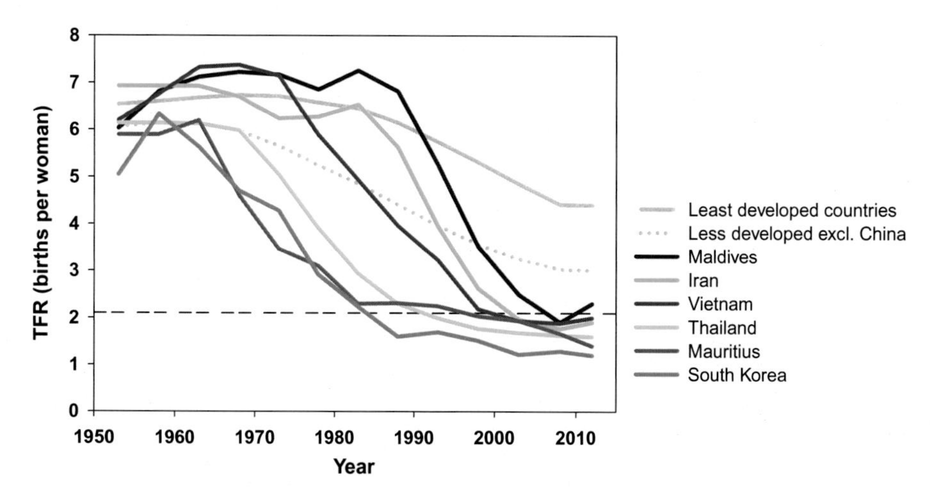


FIG. 1 Time course of total fertility rate (TFR, average births per woman) for selected countries which implemented population-focused voluntary family planning programs at differing times, showing rapid change in fertility, compared with aggregate TFR for less developed countries (excluding China), and for least developed countries.

attitudes to family size, as well as encouraging later marriage and greater autonomy for women. Different countries took different measures to address the complex barriers to family planning uptake specific to their societies; some recruited the support of religious leaders, while others provided female community health workers or mothers' clubs. Some promoted child spacing to mothers after childbirth, and some implemented school-based programs or premarital classes. However, all aimed to reduce the population growth rate, not merely to avoid unwanted births (Robinson and Ross, 2007).

Most countries in East- and South-East Asia were early and strong adopters of family planning. In the Muslim world, Tunisia led the way, particularly with measures increasing women's autonomy and reproductive rights, and good progress was made in Morocco, Egypt, Indonesia, and Bangladesh, among others. Iran came late to population concern but achieved spectacularly rapid fertility decline during the 1990s under an exemplary program focused on public education, premarital counseling, and free access to contraception, with endorsement from religious leaders (Dérer, 2019). More gradual transitions occurred in the Catholic countries of Latin America and the Philippines, several of which still have a way to go. However, in sub-Saharan Africa and parts of South and Western Asia, efforts were sparse and political will has been low. Where programs were implemented, such as in Kenya and Ghana during the 1980s and more recently in Ethiopia and Rwanda (Habumuremyi and Zenawi, 2012), fertility did fall appreciably in the communities served by the programs, but falls ended when programs ended (Ezeh et al., 2009). On a local scale, NGO projects are demonstrating considerable success, particularly those implementing integrated development interventions under the "Population, Health and Environment" (PHE) model (Oglethorpe et al., 2008; Wilson Center, 2013). However, compared with the most successful family planning countries, priority given to family planning in Africa has been low.

Since the mid-1990s, family planning has largely fallen off the international development agenda. A backlash against it was building during the 1980s through a combination of suspicion of neocolonial motives and Catholic opposition to contraception. These parties seized on the rare incidents of coercive measures within national programs, most notably China's one-child policy from 1979 and India's coerced vasectomies in the late 1970s. Coercion had never been condoned by international family planning agencies, which have always emphasized serving clients' choices safely, accessibly, and affordably, but national programs at times failed to maintain these standards. In both the above cases, coercion was the result of applying targets for local administrators to meet, with too little concern for how they were met. Such excursions into coercive measures only served to show how counterproductive they are—the outrage over the Indian program set back voluntary family planning there for decades, while in China, the rapid fertility decline achieved under its earlier voluntary program, initiated in the late 1960s, became a slower trickle-down after the one-child policy was introduced.

Despite coercive measures being anomalies which the family planning movement actively addressed, the opponents of contraception argued that any demographic motivation for family planning would naturally lead to abuses of human rights, and that family planning should be reframed as serving women's reproductive health and rights exclusively (Sinding, 2016). This reframing was consolidated at the UN's 1994 International Conference on Population and Development (ICPD) in Cairo. The Cairo statement itself (the ICPD Program of Action) (UNFPA, 1994) acknowledges the importance of minimizing population growth for poverty reduction and environmental security, but its implementers have cultivated a taboo around

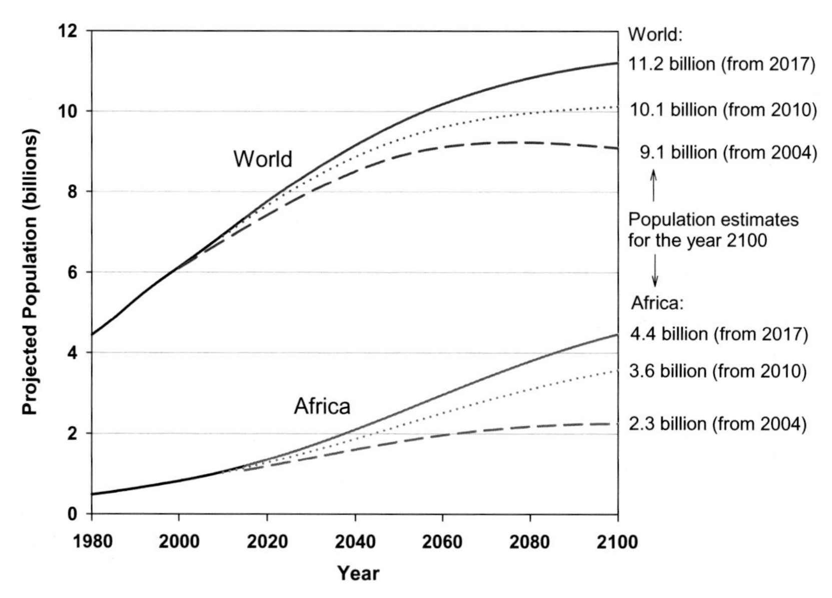


**FIG. 2** Population projections from the United Nations, showing a dramatic rise in expected outcomes since 2004.

the population issue. It is implied that any reference to population growth as a problem masks a callous agenda to suppress "people of color" or to deflect blame for global issues away from rich people in developed countries onto the poor (Monbiot, 2020). The sad irony of this perspective is that the family planning movement was instigated to empower women, deflect threats to poor communities, and close the gap between developed and underdeveloped countries (Potts, 2014).

The result of this paradigm shift has been a tragedy for the poor. International aid for family planning plummeted (Sinding, 2009), national programs were wound back, and fertility declines slowed, stalled, or even reversed in several countries such as Indonesia and Egypt (Bongaarts, 2008). The UN population projections, based on earlier trends, underestimated global growth and were repeatedly revised upward, adding 2 billion people to the global population projected for the year 2100 (Fig. 2). Almost all of this additional growth is anticipated to occur in Africa, formerly expected to level not far above 2 billion but now projected to soar past 4 billion.

When we look at the annual increment of the global population, the reversal around the year 2000 is evident (Fig. 3). The UN projections persisted in assuming fertility decline would get back on track everywhere, quickly resuming the downward trend in increment seen in the 1990s. But annual estimates of the global population, published by the Population Reference Bureau (PRB), continued to show a rising increment.

This rise cannot be explained by the factors usually cited as the main drivers of fertility decline: reducing poverty, increasing girls' education, and stemming infant mortality. All three of these aims were accelerated over the same period under the Millennium Development Goals (the overarching UN agenda between 2000 and 2015, now replaced by the Sustainable Development Goals). The decline in funding and political priority for family planning efforts is the most plausible explanation. The policy shift that was intended to

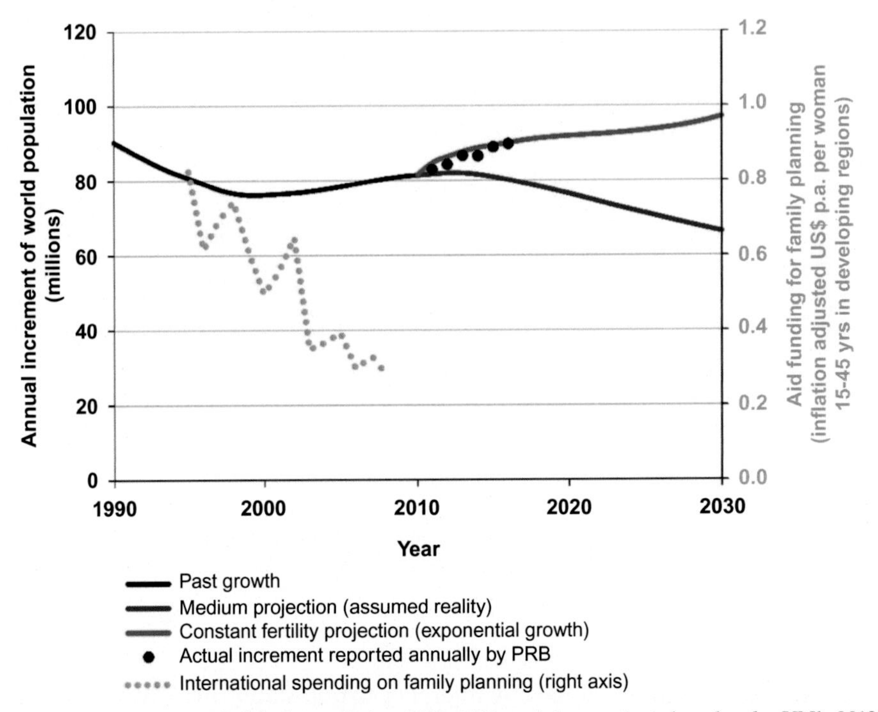


**FIG. 3** The annual increment of global population 1990–2010, and that projected under the UN's 2012 medium fertility and constant fertility projections (UNDESA, 2013). *Black dots* give estimates of actual increment reported annually in the Population Reference Bureau's "World Population Datasheets" (PRB, 2011–2016). International aid spending on family planning is plotted against the right axis (UN Economic and Social Council, 2010).

promote women's health and rights has perversely undermined women's health and rights (Sinding, 2016).

## 2.2 Myths that maintain the population taboo

To sustain the taboo on population, two myths have found a widespread following. The first is that population growth does not impede economic development as "with every mouth, God sends a pair of hands" (Robinson, 1974). It was argued that human ingenuity had always triumphed over scarcity of resources, and hence is "the ultimate resource" (Simon, 1981). While not everyone accepted the idea of "the more people, the better," economists predominantly adhered to the revisionist view that population growth was economically benign (Kelley, 2001), much to the concern of family planning practitioners, who saw the impoverishment of high fertility firsthand (Potts, 1999; Sinding, 2009).

Around the turn of the century, the increasingly evident gap between the economic performance of developing countries that lowered their fertility and those that did not led to a new theory that it was not population growth but age structure that mattered (Bloom et al., 2001). Dubbed the "demographic dividend," the theory holds that lowering the birth rate reduces the ratio of dependent children to working-age people and hence increases economic output per capita. The theory has been actively promoted by the UNFPA, allowing fertility reduction to be encouraged without invoking the taboo population growth (Herrmann, 2017). But the theory lacks direct evidence that the age structure was responsible for any part of the economic stimulus in the Asian Tiger economies, and the correlation does not hold in Africa (Garenne, 2016). Demographic dividend theory provides no incentive to get fertility below replacement level, where it needs to be to end population growth. In addition, it has the harmful effect of implying that the eventual rise in the proportion of elderly people will become an economic millstone. This "challenge of population aging" is almost universally decried, although the evidence to date is that a declining proportion of "working-age" people merely leads to less unemployment, not less employment (O'Sullivan, 2014, 2020b). Regardless of the evidence, the fear of population aging has already led some national leaders, such as Tanzania's President Magufuli, and Turkey's Prime Minister Erdoğan, to decry family planning and advocate more births (BBC, 2018; MacFarlane et al., 2016). Iran, formerly an exemplar of successful rights-based family planning, backflipped spectacularly with its 2015 "Comprehensive Population and Exaltation of Family Bill" stripping women of reproductive rights and contraception access (Kokabisaghi, 2017; Zaynab, 2018).

Throughout this discourse, the limitations of land and natural resources continue to be ignored. Even the obvious burden of needing to generate sufficient new infrastructure and other durable items each year to equip additional people has so far eluded most economic analysts (O'Sullivan, 2012). High-fertility countries find themselves scrambling up a down-escalator, and naturally get ahead more quickly when the escalator is slowed (O'Sullivan and Martin, 2016). The French economist Alfred Sauvy described this "demographic investment" as early as 1958 (Sauvy, 1958), and the Cambridge economist Austin Robinson applied Sauvy's thesis to the budget for Bangladesh's first 5-year plan. He concluded that the cost of "standing still" at the prevailing 3% per annum population growth represented around 75% of all the budgeted investment. With the planned level of investment, incomes might be raised by 30% over 20 years, but if population growth were at the European level (0.45% p.a. at that time), an income increase of 150% would be expected (Robinson, 1974). The Harvard economist Lester Thurow invoked the same idea in a 1986 opinion piece challenging the revisionists (Thurow, 1986). In a thought-experiment, he estimated that no country was likely to be able to achieve sufficient investment to get ahead if population growth exceeded 2% per annum. However, under the long shadow of the population taboo, such ideas have not regained traction.

The second myth serving the taboo is that past fertility declines have been driven by indirect influences including reducing poverty, educating girls, and lowering infant mortality rates, and that addressing these factors is the best way to solve population growth. It is indisputable that these factors tend to correlate with lower fertility, and all are goals of development in their own right. But the focus on these drivers is calculated to discredit the role of family planning programs. Literature linking female education to fertility tends to omit any mention of the presence or strength of concurrent family planning efforts, and rarely acknowledges reverse causation (the well-established effect that lower fertility enhances girls' access to education) (Lutz et al., 2014). To meet the reproductive rights of women, *access to* contraception is emphasized, but the promotion of smaller families and direct efforts to change social norms around family size are avoided. It is not that the harms of overpopulation are denied, but we are assured that when women are educated and empowered, they choose small families. While we might consider it a moral obligation to inform people of the risks of consuming alcohol or smoking tobacco, it is implied that informing them of the consequences of large families would be coercive. Any right-thinking person, the discourse maintains, knows that population growth will solve itself if we look after women and, therefore, anyone who advocates for any focus on population growth itself is acting against individual freedoms.

A brief perusal of national-level data soundly debunks both of these myths. As was illustrated in Fig. 1, national family planning programs were instrumental in every case of rapid fertility decline. The strength of family planning programs explains most of the pace of fertility decline, whereas improvement in education had little effect (de Silva and Tenreyo, 2017; Psaki et al., 2019). Importantly, these programs included both provision of contraceptives and reproductive health services, as well as mass-media campaigns to communicate the benefits of smaller families (Elkamel, 2020). Changing social norms around family size turns out to be pivotal for accelerating fertility decline (de Silva and Tenreyo, 2020). The withdrawal of advocacy components of national family planning programs after 1994 has contributed to the slowdown in the fertility transition. In their place, some excellent nongovernment initiatives have emerged, but they lack the scale and reach of national programs. These initiatives include the Population Media Center's serialized radio and television dramas produced in local languages in many high-fertility countries, whose central characters model attitudinal change to a range of social issues from domestic violence, child marriage, and female genital mutilation to contraception and the benefits of small families (Jah et al., 2014).

As Bill Ryerson, founder of Population Media Center, explained, "meeting unmet demand for contraceptives is only part of the solution. The countries that have most successfully reduced population growth have emphasized changing attitudes of the people regarding the role of women, ideal family size, age of first pregnancy, and the benefits of using modern contraceptives. ... [The] use of effective family planning methods will not result in population stabilization if desired family size is five, six or seven children" (Ryerson, 1993). Fig. 4 illustrates that desired family size correlates well with the actual fertility that countries achieve.

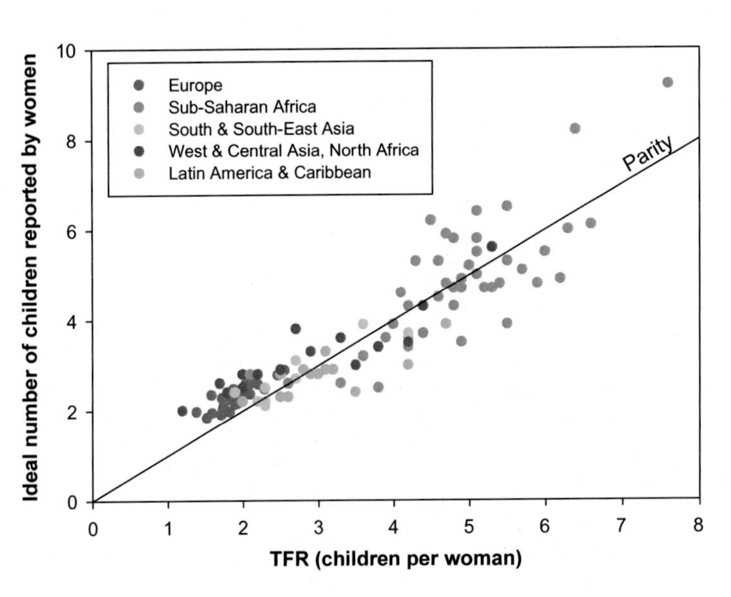


**FIG. 4** The average family size that countries achieve closely follows the number of children women say they want. *Data from DHS surveys.*

The well-recognized relationship between poverty and high fertility has driven the belief that economic development is a major driver of fertility decline. This belief is bolstered by the myth that population growth is economically neutral, as this myth dismisses the other explanation for the relationship: that population growth impoverishes, and easing it promotes economic advancement. To explore which direction of causation is more influential, Fig. 5 contrasts the rate of fertility decline as a function of the prior level of wealth and, conversely, the rate of income growth as a function of the prior level of fertility. When the pace of fertility decline is presented as a function of the level of GDP per person (Fig. 5A), there is no trend: the poorest countries could reduce fertility as rapidly as any other if they implemented appropriate programs. Conversely, in Fig. 5B, we see that the level of fertility has had a very strong influence on the likelihood of economic advancement. In countries with fertility above four children per woman, only a few oil-rich states sustained significant economic growth. Over a 20-year period, all low-fertility countries advanced economically, including those with shrinking populations. Slowing the down-escalator evidently works wonders.

An interesting phenomenon occurs when we compare countries that implemented high-profile family planning programs, such as Thailand, with those that did not, such as the Philippines (Fig. 6). In 1965, both countries had around 30 million people and fertility above six children per woman. Thailand's population will peak during the 2020s at just over 70 million, whereas the Philippines has around 110 million, on its way to over 150 million before it peaks. In 1965, the Philippines was richer, better educated, and culturally more westernized, yet its fertility transition has been slow. In Fig. 6B, we see that Thailand's GDP per person only

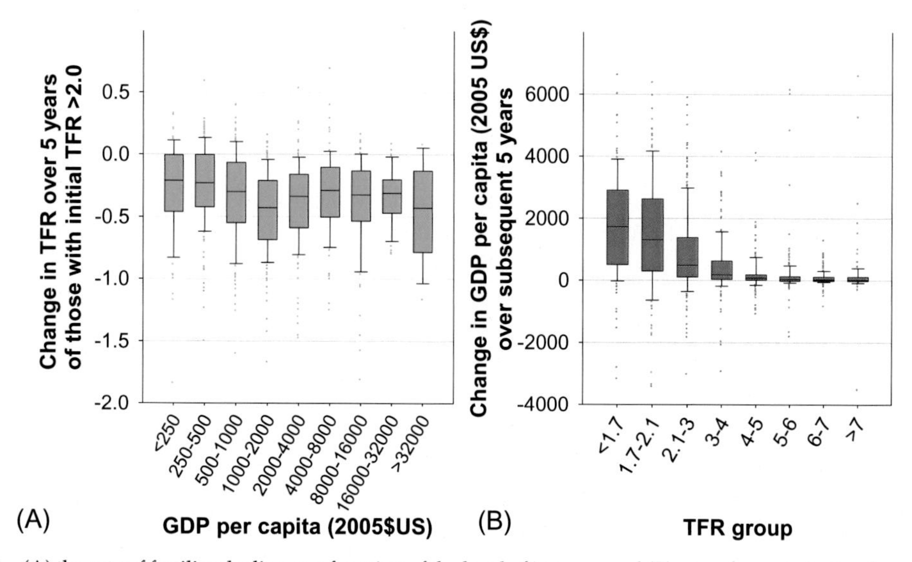


FIG. 5 (A) the rate of fertility decline as a function of the level of income, and (B) rate of economic development as a function of the level of fertility. Data points represent each country in each 5 years between 1960 and 2010. All countries and periods with available data are included. Box plots span the 25 percentile, median, and 75 percentile, and whiskers extend to the 10th and 90th percentile. GDP per capita (inflation-adjusted 2005 US$) are from the World Bank economic database and fertility data from UNDESA (2015).

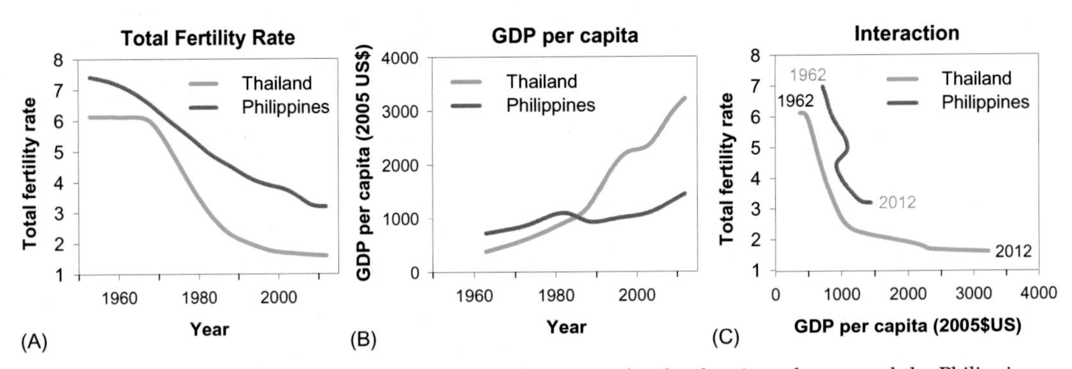


**FIG. 6** Fertility and wealth time courses for Thailand, a strong family planning adopter, and the Philippines, a weak adopter. *Left*: the change in total fertility rate (TFR, average children per woman) over 5-year intervals from 1950 to 2015; *middle*: the average GDP per capita over 5-year intervals from 1960 to 2015 (adjusted to the constant year 2005 US$); and *right*: the relationship between TFR and GDP per capita.

overtook the Philippines after its fertility had fallen well below three. In Fig. 6C, the relationship between fertility and income is steeply concave: fertility fell first before income per capita accelerated. If the adage "development is the best contraception" were true, we would see convex curves, with wealth increasing until sufficient to drive fertility down. But the most interesting thing about this chart is that both countries followed parallel paths. Thailand's family planning program merely allowed it to proceed more quickly to the preconditions that allowed economic betterment.

This pattern is repeated throughout the developing world. In Fig. 7, all the countries with fertility above five in 1950 are put into one of three groups according to the rate of their fertility transition. They are synchronized with respect to the start of their transition before averaging. Fast transition countries (achieving a fall of more than 3 units over a 20-year period), all of which promoted family planning, are approaching a peak population around 2–2.5 times the population when they began fertility reduction efforts (Fig. 7B). Those with intermediate transitions (1–3 units over two decades) have not reduced fertility as fast as the number of mothers has risen, so their population growth is not yet decelerating. The population of slow-transition countries has tripled over the same period, and still has at best another doubling in store due to demographic momentum, even if they adopt strong measures from now. In Fig. 7D, we find a remarkable commonality in the relationship between fertility and GDP per capita. It appears that, unless countries have exceptional mineral wealth, fertility decline is a prerequisite for sustained economic advancement.

## 2.3 Trends and projections

For many decades, the UN Population Division has been collating demographic data and formulating projections of future population growth. These projections are based on the past relationships between the level of fertility and the rate at which it has fallen. As we saw in Fig. 1, countries with high fertility can initially achieve rapid declines, but the rate of decline tapers when one child fewer becomes a more significant change in a family's structure.

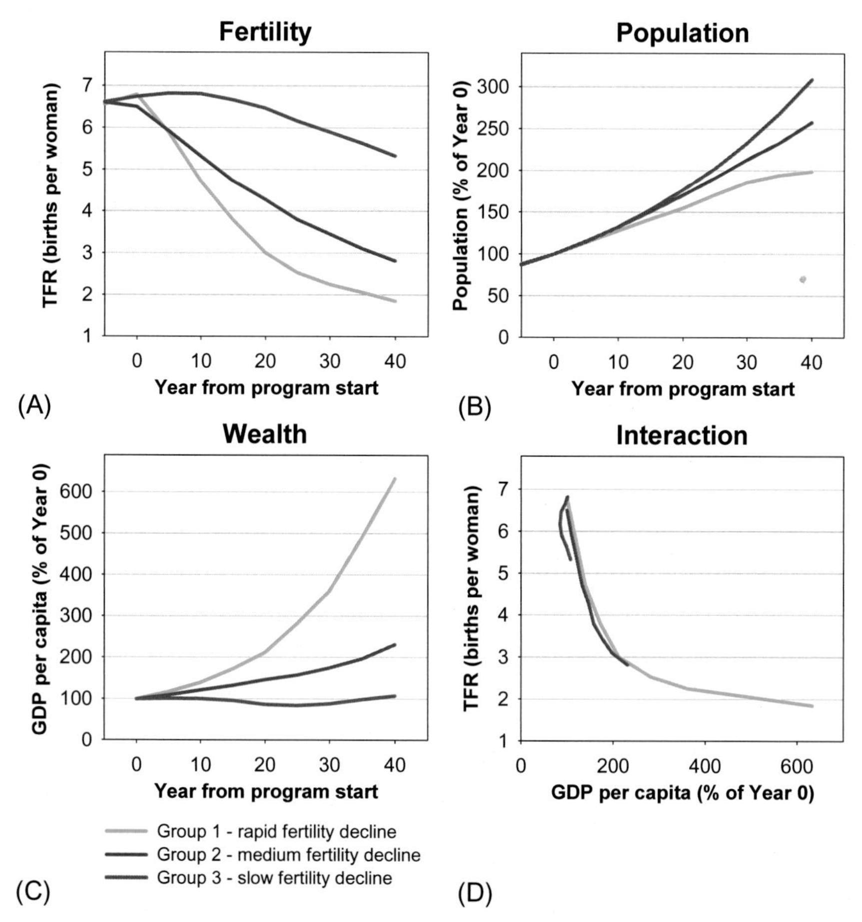


**FIG. 7** The average time-course for (A) fertility, (B) population, and (C) GDP per capita (inflation-adjusted US$), and (D) the relationship between TFR and per capita GDP for developing countries grouped according to the rate of their fertility transition. Each of the rapid transition countries (Group 1) achieved a fertility decline greater than 3 units over 20 years under family planning programs. Group 2 countries had maximum falls between 1 and 3 units over 20 years. In Group 3 countries, fertility had not fallen by more than one unit in a 20-year period, up to 2010. Year 0 is the start of the fertility transition in each country or 1970 for Group 3.

However, the UN's model has trouble predicting when a country's fertility transition will take off, and does not allow for midtransition stalls (O'Sullivan, 2016).

Modeling of future climate change mitigation scenarios has been coordinated internationally by the Intergovernmental Panel on Climate Change (IPCC) through the use of a set of "shared socioeconomic scenarios" (SSPs), which imagine different potential socio-political futures as contexts within which adaptation and mitigation actions can be applied (O'Neill et al., 2014). The population projections used in the SSPs were supplied by the Wittgenstein Centre for Demography and Global Human Capital at Austria's International Institute for Applied Systems Analysis. In contrast with the UN, they emphasize the role of human capital (particularly education attainment) in driving fertility decline (Samir and Lutz, 2017). The theory behind these projections is that future fertility declines will be more rapid than in the past because the communities and particularly the women are better educated. Hence, the central projection (SSP2) lies well below the lower bound of the UN's probable range, peaking around 9.4 billion in 2070 (Fig. 9). Because fertility is assumed to be dependent on human capital, and human capital development depends on economic progress and equity, the scenarios have higher population projections (SSP3) or lower (SSP1 and SSP5) according to whether economic development is slower or faster, respectively, than the central scenario (Jiang, 2014).

More recently, the Institute for Health Metrics and Evaluation (IHME), a University of Washington unit funded by the Bill & Melinda Gates Foundation, has produced its own projections as part of its "Global Burden of Disease" program. Like the Wittgenstein Centre, it emphasizes education as a key driver of fertility decline, and predicts a global peak population below 10 billion in the 2060s (Vollset et al., 2020). However, its methodology and findings have raised concern among demographers (Ezeh et al., 2020; Gietel-Basten and Sobotka, 2020; O'Sullivan, 2020a). The IHME has come under scrutiny as a relatively intransparent, private unit at risk of politicizing analyses (Mahajan, 2019; Schwab, 2020). The IHME initially collaborated with the World Health Organization (WHO) on its Global Burden of Disease analyses, but the WHO withdrew due to its inability to validate the IHME data (Mathers, 2020). For our purposes, we should note that the SSP2 projection has substantially underestimated population growth up to 2020 (O'Sullivan, 2019), and the IHME model is already overestimating the pace of fertility decline in sub-Saharan Africa (O'Sullivan, 2020a). It would, therefore, be rash to put faith in their projections for the purpose of estimating future greenhouse gas emissions, climate change impacts, and adaptation needs.

None of these projections examine the role of family planning efforts as drivers of fertility decline. In Fig. 8, an attempt is made to do so. Two scenarios are modeled: a business-as-usual path in which failure to promote smaller families means that fertility in high-fertility countries continues on its current slow decline; and a proactive path in which all remaining high-fertility countries provide and promote voluntary family planning, and achieve the average path of fertility decline that was achieved by voluntary family planning countries in the past (the rapid transition group in Fig. 7). The latter leads to a path similar to that of the UN's "low fertility" projection and the IPCC's SSP1/SSP5 projection, without requiring fertility to fall as unrealistically low. (SSP1 expects Africa's fertility to be below 1.3 children per woman in 2100, whereas the UN's low projection assumes the average woman has half a child fewer than the medium projection in all countries, including those that already have very low fertility.)

The key to achieving such a path is greater political will to implement voluntary family planning programs that combine the provision of services with communication campaigns to change social norms about family size and contraception acceptance, as an integrated pillar of national development plans (including climate adaptation plans). For this political will to emerge, governments and aid donors would need to rediscover the profound benefits of minimizing further growth in human numbers and acknowledge the efficacy of direct, noncoercive approaches. Recent history suggests that waiting for the indirect drivers of education, urbanization, and cultural globalization to shift social norms will be too slow. The list of advantages to having fewer future people rather than more is very long. The implications for climate change adaptation and mitigation will be discussed in Sections 4 and 5.

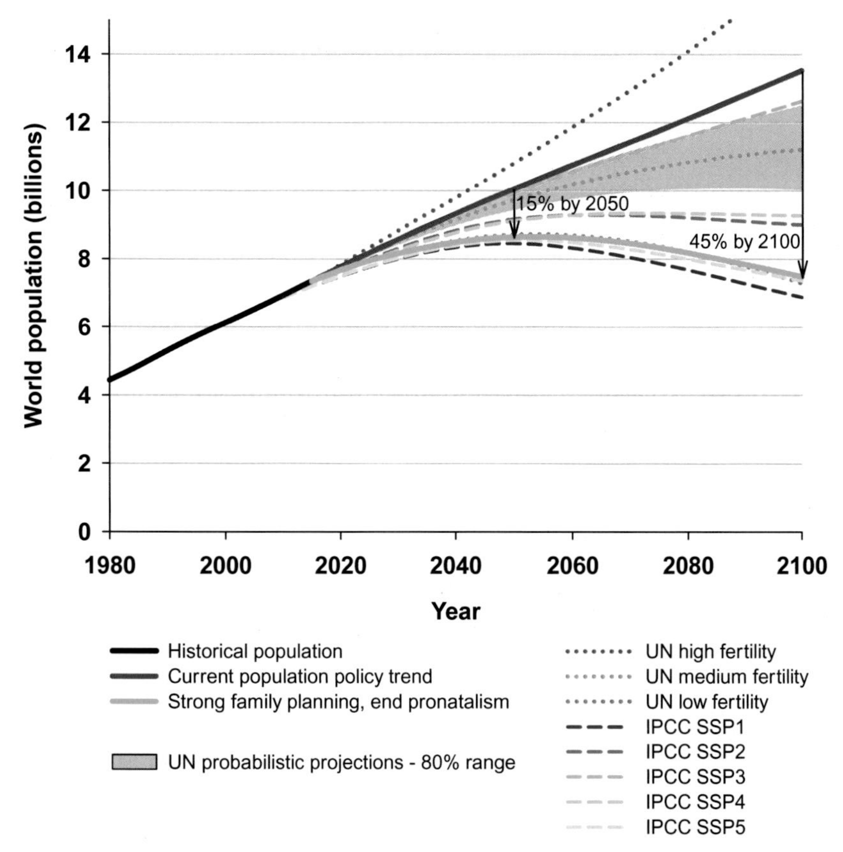


FIG. 8 Policy-based projections of the future global population, comparing outcomes if countries continue their recent trends or if remaining high-fertility countries adopt strong family planning, achieving the average path that past family planning countries achieved and assuming low-fertility countries abandon attempts to increase births. These outcomes are compared with the UN projections (UNDESA, 2015) and the IPCC's Shared Socioeconomic Pathways.

## 3 How will climate change affect demography?

Climate change will affect all aspects of our lives in the future, including our experience of human density. Here, we consider only the likely effects on human numbers and their distribution, not the welfare of those people, which will be covered in other chapters. Demographic effects will be through effects on deaths, births, and migrations.

### 3.1 Deaths

The risk of deaths from heat stress is already rising. Climate change causes a disproportionate increase in the frequency of very high temperatures in locations where they were previously very rare. This is often referred to as "shifting the bell-curve." Although the rise in

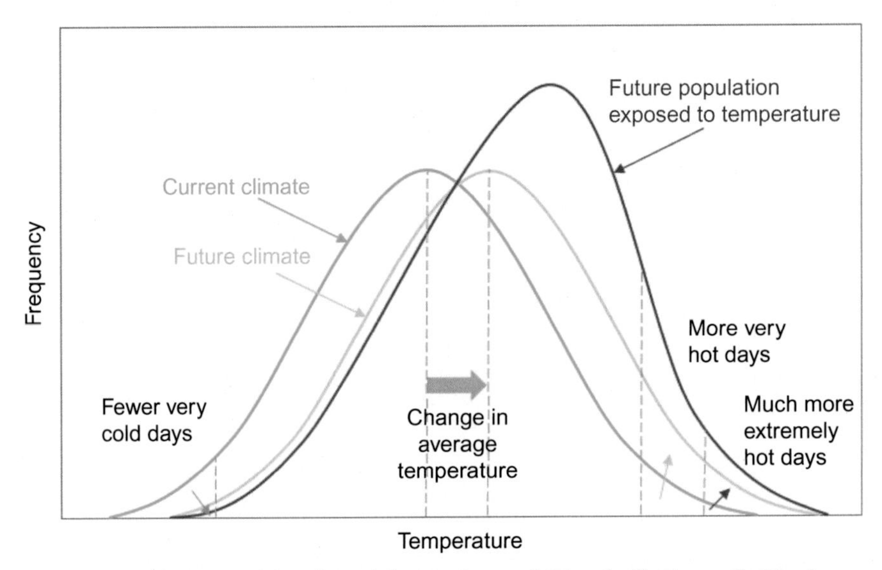


**FIG. 9** Diagrammatic illustration of the effect of climate change shifting the "bell curve" of the frequency of temperature events. In addition, population growth is occurring more in hot, humid regions, skewing the impact toward more exposure to extreme heat events.

average temperature might be small compared with the usual range of variation, the frequency of extreme heat events at the tail of the curve becomes many times its previous level (Fig. 9). The effect is exacerbated because hot locations are experiencing disproportionate amounts of population growth, increasing the numbers of people likely to be exposed to future heatwaves.

The occurrence of lethal heatwaves is commonly modeled based on the wet-bulb temperature above which a healthy person could not maintain body temperature. The wet-bulb temperature allows for evaporative cooling but is closer to the ambient temperature as humidity rises. The McKinsey Global Institute (2020) defines a lethal heatwave as "a three-day period with maximum daily wet-bulb temperatures exceeding 34°C." Without greenhouse gas mitigation, they estimate that between 0.7 billion and 1.2 billion people could live in areas exposed to lethal heatwaves by 2050 (McKinsey Global Institute, 2020). Pakistan and West Africa are two regions most vulnerable to this hazard—both regions with high rates of population growth. Similarly, Im et al. (2017) modeled the occurrence of wet-bulb temperatures above 35°C and found that "the most intense hazard from extreme future heat waves is concentrated around densely populated agricultural regions of the Ganges and Indus river basins."

Owing to the high agricultural productivity of these hot, humid regions, it is no coincidence that they also have high population densities. Rural communities are particularly vulnerable to heat exposure during agricultural work, while urban populations can suffer the additional burden of the urban heat island effect.

Xu et al. (2020) took a different approach, mapping the climatic niche of human settlement, concluding that the majority of people live where mean annual temperature (MAT) ranges from 11°C to 15°C. Currently, only a few zones in the Sahara desert, comprising 0.8% of the Earth's land surface, experience a MAT >29°C. However, without greenhouse gas mitigation and in the absence of migration, by 2070 such temperatures could apply widely across the tropical latitudes, affecting one-third of the global population. This has as much bearing on agricultural production as it does on potential heat-related mortality.

These studies have focused on conditions deemed to be unsurvivable by healthy adults. Much lower levels of heat stress can contribute to the deaths of vulnerable people, such as the elderly, ill, or obese. It has been argued that heat is a much more common contributor to deaths than is currently reported on death certificates (Longden et al., 2020). Mora et al. (2017) analyzed reported incidents of excess human mortality associated with heatwaves between 1980 and 2014 to determine a weather threshold for lethality. They estimated that 30% of the world population is already exposed to such events and this could rise to 74% without climate change mitigation, but this figure could be possibly as low as 48% if greenhouse gas emissions are rapidly reduced (Mora et al., 2017).

The lethality of heatwaves can be mitigated by changes in building design, human behaviors, and availability of air conditioning. Because of this, it is not possible to model future death rates associated with the increased incidence of heatwaves. However, it does not bode well that the regions with the greatest risk of lethal heatwaves are those suffering high rates of poverty, high population density, and high population growth rate.

Flooding and related damage, such as landslides and infrastructure collapse, have a far less predictable toll. Again, it is possible to adapt the built environment and human settlement patterns to avoid some impacts of flooding. But it is unlikely that existing infrastructure and settlements will be moved merely as a precaution against rising risk, so an increasing incidence of damage seems inevitable. Again, appropriate adaptation and crisis-response systems could minimize the direct toll of human lives.

After decades of improving nutrition levels, since 2014 the number of undernourished people in the world has been once again increasing (FAO, 2020). The possibility of lethal famines is growing as regional food production fails to keep pace with population growth (Brown, 2012). Population growth is the main reason for food insecurity, but insecurity tips toward famine in extreme weather events and these will become more extreme. Increasingly volatile weather patterns will affect food production at both local and global scales. While some regions, mainly in the northern latitudes, are likely to see increased crop yields, those in the tropics and semiarid zones are more likely to suffer net negative effects. The McKinsey Global Institute (2020) estimates that the global annual harvest currently has a 10% chance of being more than 15% below average once per decade. Without mitigating climate change, this chance could rise to 20% by 2030, and by 2050 more than a third of years could yield lower than 15% below the long-term average (McKinsey Global Institute, 2020).

However, that average will be rising, as it has done for many decades because of improvements in agricultural practices, the expansion of irrigation, and, to a lesser extent, recruitment of more land. Whether this pace of improvement can continue to outpace population growth in future decades is increasingly questioned. Globally, crop yields increased by 56% between 1965 and 1985 but only 20% between 1985 and 2005 despite a substantial, and unrepeatable, increase in the area irrigated (Foley et al., 2011). Several limits are being reached apart from climate change. Major staple crops are reaching their genetic potential, barring major breakthroughs in genetic engineering which are far from guaranteed (Grassini et al., 2013).

Freshwater is already being used in unsustainable volumes, with major aquifers depleting and rivers, from the Colorado river to the Nile and Mekong, barely reaching their deltas for much of the year due to overextraction, causing seawater to invade the valuable delta soils. Fertilizers such as phosphorus and potassium could become scarce and costly as more accessible sources are mined out. The use of nitrogen fertilizer could be constrained because of its disruption of natural ecosystems and its potential contribution to global warming (Bodirsky et al., 2014). Soils are being degraded because of overuse, or lost under urban development and infrastructure. And an increasing proportion of fisheries are also in a parlous state of overexploitation (FAO, 2020). All these challenges are symptoms of population pressure.

Mortality from contagious diseases could also increase. The WHO identifies both population growth and climate change among factors contributing to the increasing risk of disease epidemics (Global Preparedness Monitoring Board, 2019). Climate change is likely to spread the geographical range of major vector-borne diseases such as malaria (Caminade et al., 2014). Increased incidence of environmental crises leading to damaged infrastructure and displaced people create situations in which hygiene standards are compromised, leading to outbreaks of diseases, such as cholera. Heat and nutritional stresses lower human resistance to the diseases already in circulation. The increasing stress on wild animal populations could contribute to more novel zoonotic diseases passing to humans.

Mortality attributable directly to climate change is unlikely to affect local and global populations greatly. The majority of these extra deaths are likely to be among the old and frail. This does not make them less tragic but does mean that they have less impact on populations as few life-years are lost and they do not affect the number of potential parents. Between 2030 and 2050, the WHO estimates that climate change will cause approximately 250,000 additional deaths per year, from malnutrition, malaria, diarrhea, and heat stress (WHO, 2018a). This represents only around 0.3% of all deaths.

The climate change response could reduce mortality associated with air pollutants, which kill more than 7 million people per year: the largest environmental cause of ill-health globally, and the second leading cause of deaths from noncommunicable diseases (WHO, 2018b). Indoor smoke affects the respiratory health of roughly 3 billion users of biomass-fuel cooking. Cooking smoke is estimated to lead to 4.3 million premature deaths per year, particularly of women and small children (Bruce et al., 2015). Efforts to introduce improved stoves to reduce smoke exposure have been given a boost by climate finance, as this also reduces emissions of black carbon, a potent, if short-lived, climate forcing. Smog (comprising ozone, nitrogen oxides, and fine particulate matter), a rapidly increasing health hazard in burgeoning Asian cities, will also be eased by the electrification of transport. Coal-fired electricity generation is a major source of fine particulates ($<2.5\,\mu m$). However, even when the electricity to recharge vehicles is generated from coal, lower urban ozone levels yield positive net health effects (Schnell et al., 2019). Replacement of coal-fired electricity with renewables substantially reduces fine particulate pollution. Fine particulates have been associated with a range of health impacts, including an estimated 1.37 million cases of lung cancer per year (Lin et al., 2019). It is quite likely that more deaths will be averted through climate change mitigation efforts than will be caused by intensified weather events.

A cataclysmic escalation of deaths this century is a real risk, but climate change would be only an exacerbating factor in a complex of stresses generated principally by population pressure. For global population growth to be reversed through more deaths rather than fewer births, premature deaths would need to increase by more than 80 million per year—around 40 times the death rate experienced in the first 6 months of the COVID-19 pandemic, sustained for many years. The projected deaths from climate change are trivial by comparison with such a calamity. As an infinitely large population is not possible, avoiding this outcome depends on population growth ending through fewer births before environmental strains cause system collapse. Climate change adds to these strains and hence increases the urgency to minimize further population growth.

## 3.2 Births

Human fertility is less likely to be affected directly by climate change. Heat exposure has been found to reduce conception rates (Lam and Miron, 1996) and to marginally increase the risk of preterm delivery, low birth weight, and stillbirths (Bekkar et al., 2020; Grace et al., 2015). While this is detrimental to the health and well-being of individuals, such effects are unlikely to change birth rates significantly at the population level. If opportunities for conception are frequent, then the loss of a single pregnancy has little impact on parents' prospects for achieving their desired family size. Countries in West Africa have long maintained fertility levels above six children per woman despite the frequent separation of sexual partners when men undertake temporary migration for work or herding stock, and despite women farmers routinely working under conditions of heat stress. High fertility has persisted even in the presence of chronic undernutrition evident in the stunting of a high proportion of children. Where these fertility rates are declining, this is not due to health constraints but associated with changing attitudes to family size combined with improved health services and family planning.

It is possible that climate change will provide an additional incentive for couples to limit childbearing. This could be in response to specific natural disasters or gradual worsening of circumstances or the anticipation of future environmental challenges. There is some evidence that communities in Ethiopia and Bangladesh see smaller families as a means of adaptation to deteriorating environmental conditions (Rovin et al., 2013; Thompson and Sultana, 1996). Researchers in Pakistan attributed increased family planning use by flood-affected communities to women having greater contact with international agencies providing crisis health services (Sathar et al., 2018).

In developed countries, reducing births or even childlessness is increasingly discussed as a climate change response, emphasizing both the moral dilemma of bringing children into such troubling times (Overall, 2012; Conly, 2016; Hedberg, 2019), and the practical reality that having one child fewer than otherwise intended avoids more future greenhouse gas emissions than any other behavior change available to individuals (Murtaugh and Schlax, 2009; Wynes and Nicholas, 2017). However, to date, such sentiments are too rare to alter national fertility appreciably. Indeed, other cultural influences are operating in the other direction, including government promotion of births as a misconceived attempt to avoid population aging (Lee et al., 2014; Götmark et al., 2018) and an increasing tendency of celebrities to have large families (The Guardian, 2019).

More concerning is the risk of increased fertility if the availability of contraception were to deteriorate in situations of climate-related crises, or more generally if economies are strained by the burden of climate change. The COVID-19 pandemic has provided a salient example of how reproductive health care and contraception access can be disrupted in crises (Bateson et al., 2020; Makins and Arulkumaran, 2020). Two-thirds of countries surveyed by the WHO reported disruptions to family planning and contraception services due to the pandemic, and the UNFPA warned of up to 7 million additional unintended pregnancies worldwide (Associated Press, 2020). Reproductive health service provider Marie Stopes International estimated that globally the pandemic would cause an additional 1.5 million unsafe abortions and 3100 additional maternal deaths (Marie Stopes International, 2020). There are reports of increased child marriage as school closures leave youth idle and impoverished families seek to off-load daughters (Aljazeera, 2020). These experiences emphasize the importance of giving priority to women's reproductive health services throughout crisis responses.

## 3.3 Migration

The biggest demographic impact of climate change is likely to be on migration. The term "climate change refugees" has been widely used in anticipation of many millions of people having to relocate from areas affected by sea-level rise, drought, or flood. However, there is a tendency to overuse the term, applying it to any migrant whose previous life has been impacted in any way by climate change, regardless of whether this was the main cause of their migration.

A review of climate migration literature argued that both household capacity to leave and household vulnerability in staying can be affected by climate change or extreme weather events (Kaczan and Orgill-Meyer, 2020). The interplays between these influences, according to the authors, "help explain four key patterns seen in the empirical literature: (1) climate-induced migration is not necessarily more prevalent among poorer households; (2) climate-induced migration tends to be more prevalent for long-distance domestic moves than local or international moves; (3) slow-onset climate changes (such as droughts) are more likely to induce increased migration than rapid-onset changes (such as floods); and (4) the severity of climate shocks impacts migration in a nonlinear fashion, with impacts influenced by whether the capability or vulnerability channel dominates."

A World Bank study estimated that in sub-Saharan Africa, South Asia, and Latin America, by 2050 more than 140 million people could relocate within their country due to the effects of climate change (Kumari Rigaud et al., 2018). The authors suggest that strong climate change mitigation measures could reduce this flow to around 50 million. While sudden crises such as floods displace people temporarily, the slow-onset changes, including decreasing crop production, water shortages, and rising sea levels, have more lasting effects. The report suggests that if climate migration is anticipated and is embedded in development planning, it can play a role as a constructive adaptation rather than being a crisis response.

Nawrotzki et al. (2016) challenge the idea that climate migration is predominantly internal. Where communities have an established history of international migration, this can be their dominant response to climate-related crop failures (Nawrotzki et al., 2016). Lustgarten (2020) described a study using the same modeling framework as the World Bank report, which found that unmitigated climate change could increase international migration from Central

America to the United States by more than 1 million people between 2020 and 2050. However, this represented less than 5% of the total migrant flow of 30 million anticipated over that period. This flow was expected to increase each year regardless of climate change (Lustgarten, 2020).

Likewise, the 50–140 million internal climate migrants modeled by the World Bank represent a small proportion of the anticipated rural-to-urban migration due mainly to demographic pressure. UN projections suggest that well above 800 million people will move from rural to urban settings between 2020 and 2050, just in the three regions included in the World Bank study (United Nations Population Division, 2018).

Kirezci et al. (2020) modeled the impact of climate change and sea-level rise on the land area exposed to coastal flooding events. Only some of these areas would require permanent evacuation. They estimated that by 2100, under the worst-case climate scenario, the population exposed to such flooding events would increase by 52%, from 148 million currently to around 225 million, based on current population distribution (not allowing for population growth). They emphasize that these figures could be lowered by the construction of protective infrastructure such as sea walls. Again, population growth on the most vulnerable islands and river deltas, particularly in South and South-East Asia, will likely expose more people to this hazard than will climate change. Although coastal lowlands (less than 10 m above sea level) occupy only 2% of global land area, they contain 10% of the global population and more than 20% of the urban population of least developed countries—cities whose populations are doubling every few decades (McGranahan et al., 2007).

These studies demonstrate the close interactions between climate change and other drivers of migration, making it difficult to differentiate climate migration from economic migration. Baez et al. (2017) found that droughts increased the flow of youth migrants from rural to urban centers in Latin America. But such studies are unable to show whether this increases the cumulative flow, or merely shifts a portion of it from better growing seasons to worse. For instance, a larger than average exodus in a drought year might make it easier for the next year's cohort of youth to find employment locally.

Migration literature, particularly under the "new economics of labor migration" (NELM) theory, tends to ignore population growth as a driver of migration. Analyses typically present the decision of a household to send migrants as one of income diversification and self-insurance. Taylor (2002) sees rural-urban migration as a phenomenon driven by GDP growth and its implicit link with economic diversification, and suggests that constraints on local production and livelihoods are due to "market failures" such as inadequate market access, finance, and insurance systems. The presumption is that, without climate change or other exogenous factors undermining livelihoods, the economic situation would be stable or gradually improving due to development, and migration offers a means to enhance development. But nothing is stable where populations are growing. The climate migration literature does not discuss the common reality that the alternative to out-migration from rural areas is an ever-dwindling allocation of natural resources per household (arable land, water, or access to communal forest, pasture, or fishing resources) (Garedew et al., 2012), and the inevitable degradation of those resources due to overuse (Taddese, 2001). Equally absent is any recognition that such subdivisions and degradations over the past two generations have contributed to the impoverishment of households and their vulnerability to climate change.

A test for genuine climate change migration would be that the sending area should permanently decline in population, and this must be due to changed environmental circumstances attributable to climate change, rather than to other impacts of human behavior. These other impacts are often attributed to climate change, especially when they occur at a distance from the activity causing them. For instance, deforestation not only changes the hydrology of river catchments, increasing flooding and the seasonality of flows, but also reduces rainfall downwind across whole continents (Lawrence and Vandecar, 2015; Ellison et al., 2017; Pearce, 2018). Overextraction of groundwater is causing widespread land subsidence, in some places more than 2 m per decade, affecting agricultural lands and coastal cities (Herrera-García et al., 2021). In addition, groundwater depletion (transferring water from aquifers ultimately to oceans) contributed 13% of sea-level rise between 2000 and 2008 (Konikow, 2011). Climate change-related sea-level rise is often cited as the cause of saltwater intrusion into the groundwater of deltas and coastal plains (Chen and Mueller, 2018), where overextraction of groundwater and expansion of aquaculture are mostly responsible (Mabrouk et al., 2018; Chang et al., 2011). This is not to belittle the vulnerability of atolls and coastal lowlands to inundation from sea-level rise and increased storm surge. But we should be mindful that climate change is not the only, and often not the biggest, cause of loss of livelihoods due to environmental change.

An absence of population decline does not mean that a community is not impacted by climate change, but it means that the community has been able to adapt to live with that change. In the meantime, hardships caused by climate change and extreme weather events would likely have contributed to many households' decisions to leave, but their place has been filled by local population growth. Without climate change, perhaps the region could have sustained an even bigger population thanks to other advances that increase opportunities for local livelihoods. But without the population growth, the same advances would increase incomes and climate resilience. Such speculative hypotheticals are impossible to quantify. It is therefore not justifiable to claim climate change migration if the sending region's population does not decline.

We are left with the conclusion that climate change will be a contributing factor in net migration flows which are largely driven by population pressure. In the past, emerging economies such as Japan and South Korea, like Europe earlier, have also seen rural-urban migration to the extent that rural areas depopulate. This was driven by economic diversification providing more jobs in the cities, while the modernizing agricultural sector became less labor-intensive. But this was only possible after fertility levels fell, preventing those leaving from being instantly replaced. In the postindustrial, internet-connected future, it is not clear whether large cities will continue to be the focus of employment growth, or whether rural towns will also have more opportunity to diversify. Regional towns in Japan are achieving some rejuvenation by developing information technology and ecotourism industries (Matanle, 2017). But wherever fertility rates remain high, net migration to cities will continue, driven by a lack of rural opportunities rather than a surfeit of urban opportunities.

The total volume of international migrations will depend on the willingness of countries to receive migrants, more than on the factors motivating people to move. This is because those motivated to move already greatly overwhelm the capacity of destination countries to absorb them. Gallup polls regularly assess intentions to migrate across the world. In 2018, they reported that more than 750 million people would migrate if they could—around 15% of the world's adults (Esipova et al., 2018). This included a third of all adults surveyed in sub-Saharan Africa and more than a quarter of those in non-EU Europe, Latin America, and the Caribbean. A number of destination countries, including Australia, Canada, New Zealand, Singapore, and Switzerland, were named as a first preference by more would-be migrants than their current population. In reality, only around 3% of people live outside the country of their birth, and only around 1% have migrated from poor to rich countries (Abel and Nikola Sander, 2014). With such a constraint on flows, the demographic pressure for migration will likely persist for a long time even after national population growth ends. Climate change is likely to have some influence on who migrates, but little effect on the volume of international flows. Most people displaced due to climate change will move within their own country.

## 4 Demographic influences on vulnerability to climate change impacts

It is widely acknowledged that poor households in poor countries are most vulnerable to the impacts of climate change. This is partly because of their high reliance on the natural environment for their livelihood (Stern, 2006), and partly because they live predominantly in equatorial zones likely to incur the greatest negative effects of climate change, from extreme heat (Bathiany et al., 2018) and disrupted rainfall patterns to intensified cyclones and flood events. But high population densities and growth rates are also contributing to vulnerability (Das Gupta, 2013). With natural resources and human services stretched to their utmost to meet the needs of increasing numbers of people, the capacity to respond to crises is diminished and small disruptions to weather patterns can have large and cascading effects.

Much has been written on the threat of climate change to food and water security, but the role of population growth, and the potential to influence future population growth, often goes unmentioned (Beddington et al., 2011; Steiner et al., 2020), even when the focus is on reducing food demand (Bajželj et al., 2014; Tilman et al., 2011). Indeed, the population taboo is evident in the disappearance of population as a focus in food security literature, from a central theme half a century ago to relatively rare today (Tamburino et al., 2020). There is no doubt that the current food system is overreaching several planetary boundaries for sustainable impacts—by one estimate, current production and consumption patterns could sustainably provide a balanced diet for only 3.4 billion people (Gerten et al., 2020). Heroic shifts in production systems and dietary choices would be needed to allow sufficient food for our current population to be produced sustainably (Conijn et al., 2018). A study commissioned by The Lancet concluded that global food systems could provide healthy diets for up to 10 billion people by 2050 and remain within environmental boundaries, but it would take a global transformation of production systems, halving of food loss and waste, and red meat consumption reduced to about a third of current levels globally—all formidable challenges with low likelihood (Willett et al., 2019). The authors conclude that, on current trends and production systems, emissions from food production will nearly double by 2050.

The regions most vulnerable to critical shortages of food and water tend to be those with high population densities and growth rates. In these regions, population growth is a much greater driver of water and food insufficiency than climate change. Modeling by Gunasekara et al. (2013) concluded that small reductions in population growth could have large effects on the numbers of people exposed to acute water stress. Carter and Parker (2009, p. 676) evaluated threats to groundwater access in Africa, concluding, "The climate change impacts [on groundwater] are likely to be significant, though uncertain in direction and magnitude, while the direct and indirect impacts of demographic change on both water resources and water demand are not only known with far greater certainty, but are also likely to be much larger. The combined effects of urban population growth, rising food demands and energy costs, and consequent demand for fresh water represent real cause for alarm, and these dwarf the likely impacts of climate change on groundwater resources, at least over the first half of the 21st century" (Carter and Parker, 2009).

Fig. 10 demonstrates how dramatically the projected increase in population will affect African countries' ability to feed their own populations, in comparison with the modest changes in rainfall anticipated. The *dashed lines* in Fig. 10 demonstrate how much this challenge could

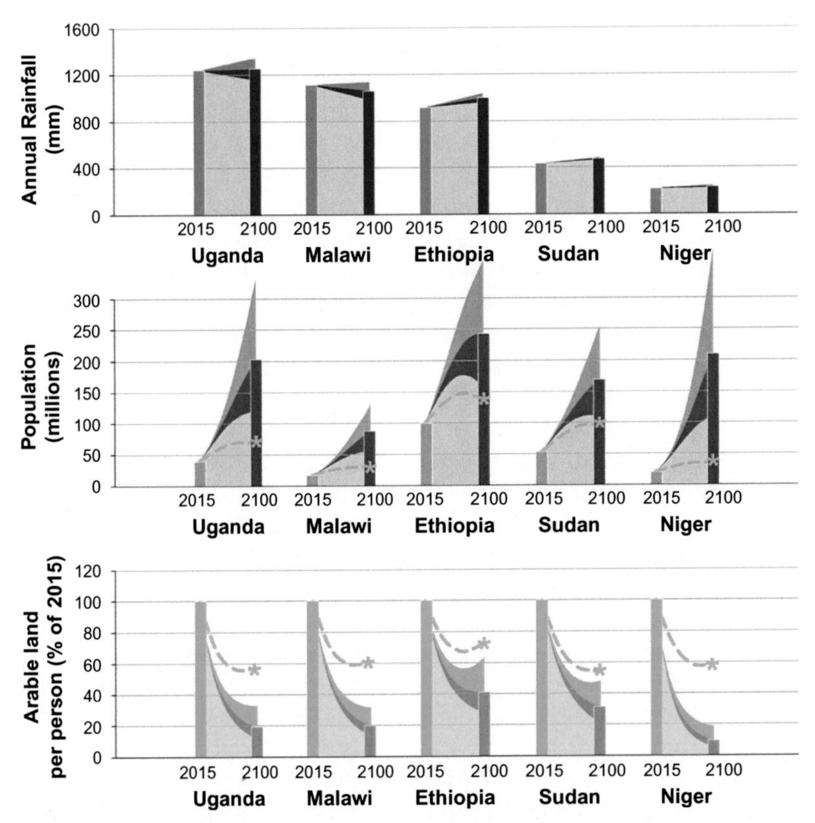


**FIG. 10** Projected change in rainfall due to climate change (from Carter and Parker, 2009) and in population (UNDESA, 2015) in five sub-Saharan African countries. Histogram bars give the central projection, and shaded areas between the bars indicate the likely range (for population, this is the 80% range using UN probabilistic projections, UNDESA, 2015). Arable land per person is calculated as a percentage of that currently available. The *dashed-line* pathways are those that would be achievable if the proactive scenario described in Fig. 8 were rapidly initiated (approximately emulating outcomes in the SSP1 and SSP5 scenarios).

be alleviated if these countries emulated the voluntary family planning successes of the past (using the same model described for Fig. 8).

Of course, these changes in annual rainfall do not do justice to the increasing unpredictability of rainfall patterns, including increasingly severe droughts and floods. Nevertheless, population growth is clearly a greater threat. Hall et al. (2017) modeled climate and population drivers of future food insufficiency, and concluded, "Very little to no difference in undernourishment projections were found when we examined future scenarios with and without the effects of climate change, suggesting population growth is the dominant driver of change." Moreland and Smith (2012) found that even a modest increase in the rate of fertility decline in Ethiopia would negate the anticipated impacts of climate change on food insecurity in that country. Thankfully, Ethiopia has since made considerable progress in extending and promoting family planning, as have Rwanda and Malawi, but most other tropical African countries are progressing more slowly.

In many areas, water scarcity will limit the further intensification of agriculture, leaving imports as the main source of increasing food supply. Food import dependence is growing in many high-fertility countries already. In terms of calories, internationally traded food has more than doubled in the past three decades (D'Odorico et al., 2014), and an increasing number of countries depend on imports for more than 25% of their cereal needs (Gardner, 2015). In years with below-average yield, the price of internationally traded cereals spikes up, placing severe stress on poor countries that depend on imported food. A flood or drought affecting major rice exporters such as Thailand and Vietnam could make food sufficiency unaffordable for hundreds of millions of West Africans: it was estimated that a 5% fall in the volume of traded rice could lead to prices rising by 17% (Bren d'Amour et al., 2016). A strong relationship exists between the global food price index and the incidence of violent unrest (Lagi et al., 2011). There is also the threat of countries banning exports in bad years to ensure domestic needs are met first (Bren d'Amour et al., 2016). Consequently, the adverse weather events caused by climate change, which might have caused ripples of hardship when most countries were self-sufficient, could turn into tidal waves of famine and violent conflict.

The relationship between natural resource scarcity and violent conflicts has been documented over many decades, in many countries (Homer-Dixon et al., 1993). Both the direct effect of population growth on the amount of land and water available per person and the indirect effect of land degradation and water depletion due to overuse contribute to this instability. Often the situation is inflamed by inequitable land titling and water access, with elites capturing the benefits of development such as irrigation infrastructure, disenfranchising communities who held traditional access. The Rwandan genocide was a case in point, escalating from severe land scarcity and power imbalances (Gasana, 2002). A 2003 study found that the risk of civil conflict was particularly elevated when land and water scarcity coincided with a youth bulge (a high proportion of those aged 15–29 in the adult population) and high rates of urban population growth (Cincotta et al., 2003). Both of these stress factors are products of rapid population growth outpacing the growth in livelihood opportunities. These researchers found that "a decline in the annual birth rate of five births per thousand people corresponded to a decline of just over 5 percent in the likelihood of civil conflict."

This nexus between population growth and climate change has already been evident in the recent unrest in the Middle East. High levels of population growth and unemployment heightened distrust between ethnic groups, laying the kindling for conflict (Friedman,

**TABLE 1** Biophysical challenges (water scarcity, peak oil, population) for selected oil-producing nations. High water and energy insecurity are indicated in bold text. Moderate water shortage is deemed to occur when availability drops below 1700 $m^3$ $cap^{-1}$ $yr^{-1}$, and severe water shortage is deemed to be below 1000 $m^3$ $cap^{-1}$ $yr^{-1}$ (Kummu et al., 2016).

| Nation | x-fold population increase 1960–2014 | 1962 Renewable freshwater per capita ($m^3$ $yr^{-1}$) | 2014 Renewable freshwater per capita ($m^3$ $yr^{-1}$) | Peak oil year | 2015 Barrels oil produced (1000s per day) | 2015 Barrels oil consumed (1000s per day) | 2015 Oil import dependence (%) |
|---|---|---|---|---|---|---|---|
| Egypt | 3.4 | **63** | **20** | 1995 | 493 | **824** | **63** |
| Iraq | 5.3 | 4587 | **998** | | 4480 | 818 | |
| Nigeria | 4.2 | 4690 | 1245 | 2005 | 1943 | 271 | |
| Saudi Arabia | 8.0 | **550** | **98** | | 12,014 | 3895 | |
| Syria | 4.1 | 1456 | **380** | 1996 | 27 | **219** | **88** |
| Yemen | 5.5 | **393** | **86** | 2002 | 22 | **168** | **88** |

*Source:* Ahmed (2017), extra data from the World Bank series EG.USE.PCAP.KG.OE.

2013). Climate change played its role in a severe drought in Syria from 2007 to 2011 (Kelley et al., 2015), which saw food imports rise steeply and prices with them. This occurred at the same time as depleted groundwater was driving many farmers from their land, and as Syria's declining oil revenue was overtaken by its oil import needs. Oil revenues have enabled countries to increase their dependence on imported food, but they have a habit of running out. Middle East analyst Nafeez Ahmed argues that the converging effects of population growth, climate change, and fossil energy depletion in these chronically water-scarce countries are setting the stage for violent upheavals and failed states (Ahmed, 2017) (Table 1).

The "Fragile States Index" scores countries on a range of political, social, and economic indicators of vulnerability to political instability. The countries that scored highest for state fragility are all experiencing high rates of population growth (Population Institute, 2015). Population density and growth rate have been described as challenge multipliers, as they exacerbate all environmental and social stresses. The stresses associated with climate change are one dimension of this convergence of vulnerability in countries suffering high demographic pressure.

Greater resilience to the impacts of climate change is often argued as a cobenefit of measures intended to promote women's reproductive rights through family planning services (De Souza, 2014). Women are particularly vulnerable to climate change impacts, and the ability to regulate their childbearing enables their greater participation in livelihoods, natural resource management, and community activities that build resilience (UNFPA, 2009). Community-level effects of smaller families are also important, such as reducing pressure on natural resources, improving child nutrition and access to education, and simply adding fewer people to the numbers exposed to environmental crises (Mogelgaard, 2018; PRB and Worldwatch, 2014).

Population, Health, and Environment (PHE) projects emphasize multisector, community-driven approaches to integrate natural resource management with livelihood diversification, gender equity, health and hygiene, and family planning (Oglethorpe et al., 2008; Gonsalves et al., 2015). PHE projects have been reported to improve climate change resilience through a number of channels, such as by strengthening social participation and cohesion, raising awareness of environmental change and conservation measures, improved hygiene and health management and income diversification, in addition to more direct effects of fewer, more widely spaced births on women's health, children's nutrition, and lessening demands on natural resources (De Souza, 2014; Hardee et al., 2018; Mohan et al., 2020). Relating family size to environmental limits has proven compelling in increasing men's support for family planning (Kock and Prost, 2017). PHE projects also tend to build male support for women's roles in livelihood enterprises and natural resource management, and greater gender equity in household decisions (Wilson Center, 2013). However, these projects have depended on short-term donor funding and rely on building relationships with existing community and sectoral organizations, creating challenges for scaling up (De Souza, 2009). Access to climate adaptation finance could magnify the impact of such projects (Mogelgaard, 2018; Patterson et al., 2019).

As part of the Least Developed Countries (LDC) Work Program of the United Nations Framework Convention on Climate Change (UNFCCC), LDCs were invited to prepare national adaptation programs of action (NAPAs) to identify priority activities for adaptation funding under the Least Developed Countries Fund (LDCF) and other mechanisms. Among the 41 NAPAs submitted by 2009, 37 highlighted population growth and density as factors increasing vulnerability. However, only two proposed projects included a population component, and none were funded (Bryant et al., 2009; Mutunga and Hardee, 2010). The sectoral structure of the framework set out for NAPAs made it difficult for a crosscutting issue such as population growth to be acknowledged. A decade later, avenues for climate adaptation funding to flow to family planning activities are yet to be opened (Patterson et al., 2019). Within the UN negotiations, inclusion of family planning is actively denounced by some agencies, with the false arguments that family planning is about control rather than choice, and that any focus on population in the climate change agenda is a denial of the responsibility of rich-world consumption patterns (CARE, 2014). Thus, the lack of attention given to population in climate adaptation activities reflects politics and ideology rather than evidence.

# 5 The influence of population change on climate change and its mitigation

Population size is a multiplier of all human impacts on the environment, including greenhouse gases. However, different people in different places have different levels of impact. In terms of greenhouse gas generation, people in developed countries and emerging economies with high fossil fuel use clearly have a much greater impact per person than the average person in India or Ethiopia. When a person in a rich country decides to have one child fewer, that decision avoids more future emissions than any other lifestyle choice they could make (Fig. 11) (Murtaugh and Schlax, 2009; Wynes and Nicholas, 2017).

As we have seen above, for people in underdeveloped countries, the decision to limit childbearing has substantial benefits for climate change adaptation, as well as for reducing poverty

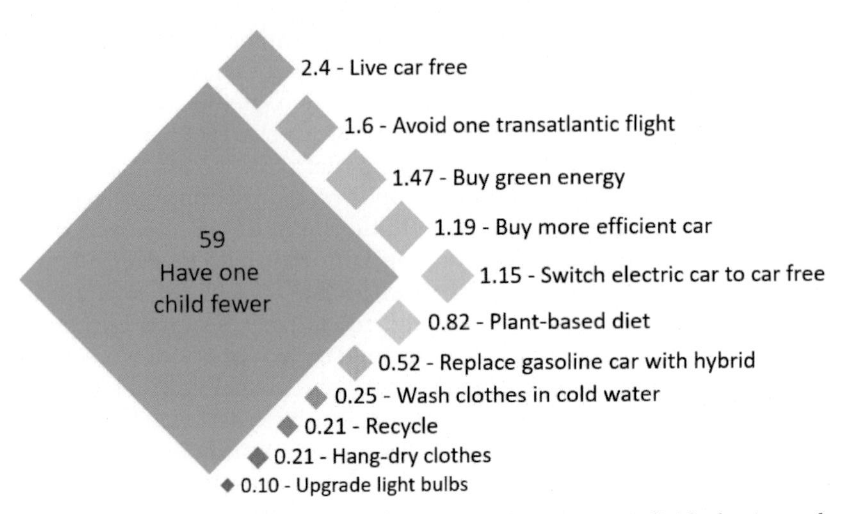


**FIG. 11** The emissions reduction (tons $CO_2e$ per year) achievable from various individual actions—the mean value of developed countries. Emissions avoided by having one child fewer assumes each parent is responsible for half the lifetime emissions of their child, a quarter of a grandchild, etc., divided among the parent's remaining years of life. *Data from Wynes, S., Nicholas, K.A., 2017. The climate mitigation gap: education and government recommendations miss the most effective individual actions. Environ. Res. Lett. 12, 074024. https://doi.org/10.1088/1748-9326/aa7541.*

and ultimately avoiding violent conflicts or famines. But the benefit for greenhouse gas emissions reduction is also significant.

Some writers have argued that people in high-fertility countries have such low per capita impact that their population growth does not matter for climate change. Advocates of this position claim that including population in the climate mitigation discourse is a means of deflecting attention from consumption behaviors of the rich, and "blaming the poor" for climate change (Monbiot, 2020). Unfortunately, they are doing the poor a disservice by reducing the political will for much-needed family planning programs. There is no trade-off between addressing fertility decline and addressing per capita emissions, as family planning programs tend to save more money than they cost. Among 16 sub-Saharan African countries, a USAID study found that fulfilling the unmet need for family planning would not only contribute materially to a range of development goals, but each dollar spent on family planning saved between two and six dollars on unneeded services for mothers and children (Moreland and Talbird, 2006). Even in the United States, where each avoided birth reduces far more emissions, a program to reduce unwanted pregnancies in teenagers and youth saved the health system around \$5.85 in perinatal care for every \$1 invested (Colorado Department of Public Health and Environment, 2015).

A study by the Global Center for Development asked whether providing much-needed additional funding for female education and family planning would be competitive with other options for climate change mitigation. Based on country-by-country estimates of emissions per person and costs per averted birth from family planning and female education, they found that "the population policy options are much less costly than almost all of the options for low-carbon energy development, including solar, wind, and nuclear power, second-generation biofuels, and carbon capture and storage. They are also cost-competitive with forest conservation and other improvements in forestry and agricultural practices" (Wheeler and Hammer, 2010). In most of the 106 developing countries studied, the cost per ton of carbon abated was less than \$10. Interestingly, although expenditure on family planning averted more births per dollar than expenditure on education, the strong synergy between these two activities meant that dividing funding between them produced the lowest cost emissions reductions.

Similarly, Wire (2009) estimated that providing family planning services to all women with an unmet need (then estimated at 215 million women globally), and thereby avoiding 75% of unintended births, would avoid 34 billion tons of carbon dioxide between 2010 and 2050 at a cost of less than \$7 per ton.

There is a fly in the ointment for these calculations, and that is the economic stimulus that reduced fertility levels are likely to promote. Extra consumption per person, particularly of fossil fuels and the products of their use, could counteract the emissions reductions from fewer births. Taking account of various ways that lower fertility might change economic growth, Casey and Galor (2017) estimated that moving from the medium to the low variant of the UN global population projection could reduce emissions from fossil fuels and industry by 10% by 2050 and 35% by 2100, despite increasing income per capita. For a similar shift in population trajectory, O'Neill et al. (2010) estimated emissions reduction around 15% by 2050 and 35%–42% by 2100, factoring in emissions from the food system as well as effects of urbanization, household size, and age structure, on a country-by-country basis, but not those resulting from economic stimulus. Neither of these studies considers the possibility that emerging economies might leapfrog fossil-fueled technologies to build postcarbon economies, and thereby avoid raising emissions per capita as incomes increase. We can conclude that providing the resources to avoid unwanted births and promote smaller families is a highly cost-effective means of reducing emissions, even when the avoided births are in communities with very low emissions per person.

Climate mitigation modeling using the SSPs has shed only indirect light on the contribution of population growth to greenhouse gas emissions. As we saw in Fig. 8, each of the five "shared socioeconomic pathways" (SSPs) has a different population projection, but it is not possible to isolate the effect of lower or higher population because many other parameters also differ between the scenarios (Samir and Lutz, 2017). Within the model, fertility decline is assumed to be driven by investments in education and health, assuming that economic development is causal on fertility but not the other way around (Jiang, 2014). Hence, the SSP framework does not lend itself to exploring the impacts of more direct investments in family planning.

However, SSP modeling does provide strong evidence that bending the population curve downward is an essential component of successful climate change mitigation. A review of scenario modeling by six separate research groups using the SSPs found that even the most extreme emissions reduction methods could not achieve less than 2°C of global warming using SSP3. This is the SSP with the highest projected population and the lowest economic development (Riahi et al., 2017). It should be noted that, since the publication of the SSPs, global population growth has most closely followed the SSP3 path.

A major contributing factor preventing strong climate change mitigation under SSP3 was the infeasibility of reversing deforestation due to increasing population pressure and demand for food (Riahi et al., 2017). Agricultural expansion is the largest driver of deforestation

(Carter et al., 2018), and forest loss is closely correlated with the increase in rural populations (FAO, 2016). In Africa, patterns of forest loss indicate that shifting agriculture, in which land was infrequently cropped and allowed to revert to forest, has seen land cropped increasingly frequently so that forest remnants shrink and become degraded before being permanently cleared (Curtis et al., 2018). Elsewhere, "commodity agriculture" predominates, but this too is largely undertaken by rural smallholders migrating into the forest frontier to produce cattle, rubber, palm oil, rice, and other cash crops, as well as illegal logging and charcoal production (Carr, 2009). Globally, population growth has been the greatest driver of the expansion of agriculture, although diet change in emerging economies (through more meat consumption) is also significant (Alexander et al., 2015; Henders et al., 2018). Thus, although only a tiny proportion of the world's population is directly involved in deforestation, regional population pressure is a push-factor, and global growth in food demand is a pull-factor, both acting to incentivize land clearing.

Designation of protected areas is less effective at preventing forest encroachment by smallholders than by industrial projects (Jayathilake et al., 2020). In regions of acute land scarcity due to population growth, such as Kenya's Chyulu Hills National Park and Tanzania's Southern Highlands, it is not politically feasible to evict squatters from conservation areas (Muriuki, 2016). In the north-western uplands of Cambodia, a 60% increase in the agricultural area occurred between 2006 and 2016 (Konga et al., 2019). In Latin America, forest frontier migrants tend to have particularly high fertility, and their children are much more likely than other rural people to become frontier migrants themselves (Lopez-Carr and Burgdorfer, 2013). The construction of new roads into forests, often associated with mining or dam projects, provides access facilitating colonization by farmers (Laurance et al., 2014). Such infrastructure development has accelerated rapidly in recent years (Laurance et al., 2015).

Greenhouse gas emissions from agriculture-related deforestation were around 1.6 Gt $CO_2$ $yr^{-1}$ in 2010–2015, and increasing rapidly in Africa (Carter et al., 2018). The rapid growth of urban populations in Africa has also greatly increased demand for charcoal, which consumes several times more wood for the same cooking fuel than direct use of wood fuel, and generates significant amounts of methane, nitrous oxide, and black carbon emissions in its production (FAO, 2017). In addition, the exposed soil continues to lose carbon, and this is exacerbated by more frequent cropping and overgrazing resulting from population growth. The combination of soil carbon loss and deforestation contributed to a net loss of 1.65 Gt of carbon from Africa's tropical zone in 2016 (Palmer et al., 2019). An estimated 116 Gt of carbon (425 Gt $CO_2$) has been released from soils over the history of agriculture (Amundson and Biardeau, 2018). Although increasing soil carbon has been widely promoted as a means of climate change mitigation, there are formidable social and logistical challenges to reversing soil carbon loss even in developed countries, and the prospects for net gains on a global scale are severely undermined by the growth in food demand (Amundson and Biardeau, 2018).

As mentioned earlier, deforestation and other reductions in vegetation also directly increase regional temperatures and alter rainfall patterns locally and over substantial distances, as well as contributing to the seasonality of river flows and the severity of flooding (Ellison et al., 2017; Lawrence and Vandecar, 2015; Mahmood et al., 2014; Pearce, 2018). Deforestation in the Amazon could reduce rainfall in the Midwest of the United States, a vital grain exporter to the world (Lawrence and Vandecar, 2015). The drying already experienced in Africa's Sahel region is due more to deforestation in East and Central Africa than to the global greenhouse effect (Zeng et al., 1999; Zheng and Eltahir, 1998).

The World Resources Institute estimated the extent to which various actions could reduce emissions from the global food system by 2050. Among the options considered was enhancing family planning to achieve replacement-rate fertility throughout Africa by 2050. This measure was estimated to contribute almost 1 Gt $CO_2e$ $yr^{-1}$ of reductions (Searchinger et al., 2018). While this was less than a tenth of their full menu of options, it was the only one offering much greater reductions, for no additional investment, in the latter half of the century. Other options, such as reducing meat consumption or lowering methane emissions from cattle, manure, and rice paddies, are one-off changes that cannot be repeated. Similarly, Project Drawdown found that providing the health and education services needed to reduce the projected global population by 1 billion could cause more emissions reduction than any other single technology deployment the project analyzed, estimated at 119 Gt between 2020 and 2050 (Hawken, 2017).

Because population growth is occurring mainly in the world's poorest countries, reducing birth rates is far from being a remedy for climate change on its own. It is nevertheless an essential component of successful climate change mitigation. Given that a dollar spent on family planning services and promotion achieves cheaper greenhouse gas emission reductions than almost all other options, and the same dollar simultaneously empowers women, saves lives, improves child nutrition and survival, reduces environmental degradation, stimulates economic development in the world's poorest communities, reduces the extent to which other climate change responses are needed, and directly saves several dollars in unneeded health services, this surely represents the climate response's low-hanging fruit.

## 6 Conclusions

Climate change is likely to have only modest impacts on the size and distribution of populations, other than in specific locations that are severely affected. In contrast, the extent of future population growth will have large impacts on both the vulnerability of communities to climate change and the emission of greenhouse gases. With or without climate change, the security and prosperity of the world's poorest countries depend on minimizing further growth in their populations.

Policies and programs are available that would accelerate the reduction in fertility in those communities where it is still high, through voluntary measures that empower women and couples to take greater control over their lives and their children's well-being. If widely supported, they could lower the global population by several billion people by the end of this century. These interventions are both highly cost-effective (generally saving more money than they cost) and widely beneficial for community development and relieving local environmental strains. The cost is modest: a doubling of current resourcing, from 1% to 2% of international aid, would make a substantial difference if supported with the political will to promote the benefits of small families and address barriers to uptake (Bongaarts, 2016). That such programs are, to date, absent from national and international climate change responses can best be explained by the prevalence of an irrational taboo against identifying population growth as a problem.

Inclusion of population issues and responses in climate change discourse is often actively denounced, and presented as morally deplorable (Monbiot, 2020). By others, it is simply dismissed as not having policy relevance, as (they argue) it is best addressed—and is being addressed—indirectly through attention to girls' education, child survival, and poverty alleviation (Rosling et al., 2018). While less accusatory, the latter commentary is similarly judgmental of population discourse as both "blaming the poor" and abetting human rights abuses in the form of involuntary population control measures. These positions are ill-informed and very damaging, not only to the prospects of communities in high-fertility countries but also to the chance of avoiding severe impacts of climate change globally.

Ending population growth is only one of many important dimensions of a successful climate change response, but it is nevertheless essential for success. If global population growth meets or exceeds the UN's current medium projection, integrated assessment models suggest no feasibility of avoiding greater than 2°C warming (Riahi et al., 2017). Lower projections, such as the IPCC's SSP2 projection, let alone SSP1, are very unlikely to be realized without substantially greater efforts to reduce desired family size and extend reliable and affordable access to contraception. The moral hazard of maintaining the population taboo has never been greater.

## References

Abel, G.J., Nikola Sander, N., 2014. Quantifying global international migration flows. Science 343, 1520. https://doi.org/10.1126/science.1248676.

Ahmed, N.M., 2017. Failing States, Collapsing Systems: Biophysical Triggers of Political Violence. Springer, p. 94. http://www.springer.com/us/book/9783319478142.

Alexander, P., Rounsevell, M.D.A., Dislich, C., et al., 2015. Drivers for global agricultural land use change: the nexus of diet, population, yield and bioenergy. Glob. Environ. Change 35, 138–147. https://doi.org/10.1016/j.gloenvcha.2015.08.011.

Aljazeera, 2020. Coronavirus despair forces girls across Asia into child marriage. Aljazeera. 2 September 2020. https://www.aljazeera.com/news/2020/09/coronavirus-despair-forces-girls-asia-child-marriage-200902071153195.html.

Amundson, R., Biardeau, L., 2018. Opinion: soil carbon sequestration is an elusive climate mitigation tool. PNAS 115 (46), 11652–11656. https://doi.org/10.1073/pnas.1815901115.

Associated Press, 2020. Millions of women lose contraceptives, abortion services amid coronavirus outbreak. LA Times. 19 August 2020. https://www.latimes.com/world-nation/story/2020-08-18/millions-of-women-lose-contraceptives-abortions-in-covid-19.

Baez, J., Caruso, G., Mueller, V., et al., 2017. Droughts augment youth migration in Northern Latin America and the Caribbean. Clim. Change 140, 423–435. https://doi.org/10.1007/s10584-016-1863-2.

Bajželj, B., Richards, K.S., Allwood, J.M., Smith, P., Dennis, J.S., Curmi, E., Gilligan, C.A., 2014. Importance of food-demand management for climate mitigation. Nat. Clim. Change 4, 924–929. http://www.nature.com/nclimate/journal/v4/n10/full/nclimate2353.html.

Bateson, D.J., Lohr, P.A., Norman, W.V., et al., 2020. The impact of COVID-19 on contraception and abortion care policy and practice: experiences from selected countries. BMJ Sex. Reprod. Health. https://doi.org/10.1136/bmjsrh-2020-200709 11 August 2020.

Bathiany, S., Dakos, V., Scheffer, M., Lenton, T.M., 2018. Climate models predict increasing temperature variability in poor countries. *Sci. Adv.* 4, eaar5809. https://doi.org/10.1126/sciadv.aar5809.

BBC, 2018. Tanzania's President Magufuli calls for end to birth control. BBC News. (10 September). https://www.bbc.com/news/world-africa-45474408.

Beddington, J., Asaduzzaman, M., Fernandez, A., Clark, M., Guillou, M., Jahn, M., Erda, L., Mamo, T., Van Bo, N., Nobre, C.A., Scholes, R., Sharma, R., Wakhungu, J., 2011. Achieving Food Security in the Face of Climate Change: Summary for Policy Makers From the Commission on Sustainable Agriculture and Climate Change. CGIAR Research Program on Climate Change, Agriculture and Food Security (CCAFS), Copenhagen, Denmark.

https://ccafs.cgiar.org/publications/achieving-food-security-face-climate-change-summary-policy-makers-commission#.X03Yq8gzbIU.

Bekkar, B., Pacheco, S., Basu, R., et al., 2020. Association of air pollution and heat exposure with preterm birth, low birth weight, and stillbirth in the US: a systematic review. JAMA Netw. Open. 3(6), e208243. https://doi.org/10.1001/jamanetworkopen.2020.8243.

Bloom, D., Canning, D., Sevilla, J., 2001. Economic Growth and the Demographic Transition. (NBER Working Paper No. 8685)National Bureau of Economic Research. https://www.nber.org/papers/w8685.

Bodirsky, B., Popp, A., Lotze-Campen, H., et al., 2014. Reactive nitrogen requirements to feed the world in 2050 and potential to mitigate nitrogen pollution. Nat. Commun. 3858 (2014), 5. https://doi.org/10.1038/ncomms4858.

Bongaarts, J., 2008. Fertility transitions in developing countries: progress or stagnation? Stud. Fam. Plan. 39 (2), 105–110. https://doi.org/10.1111/j.1728-4465.2008.00157.x.

Bongaarts, J., 2016. Slow down population growth. Nature 530, 409–412. https://www.nature.com/news/development-slow-down-population-growth-1.19415.

Borlaug, N., 1970. Nobel Lecture. 11 December 1970. https://www.nobelprize.org/prizes/peace/1970/borlaug/lecture/.

Bren d'Amour, C., Wenz, L., Kalkuhl, M., Steckel, J.C., Creutzig, F., 2016. Teleconnected food supply shocks. Environ. Res. Lett. 11, 035007. https://doi.org/10.1088/1748-9326/11/3/035007.

Brown, L.R., 2012. Full Planet, Empty Plates—The New Geopolitics of Food Scarcity. W.W. Norton & Company, USA 160 pp. ISBN-10: 0393344150.

Bruce, N., Pope, D., Rehfuess, E., et al., 2015. WHO indoor air quality guidelines on household fuel combustion: strategy implications of new evidence on interventions and exposure—risk functions. Atmos. Environ. 106, 451–457. https://doi.org/10.1016/j.atmosenv.2014.08.064.

Bryant, L., Carver, L., Butler, C.D., Anage, A., 2009. Climate change and family planning: least-developed countries define the agenda. Bull. World Health Organ. 87, 852–857. https://doi.org/10.2471/BLT.08.062562.

Caminade, C., Kovats, S., Rocklov, J., Tompkins, A.M., Morse, A.P., Colón-González, F.J., Stenlund, H., Martens, P., Lloyd, S.J., 2014. Impact of climate change on global malaria distribution. PNAS 111 (9), 3286–3291. https://doi.org/10.1073/pnas.1302089111 March 4, 2014.

CARE, 2014. Choice, Not Control: Why Limiting the Fertility of Poor Populations Will Not Solve the Climate Crisis. https://insights.careinternational.org.uk/publications/choice-not-control-why-limiting-the-fertility-of-poor-populations-will-not-solve-the-climate-crisis.

Carr, D., 2009. Population and deforestation: why rural migration matters. Prog. Hum. Geogr. 33 (3), 355–378. https://doi.org/10.1177/0309132508096031.

Carter, R.C., Parker, A., 2009. Climate change, population trends and groundwater in Africa. Hydrol. Sci. J. 54 (4), 676–689. https://doi.org/10.1623/hysj.54.4.676.

Carter, S., Herold, M., Avitabile, V., De Bruin, S., De Sy, V., Kooistra, L., Rufino, M.C., 2018. Agriculture-driven deforestation in the tropics from 1990-2015: emissions, trends and uncertainties. Environ. Res. Lett. 13, 1–13. https://doi.org/10.1088/1748-9326/aa9ea4.

Casey, G., Galor, O., 2017. Is faster economic growth compatible with reductions in carbon emissions? The role of diminished population growth. Environ. Res. Lett. 12, 014003. https://doi.org/10.1088/1748-9326/12/1/014003.

Chang, S.W., Clement, P., Simpson, M.J., Lee, K.-K., 2011. Does sea-level rise have an impact on saltwater intrusion? Adv. Water Resour. 34 (10), 1283–1291. https://doi.org/10.1016/j.advwatres.2011.06.006.

Chen, J.J., Mueller, V., 2018. Climate change is making soils saltier, forcing many farmers to find new livelihoods. The Conversation. 29 November 2018. https://theconversation.com/climate-change-is-making-soils-saltier-forcing-many-farmers-to-find-new-livelihoods-106048.

Cincotta, R.P., Engelman, R., Anastasion, D., 2003. The Security Demographic: Population and Civil Conflict After the Cold War. Population Action International, Washington, DC. https://pai.org/wp-content/uploads/2012/01/The_Security_Demographic_Population_and_Civil_Conflict_After_the_Cold_War-1.pdf.

Coale, A.J., Hoover, E.M., 1958. Population Growth and Economic Development. Princeton University Press ISBN: 9780691652672.

Colorado Department of Public Health & Environment, 2015. Preventing Unintended Pregnancies is a Smart Investment. https://web.archive.org/web/20151110092403/https://www.colorado.gov/pacific/sites/default/files/HPF_FP_UP-Cost-Avoidance-and-Medicaid.pdf.

Conijn, J.G., Bindraban, P.S., Schröder, J.J., Jongschaap, R.E.E., 2018. Can our global food system meet food demand within planetary boundaries? *Agric. Ecosyst. Environ.* 251, 244–256. https://doi.org/10.1016/j.agee.2017.06.001.
Conly, S., 2016. One Child: Do We Have a Right to Have More? Oxford University Press, Oxford, UK. https://doi.org/10.1093/acprof:oso/9780190203436.001.0001.
Curtis, P.G., Slay, C.M., Harris, N.L., Tyukavina, A., Hansen, M.C., 2018. Classifying drivers of global forest loss. Science 361, 1108–1111. https://doi.org/10.1126/science.aau3445.
D'Odorico, P., Carr, J.A., Laio, F., Ridolfi, L., Vandoni, S., 2014. Feeding humanity through global food trade. Earth's Future 2, 458–469. https://doi.org/10.1002/2014EF000250.
Das Gupta, M., 2013. Population, Poverty, and Climate Change. (World Bank Policy Research Working Paper 6631) http://documents1.worldbank.org/curated/en/116181468163465130/pdf/WPS6631.pdf.
de Silva, T., Tenreyo, S., 2017. Population control policies and fertility convergence. J. Econ. Perspect. 31 (4), 205–228. https://pubs.aeaweb.org/doi/pdf/10.1257/jep.31.4.205.
de Silva, T., Tenreyo, S., 2020. The fall in global fertility: a quantitative model. Am. Econ. J. Macroecon. 12 (3), 77–109. https://doi.org/10.1257/mac.20180296.
De Souza, R.-M., 2009. The Integration Imperative: How to Improve Development Programs by Linking Population, Health and Environment. Focus on Population, Environment and Security, Woodrow Wilson Center, Washington, DC. https://www.wilsoncenter.org/sites/default/files/media/documents/publication/Focus_19_DeSouza.pdf.
De Souza, R.-M., 2014. Resilience, integrated development and family planning: building long-term solutions. Reprod. Health Matters 22 (43), 75–83. https://doi.org/10.1016/S0968-8080(14)43773-X.
Dérer, P., 2019. The Iranian Miracle: The Most Effective Family Planning Program in History? The Overpopulation Project. 21 March 2019. https://overpopulation-project.com/the-iranian-miracle-the-most-effective-family-planning-program-in-history/.
Elkamel, F., 2020. Knowledge and social change: impact of 40 years of health and population communication in Egypt. Arab Media Soc. 6 May 2020. https://www.arabmediasociety.com/knowledge-and-social-change-impact-of-40-years-of-health-and-population-communication-in-egypt/.
Ellison, D., Morris, C.E., Locatelli, B., et al., 2017. Trees, forests and water: cool insights for a hot world. Glob. Environ. Change 43 (March), 51–61. https://doi.org/10.1016/j.gloenvcha.2017.01.002.
Esipova, N., Pugliese, A., Ray, J., 2018. More than 750 million worldwide would migrate if they could. Gallup. Published Online 10 December 2018. https://news.gallup.com/poll/245255/750-million-worldwide-migrate.aspx.
Ezeh, A.C., Mberu, B.U., Emina, J.O., 2009. Stall in fertility decline in Eastern African countries: regional analysis of patterns, determinants and implications. Philos. Trans. R. Soc. B 364, 2991–3007. https://doi.org/10.1098/rstb.2009.0166.
Ezeh, A., Kissling, F., Singer, P., 2020. Why sub-Saharan Africa might exceed its projected population size by 2100. Lancet. https://doi.org/10.1016/S0140-6736(20)31522-1 (Comment).
FAO, 2016. State of the World's Forests 2016. Forests and Agriculture: Land-Use Challenges and Opportunities. FAO, Rome. www.fao.org/publications/sofo/en/.
FAO, 2017. The Charcoal Transition: Greening the Charcoal Value Chain to Mitigate Climate Change and Improve Local Livelihoods. Food and Agriculture Organization of the United Nations, Rome. http://www.fao.org/3/a-i6935e.pdf.
FAO, 2020. The State of World Fisheries and Aquaculture 2020. Food and Agriculture Organisation of the United Nations. http://www.fao.org/state-of-fisheries-aquaculture.
FAO, IFAD, UNICEF, WFP and WHO, 2020. The State of Food Security and Nutrition in the World (SOFI) Report 2020. http://www.fao.org/publications/sofi/en/.
Foley, J.A., Ramankutty, N., Brauman, K.A., et al., 2011. Solutions for a cultivated planet. Nature 478, 337–342. https://doi.org/10.1038/nature10452.
Friedman, T.L., 2013. Tell me how this ends. New York Times. 21 May 2013. http://www.nytimes.com/2013/05/22/opinion/friedman-tell-me-how-this-ends.html.
Gardner, G., 2015. Food Trade & Self-Sufficiency. WorldWatch Institute—Vital Signs Report. https://farmlandgrab.org/24639.
Garedew, E., Sandewall, M., Soderberg, U., 2012. A dynamic simulation model of land-use, population, and rural livelihoods in the central rift valley of Ethiopia. Environ. Manag. 49, 151–162. https://doi-org.ezproxy.library.uq.edu.au/10.1007/s00267-011-9783-4.

Garenne, M., 2016. Demographic dividend in Africa: macro- and microeconomic effects. N-IUSSP. December 5, 2016. http://www.niussp.org/2016/12/05/demographic-dividend-africa-macro-micro-economic-effectsdividende-demographique-en-afrique-effets-economiques-macro-et-micro/.

Gasana, J., 2002. Remember Rwanda? World Watch Mag. 15 (5), 24–33. https://www.academia.edu/28702224/WORLD_at_BULLET_WATCH_WORLD_at_BULLET_WATCH_Remember_Rwanda_Working_for_a_Sustainable_Future.

Gerten, D., Heck, V., Jägermeyr, J., et al., 2020. Feeding ten billion people is possible within four terrestrial planetary boundaries. Nat. Sustain. 3, 200–208. https://doi.org/10.1038/s41893-019-0465-1.

Gietel-Basten, S., Sobotka, T., 2020. Uncertain population futures: critical reflections on the IHME Scenarios of future fertility, mortality, migration and population trends from 2017 to 2100. SocArXiv preprint. https://osf.io/preprints/socarxiv/5syef/.

Global Preparedness Monitoring Board, 2019. A World at Risk: Annual Report on Global Preparedness for Health Emergencies. World Health Organization, Geneva. https://apps.who.int/gpmb/assets/annual_report/GPMB_annualreport_2019.pdf.

Gonsalves, L., Donovan, S., Ryan, V., Winch, P., 2015. Integrating population, health, and environment programs with contraceptive distribution in rural Ethiopia: a qualitative case study. Stud. Fam. Plann. 46 (1), 41–54. http://www.jstor.org/stable/24642202.

Götmark, F., Cafaro, P., O'Sullivan, J., 2018. Aging human populations: good for us, good for the earth. Trends Ecol. Evol. 33 (11), 851–862. https://doi.org/10.1016/j.tree.2018.08.015.

Grace, K., Davenport, F., Hanson, H., Funk, C., Shukla, S., 2015. Linking climate change and health outcomes: examining the relationship between temperature, precipitation and birth weight in Africa. Glob. Environ. Change 35, 125–137.

Grassini, P., Eskridge, K., Cassman, K., 2013. Distinguishing between yield advances and yield plateaus in historical crop production trends. Nat. Commun. 4, 2918. https://doi.org/10.1038/ncomms3918.

Gunasekara, N.K., Kazama, S., Yamazaki, D., Oki, T., 2013. The effects of country-level population policy for enhancing adaptation to climate change. Hydrol. Earth Syst. Sci. 17, 4429–4440. https://doi.org/10.5194/hess-17-4429-2013.

Habumuremyi, P.D., Zenawi, M., 2012. Making family planning a national development priority. Lancet 380 (9837), 78–80. https://doi.org/10.1016/S0140-6736(12)60904-0.

Hall, C., Dawson, T.P., Macdiarmid, J.I., Matthews, R.B., Smith, P., 2017. The impact of population growth and climate change on food security in Africa: looking ahead to 2050. Int. J. Agric. Sustain. 15, 124–135. https://doi.org/10.1080/14735903.2017.1293929.

Hardee, K., Patterson, K., Schenck-Fontaine, A., et al., 2018. Family planning and resilience: associations found in a Population, Health and Environment (PHE) project in Western Tanzania. Popul. Environ. 40, 204–238. https://doi.org/10.1007/s11111-018-0310-x.

Hawken, P. (Ed.), 2017. Drawdown: The Most Comprehensive Plan Ever Proposed to Reverse Global Warming. Penguin Putnam Inc. 255 pp. ISBN: 9780143130444. https://drawdown.org/the-book.

Hedberg, T., 2019. The duty to reduce greenhouse gas emissions and the limits of permissible procreation. Essays Philos. 20 (1), eP1628. https://doi.org/10.7710/1526-0569.1629.

Henders, S., Ostwald, M., Verendel, V., Ibisch, P., 2018. Do national strategies under the UN biodiversity and climate conventions address agricultural commodity consumption as deforestation driver? Land Use Policy 70, 580–590. https://doi.org/10.1016/j.landusepol.2017.10.043.

Herrera-García, G., Ezquerro, P., Tomás, R., et al., 2021. Mapping the global threat of land subsidence. Science 371 (6524), 34–36. https://doi.org/10.1126/science.abb8549.

Herrmann, M., 2017. This is UNFPA: Demographic Transitions, Demographic Dividends and Poverty Reduction. (UNFPA—Presentation) https://www.un.org/development/desa/dspd/wp-content/uploads/sites/22/2017/04/DD-MH-2.pdf.

Homer-Dixon, T.F., Boutwell, J.H., Rathjens, G.W., 1993. Environmental change and violent conflict. Sci. Am. 268 (2), 38–45.

Im, E.-S., Pal, J.S., Eltahir, E.A.B., 2017. Deadly heat waves projected in the densely populated agricultural regions of South Asia. Sci. Adv. 3(8), e1603322. https://doi.org/10.1126/sciadv.1603322.

Jah, F., Connolly, S., Barker, K., Ryerson, W., 2014. Gender and reproductive outcomes: the effects of a radio serial drama in northern Nigeria. Int. J. Popul. Res. 2014, 326905. https://doi.org/10.1155/2014/326905.

Jayathilake, H.M., Prescott, G.W., Carrasco, L.R., et al., 2020. Drivers of deforestation and degradation for 28 tropical conservation landscapes. Ambio. https://doi.org/10.1007/s13280-020-01325-9.

Jiang, L., 2014. Internal consistency of demographic assumptions in the shared socioeconomic pathways. Popul. Environ. 35, 261–285. https://doi.org/10.1007/s11111-014-0206-3.

Kaczan, D.J., Orgill-Meyer, J., 2020. The impact of climate change on migration: a synthesis of recent empirical insights. Clim. Change 158, 281–300. https://doi.org/10.1007/s10584-019-02560-0.

Kelley, A.C., 2001. The population debate in historical perspective: revisionism revised. In: Birdsall, N., Kelley, A.C., Sinding, S.W. (Eds.), Population Matters: Demographic Change, Economic Growth and Poverty in the Developing World. Oxford University Press, pp. 24–54 ISBN: 978-0-19-926186-4.

Kelley, C.P., Mohtadi, S., Cane, M.A., Seager, R., Kushnir, Y., 2015. Climate change and the recent Syrian drought. PNAS. 201421533. https://doi.org/10.1073/pnas.1421533112.

Kirezci, E., Young, I.R., Ranasinghe, R., et al., 2020. Projections of global-scale extreme sea levels and resulting episodic coastal flooding over the 21st Century. Sci. Rep. 10, 11629. https://doi.org/10.1038/s41598-020-67736-6.

Kock, L., Prost, A., 2017. Family planning and the Samburu: a qualitative study exploring the thoughts of men on a population health and environment programme in rural Kenya. Int. J. Environ. Res. Public Health 14 (5), 528. https://doi.org/10.3390/ijerph14050528.

Kokabisaghi, F., 2017. Right to sexual and reproductive health in new population policies of Iran. J. Public Health Policy 38 (2), 240–256. https://doi.org/10.1057/s41271-017-0068-x.

Konga, R., Diepart, J.-C., Castella, J.-C., Lestrelin, G., Tivet, F., Belmain, E., Bégué, A., 2019. Understanding the drivers of deforestation and agricultural transformations in the Northwestern uplands of Cambodia. Appl. Geogr. 102, 84–98. https://doi.org/10.1016/j.apgeog.2018.12.006.

Konikow, L.F., 2011. Contribution of global groundwater depletion since 1900 to sea-level rise. Geophys. Res. Lett. 38, L17401. https://doi.org/10.1029/2011GL048604.

Kumari Rigaud, K., de Sherbinin, A., Jones, B., Bergmann, J., Clement, V., Ober, K., Schewe, J., Adamo, S., McCusker, B., Heuser, S., Midgley, A., 2018. Groundswell: Preparing for Internal Climate Migration. The World Bank, Washington, DC. https://www.worldbank.org/en/news/infographic/2018/03/19/groundswell—preparing-for-internal-climate-migration.

Kummu, M., Guillaume, J., de Moel, H., et al., 2016. The world's road to water scarcity: shortage and stress in the 20th century and pathways towards sustainability. Nat. Sci. Rep. 6, 38495. https://doi.org/10.1038/srep38495.

Lagi, M., Bertrand, K.Z., Bar-Yam, Y., 2011. The Food Crises and Political Instability in North Africa and the Middle East. New England Complex Systems Institute. http://arxiv.org/pdf/1108.2455.pdf.

Lam, D.A., Miron, J.A., 1996. The effects of temperature on human fertility. Demography 33 (3), 291–305.

Laurance, W.F., Sayer, J., Cassman, K.G., 2014. Agricultural expansion and its impacts on tropical nature. Trends Ecol. Evol. 29, 107–116. https://doi.org/10.1016/j.tree.2013.12.001.

Laurance, W.F., Peletier-Jellema, A., Geenen, B., et al., 2015. Reducing the global environmental impacts of rapid infrastructure expansion. Curr. Biol. 25, R259–R262. https://doi.org/10.1016/j.cub.2015.02.050.

Lawrence, D., Vandecar, K., 2015. Effects of tropical deforestation on climate and agriculture. Nat. Clim. Change 5 (1), 27–36. https://doi.org/10.1038/NCLIMATE2430.

Lee, R., Mason, A., Members of the NTA Network, 2014. Is low fertility really a problem? Population aging, dependency, and consumption. Science 346 (6206), 229–234. https://doi.org/10.1126/science.1250542.

Lin, C., Lin, R., Chen, T., et al., 2019. A global perspective on coal-fired power plants and burden of lung cancer. Environ. Health 18, 9. https://doi.org/10.1186/s12940-019-0448-8.

Longden, T., Quilty, S., Haywood, P., Hunter, A., Gruen, R., 2020. Heat-related mortality: an urgent need to recognise and record. Lancet Planet. Health 4 (5), e171. https://doi.org/10.1016/S2542-5196(20)30100-5.

Lopez-Carr, D., Burgdorfer, J., 2013. Deforestation drivers: population, migration, and tropical land use. Environ. 55 (1), 3–11. https://doi.org/10.1080/00139157.2013.748385.

Lustgarten, A., 2020. The great climate migration. N. Y. Times Mag. 23 July 2020. https://www.nytimes.com/interactive/2020/07/23/magazine/climate-migration.html.

Lutz, W., Butz, W.P., KC, S., 2014. World Population & Human Capital in the Twenty-First Century: Executive Summary. IIASA, Laxenburg, Austria. 68 pp. http://pure.iiasa.ac.at/id/eprint/11189/.

Mabrouk, M., Jonoski, A., Oude Essink, G.H.P., Uhlenbrook, S., 2018. Impacts of sea level rise and groundwater extraction scenarios on fresh groundwater resources in the Nile Delta Governorates, Egypt. Water 10 (11), 1690. https://doi.org/10.3390/w10111690.

MacFarlane, K.A., O'Neil, M.L., Tekdemire, D., et al., 2016. Politics, policies, pronatalism, and practice: availability and accessibility of abortion and reproductive health services in Turkey. Reprod. Health Matters 24 (48), 62–70. https://doi.org/10.1016/j.rhm.2016.11.002.

Mahajan, M., 2019. The IHME in the shifting landscape of global health metrics. Global Pol. 10, 110–120. https://doi.org/10.1111/1758-5899.12605.

Mahmood, R., Pielke, R.A.S., Hubbard, K.G., et al., 2014. Land cover changes and their biogeophysical effects on climate. Int. J. Climatol. 34 (4), 929–953. https://doi.org/10.1002/joc.3736.

Makins, A., Arulkumaran, S., 2020. The negative impact of COVID-19 on contraception and sexual and reproductive health: could immediate postpartum LARCs be the solution? Int. J. Gynaecol. Obstet. 150 (2), 141–143. https://doi.org/10.1002/ijgo.13237.

Marie Stopes International, 2020. Resilience, Adaptation, Action: MSI's Response to Covid-19. https://www.mariestopes.org/media/3849/resilience-adaptation-and-action.pdf.

Matanle, P., 2017. Towards an Asia–Pacific "depopulation dividend" in the 21st century: regional growth and shrinkage in Japan and New Zealand. Asia Pac. J. 15 (6), 5018. https://apjjf.org/2017/06/Matanle.html.

Mathers, C.D., 2020. History of global burden of disease assessment at the World Health Organization. Arch. Public Health 78, 77. https://doi.org/10.1186/s13690-020-00458-3.

McGranahan, G., Balk, D., Anderson, B., 2007. The rising tide: assessing the risks of climate change and human settlements in low elevation coastal zones. Environ. Urban. 19, 17–37. https://doi.org/10.1177/0956247807076960.

McKinsey Global Institute, 2020. Climate Risk and Response: Physical Hazards and Socioeconomic Impacts. 164 pphttps://www.mckinsey.com/business-functions/sustainability/our-insights/climate-risk-and-response-physical-hazards-and-socioeconomic-impacts.

Mogelgaard, K., 2018. Challenges and Opportunities for Integrating Family Planning into Adaptation Finance. Population Reference Bureau, Washington, DC.https://www.prb.org/wp-content/uploads/2018/03/Family_Planning_and_Adaptation_Finance_Full_Report_FINAL.pdf.

Mohan, V., Hardee, K., Savitzky, C., 2020. Building community resilience to climate change: the role of a population-health-environment programme in supporting the community response to cyclone Haruna in Madagascar. Jamba 12 (1), a730. https://doi.org/10.4102/jamba.v12i1.730.

Monbiot, G., 2020. Population panic lets rich people off the hook for the climate crisis they are fuelling. The Guardian. 27 August 2020. https://www.theguardian.com/commentisfree/2020/aug/26/panic-overpopulation-climate-crisis-consumption-environment.

Mora, C., Dousset, B., Caldwell, I.R., et al., 2017. Global risk of deadly heat. Nat. Clim. Change. 7. https://doi.org/10.1038/NCLIMATE3322.

Moreland, S., Smith, E., 2012. Modeling Climate Change, Food Security and Population: Pilot Testing the Model in Ethiopia. Futures Group With MEASURE Evaluation PRH. https://www.measureevaluation.org/resources/publications/sr-12-69.

Moreland, S., Talbird, S., 2006. Achieving the Millennium Development Goals: The Contribution of Fulfilling the Unmet Need for Family Planning. USAID, Washington, DC. http://pdf.usaid.gov/pdf_docs/Pnadm175.pdf.

Muriuki, G., 2016. Chyulu Hills burning reveals Kenya's squatter dilemma. The Conversation. 10/10/2016. https://theconversation.com/chyulu-hills-burning-reveals-kenyas-squatter-dilemma-65169.

Murtaugh, P.A., Schlax, M.G., 2009. Reproduction and the carbon legacies of individuals. Glob. Environ. Change. 19. https://doi.org/10.1016/j.gloenvcha.2008.10.007.

Mutunga, C., Hardee, K., 2010. Population and reproductive health in national adaptation programmes of action (NAPAs) for climate change. Afr. J. Reprod. Health 14 (4), 133–146. https://www.ajol.info/index.php/ajrh/article/view/67847.

Nawrotzki, R.J., Runfola, D.M., Hunter, L.M., et al., 2016. Domestic and international climate migration from rural Mexico. Hum. Ecol. 44, 687–699. https://doi.org/10.1007/s10745-016-9859-0.

O'Neill, B.C., Dalton, M., Fuchs, R., Jiang, L., Pachaui, S., Zigova, K., 2010. Global demographic trends and future carbon emissions. PNAS 107, 17521–17526.

O'Neill, B.C., Kriegler, E., Riahi, K., et al., 2014. A new scenario framework for climate change research: the concept of shared socioeconomic pathways. Clim. Change 122, 387–400. https://doi.org/10.1007/s10584-013-0905-2.

O'Sullivan, J., 2016. Population projections: recipes for action, or inaction? Popul. Sustain. 1 (1), 45–57. ISSN: 2398-5496. https://jpopsus.org/full_articles/population-projections-recipes-for-action-or-inaction/.

O'Sullivan, J.N., 2012. The burden of durable asset acquisition in growing populations. Econ. Aff. 32 (1), 31–37. http://onlinelibrary.wiley.com/doi/10.1111/j.1468-0270.2011.02125.x.

O'Sullivan, J., 2014. Ageing paranoia, its fictional basis and all too real costs. In: Betts, K., Goldie, J. (Eds.), Sustainable Futures: Linking Population, Resources and the Environment. CSIRO Publishing, Melbourne, pp. 47–60. ISBN: 9781486301898. https://ebooks.publish.csiro.au/content/sustainable-futures.

O'Sullivan, J., 2019. World Population Prospects, 2019—Good News or Bad? The Overpopulation Project. 26 June 2019. https://overpopulation-project.com/world-population-prospects-2019-good-news-or-bad/.

O'Sullivan, J., 2020a. Will Global Population Peak Below 10 Billion? The Overpopulation Project. 27/07/2020. https://overpopulation-project.com/will-global-population-peak-below-10-billion/.

O'Sullivan, J., 2020b. Silver tsunami or silver lining? Why we should not fear an ageing population. Discussion paper, Sustainable Population Australia. ISBN: 978-0-6487082-3-0. https://population.org.au/discussion-papers/ageing/.

O'Sullivan, J., Martin, R., 2016. The risk of misrepresenting the demographic dividend. N-IUSSP. April 18, 2016. http://www.niussp.org/2016/04/18/the-risk-of-misrepresenting-the-demographic-dividendle-risque-dune-interpretation-erronee-du-dividende-demographique/.

Oglethorpe, J., Honzak, C., Margoluis, C., 2008. Healthy People, Healthy Ecosystems: A Manual for Integrating Health and Family Planning into Conservation Projects. WWF, Washington, DC. https://www.worldwildlife.org/publications/healthy-people-healthy-ecosystems-a-manual-on-integrating-health-and-family-planning-into-conservation-projects.

Overall, C., 2012. Why Have Children? MIT Press, Cambridge, MA. https://doi.org/10.7551/mitpress/8674.001.0001.

Palmer, P.I., Feng, L., Baker, D., et al., 2019. Net carbon emissions from African biosphere dominate pan-tropical atmospheric $CO_2$ signal. Nat. Commun. 10, 3344. https://doi.org/10.1038/s41467-019-11097-w.

Patterson, K.P., Mogelgaard, K., Kabiswa, C., Ruyoka, R., 2019. Building resilience through family planning and climate adaptation finance: systematic review and opportunity analysis. Lancet Planet. Health. https://doi.org/10.1016/S2542-5196(19)30155-X.

Pearce, F., 2018. Rivers in the sky: how deforestation is affecting global water cycles. Yale Environment. 360, 24 July 2018. https://e360.yale.edu/features/how-deforestation-affecting-global-water-cycles-climate-change.

Population Institute, 2015. Demographic Vulnerability: Where Population Growth Poses the Greatest Challenges. https://www.populationinstitute.org/demovulnerability/.

Potts, M., 1999. The population policy pendulum: needs to settle near the middle—and acknowledge the importance of numbers. BMJ 319 (7215), 933–934. https://doi.org/10.1136/bmj.319.7215.933.

Potts, M., 2014. Getting family planning and population back on track. Glob. Health: Sci. Pract. 2 (2), 145–151. https://doi.org/10.9745/GHSP-D-14-00012.

PRB, 2011–2016. World Population Datasheet. Population Reference Bureau. http://www.prb.org/Publications/Datasheets/2015/2015-world-population-data-sheet.aspx.

PRB and Worldwatch, 2014. Making the Connection: Population Dynamics and Climate Compatible Development Recommendations From an Expert Working Group. Population Reference Bureau. http://www.prb.org/pdf15/population-climate-full-paper.pdf.

Psaki, S.R., Chuang, E.K., Melnikas, A.J., Wilson, D.B., Mensch, B.S., 2019. Causal effects of education on sexual and reproductive health in low and middle-income countries: a systematic review and meta-analysis. SSM—Popul. Health. 8, 100386. https://doi.org/10.1016/j.ssmph.2019.100386.

Riahi, K., van Vuuren, D.P., Kriegler, E., et al., 2017. The shared socioeconomic pathways and their energy, land use and greenhouse gas emissions implications: an overview. Glob. Environ. Change 42, 153–168. https://doi.org/10.1016/j.gloenvcha.2016.05.009.

Robinson, A., 1974. The economic development of Malthusia. Bangladesh Dev. Stud. 2 (3), 647–660. https://www.jstor.org/stable/40794163.

Robinson, W.C., Ross, J.A. (Eds.), 2007. The Global Family Planning Revolution. World Bank, Washington, DC 496 pp. ISBN-10: 0-8213-6951-2. https://openknowledge.worldbank.org/handle/10986/6788.

Rosling, H., Rosling, O., Rosling Rönnlund, A., 2018. Factfulness: Ten Reasons We're Wrong About the World—And Why Things Are Better Than You Think. Hodder & Stoughton Ltd., London 352 pp. ISBN: 9781473637498.

Rovin, K., Hardee, K., Kidanu, A., 2013. Linking population, fertility, and family planning with adaptation to climate change: perspectives from Ethiopia. Afr. J. Reprod. Health 17 (3), 15–29.

Ryerson, W.N., 1993. What's needed to solve the population problem? The Social Contract (Summer), 277–278. https://www.thesocialcontract.com/pdf/three-four/Ryerson.pdf.

Samir, K.C., Lutz, W., 2017. The human core of the shared socioeconomic pathways: population scenarios by age, sex and level of education for all countries to 2100. Glob. Environ. Change 42, 181–192. https://doi.org/10.1016/j.gloenvcha.2014.06.004.

Sathar, Z.A., Khalil, M., Hussain, S., Sadiq, M., Khan, K., 2018. Climate Change, Resilience, and Population Dynamics in Pakistan: A Case Study of the 2010 Floods in Mianwali District. Population Council, Pakistan. https://www.popcouncil.org/uploads/pdfs/2018PGY_ClimateChangePakistan.pdf.

Sauvy, A., 1958. De Malthus à Mao Tsé-Toung. Éditions Denoël, Paris. Accessed in English translation: Fertility and survival: population problems from Malthus to Mao Tse-tung (Translation by Christine Brooke-Rose, 1961). Chatto & Windus, London 252 p.

Schnell, J.L., Naik, V., Horowitz, L.W., et al., 2019. Air quality impacts from the electrification of light-duty passenger vehicles in the United States. Atmos. Environ. 208, 95–102. https://doi.org/10.1016/j.atmosenv.2019.04.003.

Schwab, T., 2020. Are Bill Gates's billions distorting public health data? The Nation. 3 December. https://www.thenation.com/article/society/gates-covid-data-ihme/.

Searchinger, T., Waite, R., Hanson, C., Ranganathan, J., Dumas, P., Matthews, E., 2018. *Creating a Sustainable Food Future: A Menu of Solutions to Feed Nearly 10 Billion People by* 2050. World Resources Institute. (Synthesis Report), December 2018. https://www.wri.org/publication/creating-sustainable-food-future.

Simon, J., 1981. The Ultimate Resource. Princeton University Press.

Sinding, S.W., 2009. Population, poverty and economic development. Philos. Trans. R. Soc. B 364, 3023–3030. https://doi.org/10.1098/rstb.2009.0145.

Sinding, S.W., 2016. Reflections on the changing nature of the population movement. J. Popul. Sustain. 1 (1), 7–14. https://jpopsus.org/full_articles/reflections-on-the-changing-nature-of-the-population-movement/.

Steiner, A., Aguilar, G., Bomba, K., et al., 2020. Actions to Transform Food Systems Under Climate Change. CGIAR Research Program on Climate Change, Agriculture and Food Security (CCAFS), Wageningen, The Netherlands. https://cgspace.cgiar.org/bitstream/handle/10568/108489/Actions%20to%20Transform%20Food%20Systems%20Under%20Climate%20Change.pdf.

Stern, N., 2006. The Economics of Climate Change. The Stern Review. Cambridge University Press, Cambridge UK, New York USA.

Taddese, G., 2001. Land degradation: a challenge to Ethiopia. Environ. Manag. 27 (6), 815–824. https://doi.org/10.1007/s002670010190.

Tamburino, L., Bravo, G., Clough, Y., Nicholas, K.A., 2020. From population to production: 50 years of scientific literature on how to feed the world. Glob. Food Sec. 24, 100346. https://doi.org/10.1016/j.gfs.2019.100346.

Taylor, J.E., 2002. The new economics of labour migration and the role of remittances in the migration process. Int. Migr. 37 (1), 63–88. https://doi.org/10.1111/1468-2435.00066.

The Guardian, 2019. Is having five children really a middle-class status symbol? The Guardian. 9 January 2019. https://www.theguardian.com/lifeandstyle/shortcuts/2019/jan/08/five-kids-club-status-symbol-celebrity-sophie-ellis-bextor.

Thompson, P.M., Sultana, P., 1996. Distributional and social impacts of flood control in Bangladesh. Geogr. J. 16 (1), 1–13.

Thurow, L., 1986. Why the ultimate size of the world's population doesn't matter. Technol. Rev. 89, 6 22 and 29.

Tilman, D., Balzer, C., Hill, J., Befort, B.L., 2011. Global food demand and the sustainable intensification of agriculture. PNAS 108 (50), 20260–20264. https://doi.org/10.1073/pnas.1116437108.

UN Economic and Social Council, 2010. Report of the Secretary-General on the Flow of Financial Resources for Assisting in the Implementation of the Programme of Action of the International Conference on Population and Development. E/CN.9/2010/5. http://www.un.org/en/development/desa/population/documents/cpd-report/index.shtml.

UNDESA, 2013. World Population Prospects, The 2012 Revision. Population Division, United Nations Department of Economic and Social Affairs. http://www.un.org/en/development/desa/publications/world-population-prospects-the-2012-revision.html.

UNDESA, 2015. World Population Prospects: the 2015 Revision. Population Division, United Nations Department of Economic and Social Affairs. https://www.un.org/en/development/desa/publications/world-population-prospects-2015-revision.html.

UNEP, 2019. Sixth Global Environmental Outlook. United Nations Environment Programme. https://www.unenvironment.org/resources/global-environment-outlook-6.

UNFPA, 1994. Programme of Action of the International Conference on Population and Development, Cairo, 5–13 September. United Nations, New York.

UNFPA, 2009. State of the World Population 2009: Facing a Changing World—Women, Population and Climate. United Nations Population Fund. https://www.unfpa.org/sites/default/files/pub-pdf/state_of_world_population_2009.pdf.

United Nations Population Division, 2018. World Urbanisation Prospects 2018. United Nations Department of Economic and Social Affairs. https://population.un.org/wup/.

Vollset, S.E., Goren, E., Yuan, C.-W., Cao, J., Smith, A.E., Hsiao, T., 2020. Fertility, mortality, migration, and population scenarios for 195 countries and territories from 2017 to 2100: a forecasting analysis for the Global Burden of Disease Study. Lancet. https://doi.org/10.1016/S0140-6736(20)30677-2 14 July 2020.

Wheeler, D., Hammer, D., 2010. The Economics of Population Policy for Carbon Emissions Reduction in Developing Countries. Center for Global Development. (Working Paper 229), November 2010. http://www.cgdev.org/publication/economics-population-policy-carbon-emissions-reduction-developing-countries-working.

WHO, 2018a. Climate Change and Health. World Health Organisation. (Fact sheet), 1 February 2018. https://www.who.int/news-room/fact-sheets/detail/climate-change-and-health.

WHO, 2018b. COP24 Special Report: Health & Climate Change. ISBN: 978-92-4-151497-2. https://www.who.int/globalchange/publications/COP24-report-health-climate-change/en/.

Willett, W., Rockström, J., Loken, B., et al., 2019. Food in the Anthropocene: the EAT-Lancet Commission on healthy diets from sustainable food systems. Lancet 393, 447–492. https://doi.org/10.1016/S0140-6736(18)31788-4.

Wilson Center, 2013. Gorillas and Family Planning: At the Crossroads of Community Development and Conservation. Environmental Change and Security Program. www.wilsoncenter.org/event/gorillas-and-family-planning-the-crossroads-community-development-andconservation.

Wire, T., 2009. Fewer Emitters, Lower Emissions, Less Cost: Reducing Future Carbon Emissions by Investing in Family Planning—A Cost/Benefit Analysis. London School of Economics. (Masters Paper) https://www.srhr-ask-us.org/publication/fewer-emitters-lower-emissions-less-cost-reducing-future-carbon-emissions-investing-family-planning/.

Wynes, S., Nicholas, K.A., 2017. The climate mitigation gap: education and government recommendations miss the most effective individual actions. Environ. Res. Lett. 12, 074024. https://doi.org/10.1088/1748-9326/aa7541.

Xu, C., Kohler, T.A., Lenton, T.M., Scenning, J.-C., Scheffer, M., 2020. Future of the human climate niche. PNAS 117 (21), 11350–11355. https://doi.org/10.1073/pnas.1910114117.

Zaynab, H., 2018. Women's bodies have become a battleground in the fight for Iran's future. Open Democracy 50, 50. 29 August 2018. https://www.opendemocracy.net/5050/zaynab-h/womens-bodies-battleground-fight-for-iran-future.

Zeng, N., Neelin, J.D., Lau, K.M., Tucker, C.J., 1999. Enhancement of interdecadal climate variability in the Sahel by vegetation interaction. Science 286, 1537–1540. https://doi.org/10.1126/science.286.5444.1537.

Zheng, X., Eltahir, E.A.B., 1998. The role of vegetation in the dynamics of West African monsoons. J. Climate 11 (8), 2078–2096. https://doi.org/10.1175/1520-0442(1998)011<2078:TROVIT>2.0.CO;2.

CHAPTER

# 13

# Assessing the social and economic impacts of sea-level rise at a global scale—State of knowledge and challenges

*A.T. Vafeidis, C. Wolff, and S. Santamaria-Aguilar*

**Coastal Risks and Sea-Level Rise Research Group, Institute of Geography, Christian-Albrechts University, Kiel, Germany**

OUTLINE



## 1 Introduction

Global mean sea level is rising at an accelerating pace, and this rise will continue for the next centuries even if large reductions in emissions are in place (Church et al., 2014). Higher sea levels will lead to increased extreme water levels encountered during storms of high tides, thus, increasing the frequency and intensity of coastal hazards, such as flooding and erosion



*The Impacts of Climate Change*
https://doi.org/10.1016/B978-0-12-822373-4.00002-1

and potentially leading to the submergence of land and the loss of wetlands (Nicholls and Cazenave, 2010). At the same time, rapid socioeconomic development in the world's coastal regions is further exacerbating the risks to population and economy, particularly in large coastal urban centers with high concentrations of people and assets (Seto et al., 2011). Assessing the potential global impacts of sea-level rise and the risks of coastal hazards for coastal population and economy has been the subject of a large number of studies in recent years. However, the results of these studies vary considerably, leading to questions and possibly some confusion regarding the extent of potential future impacts of sea-level rise (SLR). Here, we present findings of previous assessments of the potential global impacts of sea-level rise and discuss the differences in results, while critically looking at the underlying concepts, methods, and data employed; with the aim to provide a detailed overview of the available assessments of potential impacts.

## 1.1 Sea-level rise and extreme sea levels: Causes, trends, scenarios, and impacts

The global mean sea level (GMSL) (i.e., sea level averaged over a specific period of time, which is commonly from months to years) has risen at rates of ~1.56 mm $yr^{-1}$ over the 20th century (Frederikse et al., 2020) and will continue rising as a consequence of both natural and anthropogenic changes in the climate system (Oppenheimer et al., 2019). The main driving processes of GMSL rise are the thermal expansion as the ocean warms, the melting of land-based ice-sheets and glaciers (Church et al., 2014), and changes in terrestrial water storage. Over the 20th century, change in the ocean mass due to ice-mass loss from glaciers and the Greenland and Antarctic Ice Sheets, was the largest contributor to the global SLR (~1 mm $yr^{-1}$; Frederikse et al., 2020). However, the global thermosteric contribution increased over the last half of the century and especially since 2000, caused by the increase in the global mean temperatures (Frederikse et al., 2020). The contribution of the SLR-driving processes, mainly steric effects, is not spatially homogeneous, causing spatial variations in regional sea levels.

Geodynamic processes that arise from the changes in the water mass distribution between land, ice, and the ocean cause also spatial variations of SLR. Vertical land movement due to glacial-isostatic adjustment ranges from few mm $yr^{-1}$ to more than 1 cm $yr^{-1}$ in some places, but vertical land movement due to subsidence from anthropogenic groundwater extraction in delta regions can be the main driver of relative SLR (Syvitski et al., 2009; Oppenheimer et al., 2019). Therefore, the rates of relative SLR can depart by up to ±30% of the global estimates in some coastal regions.

The potential future SLR largely depends on the emission scenario, ranging from 0.43 m (0.29–0.59 m, likely range of RCP 2.6) to 0.84 m (0.61–1.10 m, likely range RCP 8.5) by 2100. Coupled climate models are used to estimate the contribution of the steric and ocean dynamic processes under the different emission scenarios over the next century. The future SLR from the ice-sheet loss is, however, estimated offline from the temperature and precipitation changes predicted by the climate models. There are large uncertainties related to the limited knowledge of some processes of ice sheet dynamics, which are especially important for estimating the Antarctic Ice Sheet contributions and can lead to larger SLR estimates than the mentioned likely ranges (Oppenheimer et al., 2019). Besides the large uncertainties of the magnitude and timing of SLR over the next century, the long thermal response of the ocean causes a long-term (millennial) commitment to elevated SLR.

Besides the rise in GMSL, changes can also be expected in extreme sea levels (ESL) arising from the combination of the tides and surges generated from low-pressure systems and strong winds. ESL show large natural variability, mainly driven by the variability of the meteorological forcing, but climate change can alter this natural variability due to changes in storminess and SLR (Vousdoukas et al., 2018b). The latter will be the main factor increasing ESL by rising the baseline sea level, but it can also induce nonlinear amplification of ESL in coastal regions by altering the water depth and thus the frictional effects (Arns et al., 2017). Projected changes in storminess are spatially heterogeneous, intensifying and reducing the meteorological component of ESL locally and regionally, and inducing changes of a lesser magnitude than SLR (Vousdoukas et al., 2017).

## 1.2 Socioeconomic development at the coast

Previous studies agree that the coastal zone, often defined in the relevant literature as the Low Elevation Coastal Zone (LECZ) (McGranahan et al., 2007) is more densely populated than the hinterland (Small and Nicholls, 2003; McGranahan et al., 2007; Neumann et al., 2015). At the same time, population growth and economic development are faster in coastal zones as a result of current global demographic changes (Neumann et al., 2015; Merkens et al., 2016). This also holds for urban land expansion and settlements, which are growing rapidly in coastal zones, with India, China, and Africa having experienced the highest changes in terms of urban land expansion (Seto et al., 2011). As a result, two-thirds of the world's urban centers with a population of more than five million are currently located in the LECZ (Brown et al., 2013).

The above trends are likely to continue, leading to even higher concentrations of people and assets at the coast. Neumann et al. (2015) present a set of scenarios where coastal population increases by the year 2060, even under the low-end forecasts. Similarly, Merkens et al. (2016) extended the Shared Socioeconomic Pathways national population projections to the coastal zone considering recent past trends in coastal population development; they showed that population in the LECZ will increase under all future scenarios, exceeding one billion people by 2050 and ranging between 830 million and 1.1 billion people by 2100, depending on the scenario.

This rapid socioeconomic development in coastal regions leads to increased exposure and exacerbates the risks of coastal hazards if no adaptation is undertaken. Combined with sea-level rise, the potential impacts can have devastating effects on coastal population and economies, particularly in densely populated areas experiencing accelerated subsidence as a result of anthropogenic activities (Syvitski et al., 2009). Adaptation, in some form, is therefore essential for reducing the potential impacts of sea-level rise and associated hazards to coastal communities. To understand adaptation needs and to plan and guide adaptation efforts at the international level, global assessments of potential impacts under a range of physical, socioeconomic and adaptation scenarios are necessary.

## 2 Global vulnerability, impact and risk assessments—Methods and data

Global-scale understanding of risk is necessary for guiding disaster risk reduction (DRR) efforts and policies (Jongman et al., 2014; Ward et al., 2020), informing mitigation targets, assessing financial and other needs for adaptation; and loss and damage estimates (Vafeidis et al., 2019). Global-scale assessments are important for identifying regions most at risk, where further, more detailed analysis may be required; providing science-based information for designing policies, such as the European Disaster Risk Reduction Strategy (COM 2008, 130) and the EU Floods Directive (Directive 2007/60/EC); for distributing adaptation funds; and for assessing the effectiveness of potential strategies (Ward et al., 2018).

During the last two decades, following advances in data acquisition and processing as well as in computational power, there has been significant progress in the development of data, methods, and tools for conducting global-scale assessments of the potential impacts and risks related to sea-level rise. The majority of relevant studies has focused on coastal flooding, which is expected to be the main source of damages in the 21st century (Hinkel et al., 2014), while fewer studies have looked at erosion, wetland loss, and saltwater intrusion (Hinkel et al., 2013; Schuerch et al., 2018). In most cases, assessments start with the identification of the land area that lies below a certain elevation (or water level) threshold. This estimation (often termed in the literature as the "bathub" model) is based on global digital elevation models (DEM) and assumes that all areas below a certain water level are submerged. To overcome the problem of erroneously including areas that lie further inland and are not connected to the ocean, hydrological connectivity (Poulter and Halpin, 2008) is considered. For the definition of the potential floodplain, most studies have employed arbitrary water-level values or definitions of water levels of specific return periods (e.g., the 1 in 100-year extreme water level). Once the potentially flooded area has been defined, global spatial population datasets are overlaid, e.g., in a Geographic Information System (GIS), to estimate the number of people located in the coastal floodplain. Assessments of economic exposure employ further information on per capita gross domestic product (GDP) or spatial land use information. The results of this process represent the current exposure of land, population, and assets to sea-level rise and are a measure of maximum potential impacts as they usually do not account for adaptation. As such, they provide a rather static view of potential impacts.

In order to assess future potential impacts and address uncertainty related to the total amount of rising, recent studies have employed physical scenarios of sea-level rise, either arbitrary or based on published scenarios such as the Special Report on Emission Scenarios (SRES) of the Intergovernmental Panel on Climate Change (IPCC) (Nakicenovic et al., 2000) or the Representative Concentration Pathways (RCP) that involve climate model outputs (Van Vuuren et al., 2011). Socioeconomic development has not always been accounted for in future estimates, however, some studies have employed the SRES scenarios or more recently the Shared Socioeconomic Pathways (SSP) (O'Neill et al., 2017) to estimate the contribution of socio-economic development in the impacts and the associated uncertainties. This work has involved the development of narratives or extending existing narratives, which are then translated into population growth rates that are employed for the future projections.

Assessments based on the methods described above have produced estimates of the distribution of land, people, or assets per elevation, and provide a measure of exposure of people living below a certain water level or within a zone starting at the coastline and defined by distance or elevation (e.g., LECZ) or both (Small and Nicholls, 2003; Reimann et al., 2018). However, these studies do not account for the fact that certain coastal areas and populations are protected from ESL or will respond to this hazard, in some form, during the course of the century. This response, classified into four general adaptation categories, namely protection, accommodation, retreat, and nature-based solutions (Oppenheimer et al., 2019) will define to a large extent the future impacts of sea-level rise.

So far, only a few broad-scale studies have managed to incorporate adaptation into assessments of potential SLR impacts. This is due to a large number of combinations of physical, socioeconomic, and adaptation scenarios that are required in adaptation analysis; as well as to the lack of information on current adaptation measures and practices of coastal regions (Vousdoukas et al., 2020). One of the most commonly used methods is the Dynamic and Interactive Vulnerability Assessment (DIVA) modeling framework (Hinkel et al., 2014), which incorporates a global coastal database (Vafeidis et al., 2008) and evaluates through a series of indicators the main impacts of sea-level rise while also accounting for adaptation in assessments of future impacts. DIVA incorporates two main adaptation options, namely hard protection (in the form of dikes) for flooding and beach nourishment for beach erosion. Recently, new broad-scale modeling efforts have been presented that employ new global or regional datasets on hard coastal protection measures and account for adaptation in the form of dike building (Tiggeloven et al., 2020; Vousdoukas et al., 2020). Further, new studies have incorporated additional types of adaptation, for example, setback zones (Lincke et al., 2020); and have proposed methods to evaluate the potential effects of nature-based protection in reducing impacts at global scales (Van Coppenolle and Temmerman, 2020; Vafeidis et al., 2019) or have produced assessments of the benefits of certain types of nature-based protection (e.g., Beck et al., 2018).

# 3 Results of impact studies

## 3.1 Current exposure

Estimates of the actual number of people that live at the coast, although in the same order of magnitude, differ considerably. Small and Nicholls (2003) estimated population exposure to 1200 million people within 100 km of the shoreline (termed as the near-coastal population in the study). McGranahan et al. (2007) introduced the LECZ, which is a spatially more limited definition of the coastal zone and includes all land area below 10 m of elevation that is hydrologically connected to the ocean. They showed that the LECZ covers 2% of the world's land area (2.64 million $km^2$) and contained 634 million people in 2000 (10% of the global population). Nicholls et al. (2008) found approximately 400 million people to be exposed to a 5-m SLR scenario due to the collapse of the West Antarctica Ice Shield (WAIS), also considering variations due to tides. Importantly, this study was one of the first to estimate the impacts of a high-end sea-level rise scenario while accounting for adaptation and projecting impacts up to the year 2300 for different socioeconomic scenarios. The study also pointed out large differences in coastal population exposure related to the use of different population datasets. To evaluate these differences, Lichter et al. (2011) compared different population and elevation datasets to calculate area and population in the LECZ. They calculated the LECZ population to range between 557 million people and 709 million people depending on the combination of datasets, and found that population estimates were up to 20% higher depending on the DEM employed for defining the LECZ and approximately 10% depending on the population dataset employed.

An interesting finding was that offsets between DEMs and population data, for instance, due to differences in the coastlines employed for distinguishing land from ocean, could lead to large systematic errors in the assessment of exposure (Lichter et al., 2011). The study of Mondal and Tatem (2012) employed two population datasets and estimated the LECZ population between 695 million and 726 million. Further, they provided insights into regional differences between LECZ population estimates and found that small island nations showed the most substantial differences in the estimation of the LECZ population due to differences in the base data. In Europe differences were small, whereas in African countries much higher differences between the population datasets and LECZ population estimates existed due to issues related to input data resolution and quality and modeling approaches. Most of these studies have been based on DEMs such as GTOPO30 and in recent years on the Shuttle Radar Topography Mission (SRTM) DEM (or others that use SRTM as their basis), which is a surface model and does not always depict terrain elevation (Hirt, 2018). A recent study by Kulp and Strauss (2019) challenged previous findings by reporting much higher estimates of population exposure using a new DEM, which is based on SRTM but implements corrections developed specifically for coastal areas. They calculated the 2010 LECZ population to be 780 million people but found much higher numbers (1.04 million) when using the CoastalDEM, which is based on the SRTM, raising the issue of potential underestimation of exposure in previous studies.

Finally, some studies have used a different definition of the exposed area and have looked at the population living below the level of a specific return water level (Muis et al., 2017; Arns et al., 2020). Muis et al. (2017) assessed coastal flood impacts using two different extreme water level datasets and estimated the people living in the 100-year floodplain to range between 99 million and 157 million people, depending on the extreme water level dataset employed. Arns et al. (2020) conducted a similar analysis when exploring the effects of tide-surge interactions in exposure estimates and calculated 133 million people in the 100-year floodplain.

## 3.2 Future exposure

Several studies have tried in recent years to develop projections of the future population. Neumann et al. (2015) estimated that the LECZ in the year 2000 comprised 2.3% (2.6 million $km^2$) of the total land area but contained 10.9% (625 million) of the population, an estimate which is very close to that of McGranahan et al. (2007). The study of Neumann et al. (2015) also developed four different coastal population scenarios based on the Foresight project (Government Office for Science, 2011) storylines estimating that the population in the

LECZ will increase to between 879 million (scenario B; global population: 7.8 billion) and 949 million people (scenario C; global population: 8.7 billion) in the year 2030. By 2060, they found that the LECZ population is likely to approach up to 1.4 billion people (534 people per $km^2$) under the highest-end growth assumption (scenario C). More recent work on this topic involves the work of Jones and O'Neill (2016) who used the five SSPs (O'Neill et al., 2017) to develop spatial population projections until 2100. They employed the LECZ and calculated exposure to SLR to be 702 million people in 2010 (base year) and to range between 492 million and 1.15 billion people in 2100. They also showed that relative to base-year levels the largest projected global increase in LECZ population occurs in SSP3, with high regional variations depending on the scenario (e.g., exposure is highest in SSP5 in Europe, North America, and Oceania, which is the highest growth scenario in these regions; whereas exposure is greatest in SSP3 for Latin America, Asia, and Africa). Similarly, Merkens et al. (2016) developed spatial population projections for the SSPs that accounted for recent population growth trends that indicated a faster growth in coastal regions. They extended the SSP narratives using additional elements for coastal population growth and projected increased population (up to 1.1 billion people) in the LECZ for 2050 under all scenarios. By 2100, coastal population declines under some scenarios, ranging between 830 million (SSP4) and 1.18 billion people (SSP3), which is marginally higher than the estimates of Jones and O'Neill (2016) who did not specifically consider coastal growth.

## 3.3 Impacts with adaptation

Attempts to estimate the impacts of SLR while accounting for adaptation are limited. Some of the first studies that tried to introduce adaptation, in the form of hard protection (dikes) include the study of Nicholls (2004) and Nicholls et al. (2008). Nicholls (2004) used GDP per capita as a direct measure of adaptive capacity to address the lack of global databases on flood protection levels. He found the population in the hazard zone (defined as the area below the 1 in 1000-year extreme sea level) to reach up to 840 million in 2080, also considering basic scenarios of anthropogenic subsidence. He further found adaptation to reduce impacts by up to two orders of magnitude. Nicholls et al. (2008) employed the Climate Framework for Uncertainty, Negotiation, and Distribution (FUND) model to assess the direct impacts of extreme (5m) SLR, considering the interaction of (dry) land loss, wetland loss, protection costs, and human displacement, assuming perfect adaptation based on cost-benefit analysis. They found that the numbers of people displaced and the potential costs depended largely on the rate of the rise but adaptation would be efficient in reducing impacts (albeit at high monetary costs). One of the most comprehensive studies on uncertainties related to the calculation of potential impacts, with and without adaptation, is the study of Hinkel et al. (2014) who analyzed different sources of uncertainty in impacts assessments. According to their results population in the 100-year floodplain ranges between 93 million and 310 million people (depending on the dataset used). By 2100, up to 4.6% of the global population is expected to be flooded annually; however, this number reduces by two orders of magnitude if protection is implemented. In terms of economic impacts, results of previous studies are very difficult to compare due to issues that include, among others, the way adaptation costs are calculated, use of different damage curves for assessing damages to assets, discounting of future damages and adaptation costs, and differences in the implementation of adaptation measures (Diaz, 2016; Hinkel et al., 2014; Jongman et al., 2014; Nicholls, 2004; Vousdoukas et al., 2020).

## 3.4 Overview of results

Differences in the assessments discussed above are due to a number of limitations and uncertainties that range from the type of data employed for the analyses to methods and definitions of concepts such as vulnerability or risk. When considering these differences, some characteristic trends can be identified and conclusions can be drawn from the results.

In terms of current and future population and land exposure to SLR and coastal hazards, studies seem to converge toward the use of two main indicators, namely the number of people and the total land area for the LECZ and the 1 in the 100-year floodplain. In the case of the LECZ, most studies indicate current (defined as the year 2010 or 2015, depending on the underlying data used) population exposure to be in the order of 700 million people (Jones and O'Neill, 2016; Merkens et al., 2016, 2018). Although larger values have recently been presented (Kulp and Strauss, 2019) based on new approaches for correcting elevation data, suggesting that population and area exposure to SLR may have been underestimated, these still remain to be confirmed by further studies. Future exposure in the LECZ, which depends on socioeconomic scenarios, appears to peak toward the middle of the century at up to 1.4 billion people, falling below 1.2 billion by the end of the century. Accordingly, values for people living in the 1–100-year floodplain range between 100 and 160 million at the end of the 21st century. Besides the global values, studies seem to agree on regional differences in exposure, with Asia being the most exposed continent and Africa the continent with the fastest rise in exposure. The greatest uncertainties regarding results can be found in small island nations and Africa, where lack of population data or differences in the quality of census population constitute the most important issues (Mondal and Tatem, 2012).

Studies of impacts that consider adaptation, report results on various indicators that range from monetary costs (e.g., annual or total damages, adaptation building, and maintenance) to numbers of people or estimates of the area exposed to, or at risk from, flooding or erosion. These values are difficult to compare as they can vary considerably depending on, besides the underlying data and methods, the scenarios (or combinations thereof) employed in the analysis. Those studies, however, agree that proactive adaptation can reduce impacts, in some cases by several orders of magnitude, and that adaptation is in general cost-efficient and affordable. Importantly, the results of impact studies appear to be most sensitive to the choice of adaptation scenario (Hinkel et al., 2014).

## 3.5 Limitations of exposure and impact studies

Although the findings of global studies provide useful insights on exposure and impacts of SLR, as well as on the potential benefits of adaptation, they are also subject to limitations and uncertainties that need to be considered when evaluating the results and employing them for supporting policy decisions.

One of the main limitations of exposure and impact assessments is related to the underlying data, mainly the elevation and population datasets employed in the analyses. Despite significant advances in data collection and analysis, providing high-accuracy elevation data of the earth's surface still constitutes a major challenge. Most of the above studies have been based on elevation data that usually have a vertical resolution of 1 m (some new datasets, such as the Multi-Error-Removed Improved-Terrain (MERIT) DEM now report values in cm) and accuracy, which is spatially variable and ranges from a few meters to up to 16 m. Although the relative accuracy is usually higher (Hinkel et al., 2014), low absolute accuracy is still a major obstacle that can introduce large errors in the calculations (Gesch, 2018). More detailed elevation data at high horizontal resolution do exist for some regions in the world. However, these data are usually not freely available and cannot be employed in global studies due to the sheer volume of the datasets and limitations in computing power. Similarly, population data are plagued by large inconsistencies in census data quality between different regions. Modeling techniques are nevertheless being developed (Lloyd et al., 2019) to address such issues. Importantly, there exist often spatial mismatches between population and elevation datasets, which are more pronounced at the coastal boundary between land and ocean. These mismatches can significantly affect estimates of exposure to SLR (Lichter et al., 2011) and need to be addressed and documented in the initial steps of the analysis (e.g., see Wolff et al., 2018).

While the LECZ has been commonly employed for providing measures of exposure, recent studies are increasingly using the 100-year floodplain as a zone that is more indicative of the true exposure of population for specific regions. The calculation of 100-year floodplain relies on the estimation of the surge level with a 100-year return period. To this, SLR and subsidence estimates are linearly added to estimate future ESL. Uncertainties, however, in the calculation of the 100-year water level; limited information on natural and anthropogenic subsidence rates; regional differences in sea level; nonlinear interactions between SLR and surges; and the omission of the contribution of waves can lead to errors in the estimation of current and future ESL and, thus, the 100-year floodplain.

Accounting for adaptation has so far only been achieved to a limited extent. The lack of global datasets on coastal protection measures has been one major drawback (Vousdoukas et al., 2018a), with different methodological tools implementing workarounds to solve this problem (e.g., the DIVA modeling framework, which models protection). Furthermore, the evaluation of the effectiveness of adaptation has primarily been based on monetary costs and specifically cost-benefit analysis and no other decision-making frameworks have been explored at this scale. Moreover, the main adaptation option that has been researched is hard protection, usually in the form of dikes. However, dikes are not appropriate or desirable as an adaptation for many coastal types and regions as they can have numerous detrimental effects on the coastal environment. Beach nourishment has also been employed as an adaptation measure (Hinkel et al., 2013) but mainly as a measure against coastal erosion. Further options, such as nature-based solutions, retreat, or accommodation have yet to be implemented in impact assessments.

Last, the vulnerability has only been considered in the assessment of damages to assets, mainly in the form of damage functions for specific types of land use (Vousdoukas et al., 2020; Tiggeloven et al., 2020). No global studies have so far looked at the vulnerability of specific population groups or have directly accounted for the adaptive capacity of nations, regions, or impacted communities.

## 4 Conclusions and ways forward

Global assessments of the potential impacts of sea-level rise have come a long way since the Global Vulnerability Assessment study of Hoozemans et al. (1993). Significant data and methodological advances in the assessment of exposure and potential impacts have been made in recent years, allowing us to develop a clear view of the current and future magnitude of the problem. In this context, current exposure is fairly well known, not only in absolute numbers but also in terms of its spatial variation globally. Further, future exposure and potential impacts have been assessed for a wide range of physical and socioeconomic scenarios, which have helped to bound uncertainty and comprehensively explore the solution space. Studies accounting for adaptation have demonstrated the need to adapt proactively, going as far as suggesting that hard protection will be economically robust for a large part of the global coastline in the 21st century, particularly for densely populated regions with high concentrations of assets (Lincke and Hinkel, 2018), even under extreme rises in sea levels (Nicholls et al., 2008). The information on the potential future impacts of SLR has, therefore, explicitly specified the extent of the problem and has shown which areas will be most affected, as well as which regions or nations will be better able to cope with SLR. These findings can support the planning of responses to SLR and directing of global adaptation efforts and funds to those regions where they most required.

However, several questions remain unanswered and will be the focus of future research on the topic. Increasing the accuracy of these assessments is a major task, which will be supported by advances in data, methodological approaches, and computational advances. For example, methods for improving the quality of elevation data (e.g., Kulp and Strauss, 2019) as well as new datasets on elevation that will become available in the next years can lead to large improvements in identifying area exposure. To this end, information on anthropogenic subsidence is essential for understanding the true rates of sea-level change in particular locations.

Similarly, referring to the elevation data, improving the spatial representation of the population can provide a much more realistic view of the exposed population and the associated risks (Merkens and Vafeidis, 2018), particularly in combination with new, higher-resolution, population datasets (Lloyd et al., 2019). Further, recent advances in computational power have in recent years allowed the use of more complex hydrodynamic models for identifying the coastal floodplain, for modeling the nonlinear interactions between, e.g., tides and surges or including the contribution of waves (Melet et al., 2020; Vousdoukas et al., 2017). At the same time, simpler methods are being developed for the same purposes, which can provide more flexibility in conducting the large numbers of runs required for adaptation analysis.

One of the main requirements for effectively informing policies, supporting decisions, and planning pathways to adapt to SLR (Haasnoot et al., 2019) is information on the potential benefits of different adaptation options and their combinations. Besides evaluating the effects of hard protection, (and in some cases soft protection) recent work on the potential of nature-based solutions (Beck et al., 2018; van Coppenolle and Temmerman, 2019) needs to be extended and incorporated in global assessments. At the same, efforts to evaluate the effects of setback zones (Lincke et al., 2020) and retreat (Reimann et al., in preparation) can, if included in global models, provide important information for supporting policy decisions on these extensively debated topics. Such decisions can also benefit immensely from new studies of assessing existing coastal protection measures (Scussolini et al., 2016) or from information on the vulnerability of the population that is exposed or impacted. For example, new data on population characteristics that are becoming available (e.g., Hauer, 2019) can be employed to provide valuable insights regarding the current and future adaptation needs.

Lastly, it is important to note that most SLR exposure, impact, and vulnerability assessments have focused on the direct impacts of flooding and erosion. Other SLR impacts, such as wetland loss (Schuerch et al., 2018) or saltwater intrusion (Polemio and Walraevens, 2019) have received considerably less attention while indirect impacts of SLR have only been assessed in a very limited number of studies (e.g., Bosello et al., 2012). Such work can contribute significantly to highlighting the true extent of potential SLR impacts, thus, strengthening existing policy initiatives on adaptation disaster risk reduction as well as informing international funding mechanisms for adapting to climate change.

Despite the recent advancements in global studies, there is still ample space for improvements in assessing the future impacts of SLR. Nevertheless, existing knowledge gives a clear view of the potential magnitude of impacts, points out explicitly the future needs and priorities for addressing and coping with SLR impacts, and provides sufficient justification for decisions related to adaptation and distribution of funding for addressing SLR.

## References

Arns, A., Dangendorf, S., Jensen, J., Talke, S., Bender, J., Pattiaratchi, C., 2017. Sea-level rise induced amplification of coastal protection design heights. Sci. Rep. 7, 40171.

Arns, A., Wahl, T., Wolff, C., Vafeidis, A.T., Haigh, I.D., Woodworth, P., Niehuser, S., Jensen, J., 2020. Non-linear interaction modulates global extreme sea levels, coastal flood exposure, and impacts. Nat. Commun. 11.

Beck, M.W., Losada, I.J., Menendez, P., Reguero, B.G., Diaz-Simal, P., Fernandez, F., 2018. The global flood protection savings provided by coral reefs. Nat. Commun. 9.

Bosello, F., Nicholls, R.J., Richards, J., Roson, R., Tol, R.S., 2012. Economic impacts of climate change in Europe: sea-level rise. Clim. Change 112, 63–81.

Brown, S., Nicholls, R.J., Woodroffe, C.D., Hanson, S., Hinkel, J., Kebede, A.S., Neumann, B., Vafeidis, A.T., 2013. Sea-level rise impacts and responses: a global perspective. In: Coastal Hazards. Springer.

Church, J.A., Clark, P.U., Cazenave, A., Gregory, J.M., Jevrejeva, S., Levermann, A., Merrifield, M.A., Milne, G.A., Nerem, R.S., Nunn, P.D., Payne, A.J., Pfeffer, W.T., Stammer, D., Unnikrishnan, A.S., Bahr, D., Box, J.E., Bromwich, D.H., Carson, M., Collins, W., Fettweis, X., Forster, P., Gardner, A., Gehrels, W.R., Giesen, R., Gleckler, P.J., Good, P., Graversen, R.G., Greve, R., Griffies, S., Hanna, E., Hemer, M., Hock, R., Holgate, S.J., Hunter, J., Huybrechts, P., Johnson, G., Joughin, I., Kaser, G., Katsman, C., Konikow, L., Krinner, G., LE Brocq, A., Lenaerts, J., Ligtenberg, S., Little, C.M., Marzeion, B., Mcinnes, K.L., Mernild, S.H., Monselesan, D., Mottram, R., Murray, T., Myhre, G., Nicholas, J.P., Nick, F., Perrette, M., Pollard, D., Radic, V., Rae, J., Rummukainen, M., Schoof, C., Slangen, A., Van Angelen, J.H., Van De Berg, W.J., Van Den Broeke, M., Vizcano, M., Wada, Y., White, N.J., Winkelmann, R., Yin, J.J., Yoshimori, M., Zickfeld, K., 2014. Sea level change. In: Climate Change 2013: The Physical Science Basis, pp. 1137–1216.

Diaz, D.B., 2016. Estimating global damages from sea level rise with the Coastal Impact and Adaptation Model (CIAM). Clim. Change 137, 143–156.

Frederikse, T., Landerer, F., Caron, L., Adhikari, S., Parkes, D., Humphrey, V.W., Dangendorf, S., Hogarth, P., Zanna, L., Cheng, L., Wu, Y.H., 2020. The causes of sea-level rise since 1900. Nature 584, 393–397.

Gesch, D.B., 2018. Best practices for elevation-based assessments of sea-level rise and coastal flooding exposure. Front. Earth Sci. 6.

Government Office for Science, 2011. Foresight: Migration and Global Environmental Change Final Project Report. Available at http://go.nature.com/somswg.

Haasnoot, M., Brown, S., Scussolini, P., Jimenez, J.A., Vafeidis, A.T., Nicholls, R.J., 2019. Generic adaptation pathways for coastal archetypes under uncertain sea-level rise. Environ. Res. Commun. 1, 071006.
Hauer, M.E., 2019. Population projections for US counties by age, sex, and race controlled to shared socioeconomic pathway. Sci. Data 6, 190005.
Hinkel, J., Nicholls, R.J., Tol, R.S.J., Wang, Z.B., Hamilton, J.M., Boot, G., Vafeidis, A.T., McFadden, L., Ganopolski, A., Klein, R.J.T., 2013. A global analysis of erosion of sandy beaches and sea-level rise: an application of DIVA. Global Planet. Change 111, 150–158.
Hinkel, J., Lincke, D., Vafeidis, A.T., Perrette, M., Nicholls, R.J., Tol, R.S.J., Marzeion, B., Fettweis, X., Ionescu, C., Levermann, A., 2014. Coastal flood damage and adaptation costs under 21st century sea-level rise. Proc. Natl. Acad. Sci. U. S. A. 111, 3292–3297.
Hirt, C., 2018. Artefact detection in global digital elevation models (DEMs): the maximum slope approach and its application for complete screening of the SRTM v4. 1 and MERIT DEMs. Remote Sens. Environ. 207, 27–41.
Hoozemans, F., Marchand, M., Pennekamp, H., 1993. A Global Vulnerability Analysis: Vulnerability Assessment for Population, Coastal Wetlands and Rice Production on a Global Scale. Delft Hydraulics, The Netherlands.
Jones, B., O'Neill, B.C., 2016. Spatially explicit global population scenarios consistent with the shared socioeconomic pathways. Environ. Res. Lett. 11.
Jongman, B., Hochrainer-Stigler, S., Feyen, L., Aerts, J.C.J.H., Mechler, R., Botzen, W.J.W., Bouwer, L.M., Pflug, G., Rojas, R., Ward, P.J., 2014. Increasing stress on disaster-risk finance due to large floods. Nat. Clim. Change 4, 264–268.
Kulp, S.A., Strauss, B.H., 2019. New elevation data triple estimates of global vulnerability to sea-level rise and coastal flooding. Nat. Commun. 10.
Lichter, M., Vafeidis, A.T., Nicholls, R.J., Kaiser, G., 2011. Exploring data-related uncertainties in analyses of land area and population in the "Low-Elevation Coastal Zone" (LECZ). J. Coast. Res. 27, 757–768.
Lincke, D., Hinkel, J., 2018. Economically robust protection against 21st century sea-level rise. Global Environ. Change 51, 67–73.
Lincke, D., Wolff, C., Hinkel, J., Vafeidis, A., Blickensdörfer, L., Povh Skugor, D., 2020. The effectiveness of setback zones for adapting to sea-level rise in Croatia. Reg. Environ. Change 20, 1–12.
Lloyd, C.T., Chamberlain, H., Kerr, D., Yetman, G., Pistolesi, L., Stevens, F.R., Gaughan, A.E., Nieves, J.J., Hornby, G., MacManus, K., 2019. Global spatio-temporally harmonised datasets for producing high-resolution gridded population distribution datasets. Big Earth Data 3, 108–139.
Mcgranahan, G., Balk, D., Anderson, B., 2007. The rising tide: assessing the risks of climate change and human settlements in low elevation coastal zones. Environ. Urban. 19, 17–37.
Melet, A., Almar, R., Hemer, M., LE Cozannet, G., Meyssignac, B., Ruggiero, P., 2020. Contribution of wave setup to projected coastal sea level changes. J. Geophys. Res. Oceans 125, e2020JC016078.
Merkens, J.L., Vafeidis, A.T., 2018. Using Information on settlement patterns to improve the spatial distribution of population in coastal impact assessments. Sustainability 10.
Merkens, J.L., Reimann, L., Hinkel, J., Vafeidis, A.T., 2016. Gridded population projections for the coastal zone under the Shared Socioeconomic Pathways. Global Planet. Change 145, 57–66.
Merkens, J.L., Lincke, D., Hinkel, J., Brown, S., Vafeidis, A.T., 2018. Regionalisation of population growth projections in coastal exposure analysis. Clim. Change 151, 413–426.
Mondal, P., Tatem, A.J., 2012. Uncertainties in measuring populations potentially impacted by sea level rise and coastal flooding. PLoS One 7.
Muis, S., Verlaan, M., Nicholls, R.J., Brown, S., Hinkel, J., Lincke, D., Vafeidis, A.T., Scussolini, P., Winsemius, H.C., Ward, P.J., 2017. A comparison of two global datasets of extreme sea levels and resulting flood exposure. Earth's Future 5, 379–392.
Nakicenovic, N., Alcamo, J., Grubler, A., Riahi, K., Roehrl, R., Rogner, H.-H., Victor, N., 2000. Special Report on Emissions Scenarios (SRES), A Special Report of Working Group III of the Intergovernmental Panel on Climate Change. Cambridge University Press.
Neumann, B., Vafeidis, A.T., Zimmermann, J., Nicholls, R.J., 2015. Future coastal population growth and exposure to sea-level rise and coastal flooding—a global assessment. PLoS One 10.
Nicholls, R.J., 2004. Coastal flooding and wetland loss in the 21st century: changes under the SRES climate and socio-economic scenarios. Global Environ. Change 14, 69–86.
Nicholls, R.J., Cazenave, A., 2010. Sea-level rise and its impact on coastal zones. Science 328, 1517–1520.
Nicholls, R.J., Tol, R.S.J., Vafeidis, A.T., 2008. Global estimates of the impact of a collapse of the West Antarctic ice sheet: an application of FUND. Clim. Change 91, 171–191.

O'Neill, B.C., Kriegler, E., Ebi, K.L., Kemp-Benedict, E., Riahi, K., Rothman, D.S., Van Ruijven, B.J., Van Vuuren, D.P., Birkmann, J., Kok, K., Levy, M., Solecki, W., 2017. The roads ahead: narratives for shared socioeconomic pathways describing world futures in the 21st century. Global Environ. Change 42, 169–180.

Oppenheimer, M., Glavovic, B., Hinkel, J., Van De Wal, R., Magnan, A.K., Abd-Elgawad, A., Cai, R., Cifuentes-Jara, M., Deconto, R.M., Ghosh, T., Hay, J., Isla, F., Marzeion, B., Meyssignac, B., Sebesvari, Z., 2019. Sea level rise and implications for low lying islands, coasts and communities. In: IPCC Special Report on the Ocean and Cryosphere in a Changing Climate. vol. 355, pp. 126–129.

Polemio, M., Walraevens, K., 2019. Recent Research Results on Groundwater Resources and Saltwater Intrusion in a Changing Environment. Multidisciplinary Digital Publishing Institute.

Poulter, B., Halpin, P.N., 2008. Raster modelling of coastal flooding from sea-level rise. Int. J. Geogr. Inf. Sci. 22, 167–182.

Reimann, L., Merkens, J.L., Vafeidis, A.T., 2018. Regionalized shared socioeconomic pathways: narratives and spatial population projections for the Mediterranean coastal zone. Reg. Environ. Change 18, 235–245.

Schuerch, M., Spencer, T., Temmerman, S., Kirwan, M.L., Wolff, C., Lincke, D., Mcowen, C.J., Pickering, M.D., Reef, R., Vafeidis, A.T., Hinkel, J., Nicholls, R.J., Brown, S., 2018. Future response of global coastal wetlands to sea-level rise. Nature 561, 231.

Scussolini, P., Aerts, J.C., Jongman, B., Bouwer, L.M., Winsemius, H.C., De Moel, H., Ward, P.J., 2016. FLOPROS: an evolving global database of flood protection standards. Nat. Hazards Earth Syst. Sci. 16.

Seto, K.C., Fragkias, M., Guneralp, B., Reilly, M.K., 2011. A meta-analysis of global urban land expansion. PLoS One 6.

Small, C., Nicholls, R.J., 2003. A global analysis of human settlement in coastal zones. J. Coast. Res. 19, 584–599.

Syvitski, J.P.M., Kettner, A.J., Overeem, I., Hutton, E.W.H., Hannon, M.T., Brakenridge, G.R., Day, J., Vorosmarty, C., Saito, Y., Giosan, L., Nicholls, R.J., 2009. Sinking deltas due to human activities. Nat. Geosci. 2, 681–686.

Tiggeloven, T., De Moel, H., Winsemius, H.C., Eilander, D., Erkens, G., Gebremedhin, E., Loaiza, A.D., Kuzma, S., Luo, T.Y., Iceland, C., Bouwman, A., Van Huijstee, J., Ligtvoet, W., Ward, P.J., 2020. Global-scale benefit-cost analysis of coastal flood adaptation to different flood risk drivers using structural measures. Nat. Hazards Earth Syst. Sci. 20, 1025–1044.

Vafeidis, A.T., Nicholls, R.J., McFadden, L., Tol, R.S.J., Hinkel, J., Spencer, T., Grashoff, P.S., Boot, G., Klein, R.J.T., 2008. A new global coastal database for impact and vulnerability analysis to sea-level rise. J. Coast. Res. 24, 917–924.

Vafeidis, A.T., Schuerch, M., Wolff, C., Spencer, T., Merkens, J.L., Hinkel, J., Lincke, D., Brown, S., Nicholls, R.J., 2019. Water-level attenuation in global-scale assessments of exposure to coastal flooding: a sensitivity analysis. Nat. Hazards Earth Syst. Sci. 19, 973–984.

van Coppenolle, R., Temmerman, S., 2019. A global exploration of tidal wetland creation for nature-based flood risk mitigation in coastal cities. Estuar. Coast. Shelf Sci. 226, 106262.

Van Coppenolle, R., Temmerman, S., 2020. Identifying ecosystem surface areas available for nature-based flood risk mitigation in coastal cities around the world. Estuaries Coast 43, 1335–1344.

Van Vuuren, D.P., Edmonds, J., Kainuma, M., Riahi, K., Thomson, A., Hibbard, K., Hurtt, G.C., Kram, T., Krey, V., Lamarque, J.-F., 2011. The representative concentration pathways: an overview. Clim. Change 109, 5.

Vousdoukas, M.I., Mentaschi, L., Voukouvalas, E., Verlaan, M., Feyen, L., 2017. Extreme sea levels on the rise along Europe's coasts. Earth's Future 5, 304–323.

Vousdoukas, M.I., Mentaschi, L., Voukouvalas, E., Bianchi, A., Dottori, F., Feyen, L., 2018a. Climatic and socioeconomic controls of future coastal flood risk in Europe. Nat. Clim. Change 8, 776.

Vousdoukas, M.I., Mentaschi, L., Voukouvalas, E., Verlaan, M., Jevrejeva, S., Jackson, L.P., Feyen, L., 2018b. Global probabilistic projections of extreme sea levels show intensification of coastal flood hazard. Nat. Commun. 9, 1–12.

Vousdoukas, M.I., Mentaschi, L., Hinkel, J., Ward, P.J., Mongelli, I., Ciscar, J.C., Feyen, L., 2020. Economic motivation for raising coastal flood defenses in Europe. Nat. Commun. 11.

Ward, P.J., De Perez, E.C., Dottori, F., Jongman, B., Luo, T., Safaie, S., Uhlemann-Elmer, S., 2018. The need for mapping, modeling, and predicting flood hazard and risk at the global scale. In: Global Flood Hazard: Applications in Modeling, Mapping, and Forecasting, pp. 1–15.

Ward, P.J., Blauhut, V., Bloemendaal, N., Daniell, J.E., De Ruiter, M.C., Duncan, M.J., Emberson, R., Jenkins, S.F., Kirschbaum, D., Kunz, M., Mohr, S., Muis, S., Riddell, G.A., Schafer, A., Stanley, T., Veldkamp, T.I.E., Winsemius, H.C., 2020. Review article: natural hazard risk assessments at the global scale. Nat. Hazards Earth Syst. Sci. 20, 1069–1096.

Wolff, C., Vafeidis, A.T., Muis, S., Lincke, D., Satta, A., Lionello, P., Jimenez, J.A., Conte, D., Hinkel, J., 2018. A Mediterranean coastal database for assessing the impacts of sea-level rise and associated hazards. Sci. Data 5.

CHAPTER

14

# Societal adaptation to climate change

*Julie L. Drolet*

Faculty of Social Work, University of Calgary, Edmonton, AB, Canada

OUTLINE



## 1 Adaptation to climate change

Climate change is one of the greatest social and ecological challenges of the 21st century (Dietz et al., 2020). Scientific knowledge, and the realities of climate change, has contributed to new perspectives and understandings that have shifted over time. Debates focused on whether climate change was occurring or will occur at all to the need for climate change adaptation and mitigation strategies. With the release of the Fourth Assessment Report of the Intergovernmental Panel on Climate Change (IPCC) in 2007, more public and political discussions affirmed that there is and will be human-induced climate change (Frommer, 2013). The Intergovernmental Panel on Climate Change (IPCC) Fifth Assessment Report published in 2014 provides an integrated view of climate change, climate change adaptation and

*The Impacts of Climate Change*
https://doi.org/10.1016/B978-0-12-822373-4.00011-2

mitigation, and future pathways for adaptation, mitigation, and sustainable development. Since the preindustrial era, anthropogenic greenhouse gas (GHG) emissions have driven large increases in the atmospheric concentrations of carbon dioxide ($CO_2$), methane ($CH_4$), and nitrous oxide ($N_2O$), due largely to economic growth and population growth. The unprecedented nature of these changes in the last 800,000 years, and their effects in the climate system, are recognized as the extremely likely cause of global warming since the mid-20th century (IPCC, 2014, p. 4). Specifically, the Earth's average temperature increased by 1°C above preindustrial levels due to GHG emissions (IPCC, 2018). This chapter acknowledges the scientific evidence of climate change as "fact" and will focus on the need for societal adaptation to climate change given this reality.

Article 1 of the United Nations Framework Convention on Climate Change (UNFCC) defines climate change as "a change of climate, which is attributed directly or indirectly to human activity that alters the composition of the global atmosphere and which is in addition to natural climate variability observed over comparable time periods" (UNFCC, 1992, p. 4).

Adaptation refers to the process of adjustment to actual or expected climate and its effects in order to either lessen or avoid harm or exploit beneficial opportunities (IPCC, 2014, p. 118). In human systems, adaptation seeks to moderate or avoid harm or exploit beneficial opportunities (IPCC, 2014, p. 118). Adaptation measures can reduce the risks of climate change impacts, but there are limits to its effectiveness, especially with greater magnitudes and rates of climate change. It is necessary to adopt a long-term perspective, in the context of sustainable development, in order to increase the likelihood that more immediate adaptation actions will also enhance future options and preparedness (IPCC, 2014, p. 19).

Adaptation is place- and context-specific, and there is no "one-size-fits-all" approach to reduce risks in all communities and settings. Locally, the effects of climate change vary, depending upon the region, with communities experiencing the impacts of climate change differently and at various degrees (Drolet and Sampson, 2014). Lebel (2012) defines local adaptation as the "decision-making processes and actions undertaken to address current or maintain capacities to deal with future change or disturbances to a local social-ecological system arising from climate change" (p. 1058). Societal adaptation to climate change is needed to contribute to the well-being of people, infrastructures, and services now and in the future. However, too often, climate change is associated with loss—depletion, destruction, dispossession, disappearance, and collapse (Elliott, 2018). For example, "loss of place" occurs when a community disappears due to sea level rise in coastal regions worldwide. Alston et al. (2018) discuss the need for social workers and others to be aware of tangible and intangible losses in disaster recovery efforts. Adaptation research since the IPCC Fourth Assessment Report has evolved from engineering and technological adaptation pathways to include more ecosystem-based, institutional, and social measures, and the range of specific adaptation measures has also expanded, along with the links to sustainable development (IPCC, 2014).

Societal adaptation to climate change involves planning and implementation that can be enhanced through complementary actions from individuals to governments. National governments serve an important role in coordinating adaptation efforts, by protecting vulnerable groups, by supporting economic diversification and by providing information, policy and legal frameworks and financial support (IPCC, 2014). Local government efforts are critical to advance progress in adaptation given their roles in scaling up adaptation of communities and households and in managing risks. Adaptation must address the underlying factors that determine chronic poverty, vulnerability, and adaptive capacity—the ability to undertake adaptations or system changes (Tschakert and Dietrich, 2010).

## 2 Social development

Social development is a process that involves all levels of institutions, from national governments to diverse civil society organizations, to build an equitable and just society through the apportion of economic opportunities and social services while addressing power imbalance (United Nations, 2005). Social development utilizes and/or changes the processes of societal institutions and systems, through policies and programs, to strengthen the capabilities and capacities of individuals, families, and communities (Drolet and Sampson, 2014). Integrating a social development perspective into climate change adaptation and mitigation programs can improve the design and implementation of climate change response measures while promoting social development goals (Pozarny, 2016). It is critical to adopt a social analysis lens to emphasize issues of equity, social justice, and engagement among countries and among marginalized and vulnerable populations. By promoting sustainable development and climate change adaptation through the social development approach, it possible to recognize that social and ecological systems are interconnected and intrinsically linked to human rights, social justice, and environmental justice (Drolet and Sampson, 2014; Miller et al., 2012).

Climate change adaptation strategies need to acknowledge the different capacities of men and women to cope with the adverse effects and to acknowledge their ability to become actively engaged in community efforts to promote social development (Drolet and Sampson, 2014). Affected community members revealed an interest in sharing local knowledge and practices in support of sustainable social development and recognized the interdependency of social, economic, and environmental considerations (Drolet and Sampson, 2014). Drolet (2012) discussed how community actions to implement practical solutions are already being made, but there is a need to support long-term sustainable development in collaboration with other social, economic, gender, and health considerations to promote adaptation, mitigation, and community resilience.

## 3 Indigenous knowledge systems and practices

Indigenous, local, and traditional knowledge systems and practices, including indigenous peoples' holistic view of community and environment, are a major resource for adapting to climate change, but these have not been used consistently in existing adaptation efforts (IPCC, 2014). Indigenous knowledge, observations, and interpretations are significant for understanding livelihoods, security, and well-being, which is essential for adaptation (Raygorodetsky, 2011). Integrating indigenous and traditional knowledge with existing practices increases the effectiveness of adaptation (IPCC, 2014, p. 19). Indigenous peoples are becoming recognized as "agents of change" in achieving strong and meaningful climate action (ILO, 2019).

## 4 Adaptation and mitigation

Adaptation and mitigation are considered two complementary strategies for responding to climate change. Mitigation consists in addressing the causes of climate change and adaptation refers to coping with the ill-effects of climate change (Jagers and Duus-Otterström, 2008). Mitigation is the process of reducing emissions or enhancing sinks[a] of GHGs in order to limit future climate change. Mitigation is necessary to reduce emissions over the next few decades in order to reduce climate risks in the future. Both adaptation and mitigation can reduce and manage the risks of climate change impacts. Yet adaptation and mitigation can also create other risks, as well as benefits, which will be further discussed later in this chapter.

## 5 Impacts of climate change

In recent decades, changes in climate have caused impacts on natural and human systems on all continents and across the oceans. Impacts are due to observed climate change and the sensitivity of natural and human systems to changing climate (IPCC, 2014, p. 6).

Climate change impacts, such as sea-level rise and changes in temperature and precipitation, will change life on earth. To limit global warming requires rapid societal transitions in energy, land, urban, and key infrastructure and industrial systems.

Today, some ecosystems and cultures are already at risk from climate change, and additional warming will continue to increase the threats and risks. With increasing warming, some physical and ecological systems are at risk of abrupt and/or irreversible changes. Systems with limited adaptive capacity, such as Artic sea ice and coral reefs, are subject to high risks with additional warming. Coral reefs are one of the most threatened ecosystems on Earth due to temperature changes causing coral bleaching. Arctic ecosystems are already experiencing irreversible shifts, as melting glaciers and ice sheets are adding water to the ocean, contributing to rising seas that directly threaten lives and livelihoods. Terrestrial species are also sensitive to the rate of warming, and marine species to the rate and degree of ocean acidification and coastal systems to sea level rise. For example, Pacific salmon from the Mexican border to the Canadian border are facing increasing pressure due to climate change (Berwyn, 2019).

## 6 Extreme weather events

Changes in many extreme weather and climate events have been observed since about 1950. Some of these changes have been linked to human influences, including a decrease in cold temperature extremes, an increase in warm temperature extremes, an increase in extreme high sea levels and an increase in the number of heavy precipitation events in a number of regions (IPCC, 2014, p. 7). Many extreme weather events have and will become more frequent and more intense due to climate change.

[a] A sink refers to any process, activity or mechanism that removes a greenhouse gas, an aerosol or a precursor of GHG or aerosol from the atmosphere (IPCC, 2014, p. 127).

The number of climate-related disasters has tripled in the last 30 years (Oxfam, 2020). In 2019, Cyclone Idai and Kenneth affected millions in Southern Africa, leaving populations in Zimbabwe, Malawi, and Mozambique without food or basic services (OCHA, 2019). The 2020 Australian bushfires destroyed more than 10 million hectares, burning homes and entire communities, and affecting millions in a smoke haze (Cave, 2020). Such catastrophic events are forcing populations to image an entirely new way of life as a result of "the unfolding wings of climate change" (Cave, 2020).

The cumulative effect of disasters produces a significant personal, material, and economic strain on individuals, communities, and the fiscal capacity of all levels of governments (Drolet, 2019). In Canada, extreme weather and climate events have contributed to the 2013 floods and the 2016 wildfires in the province of Alberta, which are recognized as among the worst disasters in Canadian history (Brown et al., 2019; Lalani and Drolet, 2019; Drolet et al., 2020). The 2013 flood was the result of heavy rainfall and thunderstorms, coupled with rapid mountain snowmelt, which contributed to the overflow of rivers and flood conditions in Southern Alberta. The flood led to severe disruptions, and many diverse populations faced socioeconomic impacts, trauma, job loss, loss and grief, and housing instability as a result of the flood and its aftermath (Drolet et al., 2020). The 2016 wildfire spread quickly, resulting in a mandatory evacuation of all 88,000 residents in Fort McMurray. Research on the mental health effects of the 2016 wildfire on children and youth found significant posttraumatic stress disorder and other negative mental health effects (e.g., depression, anxiety, alcohol/substance use) 18 months after the wildfire (Brown et al., 2019).

In Western Canada, warmer winter temperatures contributed to the Mountain Pine Beetle epidemic in British Columbia (BC) with severe impacts at the community level (Drolet, 2012). In the past, the larvae of the beetle were killed during the coldest winter periods of −30°C temperatures or colder. The rate of spread and severity of the beetle outbreak is the largest recorded in the history of Canada, infesting over 20 million hectares of pine forest in Western Canada (Dhar et al., 2016). Climate change contributed to the outbreak in BC with warmer winters that allowed the beetle to reproduce at higher rates with far reaching and severe consequences for the forest industry (Dhar et al., 2016).

According to the IPCC (2014), climate change-related risks from extreme events, such as heat waves, heavy precipitation, and coastal flooding, are already moderate; risks associated with some types of extreme events increase progressively with further warming. Tschumi and Zscheischler (2019) found that a link between recorded climate-related disasters and observed climate data, which is a link between climate and impacts communities. For example, in 2019, many countries in South Asia were affected by massive floods and landslides that forced 12 million people from their homes in India, Nepal and Bangladesh in 2019 (Oxfam, 2020). Approximately, one third of Bangladesh was underwater in 2018 and in 2020 due to flooding in the monsoon season that was intensified by rising sea surface temperatures in South Asia that contributed to more moisture in the atmosphere and torrential rainfall (Boyle, 2020).

But it is not only the increase in extreme weather events that imperils socioeconomic and socioecological systems. Also, the gradual or "incremental" change in climate conditions such as the raise in temperature and the change in precipitation patterns will affect society, the economy, and environment in new ways. In sub-Saharan Africa, longer droughts, diminished groundwater supply, and soil degradation will have increasingly negative impacts on crops and livestock. The severe droughts of 2011, 2017, and 2019 wiped out crops and livestock, forcing people from their homes due to acute food and water shortages. The Food and Agriculture Organization (FAO) reported that agriculture is particularly vulnerable to disasters because of its' heavy reliance on weather, climate, land, and water, as disasters reduce production and direct economic losses for farmers, with cascading effects on rural livelihoods (FAO, 2017).

## 7 Vulnerability to climate risks

Climate risks are unevenly distributed between groups of people and between regions, affecting disadvantaged people and disadvantaged communities in greater ways. This is particularly important in the social work profession as vulnerable and marginalized populations and communities are increasingly affected by extreme weather events and climate impacts. The human, social, and structural dimensions are central in addressing vulnerabilities, as people's health and well-being suffer as a result of inequalities, poverty, and unsustainable development practices (Drolet et al., 2013). Differences in vulnerability and exposure arise from nonclimatic factors and from multidimensional inequalities often produced by uneven development processes. These differences shape differential risks from climate change. People who are socially, economically, culturally, politically, institutionally, or otherwise marginalized are especially vulnerable to climate change and also to some adaptation and mitigation responses. This heightened vulnerability is rarely due to a single cause but intersecting social processes that result in inequalities in socioeconomic status and income, as well as in exposure. Such social processes include, for example, discrimination on the basis of gender, class, ethnicity, age, and (dis)ability (IPCC, 2014, p. 54).

The character and severity of impacts from climate change and extreme events emerge from risk that depends not only on climate-related hazards but also on exposure (people and assets at risk) and vulnerability (susceptibility to harm) of human and natural systems (IPCC, 2014, p. 54). Exposure and vulnerability are influenced by a wide range of social, economic, and cultural factors and processes including wealth and its distribution across society, demographics, migration, access to technology and information, employment patterns, the quality of adaptive responses, societal values, governance structures, and institutions (IPCC, 2014, p. 54).

Societal adaptation to climate change also needs to consider human health, security, and livelihoods. Adaptation options that focus on strengthening existing delivery systems and institutions, as well as insurance and social protection strategies, can improve health, security and livelihoods. The most effective way to reduce vulnerability in health is to implement and improve basic public health measures, provide essential health care, increase capacity for disaster preparedness and response, and alleviate poverty (IPCC, 2014). Insurance programs, social protection measures, and disaster risk management may enhance long-term livelihood resilience among the poor and marginalized people, if policies address multidimensional poverty (IPCC, 2014, p. 97). The COVID-19 pandemic has brought to light the importance of addressing collective problems in a collective manner and the need for collaboration and cooperation.

Many adaptation and mitigation options can help address climate change, but no single option is sufficient by itself. Effective implementation depends on policies and cooperation at all scales and can be enhanced through integrated responses that link mitigation and adaptation with other societal objectives (IPCC, 2014, p. 94). Many governments favor climate change and disaster policies that promote "community resilience." Community resilience is defined as a community or region's capability to prepare for, respond to, and recover from significant multihazard threats with minimum damage to public safety and health, the economy, and national security (Colten et al., 2008, p. 38). Benefits from adaptation can already be realized in addressing current risks, which can be realized in the future for addressing emerging risks (IPCC, 2014, p. 18).

Table 1 presents adaptation approaches for managing the risks of climate change.

**TABLE 1** Approaches for managing the risks of climate change through adaptation.

| Category | Examples |
|---|---|
| Human development | Improved access to education, nutrition, health facilities, energy, safe housing and settlement structures, and social support structures; reduced gender inequality and marginalization in other forms |
| Poverty alleviation | Improved access to and control of local resources; land tenure; disaster risk reduction; social safety nets and social protection; insurance schemes |
| Livelihood security | Income, asset, and livelihood diversification; improved infrastructure; access to technology and decision-making fora; increased decision-making power; changed cropping, livestock and aquaculture practices; reliance on social networks |
| Disaster risk management | Early warning systems; hazard and vulnerability mapping; diversifying water resources; improved drainage; flood and cyclone shelters; building codes and practices; storm and wastewater management; transport and road infrastructure; disaster risk reduction |
| Ecosystem management | Maintaining wetlands and urban green spaces; coastal afforestation; watershed and reservoir management; reduction of other stressors on ecosystems and of habitat fragmentation; maintenance of genetic diversity; manipulation of disturbance regimes; community-based natural resource management; forestry management |
| Spatial or land-use planning | Provisioning of adequate housing, infrastructure, and services; managing development in flood prone and other high-risk areas; urban planning and upgrading programs; land zoning laws; easements; protected areas |
| Structural/physical | Engineered and built-environment options: Sea walls and coastal protection structures; flood levees; water storage; improved drainage; flood and cyclone shelters; building codes and practices; storm and wastewater management; transport and road infrastructure improvements; floating houses; power plant and electricity grid adjustments<br>Technological options: New crop and animal varieties; indigenous, traditional and local knowledge, technologies and methods; efficient irrigation; water-saving technologies; desalinization; conservation agriculture; food storage and preservation facilities; hazard and vulnerability mapping and monitoring; early warning systems; building insulation; mechanical and passive cooling; technology development, transfer and diffusion<br>Ecosystem-based options: Ecological restoration; soil conservation; afforestation and reforestation; mangrove conservation and replanting; green infrastructure (e.g., shade trees, green roofs); controlling overfishing; fisheries comanagement; assisted species |

*Continued*

TABLE 1 Approaches for managing the risks of climate change through adaptation—cont'd

| Category | Examples |
|---|---|
| | migration and dispersal; ecological corridors; seed banks, gene banks and other ex situ conservation; community-based natural resource management<br>Services: Social safety nets and social protection; food banks and distribution of food surplus; municipal services including water and sanitation; vaccination programs; essential public health services; enhanced emergency medical services; provision of social services |
| Institutional | Economic options: Financial incentives; insurance; catastrophe bonds; payments for ecosystem services; pricing water to encourage universal provision and careful use; microfinance; disaster contingency funds; cash transfers; public-private partnerships<br>Laws & regulations: Land zoning laws; building standards and practices; easements; water regulations and agreements; laws to support disaster risk reduction; laws to encourage insurance purchasing; defined property rights and land tenure security; protected areas; fishing quotas; patent pools and technology transfer<br>National & government policies & programs: National and regional adaptation plans including mainstreaming; subnational and local adaptation plans; economic diversification; urban upgrading programs; municipal water management programs; disaster planning and preparedness; integrated water resource management; integrated coastal zone management; ecosystem-based management; community-based adaptation |
| Social | Educational options: Awareness raising and integrating into education; gender equity in education; extension services; sharing indigenous, traditional, and local knowledge; participatory action research and social learning; knowledge-sharing and learning platforms<br>Informational options: Hazard and vulnerability mapping; early warning and response systems; systematic monitoring and remote sensing; climate services; use of indigenous climate observations; participatory scenario development; integrated assessments<br>Behavioral options: Household preparation and evacuation planning; migration; soil and water conservation; storm drain clearance; livelihood diversification; changed cropping, livestock, and aquaculture practices; reliance on social networks |
| Spheres of change | Practical: Social and technical innovations, behavioral shifts, or institutional and managerial changes that produce substantial shifts in outcomes<br>Political: Political, social, cultural, and ecological decisions and actions consistent with reducing vulnerability and risk and supporting adaptation, mitigation, and sustainable development<br>Personal: Individual and collective assumptions, beliefs, values, and worldviews influencing climate-change responses |

*Adapted from IPCC, 2014. In: Core Writing Team, Pachauri, R.K., Meyer, L.A. (Eds.), Climate Change 2014: Synthesis Report. Contribution of Working Groups I, II and III to the Fifth Assessment Report of the Intergovernmental Panel on Climate Change. Geneva, Switzerland, p. 27.*

Scientists have documented that it will be difficult to limit total global warming to less than 2°C, and extremely difficult to reach the goal of 1.5°C, which is regarded as an upper limit to avoid great risk of harm to the economy, human health, and well-being, and Earth's ecosystems. It is very likely that heat waves will occur more often and last longer, and that extreme precipitation events will become more intense and frequent in many regions. The ocean will continue to warm and acidify, and global mean sea level will rise (IPCC, 2014).

The emission of GHSs will continue to cause further global warming and long-lasting changes in all aspects of the climate system, increasing the likelihood of severe, pervasive and irreversible impacts for people and ecosystems (IPCC, 2014). Article 8 of the 2015 Paris Agreement recognizes "the importance of averting, minimizing and addressing loss and damage associated with the adverse effects of climate change including extreme weather events and slow onset events, and the role of sustainable development in reducing the risk of loss and damage" (UNFCCC, 2015, p. 12).

Many aspects of climate change and associated impacts will continue for centuries, even if anthropogenic emissions of GHGs are stopped. The risks of abrupt or irreversible changes increase as the magnitude of the warming increases (IPCC, 2014, p. 16). Given the situation, there is an urgent need for societal adaptation to climate change in all dimensions and levels, spanning the micro to macro spectrum. Macro-level research has found that population size, affluence, and the structure of the political economy are important drivers of GHG emissions. At the meso-level, studies of how various organizations and sectors exert influence on climate change responses are an active area of research. Actions to address climate change vary from climate skepticism to active engagement to mitigate and adapt, but they are often constrained by institutional, political, and economic contexts and processes. At the micro-level, researchers have found that both social structural factors (gender, political ideology, education) and social psychological factors (values, beliefs, norms, trust, identity) predict climate change public opinion and individual responses to climate change (voting, policy support, household behaviors) (Dietz et al., 2020).

Although scholars have explored the intersection of race and class in climate justice, other key dimensions of identity and social status have been neglected. These include the needs of indigenous communities; queer and gender-diverse people; people with disabilities or those needing special accommodations, such as older adults, pregnant individuals, and children; and immigrant and undocumented communities.

## 8 Neoliberalism and climate denial

For at least two decades, governments have adopted neoliberalism by giving priority to market forces over social benefits, resulting in reduced government interventions, a decline in infrastructure, and an expectation that people and communities will become more self-reliant (Alston and Kent, 2009). The environmental crisis has called into question contemporary social and economic systems, and related production and consumption patterns that depend on the unsustainable exploitation of natural resources (UNRISD, 2015). Yet political ideology continues to influence how social-economic-environment relationships are imagined.

The growing scientific consensus on climate change and efforts to formulate policy has been met with active campaigns by conservative movements to undermine climate science and policy in the United States (McCright and Dunlap, 2010). Campaigns of climate denial have contributed to the polarization on climate change in the United States, with conservatives far less willing than liberals to support climate policy (Ballew et al., 2019).

Studies demonstrate that beyond political ideology, racial bias is related to a lack of concern about climate change, contributing to polarization in the United States (Benegal, 2018). Benegal (2018) refers to the "spillover of racialization" on policy matters in areas such as health care reform and climate change. The COVID-19 pandemic in 2020 has brought to the forefront how racial bias is present at all levels in society, with the "Black Lives Matter" movement and antiracism demonstrations in many parts of the world. These interconnected crises—climate change, COVID-19, systemic racism—are severely affecting racialized and indigenous communities, calling for transformative change and justice (Cooper, 2020). In considering societal adaptation to climate change, it is necessary to discuss environmental justice issues related to climate justice.

## 9 Environmental and climate justice

Environmental justice is a social movement and discourse that influenced how climate justice has been conceptualized and understood (Schlosberg and Collins, 2014). Climate justice is local and global, concerned with the causes and consequences of climate change, and the inequitable impacts of climate change. Climate change raises many ethical challenges, especially on the issue of climate justice. Environmental disasters, such as Hurricane Katrina in 2005 revealed how preexisting injustices like poverty, racial discrimination, segregation, a poor education system, substandard housing, and a lack of community preparation, were exacerbated by extreme weather events (Schlosberg and Collins, 2014).

Protests and marches by school children throughout the world demonstrate the intergenerational injustice of climate change. Greta Thunberg, a young Swedish climate activist, delivered the famous "how dare you" speech to the United Nations Assembly (Rowlatt, 2020). School climate strikes were organized to bring attention to climate justice with messages to world leaders to act to save the planet and the future from climate emergency (United Nations, 2019a).

At a macro-level, political economy debates continue on forms of Marxist theory that assert that a capitalist political economy requires growth that will be antithetical to protecting the environment and alternative views, such as ecological modernization theory, that hold that capitalism can be substantially reformed to reduce environmental impact (Dietz et al., 2020). For example, Naomi Klein discusses how capitalism contributes to climate change inequities and injustices, calling for transformative economic structural change (Klein, 2014, 2019). There is a need for re-visioning the relationship between people and the planet.

## 10 New vision of "development"

The Global Sustainable Development Report 2019 (GSDR) states that unless there is a fundamental—and urgent—change in the relationship between people and nature, and a significant reduction in inequalities between and inside countries, any "development" progress of the last two decades risks being undone (United Nations, 2019b). The GSDR 2019 stresses the need to transform key areas of human activity, which could otherwise lead to systems failure—including food, energy, consumption, and production, and cities—and increase resilience to economic shocks and disasters caused by natural and man-made hazards, through active implementation of the Sendai Framework (United Nations, 2019b). The Sendai Framework was adopted in 2015 at the Third UN World Conference on Disaster Risk Reduction. It aims to reduce disaster risks and losses in lives, livelihoods, and to enhance disaster preparedness by "building back better" in recovery, rehabilitation, and reconstruction (UNDRR, 2015).

The goal is to shift from an unsustainable development model to a risk-informed one that restores and regenerates natural systems to ensure that "no ecosystem is left behind." The resilience and stability of natural ecosystems, their restoration and regeneration are of paramount importance for systemic risks to be managed effectively. Within this context, the vision of resilient society gains relevance as the concept of resilience comprises three components important to cope with (climate) change. These are the capacity (1) to resist and (2) to recover from disturbances and shocks and (3) to adjust functioning—prior to or following—changes and disturbances (Frommer, 2013).

Adaptations to climate change will have transformational effects on humans and societies at every level from the local to the global.

## 11 Transformative change

It is widely recognized that transformative change is needed for societal adaptation to climate change. For example, transformations in economic, social, technological, and political decisions and actions can enhance adaptation and promote sustainable development. At the national level, transformation is considered most effective when it reflects a country's own visions and approaches to achieving sustainable development in accordance with its national circumstances and priorities (IPCC, 2014). Restricting adaptation responses to incremental changes to existing systems and structures, without considering transformational change, may increase costs and losses and miss opportunities (IPCC, 2014). Planning and implementation of transformational adaptation may reflect strengthened, altered or aligned paradigms and may place new and increased demands on governance structures to reconcile different goals and visions for the future and to address equity and ethical implications (IPCC, 2014).

## 12 Conclusion

Many adaptation and mitigation options can help address climate change, but no single option is sufficient by itself. Effective implementation depends on policies and cooperation at all levels and can be enhanced through integrated responses that link adaptation and mitigation with other societal objectives (IPCC, 2014, p. 26). Vulnerability to climate change, GHG emissions, and the capacity for adaptation and mitigation is strongly influenced by livelihoods, lifestyles, behavior, and culture (IPCC, 2014). Effective adaptation and mitigation responses will depend on policies and measures across multiple levels: international, regional, national, and subnational.

International cooperation is critical for effective climate change mitigation, with local cobenefits. Adaptation focuses primarily on local to national scale outcomes, but its effectiveness can be enhanced through coordination. The United Nations Framework Convention on

Climate Change (UNFCCC) is the main multilateral forum focused on addressing climate change. The Kyoto Protocol offers lessons toward achieving the objective of the UNFCCC, with respect to participation, implementation, flexibility mechanisms, and environmental effectiveness. International cooperation for supporting adaptation planning and implementation has received less attention than mitigation but is increasing and has assisted in the creation of adaptation strategies, plans, and actions at the national, subnational, and local level (IPCC, 2014).

Governments at various levels have begun to develop adaptation plans and policies and integrate climate change considerations into broader development plans. Examples of adaptation are now available from all regions of the world (IPPC, 2014, p. 54), and can be found in this book.

## References

Alston, M., Hargreaves, D., Hazeleger, T., 2018. Postdisaster social work: reflections on the nature of place and loss. Aust. Soc. Work 71 (4), 405–416.

Alston, M., Kent, J., 2009. Generation X-pendable: the social exclusion of rural and remote young people. J. Sociol. 45 (1), 89–107.

Ballew, M.T., Leiserowitz, A., Roser-Renouf, C., Rosenthal, S.A., Kotcher, J.E., et al., 2019. Climate change in the American mind: data, tools, and trends. Environ. Sci. Policy Sustain. Dev. 61 (3), 4–18.

Benegal, S.D., 2018. The spillover of race and racial attitudes into public opinion about climate change. Environ. Polit. 27 (4), 733–756.

Berwyn, B., 2019. Global Warming Is Pushing Pacific Salmon to the Brink, Federal Scientists Warn. Inside Climate News. https://insideclimatenews.org/news/29072019/pacific-salmon-climate-change-threat-endangered-columbia-river-california-idaho-oregon-study.

Brown, M., Agyapong, V., Greenshaw, A.J., Cribben, I., Brett-MacLean, P., Drolet, J., McDonald-Harker, C., Omeje, J., Mankowski, M., Noble, S., Kitching, D., Silverstone, P., 2019. Significant PTSD and other mental health effects present 18 months after the Fort McMurray wildfire: findings from 3,070 grade 7–12 students. Front. Psych. 10, 623. section Children and adolescent psychiatry.

Boyle, L., 2020, July 31. 'Fingerprint' of the climate crisis in flooding that has left a third of Bangladesh under water. In: Independent. https://www.independent.co.uk/environment/climate-crisis-floods-bangladesh-monsoon-cyclone-amphan-a9648401.html.

Cooper, R., 2020. On Climate, COVID, and Race. Psychiatric Times. https://www.psychiatrictimes.com/view/on-climate-covid-and-race.

Cave, D., 2020. The End of Australia as We Know It. The New York Times. https://www.nytimes.com/2020/02/15/world/australia/fires-climate-change.html.

Colten, C.E., Kates, R.W., Laska, S.B., 2008. Three years: lessons for community resilience. Environment 50, 36–47.

Dhar, A., Parrott, L., Hawkins, C.D.B., 2016. Aftermath of mountain pine beetle outbreak in British Columbia: stand dynamics, management response and ecosystem resilience. Forests 7 (171), 1–19.

Dietz, T., Shwom, R.L., Whitley, C.T., 2020. Climate change and society. Annu. Rev. Sociol. 46, 135–158.

Drolet, J., McDonald-Harker, C., Lalani, N., Tran, J., 2020. Impacts of the 2013 flood on immigrant children, youth and families in Alberta, Canada. Int. J. Soc. Work 7 (1), 56–74.

Drolet, J. (Ed.), 2019. Rebuilding Lives Post-Disaster. Oxford University Press, New York, NY.

Drolet, J.L., Sampson, T., 2014. Addressing climate change from a social development approach: small cities and rural communities' adaptation and response to climate change in British Columbia, Canada. Int. Soc. Work 60 (1), 61–73.

Drolet, J., Sampson, T., Jebaraj, D., Richard, L., 2013. Social work and environmentally induced displacement. A commentary. Refug. Can. J. Refug. 29, 55–62.

Drolet, J., 2012. Climate change, food security, and sustainable development: a study on community-based responses and adaptations in British Columbia, Canada. Community Dev. 43 (5), 630–644.

Elliott, R., 2018. The sociology of climate change as a sociology of loss. Eur. J. Sociol. 59 (3), 301–337.

Food and Agriculture Organization of the United Nations (FAO), 2017. The Impact of Disasters and Crises on Agriculture and Food Security. Rome, Italy. http://www.fao.org/3/I8656EN/i8656en.pdf.

Frommer, B., 2013. Climate change and the resilient society: utopia or realistic option for German regions? Nat. Hazards 67, 99–115.

International Labour Organization (ILO), 2019. Indigenous Peoples and Climate Change: Emerging Research on Traditional Knowledge and Livelihoods. Geneva, Switzerland. https://www.ilo.org/wcmsp5/groups/public/—ed_protect/—protrav/—ilo_aids/documents/publication/wcms_686780.pdf.

IPCC, 2018. Summary for Policymakers of IPCC Special Report on Global Warming of 1.5°C Approved by Governments. https://www.ipcc.ch/2018/10/08/summary-for-policymakers-of-ipcc-special-report-on-global-warming-of-1-5c-approved-by-governments/.

IPCC, 2014. Core Writing Team, , Pachauri, R.K., Meyer, L.A. (Eds.), Climate Change 2014: Synthesis Report. Contribution of Working Groups I, II and III to the Fifth Assessment Report of the Intergovernmental Panel on Climate Change. Geneva, Switzerland.

Jagers, S.C., Duus-Otterström, G., 2008. Dual climate change responsibility: on moral divergences between mitigation and adaptation. Environ. Polit. 17 (4), 576–591.

Klein, N., 2019. On Fire: The (Burning) Case for a Green New Deal. Simon & Schuster, New York.

Klein, N., 2014. This Changes Everything: Capitalism vs. the Climate. Simon & Schuster, New York.

Lalani, N., Drolet, J., 2019. Impacts of the 2013 floods on families' mental health in Alberta: perspectives of community influencers and service providers in rural communities. Best Pract. Ment. Health 15 (2), 74–92.

Lebel, L., 2012. Local knowledge and adaptation to climate change in natural resource-based societies of the Asia-Pacific. Mitig. Adapt. Strat. Glob. Chang. 18 (7), 1057–1076.

McCright, A.M., Dunlap, R.E., 2010. Anti-reflexivity. Theory Cult. Soc. 27 (2/3), 100–133.

Miller, S.E., Hayward, R.A., Shaw, T.V., 2012. Environmental shifts for social work: a principles approach. Int. J. Soc. Welf. 21 (3), 270–277.

Oxfam, 2020. 5 Natural Disasters That Beg for Climate Action. https://www.oxfam.org/en/5-natural-disasters-beg-climate-action.

Pozarny, P., 2016. Understanding Climate Change as a Social Development Issue. https://gsdrc.org/topic-guides/climate-change-and-social-development/understanding-climate-change-as-a-social-development-issue/.

Raygorodetsky, G., 2011. Why Traditional Knowledge Holds the Key to Climate Change. United Nations University. https://unu.edu/publications/articles/why-traditional-knowledge-holds-the-key-to-climate-change.html.

Rowlatt, J., 2020. Greta Thunberg: Climate Change 'as Urgent' as Coronavirus. BBC News. https://www.bbc.com/news/science-environment-53100800.

Schlosberg, D., Collins, L.B., 2014. From environmental to climate justice: climate change and the discourse of environmental justice. Wiley Interdiscip. Rev. Clim. Chang. 5 (3), 359–374.

Tschakert, P., Dietrich, K.A., 2010. Anticipatory learning for climate change adaptation and resilience. Ecol. Soc. 15 (2), 11.

Tschumi, E., Zscheischler, J., 2019. Countrywide climate features during recorded climate-related disasters. Clim. Change 158, 593–609.

United Nations, 2019a. Climate Justice. Sustainable Development Goals. https://www.un.org/sustainabledevelopment/blog/2019/05/climate-justice/.

United Nations, 2019b. Global Sustainable Development Report. https://sustainabledevelopment.un.org/content/documents/24797GSDR_report_2019.pdf.

United Nations, 2005. The Social Summit Ten Years Later. Department of Economic and Social Affairs. www.un.org/esa/socdev/publications/SocialSummit-10YearsLater.pdf.

UNFCCC, 2015. Adoption of the Paris Agreement, Conference of the Parties on Its Twenty-first Session, FCCC/CP/2015/L.9. https://unfccc.int/process-and-meetings/the-paris-agreement/the-paris-agreement.

United Nations Framework Convention on Climate Change, 1992. United Nations Framework Convention on Climate Change. https://treaties.un.org/doc/Treaties/1994/03/19940321%2004-56%20AM/Ch_XXVII_07p.pdf.

United Nations Office for the Coordination of Humanitarian Affairs (OCHA), 2019. Cyclones Idai and Kenneth. https://www.unocha.org/southern-and-eastern-africa-rosea/cyclones-idai-and-kenneth.

United Nations Office for Disaster Risk Reduction (UNDRR), 2015. Sendai Framework for Disaster Risk reduction 2015–2030. https://www.undrr.org/publication/sendai-framework-disaster-risk-reduction-2015-2030.

United Nations Research Institute for Social Development (UNRISD), 2015. Research for Social Change: Transformations to Equity and Sustainability. UNRISD, Geneva.

CHAPTER

# 15

# Managing urban climate change risks: Prospects for using green infrastructure to increase urban resilience to floods

*Juliana Reu Junqueira*[a], *Silvia Serrao-Neumann*[a,b], *and Iain White*[a]

[a]Environmental Planning Programme, School of Social Sciences, University of Waikato, Hamilton, New Zealand [b]Cities Research Institute, Griffith University, Brisbane, QLD, Australia

OUTLINE



## 1 Introduction

Changes in extreme weather events have been observed since the 1950s (IPCC, 2014) and have become more frequent during the last decades (Cheng et al., 2017). These extremes include an increase in warmer temperatures, storminess, and sea level rise, and an escalation in the occurrence and severity of heavy rainfall (Hettiarachchi et al., 2018; IPCC, 2019; Liu and Cheng, 2014). Although the impacts of climate change tend to adversely affect cities due to their concentration of capital, urban areas can also contribute to achieve climate change


https://doi.org/10.1016/B978-0-12-822373-4.00013-6

mitigation and adaptation and increasing urban resilience, especially through effective urban planning (Grafakos et al., 2020; IPCC, 2014).

With respect to climate change, a key challenge confronting urban areas is the risk of precipitation extremes (Drosou et al., 2019; Hettiarachchi et al., 2018; White, 2010). Typically, the spread of urbanization and ineffective policies and flood management strategies have increased exposure to flood risk (Batica, 2015; Cheng et al., 2017; Galderisi and Treccozzi, 2017; Sörensen, 2018). In particular, problems can emerge from the weakening of water-retaining function of soils in urbanized areas (McFarland et al., 2019). For example, when buildings and roads replace natural ground cover, the flow of water is severely modified, not allowing water runoff to be absorbed into soil. So, stormwater runoff flows on the surface and increases the chances of erosion and flooding of urban streams, causing damage to surrounding habitats, properties, and infrastructure (Kazemi, 2014). Urbanization can also reduce the area for river overflowing ( Jha, 2012), increase water flows and speed leading to downstream flooding of built-up areas (Liao, 2012), increase the demands on existing drainage infrastructure, and increase people and assets exposure to flood risks (Struck and Lichten, 2010).

Although the concept of resilience can be understood from many differing lenses (Davidson et al., 2016), urban resilience is defined here as the capacity of an urban area to overcome a diverse variety of disturbances and stresses, and adapt to new siloed (Béné et al., 2018; Doyon, 2016; Meerow et al., 2016). Resilient urban areas are also capable of enacting systemic change, therefore reducing their exposure to shocks (White and O'Hare, 2014). Improving urban resilience requires a better understanding, and design, of interventions aimed at reducing the root causes of vulnerabilities (Irajifar, 2016; Lamond et al., 2015). In practice, climate risks are not equally distributed and may inflict higher impacts on underprivileged people and communities (Bird et al., 2013; O'Hare and White, 2018) which, in turn, contribute to increased socioeconomic disparities (Albrechts et al., 2020). According to Handmer et al. (2012) a lack of effective urban planning, limited protective infrastructure, and capacity constraints that affect risk reduction, preparedness, and recovery are some of the factors that can increase urban vulnerability to climate change. In addition to directly addressing hazards, improving urban resilience can also be improved indirectly, such as by improving spatial planning practices (Folke et al., 2010; UNSDR, 2007), decision making that can better consider long-term pressures and threats, and development that links the natural and built environments (Hughes and Sharman, 2015).

Green infrastructure has emerged as a promising measure to mitigate some of the negative impacts of climate change in urban areas (Salata and Yiannakou, 2016), especially as it can deliver multiple benefits for people, places, and ecosystems (Demuzere et al., 2014). As green infrastructure connects natural and social systems, it can potentially contribute to both specifically improve urban flood resilience and more generally build resilience to climate change (Cole et al., 2017; Salata and Yiannakou, 2016). But this connection must be effectively planned, designed, and implemented, as climate change adaptation measures should consider multiple spatial scales, and the specific context will determine precisely what type, combination, and magnitude of green infrastructure are required. In particular, green infrastructure may need to be integrated into urban planning practices as part of a system perspective. This means it is able to consider the socioeconomic, technical, and biophysical characteristics of urban areas, as well as the trade-offs inherent within and between these

(Salata and Yiannakou, 2016). This means that in practice, the use of green infrastructure as an approach to address climate change impacts and flood risks needs to be seen in the context of solving interrelated spatial, engineering, environmental, and sociopolitical challenges. For example, the use of natural features, such as rain gardens, bioretention cells, green roofs, and urban forests as "infrastructure" demands new evidence, processes, and approaches beyond those found in traditional flood risk management practices (Benedict and McMahon, 2006).

There are also uncertainties regarding the efficiency of green infrastructure when compared with that of gray infrastructure (Thorne et al., 2018). Few studies have attempted to evaluate hydrological efficiency of green infrastructure by applying different models to assess their impact at the catchment scale (Juan et al., 2016). In reality, the effectiveness of green infrastructure is frequently assessed from a specific managerially convenient viewpoint (such as stormwater management) rather than having a systemic view of the entire network of benefits (Alves et al., 2019). Secondary benefits, such as reduced urban heat island effects, improved water quality, improved human health and scenic amenities are often underestimated and monetizing these secondary benefits are not easy. More generally, there is also lack of confidence in public acceptance of green infrastructure alternatives, which limits their wide adoption (Thorne et al., 2018). In practice, the lack of a coherent framework relating to green infrastructure planning, design, and implementation could lead to misunderstandings and inconsistencies in the form they are rolled out in urban areas (Salata and Yiannakou, 2016).

In simple terms, for green infrastructure to be widely implemented, in many areas a review of current urban planning practices is required and further attention should be given to the role that planning has to play in promoting natural connectivity within the urban landscape (Lemes de Oliveira, 2019; Sinnett et al., 2019). This chapter seeks to reconcile this impasse by reviewing literature relating to the use of green infrastructure into urban planning as a means to increase urban flood resilience as a result of climate change, identifying the key concepts, highlighting the main debates, and focusing on the key questions that need to be resolved.

## 2 The potential of green infrastructure for increasing urban resilience to floods

### 2.1 Key concepts and terminologies

Green infrastructure comprises an integrated system of natural areas and other open spaces that aim to conserve the principles and roles of natural environments and associated ecosystem services (Benedict and McMahon, 2006; European Union, 2013). It has been defined in multiple ways due to its multifunctional character (Mell et al., 2017) and is seen to encapsulate spatially and strategically managed networks of natural and seminatural areas (Cole et al., 2017; Federal Agency for Nature Conservation (BfN), 2017; Salata and Yiannakou, 2016). Originally linked to both landscape architecture and landscape ecology (Fletcher et al., 2015), green infrastructure is now recognized for its functionality, particularly as it provides a broad variety of environmental services and other benefits, such as enhanced biodiversity, improving water quality, carbon reduction and sequestration, groundwater recharge, flood mitigation, and potential natural landscape scenario (Ashley et al., 2018).

Fletcher et al. (2015) have identified a range of terms used to describe green infrastructure strategies based on different national contexts and disciplinary traditions, including water-sensitive urban design, low impact development, and nature-based solutions (see Table 1). The different terminologies used to describe these strategies reflect the emergent nature of the field as well as the issues it is deemed to address. For instance, water-sensitive urban design, low impact development, and blue-green infrastructure are primarily focused on climate change adaptation and flood risk management, whereas nature-based solutions aim

**TABLE 1** Different terminologies used for green infrastructure strategies.

| Terminology | Concept | Countries/regions |
|---|---|---|
| Water sensitive urban design | Water sensitive urban design is a cross-disciplinary collaboration between water management, urban landscape, and planning (Hoyer et al., 2011; van Roon, 2017). Water sensitive urban design can combine sustainable stormwater management challenges with urban planning requirements and can also create holistic solutions to environmental, economic, social, and cultural preservation (Hoyer et al., 2011) | Australia (Fletcher et al., 2015) |
| Blue-green infrastructure | Blue-green infrastructure is a principle in landscape design that combines design principles such as green spaces and ecological systems (Ghofrani et al., 2017). Blue-green infrastructure focuses on biophysical processes, such as retention, storage and absorption, and control of stormwater volume and quality (Liao et al., 2017) | United States, The Netherlands, Belgium, Japan, India (Ghofrani et al., 2017) |
| Low impact design/low impact development—LID/low impact urban design and development—LIUDD | The low impact design approach is primarily used to manage urban water runoff (van Roon, 2012). It manages stormwater by small, cost-effective landscape elements rather than managing stormwater in extensive, unnatural and costly facilities (Dredge, 2015) | North America and New Zealand (Fletcher et al., 2015) |
| Nature-based solutions/nature-based flood solution | Nature-based solutions consider nature-based resources (Emilsson and Sang, 2017). Nature-based solutions offer natural, social and economic advantages to urban resilience through the preservation, improvement or regeneration of biodiversity and habitats to address different issues at the same time (Depietri and McPhearson, 2017; Mabon, 2019) | Europe (Pauleit et al., 2017) |

to increase environmental, social and economic benefits by strengthening biodiversity and ecosystems (Fletcher et al., 2015). However, despite some minor specificities, all terminologies aim to use, or work with, nature to reduce urbanization impacts on, for example, waterways, rather than trying to control natural processes.

## 2.2 The different types of green infrastructure and their differing potential to increase urban resilience to floods

The multifunctional character of green infrastructure has led to the emergence of a range of types and design options (see Table 2). It is important to note that, as with traditional infrastructure, these have differing suitability and applications. Some may be more cost-effective, others may require considerable resources and planning to be effectively implemented. For example, rain gardens, bioretention cells, planter boxes, and permeable pavements can be more easily retrofitted in public areas within cities because they do not significantly affect private properties (Hu et al., 2019; Lamond et al., 2015). On the contrary, green roofs and large-scale rainwater harvesting schemes may be economically prohibitive as a retrofit alternative, but may be easier to implement in greenfield developments (Dover, 2015; Staddon et al., 2018).

**TABLE 2** Suitability of green infrastructure types for increasing urban resilience to floods.

| Type of green infrastructure | Definition | Role in increasing urban resilience to floods | Advantages | Limitations |
|---|---|---|---|---|
| Bioretention cells | Similar to rain gardens but primarily designed to contain a certain quantity of runoff from a wide impervious area such as a carpark or a road (Wang et al., 2019) | • Bioretention cells minimize flood risks because they capture, and absorb rainfall runoff as it flows downstream (National Association of City Transportation Officials, 2019; Soil Science Society of America (SSSA), 2019) | • Can increase evapotranspiration and aid urban cooling (Ahiablame and Shakya, 2016)<br>• Efficient in improving water quality (Ahiablame and Shakya, 2016)<br>• Require minimum space to be implemented (Lin et al., 2018) | • Maintenance includes several variables such as irrigation, sediment removal, plant management, fertilizing and pest control (Chui et al., 2016)<br>• Only effective to alleviate risk of small to medium floods (Akhter and Hewa, 2016)<br>• Only efficient to reduce runoff volume and peak discharge in short return period rainfall events (Wang et al., 2019) |

*Continued*

TABLE 2 Suitability of green infrastructure types for increasing urban resilience to floods—Cont'd

| Type of green infrastructure | Definition | Role in increasing urban resilience to floods | Advantages | Limitations |
|---|---|---|---|---|
| Downspouts disconnection | Reconfiguration of pipes that carry roof runoffs (Kazemi, 2014) | • Downspouts disconnection can minimize risk of contamination from floods by preventing stormwater from reaching sewer systems (US EPA, 2020) | • Low-cost to be implemented (Kazemi, 2014) | • Dependent on other alternatives such as rainwater harvesting, cisterns or rain gardens to convey rooftop runoff (Staddon et al., 2018) |
| Green roofs | Green roofs, or living roofs, are covered—entirely or partially—with vegetation and a growing media, planted over a waterproof membrane (Dover, 2015) | • Green roofs help reduce rooftop temperatures and provide insulation to buildings (Koc et al., 2016)<br>• They can aid urban stormwater management, as they improve the retention capacity of rainfall runoff (Cipolla et al., 2016) | • Can restore runoff rates after urbanization to predevelopment ones (Cipolla et al., 2016)<br>• Low cost to be implemented in new buildings (estimated at 0.5% of the total building costs) (Dover, 2015)<br>• Can be designed for specific functions such as attenuation of peak flows and reduction of stormwater runoff (Dover, 2015) | • More efficient for frequent storms of smaller magnitude (Ercolani et al., 2018)<br>• Require trained contractors to do green roof's installation (Dover, 2015)<br>• Restrict to be used as a retrofit alternative due to structural loading limitations (Cipolla et al., 2016; Dover, 2015) |
| Planter box | Similar to rain gardens, but in a smaller dimension, planter boxes are very suitable for space-limited dense urban areas (Kazemi, 2014) | • Planter boxes can minimize flood risks as they can detain, filter and infiltrate rainfall runoff (Kazemi, 2014) | • Ideal for space-limited dense urban areas (US EPA, 2020)<br>• Can be used as vertical building walls (Staddon et al., 2018)<br>• Can improve water quality and reduce stormwater runoff and peak flow rates (US EPA, 2020) | • Only effective to reduce small flood risks (US EPA, 2020) |

**TABLE 2** Suitability of green infrastructure types for increasing urban resilience to floods—Cont'd

| Type of green infrastructure | Definition | Role in increasing urban resilience to floods | Advantages | Limitations |
|---|---|---|---|---|
| Permeable pavements | Pavement surfaces that allow stormwater runoff to infiltrate into the soil through several layers of permeable materials (United States Environmental Protection Agency, 2014) | • Permeable pavements can increase collection and infiltration of rainfall and stormwater runoff; recharge groundwater, and increase water quality (Dover, 2015) | • More cost-effective for frequent storms (Wang et al., 2019)<br>• Low construction and maintenance costs (Dover, 2015; Zhu and Chen, 2017)<br>• Can be incorporated through the modification of surfaces that would normally be impermeable (Dover, 2015) | • Effective for reducing impacts from rainfall events with short return period, but not for extreme downpours with prolonged duration (Wang et al., 2019) |
| Rain gardens | It is a garden in a small depression, generally built on natural slopes ( Jia et al., 2016) | • Because rain gardens momentarily retain and soak rainwater runoff from impervious surfaces they contribute to the reduction of flood risks (Groundwater Foundation, 2019) | • Require minimum space to be implemented (US EPA, 2020)<br>• Help to recharge aquifers (Zhang et al., 2019)<br>• Efficient in reducing peak flow rates and total volume of stormwater runoff through retention and re-rotation processes (Zhang et al., 2019) | • Designed to store a predefined volume of stormwater runoff ( Jia et al., 2016)<br>• Additional runoff is discharged into urban stormwater drainage systems ( Jia et al., 2016)<br>• Reduced capability of high-infiltration after some years (Guo, 2012)<br>• Only efficient to alleviate risk of small to medium floods (Akhter and Hewa, 2016) |
| Rainwater harvesting | Rainwater harvesting systems are usually used with downspout disconnection to capture the runoff in rain barrels, rain | • It can slow and reduce stormwater runoff decreasing flood risk (Staddon et al., 2018) | • Water collected during a rain event can be used for other purposes such as outdoor irrigation and indoor secondary | • Limited capacity for rainwater storage (according to the size of the storage tank) (Damodaram and Zechman, 2013) |

*Continued*

TABLE 2 Suitability of green infrastructure types for increasing urban resilience to floods—Cont'd

| Type of green infrastructure | Definition | Role in increasing urban resilience to floods | Advantages | Limitations |
|---|---|---|---|---|
| | tanks or cisterns reducing rainfall runoff (Bartos et al., 2018) | | water uses (Damodaram and Zechman, 2013)<br>• Prevent sewer overflows (Bartos et al., 2018) | • Limited capacity to contain precipitation extremes (Akhter and Hewa, 2016) |
| Urban forestry | Also known as urban tree canopy, urban forestry is the designed implementation and maintenance of trees in an urban landscape (Naturally Resilient Communities, 2019) | • Urban forestry can decrease stormwater runoff by retrieving rainwater on branches and leaves (Naturally Resilient Communities, 2019; Staddon et al., 2018) | • Can enhance the value of adjacent properties (Dover, 2015)<br>• Can have cooling effects, improve air quality and reduce stormwater runoff (Escobedo et al., 2011) | • Expensive maintenance (Dover, 2015)<br>• Not suitable for small areas because the size of urban forestries is directly related to the quantity of ecosystem services provided (Escobedo et al., 2011)<br>• Causal agent of damage to buildings and sewers (Dover, 2015) |

Overall, rain gardens, bioretention cells, and permeable pavements are easier to be implemented in urbanized areas. In particular, they are especially suitable to be retrofitted into existing built up areas because they neither need large extensions to be implemented nor are dependent on structural loading capacity of buildings (such as in the case of green roofs). For example, rain gardens can be inserted in small areas along parks and in private gardens if promoted by local planning policies. Bioretention cells can be introduced along sidewalks, and permeable pavements could replace cover in driveways and parking lots. However, the main drawback of these alternatives concerns their reduced efficiency to deal with heavy storms (Akhter and Hewa, 2016; Jia et al., 2016). In particular, once they reach capacity their effectiveness in reducing stormwater runoff starts to decrease (Webber et al., 2019). Their performance is also affected by design patterns, percentage of impervious surface, surface roughness, and infiltration rate (Goncalves et al., 2018; Gulbaz and Kazezyilmaz-Alhan, 2017). Thus, the effectiveness of green infrastructure in minimizing the impacts of rainfall events, especially precipitation extremes, is highly context specific and demands further analysis from urban managers, which can be difficult to achieve depending on their internal technical, organizational capacity.

On the contrary, the implementation of combined green infrastructure alternatives may provide higher storage capacity and conveyance systems to capture intense runoffs which, in turn, better deal with precipitation extremes (Goncalves et al., 2018; Webber et al., 2019). For example, alternatives, such as permeable pavement, rain gardens or green roofs, may be paired with storage tanks (or other rainwater harvesting systems) to reduce flood volume. In reducing flood volume, many measures can also slow down the flow, which reduces the peak load on infrastructure and can also complement more traditional flood resilience measures (Webber et al., 2019) such as early warning systems and evacuation of at risk communities (IPCC, 2014). However, the integration of storage tanks into green infrastructure alternatives tends to require greater investments and may be impeded by the limited financial capacity of local territorial authorities.

This initial discussion emphasizes that for green infrastructure to be capable of dealing with severe rainfall events, they need to be implemented in a strategic manner to maximize their individual and complementary efficiency. For example, a feasible scenario may be the incorporation of one alternative to slow down stormwater runoff (e.g., permeable pavement), another to prevent sewer overflow by providing temporary storage (e.g., rain garden), and another to absorb or re-rotate stormwater runoff (e.g., bio-retention cells). As a retrofitting option, small-scale green infrastructure investment seems to have limited capacity to mitigate precipitation extremes associated with climate change, but strategically planned it can have a more significant affect. Yet, this scale of intervention can be difficult in existing urban areas. In comparison, implementing combined green infrastructure alternatives in greenfield developments to improve flood resilience may be more feasible because it can be master-planned from the start for maximum benefit (Webber et al., 2019). Hence, reaching a significant level of green infrastructure implementation in urban areas may take time and require the efforts of and collaboration between many stakeholders (Burns et al., 2015). Table 2 summarizes the suitability of green infrastructure types as a measure to manage urban flooding.

## 3 Barriers to incorporating green infrastructure into urban planning

Although different policy sectors have been including green infrastructure principles in their practice (e.g., land use planning and spatial development, water management, climate change adaptation, and environmental protection), the integration between land-use planning and urban water management is still limited due to several barriers (Slätmo et al., 2019). This section summaries these issues together into four main areas.

The first barrier stems from uncertainties seen by many flood risk management professionals regarding the effectiveness of green infrastructure when compared with gray infrastructure (Thorne et al., 2018). For example, O'Donnell et al. (2017) conducted a study in Newcastle to recognize obstacles to the introduction of green infrastructure and identified that the majority of respondents perceived a hesitancy to embrace new solutions, specifically shifting from conventional engineering solutions to more environmentally sustainable solutions. More broadly, the lack of knowledge, awareness, and recognition of green infrastructure potential has also been identified as a major obstacle to winning support from local governments and communities (Mell and Lemes de Oliveira, 2019; O'Donnell et al., 2017).

The second barrier relates to the need to adopt an integrated approach to spatial planning which is able to support a wide variety of benefits for the community and enhance ecological cohesion between protected and unprotected areas (Lafortezza et al., 2013). This issue is because green infrastructure differs from other landscape planning approaches as it requires the combination of ecosystem-based approaches (Yiannakou and Salata, 2017) in parallel to the more social and economic aspects associated with urban development (Lafortezza et al., 2013; Lawson et al., 2014). The latter is particularly important if green infrastructure is to be used to also offset the lack of amenities or disparity of impacts floods can cause for different parts of a city, or neighborhoods already experiencing socioeconomic disadvantages.

A third barrier relates to the spatial scale in which green infrastructure is implemented, which can range from the whole catchment to single waterways or sites. This aspect is therefore determined and influenced by a range of policies, plans, and other activities carried out at what may be distinct spatial scales by differing agencies (Lafortezza et al., 2013). Waterways are linked with one another and with the wider urban system, so the advantages of green infrastructure implementation in mitigating the effects of extreme rainfall events should not be measured in isolation. However, if green infrastructure is primarily being implemented along single waterways, it is challenging to capture a more comprehensive picture of their role in mitigating the effects of intense rainfall events on a broader spatial scale.

Finally, a key barrier concerns economic aspects. Financial benefits produced by green infrastructure implementation are not easily measurable (Mell et al., 2013; Nordman et al., 2018). Furthermore, benefits of green infrastructure implementation often emerge as indirect impacts, which mean that those who pay for the service are not always the ones who benefit directly, especially for cultural and regulatory services including health enhancement and flood risk management (Hislop et al., 2019). Added to that are the costs of maintaining green infrastructure, so it continues to perform its multiple functions effectively. In sum, green infrastructure benefits are often tricky to determine and gravitate toward quantifiable indicators that may only partially encompass their value. There are also temporal issues. For instance, there is a desire for the development process to bring short term profits, whereas the benefits from green infrastructure may unfold over the longer term as part of a wider enhancement of social-ecosystem resilience (Hislop et al., 2019; Mell et al., 2013).

## 4 Going forward

It is clear from the discussion thus far that there are many challenges to implementing green infrastructure despite its potential as a means to improve urban flood resilience. For this situation to be addressed, there are two key areas that need attention. First, urban planning needs to connect diverse principles (e.g., political, ecological, and socioeconomic) in a way that can recognize and capture diverse benefits (Lemes de Oliveira, 2019). Second, spatial planning needs to be guided by a strategic green infrastructure approach which integrates multiple measures, performing multiple functions, across an urban network (Marot et al., 2015; Pauleit et al., 2017).

Urban adaptation to climate change implies analyzing, designing, planning, and making decisions as a society (Toimil et al., 2020; White, 2010). However, there is considerable evidence that adaptation issues and challenges facing cities and regions cannot be adequately addressed and handled by "conventional" urban planning (Albrechts et al., 2020; Lemes de Oliveira, 2019). Although climate change is commonly considered within urban planning, the significant uncertainty in future climate and the long-time frames associated with climate scenarios present a challenge to this task (IPCC, 2014; Woodward et al., 2014). For example, future risk and decision uncertainty increase over longer temporal scales, which bring problems for urban planning processes and political systems that typically have a "presentist bias" (White and Haughton, 2017).

In order to address this impasse and the above-mentioned barriers and limitations associated with different green infrastructure strategies, Scott (2019) argues that opportunities can be created by reconnecting diverse perspectives into common values. In particular, these opportunities can be optimized by involving a wider variety of stakeholders in risk assessment and decision-making processes, and by improving techniques to allow complementary approaches for action on climate change to be incorporated by other groups (Grafakos et al., 2020). For example, computational models can be used to increase efficiency, including stratified sampling methods or hybrid approaches, therefore, lowering the quantity of simulations necessary to assess uncertainty (Toimil et al., 2020). These models also need to be linked to development pressures and political issues. As such, methods such as scenario planning can help address future uncertainty by discussing trajectories, emergent threats, and possible management decisions to manage uncertainty and highlight with unintended outcomes (Brewington et al., 2019). Spatially-explicit scenario analyses have been used to forecast future urban growth in multiple social, cultural, environmental, and political contexts, advice sustainable agroforestry practices, and model possible outcomes of resource policies (Brewington et al., 2019). However, there are difficulties involved in the use of this type of analysis that discourages their application, in particular their fit with traditional siloed norms of decision making (Carter and White, 2012).

Opportunities for green infrastructure incorporation can also arise when core environmental principles are transformed into terms and ideas that are used and prioritized by socioeconomic and urban planning objectives (Lemes de Oliveira, 2019). For example, green infrastructure alternatives can work simultaneously to reduce stormwater runoff and improve the accessibility of greenspace areas around the city. Additionally, environmental concepts (e.g., flood minimization, heat islands mitigation, enhanced water quality) need to be translated into spatial planning policies and decisions to start collaborative spaces for dialog to create new avenues and opportunities for the incorporation of green infrastructure (Scott, 2019).

Resilience scholarship emphasizes how urban spaces can be planned and reorganized over time in accordance with land suitability and other recognized land-use planning principles so that they can withstand unexpected shocks and inform future activities within it (Herath and Wijesekera, 2019; Ran and Nedovic-Budic, 2016). For example, urban spaces can be redesigned to incorporate green infrastructure solutions that create more space for water (Busscher et al., 2019), such as leaving flood-prone areas undeveloped to minimize flood damage on public and private assets (Cilliers, 2019). These urban reconfigurations, however, need to be guided by a systemic approach underpinning plan making and implementation processes which consider social and political aspects, a wider spatial and temporal scale (including to deal with climate change impacts), and regulatory frameworks able to influence landowners or affected parties (Ran and Nedovic-Budic, 2016).

Similarly, urban flood risks cannot be managed in isolation and should be considered at city and catchment scales to improve urban resilience (Jayawardena, 2018; Jha, 2012), particularly as the frequency and magnitude of flood peaks in urban areas is likely to increase due to climate change (Emilsson and Sang, 2017; Suarez et al., 2005). Hence, improving urban flood resilience requires effective links between policy, practice, or design, and balance between potentially competing political objectives associated with urban density, efficient use of land, and rapid decision making (Hettiarachchi et al., 2018). Adopting a systemic approach to both urban planning and flood risk management processes is a complex task which further compounds their integration (Francesch-Huidobro et al., 2017; Meng et al., 2019).

The literature suggests that such a systemic approach would need to be underpinned by the following key principles: (i) connectivity; (ii) quantity and size; (iii) proximity; (iv) distribution and accessibility; (v) multiscale approach; and (vi) multifunctionality (Lemes de Oliveira, 2019). To expand further, this means that the integration of green infrastructure into urban planning should explore the implementation of a cohesive network of green infrastructure (*connectivity*) underpinned by ecological functions so as to maximize its benefits, such as temperature regulation, water quality, and runoff reduction (Lafortezza et al., 2013; Lemes de Oliveira, 2019). In addition, strategies and policies should promote uniform distribution of green infrastructure areas throughout the city and vary both in terms of function and design, in order to promote stronger proximity between people and green infrastructure areas (*quantity and size*) and reduce access inequality (Natural England, 2010). The more people have access to urban areas with green infrastructure (*proximity*), the more likely green infrastructure can contribute to building urban resilience associated with people's well-being (Davies and Lafortezza, 2019; Lemes de Oliveira, 2019). Green infrastructure networks should provide a well-defined hierarchical network (such as local use, suburb, city, and regional level) when it applies to urban planning (*multi-scale*) (Lemes de Oliveira, 2019; Natural England, 2010). Finally, a variety of functions should be established to address different issues (*multifunctionality*) (Sinnett et al., 2015). The more multifunctional urban planning responses address climate change effects, the more they can contribute to urban resilience (Davies and Lafortezza, 2019). Hence, urban planning can incorporate green infrastructure into cities and help to alleviate a variety of environmental and social issues and produce several benefits, such as increasing urban resilience to climate change (Mell, 2020). Nonetheless, the outcomes of the implementation of these principles rely on effective collaborations and public engagement, design quality, policy context, and equality (Dhakal, 2017; Hislop et al., 2019; Lemes de Oliveira, 2019).

## 5 Conclusion

Climate change impacts have become more frequent during the last decades, including the increased frequency and intensity of heavy precipitation events leading to intensified flood risks. As a result, planners, designers, and flood managers face new challenges in flood control and adaptation of urban stormwater drainage systems to both current and future climate change risks.

Green infrastructure has emerged as a valuable measure to address the negative impacts of climate change in urban areas, particularly as it can mitigate flooding by balancing water flows and reduce the burden on aging stormwater infrastructure systems, but also due to its wider benefits, such as by shading vegetation providing thermal comfort. The concept of green infrastructure emphasizes the need to manage runoff at source, unlike traditional stormwater strategies designed to capture, transport, and mitigate stormwater in a downstream location. The chapter has highlighted, however, a range of obstacles to the incorporation of green infrastructure in urban areas, such as uncertainty about its efficiency in flood risk management, lack of integration with spatial planning, lack of a comprehensive understanding of its potential, and difficulties in measuring economic advantages.

To address these challenges, we argue for a comprehensive rethink of existing policies and practices involving urban planning and flood management at various scales. For green infrastructure to fulfill its potential as a natural solution for urban climate change adaptation and mitigation, planning needs to be conducted in a more systematic and considered fashion. Formal policy and regulations, as well as more informal politics and incentives can all play a part in promoting green infrastructure uptake. A starting step would be to provide more quality information on their suitability and application to allow the variety of urban stakeholders (such as planners, engineers, private land-owners, or political decision-makers) to better understand current and future risks and the potential of green infrastructure to enhance resilience to climate change. More work is yet to be done on compiling the right information in the right format, however, particularly given the cross-departmental nature of implementing green infrastructure. Finally, to retrofit the slow incorporation of green infrastructure into urban areas, a more collaborative approach is required that acknowledges scientific limitations, aspirational visions of future urban form, and differing stakeholders' needs and values.

## References

Ahiablame, L., Shakya, R., 2016. Modeling flood reduction effects of low impact development at a watershed scale. J. Environ. Manage. 171, 81–91. https://doi.org/10.1016/j.jenvman.2016.01.036.

Akhter, M., Hewa, G., 2016. The use of PCSWMM for assessing the impacts of land use changes on hydrological responses and performance of WSUD in managing the impacts at Myponga Catchment, South Australia. Water 8 (11), 511. https://doi.org/10.3390/w8110511.

Albrechts, L., Barbanente, A., Monno, V., 2020. Practicing transformative planning: the territory-landscape plan as a catalyst for change. City Territory Archit. 7(1). https://doi.org/10.1186/s40410-019-0111-2.

Alves, A., Gersonius, B., Kapelan, Z., Vojinovic, Z., Sanchez, A., 2019. Assessing the co-benefits of green-blue-grey infrastructure for sustainable urban flood risk management. J. Environ. Manage. 239, 244–254. https://doi.org/10.1016/j.jenvman.2019.03.036.

Ashley, R., Gersonius, B., Digman, C., Horton, B., Smith, B., Shaffer, P., 2018. Including uncertainty in valuing blue and green infrastructure for stormwater management. Ecosyst. Serv. 33, 237–246. https://doi.org/10.1016/j.ecoser.2018.08.011.

Bartos, M., Wong, B., Kerkez, B., 2018. Open storm: a complete framework for sensing and control of urban watersheds. Environ. Sci.: Water Res. Technol. 4 (3), 346–358. https://doi.org/10.1039/c7ew00374a.

Batica, J., 2015. Methodology for Flood Resilience Assessment in Urban Environments and Mitigation Strategy Development. (PhD thesis)Université Nice Sophia Antipolis, Nice, France.

Béné, C., Mehta, L., McGranahan, G., Cannon, T., Gupte, J., Tanner, T., 2018. Resilience as a policy narrative: potentials and limits in the context of urban planning. Clim. Dev. 10 (2), 116–133. https://doi.org/10.1080/17565529.2017.1301868.

Benedict, M.A., McMahon, E.T., 2006. Green Infrastructure: Linking Landscapes and Communities. Island Press, Washington.

Bird, D., King, D., Haynes, K., Box, P., Okada, T., Nairn, K., 2013. Impact of the 2010–11 Floods and the Factors That Inhibit and Enable Household Adaptation Strategies. National Climate Change Adaptation Research Facility, Gold Coast.

Brewington, L., Keener, V., Mair, A., 2019. Simulating land cover change impacts on groundwater recharge under selected climate projections, Maui, Hawai. Remote Sens. 11(24). https://doi.org/10.3390/rs11243048.

Burns, M.J., Wallis, E., Matic, V., 2015. Building capacity in low-impact drainage management through research collaboration. Freshw. Sci. 34 (3), 1176–1185. https://doi.org/10.1086/682565.

Busscher, T., van den Brink, M., Verweij, S., 2019. Strategies for integrating water management and spatial planning: organising for spatial quality in the Dutch "Room for the River" program. J. Flood Risk Manag. 12(1). https://doi.org/10.1111/jfr3.12448.

Carter, J.G., White, I., 2012. Environmental planning and management in an age of uncertainty: the case of the water framework directive. J. Environ. Manage. 113, 228–236. https://doi.org/10.1016/j.jenvman.2012.05.034.

Cheng, C., Yang, Y.C.E., Ryan, R., Yu, Q., Brabec, E., 2017. Assessing climate change-induced flooding mitigation for adaptation in Boston's Charles River watershed, USA. Landsc. Urban Plan. 167 (C), 25–36. https://doi.org/10.1016/j.landurbplan.2017.05.019.

Chui, T.F.M., Liu, X., Zhan, W.T., 2016. Assessing cost-effectiveness of specific LID practice designs in response to large storm events. J. Hydrol. 533, 353–364. https://doi.org/10.1016/j.jhydrol.2015.12.011.

Cilliers, D.P., 2019. Considering flood risk in spatial development planning: a land use conflict analysis approach. Jamba J. Disaster Risk Stud. 11 (1), 537. https://doi.org/10.4102/jamba.v11i1.537.

Cipolla, S.S., Maglionico, M., Stojkov, I., 2016. A long-term hydrological modelling of an extensive green roof by means of SWMM. Ecol. Eng. 95, 876–887. https://doi.org/10.1016/j.ecoleng.2016.07.009.

Cole, L.B., McPhearson, T., Herzog, C.P., Russ, A., Russ, A., Krasny, M.E., 2017. Green infrastructure. In: U.P.S. Online, (Ed.), Urban Environmental Education Review. University of Cornell University Press, p. 8. https://doi.org/10.7591/cornell/9781501705823.003.0028.

Damodaram, C., Zechman, E.M., 2013. Simulation-optimization approach to design low impact development for managing peak flow alterations in urbanizing watersheds. J. Water Resour. Plan. Manag. 139 (3), 290–298. https://doi.org/10.1061/(Asce)Wr.1943-5452.0000251.

Davidson, J.L., Jacobson, C., Lyth, A., Dedekorkut-Howes, A., Baldwin, C.L., Ellison, J.C., et al., 2016. Interrogating resilience: toward a typology to improve its operationalization. Ecol. Soc. 21 (2), 15. https://doi.org/10.5751/es-08450-210227.

Davies, C., Lafortezza, R., 2019. Transitional path to the adoption of nature-based solutions. Land Use Policy 80, 406–409. https://doi.org/10.1016/j.landusepol.2018.09.020.

Demuzere, M., Orru, K., Heidrich, O., Olazabal, E., Geneletti, D., Orru, H., et al., 2014. Mitigating and adapting to climate change: multi-functional and multi-scale assessment of green urban infrastructure. J. Environ. Manage. 146, 107–115. https://doi.org/10.1016/j.jenvman.2014.07.025.

Depietri, Y., McPhearson, T., 2017. Integrating the grey, green, and blue in cities: nature-based solutions for climate change adaptation and risk reduction. In: Kabisch, N., Korn, H., Stadler, J., Bonn, A. (Eds.), Nature-Based Solutions to Climate Change: Adaptation in Urban Areas. Springer International Publishing, Cham, Switzerland, pp. 91–109. https://doi.org/10.1007/978-3-319-56091-5_6.

Dhakal, K.P., 2017. Climate Change Impact on Urban Stormwater System and Use of Green Infrastructure for Adaptation: An Investigation on Technology, Policy, and Governance. (PhD thesis)Southern Illinois University, Carbondale, IL.

Dover, J.W., 2015. Green Infrastructure: Incorporating Plants and Enhancing Biodiversity in Buildings and Urban Environments. Routledge, Taylor & Francis Group, Earthscan from Routledge, New York.

Doyon, A., 2016. An Investigation Into Planning for Urban Resilience Through Niche Interventions. (PhD thesis)The University of Melbourne, Melbourne, Australia.

Dredge, M., 2015. An Overview of Natural Hazards for the Hamilton City Council. Waikato Regional Council. https://www.waikatoregion.govt.nz/assets/PageFiles/37500/TR201404.pdf.

Drosou, N., Soetanto, R., Hermawan, F., Chmutina, K., Bosher, L., Hatmoko, J.U.D., 2019. Key factors influencing wider adoption of blue-green infrastructure in developing cities. Water. 11(6). https://doi.org/10.3390/w11061234.

Emilsson, T., Sang, Å.O., 2017. Impacts of climate change on urban areas and nature-based solutions for adaptation. In: Kabisch, N., Korn, H., Stadler, J., Bonn, A. (Eds.), Nature-Based Solutions to Climate Change Adaptation in Urban Areas. Springer International Publishing, Cham, Switzerland, pp. 15–27.

Ercolani, G., Chiaradia, E.A., Gandolfi, C., Castelli, F., Masseroni, D., 2018. Evaluating performances of green roofs for stormwater runoff mitigation in a high flood risk urban catchment. J. Hydrol. 566, 830–845. https://doi.org/10.1016/j.jhydrol.2018.09.050.

Escobedo, F.J., Kroeger, T., Wagner, J.E., 2011. Urban forests and pollution mitigation: analyzing ecosystem services and disservices. Environ. Pollut. 159 (8–9), 2078–2087. https://doi.org/10.1016/j.envpol.2011.01.010.

European Union, 2013. Building a Green Infrastructure for Europe. European Commission, Belgium.https://www.greengrowthknowledge.org/resource/building-green-infrastructure-europe.

Federal Agency for Nature Conservation (BfN), 2017. Federal Green Infrastructure Concept. Nature Conservation Foundations for Plans Adopted by the German Federation. Bundesamt für Naturschutz.www.bfn.de/bkgi.html.

Fletcher, T.D., Shuster, W., Hunt, W.F., Ashley, R., Butler, D., Arthur, S., et al., 2015. SUDS, LID, BMPs, WSUD and more—the evolution and application of terminology surrounding urban drainage. Urban Water J. 12 (7), 525–542. https://doi.org/10.1080/1573062X.2014.916314.

Folke, C., Carpenter, S.R., Walker, B., Scheffer, M., Chapin, T., Rockstrom, J., 2010. Resilience thinking: integrating resilience, adaptability and transformability. Ecol. Soc. 15 (4), 9.

Francesch-Huidobro, M., Dabrowski, M., Tai, Y., Chan, F., Stead, D., 2017. Governance challenges of flood-prone delta cities: integrating flood risk management and climate change in spatial planning. Prog. Plan. 114, 1–27. https://doi.org/10.1016/j.progress.2015.11.001.

Galderisi, A., Treccozzi, E., 2017. Green strategies for flood resilient cities: the Benevento case study. Green Urban. 37, 655–666. https://doi.org/10.1016/j.proenv.2017.03.052.

Ghofrani, Z., Sposito, V., Faggian, R., 2017. A comprehensive review of blue-green infrastructure concepts. Int. J. Environ. Sustain. 6 (1), 15–36.

Goncalves, M.L.R., Zischg, J., Rau, S., Sitzmann, M., Rauch, W., Kleidorfer, M., 2018. Modeling the effects of introducing low impact development in a tropical city: a case study from Joinville, Brazil. Sustainability. 10(3). https://doi.org/10.3390/su10030728.

Grafakos, S., Viero, G., Reckien, D., Trigg, K., Viguie, V., Sudmant, A., et al., 2020. Integration of mitigation and adaptation in urban climate change action plans in Europe: a systematic assessment. Renew. Sust. Energ. Rev. 121. https://doi.org/10.1016/j.rser.2019.109623.

Groundwater Foundation, 2019. From: https://www.groundwater.org/action/home/raingardens.html.

Gulbaz, S., Kazezyilmaz-Alhan, C.M., 2017. An evaluation of hydrologic modeling performance of EPA SWMM for bioretention. Water Sci. Technol. 76 (11–12), 3035–3043. https://doi.org/10.2166/wst.2017.464.

Guo, J.C.Y., 2012. Cap-orifice as a flow regulator for rain garden design. J. Irrig. Drain. Eng. 138 (2), 198–202. https://doi.org/10.1061/(ASCE)IR.1943-4774.0000399.

Handmer, J., Honda, Y., Kundzewicz, Z.W., Arnell, N., Benito, G., Hatfield, J., et al., 2012. Changes in impacts of climate extremes: human systems and ecosystems. Managing the Risks of Extreme Events and Disasters to Advance Climate Change Adaptation: Special Report of the Intergovernmental Panel on Climate Change. Cambridge University Press, pp. 231–290.

Herath, H.M.M., Wijesekera, N.T.S., 2019. A state-of-the-art review of flood risk assessment in urban area. In: IOP Conference Series: Earth and Environmental Science. https://doi.org/10.1088/1755-1315/281/1/012029.

Hettiarachchi, S., Wasko, C., Sharma, A., 2018. Increase in flood risk resulting from climate change in a developed urban watershed—the role of storm temporal patterns. Hydrol. Earth Syst. Sci. 22 (3), 2041–2056. https://doi.org/10.5194/hess-22-2041-2018.

Hislop, M., Scott, A.J., Corbett, A., 2019. What does good green infrastructure planning policy look like? Developing and testing a policy assessment tool within Central Scotland UK. Plan. Theory Pract. 20 (5), 633–655. https://doi.org/10.1080/14649357.2019.1678667.

Hoyer, J., Dickhaut, W., Kronawitter, L., Weber, B., 2011. Water Sensitive Urban Design. Principles and Inspiration for Sustainable Stormwater Management in the City of the Future. Jovis, Hamburg, Germany. http://www.switchurbanwater.eu/outputs/pdfs/W5-1_GEN_MAN_D5.1.5_Manual_on_WSUD.pdf.

Hu, M.C., Zhang, X.Q., Li, Y., Yang, H., Tanaka, K., 2019. Flood mitigation performance of low impact development technologies under different storms for retrofitting an urbanized area. J. Clean. Prod. 222, 373–380. https://doi.org/10.1016/j.jclepro.2019.03.044.

Hughes, J., Sharman, B., 2015. Flood resilient communities: a framework and case studies. In: Asia Pacific Stormwater Conference, 23.

IPCC, 2014. Climate change 2014: synthesis report. In: Contribution of Working Groups I, II and III to the Fifth Assessment Report of the Intergovernmental Panel on Climate Change. Intergovernmental Panel on Climate Change, Geneva, Switzerland.

IPCC, 2019. IPCC Special Report on the Ocean and Cryosphere in a Changing Climate. .

Irajifar, L., 2016. Development and Validation of a Neighbourhood Disaster Resilience Index a Case Study From Australia. (PhD thesis) Griffith University, Brisbane, Australia.

Jayawardena, H.M.I., 2018. A Research Approach to Develop an Ecologically Inspired Water Sensitive Planning and Design Framework for Delivering Climate Resilient Cities: A Case Study of Colombo. (Sri Lanka thesis)University of Auckland.

Jha, A.K., 2012. Cities and Flooding a Guide to Integrated Urban Flood Risk Management for the 21st Century. World Bank, Washington, DC.

Jia, Z., Tang, S., Luo, W., Li, S., Zhou, M., 2016. Small scale green infrastructure design to meet different urban hydrological criteria. J. Environ. Manage. 171, 92–100. https://doi.org/10.1016/j.jenvman.2016.01.016.

Juan, A., Hughes, C., Fang, Z., Bedient, P., 2016. Hydrologic performance of watershed-scale low-impact development in a high-intensity rainfall region. J. Irrig. Drain. Eng. 143 (4), 04016083.

Kazemi, H., 2014. Evaluating the Effectiveness and Hydrological Performance of Green Infrastructure Stormwater Control Measures. (PhD thesis)University of Louisville.

Koc, C.B., Osmond, P., Peters, A., 2016. A green infrastructure typology matrix to support urban microclimate studies. Process. Eng. 169, 183–190. https://doi.org/10.1016/j.proeng.2016.10.022.

Lafortezza, R., Davies, C., Sanesi, G., Konijnendijk, C.C., 2013. Green infrastructure as a tool to support spatial planning in European urban regions. iForest 6, 102–108. https://doi.org/10.3832/ifor0723-006.

Lamond, J.E., Rose, C.B., Booth, C.A., 2015. Evidence for improved urban flood resilience by sustainable drainage retrofit. Proc. Inst. Civ. Eng. Urban Des. Pla. 168 (2), 101–111. https://doi.org/10.1680/udap.13.00022.

Lawson, E., Thorne, C., Ahilan, S., Allen, D., Arthur, S., Everett, G., et al., 2014. Delivering and evaluating the multiple flood risk benefits in blue-green cities: an interdisciplinary approach. WIT Trans. Ecol. Environ. 184, 113–124. https://doi.org/10.2495/FRIAR140101.

Lemes de Oliveira, F., 2019. Towards a spatial planning framework for the re-naturing of cities. In: Newman, P., Desha, C., Mell, I. (Eds.), Planning Cities With Nature. Springer International Publishing, Cham, Switzerland, pp. 81–95.

Liao, K.H., 2012. A theory on urban resilience to floods-a basis for alternative planning practices. Ecol. Soc. 17 (4), 48. https://doi.org/10.5751/Es-05231-170448.

Liao, K.-H., Deng, S., Tan, P.Y., 2017. Blue-green infrastructure: new frontier for sustainable urban stormwater management. In: Tan, P.Y., Jim, C.Y. (Eds.), Greening Cities. Springer International Publishing, Singapore, pp. 203–226. https://doi.org/10.1007/978-981-10-4113-6.

Lin, J.Y., Chen, C.F., Ho, C.C., 2018. Evaluating the effectiveness of green roads for runoff control. J. Sustain. Water Built Environ. 4(2). https://doi.org/10.1061/Jswbay.0000847.

Liu, Y.C., Cheng, C.L., 2014. A solution for flood control in urban area: using street block and raft foundation space operation model. Water Resour. Manag. 28 (14), 4985–4998. https://doi.org/10.1007/s11269-014-0783-z.

Mabon, L., 2019. Enhancing post-disaster resilience by 'building back greener': evaluating the contribution of nature-based solutions to recovery planning in Futaba County, Fukushima prefecture, Japan. Landsc. Urban Plan. 187, 105–118. https://doi.org/10.1016/j.landurbplan.2019.03.013.

Marot, N., Golobič, M., Müller, B., 2015. Green infrastructure in Central, Eastern, and South-Eastern Europe: is there a universal solution to environmental and spatial challenges? Urbani Izziv 26 (Supplement), S1–S12. https://doi.org/10.5379/urbani-izziv-en-2015-26-supplement-000.

McFarland, A.R., Larsen, L., Yeshitela, K., Engida, A.N., Love, N.G., 2019. Guide for using green infrastructure in urban environments for stormwater management. Environ. Sci.: Water Res. Technol. 5 (4), 643–659. https://doi.org/10.1039/c8ew00498f.

Meerow, S., Newell, J.P., Stults, M., 2016. Defining urban resilience: a review. Landsc. Urban Plan. 147 (2016), 38–49. https://doi.org/10.1016/j.landurbplan.2015.11.011.

Mell, I., 2020. What future for green infrastructure planning? Evaluating the changing environment for green infrastructure planning following the revocation of regional planning policy in England. Plan. Pract. Res. 35 (1), 18–50. https://doi.org/10.1080/02697459.2020.1714271.

Mell, I., Lemes de Oliveira, F., 2019. Re-naturing our future cities. In: Newman, P., Desha, C., de Lemes Oliveira, F., Mell, I. (Eds.), Planning Cities With Nature. Springer International Publishing, Cham, Switzerland, pp. 281–285.

Mell, I.C., Henneberry, J., Hehl-Lange, S., Keskin, B., 2013. Promoting urban greening: valuing the development of green infrastructure investments in the urban core of Manchester, UK. Urban For. Urban Green. 12 (3), 296–306. https://doi.org/10.1016/j.ufug.2013.04.006.

Mell, I., Allin, S., Reimer, M., Wilker, J., 2017. Strategic green infrastructure planning in Germany and the UK: a transnational evaluation of the evolution of urban greening policy and practice. Int. Plan. Stud. 22 (4), 333–349. https://doi.org/10.1080/13563475.2017.1291334.

Meng, M., Dąbrowski, M., Tai, Y., Stead, D., Chan, F., 2019. Collaborative spatial planning in the face of flood risk in delta cities: a policy framing perspective. Environ Sci Policy 96, 95–104. https://doi.org/10.1016/j.envsci.2019.03.006.

National Association of City Transportation Officials, 2019. Urban Street Design Guide. From: https://nacto.org/publication/urban-street-design-guide/street-design-elements/stormwater-management/bioswales/.

Natural England, 2010. Nature Nearby: Accessible Natural Greenspace Guidance. Natural England, Peterborough.

Naturally Resilient Communities, 2019. Urban Trees + Forests. From: http://nrcsolutions.org/urban-forests-trees/.

Nordman, E.E., Isely, E., Isely, P., Denning, R., 2018. Benefit-cost analysis of stormwater green infrastructure practices for Grand Rapids, Michigan, USA. J. Clean. Prod. 200, 501–510. https://doi.org/10.1016/j.jclepro.2018.07.152.

O'Donnell, E.C., Lamond, J.E., Thorne, C.R., 2017. Recognising barriers to implementation of blue-green infrastructure: a Newcastle case study. Urban Water J. 14 (9), 964–971. https://doi.org/10.1080/1573062x.2017.1279190.

O'Hare, P., White, I., 2018. Beyond 'just' flood risk management: the potential for-and limits to-alleviating flood disadvantage. Reg. Environ. Chang. 18 (2), 385–396. https://doi.org/10.1007/s10113-017-1216-3.

Pauleit, S., Zölch, T., Hansen, R., Randrup, T.B., van den Bosch, C.K., 2017. Nature-based solutions and climate change—four shades of green. In: Kabisch, N., Korn, H., Stadler, J., Bonn, A. (Eds.), Nature-Based Solutions to Climate Change Adaptation in Urban Areas. Springer International Publishing, Cham, Switzerland, pp. 29–49.

Ran, J., Nedovic-Budic, Z., 2016. Integrating spatial planning and flood risk management: a new conceptual framework for the spatially integrated policy infrastructure. Comput. Environ. Urban. Syst. 57, 68–79. https://doi.org/10.1016/j.compenvurbsys.2016.01.008.

Salata, K., Yiannakou, A., 2016. Green infrastructure and climate change adaptation. Tema 9 (1), 10–27. https://doi.org/10.6092/1970-9870/3723.

Scott, A., 2019. Mainstreaming the environment in planning policy and decision making. In: Davoudi, S., Cowell, R., White, I., Blanco, H. (Eds.), The Routledge Companion to Environmental Planning. Taylor & Francis Group, Milton, pp. 420–433.

Sinnett, D., Smith, N., Burgess, S., 2015. Handbook on Green Infrastructure: Planning, Design and Implementation, 6th ed. 3. Edward Elgar Publishing, Beaverton, United States.

Sinnett, D., Calvert, T., Smith, N., 2019. Do built environment assessment systems include high-quality green infrastructure? In: Newman, P., Desha, C., de Lemes Oliveira, F., Mell, I. (Eds.), Planning Cities With Nature. Springer International Publishing, Cham, Switzerland, pp. 169–186.

Slätmo, E., Nilsson, K., Turunen, E., 2019. Implementing green infrastructure in spatial planning in Europe. Land. 8(4). https://doi.org/10.3390/land8040062.

Soil Science Society of America (SSSA), 2019. Soils Sustain Live. From: https://www.soils.org/discover-soils/soils-in-the-city/green-infrastructure/important-terms/rain-gardens-bioswales.

Sörensen, J., 2018. Urban, Pluvial Flooding: Blue-Green Infrastructure as a Strategy for Resilience. (PhD thesis). Lund University, Sweden.

Staddon, C., Ward, S., De Vito, L., Zuniga-Teran, A., Gerlak, A.K., Schoeman, Y., et al., 2018. Contributions of green infrastructure to enhancing urban resilience. Environ. Syst. Decis. 38 (3), 330–338. https://doi.org/10.1007/s10669-018-9702-9.

Struck, S., Lichten, K.H., 2010. Low Impact Development 2010: Redefining Water in the City. American Society of Civil Engineers, Reston, United States.

Suarez, P., Anderson, W., Mahal, V., Lakshmanan, T.R., 2005. Impacts of flooding and climate change on urban transportation: a systemwide performance assessment of the Boston Metro Area. Transp. Res. Part D: Transp. Environ. 10 (3), 231–244. https://doi.org/10.1016/j.trd.2005.04.007.

Thorne, C.R., Lawson, E.C., Ozawa, C., Hamlin, S.L., Smith, L.A., 2018. Overcoming uncertainty and barriers to adoption of blue-green infrastructure for urban flood risk management. J. Flood Risk Manag. 11 (S2), S960–S972. https://doi.org/10.1111/jfr3.12218.

Toimil, A., Losada, I.J., Nicholls, R.J., Dalrymple, R.A., Stive, M.J.F., 2020. Addressing the challenges of climate change risks and adaptation in coastal areas: a review. Coast. Eng. 156. https://doi.org/10.1016/j.coastaleng.2019.103611.

United States Environmental Protection Agency, 2014. Using Green Infrastructure to Mitigate Flooding in La Crosse. Assessment of Climate Change Impacts and System-Wide Benefits. United States Environmental Protection Agency, La Crosse. https://www.epa.gov/sites/production/files/2015-10/documents/lacrosse_tech_assistance.pdf.

United States Environmental Protection Agency, 2020. Green Infrastructure. From: https://www.epa.gov/green-infrastructure.

UNSDR, 2007. International Strategy for Disaster Reduction. www.unisdr.org/unisdr.

van Roon, M.R., 2012. Wetlands in the Netherlands and New Zealand: optimising biodiversity and carbon sequestration during urbanisation. J. Environ. Manage. 101, 143–150. https://doi.org/10.1016/j.jenvman.2011.08.026.

van Roon, M.R., 2017. Justifying water sensitive development: science informing policy and practice. Eur. J. Sustain. Dev. 6 (4), 21–31. https://doi.org/10.14207/ejsd.2017.v6n4p21.

Wang, M., Zhang, D., Cheng, Y., Tan, S.K., 2019. Assessing performance of porous pavements and bioretention cells for stormwater management in response to probable climatic changes. J. Environ. Manage. 243, 157–167. https://doi.org/10.1016/j.jenvman.2019.05.012.

Webber, J.L., Fletcher, T.D., Cunningham, L., Fu, G., Butler, D., Burns, M.J., 2019. Is green infrastructure a viable strategy for managing urban surface water flooding? Urban Water J. 1–11. https://doi.org/10.1080/1573062X.2019.1700286.

White, I., 2010. Water and the City: Risk, Resilience and Planning for a Sustainable Future. Routledge, New York.

White, I., Haughton, G., 2017. Risky times: hazard management and the tyranny of the present. Int. J. Disaster Risk Reduct. 22, 412–419. https://doi.org/10.1016/j.ijdrr.2017.01.018.

White, I., O'Hare, P., 2014. From rhetoric to reality: which resilience, why resilience, and whose resilience in spatial planning? Eviron. Plann. C. Gov. Policy 32 (5), 934–950. https://doi.org/10.1068/c12117.

Woodward, M., Kapelan, Z., Gouldby, B., 2014. Adaptive flood risk management under climate change uncertainty using real options and optimization. Risk Anal. 34 (1), 75–92. https://doi.org/10.1111/risa.12088.

Yiannakou, A., Salata, K.-D., 2017. Adaptation to climate change through spatial planning in compact urban areas: a case study in the city of Thessaloniki. Sustainability 9 (2), 271.

Zhang, L., Oyake, Y., Morimoto, Y., Niwa, H., Shibata, S., 2019. Rainwater storage/infiltration function of rain gardens for management of urban storm runoff in Japan. Landsc. Ecol. Eng. 15 (4), 421–435. https://doi.org/10.1007/s11355-019-00391-w.

Zhu, Z.H., Chen, X.H., 2017. Evaluating the effects of low impact development practices on urban flooding under different rainfall intensities. Water. 9(7). https://doi.org/10.3390/w9070548.

# CHAPTER 16

# Effect of climate change on the insurance sector

*Adam D. Krauss*

Traub Lieberman Straus & Shrewsberry LLP, Hawthorne, NY, United States

OUTLINE



## 1 Introduction

The issue of climate change is increasingly permeating virtually every topic of discussion and the concern is unlikely to dissipate anytime soon. Rightly so, as many of the scientific models present ongoing catastrophic damage scenarios affecting persons, property,



*The Impacts of Climate Change*
https://doi.org/10.1016/B978-0-12-822373-4.00014-8

businesses governments, economies, ecosystems, and natural resources, to name but a few. The scope and scale of estimated damage from climate change is unprecedented and the costs to mitigate the risk no less daunting. Insurers and their policyholders face high exposure risk from climate change on many fronts, including general liability claims for third-party bodily injury and property damage, Directors and Officers ("D&O") claims for a Company's failure to properly disclose climate-related risk to its business and/or failure to "align its business model with a low-carbon future" and first-party loss, including business interruption. Although, to date, there have been minimal lawsuits seeking insurance coverage relating to climate change damages, we expect that to change, given the increasing number of underlying lawsuits and related activity, coupled with the staggering liability that is at stake.

The litigation landscape has been changing and this in turn has implications for the (re)insurance industry. Climate change is not only about the liability side of the balance sheet. (Re) insurers, as investors, need to appraise existing investment strategy including fossil fuel and renewable energy companies to help mitigate the projected impact of climate change.

## 2 Climate change science—The basics

Natural gases, alternatively known as greenhouse gases ("GHGs"), in the Earth's atmosphere impact the levels of radiation from the Sun. Human activity, most notably the burning of fossil fuels, has increased the quantum of these natural gases and has caused changes in climate. These GHGs vary in life span and global warming potential ("GWP") but are universal since they are well mixed and do not respect national boundaries.

### 2.1 What is "climate change"?

"Climate" is how the atmosphere behaves over relatively long periods of time, whereas "weather" refers to conditions of the atmosphere over a short period of time. Climate is the aggregated patterns over time of weather, meaning averages, extremes, timing, and spatial distribution of weather events. Relatively, small temperature changes can result in big alterations to weather patterns.[a]

### 2.2 The "greenhouse effect"

As depicted in Fig. 1[b], the Greenhouse Effect describes how natural gases in the Earth's atmosphere reduce the amount of heat escaping from the Earth and how increased levels of these GHGs from human activities can amplify global warming through an "enhanced" greenhouse effect.

[a] https://oceanservice.noaa.gov/facts/weather_climate.html.

[b] http://css.umich.edu/factsheets/climate-change-science-and-impacts-.

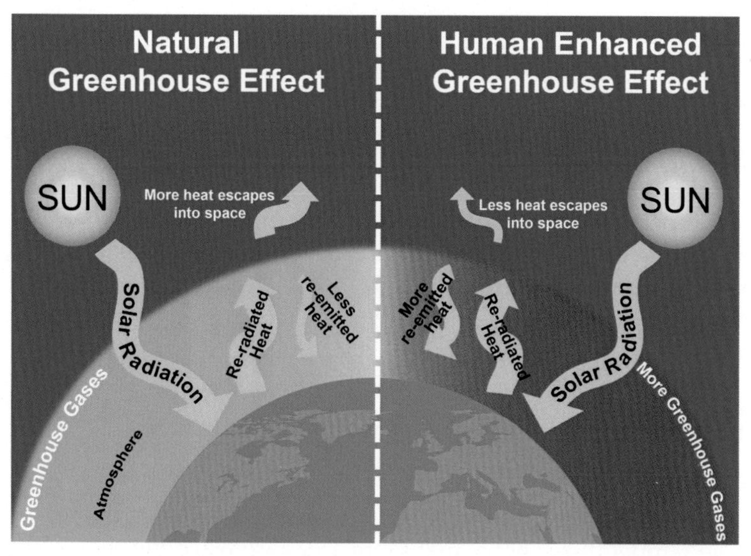


**FIG. 1** The greenhouse effect. *From Center for Sustainable Systems, University of Michigan. 2020. Climate Change: Science and Impacts Factsheet. Pub. No. CSS05-19; Will Elder/NPS.*

The Earth's atmosphere traps outgoing radiation (the Greenhouse Effect). It's this equilibrium of incoming and outgoing radiation that makes the Earth habitable. Thus, although global warming is often called the "Greenhouse Effect," it is more properly termed the "Enhanced Greenhouse Effect."

The GHGs are water vapor ($H_2O$), carbon dioxide ($CO_2$), nitrous oxide ($N_2O$), chlorofluorocarbons ("CFCs"), methane ($CH_4$), and ozone.[c] Various human activities release gases and contribute to the Greenhouse Effect. Some of the largest sources of these GHGs include: the burning of fossil fuels (carbon dioxide); aerosol sprays and refrigerants (CFCs); agricultural and industrial activities (nitrous oxide); livestock and other agricultural practices and decay of organic waste (methane).[d]

Each of these gases can remain in the atmosphere for different amounts of time, ranging from a few years to thousands of years.[e] Importantly, all of these gases remain in the atmosphere long enough to become well mixed, meaning that the amount that is measured in the atmosphere is roughly the same all over the world, regardless of the source of the emissions.[f]

Some gases are more effective than others at making the planet warmer. For each GHG, a Global Warming Potential ("GWP") has been determined to reflect how long it remains in the atmosphere, on average, and how strongly it absorbs energy. Gases with a higher GWP absorb more energy, per pound, than gases with a lower GWP, and thus contribute more to warming the Earth.

[c] http://www.epa.gov/ghgemissions/overview-greenhouse-gases.

[d] Id.

[e] Id.

[f] Id.

# 3 Climate change data and accountability

There is overwhelming scientific agreement that human activities are changing the global climate system and these changes are already affecting human and natural systems. Other observed changes include rising sea levels, ocean warming and acidification, melting sea ice, thawing permafrost, increases in the frequency and severity of extreme events, and a variety of impacts on people, communities, and ecosystems.[g]

Measures of climate change include global surface temperatures and $CO_2$ levels which have increased at an unprecedented pace. Since 1993 average sea levels have risen at twice the rate of the long-term trend.[h]

## 3.1 Key trends: Increasing surface and water temperatures

### 3.1.1 *Surface temperatures*

Though surface warming has not been uniform across the planet, the upward trend in the globally averaged temperature shows that more areas are warming than cooling. According to the NOAA 2019 Global Climate Summary, the combined land and ocean temperature has increased at an average rate of 0.07°C (0.13°F) per decade since 1880; however, the average rate of increase since 1981 (0.18°C/0.32°F) is more than twice as great (Fig. 2).[i]

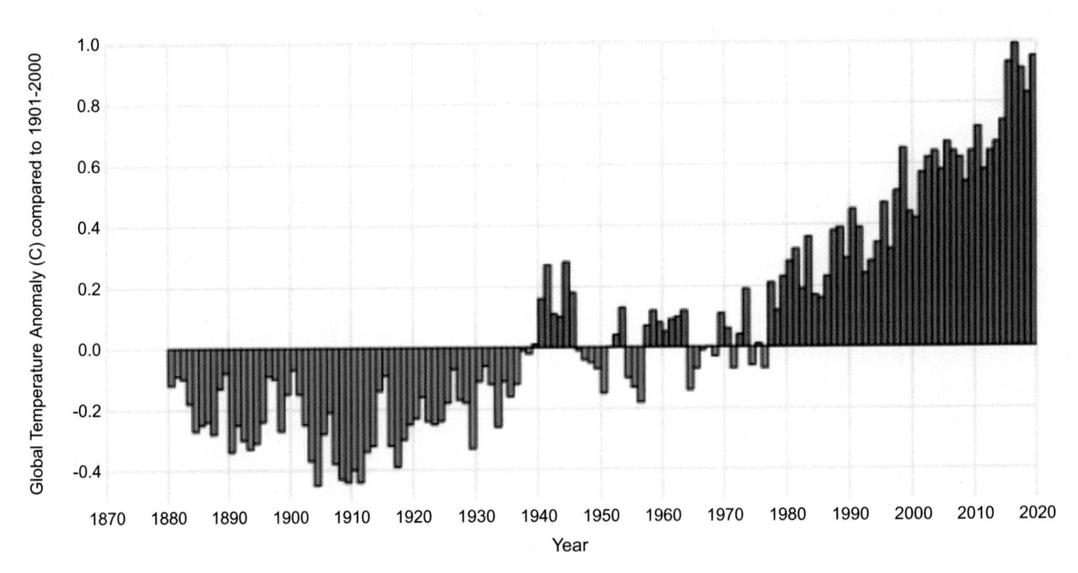


**FIG. 2** History of global surface temperature since 1880 (Id.).

[g]https://climate.nasa.gov/evidence/.

[h]http://www.epa.gov/climate-indicators/climate-change-indicators-sea-level.

[i]https://www.climate.gov/news-features/understanding-climate/climate-change-global-temperature.

A total of 19 of the 20 warmest years all have occurred since 2001, with the exception of 1998.[j] 2010–19 was the hottest decade ever recorded and the World Meteorological Organization ranked 2019 second warmest for the globe.[k]

### 3.1.2 *Ocean temperatures*

More than 90% of the excess heat is stored within the world's oceans, where it accumulates and causes increases in ocean temperature. Because the oceans are the main repository of the Earth's energy imbalance, measuring ocean heat content is one of the best ways to quantify the rate of global warming. Current data reveal that the world's oceans (especially the upper 2000m) in 2019 were the warmest in recorded human history.[l]

The global ocean-only temperature departure from average for August 2019 was the highest on record at 0.84°C (1.51°F) above the twentieth century average of 16.4°C (61.4°F). The ten highest monthly global ocean surface temperature departures from average have occurred since September 2015.[m]

## 3.2 Key trends: Sea level rise

The two major causes of global sea level rise are: (1) thermal expansion caused by warming of the ocean (since water expands as it warms) and (2) increased melting of land-based ice, such as glaciers and ice sheets. The oceans are absorbing more than 90% of the increased atmospheric heat associated with emissions from human activity.[n]

Global mean sea level has risen about 8–9 in. (21–24 cm) since 1880, with about a third of that coming in just the last two and a half decades. In 2019, global mean sea level was 3.4 in. (87.6 mm) above the 1993 average—the highest annual average in the satellite record (1993–present). From 2018 to 2019, global sea level rose 0.24 in. (6.1 mm) (Fig. 3).[o]

## 3.3 Key trends: Increasing $CO_2$

According to measurements taken since 1960 by the NOAA Earth System Research Laboratory at the Mauna Loa Observatory in Hawaii, atmospheric carbon dioxide ("$CO_2$") has been steadily increasing with readings in September 2020 approaching 412 parts per million ("ppm"), as shown in Fig. 4.

$CO_2$ concentrations have increased by more than 40% to levels not seen in at least 3 million years.[p]

[j]https://climate.nasa.gov/vital-signs/global-temperature/.

[k]https://public.wmo.int/en/media/press-release/wmo-confirms-2019-second-hottest-year-record.

[l]https://link.springer.com/content/pdf/10.1007%2Fs00376-020-9283-7.pdf.

[m]https://www.ncdc.noaa.gov/sotc/global/201908.

[n]https://oceanservice.noaa.gov/facts/sealevel.html.

[o]https://www.climate.gov/news-features/understanding-climate/climate-change-global-sea-level.

[p]WMO Greenhouse Gas Bulletin, No. 13 (Oct. 30, 2017).

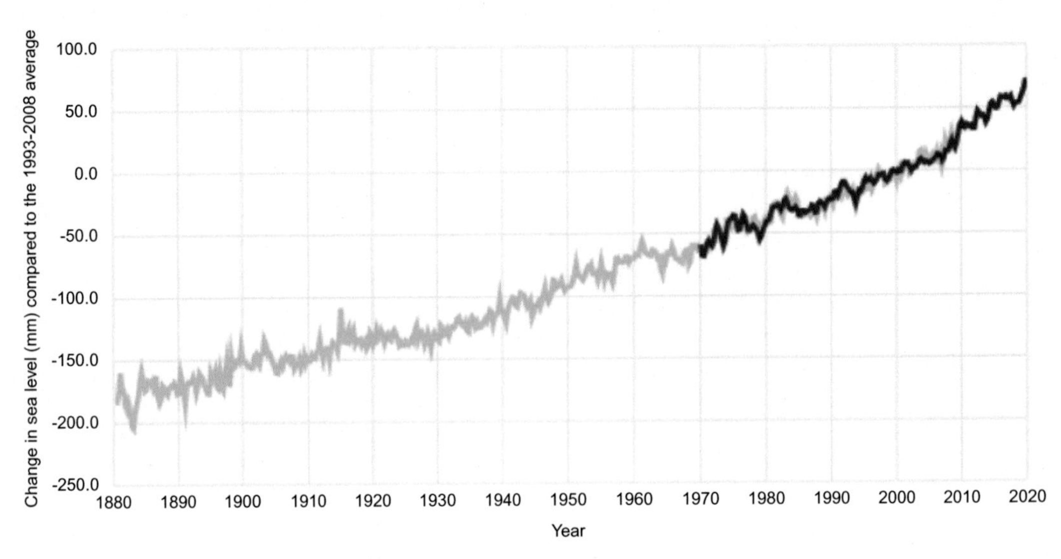


**FIG. 3** Global sea level since 1880 (Id.).

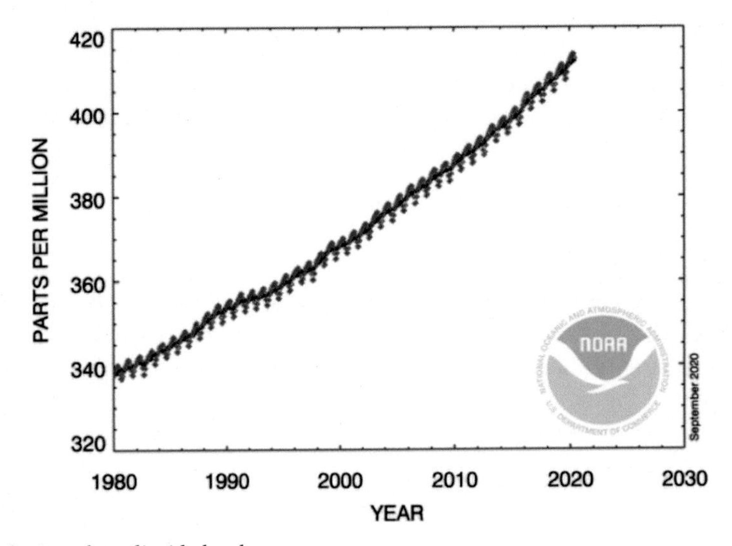


**FIG. 4** Atmospheric carbon dioxide levels

The world's countries emit vastly different amounts of GHGs into the atmosphere. Fig. 5 reflects data compiled by the International Energy Agency, which estimates each region's $CO_2$ emissions from the combustion of coal, natural gas, oil, and other fuels, including industrial waste and nonrenewable municipal waste:

Largely as a result of the COVID-19 pandemic, global $CO_2$ emissions declined by approximately 8% in 2020. Such a year-on-year reduction would be the largest ever, six times larger

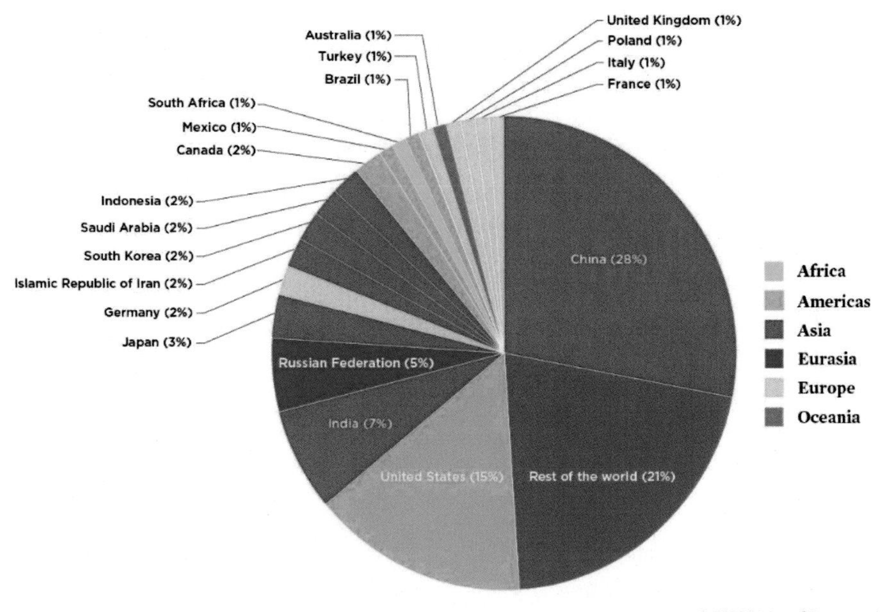


**FIG. 5** Breakdown of $CO_2$ emissions by Country (2020) (https://www.ucsusa.org/resources/each-countrys-share-co2-emissions).

than the previous record reduction of 0.4 Gt in 2009—caused by the global financial crisis—and twice as large as the combined total of all previous reductions since the end of World War II (Fig. 6).[q]

Similarly, energy demand declined in all major regions in 2020. Demand in China was projected to decline in 2020 by more than 4%, a reversal from average annual demand growth of nearly 3% between 2010 and 2019 (Fig. 7).[r]

Despite this predicted short-term fall in $CO_2$ emissions, atmospheric $CO_2$ measurements remain at record high levels due to contributing anthropogenic emissions.

## 3.4 2013 and 2017 climate change accountability studies

In 2013, the Climate Accountability Institute Study quantified total historical carbon dioxide and methane emissions of the top 90 fossil fuel companies and other carbon producers for the period 1854–2010.[s] A 2017 study builds upon the earlier study to conclude that since 1880, 90 companies are responsible for up to 50% of global temperature rise, 57% of the increase in atmospheric $CO_2$, and between 26% and 32% of sea level rise.[t]

[q]https://www.iea.org/world.

[r]IEA (2020), Global Energy Review 2020, IEA, Paris https://www.iea.org/reports/global-energy-review-2020.

[s]https://link.springer.com/article/10.1007/s10584-013-0986-y.

[t]https://link.springer.com/article/10.1007%2Fs10584-017-1978-0.

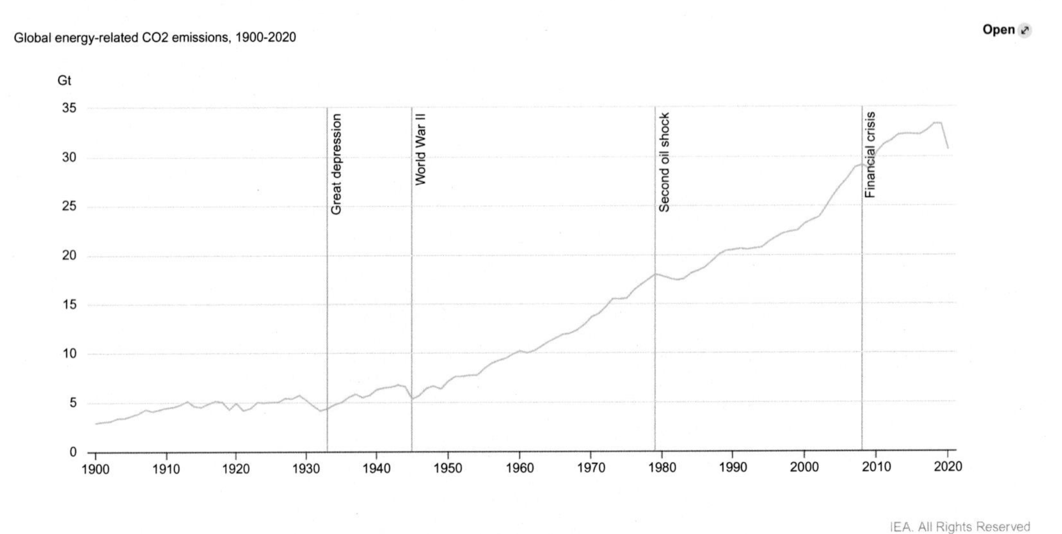


**FIG. 6** Global energy related $CO_2$ emissions (1900–2020) (Id.).

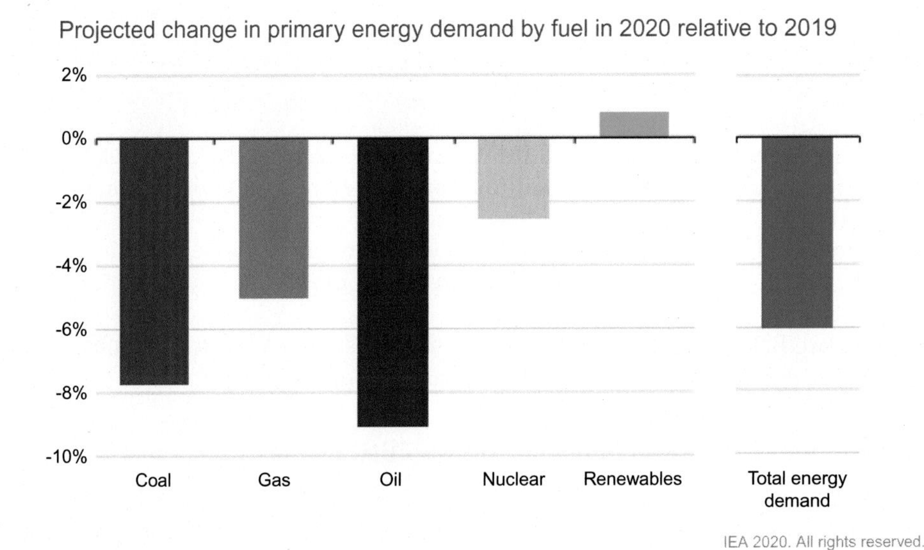


**FIG. 7** Projected change in primary energy demand by fuel type in 2020 versus 2019 (Id.).

As depicted in Figs. 8 and 9, the 2017 study rankings include both independent companies and sovereign nations.

## 3.5 Attribution science and climate change

The evolving field of climate change attribution science is playing a key role in shaping our understanding of how humans are affecting the global climate system, and in informing

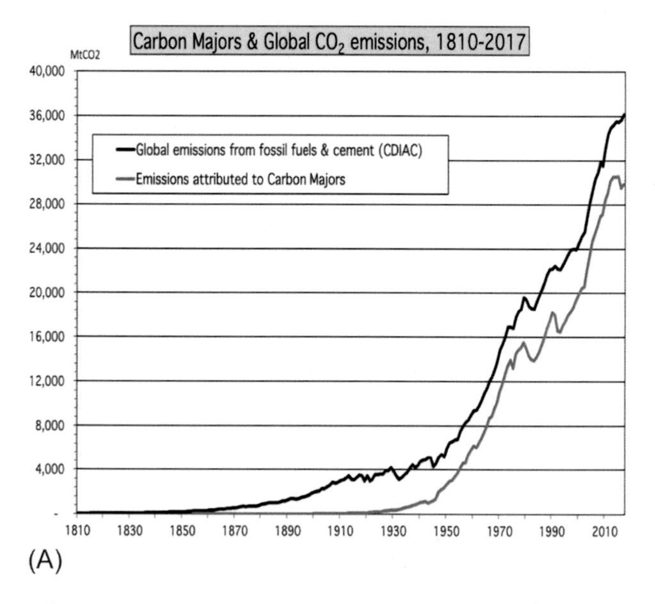


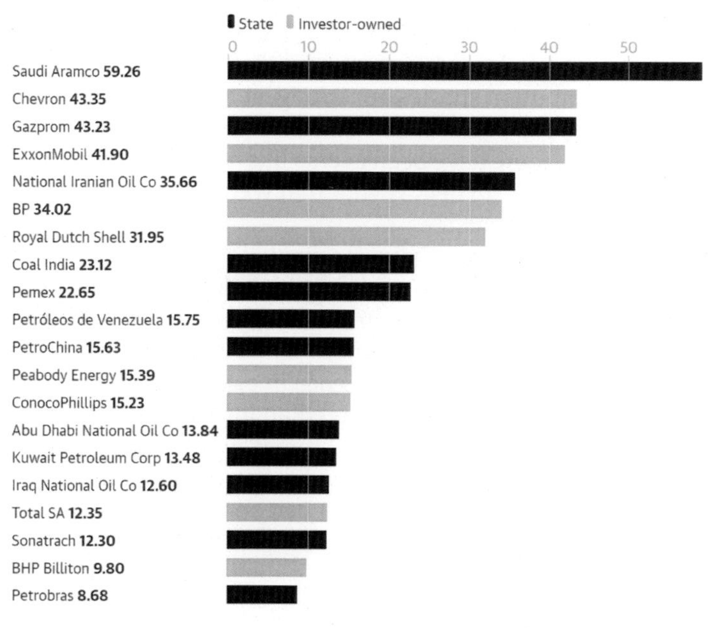


**FIG. 8** (A, B) "Carbon majors" and Global $CO_2$ emissions (1810–2017).

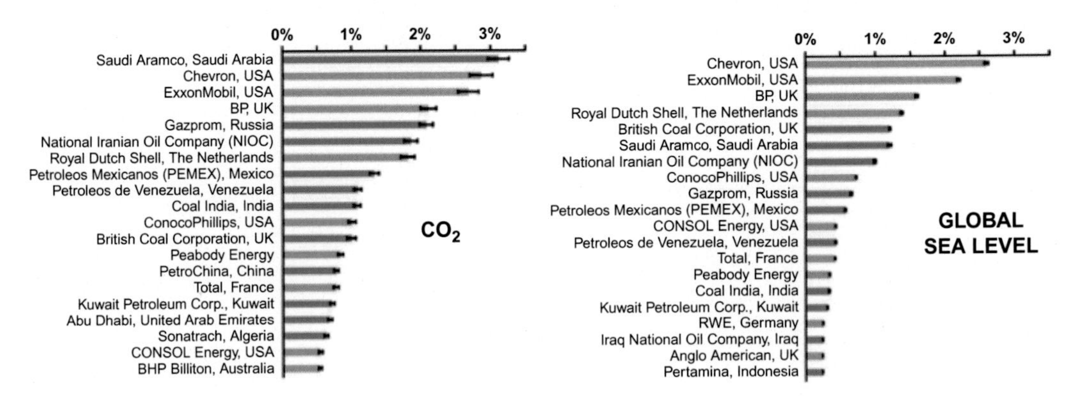


**FIG. 9** "Carbon majors" percent contribution of historical global rise in $CO_2$ and seal level (1880–2010) (https://link.springer.com/article/10.1007/s10584-017-1978-0).

discussions about responsibility for climate change impacts. New research and more powerful computer models are advancing scientists' ability to dissect the forces that can worsen extreme weather. Attribution studies generally try to determine how the intensity or frequency of an event was altered by climate change or other factors, such as weather patterns like El Niño. Often, they take the details of an event, put them into a computer model and then use the model to show what would have happened without certain variables, such as rising temperatures caused by increased greenhouse gases. Some use models to estimate the probability of an event occurring with or without changes, then use the probabilities to estimate the fraction of attributable risk. In a 2016 report by the National Academy of Sciences, the committee stresses that attribution studies are strongest for events that have a long-term historical record of observations and can be simulated in climate models.[u]

Within the past few years, there have been events like a 2017 marine heat wave off the coast of Australia[v] or Japan's deadly 2018 heat wave[w] that were found to be almost entirely attributable to human-caused climate change.

The consequences of climate change have received increasing attention in recent years, as communities around the world have been hit hard by climate-related natural disasters. As discussed below, attribution science is central to the recent wave of climate litigation in the United States ("U.S.") and abroad, as it informs discussions of responsibility for climate change. For example, the role of attribution science in establishing legal "standing," is being used to determine whether plaintiffs have suffered an injury, or risk of an injury, that can be linked to anthropogenic climate change, and therefore linked to emissions that were generated by a private entity or inadequately regulated by a government entity. Attribution data are a valuable adjunct to impact projections as it can be used to establish an existing injury while also lending credibility to projections of future harm.[x]

[u]https://www.nap.edu/catalog/21852/attribution-of-extreme-weather-events-in-the-context-of-climate-change.

[v]http://www.ametsoc.net/eee/2017a/ch20_EEEof2017_Perkins.pdf.

[w]https://www.jstage.jst.go.jp/article/sola/advpub/0/advpub_15A-002/_pdf/-char/ja.

[x]https://journals.library.columbia.edu/index.php/cjel/article/view/4730/2118.

# 4 Climate change litigation

Climate litigation is increasing in the U.S. and abroad.

In total, 1587 cases of climate litigation have been identified as being brought between 1986 and the end of May 2020[y]: 1213 cases in the U.S. and 374 cases in 36 other countries and eight regional or international jurisdictions. Outside the U.S., the majority of cases have been brought in Australia (98 cases), the U.K. (62), and the EU bodies and courts (57).[z] Traditionally, these cases have been brought against governments, but beginning in 2017, there has been a steep rise in climate lawsuits brought directly against companies—particularly in the U.S.

In the landmark case of *Massachusetts v. EPA*, 549 U.S. 497 (2007), the U.S. Supreme Court determined that carbon dioxide was a "pollutant" under the Clean Air Act ("CAA") and that the United States Environmental Protection Agency ("USEPA") was remiss in failing to promulgate regulations governing GHG emissions.[aa]

Thereafter, the U.S. Supreme Court issued another watershed ruling in *American Electric Power, et al. v. Connecticut*, et al., 564 U.S. 410 (2011), which involved eight states, New York City, and various land trusts which sued power companies that operated plants, seeking to cap-and-abate the companies GHG emissions due to alleged ongoing contributions to global warming. The states sought a decree setting carbon-dioxide emissions for each company. The U.S. Supreme Court held that per *Massachusetts v. EPA* precedent, the authority of the USEPA to regulate GHGs delegated by the CAA displaces any federal common law right of state, city, and private parties to seek abatement of carbon-dioxide emissions from fossil-fuel fired power plants. The court went on to find that *Massachusetts v. EPA* established that emissions of carbon dioxide qualify as air pollution subject to regulation under the CAA and this "speaks directly" to the emissions of carbon dioxide from the companies' power plants.

Since that time, there have been a handful of U.S. climate change suits alleging public nuisance, all of which were dismissed on justiciability, displacement, preemption, and/or standing grounds, citing to the legal precedent of *Massachusetts v. EPA* and its progeny and holding that the CAA supplants any private cause of action for common law nuisance and it is for the USEPA to regulate GHG's, not the courts.[ab]

[y] A fairly narrow definition of "climate litigation" was adopted to arrive at this figure, which focuses on judicial cases and targeted adjudications involving climate change presented to administrative entities (including courts) and a few international bodies. Commercial disputes, which are increasingly administered by dispute resolution bodies, were not included. https://www.lse.ac.uk/granthaminstitute/wp-content/uploads/2020/07/Global-trends-in-climate-change-litigation_2020-snapshot.pdf.

[z] Id.

[aa] With the recent passing of U.S. Supreme Court Justice Ruth Bader Ginsburg, legal experts say that the addition of a sixth conservative justice to the high court could lock in opposition to expansive readings of the Clean Air Act that encompass greenhouse gas emissions or trigger a reexamination of the landmark 2007 climate case Massachusetts v. EPA. https://www.eenews.net/stories/1063714329?utm_source=Energy+News+Network+daily+email+digests&utm_campaign=9e62c4b5e2-EMAIL_CAMPAIGN_2020_05_11_11_46_COPY_01&utm_medium=email&utm_term=0_724b1f01f5-9e62c4b5e2-89268843.

[ab] See, *Korsinsky v. EPA*, 2006 U.S. App. LEXIS 21024 (2nd Cir. 2006); *Comer v. Murphy Oil USA*, 585 F.3d 855 (5th Cir. 2009); *Native Village of Kivalina v. ExxonMobil Corp.*, 696 F.3d 849 (9th Cir. 2012); *People of the State of Calif. v. General Motors, et al.* – Northern Dist. Of Calif., No. 3:06-cv-05755 (2007).

However, in the last few years, there have been a proliferation of new climate change suits by municipalities across the U.S., seeking to hold the fossil fuel companies liable for their contribution to climate change and to recover damages for the cost of adapting to climate change and mitigating its effects. There have been no substantive rulings in any of these cases on the merits raised by the plaintiffs. All of these cases were filed in state court and removed to federal court by the defendants to take advantage of *Massachusetts v. EPA* and similar federal precedent. However, as of this writing, all appellate courts that have ruled on the issue, have determined that the cases should be remanded back to the state court in which they were originally filed. As discussed below, this is an extremely significant development that has and will likely continue to precipitate additional lawsuit filings of a similar nature. Moreover, it is important to note that these municipality lawsuits all make clear in their initial complaints that plaintiffs are not seeking to have the respective courts regulate the fossil fuel companies GHG emissions, but rather to compensate them for the costs to mitigate against and adapt to the effects of climate change. This arguably serves to distinguish these cases from the precedent of *Massachusetts v. EPA* and *American Electric Power, et al. v. Connecticut.*

The major pending U.S. municipality climate change lawsuits are:

- County of San Mateo[ac]
- County of Marin[ad]
- City of Santa Cruz[ae]
- City of Richmond[af]
- City of Imperial Beach[ag]
- City and County of San Francisco and City of Oakland[ah]
- City of New York[ai]
- City of Boulder[aj]
- King County[ak]
- Rhode Island[al]
- Mayor and City Council of Baltimore[am]
- City of Honolulu[an]
- District of Columbia[ao]

[ac] http://climatecasechart.com/case/county-san-mateo-v-chevron-corp/.

[ad] https://www.marincounty.org/-/media/files/departments/ad/press-releases/2017/20170717-marin-co-sea-level-rise-complaint-final-endorsed.pdf?la=en.

[ae] http://climatecasechart.com/case/county-santa-cruz-v-chevron-corp/.

[af] http://climatecasechart.com/case/county-santa-cruz-v-chevron-corp/.

[ag] http://climatecasechart.com/case/county-san-mateo-v-chevron-corp/.

[ah] http://climatecasechart.com/case/people-state-california-v-bp-plc-oakland/.

[ai] http://climatecasechart.com/case/city-new-york-v-bp-plc/.

[aj] http://climatecasechart.com/case/board-of-county-commissioners-of-boulder-county-v-suncor-energy-usa-inc/.

[ak] http://climatecasechart.com/case/king-county-v-bp-plc/.

[al] http://climatecasechart.com/case/rhode-island-v-chevron-corp/.

[am] http://climatecasechart.com/case/mayor-city-council-of-baltimore-v-bp-plc/.

[an] http://climatecasechart.com/case/city-county-of-honolulu-v-sunoco-lp/.

[ao] http://climatecasechart.com/case/district-of-columbia-v-exxon-mobil-corp/.

- City of Hoboken[ap]
- Minnesota[aq]
- Connecticut[ar]
- Delaware[as]
- City of Charleston[at]
- Maui, Hawaii[au]
- City of Annapolis[au1]

These municipality suits all contain one or more of the following causes of action and prayers for relief, each of which must be analyzed by insurers which may be asked by their respective insured to provide a defense and/or indemnity:

- Damage to the municipality's property, as well as to the public at large.
- Nuisance due to sea level rise, increased flooding, and intensified storms.
- Trespass due to sea level rise and increased flooding onto property.
- Defendants' historical knowledge of global warming, sea level rise, and other climate change.
- Strict liability (failure to warn and design defect) and negligent failure to warn.
- Unjust enrichment and deceptive trade practices.
- Climate change "data" regarding impact of sea level rise particular to Plaintiff's geographic location.
- Compensatory damages, abatement of the alleged nuisance, punitive/treble damages, and disgorgement of profits.
- An order requiring the defendants to abate the nuisance by funding a "climate adaptation program" to build sea walls and other infrastructure necessary to protect public and private property from sea level rise and other climate impacts.
- Loss of income from reduced agricultural productivity.

All of these municipality suits were removed by defendant fossil fuel companies to federal court—except for the recently filed *City of Annapolis* and *Charleston, SC* suits and the *NYC* suit which was originally brought in federal court. Motions to remand the cases back to state court were filed by the plaintiffs in all instances, with the majority of federal trial court's ruling in favor of remand. Moreover, the Ninth, Tenth, and Fourth federal circuit courts of appeal, recently issued rulings affirming the remand, respectively, in the seven *California*[av] cases, as

[ap]http://climatecasechart.com/case/city-of-hoboken-v-exxon-mobil-corp/.

[aq]http://climatecasechart.com/case/state-v-american-petroleum-institute/.

[ar]http://climatecasechart.com/case/state-v-exxon-mobil-corp/.

[as]http://climatecasechart.com/case/state-v-bp-america-inc/.

[at]http://climatecasechart.com/case/city-of-charleston-v-brabham-oil-co/.

[au]https://www.mauicounty.gov/DocumentCenter/View/124390/Maui-County-Climate-Change-Litigation-Complaint?bidId=.

[au1]http://climatecasechart.com/case/city-of-annapolis-v-bp-plc/.

[av]https://www.climatedocket.com/2020/08/27/ninth-circuit-california-climate-suits-stay/#more-9297.

well as the *Boulder*[aw] and *Baltimore*[ax] cases—although a Petition for a Writ of *Certiorari* seeking U.S. Supreme Court review was filed by the fossil fuel defendants in the *Baltimore* case and granted by the Court and similar Writs are likely to be filed in all other cases.

In its ruling that the *San Francisco* and *Oakland* municipality climate change cases should be remanded and proceed in state court, the Ninth Circuit Court of Appeals, found that adjudicating the state-law claim for public nuisance does not require the resolution of a substantial question of federal law or the interpretation of federal statutes, stating that "[t]he question whether the 'Energy Companies' can be held liable for public nuisance based on production and promotion of the use of fossil fuels and be required to spend billions of dollars on abatement is no doubt an important policy question, but it does not raise a substantial question of federal law for the purpose of determining whether there is jurisdiction under § 1331."[ay]

Most notably, the Ninth Circuit similarly rejected the "Energy Companies" assertion that the CAA completely preempts the state-law nuisance claims. In support of this most significant conclusion, the Ninth Circuit explained (1) the Supreme Court has not determined that the Clean Air Act completely preempts state law, (2) the statutory language of the Clean Air Act does not indicate that Congress intended to preempt every state law cause of action, and (3) the Clean Air Act does not provide the plaintiffs with a "substitute" cause of action.[az]

As of this writing and as reflected in Fig. 10, there are six federal appellate courts that have/will be simultaneously addressing essentially the same jurisdictional question, which could result in inconsistent and conflicting rulings:

- **First Circuit** (Rhode Island)
- **Second Circuit** (New York)
- **Fourth Circuit** (Baltimore) (Affirmed Remand to State Court on 3/6/20—Petition for Cert. to U.S. Supreme Ct. pending)
- **Ninth Circuit** (addressing split in trial court rulings of the California cases) (Affirmed Remand to State Court on 5/26/20).
- **Tenth Circuit** (City of Boulder) (Affirmed Remand to State Court on 7/7/20)
- **Eighth Circuit** (Minnesota) (Suit filed June 2020 in state court and removed in July 2020: Appeal likely)

Although the subject of some debate in the pending appeals, appellate review of federal remand orders is "substantially limited" arguably only to cases involving issues of "federal officer removal" or civil rights.[ba]

Although the fossil fuel defendants have asserted in many of these municipality suits that the "federal officer" ground for review applies because they operate under permits issued by federal officials, the trial and appellate courts granting remand so far have rejected such ground, therefore finding the federal district courts had no subject matter jurisdiction.[bb]

[aw] http://climatecasechart.com/case/board-of-county-commissioners-of-boulder-county-v-suncor-energy-usa-inc/.

[ax] http://climatecasechart.com/case/mayor-city-council-of-baltimore-v-bp-plc/.

[ay] http://climatecasechart.com/case/people-state-california-v-bp-plc-oakland/.

[az] Id.

[ba] Id.

[bb] Id.

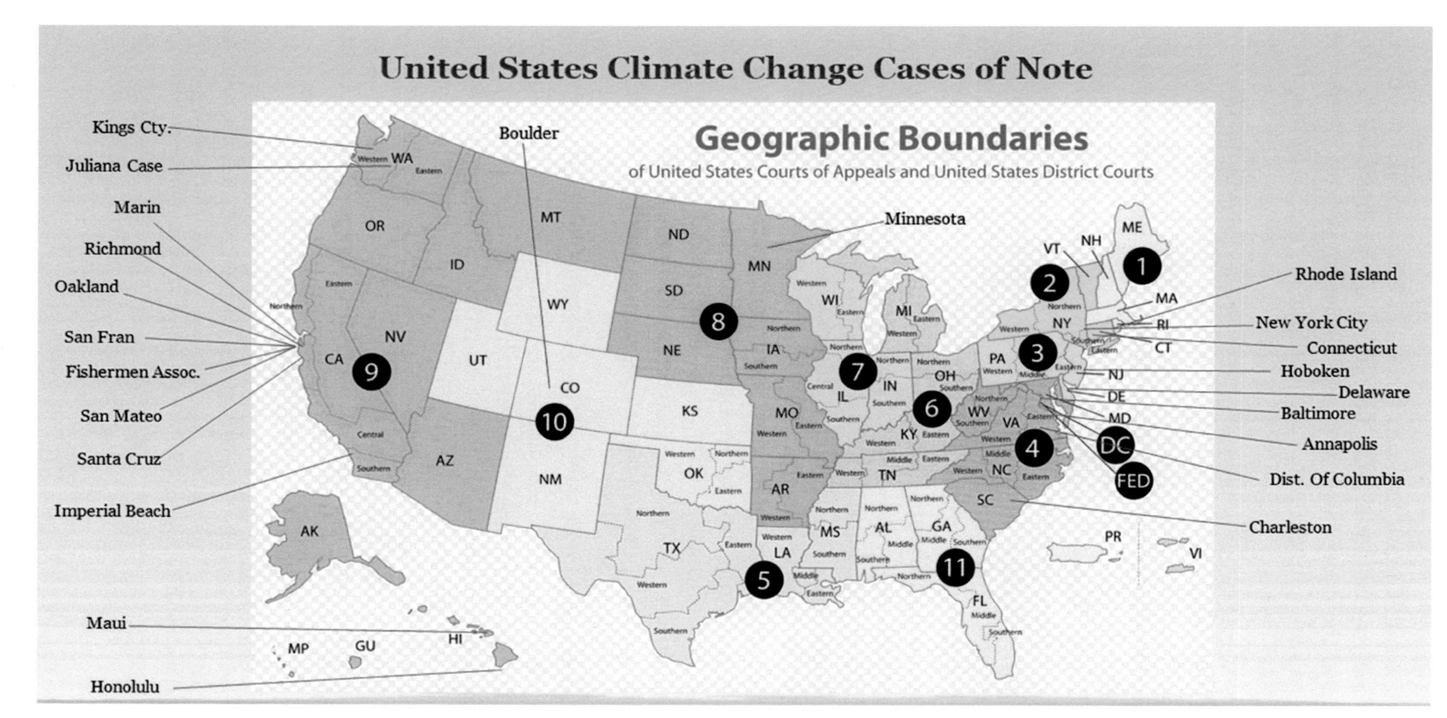


FIG. 10 Pending U.S. "municipality" climate change cases by federal circuit court region.

For example, the Ninth Circuit found in the California appeals that the contracts entered into between the Energy Companies and the federal government were "typical of any commercial contract" and the companies' activities under those agreements are not "so closely related to the government's function that the [companies] face a significant risk of state-court prejudice."[bc]

Accordingly, there is a greater likelihood that at least certain of the cases remanded to state court will remain there—creating increased chance for success on the merits. At minimum, the remand rulings move the cases closer to institution of the discovery process, requiring disclosure of documents and conduct of depositions, although motions to dismiss will likely be filed before discovery begins in earnest.

Notably, following the federal circuit court rulings discussed above remanding these municipality cases back to their respective state courts, there has been increased filings of similar new suits (e.g., *Annapolis, Connecticut, Charleston, Delaware, District of Columbia, Minnesota, and Maui*). It is likely that the municipalities involved in filing these latest suits were motivated by the appellate rulings granting remand and that additional similar suits will follow.

***Additional Noteworthy Climate Change Cases***

Aside from the municipality cases discussed above, there are several additional notable cases involving Climate Change, both in the U.S. and abroad:

***Pacific Coast Federation of Fishermen's Associations, Inc. v. Chevron Corp., et al.***, No. 3:18-cv-07477, N.D. Calif. (as removed).

In November 2018, as the first of its kind, a commercial fishing industry trade group filed suit seeking to hold fossil fuel companies liable for adverse climate "change impacts to the ocean off the coasts of California and Oregon, asserting that it resulted in 'prolonged closures' of Dungeness crab fisheries."[bd] The case was removed by the fossil fuel defendants to federal court and has been stayed pending appellate rulings in the *California* municipality suits discussed above.

***Juliana v. United States***, No. 18-36082 is climate-change lawsuit brought in Oregon federal court by 21 young people (ages 11–22) against the United States. The plaintiffs allege that the government has violated their constitutional rights to life, liberty and property by failing to prevent dangerous climate change. They seek an order compelling the federal government to prepare a plan that will ensure the level of carbon dioxide in the atmosphere falls below 350 ppm by 2100, down from an average of 405 ppm in 2017. On January 17, 2020, the Ninth Circuit Court of Appeals issued a 2–1 decision dismissing the Youth Climate Change Suit, although recognizing the harms brought by climate change and the contribution of fossil fuels:

> In the mid-1960s, a popular song warned that we were 'on the eve of destruction.' The plaintiffs in this case have presented compelling evidence that climate change has brought that eve nearer. A substantial evidentiary record documents that the federal government has long promoted fossil fuel use despite knowing that it can cause catastrophic climate change, and that failure to change existing policy may hasten an environmental apocalypse...[be]

[bc] Id.

[bd] http://climatecasechart.com/case/pacific-coast-federation-of-fishermens-associations-inc-v-chevron-corp/.

[be] http://climatecasechart.com/case/juliana-v-united-states/.

The *Juliana* court found that Plaintiffs had established concrete and particularized injuries, such as being forced to leave their homes because of water security or flooding and that the causation requirement was satisfied because the injuries were caused by carbon emissions from fossil fuel production, extraction, and transportation. The court also held that there was at least a genuine factual dispute as to whether the government's policies, from subsidizing fossil fuel production to offering drilling permits, constituted a substantial factor in causing the carbon emissions, which, in turn, caused the plaintiffs' injuries.[bf] However, the court went on to hold:

> 'Reluctantly, we conclude that such relief [sought by plaintiffs] is beyond our constitutional power.' Reducing 'the global consequences of climate change ... calls for no less than a fundamental transformation of this country's energy system, if not that of the industrialized world.' As a result, 'any effective plan would necessarily require a host of complex policy decisions entrusted, for better or worse, to the wisdom and discretion of the executive and legislative branches.'[bg]

A Motion for Rehearing was denied by the Ninth Circuit in the *Juliana* case, however, it is likely that a Writ of Certiorari will be filed seeking appeal before the United States Supreme Court.

***WildEarth Guardians v. Zinke***, marks an important decision prompting the U.S. Bureau of Land Management ("BLM") to seriously consider GHG emissions when performing environmental assessments for oil and gas leasing.[bh] On March 19, 2019, a Colorado federal district court invalidated drilling leases for more than 300,000 acres of federal land, ruling that the Department of Interior ("DOI") and the BLM's authorization of oil and gas leasing in Wyoming failed to adequately consider climate change in its environmental impact statement ("EIS") under the National Environmental Policy Act ("NEPA").[bi]

BLM stated in the EIS that the leases would not have a "measurable effect" on national or global emissions. The court disagreed, emphasizing that "[t]he leasing stage... is the point of no return with respect to emissions." The court went on to hold that BLM must provide documents backing its claim that the leases would not affect emissions and after such documentation is provided, the plaintiffs may again challenge the EIS. Until such time, the leases in question were declared invalid. The court urged BLM to take the responsibility to assess the leases' impact on the environment seriously, stating "[c]ompliance with NEPA cannot be reduced to a bureaucratic formality, and the court expects [BLM] not to treat remand as an exercise in filling out the proper paperwork."[bj] Notably, a DOI study found that for the period 2005–14, GHG emissions from public lands accounted for 25% of overall U.S. emissions.[bk]

In ***Urgenda Foundation v. State of the Netherlands***, a Dutch environmental group and 900 Dutch citizens sued the Dutch government to require it to do more to prevent global climate change. The court in the Hague ordered the Dutch state to limit GHG emissions to 25% below

[bf] Id.

[bg] Id.

[bh] *WildEarth Guardians v. Zinke*, 368 F. Supp. 3d 41 (D.D.C. 2019).

[bi] Id.

[bj] Id.

[bk] https://pubs.usgs.gov/sir/2018/5131/sir20185131.pdf.

1990 levels by 2020, finding the government's existing pledge to reduce emissions by 17% insufficient to meet the state's fair contribution toward the UN goal of keeping global temperature increases within 2°C of preindustrial conditions. The court concluded that the state has a duty to take climate change mitigation measures due to the "severity of the consequences of climate change and the great risk of climate change occurring." In reaching this conclusion, the court cited (without directly applying) Article 21 of the Dutch Constitution and EU emissions reduction targets, as well as various principles and doctrines. The court did not specify how the government should meet the reduction mandate, but offered several suggestions, including emissions trading or tax measures. This is the first decision by any court in the world ordering states to limit greenhouse gas emissions for reasons other than statutory mandates.[bl]

On October 9, 2018, the Hague Court of Appeal upheld the District Court's ruling, concluding that by failing to reduce greenhouse gas emissions by at least 25% by end-2020, the Dutch government is acting unlawfully in contravention of its duty of care under Articles 2 and 8 of the European Convention on Human Rights ("ECHR").[bm]

On December 20, 2019, the Supreme Court of the Netherlands upheld the decision under Articles 2 and 8 of the ECHR.[bn]

In ***Milieudefensie v. Royal Dutch Shell***, seven environmental and human rights groups in the Netherlands filed a 2019 suit in the Hague Court of Appeals against Royal Dutch Shell for failing to align its business model with the goals of the Paris Climate Agreement.[bo] Plaintiffs are not seeking monetary compensation, but are demanding that Shell adjust its business model in order to keep global temperature rise below 1.5°C (2.7°F), and specifically that Shell must reduce its $CO_2$ emissions by 45% by 2030 compared to 2010 levels and to zero by 2050, in line with the Paris Climate Agreement and recommendations by the Intergovernmental Panel on Climate Change ("IPCC").[bp] The court heard argument in December 2020 with respect to issues specified by the court, with a decision expected in late-May 2021.

On March 6, 2020, in ***Smith v. Fronterra Co-Operative Group Limited, et al.***, CIV-2019-404-001730 ([2020] NZHC 419), the High Court Of New Zealand (Auckland Registry) rejected two claims brought against major GHG emitters alleging that the defendants' actions constituted public nuisance and negligence, but allowed a common law claim to proceed.[bq]

Plaintiff asserts that he is of Ngāpuhi and Ngāti Kahu descent and that he is the "climate change spokesman" for the Iwi Chairs' Forum. He claims customary interests in lands and other resources situated in or around Mahinepua in Northland and asserts that various sites of customary, cultural, historical, nutritional, and spiritual significance to him are close to the coast, on low-lying land or are in the sea. Plaintiff brought suit against several defendants that operate facilities that emit greenhouse gas emissions, including dairy farms, a power station,

[bl] http://climatecasechart.com/non-us-case/urgenda-foundation-v-kingdom-of-the-netherlands/.

[bm] Id.

[bn] Id.

[bo] http://blogs2.law.columbia.edu/climate-change-litigation/wp-content/uploads/sites/16/non-us-case-documents/2019/20190405_8918_summons.pdf.

[bp] Id.

[bq] http://blogs2.law.columbia.edu/climate-change-litigation/wp-content/uploads/sites/16/non-us-case-documents/2020/20200306_2020-NZHC-419_opinion-1.pdf.

and an oil refinery. Plaintiff alleges that the defendants' contributions to climate change constitute a public nuisance, negligence, and breach of a duty cognizable at law to cease contributing to climate change.[br]

The High Court of New Zealand dismissed the first two claims, finding that Plaintiff could not demonstrate public nuisance because the damage claimed was neither particular to him, nor the direct consequence of the defendants' actions. The court further reasoned that showing a public nuisance was difficult given that the defendants are complying with all relevant statutory and regulatory requirements. The court determined that Plaintiff's negligence claim must fail because he was unable to show that the defendants owed him a duty of care, concluding that the damage claimed was not reasonably foreseeable or proximately caused by their actions.[bs]

However, the High Court of New Zealand declined to strike the third cause of action, which alleged that the defendants have a duty to cease contributing to climate change. The court found that there were "significant hurdles" for Smith in persuading the court that this new duty should be recognized, but determined that the relevant issues should be explored at a trial. The court explained that "[i]t may, for example, be that the special damage rule in public nuisance could be modified; it may be that climate change science will lead to an increased ability to model the possible effects of emissions." The court warned, however, that it would likely be unable to provide the injunctive relief that Smith seeks, which would require a "bespoke emission reduction scheme."[bt]

## 5 Climate change damages

The costs of climate change are increasing and are substantial. 2017 was the costliest year on record for natural catastrophe events, with US$344 billion in global economic loss, of which 97% was due to weather-related events.[bu] Insured loss estimates from natural catastrophes totaled $140 billion in 2017.[bv] Insurance industry losses from natural catastrophes and man-made disasters globally amounted to US$83 billion in 2020, according to Swiss Re Institute's preliminary sigma estimates. This makes it the fifth-costliest year for the industry since 1970.[bv1] The 2018 IPCC Special Report estimated that global economic damages by 2100 would reach $54 trillion with 1.5°C (2.7°F) of warming, $69 trillion with 2°C (3.7°F) warming and $551 trillion with 3.7°C (6.7°F) of warming above preindustrial levels.[bw]

The insurance industry has fundamental exposure to climate risks through physical and transition risks. Physical risks are actual climate change effects, such as storm surge as a result

[br] Id.

[bs] Id.

[bt] Id.

[bu] http://thoughtleadership.aonbenfield.com/Documents/20180124-ab-if-.

[bv] http://www.munichre.com/en/media-relations/publications/press-.

[bv1] https://www.swissre.com/media/news-releases/nr-20201215-sigma-full-year-2020-preliminary-natcat-loss-estimates.html.

[bw] https://www.ipcc.ch/sr15/.

of a hurricane. Transition risks include policy changes, reputational impacts, and shifts in market preferences, norms, and technology, as the world moves toward a zero-carbon economy. Since 1980, the majority of rising losses associated with weather events has been due to exposure accumulation that comes with economic growth and urbanization. The concentration of assets (human and physical), particularly in urban areas such as low-lying coastal regions that are vulnerable to adverse weather conditions, inflates the loss potential when a severe weather event strikes. Other socioeconomic factors account for most of the remainder of the trend of rising losses over time.[bx]

According to Swiss Re's recent Sigma 2/2020 Report, many of today's catastrophe models are rooted in the past and do not fully account for rising exposure from increased value concentration in a rapidly urbanizing and, at times, more vulnerable world, especially when sprawling into higher hazard regions.[by]

[Re]insurance companies face climate-change risks on both sides of their balance sheets, which can have adverse effects on underwriting profitability and solvency in the long term. On the liability side, the main risk is underestimating insurance risk premiums due to reliance on historical loss data or incomplete/outdated models. On the asset side, the exposure derives from the impact of physical and transition risks on invested assets, including infrastructure funds and corporate bond holdings.[bz]

In the first wide-ranging federal government study focused on the specific impacts of climate change on Wall Street, a September 2020 report commissioned by U.S. federal regulators overseeing the commodities markets concluded that climate change presents significant threats to the U.S. financial markets, as the costs of wildfires, storms, droughts, and floods spread through insurance and mortgage markets, pension funds, and other financial institutions.[ca]

The full extent of damages and associated costs from climate change are far reaching and include:

- Direct damage to real and personal property
- Costs to mitigate against sea level rise (dams, levees, raising/strengthening buildings and structures, relocation, etc.)
- Impacts of climate change on physical systems, such as oceans, lakes, and snowpack
- Impacts of climate change on biological systems—humans, vegetation, and wildlife
- Loss of natural resources (e.g., forest wildfires, complete submergence of sensitive wetlands, coral reefs, impacts to fish and wildlife)
- Alteration of marine and terrestrial animal indigenous habitat/range
- Bodily injury
- Increased water-borne pests and associated disease (mosquitos, etc.)
- Damage to crops, agricultural productivity, growing season, and soil erosion

[bx] https://www.swissre.com/dam/jcr:85598d6e-b5b5-4d4b-971e-5fc9eee143fb/sigma%202%202020_EN.pdf.

[by] Id.

[bz] Id.

[ca] https://www.cftc.gov/sites/default/files/2020-09/9-9-20%20Report%20of%20the%20Subcommittee%20on%20Climate-Related%20Market%20Risk%20-%20Managing%20Climate%20Risk%20in%20the%20U.S.%20Financial%20System%20for%20posting.pdf.

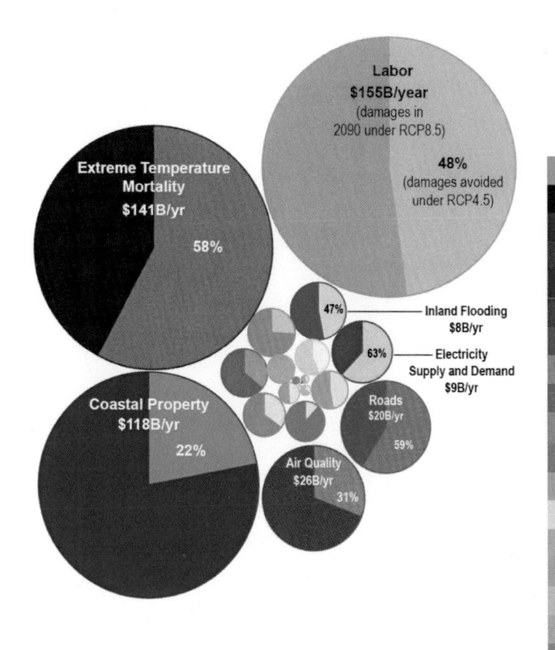


| Annual Economic Damages in 2090 | | |
|---|---|---|
| Sector | Annual damages under RCP8.5 | Damages avoided under RCP4.5 |
| Labor | $155B | 48% |
| Extreme Temperature Mortality◊ | $141B | 58% |
| Coastal Property◊ | $118B | 22% |
| Air Quality | $26B | 31% |
| Roads◊ | $20B | 59% |
| Electricity Supply and Demand | $9B | 63% |
| Inland Flooding | $8B | 47% |
| Urban Drainage | $6B | 26% |
| Rail◊ | $6B | 36% |
| Water Quality | $5B | 35% |
| Coral Reefs | $4B | 12% |
| West Nile Virus | $3B | 47% |
| Freshwater Fish | $3B | 44% |
| Winter Recreation | $2B | 107% |
| Bridges | $1B | 48% |
| Munic. and Industrial Water Supply | $316M | 33% |
| Harmful Algal Blooms | $199M | 45% |
| Alaska Infrastructure◊ | $174M | 53% |
| Shellfish* | $23M | 57% |
| Agriculture* | $12M | 11% |
| Aeroallergens* | $1M | 57% |
| Wildfire | -$106M | -134% |

**FIG. 11** Estimated annual U.S. economic damages in 2090 under RCP8.5 and RCP4.5 scenarios.

- Salinization and other contamination of potable of water supply
- Business interruption (including supply chain) and loss of income
- Lost worker productivity
- Diminished property value
- Impacts to national security

As depicted in Fig. 11,[cb] the U.S. Fourth National Climate Assessment (2018) ("NCA4") cautions that climate change could cost the U.S. hundreds of billions of dollars *annually* by century's end[cc]:

[cb]https://nca2018.globalchange.gov/. The total area of each circle in Fig. 11 reflects projected damages under Representative Concentration Pathways ("RCP") 8.5, whereas the lighter shaded area reflects estimated damages under RCP4.5. NCA4 utilizes RCP8.5 as a "higher" scenario, associated with more warming, and RCP4.5 as a "lower" scenario with less warming. These RCPs represent the most recent set of climate projections developed by the international scientific community and have been adopted by the IPCC in its latest reports as a basis for climate predictions and projections.

[cc]Development of the Fifth National Climate Assessment (NCA5) is currently underway, with anticipated delivery in 2023. https://www.globalchange.gov/nca5.

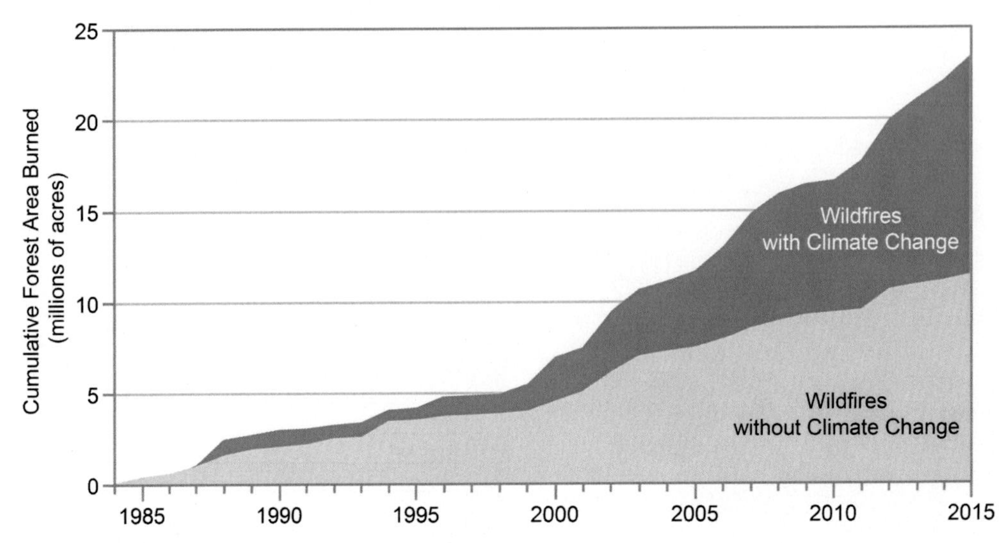


**FIG. 12** Climate change comparison, estimated U.S. western forest areas burned.

According to the 2018 NCA4, human-caused warming has increased the area burned by wildfire in the Western United States, "particularly by drying forests and making them more susceptible to burning."[cd] By 2050, according to the NCA4, the area that burns yearly in the West could be two to six times larger than today. The NCA4 estimated the total acres burned in western forests under current climate conditions and in a model without human-caused warming. As shown in Fig. 12, it found that half as much forest area would have burned between 1984 and 2015 in a world not warmed by climate change.

According to the 2018 NCA4, coral reefs in the U.S. Caribbean, Hawaii, Florida, and the U.S. Pacific islands are already being affected by bleaching and disease caused in part by climate change. The NCA4 finds that if $CO_2$ levels reach 450–500 ppm, reef erosion could exceed calcification, meaning that reef structure is likely to erode and coral cover is likely to decline dramatically. Should $CO_2$ levels exceed 500 ppm, corals are not expected to survive. The NCA4 concludes that loss of recreational benefits alone from coral reefs in the U.S. could reach $140 billion by 2100.[ce]

The NCA4 determined that as of 2013, U.S. coastal shoreline counties were home to 133.2 million people, or 42% of the population.[cf] A June 2018 study by the Union of Concerned Scientists found that high-tide flooding in the U.S. could put over 300,000 coastal homes and commercial properties in the lower 48 states with a collective market value of about $136 billion in 2018 dollars at risk within the next 30 years. By the end of the century, over 2.5

[cd] https://nca2018.globalchange.gov/.

[ce] Id.

[cf] Id.

million homes and commercial properties currently worth more than $1 trillion altogether could be at risk.[cg]

Similarly, a 2017 Report from Zillow determined[ch]:

- By 2050, more than 386,000 existing homes on the U.S. coastal areas, worth $209.6 billion (2018 dollars), are likely to be at risk of permanent inundation from sea level rise alone.
- If sea levels rise as predicted by the year 2100, almost 300 U.S. cities would lose at least half their homes, and 36 U.S. cities would be completely lost.
- One in eight Florida homes would be under water, accounting for nearly half of the lost housing value nationwide.
- Nationwide, almost 2.5 million homes—worth a combined $1.3 trillion—are at risk of being underwater by 2100.

According to a recent study, by 2040, more than 67,000 (64%) of the septic tanks in Miami-Dade County, Florida could have yearly issues due to sea level rise—impacting not only the tank owners, but the potable water supply and the health of residents and the ecosystem. The building code has been revised to double the amount of clean fill that needs to be placed under the septic, as groundwater level continues to rise. Officials have considered excavating/abandoning all the septic tanks and connecting homes to the county's sewer lines, at an estimated cost of $3.3 billion.[ci]

The City of Rotterdam, 90% of which lies below sea level, has decentralized many aspects of water management, with flood protection now the responsibility of regional water management boards. They have also bolstered defenses, including a 3700 km network of dikes, dams, and seawalls, to protect against worst case sea level rise scenarios.[cj]

In a 2019 study assessing the economic cost of climate change in Europe, the authors estimated that sea-level rise under various RCP scenarios, could result in annual flooding impacting 1.8 million people under RCP2.6 to 2.9 million people under RCP8.5 by the 2050s and, potentially, 4.7 million people under RCP2.6 to 9.6 million people under RCP8.5 by the 2080s, if there is no investment in adaptation.[ck] The study found that this flooding, along with other impacts of sea level rise (e.g., erosion), will lead to *annual* estimated damage costs in Europe of €135 billion to €145 billion (mid estimates) for RCP2.6 and RCP4.5 respectively for the 2050s, rising to €450 billion to €650 billion by the 2080s.[cl]

[cg] https://www.ucsusa.org/sites/default/files/attach/2018/06/underwater-analysis-full-report.pdf.

[ch] https://www.zillow.com/research/ocean-at-the-door-21931/.

[ci] https://www.miamidade.gov/green/library/vulnerability-septic-systems-sea-level-rise.pdf.

[cj] https://www.weforum.org/agenda/2019/01/the-world-s-coastal-cities-are-going-under-here-is-how-some-are-fighting-back/.

[ck] https://www.coacch.eu/wp-content/uploads/2019/11/COACCH-Sector-Impact-Economic-Cost-Results-22-Nov-2019-Web.pdf.

[cl] Id.

## 6 The insurance coverage implications of climate change

To date, there has been a dearth of coverage actions and decisional law relating to insurance for climate change liability. However, this will likely change soon, given the rising prominence of the issue, the mounting scientific evidence, the substantial costs involved and the increased litigation activity by municipalities and private parties against fossil fuel companies and other target defendants, as well as certain pro-plaintiff rulings discussed elsewhere in this chapter. Commercial General Liability, Directors and Officers ("D&O"), and Property Insurance are all expected targets of claims stemming from climate change litigation.

To date, *AES Corp. v. Steadfast Ins. Co.*, 725 S.E. 2d 532 (Va. 2012), is the only reported U.S. decision involving insurance coverage for climate change liabilities, where the Virginia Supreme Court held that the insurer had no obligation to provide a defense or coverage for the insured's potential climate change-related liabilities arising from the *Native Village of Kivalina* suit.[cm] However, the *AES Corp.* case was disposed by summary judgment solely on the lack of an "occurrence" issue. More specifically, the Supreme Court of Virginia found that the underlying allegations asserting that the insured intentionally released tons of carbon dioxide and GHGs into the atmosphere as part of its business operations did not constitute an "occurrence" within the terms of the insurance policies. Notably, even though the underlying Complaint alleged both negligent and intentional conduct of the insured AES Corp., the court held that "whether or not AES's intentional act constitutes negligence, the natural or probable consequence of that intentional act is not covered."

Below is a brief overview of the most significant issues insurers will need to address under a commercial general liability ("CGL") policy when faced with a claim for coverage arising from climate change liabilities[cn]:

**Choice of law—What law is to apply to the insurance coverage dispute?** Choice of law is the first step in resolving any substantive legal issue. State courts must choose when a conflict exists between substantive law of two or more states relevant to the insurance coverage issues. Which State's law is chosen can often be dispositive in the coverage action, as a States frequently have diametrically conflicting views on certain of the coverage issues and policy language. To the extent the underlying climate change suit by a particular municipality, for example, is for alleged injury to property and persons solely within its geographic borders, the coverage court could simply apply that state's law, particularly if it is also the forum state. However, given the transient nature of GHGs and the global impact of climate change, a choice of law analysis could likely become exceedingly complex.

**Do the Climate Change Suits Against the Insureds Seek "Damages"?** The term "damages" is not defined under most CGL policies. Insurers typically assert the term is limited to "legal" damages and does not include equitable relief. The majority of U.S. states have ruled that environmental response costs are "damages" and are covered under the CGL

[cm] *Native Village of Kivalina v. ExxonMobil Corp.*, 696 F.3d 849 (9th Cir. 2012). The underlying *Kivalina* suit alleged damages to the residents' native village in Alaska due to rising sea level from global warming/GHGs.

[cn] CGL insurance policies generally provide defense and indemnity coverage when a third-party sues the insured for money damages due to bodily injury and/or property damage that took place during the policy period, which arose from an "occurrence."

policy. Accordingly, monetary relief as compensatory damages sought in the climate change suits should qualify. However, insurers will likely argue that the injunctive relief to abate the nuisance does not qualify as "damages." Certain of the plaintiffs in the pending municipality climate change suits seek an order requiring the companies to pay monies into a "Climate Change Abatement Fund" for future perceived harm, which raises additional issues, particularly if there has been no present finding of "property damage" or "bodily injury." Declaratory and various types of equitable relief sought may also create coverage disputes.

**Do the Climate Change Suits Against the Insureds Involve "Property Damage"?** "Property damage" is generally defined in most CGL policies as: "Physical injury to tangible property, or loss of use of that same physically injured tangible property." Some CGL policies also include within the "property damage" definition, the "loss of use of tangible property that is not physically injured." To determine "physical injury," U.S. courts often look at whether the tangible property was altered in appearance, shape, color, or in another material dimension. Generally environmental damage to property has been found by courts to constitute physical injury to tangible property. To the extent the climate suits allege water damage to real property, buildings and structures from sea level rise; they may qualify as "property damage." However, mitigative and preventative efforts to curtail or avert "property damage" (e.g., dams, dikes and raising or relocating buildings) will likely raise disputes. Courts have found coverage for mitigative and prophylactic costs, especially where "property damage" is present and the mitigation is to avoid further damage.[co]

Economic loss alone, without any accompanying damage to or loss of use of tangible property, is not covered property damage. Accordingly, insurers would likely argue that coastal property which has decreased in value due to rising sea level is not covered, unless there is an accompanying damage or loss of use.

Alleged damages in climate change suits resulting from a decrease in crop yields may not be covered. U.S. courts have sometimes found coverage in other contexts if there was physical damage to the crops. However, coverage denials have been upheld for costs arising from crop failures due to the seeds failing to germinate.[cp]

In *Concord Gen Mut. Ins. Co. v. Green & Co. Bldg. & Dev. Corp.*[cq], the Supreme Court of New Hampshire held that there was no requisite physical injury to tangible property, where $CO_2$ was leaking from insured's chimney, as the gases did not physically alter the property and the homeowners were able to continue living in their house, although they could not use their chimney.

[co]See, *Oak Ford Owners Ass'n v. Auto-Owners Ins. Co.*, 510 F. Supp. 2d 812 (M.D. Fla. 2007) (Court held insured's dredging of a creek to make it deeper and wider without proper permit and damage to wetlands constituted damage to tangible property. Court stated "injury has occurred even though the effects such as flooding and erosion would increase over time").

[cp]See, *Farm Bureau Mut. Ins. Co. v. Earthsoils*, 812 N.W.2d 873 (Minn. Ct. App. 2012) (Farmer had only half the crop yield due to fertilizer with insufficient nitrogen content. Court ruled no coverage because less than anticipated crop yield did not result from physical injury to the crop itself ). But see, *W. Heritage Ins. Co. v. Green*, 54 P.3d 948 (Idaho 2002) (Alleged misapplication of fertilizer caused some of the potato plants to form yellow foliage, poor root systems and misshapen potatoes. Court found physical injury to plants was a covered loss).

[cq]8 A.3d 399 (N.H. 2010).

**Is There "Property Damage" During the Policy Period?** This will undoubtedly be a disputed issue in a climate suit context and often involve a "battle of the experts." If "property damage" has happened, in which year(s) did it take place? Most large target companies named as defendants in these suits have "legacy" liability insurance policies dating back to the 1940s or earlier. Accordingly, nearly every major insurance company will be implicated if the "property damage" is deemed to have occurred from the 1940s through present.

**Do the Climate Change suits against the insureds involve "Property Damage" arising from an "Occurrence"?** "Occurrence" is generally defined as:

> An accident, including continuous or repeated exposure to substantially the same general harmful conditions which results in injury or damage which is neither expected nor intended from the standpoint of the insured.

"Occurrence" is not the trigger of coverage. Rather, it is the act of the insured (the accident, event or conditions) that results in injury—the cause. It is the resulting injury/damage during the policy period that triggers coverage—the effect. All of the pending U.S. municipality climate change suits allege intentional and knowing conduct on the part of the fossil fuel defendants dating back to at least the 1960s. Such allegations may support a finding of no "occurrence."[cr]

**How Many Occurrences Are There?** The answer to this question could have huge monetary implications on available policy limits and exhaustion of coverage benefits. The analysis of this issue could prove extremely complex in climate change suits, where the alleged damages involve both traditional concepts of property damage and bodily injury, as well as injury to ecosystems, marine life, and natural resources, etc. separated by time and place.

Typical limits of liability language in a CGL policy states for the "purpose of determining the limits of the Company's liability, all injury or damage arising out of continuous or repeated exposure to substantially the same general harmful conditions shall be considered as arising out of one occurrence."

The U.S. courts generally apply either the "cause test" or the "effects test" in determining the number of occurrences. Under the "cause" test, the inquiry is whether the diverse injuries or claims share a common, uninterrupted proximate cause? This often results in a one occurrence finding. In contrast, under the "effects test" the focus is on the point at which people or property are damaged by insured's act or omission, which militates in favor of a multiple occurrence finding, should the facts permit. Two possible outcomes in a climate change coverage action would be: one "occurrence"—the insured's decision to manufacture and supply a "defective" product (fossil fuels which, when burned, release persistent GHGs), or multiple "occurrences"—any isolated discrete injuries separated in place and time.

**Does the claim fall within the Products/Completed Operations Hazards?** Many CGL policies only contain aggregate policy limits for products/completed operations hazards (as defined). The assertion of strict liability and other "defective product" allegations in the climate change Complaints could implicate this aggregate limitation. Depending on the number of occurrences outcome, the applicability of the products hazard definition could have a significant impact on available policy limits.

[cr] See, *AES Corp. v. Steadfast Ins. Co.*, supra.

**What is the appropriate "Trigger of Coverage"?** Trigger of coverage refers to what must occur during the policy period to give rise to potential coverage under the specific terms of the policy. There are four main GL trigger theories which could be applied to these climate change suits, the selection of which could have a significant impact on the number of policy years implicated: (1) Injury in fact; (2) Exposure; (3) Manifestation; and (4) Continuous.[cs]

**Does a Pollution Exclusion Apply?** The three main types of pollution exclusions likely to be encountered in climate change coverage actions are: (1) Sudden and Accidental (1973–85); (2) Absolute (1986–); and Total (1988–). All three of these variants, exclude coverage for, *inter alia*, "property damage" arising out of the discharge of "pollutants…" The term "pollutant" is most commonly defined in a CGL policy as: "Any solid, liquid, gaseous or thermal irritant or contaminant, including smoke, vapor, soot, fumes, acids, alkalis, chemicals and waste. Waste includes materials to be recycled, reconditioned or reclaimed."

U.S. courts generally apply either a "traditional environmental pollution" approach or a broader, literal interpretation to the exclusions. Under the "traditional" approach, courts interpret the exclusion to preclude coverage only for those claims that are commonly considered to arise from "traditional" environmental pollution (e.g., dumping waste at a landfill). Under the "literal" approach, courts focus on the plain language of the policies and apply the exclusion to all claims arising from contaminants or irritants that cause damage, regardless of whether the claims involve traditionally understood contamination. Importantly, the U.S. Supreme Court has on multiple occasions held that greenhouse gases (including carbon dioxide and methane) fall within the CAA's definition of "air pollutant."[ct]

## 6.1 Is climate change liability a D&O issue?

According to a May 2018 Zurich Quarterly Claim Journal, climate change liability presents significant D&O exposure:

> From a D&O perspective it is more than likely that the industry will see an increase in claims in the future as a result of companies failing to adequately manage the risk of climate change on their business and to disclose these risks to investors. With respect to Financial Lines, it is most likely that D&O insurance will take the brunt of the Impact.[cu]

Zurich further reported in May 2019, that:

> … [W]e may shortly arrive at a time where the use of fossil fuels is severely restricted. There is therefore an argument that the fossil fuel reserves that currently exist will never be used. The concern is that energy companies and their directors are aware of this risk, however have not taken this into account when stating their reserves, thus massively overstating the value of their business and leaving them open to the risk of actions against them. This may also have a knock on effect to their advisors, e.g. actions against their auditors and investment banks.[cv]

[cs]Allocation of damages goes hand-in-hand with trigger and could result in the insured being able to select any triggered policy to pay "all sums" (subject to that selected insurer's contribution rights) or each triggered policy solely obligated to pay a pro-rata share of the damage/injury that took place during its policy period.

[ct]See, e.g., *Am. Elec. Power Co. v. Connecticut*, 564 U.S. 410 (2011).

[cu]https://insider.zurich.co.uk/app/uploads/2018/05/Zurich-Claims-Quarterly- Journal-Spring-2018.pdf.

[cv]https://insider.zurich.co.uk/risk-mitigation/predicting-the-future-climate-change-and-do-insurance/.

According to the 2019 Zurich Report, D&O liability could revolve around the following three scenarios:

**(1)** Companies who fail to disclose how climate change may affect their business in the future.
**(2)** Investors and shareholders suing investment and pension funds for investing in businesses adversely affected by climate change.
**(3)** Companies who have allegedly directly contributed to the rise of Climate Change internationally and its affects locally having a liability to those affected by the consequences of climate change.[cw]

It is important to recognize that in 2010, the U.S. Securities Exchange Commission ("S.E. C.") issued a 29 page "interpretive guidance" (not a new rule) on existing disclosure requirements regarding how companies are to address the risks posed by climate change in their securities filings.[cx] The "Guidance" stressed that "[t]his interpretive release is intended to remind companies of their obligations under existing federal securities laws and regulations to consider climate change and its consequences as they prepare disclosure documents to be filed with us and provided to investors."[cy]

After largely a decade of silence, in January 2020, the S.E.C. Chairman and Commissioner issued the following joint statements with respect to climate change and financial disclosure concerns:

> Much has changed in the last decade with respect to what we know about climate change and the financial risks it creates for global markets. The science is largely undisputed and the effects increasingly visible and dire; the looming economic threat to markets worldwide is more and more apparent; investors have increased their demands on companies and regulators for consistent, reliable, and comparable disclosures. [Climate Change] may be the single most momentous risk to face markets since the financial crisis. ***(SEC Commissioner Allison Herren Lee)***

> [T]he landscape around these issues is, and I expect will continue to be, complex, uncertain, multinational/jurisdictional and dynamic. In 2010, the Commission issued an interpretive release to provide guidance to public companies regarding the Commission's existing disclosure requirements as they apply to environmental and climate-related matters. Since then, SEC staff has continued to consider these matters, including, as part of regular reviews of annual and periodic reports and other company filings by the Division of Corporation Finance. Our staff in the Office of Compliance Inspections and Examinations is reviewing disclosures of investment advisers and other issuers regarding funds and other products that pursue environmental or climate-related investment mandates to ensure that investors are receiving accurate and adequate information about the material aspects of those strategies.[cz] ***(SEC Chairman Jay Clayton)***

Following the election of U.S. President Biden, on February 24, 2021, Acting SEC Chair Lee directed the Division of Corporation Finance to scrutinize disclosures for adherence to the

[cw] Id.

[cx] The United States Supreme Court has ruled that "information is material if there is a substantial likelihood that a reasonable investor would consider it important in deciding how to vote or make an investment decision, or, put another way, if the information would alter the total mix of available information." *TSC Indust. v. Northway, Inc.*, 426 U.S. 438 (1976).

[cy] http://www.sec.gov/rules/interp/2010/33-9106.pdf.

[cz] https://www.sec.gov/news/public-statement/clayton-mda-2020-01-30
https://www.sec.gov/news/public-statement/lee-mda-2020-01-30

SEC's 2010 guidance on climate change-related disclosures and later announced that in the context of inspections, "emerging risks, including those relating to climate and ESG," will be a priority. Further, in March 2021, the SEC sent a very clear signal about one of its chief enforcement priorities by announcing the creation of a Climate and Environmental, Social, and Governance ("ESG") Task Force within the Division of Enforcement.[cz1]

In 2015, the International Financial Stability Board established the Task Force on Climate-Related Financial Disclosures ("TCFD") to develop guidance for companies in disclosing clear, comparable and consistent information on the financial risks and opportunities presented by climate change.[da] The final recommendations, released in June 2017, were designed to bring consideration of climate risk into the forefront of business and investment decision-making to facilitate efficient allocation of capital and to enable a smooth transition to a low-carbon economy. According to the May 2019 TCFD Status Report, the recommendations of the TCFD had received widespread business support, including over 785 companies and other organizations, and 340 investors with approximately $34 trillion in assets under management ("AUM").[db] The prior TCFD 2018 Status Report cautions that the expected transition to a lower-carbon economy is estimated to require around $3.5 trillion, on average, in energy sector investments per year for the foreseeable future, generating new investment opportunities. However, the risk-return profile of companies exposed to climate-related risks may change significantly because of physical impacts of climate change, climate policy, or new technologies. TCFD points to one study which estimated the value at risk to the total global stock of manageable assets because of climate change ranges from $4.2 trillion to $43 trillion between now and the end of the century.[dc]

In its 2019 white paper, FM Global focused on why CFO's must initiate natural catastrophe preparedness, noting that "[i]f the CFO doesn't lead the charge to invest in reducing … [threats from natural catastrophe – including climate change], they will be the ones that stakeholders hold accountable for not properly addressing the risks."[dd] Based on its analysis of over 10,000 wind and flood related investments by 1800 clients, FM Global determined that for every US$1 a company spends to protect structures from hurricane, wind and flood damage, estimated loss exposures decrease by an average US$105, in relation to those companies' associated reductions in property loss and business interruption exposures.[de] FM Global's review of nearly 100 10-K filings of public companies that experienced property damage and/or business interruption from Hurricane's Harvey, Irma or Maria, provides insightful and unsettling information on the breadth and extent of losses across a wide array of business sectors and the impact deep into these companies supply chains.[df]

[cz1] https://www.sec.gov/news/press-release/2021-42; https://www.sec.gov/news/public-statement/lee-statement-review-climate-related-disclosure?utm_medium=email&utm_source=govdelivery.

[da] http://www.fsb-tcfd.org/about/.

[db] https://www.fsb-tcfd.org/publications/tcfd-2019-status-report/.

[dc] https://www.fsb-tcfd.org/publications/tcfd-2018-status-report/.

[dd] http://cms.ipressroom.com.s3.amazonaws.com/240/files/20190/Master+the+disaster+-+CFO+natural+disaster+preparedness+in+2019+and+beyond.pdf.

[de] Id.

[df] Id.

Laurence D. Fink, the founder and chief executive of BlackRock, announced in January 2020 that his firm would make investment decisions with environmental sustainability as a core goal. BlackRock is the world's largest asset manager with nearly $7 trillion in investments, and this move will fundamentally shift its investing policy, causing other large money managers to follow suit. Mr. Fink's annual letter to the chief executives of the world's largest companies is closely watched, and in the 2020 edition he said BlackRock would begin to exit certain investments that "present a high sustainability-related risk," such as those in coal producers. His intent is to encourage every company, not just energy firms, to rethink their carbon footprints.[dg]

In his influential annual letter, Mr. Fink made clear that a company's potential liabilities from climate change represents a core investing concern and Blackrock will hold board members accountable where they feel a company is not properly addressing the issue:

> Climate change has become a defining factor in companies' long-term prospects... The evidence on climate risk is compelling investors to reassess core assumptions about modern finance... Investors are increasingly reckoning with these questions and recognizing that climate risk is investment risk.... Indeed, climate change is almost invariably the top issue that clients around the world raise with BlackRock... We believe that when a company is not effectively addressing a material issue, its directors should be held accountable...Where we feel companies and boards are not producing effective sustainability disclosures or implementing frameworks for managing these issues, we will hold board members accountable.[dh]

Indeed, shareholders are increasingly demanding more climate related disclosures and seeking to hold companies accountable for failure to do so.

In the pending lawsuit of *Ramirez v. Exxon Mobil Corp.*, No. 3:16-cv-311, N.D. Tex., investors filed a securities class action against Exxon Mobil for failure to disclose climate risks. The complaint alleges that Exxon's public statements during the defined class period were materially false and misleading because they failed to disclose that internally generated reports concerning climate change recognized the environmental risks caused by global warming and climate change; that due to risk associated with climate change Exxon would not be able to extract existing hydrocarbon reserves it claimed to have; and that Exxon had used an inaccurate price of carbon to calculate the value of certain oil and gas prospects. The complaint alleged that as a result of positive statements Exxon made during the class period, the common stock price was artificially inflated, and that Exxon's release of its third quarter financial results on October 28, 2016, in which it disclosed it might have to write down 20% of its "stranded" oil and gas assets, resulted in the stock price falling by more than $2 per share.[di]

In turn public companies are beginning to acknowledge in 2019 10-k annual reports that potential climate change liabilities threaten their bottom line:

> **Arch Coal**—"Increasing attention to global climate change has resulted in an increased possibility of governmental investigations and, potentially, private litigation against us and our customers. For example, claims have been made against certain energy companies

[dg]https://www.blackrock.com/corporate/investor-relations/larry-fink-ceo-letter.

[dh]Id.

[di]http://blogs2.law.columbia.edu/climate-change-litigation/wp-content/uploads/sites/16/case-documents/2016/20161107_docket-316-cv-3111_complaint.pdf.

alleging that greenhouse gas emissions constitute a public nuisance. While the United States Supreme Court held that federal common law provides no basis for public nuisance claims against energy companies, state law tort claims remain a possibility and a source of concern ... Moreover, the proliferation of successful climate change litigation could adversely impact demand for coal and ultimately have a material adverse effect on our business, financial condition and results of operations."[dj]
**Alliance Resource Partners** (second largest coal producer in eastern U·S)—"The United States Supreme Court did not ... decide whether similar [Climate Change] claims can be brought under state common law... As a result ... tort-type liabilities remain a concern."[dk]
**ConocoPhillips**—"Increasing attention to global climate change has resulted in an increased likelihood of governmental investigations and private litigation, which could increase our costs or otherwise adversely affect our business... The ultimate outcome and impact to us cannot be predicted with certainty, and we could incur substantial legal costs associated with defending these and similar lawsuits in the future."[dl]
**Devon Energy**—"These and other regulatory, social and market risks relating to climate change described above could result in unexpected costs, increase our operating expense and reduce the demand for our products, which in turn could lower the value of our reserves and have a material adverse effect on our profitability, financial condition and liquidity."[dm]

On October 24, 2019, the Massachusetts Attorney General ("AG") filed suit in state court seeking a ruling that Exxon Mobil is violating the state's Consumer Protection Act and an order that Exxon pay civil penalties and perform injunctive relief. The suit asserts that Exxon allegedly deceived investors by failing to divulge potential climate change related risks to their investments and that Exxon violated Massachusetts consumer protection laws by misleading consumers on the impact of its products on climate change.[dn] Exxon removed to federal court and in March 2020 a motion to remand was granted.

In this age of uncertainty as to potential climate-change liability, Zurich has offered a D&O liability policy which includes coverage for liabilities "in connection with misrepresenting or failing to disclose information related to greenhouse gases or actual or alleged global warming or climate changes."[do]

[dj]https://www.sec.gov/Archives/edgar/data/1037676/000162828020001344/aci-20191231x10k.htm.

[dk]https://www.snl.com/Cache/IRCache/cf1bcdd44-d65d-0f67-0451-a952e854f3e6.html.

[dl]https://conocophillips.gcs-web.com/node/9221/html.

[dm]http://d18rn0p25nwr6d.cloudfront.net/CIK-0001090012/314baf25-9418-4773-bcc9-eb3152d791c7.pdf.

[dn]http://blogs2.law.columbia.edu/climate-change-litigation/wp-content/uploads/sites/16/case-documents/2019/20191024_docket-1984CV03333-_complaint.pdf.

[do]https://www.zurich.com.au/content/dam/au-documents/business-insurance/financial-lines/directors-and-officers/directors-and-officers-liability-policy.pdf.

## 6.2 First-party property insurance

The property insurance market is likewise in the cross-hairs of climate change-related losses.

As losses from hurricanes, wildfires, and other natural disasters mushroom, insurance companies have retreated from risky areas, leaving homeowners to rely on subsidized state programs, which are struggling to stay financially sustainable. At the same time, mortgage lenders making loans to homebuyers in some high-risk areas are increasingly selling those riskier loans to Fannie Mae and Freddie Mac ("Fannie & Freddie"), which pool the U.S. mortgages into financial assets. If government-backed insurance programs and mortgages fail, it could result in demand for billions of dollars of taxpayer money for bailouts. Indeed, in a January 2020 letter directed to Fannie & Freddie, the U.S. Senate expressed concern that they were not adequately prepared to adapt to the impact of climate change, stating "it is critical that Fannie Mae is prepared for these increasing risks so that it can continue fulfilling its Congressionally-directed mission to facilitate access to homeownership throughout the nation …" The Senate letter goes on to request that Fannie & Freddie answer a number of detailed questions concerning their preparedness to climate change.[dp]

### 6.2.1 Flood-risk

A working paper published by the National Bureau of Economic Research found that homes at risk of flooding in the U.S. are currently overvalued by $34 billion, pointing to a potential real estate bubble fueled by climate threats.[dq] Recent research by consulting firm McKinsey Global Institute ("MGI") likewise concluded that coastal homes in Florida could lose 15%–35% of their value by 2050.[dr] By 2050, an analysis by the First Street Foundation shows Monroe County, Florida alone could see 35% of its properties (over 12,000 homes) flooded 50 or more times a year. At that point, the study assumes, they would lose all their value.[ds]

As floods and powerful hurricanes become more common, insurance premiums go up. Flood prone communities in the U.S. are expected to be hit with a serious price hike next year when the National Flood Insurance Program (NFIP) recalculates its rates. The MGI report found that if prices track with rising risk, average annual flood premiums could increase by about 50% from $800 to $1200, with bigger jumps in higher-risk properties.[dt] Very few

[dp] https://www.banking.senate.gov/imo/media/doc/Fanine%20Freddie%20Letters%20Climate%20Risks.pdf.

[dq] https://www.nber.org/papers/w26807.

[dr] https://www.mckinsey.com/~/media/McKinsey/Business%20Functions/Sustainability/Our%20Insights/Will%20mortgages%20and%20markets%20stay%20afloat%20in%20Florida/MGI_Climate%20Risk_Case%20Studies_Florida_May2020.pdf.

[ds] https://wusfnews.wusf.usf.edu/environment/2020-05-04/sea-rise-wont-sink-all-of-floridas-real-estate-market-experts-say-just-parts-of-it. Globally, since 2000 the number of people living in the low-elevation coastal zone which are exposed to storm surge events has grown by about 1.3% annually, or 0.8% faster than the overall population, with such growth particularly notable in Asia and Africa. https://www.swissre.com/dam/jcr:85598d6e-b5b5-4d4b-971e-5fc9eee143fb/sigma%202%202020%20_EN.pdf.

[dt] https://www.mckinsey.com/~/media/McKinsey/Business%20Functions/Sustainability/Our%20Insights/Will%20mortgages%20and%20markets%20stay%20afloat%20in%20Florida/MGI_Climate%20Risk_Case%20Studies_Florida_May2020.pdf.

home buyers in the U.S. would likely learn about a property's flood risk before they make their offer, as just a few states currently require sellers to disclose how much they pay in flood insurance.[du]

The NFIP was established in 1968 in response to a lack of affordable private flood insurance. Since the end of 2017, 15 short-term NFIP reauthorizations have been enacted. The NFIP is currently authorized until September 30, 2021. The NFIP is the primary source of flood insurance coverage for residential properties in the U.S. The NFIP has more than 5 million flood insurance policies providing over $1.3 trillion in coverage, with over 22,400 communities in 56 states and jurisdictions participating. The program collects about $4 billion in annual premium revenue and fees.[dv] In 2017, the NFIP took in approximately $3.6 billion in annual premium and paid out $8.7 billion.[dw] Approximately 3.8% of NFIP policyholders have filed for repetitive losses, accounting for a disproportionate 35.5% of flood loss claims and 30.5% of claim payments. Of those serial recipients, the Federal Emergency Management Agency ("FEMA") estimates that 90% pay grandfathered below-market rates.[dx] Instead of charging insurance premiums that cover the expected cost of floods, FEMA offers partly subsidized insurance. The NFIP was originally designed to generate enough revenue for a "historical average loss year," which is the mean of the annual losses over the life of the program. However, this method of calculating needed revenue underrepresents losses from catastrophic storms. The NFIP is over $20 billion in debt as of December 2019.[dy]

The NFIP currently covers up to $250,000 for residential buildings and $500,000 for nonresidential buildings damaged by flooding. Coverage for certain commonly owned contents, such as furniture and electronic equipment, is also available for purchase up to $100,000 for residential properties and $500,000 for nonresidential. As of September 2019, single-family homes constitute 69% of the properties covered by the NFIP.[dz]

FEMA has confirmed that it is switching the NFIP to risk-based pricing, which would end the subsidies most coastal communities enjoy on their flood insurance premiums and show the true dollar cost of living in areas repeatedly pounded by hurricanes and drenched with floods. Such a paradigm shift could have a significant effect on the fair market value of flood-prone properties. Although the new risk rating structure was due to take effect in October 2020, it has been deferred to October 2021. FEMA states that the new risk rating (dubbed "Risk Rating 2.0") "will deliver rates that are fair, make sense, are easier to understand and better reflect a property's unique flood risk."[ea]

A January 2021 Report by the University of Pennsylvania's Wharton Risk Management and Decision Processes Center states that the U.S. federal government's flood insurance

[du] https://www.nber.org/papers/w26807.

[dv] https://fas.org/sgp/crs/homesec/IN10835.pdf; https://www.fema.gov/flood-insurance/rules-legislation/congressional-reauthorization.

[dw] https://www.washingtonpost.com/opinions/flood-insurance-reform-wont-be-pleasant-but-its-necessary/2019/03/30/8f07f198-4a72-11e9-93d0-64dbcf38ba41_story.html.

[dx] https://www.latimes.com/opinion/op-ed/la-oe-welch-flood-insurance-20170918-story.html.

[dy] https://fas.org/sgp/crs/homesec/R44593.pdf.

[dz] https://www.insurance.com/home-and-renters-insurance/natural-disasters/flood-insurance.html; https://www.pgpf.org/budget-basics/the-national-flood-insurance-program.

[ea] https://www.fema.gov/flood-insurance/work-with-nfip/risk-rating.

program is "so complicated" that homeowners and insurance agents do not understand how to obtain lower premiums. Although Wharton focused only on Portland, OR, the study says its findings are far-reaching, noting that "the complexities of current NFIP pricing are universal, so these errors and missed opportunities likely exist in other communities, as well." The Wharton report further suggests that states should require insurance agents who write flood policies to undergo mandatory training every year, so they better understand NFIP and also urge the FEMA to make policy pricing "easier and more transparent."[ea1]

### 6.2.2 Wildfire-risk

The central overriding message of a 2020 Report by an advisory panel to the U.S Commodities Futures Trading Commission is that U.S. financial regulators must recognize that climate change poses serious emerging risks to the U.S. financial system, and they should move urgently and decisively to measure, understand, and address these risks.[eb] According to this Report, increased wildfire activity caused by higher temperatures and drier conditions across the U.S. West are among the sparks from climate change that could ignite a U.S. financial crisis by damaging home values, state tourism and local government budgets.[ec] In turn, the increased frequency of wildfires leads to the release of more carbon into the atmosphere, which further exacerbates the problem of global warming by effectively turning carbon sinks (forest) into carbon sources.

CalFire, California's fire-fighting agency, states that approximately 3 million of the state's 12 million homes are at high risk from wildfires.[ed] In many coastal and wildfire-prone regions, insurers are retreating, finding that the potential losses just aren't worth the business. In cases where the industry is willing to offer policies, premiums are rising. California homeowners living in areas at high risk for wildfires, for example, have seen their premiums rise by as much as 500%.[ee] Insurance companies in California have declined to renew nearly 350,000 policies since 2015 in areas at high risk for wildfires.[ef] As a result, homeowners who need fire, windstorm, and hail insurance are increasingly turning to subsidized, state-backed programs. Typically called FAIR—or Fair Access to Insurance Requirements plans—about 30 U.S. states have an insurance program of last resort for homeowners unable to find insurance on the private market. These programs have ballooned in recent decades. In 1990, FAIR programs held about 780,000 insurance policies. By 2014, that figure had grown to about 2.1 million.[eg]

[ea1] https://riskcenter.wharton.upenn.edu/wp-content/uploads/2021/01/Improving-LMI-Household-Flood-Insurance-Options_Issue-Brief.pdf.

[eb] https://www.cftc.gov/sites/default/files/2020-09/9-9-20%20Report%20of%20the%20Subcommittee%20on%20Climate-Related%20Market%20Risk%20-%20Managing%20Climate%20Risk%20in%20the%20U.S.%20Financial%20System%20for%20posting.pdf.

[ec] Id.

[ed] https://www.sacbee.com/news/california/article233142921.html.

[ee] https://www.hcn.org/articles/wildfire-in-california-more-than-340000-lose-wildfire-insurance.

[ef] https://www.claimsjournal.com/news/west/2020/02/21/295625.htm.

[eg] https://grist.org/climate/insurance-companies-and-lenders-are-responding-to-climate-change-by-shifting-risk-to-taxpayers/.

A 2019 white paper by reinsurance giant Munich Re focused on climate change and California wildfires, but also observed that there are many regions across the globe where the environmental conditions favoring a severe local fire season have intensified significantly over the last few decades as well. These areas include: Mexico, Brazil, the Mediterranean region of southern Europe, large parts of sub-Saharan Africa, the east of Australia, eastern China and Korea.[eh]

## 7 Insurance market reaction and preparedness

Insurers typically provide 1-year policies, and their underwriting decisions tend to be made using retrospective models with short time horizons. Thus, although this short time horizon allows for repricing risk and therefore some insulation from climate change exposure, there is concern that they, their reinsurers, and their regulators could neglect to account for climate change-related shifts in the frequency or intensity of catastrophic events that unfold over multiple years or decades.

In fact, the 2019 Insurance Regulator State of Climate Risks Survey, conducted by the Deloitte Center for Financial Services, found that a majority of U.S. state insurance regulators expect all types of insurance companies' climate change risks to increase over the medium to long term, including physical risks, liability risks, and transition risks. More than half of the regulators surveyed also indicated that climate change was likely to have a high impact or an extremely high impact on coverage availability and underwriting assumptions.[ei]

In the 2019 study of "Loss and Damage From Climate Change", the authors raise warning that climate change may make some risks uninsurable and urge insurers to adjust their underwriting practices that are typically based on recent past loss experience.[ej]

Regulators in some jurisdictions are experimenting with climate risk stress testing. For example, the Bank of England in 2019 announced plans to conduct climate risk stress tests of major U.K. banks and insurers. Also, as part of the Bank's Biennial Exploratory Scenario (BES), scheduled to start in 2021, it will ask major U.K. banks and insurers to estimate the size of climate change risks in three scenarios over a 30-year time horizon and consider how they would adjust their business models under each scenario. Similarly, the Bank of France, the Australian Prudential Regulation Authority, and the Bank of the Netherlands have completed or are in the process of launching climate risk stress tests for banks and insurers.[ek]

### 7.1 Divestment from the coal industry

In May 2018, Allianz announced that effective immediately, it will no longer insure both single fired coal power plants and all planned and operating coal mines. Allianz will also no

[eh]https://www.munichre.com/content/dam/munichre/global/content-pieces/documents/Whitepaper%20wildfires%20and%20climate%20change_2019_04_02.pdf.

[ei]https://www2.deloitte.com/us/en/pages/financial-services/articles/insurance-companies-climate-change-risk.html.

[ej]https://link.springer.com/content/pdf/10.1007%2F978-3-319-72026-5.pdf.

[ek]https://cleantechnica.com/files/2020/09/Managing-Climate-Risk-in-the-U.S.-Financial-System.png.

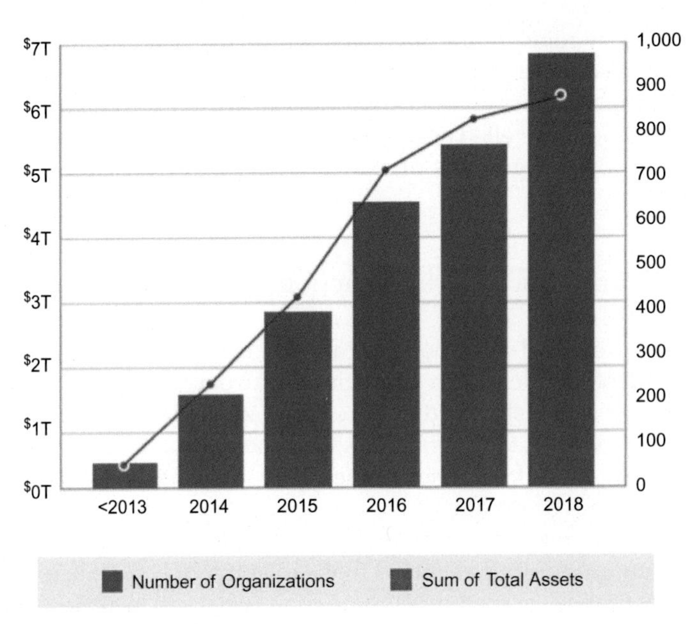


**FIG. 13** Growth in divestment commitments.

longer invest in energy companies that put the 2°C Paris Treaty temperature reduction target at risk, by extensively building coal-fired power plants.[el] Thereafter, a number of other insurers have followed suit or indicated intent to do so. International insurers Swiss Re, Munich Re, AXA and Zurich have all opted to limit their insurance dealings with coal.[em] In the last 6 months of 2019, U.S. insurance companies Chubb, AXIS Capital, Liberty Mutual and The Hartford took steps to limit coal insurance. Coal exit policies have been announced by at least 17 of the World's biggest insurers which control 46% of the global reinsurance market and 9.5% of the primary insurance market and account for more than $9 trillion of insurers' assets. Most refuse to insure new mines and power plants, while companies such as Swiss Re and Zurich, also ended coverage for existing coal projects and the companies that operate them, and have adopted similar policies for tar sands.[en]

According to a 2018 Report by Arabella Advisors, nearly 1000 institutional investors with $6.2 trillion in assets have committed to divest from fossil fuels, exhibiting a striking upward trend, as shown in Fig. 13. The insurance sector is estimated to account for $3 trillion of the $6.2 trillion in divestitures.[eo]

[el] https://www.insurancebusinessmag.com/asia/news/breaking-news/allianz-bows-to-pressure-announces-it-will-stop-insuring-coal-firm-99777.aspx.

[em] https://dailycaller.com/2018/11/20/insurance-firms-coal/.

[en] https://www.insureourfuture.us/updates/2020/1/27/2019-a-record-breaking-year-for-insurance-policies; https://theclimatecenter.org/us-insurance-industry-lags-in-divesting-from-fossil-fuels/.

[eo] https://www.arabellaadvisors.com/wp-content/uploads/2018/09/Global-Divestment-Report-2018.pdf.

Boulder, Colorado recently passed a resolution and became the first county in the United States to call on insurance companies to stop insuring and investing in fossil fuels, and to announce that it will screen potential county insurance providers for those who continue to support fossil fuels without phase-out plans.[ep]

## 7.2 Catastrophe bonds and insurance-linked securities

Since the early 2000s, much of the commercial insurance industry has experienced a soft market. Prices have remained relatively low despite numerous natural disasters that have been costly for insurers.[eq] Why has the insurance market remained soft despite all the costly disasters? The answer lies in part that insurers have easy access to capital from nontraditional sources. After Hurricane Andrew in 1992, insurers needed a new source of capital. Hedge funds, mutual funds, pension funds, and other investors responded by directing money into catastrophe bonds ("Cat Bonds"), and other types of "alternative capital" or "insurance linked securities" ("ILS").[er] The ILS market has grown significantly since hurricane Katrina in 2005.[es]

Much of the alternative capital has been concentrated in the catastrophe business, protecting insurers from natural disasters. The Cat Bond's purpose is to protect insurers from "catastrophic" costs tied to damage from hurricanes, floods or other natural disasters.

Moody's recently reported that the role of ILS and alternative reinsurance capital in paying catastrophe claims related to climate change is set to increase, as the insurance and reinsurance sector becomes increasingly aware of the climate risk it faces and turns to efficient capacity to help it offset them.[et] In fact, the Cat Bond market has been extremely active, with the record issuance of more than $11 billion in 2018 and 2017 and about $100 billion invested to date.[eu] Artemis and A.M. Best report that 2020 could post a new record for Cat Bond issuance, with $7.5 billion having been issued as of September 2020.[ev]

Despite the recent trend of record natural disasters, Cat Bonds have been posting significant positive returns for investors.[ew] Indeed, the use of Cat Bonds to hedge against climate change liabilities may be a prudent strategy. This is especially because the typical Cat Bond is "triggered" on very specific terms. For example, a Cat Bond may only cover wind damage for a named Hurricane in a particular geographic area, but not flooding.[ex] The Cat Bonds are

[ep] https://www.insurancebusinessmag.com/us/news/environmental/co-county-to-screen-insurance-contracts-based-on-fossil-fuel-policies-213753.aspx. Boulder is also a plaintiff in a pending climate change suit against various fossil fuel defendants.

[eq] https://www.naic.org/capital_markets_archive/primer_180705.pdf.

[er] https://seekingalpha.com/article/4151617-insurance-linked-securities-embracing-catastrophic-risks.

[es] Id.

[et] https://www.artemis.bm/news/ils-role-in-paying-climate-related-catastrophe-claims-to-increase-moodys/.

[eu] https://www.reinsurancene.ws/total-cat-bond-issuance-surpasses-100bn-aon-securities/.

[ev] https://www.artemis.bm/news/catastrophe-bond-market-may-surpass-records-in-2020-a-m-best/.

[ew] https://www.insurancejournal.com/news/international/2018/09/11/500669.htm.

[ex] Id.

also not correlated to the stock or other capital market.[ey] Moreover, the majority of the Cat Bonds is effective for only 1–3 years in duration and therefore are priced for the short term, and as a result, are not exposed to climate change uncertainty over a longer multidecade horizon.[ez] In fact, in a first of its kind, FEMA recently purchased a $500 million Cat Bond to reinsure a portion of its potential exposure under the ailing National Flood Insurance Program.[fa]

## 8 Closing thoughts

It is beyond reproach that the Earth's climate is changing in ways that will undoubtedly present negative impacts that will be borne on some level by every person and business. Many of the scientific models present catastrophic damage scenarios occurring relatively soon—within the lifespan of our children. The monetary cost of climate change—be it for adaptive actions or failure to adapt, will be enormous, eclipsing the GDP of many developed countries for decades to come.

Given the foregoing, lawsuits against fossil fuel companies and other carbon producers seeking to hold them responsible for the effects of climate change, will continue to grow. Similarly, suits by stakeholders against public companies and their directors and officers will likely proliferate.

Coverage actions and decisional law relating to insurance for climate change liability are virtually nonexistent, but that is likely to change soon.

In the last few years, there has been a proliferation of new climate change suits by municipalities across the U.S., seeking to hold the fossil fuel companies accountable for the past and future costs arising from climate change. At present, there have been no substantive rulings in any of these cases on the merits raised by the plaintiffs. Should any of these suits survive motions to dismiss and result in successful judgments, the damages are virtually limitless.

We expect to see financial institutions continuing to divest from fossil fuel related investments, with insurers occupying a big part of that role. Insurance linked securities, including Cat bond issuance, will almost certainly continue to grow as a vehicle to hedge against uncertain climate-related liability.

A July 2020 report by The Institute and Faculty of Actuaries sets out a sobering and comprehensive accounting of key implications of climate change for the insurance industry:

- Changing weather patterns and a changing climate will impact property- and agriculture-related losses through changes in frequency and severity of flood, wind, drought, hail and other climate-related events. These changes will need to be modeled and allowed for in all aspects of the business, i.e., pricing, reserving and capital modeling.

[ey] Id.

[ez] https://www.schroders.com/en/sysglobalassets/digital/insights/pdfs/climate-change-threat-to-ils.pdf; https://www.fcm.com/insurance-linked-securities-climate-change-and-esg-considerations.html.

[fa] https://www.artemis.bm/news/fema-secures-500m-nfip-cat-bond-with-backing-of-over-35-investors/.

- Insurers typically charge higher prices for risks where there is greater uncertainty around their scale, nature and frequency. For some types of insurance, the uncertainty around climate risks could lead to prices that few customers or businesses could afford and lower rates of insurance penetration. It could also widen the protection gap, i.e., between those who can afford protection and those who cannot.
- Exposed coastal properties and coal-related activities are examples of risks that might soon become uninsurable. To support the Paris Agreement commitment to keep climate change below 2°C, some insurers and reinsurers have divested from coal companies while others refuse to underwrite new coal projects. Insurers and reinsurers face the risk of a shrinking market for certain products.
- The risk of over-exposure to a single (climate-related) event may increase as significant climate events become more common and/or more severe, e.g., increased frequency of tropical cyclones, tornado, hail, drought, flood, famine, etc. Insurers and reinsurers should ensure that accumulations are managed in areas that become more susceptible to these developing risks, e.g., increased density of coastal property coverage as sea levels rise.
- Events that are usually uncorrelated may become more correlated because of climate change, e.g., correlation of political risk with droughts or floods. These correlated risks are difficult to quantify and manage and contribute to a greater accumulation of risks.
- At some point in the future, liability claims relating to climate change could emerge with some latency. There are already many examples of legal proceedings relating to climate-change both inside and outside of the U.S. Lawsuits where the negative impact of carbon emissions is central to the claim are increasing.
- There is a risk that the different models used by actuaries to calculate premiums, reserves and capital do not adequately represent the reality of a world impacted by climate change (and if they do now, they may not in the future). In particular, actuaries should consider how sensitive their models are to assumptions and data that could be impacted by climate change.
- Those insurers who do not adequately account for climate risks in their pricing models may be more susceptible to adverse selection by policyholders, e.g., because they unwittingly offer cheaper premiums to customers than competitors who have adequately accounted for climate risks.
- Climate change could lead to greater capital requirements for insurance companies because of the increased frequency and severity of extreme events combined with the risk climate change poses to assets.
- There is a move toward standardized reporting requirements, and a growing number of jurisdictions have introduced requirements for insurance companies. Rating agencies have also flagged that climate risks need to be incorporated in credit ratings.[fb]

According to a 2019 CRO Forum Report, as hazard modeling becomes ever more precise, certain local peak risks may exceed capacity or become unaffordable to insure. Certain coastal or forest-fringe properties in U.S. are already on the edge of insurability. Governments can overcome some of the issues through pooling mechanisms that share the peak risk across a wider pool of participants; however, unless these are designed very carefully, they can make

[fb]https://www.actuaries.org.uk/system/files/field/document/Climate-change-report-29072020.pdf.

the problem worse by incentivizing unsustainable development. However, in the more extreme warming scenario of >5°C, severe damage and disruption could become so frequent later in the century that many risks may be uninsurable, with a profound impact on the economy and on society.[fc]

Since GHG emissions do not obey political boundaries, climate change will firmly remain an issue to be grappled with on local, state, national and international levels—and an issue that will remain at the forefront for decades to come.

[fc]https://www.thecroforum.org/wp-content/uploads/2019/01/CROF-ERI-2019-The-heat-is-on-Position-paper-1.pdf.

# CHAPTER 17

# Game theory and climate change

*David Mond*

**Mathematics Institute, University of Warwick, Coventry, United Kingdom**

OUTLINE



## 1 Introduction

The Nobel Prize in Economics has been awarded to 84 people. In all, 17 have won for their contributions to game theory—the study of the incentives operating on individuals (which may be people, animals, species, countries, etc.), and the strategies that they can adopt, in situations ("games") where different levels of benefit (or "utility") are available to the participants, depending on the strategies that they choose. Game theory has been most widely applied in economics, and often is viewed as a part of that subject. But in its abstraction, which sees the same pattern in very different kinds of events, it is more a branch of mathematics; its need to invoke and critique human rationality in order to understand human responses to the incentives it identifies, means that it also draws on psychology.

The meanings of the word "game" in everyday use are hard to circumscribe; the same is true in game theory. Two common features are, first, that a game has rules, and, second, that a game is an activity carried out for fun—in other words, in which the serious considerations governing our lives are temporarily set aside. The term "game theory" reflects the former meaning rather than the latter.

Below I introduce two abstract model situations described by game theory ("games," for short) with implications for our efforts to prevent climate change. I do this in a resolutely

*The Impacts of Climate Change*
**https://doi.org/10.1016/B978-0-12-822373-4.00005-7**

nonmathematical way. A qualitative description is enough to make clear their relevance. The only point where we use any mathematical notation is in our description of the Prisoner's Dilemma in Section 2.1. The reader who is put off by mathematical notation can safely skip Section 2.2.1.

In applications of game theory in Economics and Business, the nature of the game is usually out in the open. Generally, there is a situation of open competition, and game theory is brought to bear on the different strategies available, and their likely pay off, or "utility." In the context of climate change, the situation is somewhat different. There is a shared problem: our energy production and agriculture are releasing gases into the atmosphere which are causing global warming, which threatens major disruption and worse, and so we—all of humanity—need to change our behavior. The situation is not at first sight competitive. Technological developments which will, or would, make this change possible, are developing rapidly, and there is a consensus that the problem can be solved. The changes we need could be made, in time to avoid catastrophe. But they are not being made. Since the problem first came to the attention of politicians, they have spent nearly 50 years postponing resolute action.

Fig. 1 shows the Keeling Curve, which shows the atmospheric concentration of carbon dioxide from 1956 to the present day, measured at the top of the Mauna Loa Volcano on the

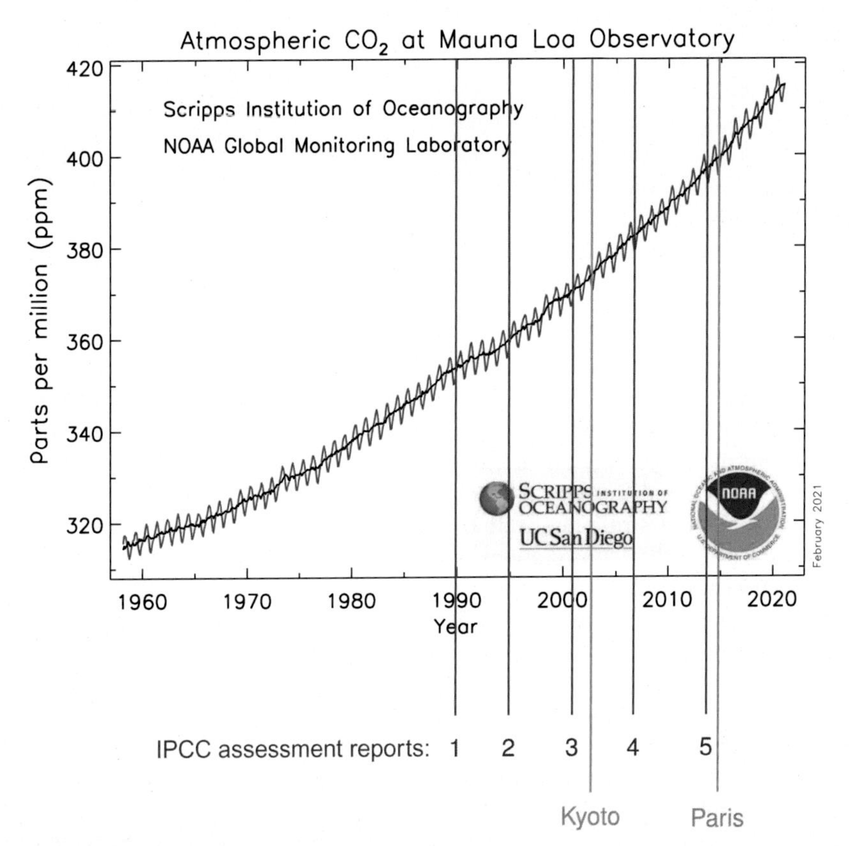


**FIG. 1** The Keeling Curve, and the dates of the IPCC reports and the Kyoto and Paris agreements. *Courtesy of NOAA and Scripps Institution of Oceanography at UC Dan Diego.*

island of Hawaii, far from the major sources of carbon dioxide. The small ripples are the annual fluctuations due to the uptake of atmospheric carbon by plants during the northern hemisphere summer—there is more land, and therefore more plants, in the northern hemisphere than in the southern. Over this diagram, I have overlaid the dates of the five Assessment Reports of the Intergovernmental Panel on Climate Change (IPCC) and the dates of two international agreements which, it was (is?) hoped, would curb carbon emissions. If there is any deviation from the steady increase as a result of the warnings of the world's scientists, it is hard to discern. We are not acting appropriately to solve the problem.

What is behind this failure to act? Some of the reasons are rather obvious. It is difficult and painful to give up the technologies and comforts that we have grown used to. The developed economies depend on a high level of consumption, which the developing economies aspire to. Agreeing who should make the sacrifices has proved extremely contentious and ultimately impossible. The fossil fuel lobby has spent billions of dollars disputing the science and encouraging doubt, as documented by Oreskes and Conway (2012). The contribution of game theory is merely to point out the perverse incentives which prevent ordinary individuals from changing our behavior, and lead us unwittingly, as nations, to a stalemate in the international negotiations needed to agree on reductions in carbon emissions. The games at work influence the behavior of individuals, and nations, without us quite understanding why this is happening. In this essay, I argue that what is needed, above all, is information. We need awareness at many levels. First and foremost, we need awareness of the science itself. But this must be coupled with awareness of the reasons for our failure to agree on emissions reductions. I hope that a small dose of game theory will contribute to this.

## 2 Model games and climate change

### 2.1 The Prisoner's Dilemma

The Prisoner's Dilemma is probably the most famous of the model games. Though its original setting, described here, is artificial, many instance of the same structure arise naturally.

Two suspects, Al and Bob, are arrested for an attempted robbery, and interrogated by the police in separate rooms. Neither knows what his accomplice is saying. Each has two options—to confess, in the process implicating his accomplice, or to keep silent. The authorities are anxious to obtain a conviction, so they design a set of penalties to secure one, and explain them to Al and Bob. To display them, we use a $2 \times 2$ grid. Since each suspect has two options, Confession (C) or Silence (S), there are $2 \times 2 = 4$ possibilities. In the grid, shown in Fig. 2, the top-left square shows the penalties for (C, C): Al gets 4 years and so does Bob; the top-right square shows the penalties for (C, S): Al gets 1 year and Bob gets 6. One can make up a plausible rationale for these penalties: if Bob confesses but Al keeps quiet (the top-right square), then Bob gets a reduced sentence of 1 year in return for his collaboration, and for testifying against Al, who gets a heavy sentence of 6 years for holding out. In the bottom-left square, the choices, and the penalties, are reversed. If neither suspect confesses, they will be convicted of a less serious crime—perhaps they parked the getaway car on a double yellow line. If both confess, neither is rewarded for betraying his accomplice, and both face a stiff term, but not as stiff as for the one who is silent while the other confesses.

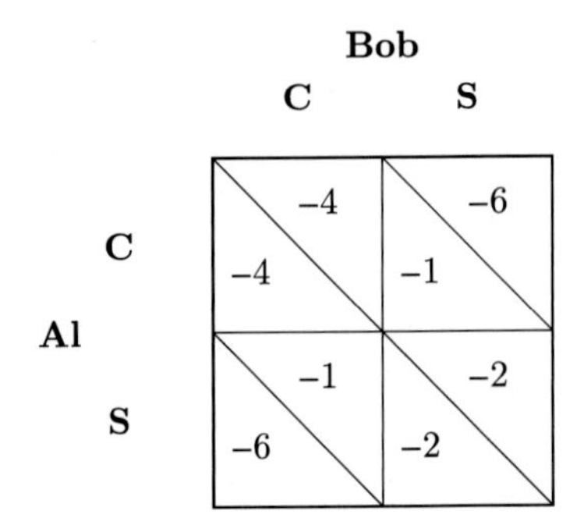


FIG. 2 Utilities in years[1] in Prisoner's Dilemma.

So what do they do? We make two additional assumptions:

*PD1*: There is no honor among thieves. Al and Bob are concerned only for their own well-being, and do not care about one another. And each knows this of the other.
*PD2*: Both Bob and Al are rational, in the sense that they can think through their situation, and make decisions which serve their aims.

To understand what they do, we first look at things from Bob's perspective. He knows that either Al will confess, or will keep quiet, but does not know which he will do. We examine each of these two cases in turn.

- If Al keeps quiet, we are in the bottom row of the $2 \times 2$ table in Fig. 2. If now Bob confesses, he gets 1 year in jail, preferable to the 2 years he gets if he keeps quiet. So,
  If Al keeps quiet, then Bob does better to confess
- Suppose instead that Al confesses. If now Bob confesses, he gets 4 years in jail, rather than the 6 he would get if he kept quiet. So,
  If Al confesses, then Bob does better to confess

Bob does not know whether Al confesses or keeps quiet, but this reasoning shows that he does not need to know. In the rational pursuit of his own interests, he should confess. Confessing is a "dominant strategy."

Interchanging the roles of Al and Bob, the same reasoning shows that Al should confess. So, the outcome is that both confess, and both get 4 years in jail.

This is a disturbing result, for, motivated exclusively by their own interests, and following impeccable logic, both end up going to jail for 4 years instead of the 2 years that they would have got if they had both remained silent.

The Prisoner's Dilemma poses a conundrum that still occupies the attention of political theorists and psychologists. My Google Scholar search for "Prisoner's Dilemma" in 2020 returned 128,000 results. Elinor Ostrom (Nobel Prize for Economics, 2009) in her classic study "Governing the Commons" (Ostrom, 2015), quotes Campbell (1985), on the attraction of dilemmas of this kind:

> Quite simply, these paradoxes cast in doubt our understanding of rationality and, in the case of the Prisoner's Dilemma suggest that it is impossible for rational creatures to collaborate. Thus, they bear directly on fundamental issues in ethics and political philosophy and threaten the foundations of the social sciences. It is

[1] A penalty is a negative utility; a sentence is a loss of years of freedom.

the scope of these consequences that explains why these paradoxes have drawn so much attention and why they command such a central place in philosophical discussion.

### *2.1.1 Robustness of the model*

The assumption of symmetry (the table is unchanged if we interchange the roles and outcomes, for Al and Bob) is not especially unreasonable in the Prisoner's Dilemma, since in any case the whole setup is artificial. But it is not necessary for the game to work. If we allow different sentences for Al and Bob, then provided both obey the inequalities shown in line (4) below, the outcome will be the same.

## 2.2 Prisoners in the real world

Though the setup here is artificial, the very same structure can be observed in many natural situations.

**Example 2.1**

1. Car companies Anvil and Bellows compete in the production and sale of unnecessarily large vehicles (UAVs). If they collaborate, they can both maintain high prices (at the expense of the consumer, of course) and high profits. If one of them defects and lowers its price, it will corner the market, unless the other follows suit. Having both lowered their prices, the two companies make less profit than in the earlier period of their price-fixing collaboration. The incentives here operate in exactly the same way as in the Prisoner's Dilemma, this time to the advantage of the consumer (though not, perhaps, of the rest of us). In this example, the role of the assumption that the prisoners may not communicate is played by the law prohibiting price-fixing cartels, which make it illegal for them to share their pricing intentions with one another.
2. The Prisoner's Dilemma can be seen at work in climate negotiations, in the form of the Polluter's Dilemma. We imagine two blocks, say China and the United States, negotiating to limit emissions. If both are willing to do so, all will benefit. But the same structures as in the classic Prisoner's Dilemma militate against this outcome. If China does comply, then the United States will gain economically by not complying—the unfettered burning of fossil fuels will give it a competitive advantage over China. If China does not comply, the United States would be foolish to allow China the very same competitive advantage. In either case, there is an incentive to defect. As soon as one side defects, the other must follow suit. Now, the two are once again in a situation of parity, but considerably worse off because the attempt to reduce emissions has failed. The story explaining the different utilities is different from the original Prisoner's Dilemma, but the inequalities between them are the same.
3. Campbell (1985, p. 7) offers the example of superpowers who have signed a nuclear arms limitation treaty but distrust their adversary's adherence, and are therefore inclined to defect.

### *2.2.1 Abstraction and generalization*

Consider a two-player game where both players, Al and Bob, have two options, $P$ and $Q$, with symmetric utility table.

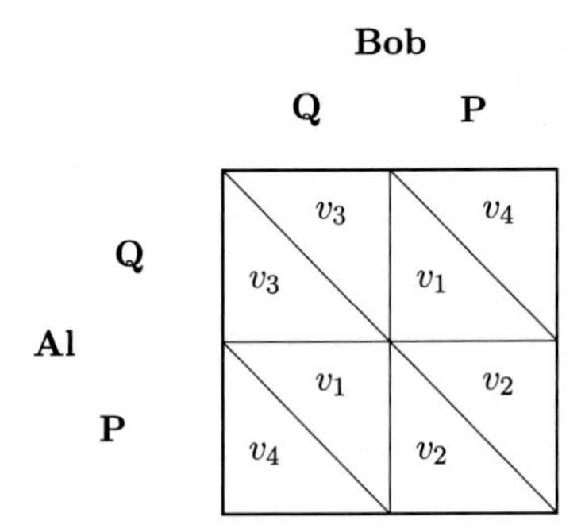


**FIG. 3** Utility table in two-person two-strategy game.

The term "utility" suggests something desirable, so we assume that given a choice, players will prefer a higher utility.

The striking feature of the Prisoner's Dilemma is that the two prisoners are led by the incentives to choose a suboptimal outcome; they end up with 4 years each instead of the 2 years they would have got if they had kept quiet. What inequalities in the utilities in Fig. 3 will yield a similar result? Let us first assume that

$$v_2 > v_3 \tag{1}$$

so that the top-left square, corresponding to (Q, Q), is less desirable for both than the bottom-right square. How can we drive the two players to choose (Q, Q) instead of (P, P)? We use the same reasoning as in the Prisoner's Dilemma, beginning by looking from Bob's point of view. Either Al chooses P, or he chooses Q.

If Al chooses P, then in order that Bob should prefer Q, we must set

$$v_1 > v_2 \tag{2}$$

If Al chooses Q, then in order that Bob should prefer Q, we must set

$$v_3 > v_4 \tag{3}$$

The same reasoning shows that these inequalities drive Al to choose Q.

Combining the inequalities (1)–(3), we conclude that the two players choose (Q, Q) in preference to (P, P), and this is a worse outcome, if

$$v_1 > v_2 > v_3 > v_4 \tag{4}$$

So it was not all that hard to devise penalties in the Prisoner's Dilemma to obtain confessions.

There are 24 different orderings for the quantities $v_1$, $v_2$, $v_3$, and $v_4$, so the set

$$S_{1234} := \{(v_1, v_2, v_3, v_4) \in \mathbb{R}^4 \; : \; v_1 > v_2 > v_3 > v_4\} \tag{5}$$

occupies 1/24th of the space of all[2] possible utility assignments,

[2]We ignore the possibility that some of them are equal, since the set where this happens has zero 4-dimensional volume.

$$V = \{(v_1, v_2, v_3, v_4) : v_1, v_2, v_3, v_4 \in \mathbb{R}\} \tag{6}$$

The opposite order, $v_4 > v_3 > v_2 > v_1$, pushes both Al and Bob toward (P, P), and makes this a suboptimal outcome, since the utility for (P, P) is lower than the utility for (Q, Q). Together, $S_{1234}$ and $S_{4321}$ occupy 1/12th of the space $V$. So, 1 in 12 random assignments of utilities result in Al and Bob being pushed, by their own choices, into adopting a strategy that is to their disadvantage.

Although nature is not governed by random chance, this simple analysis suggests we should expect to see many instances of Prisoner's Dilemma-type perverse incentives.

### 2.2.2 Nash equilibrium

The mathematician John Nash was awarded the Nobel Prize in Economics in 1994 for contributions to game theory. A central part of his contribution is a notion of equilibrium, now known as a Nash equilibrium, in a competitive game. A collection of strategies, one for each player, is known as a *Nash equilibrium* if, once it is adopted, no player can improve his own outcome *by unilateral action alone*. In the Prisoner's Dilemma, an agreement to keep quiet would lead to a shorter sentence, but agreement is not "unilateral action." The pair (C, C) is a Nash equilibrium. There is a related notion, of *Pareto optimum*, or *Pareto equilibrium*, defined as a collection of strategies in which no player's outcome can be improved without worsening the outcome for at least one of the other players. In the Prisoner's Dilemma, (S, S) is a Pareto equilibrium. In general, a Pareto optimum is what one would choose if given the power to maximize utility in a fair way, and is often the outcome in a negotiation where there is an arbitrator. In competitive games with no arbitrator, Nash equilibrium is the more important notion, because generally speaking, a Nash equilibrium is where the players end up. They may arrive there through rational evaluation of the available strategies, as in our description of the Prisoner's Dilemma. But in repeated games, they may reach a Nash equilibrium through trial and error—or just plain error—and it may be hard to escape from. Evolutionary Game Theory studies how this occurs—and how "players" (societies, species, political parties, competing companies, etc.) can sometimes avoid falling into a damaging Nash equilibrium, as in the Prisoner's Dilemma, for example, by evolving altruism.

## 2.3 The tragedy of the commons: Keep the gain, share the pain

Garrett Hardin introduced the term "The Tragedy of the Commons" in a 1968 paper of that name (Hardin, 1968). His observation was that the users of a shared resource, such as the common land on which villagers once grazed their animals, have an incentive to overconsume. Suppose that the butcher pays a sum of $B$ for a cow ready for slaughter, while the cost in resources of raising the cow is $C$. If $B > C$, livestock raising is a profitable enterprise. But now suppose that the cows graze on common land, but are still privately owned. Now the cost, $C$, is shared among all the villagers, while the benefit of the sale to the butcher, $B$, all goes to the owner of the cow. Immediately, the profitability of rearing livestock increases. The added incentive may lead to the raising of too many cows and the consequent exhaustion of the pastures, and thus the loss of the shared resource. Since it is the attractive idealism embodied in the shared ownership of the resource that has led to its destruction, this is indeed a "tragedy."

A mathematical derivation of the Nash equilibrium and the Pareto optimum for a community of goat herders sharing grazing land is given in Gibbons (1992, pp. 27–29). Here, the strategy for each herder is simply the number of goats that he chooses to graze on the common land. Under mild assumptions, it shows that the Nash equilibrium is both less profitable for each herder, and more damaging to the shared resource, than the Pareto optimum.

Scholarly work on this issue typically refers to the shared resource as a *common pool resource*, or CPS. Those who rely on the efforts of others in a collaborative enterprise (such as, for example, reducing our carbon emissions), without contributing themselves, are referred to as "free riders."

There are many instances in which the strategy of privatizing the gain and sharing the pain functions to the advantage of its users. The common pool resource may be a fishery, a functioning market economy, or, in the case of climate change, the world's atmosphere.

**Example 2.2**

1. A company which automates and downsizes to increase its productivity will become more profitable. But there is a social cost: the purchasing power of the workers it has laid off will be reduced, and this reduction will harm the economy. Here again, the company alone will reap the benefit of the downsizing, but the cost of the resulting loss of customer purchasing power will be shared among all companies. In many circumstances, the improvements in productivity and the freeing up of surplus labor will lead to economic gains, and the belief that this is always the case is an axiom for most modern economists. But this is contingent on many other factors, and is by no means guaranteed, nor, necessarily, sustainable. It is interesting to note that employers generally object to measures like the introduction of a minimum wage, and the unionization of the workforce, which seek to counter the effects of too much "freeing up of labor." They see these measures as damaging their profitability, even though in the long run the resulting increase in the purchasing power of the workforce may lead to a general increase in economic activity and a wealthier society. There is an epistemological issue here: the costs to the employer of a union or a minimum wage are immediately obvious, less so the long-term economic benefit of a higher-earning public.
2. The "Diner's Dilemma" is a disconcerting instance of the same principle, which many of us who are neither employers nor cattle-herding villagers personally experience. A group of, let us say, four friends eat together in a restaurant and decide to split the bill, in a profession of goodwill and financial unconcern. For the sake of simplicity, we imagine that two meals are available, one moderate and one large and expensive. Denote the costs of the two meals by $e$ and $E$, respectively, and the pleasure they yield by $p$ and $P$. Although pleasure is not usually measured in the same units as cost, some comparison of these quantities underlies a diner's decision of which meal to choose, so in the following we reason using simple numerical comparisons.

   We assume that $e < p$, so that at the very least each diner would be willing to pick the smaller meal. If also $P < E$, then each diner, paying his own bill, would reject the larger meal. However, when the bill is shared between four, then the basis for the comparison is different. In choosing the larger meal over the smaller, the diner will gain $P - p$ in pleasure, but will pay only $\frac{1}{4}(E - e)$ extra. So if $P - p > \frac{1}{4}(E - e)$, each diner chooses the larger meal. Not surprisingly, the socialization of the cost has increased our willingness to spend. Unfortunately for each diner, all of the other diners reason in same way, so in the end all pay the full cost of the expensive meal, which none of them wanted to do while on their way to the restaurant. As they disperse at the end of the meal, everyone feels a fool.

The inconsistency of our behavior in response to the changing incentives brings home the degree to which we are vulnerable to incentives of this kind. Prior to arriving at the restaurant, we have an idea of how much we want to spend. When we order, after agreeing to split the bill, the regime is different, and we all pick the expensive meal. After we go our separate ways, the regime reverts to what it was before, and we reproach ourselves for our greedy self-indulgence. This behavior has been studied by Gneezy et al. (2004), an interesting example of the intersection of game theory and experimental psychology.

3. Climate change provides an extreme example of the tragedy of the commons. The damage caused by burning fossil fuels, and rearing methane-producing livestock, is shared across the whole world. The benefit from these two activities goes to the perpetrators. The same principle, in reverse (now "keep the pain, share the gain"), applies to measures to reduce emissions. The cost of the measures is paid by their author, while the benefit is shared with the rest of the world. Once again, the incentives work against the climate. In this connection, it is worth noting that China, with one-sixth of the world's population, therefore, stands to reap one-sixth of the benefit of any action it takes to limit emissions (in addition, or course, to the local benefits in cleaner air, etc.), whereas for the United States, with only 1/24th, any reduction in emissions will be of necessity more altruistic. Not to mention minnows like the United Kingdom (1/100th of the world's population) or Australia (1/300th). The EU combines its relatively large size (1/15th of the world's population) with a highly developed awareness of the dangers of climate change—it is host to some of the most important centers of research on climate change, the Potsdam Institute, Climate Analytics and NewClimate in Germany, and, until recently, the UK Meteorological Office's Hadley Centre in Exeter and the Grantham Institute in Imperial College London. Both factors ought to lead to a greater willingness to reduce emissions. But in general, the division of the world into competing nations place us in a supremely dangerous Nash equilibrium. Our own efforts, made in isolation, will bring us insufficient benefit to justify their expense, so by unilateral effort we cannot improve our situation. The remedy may be to form climate coalitions, or climate clubs, in the words of William Nordhaus (Nobel Prize in Economics, 2019): groups of nations which agree to account for costs and benefits jointly (Nordhaus, 2015).

Finding ways of correcting the perverse incentive due to the tragedy of the commons in climate change is a central preoccupation of climate economists.

Although Hardin's theoretical analysis seems hard to fault, there in fact are many instances where common pool resources have been managed sustainably, sometimes for centuries. In her classic study *Governing the Commons* (Ostrom, 2015), Elinor Ostrom documents some of them. She also identifies two currents in the thinking of those who accept Hardin's argument that an unregulated commons will fail. One argues in favor of a central authority which will impose penalties on those who overexploit the shared resource, thereby correcting the perverse incentives. The other calls for the privatization of the resource, for example, by sharing out the common land into smallholdings, whose owners, one might expect, have every incentive to use it sustainably.

In an important category of cases, which include the climate, the second is not an option, and regulation of some kind may appear to be the only remedy. Privatizers are at this point inclined to deny that there is a problem, or to argue that the free market will provide the solution. The two currents are clearly visible in the following two comments on the 2015 Paris Accord. From James Hansen, the former US Government scientist and veteran climate campaigner, quoted in Milman (2015a):

> It's a fraud really, a fake. It's just bullshit for them to say: "We'll have a 2C warming target and then try to do a little better every five years." It's just worthless words. There is no action, just promises. As long as fossil fuels appear to be the cheapest fuels out there, they will be continued to be burned.

From John Kerry, Secretary of State under President Obama, and former Democrat candidate for the US presidency, quoted in Milman (2015b) in response to Hansen's comment:

> Look, I have great respect for Jim Hansen [...] I understand the criticisms of the agreement because it doesn't have a mandatory scheme and it doesn't have a compliance enforcement mechanism. That's true. What we're doing is sending the marketplace an extraordinary signal—that those 186 countries are really committed—and that helps the private sector to move capital into that, knowing there's a future that is committed to this sustainable path. The result will be a very clear signal to the marketplace of the world that people are moving into low carbon, no carbon, alternative renewable energy. And I think it's going to create millions of jobs, enormous new investment in R&D, and that R&D is going to produce the solutions, not government.

To economists, the damage to the climate caused by fossil fuels and agriculture is an "externality"—it does not enter into the accounting which decides on the profitability, or otherwise, of these activities. Where privatization is not an option, the standard approach of free market economics to remove the inefficiency of externalities is to seek ways of including their costs into the accounting. This can be done by the introduction of carbon taxes, as in Sweden, or cap-and-trade schemes, such as the European Emissions Trading Scheme. Calculating the correct price to charge for carbon emissions is a thorny problem for economists, since the discount rate, the rate at which we discount future benefits in relation to present costs, is itself the subject of much disagreement. The late Harvard economist Martin Weitzmann spoke eloquently about this in a lecture at the European University Institute in 2015, available to view on YouTube, at (Weitzmann, 2015):

> If you take a discount rate of 6% or 7%, which corresponds to the rate of return on capital and stock markets over long terms, the social cost of carbon, if you calculate it according to US methodology, it comes out to about \$1; if you take a 1% discount rate, the social cost of carbon is something like \$600 or \$700. Depending on what discount rate you choose, you can get almost any answer.

The inability of economics alone to arrive at a reliable estimate for the social cost of carbon reflects the fact that behind the choice of the discount rate there is a moral judgment. The importance attached to the welfare of future generations is not a purely economic issue. Confidence in the steady march of progress leads some economists to argue that we should not sacrifice current welfare in favor of the welfare of future generations who will in any case be wealthier than us; others, including Weitzmann himself, believe that there is a significant likelihood of a catastrophic outcome.

Externalities can also be internalized, at least for the purpose of policy discussion, by a change in accounting practices: some economists at the International Monetary Fund now include the (unpaid) cost of the damage to the climate caused by fossil fuel use in their estimate of the value of the subsidies[3] to fossil fuel use. In Coady et al. (2019), these are now valued at 5.2 trillion dollars annually, an amount equal to about one quarter of the GDP of the United States. See also Irfan (2019) for a journalistic account.

#### 2.3.1 *Brief pessimistic conclusion*

The games presented here paint a rather gloomy picture. All suggest that the regime of incentives driving our behavior is stacked against our being able to restrain ourselves, and one another, from continuing to pump carbon into the atmosphere. And the failure of 30 years of international climate negotiations and treaties to have any measurable effect on global carbon emissions seems to bear this out.

### 2.4 The importance of knowledge

In all of the games described previously, a lack of information plays an important role.

- This is obvious in the case of the Prisoner's Dilemma: the pressure to confess relies on neither prisoner knowing what the other is doing. This is what ensures that the prisoners end up in the Nash equilibrium rather than the Pareto equilibrium.
- In Example 2.1, fictional car companies are obliged to "defect" from any agreement to maintain high prices, by legislation which prevents them from sharing their pricing intentions.
- In the Diner's Dilemma, it is, apparently, the freedom to make a decision individually that leads to overconsumption. However, this relies on the fact that individuals have privileged access to their own intentions. If the diners were willing to raise the issue in discussion rather than reasoning in private, it is possible that together they could resist the temptation. The concepts of "Trust" and "Information" are closely linked. I distrust you because I do not know your intentions.
- Hardin's Tragedy of the Commons once again relies on ignorance of the other's intentions. In the real world, as empirical work of Elinor Ostrom and others has shown (see Ostrom, 2015; Ostrom et al., 2010), shared resources are often used sustainably when there is a suitable mechanism for ensuring that everyone knows what everyone else is doing. This operates on two levels: the threat of social disapproval can prevent excessive exploitation of the shared resource and the awareness of others' actions plays a role in influencing one's own. Awareness of our own intentions also plays a role in our prediction of what others

[3]It may be disconcerting that what was previously accepted and ignored has become a "subsidy." Something similar was attempted by the Ecuadorian Government in 2010. They argued that by abstaining from the exploitation of oil resources in the Amazon region in order to preserve the rain forest and hence the climate, they were in effect subsidizing the rest of the world's climate stability, and requested compensation for this subsidy, see Vidal and Carroll (2010). The scheme was later abandoned under the pressure of an increase in Ecuador's foreign debt and the unwillingness of the creditor nations to accept the Ecuadorian's redefinition, see Fernholz (2013).

will do. In Ostrom (2009), Ostrom urged the adoption of a range of such mechanisms, at different levels, to cope with climate change.

We do not face these dilemmas because of "human nature." They are epistemological. Any society made up of separate beings, human or extraterrestrial, with privileged access to their own intentions, would face the same problems. It has been hypothesized (e.g., in Byers and Peacock, 2019; Frank et al., 2018) that the failure of technologically advanced alien civilizations to overcome these dilemmas, and their catastrophic environmental consequences, explains the Fermi Paradox—the cosmic silence that greets our search for extraterrestrial intelligence.

## 2.5 The failure of politics

Collective action is required, but prevented by the Nash equilibrium of the Polluter's Dilemma, and what OECD Secretary General Angel Gurría, in a reference to the Tragedy of the Commons, calls the "Tragedy of the National Horizons—an insistence on fashioning policies designed to address climate change primarily, if not exclusively, from a national perspective," in Gurría (2017). The kind of international collaboration needed to overcome these game-theoretic dilemmas has become harder, and not easier, given the recent tendency to the aggressive pursuit of "national interests" by some governments, and the distrust of international institutions that seem to be a feature of much populist thinking. A recent opinion poll (eupinions, 2020) finds that on average, 58% of Europeans would like EU countries to reduce their carbon emissions to no excess emissions by 2030. The current Commission target of EU carbon neutrality by 2050 received the support of only 8% of the young and 10% of those aged 30–49 and 50–69. But the democratic process does not seem to offer a solution, since politicians expect too many voters to reject the sacrifice of immediate self-interest implicit in measures to limit carbon emissions, as predicted earlier. The same opinion poll finds that 53% of young Europeans place more confidence in authoritarian states than democracies when it comes to addressing the climate crisis. The ability of the authoritarian government of China to act, where the democratic nations fail to do so, is the premise of the brilliant piece of climate science fiction "The collapse of western civilization: a view from the future" of Oreskes and Conway (2014). But a quick inspection of the climate policies of present-day authoritarian governments does not support this view. Indeed, China's pursuit of the rapid improvement in living standards for the mass of its population in part reflects the need of its authoritarian government to legitimate its power and control, and so the same incentives operate there as in countries with electoral cycles.

It seems that we are in a very dangerous situation.

The response of politicians to the challenge of changing behavior on the scale required to avert catastrophic climate change has been, in general, to downplay the problem.

> When I analyzed the words that politicians use when speaking about climate in public debates, a key finding was that there is a strong tendency to dilute the message. Politicians were reluctant to speak out about the severity of the climate problem, or the far-reaching changes to our society and economy that will be required. In nearly a hundred thousand words of debate about climate change in 2009, there were only three mentions of abrupt or irreversible impacts, often called "tipping points" even though these are widely discussed in the scientific literature. The politicians appeared to be presenting climate change as a relatively unthreatening, manageable problem. ***From page 58 of* Too hot to handle*, by Willis (2020); see also Willis (2019).***

It seems that they calculate that the electorate will not be willing to make the sacrifice of immediate self-interest that resolute action would require, and therefore try to convince the voters that timid measures which will help them in other ways, are all that can be delivered.

> It was fall 1985, and Curtis Moore, a Republican staff member on the Committee on Environment and Public Works, was telling Rafe Pomerance [a veteran climate campaigner] that the greenhouse effect wasn't a problem.
>
> With his last ounce of patience, Pomerance begged to disagree.
>
> Yes, Moore clarified, of course it was an existential problem—the fate of civilisation depended on it, the oceans would boil, all of that. But it wasn't a *political* problem. Know how you could tell? Political problems had solutions. And the climate issue had none. Without a solution—an obvious, attainable one—any policy could only fail. ***From page 107 of* Losing Earth—the decade we could have stopped climate change, *by Rich (2020).***

Politicians facing reelection are unwilling to grapple with the climate emergency for many reasons. These can be bundled together, as Curtis Moore does in the passage just quoted, as an unwillingness to be seen to fail. A corollary of this is the downplaying of the issue that Rebecca Willis comments on. It should be clear that this is precisely the opposite of what they should be doing. What we need at every stage in our perception of the problem is more information and not less. But this requires politicians to put the long-term interests of their community above their short-term desire to be reelected. In other words, to stop playing the game of politics.

## 2.6 Altruism as a survival strategy

Experimental psychology suggests that humans are endowed with an in-built tendency to altruism, which sometimes overrules our individual preference for self-preservation. Society tries to complement this by honoring those who make sacrifices and expose themselves to danger for the common good. In the United States, the Civil Rights movement of the 1960s and 1970s is an inspiring recent example, as witness Barack Obama's eulogy to the late civil rights activist and congressman John Lewis (The New York Times, 2020). It seems to me that, in view of the obstacles to agreement that have all too clearly stymied international efforts, our only hope is to mobilize an awareness of the malign Nash Equilibrium that we are in, and a willingness to act locally on behalf of the climate, even though this will bring us few direct benefits, in the hope that this will encourage others to do the same.

Altruism can be a survival strategy when it stimulates altruism in others. The civil rights protestors who refused to move from whites only lunch counters, or were clubbed by police in Alabama and Mississippi, in the main improved their own lives, though at considerable risk, because thousands of others were inspired to join them. Once again, information, in the form of publicity, is a vital component of this strategy, and thus it is likely to succeed only where information circulates widely and freely. Modern technology would seem to make this more possible now than ever before.

I end by citing one less dramatic example. The elected members of Warwick District Council, responsible for local administration in a district of 140,000 people in the center of England, have voted unanimously to make the district carbon neutral by 2030, a far more ambitious goal than the UK Government's aim of carbon neutrality by 2050. This will necessitate a local tax rise, amounting to about £57 per year for an average household, and the relevant legislation obliges the council to hold a local referendum on the tax rise, to be held in May 2021. It will be the first such referendum in the United Kingdom. It may be that by the time you are reading this, the results will be known. The hope is that approval of the tax rise and of the ambitious plans for a rapid reduction in carbon emissions would lead to other local councils following suit, and could tip the United Kingdom into efforts concomitant with the gravity of the situation.

An interesting feature of the situation is that no one political party is in control of Warwick District Council. The elected councilors are members of the Conservative, Labor, Liberal Democrat and Green Parties, and of a local residents' group. The largest group is the Conservative, and the leader of the council is from this party, but decisions need multiparty support. I suspect that it is the need to work together across party boundaries that has made a unanimous agreement possible. Free riding, in this situation, would be to oppose the tax increase, to gain cheap electoral advantage. That this has not happened is no doubt a sign of the good sense of the councilors, but may well have something to do with the sharing of power.

Whether it is through the dramatic confrontation of protestors with police, or the patient building of a consensus at all levels, from local to international, the crucial feature of effective action to save the climate is the sharing of information (Fig. 4). This should include an understanding of the trap of a suboptimal Nash equilibrium and of the tragedy of the commons.

**FIG. 4** Police surround a boat parked by climate protestors to disrupt traffic in the center of London, April 2019. *From photo of Extinction Rebellion/Lola Perrin.*

## References

Byers, M., Peacock, K., 2019. Did climate change destroy the aliens? Bull. At. Sci. https://thebulletin.org/2019/07/did-climate-change-destroy-the-aliens/.

Campbell, R., 1985. Background for the uninitiated. In: Campbell, R., Sowden, L. (Eds.), Paradoxes of Rationality and Cooperation: Prisoner's Dilemma and Newcomb's Problem. University of British Columbia Press.

Coady, D., Parry, I., Le, N.P., Shang, B., 2019. Global fossil fuel subsidies remain large. An update based on country-level estimates. IMF Working Paper. May.

eupinions, 2020. https://eupinions.eu/de/text/in-crisis-europeans-support-radical-positions.

Fernholz, T., 2013. Ecuador abandons rain-forest protection to pay its China debts. Quartz. https://qz.com/116321/ecuador-terminates-a-3-6-billion-plan-to-protect-the-amazon-from-oil-drilling/.

Frank, A., Carroll-Nellenback, J., Alberti, M., Kleidon, A., 2018. The anthropocene generalized: evolution of exo-civilizations and their planetary feedback. Astrobiology 18. https://doi.org/10.1089/ast.2017.1671.

Gibbons, R., 1992. A Primer in Game Theory. Harvester Wheatsheaf.

Gneezy, U., Haruvy, E., Yafe, H., 2004. The inefficiency of splitting the bill. Econ. J. 114 (April), 265–280.

Gurría, A., 2017. Climate action: time for implementation? OECD Secretary-General Angel Gurría, speech at the Munk School of Global Affairs and Public Policy. https://www.youtube.com/watch?v=wI227Dgt6pE

Hardin, G., 1968. The tragedy of the commons. Science 162, 1243–1248 Available online from Creative Commons.

Irfan, U., 2019. Fossil fuels are underpriced by a whopping $5.2 trillion: we can't take on climate change without properly pricing coal, oil, and natural gas. Vox, May 17. https://www.vox.com/2019/5/17/18624740/fossil-fuel-subsidies-climate-imf.

Milman, O., 2015a. Father of climate change awareness, calls Paris talks 'a fraud'. The Guardian, 13-12-2015. https://www.theguardian.com/environment/2015/dec/12/james-hansen-climate-change-paris-talks-fraud

Milman, O., 2015b. https://www.theguardian.com/environment/2015/dec/13/john-kerry-james-hansen-climate-change-paris-talks-fraud.

Nordhaus, W., 2015. Climate clubs: overcoming free-riding in international climate policy. Am. Econ. Rev. 105 (April), 1339–1370.

Oreskes, N., Conway, E., 2012. Merchants of Doubt, Paperback ed. Bloomsbury Press.

Oreskes, N., Conway, E., 2014. The collapse of Western Civilisation: a view from the future. Dædalus J. Am. Acad. Arts Sci. (Winter 2013), 40–58.

Ostrom, E., 2009. A polycentric approach for coping with climate change, Policy Research Working Paper 5095. World Bank.https://openknowledge.worldbank.org/bitstream/handle/10986/9034/WPS5095_WDR2010_0021.pdf.

Ostrom, E., 2015. Governing the Commons, The Evolution of Institutions for Collective Action. Canto Classics. First published by Cambridge University Press, 1991.

Ostrom, E., Poteete, A., Janssen, M., 2010. Working Together: Collective Action, the Commons and Multiple Methods in Action. Princeton University Press.

Rich, N., 2020. Losing earth—the decade we could have stopped climate change. Picador.

Vidal, J., Carroll, R., 2010. Ecuador signs $3.6bn deal not to exploit oil-rich Amazon reserve. The Guardian, 4-8-2010. https://www.theguardian.com/environment/2010/aug/04/ecuador-oil-drilling-deal-un

The New York Times, 2020. Eulogy of Barack Obama to John Lewis. The New York Times.https://www.nytimes.com/2020/07/30/us/obama-eulogy-john-lewis-full-transcript.html.

Weitzmann, M., 2015. Why is the economics of climate change so difficult and controversial?, Max Weber Lecture, European University Institute. https://www.youtube.com/watch?v=pmtpKPEVSPU.

Willis, R., 2019. Taming the climate. Corpus analysis of politicians speech on climate change. Environ. Politics 26, 212–231. https://doi.org/10.1080/09644016.2016.1274504.

Willis, R., 2020. Too Hot to Handle? The Democratic Challenge of Climate Change. Bristol University Press.

# CHAPTER 18

# Urban life and climate change

*Tobias Emilsson*

**Department of Landscape Architecture, Planning and Management, Swedish University of Agricultural Sciences, Alnarp, Sweden**

OUTLINE



## 1 Introduction

Urban environments are constantly expanding and have been expected to continue to increase in the pre-COVID projections. The current proportion of the global population living in urban centers has been estimated to above 50% and the prognosis is that it will reach close to 70% by 2050 (United Nations et al., 2019). Thus, creating livable urban environments is paramount for a very large number of people.

Local environments and local processes are challenged by degradation due to the ongoing urban development and activities. Increasing population densities inevitably lead to increasing building densities, the development of infrastructure for transport of goods, people, and waste as well as release of different substances to both land, air, and water (Grimm et al., 2008).

As more people are expected to live their entire lives within urban environments, the physical and psychological framework of urban areas is becoming increasingly important. In the light of climate change, cities and urban centers are stressed environments that have been



*The Impacts of Climate Change*
**https://doi.org/10.1016/B978-0-12-822373-4.00009-4**

pointed out as being especially vulnerable and where catastrophes will impact large numbers of people (Revi et al., 2014). The urban development is driving the urban climate in a particular direction but climate change is also influencing the physical conditions with increasing temperatures but the two are also closely connected through multiple feedback loops with implications for human wellbeing.

This chapter will (1) give an introduction to the characteristics of urban climate, its drivers, and how it is affected by climate change, and (2) review potential tools and strategies for climate change mitigation.

## 2 Urban climate

Urban climate is in large dependent not only on regional factors, such as latitude, topography, and distance to large water bodies (Grimmond, 2004) but also on particular effects of the urbanization per se. The basic foundation for the urban climate can be related to both a change in physical characters of the landscape as compared to the rural surrounding but also the activities and processes taking place within the urban zone. The urban environment is characterized by high building densities, high soil sealing rates, and high levels of dark colored materials. Building materials, such as concrete structures, paving, and roof sealing membranes have rater low albedo, that is, only a minor part of the incoming radiation will be reflected. The low reflection means that the heat will be absorbed and temperature will increase. The albedo effect has been directly linked to increasing both day- and night-time temperature (Kolokotroni and Giridharan, 2008). By definition, albedo can vary from 0 where all the energy is absorbed and 1 when all incoming radiation is reflected. Building materials, such as asphalt paving or roof membranes, can have an albedo as low as 0.10 (Taha et al., 1988). This can be compared to natural components such as vegetation, and canopies falling between 0.15 and 0.30. Modern building materials, such as steel, concrete, and pavements also have a high heat capacity, that is, they have a capacity to store much more heat as compared to natural materials, such as soil or sand. Urban areas might store as much as double the amount of heat as compared to nonurbanized surrounding areas (Christen and Vogt, 2004). Thus, urban areas tend to have a warmer climate as compared to the less dense rural surrounding areas; a phenomenon that has been the subject of intense scrutiny for more than 50 years (Oke, 1973). The phenomenon is termed "the urban heat island effect" (UHI) and describes the link between urbanization and increasing surface and atmospheric temperatures along a rural to urban gradient (Voogt and Oke, 2003). In general, the term UHI is most often used to describe the difference between air temperatures in the canopy layer between the urban and nonurbanized surrounding areas but other definitions exist (Voogt and Oke, 2003). The strength of the UHI is most pronounced during calm clear nights.

The increase in building densities and soil sealing is also directly linked to a reduction of vegetated surfaces. Unvegetated sealed surfaces as roads or buildings will have low capacity to hold water and precipitation will rapidly be transported from the sealed surfaces and directed to underground pipes. The small amount of water available for evaporative cooling will only last for a short time meaning that surfaces rapidly heat up as only limited energy will be converted to latent heat (Founda and Santamouris, 2017). The reduction in vegetated surfaces also mean that there will only be limited transpiration from plant surfaces.

The urban climate will not only be affected by the surface materials but also by the organization and urban form. Using urban form to counteract the negative implications of climate has been explored in many cities throughout the centuries, for example, in respect to reduction in wind chill and prevention of heat absorption in hot climates, but urban form might also be part of the problem with urban climates (Pattacini, 2012). The connection between form and temperature is complex and multidimensional. Form will have an impact on reflection and transfer of heat but also on shading, wind speed, wind direction, and consequently ventilation. During summer, building density is important but the buildup of higher temperatures is driven by the vertical organization of buildings where tall buildings increase solar absorption during the day and reduce the nocturnal heat loss (Salvati et al., 2019). Even without particular design, the traditional layout of urban centers with high buildings and narrow streets will cause an increase of urban temperatures. Tall buildings will trap the incoming solar radiation and energy through reflection to surrounding tall buildings causing rising temperatures in street canyons. The reduction in nighttime temperatures due to outgoing longwave radiation is dependent on the sky view factor which is also reduced by urban morphologies with tall buildings (Salvati et al., 2019).

On top of the physical characteristics of the urban fabric is the effect of internal energy use and release of low valued heat from, for example, combustion engines, industrial production, and air conditioning, further increasing the problems derived from urban form and surface materials. These feedback loops where increasing temperatures drive up increased air conditioning use, further increase urban temperatures. UHI might increase the cooling energy need from 10% to 120% during summer. On the contrary, winter heating demand might be reduced from 3% to 45% (Li et al., 2019). Internal energy use in industrial production, transportation, and in particular combustion will also increase the air pollution and the presence of airborne particles. Air pollution have dual effects on the urban climate as it both increase reflection of incoming radiation as well as increasing the absorption and re-emission of longwave radiation (Haywood and Boucher, 2000; Li et al., 2018).

In general, climate change is projected to increase intensity and duration of heat waves as well as the occurrence and the intensity of cloud bursts and this will also have consequences for the urban climate (Seneviratne et al., 2012). Modeling studies have shown that heat waves might become more common and have an extended duration under several climate change scenarios (Lemonsu et al., 2014). This increase in heat wave intensity and duration is especially troublesome in an UHI perspective as the effects of urban heat will be even further enhanced. Heat waves are particularly troublesome as they often exhibit characters that worsen the UHI effect, for example, low wind speed and reduced ventilation, as well as long drought periods were many water reservoirs, temporary stored soil water and plant systems become exhausted of their water supply leading to drastically reduced evaporation rates. Thus, the difference between the urban and the rural climate is further enhanced during heat waves (Li and Bou-Zeid, 2013). The climate change and urbanization process are in many places working in the same directions, reinforcing the effects. Increasing urban temperatures are indeed troublesome and the overall increase in urban heat waves can be linked to decreasing well-being and increasing mortality rates in the urban population (Gabriel and Endlicher, 2011; Guo et al., 2018; Paravantis et al., 2017). Even nonmortal effects can be dramatic, especially among vulnerable population groups, such as poor and elderly, who experience decreased quality of life during longer summer periods (Gabriel and Endlicher, 2011; Kovats and Hajat, 2008). Stress effect of excessive heat might manifest in a range of syndromes ranging from lack of concentration to exhaustion and circulatory disorders (Kovats and Hajat, 2008; Scherer et al., 2013).

## 3 Urban climate and water

The same factors that drive urban temperatures will also have profound effect on urban water flow. The conversion of vegetated and open surfaces to buildings, pavements, or roads using material that are impervious and have low water holding capacity will change both runoff volume and the runoff dynamic. Runoff in an urban setting is much faster and larger in volume as compared to runoff volumes produced by a rainfall event in a natural landscape. As rain hits urban surfaces it will rapidly runoff and be directed toward underground channels or pipes, to prevent undesired flooding and damages on buildings. This means that a large amount of the rain volume must be collected and transported within a short time period from the urbanized areas to the recipient, resulting in increasing storm water volumes and a short time of concentration as compared to undeveloped areas (Saghafian et al., 2008; Xu and Zhao, 2016). Urban streams and pipes will have high flows during and directly after rain events, which that also might cause erosion and destruction of riparian habitats (Walsh et al., 2004). Furthermore, these habitats will be much drier as compared to the conditions in more natural areas between rain events, and smaller creeks and tributaries might be left completely dry resulting in reduced habitat value. Lack of vegetated surfaces means that there is both comparably less infiltration and a dramatic reduction of evapotranspiration. Moreover, a reduction of vegetated surfaces leads to a reduction in rainfall interception, a component that can store several millimeter of rain, resulting in an overall reduction of runoff close to 20% of total rain fall (Kermavnar and Vilhar, 2017).

The effects of climate change will come as an additional stress on top of a system already under pressure due to ongoing urbanization (Skougaard Kaspersen et al., 2017). Climate change is projected to lead to an increase in high intensity cloud bursts resulting in severe storms with large rain volumes becoming more frequent and this will consequently lead to more flash floods and destruction of property value and even loss of life (Henstra et al., 2020; Revi et al., 2014). Indeed, more intense storms will increase the probability of pluvial flooding as precipitation rates over different time scales more often will exceed both infiltration and drainage capacity of the soil, and the capacity of the urban drainage system to handle the particular rain volume (Blanc et al., 2012). Sea level rise might also influence the capacity of the drainage system to handle large rain events, making pluvial flooding events more severe or more common even in northern parts of Europe (Bevacqua et al., 2019).

There is an ongoing struggle to increase the resilience of the urban storm water system but many cities are still seeing continued growth of impervious surfaces since the 1990s with absolute numbers reaching more than 10% (Skougaard Kaspersen et al., 2017). This process is ongoing and adds to the complexity of climate change (Semadeni-Davies et al., 2008). Even if urbanization and climate change are driving the system in the same direction, there are different approaches that are needed to solve the particular problems arising from the two processes. The effect of urbanization on flood risk is comparably larger in areas with coarse soils and high infiltration rates as compared to areas with lower infiltrations rates. Thus, areas where the natural conditions for infiltration are good will be more negatively impacted by increasing soil sealing compared to areas where the natural conditions for infiltration are already low and surface runoff is already a problem. Updating the urban drainage system to cope with development processes might be easier as compared to making it ready to handle climate change driven extreme events with longer return periods (Skougaard Kaspersen et al., 2017).

## 4 Mitigating the negative effects of climate change

Finding strategies for mitigating climate change and handling the negative effects of continuing urbanization is crucial for achieving livable urban environments. There are several approaches that include changing the urban energy exchange pattern and hydrology through alterations of surface materials; making the urban area behave more like its predevelopment state. Most strategies focus on increasing the amount of vegetation within the urban area and specifically on the development of a network of green and blue vegetated surfaces interconnected in strategically planned green infrastructure (GI) network. This network of connected green areas, may contribute to handling storm water volumes but on the same time impact water quality, urban heat, biodiversity, and aesthetic values (Gaffin et al., 2012). By definition GI is multifunctional and focused on the delivery of ecosystem services that are connected to different types of surfaces (European Commission, 2013). In an urban setting, they can be comprised of more familiar vegetated elements, such as urban parks, remaining woodlands, or allotment gardens but also more novel solutions such as green walls, green roofs, or rain gardens.

Urban heat mitigation strategies can be developed with different focuses, for example, focusing on reducing the overall heat load on a larger scale or on increasing the climate comfort at a more local at street level. The strategies and choice of intervention will vary depending on focus. High temperatures at street level and local temperature discomfort can be addressed by, for example, installation of street tree plantations that will contribute with shade and evapotranspirative cooling or green facades that will deliver local evaporative cooling (Zölch et al., 2016). Trees with large canopy areas and closed canopies might have an increased potential in providing better street level microclimate during the day in relation to air temperature, solar radiation, and relative humidity. The climatic effect is species specific as it is related to the tree species ability to maintain high transpiration rates during heat and drought as well as having a high leaf area density (Gillner et al., 2015; Lindén et al., 2016). On the contrary, the dense canopy might have negative effects on nocturnal cooling in that the large canopy reduces the sky view factor (Sanusi et al., 2017).

More complex vegetated system, such as parks and remaining patches of original vegetation, can function as "cold spots" or low temperature refugees within an urban area, particularly for vulnerable groups (Brown et al., 2015). Vegetation structure including multiple layers with shrubs, trees, and ground cover combined with adequate maintenance and in particular, park size is important to create sufficient temperature reduction (Bowler et al., 2010; Vieira et al., 2018).

Green roofs can have a role in decreasing urban temperatures but less so on street level temperatures (Norton et al., 2015). Green roofs will increase reflection and have a potential to contribute with evaporative cooling as long as they are not completely dry. In many environments this will require irrigation to maintain sufficient cooling (Gomes et al., 2019). Thicker substrate layers and a denser vegetation has a higher capacity as compared to low growing ground covering species (Alcazar et al., 2016). Still, green roofs will influence the temperatures adjacent but to increase urban comfort they need to be combined with vegetation directly adjacent to walkways and pedestrian zones (Alcazar et al., 2016). The effect of green roofs and green walls on urban temperature is most pronounced in dry and hot climates where simulation studies have pointed to a close to 10°C reduction in average maximum air temperature. Again, the effect of green walls is more important for street level comfort as compared to roof-based systems (Alexandri and Jones, 2008).

However, there are also alternatives to exclusively green solutions including both GI and physical alterations of buildings or building material. Sometimes these solutions will be more cost effective in solving a single focused problem as compared to installing multifunctional green spaces contributing to a range of factors, such as temperature reduction, amenity value, storm water reductions, and biodiversity (Carvalho et al., 2017). The effect of green systems on urban heat is, for example, dependent on rain fall patterns and evaporative cooling meaning that high reflective roofs might be the preferred choice over green roofs in dry locations where only temperature reduction on the roof is the prioritized value (Carvalho et al., 2017). Reflective roofs or buildings have no benefit to pedestrian climate comfort but can decrease the buildup of heat within a building (Falasca et al., 2019).

Optimization of urban cooling for the wellbeing of urban inhabitants is not only about selection of the proper physical structure. Norton et al. (2015) argue that achieving urban cooling and mitigating increasing urban heat can be seen as a five-step process involving (1) neighborhood prioritization, (2) mapping of neighborhood qualities, (3) utilizing and improving existing green network through changed management, (4) local scale optimization in relation to street orientation and design, and finally (5) actual placement of GI initiatives. A strategical approach is needed to actually focus the installation of green spaces in communities where they are really needed. Changed management in the form of irrigation can also be an efficient way to promote functions from surfaces without making new installations (Shashua-Bar et al., 2011). Similar models have been developed in Detroit that aims to take an overarching view on not only the just placement of green space contributing to climate mitigation and cooling but also for using the green spaces to connect neighborhoods, support development of local biodiversity. Looking on several different values within the same model also makes it clear that there are not always win-win situations but that there is a need for finding solution that balance different interests and that the different tradeoffs between different values into account (Meerow and Newell, 2017).

GI not only has a potential to reduce and balance urban temperatures but has also been proposed to be an efficient way to handle local stormwater problems and move toward urban hydrological systems that have runoff dynamics that more resemble natural catchments. Stormwater must be handled on many levels and GI has a potential to reduce the severity of floods and reduce the occurrence of some events. There are a wide arrange of solutions that can be installed to reduce the negative effects of climate change and urbanization and making the urban system more similar to the predevelopment analog. Over time the focus on storm water reduction and climate change mitigation has grown in scope and ambition moving away from a simple-minded focus on volume control to system that is true multifunctional where the term GI has also entered the storm water vocabulary (Fletcher et al., 2015). The most commonly discussed types of GI system with potential stormwater function are rain gardens, bioswales, constructed wetlands, retention and detention basins, and green roofs (Liao et al., 2017). Street trees using structural soil has been increasingly used during the last years as for stormwater retention (Grey et al., 2018). GI has been seen as a way forward to achieve increased sustainability of stormwater systems but there has been a general lack of support for their efficiency from real-world installations. A case study from Augustenborg using insurance data as a proxy for flooding effect shows that green blue refurbishment of an entire neighborhood can be successful in reducing the probability of local flooding and for reducing the societal costs resulting from damaged properties following flooding (Sörensen and Emilsson, 2019).

## 5 Conclusion

Urbanization is still an ongoing process in many parts of the world resulting in environmental challenges and decreased quality of life for the ever-growing global urban population. Focusing on adaptation of current building practices to decrease the negative impact on urban heat and stormwater runoff is crucial. This involves integration of more soft pervious surfaces that reflect incoming radiation and that can contribute to continued evapotranspiration. The effects of changed building practices are certain and they will decrease the negative effects of urbanization. Additional work is required to mitigate the more uncertain effects of a future changing climate.

## References

Alcazar, S.S., Olivieri, F., Neila, J., 2016. Green roofs: experimental and analytical study of its potential for urban microclimate regulation in Mediterranean–continental climates. Urban Clim. 17, 304–317. https://doi.org/10.1016/j.uclim.2016.02.004.

Alexandri, E., Jones, P., 2008. Temperature decreases in an urban canyon due to green walls and green roofs in diverse climates. Build. Environ. 43 (4), 480–493 (Scopus).

Bevacqua, E., Maraun, D., Vousdoukas, M.I., Voukouvalas, E., Vrac, M., Mentaschi, L., Widmann, M., 2019. Higher probability of compound flooding from precipitation and storm surge in Europe under anthropogenic climate change. Sci. Adv. 5 (9), eaaw5531. https://doi.org/10.1126/sciadv.aaw5531.

Blanc, J., Hall, J.W., Roche, N., Dawson, R.J., Cesses, Y., Burton, A., Kilsby, C.G., 2012. Enhanced efficiency of pluvial flood risk estimation in urban areas using spatial–temporal rainfall simulations. J. Flood Risk Manag. 5 (2), 143–152. https://doi.org/10.1111/j.1753-318X.2012.01135.x.

Bowler, D.E., Buyung-Ali, L., Knight, T.M., Pullin, A.S., 2010. Urban greening to cool towns and cities: a systematic review of the empirical evidence. Landsc. Urban Plan. 97 (3), 147–155. (Scopus). https://doi.org/10.1016/j.landurbplan.2010.05.006.

Brown, R.D., Vanos, J., Kenny, N., Lenzholzer, S., 2015. Designing urban parks that ameliorate the effects of climate change. Landsc. Urban Plan. 138, 118–131. https://doi.org/10.1016/j.landurbplan.2015.02.006.

Carvalho, D., Martins, H., Marta-Almeida, M., Rocha, A., Borrego, C., 2017. Urban resilience to future urban heat waves under a climate change scenario: a case study for Porto urban area (Portugal). Urban Clim. 19, 1–27. https://doi.org/10.1016/j.uclim.2016.11.005.

Christen, A., Vogt, R., 2004. Energy and radiation balance of a central European city. Int. J. Climatol. 24 (11), 1395–1421. https://doi.org/10.1002/joc.1074.

European Commission, (Ed.), 2013. Building a Green Infrastructure for Europe. Publ. Office of the European Union. https://doi.org/10.2779/54125.

Falasca, S., Ciancio, V., Salata, F., Golasi, I., Rosso, F., Curci, G., 2019. High albedo materials to counteract heat waves in cities: an assessment of meteorology, buildings energy needs and pedestrian thermal comfort. Build. Environ. 163. https://doi.org/10.1016/j.buildenv.2019.106242 (Scopus).

Fletcher, T.D., Shuster, W., Hunt, W.F., Ashley, R., Butler, D., Arthur, S., Trowsdale, S., Barraud, S., Semadeni-Davies, A., Bertrand-Krajewski, J.-L., Mikkelsen, P.S., Rivard, G., Uhl, M., Dagenais, D., Viklander, M., 2015. SUDS, LID, BMPs, WSUD and more—the evolution and application of terminology surrounding urban drainage. Urban Water J. 12 (7), 525–542. https://doi.org/10.1080/1573062X.2014.916314.

Founda, D., Santamouris, M., 2017. Synergies between Urban Heat Island and Heat Waves in Athens (Greece), during an extremely hot summer (2012). Sci. Rep. 7. https://doi.org/10.1038/s41598-017-11407-6.

Gabriel, K.M.A., Endlicher, W.R., 2011. Urban and rural mortality rates during heat waves in Berlin and Brandenburg, Germany. Environ. Pollut. 159 (8), 2044–2050. https://doi.org/10.1016/j.envpol.2011.01.016.

Gaffin, S.R., Rosenzweig, C., Kong, A.Y.Y., 2012. Adapting to climate change through urban green infrastructure. Nat. Clim. Chang. 2 (10), 704. https://doi.org/10.1038/nclimate1685.

Gillner, S., Vogt, J., Tharang, A., Dettmann, S., Roloff, A., 2015. Role of street trees in mitigating effects of heat and drought at highly sealed urban sites. Landsc. Urban Plan. 143, 33–42. (Scopus). https://doi.org/10.1016/j.landurbplan.2015.06.005.

Gomes, M.G., Silva, C.M., Valadas, A.S., Silva, M., 2019. Impact of vegetation, substrate, and irrigation on the energy performance of green roofs in a Mediterranean climate. Water 11 (10), 2016. https://doi.org/10.3390/w11102016.

Grey, V., Livesley, S.J., Fletcher, T.D., Szota, C., 2018. Tree pits to help mitigate runoff in dense urban areas. J. Hydrol. 565, 400–410.

Grimm, N.B., Faeth, S.H., Golubiewski, N.E., Redman, C.L., Wu, J., Bai, X., Briggs, J.M., 2008. Global change and the ecology of cities. Science 319 (5864), 756–760. https://doi.org/10.1126/science.1150195.

Grimmond, S., 2004. Understanding urban climates. In: Janelle, D.G., Warf, B., Hansen, K. (Eds.), WorldMinds: Geographical Perspectives on 100 Problems. Springer, Netherlands, pp. 481–486. https://doi.org/10.1007/978-1-4020-2352-1_78.

Guo, Y., Gasparrini, A., Li, S., Sera, F., Vicedo-Cabrera, A.M., de Coelho, M.S.Z.S., Saldiva, P.H.N., Lavigne, E., Tawatsupa, B., Punnasiri, K., Overcenco, A., Correa, P.M., Ortega, N.V., Kan, H., Osorio, S., Jaakkola, J.J.K., Ryti, N.R.I., Goodman, P.G., Zeka, A., … Tong, S., 2018. Quantifying excess deaths related to heatwaves under climate change scenarios: a multicountry time series modelling study. PLoS Med. 15 (7). https://doi.org/10.1371/journal.pmed.1002629 e1002629.

Haywood, J., Boucher, O., 2000. Estimates of the direct and indirect radiative forcing due to tropospheric aerosols: a review. Rev. Geophys. 38 (4), 513–543. https://doi.org/10.1029/1999RG000078.

Henstra, D., Thistlethwaite, J., Vanhooren, S., 2020. The governance of climate change adaptation: stormwater management policy and practice. J. Environ. Plan. Manag. 63 (6), 1077–1096. https://doi.org/10.1080/09640568.2019.1634015.

Kermavnar, J., Vilhar, U., 2017. Canopy precipitation interception in urban forests in relation to stand structure. Urban Ecosyst. 20 (6), 1373–1387. https://doi.org/10.1007/s11252-017-0689-7.

Kolokotroni, M., Giridharan, R., 2008. Urban heat island intensity in London: an investigation of the impact of physical characteristics on changes in outdoor air temperature during summer. Sol. Energy 82 (11), 986–998. https://doi.org/10.1016/j.solener.2008.05.004.

Kovats, R.S., Hajat, S., 2008. Heat stress and public health: a critical review. Annu. Rev. Public Health 29, 41–55. https://doi.org/10.1146/annurev.publhealth.29.020907.090843.

Lemonsu, A., Beaulant, A.L., Somot, S., Masson, V., 2014. Evolution of heat wave occurrence over the Paris basin (France) in the 21st century. Clim. Res. 61 (1), 75–91. https://doi.org/10.3354/cr01235.

Li, D., Bou-Zeid, E., 2013. Synergistic interactions between Urban Heat Islands and Heat Waves: the impact in cities is larger than the sum of its parts. J. Appl. Meteorol. Climatol. 52 (9), 2051–2064. https://doi.org/10.1175/JAMC-D-13-02.1.

Li, H., Meier, F., Lee, X., Chakraborty, T., Liu, J., Schaap, M., Sodoudi, S., 2018. Interaction between urban heat island and urban pollution island during summer in Berlin. Sci. Total Environ. 636, 818–828. https://doi.org/10.1016/j.scitotenv.2018.04.254.

Li, X., Zhou, Y., Yu, S., Jia, G., Li, H., Li, W., 2019. Urban heat island impacts on building energy consumption: a review of approaches and findings. Energy 174, 407–419. https://doi.org/10.1016/j.energy.2019.02.183.

Liao, K.-H., Deng, S., Tan, P.Y., 2017. Blue-green infrastructure: new frontier for sustainable urban stormwater management. In: Tan, P.Y., Jim, C.Y. (Eds.), Greening Cities: Forms and Functions. Springer, pp. 203–226. https://doi.org/10.1007/978-981-10-4113-6_10.

Lindén, J., Fonti, P., Esper, J., 2016. Temporal variations in microclimate cooling induced by urban trees in Mainz, Germany. Urban For. Urban Green. 20, 198–209. https://doi.org/10.1016/j.ufug.2016.09.001.

Meerow, S., Newell, J.P., 2017. Spatial planning for multifunctional green infrastructure: growing resilience in Detroit. Landsc. Urban Plan. 159, 62–75. https://doi.org/10.1016/j.landurbplan.2016.10.005.

Norton, B.A., Coutts, A.M., Livesley, S.J., Harris, R.J., Hunter, A.M., Williams, N.S.G., 2015. Planning for cooler cities: a framework to prioritise green infrastructure to mitigate high temperatures in urban landscapes. Landsc. Urban Plan. 134, 127–138. https://doi.org/10.1016/j.landurbplan.2014.10.018.

Oke, T.R., 1973. City size and the urban heat island. Atmos. Environ. (1967) 7 (8), 769–779. https://doi.org/10.1016/0004-6981(73)90140-6.

Paravantis, J., Santamouris, M., Cartalis, C., Efthymiou, C., Kontoulis, N., 2017. Mortality associated with high ambient temperatures, heatwaves, and the Urban Heat Island in Athens, Greece. Sustainability 9 (4), 606. https://doi.org/10.3390/su9040606.

Pattacini, L., 2012. Climate and urban form. Urban Des. Int. 17 (2), 106–114. https://doi.org/10.1057/udi.2012.2.

Revi, A., Satterthwaite, D.E., Aragón-Durand, F., Corfee-Morlot, J., Kiunsi, R.B.R., Pelling, M., Roberts, D.C., Solecki, W., 2014. Urban areas. In: Field, C.B., Barros, V.R. (Eds.), Climate Change 2014—Impacts, Adaptation and Vulnerability: Global and Sectoral Aspects. Cambridge University Press, New York, NY, pp. 535–612.

Saghafian, B., Farazjoo, H., Bozorgy, B., Yazdandoost, F., 2008. Flood intensification due to changes in land use. Water Resour. Manag. 22 (8), 1051–1067. https://doi.org/10.1007/s11269-007-9210-z.

Salvati, A., Monti, P., Coch Roura, H., Cecere, C., 2019. Climatic performance of urban textures: analysis tools for a Mediterranean urban context. Energ. Buildings 185, 162–179. https://doi.org/10.1016/j.enbuild.2018.12.024.

Sanusi, R., Johnstone, D., May, P., Livesley, S.J., 2017. Microclimate benefits that different street tree species provide to sidewalk pedestrians relate to differences in plant area index. Landsc. Urban Plan. 157, 502–511. https://doi.org/10.1016/j.landurbplan.2016.08.010.

Scherer, D., Fehrenbach, U., Lakes, T., Lauf, S., Meier, F., Schuster, C., 2013. Quantification of heat-stress related mortality hazard, vulnerability and risk in Berlin, Germany. J. Geogr. Soc. Berlin 144 (3–4), 238–259. https://doi.org/10.12854/erde-144-17.

Semadeni-Davies, A., Hernebring, C., Svensson, G., Gustafsson, L.G., 2008. The impacts of climate change and urbanisation on drainage in Helsingborg, Sweden: combined sewer system. J. Hydrol. 350 (1–2), 100–113.

Seneviratne, S., Nicholls, N., Easterling, D., Goodess, C., Kanae, S., Kossin, J., Luo, Y., Marengo, J., McInnes, K., Rahimi, M., Reichstein, M., Sorteberg, A., Vera, C., Zhang, X., Alexander, L.V., Allen, S., Benito, G., Cavazos, T., Clague, J., …Zwiers, F.W., 2012. Christopher B. Field & Intergovernmental Panel on Climate Change, (Ed.), Changes in Climate Extremes and Their Impacts on the Natural Physical Environment. Cambridge University Press, pp. 109–230. https://doi.org/10.7916/d8-6nbt-s431.

Shashua-Bar, L., Pearlmutter, D., Erell, E., 2011. The influence of trees and grass on outdoor thermal comfort in a hot-arid environment. Int. J. Climatol. 31 (10), 1498–1506. (Scopus). https://doi.org/10.1002/joc.2177.

Skougaard Kaspersen, P., Høegh Ravn, N., Arnbjerg-Nielsen, K., Madsen, H., Drews, M., 2017. Comparison of the impacts of urban development and climate change on exposing European cities to pluvial flooding. Hydrol. Earth Syst. Sci. 21 (8), 4131–4147. https://doi.org/10.5194/hess-21-4131-2017.

Sörensen, J., Emilsson, T., 2019. Evaluating flood risk reduction by urban blue-green infrastructure using insurance data. J. Water Resour. Plan. Manag. 145 (2). https://doi.org/10.1061/(ASCE)WR.1943-5452.0001037 (Scopus).

Taha, H., Akbari, H., Rosenfeld, A., Huang, J., 1988. Residential cooling loads and the urban heat island—the effects of albedo. Build. Environ. 23 (4), 271–283. https://doi.org/10.1016/0360-1323(88)90033-9.

United Nations, Department of Economic and Social Affairs, Population Division, 2019. World Urbanization Prospects: The 2018 Revision (ST/ESA/SER.A/420). United Nations, New York.

Vieira, J., Matos, P., Mexia, T., Silva, P., Lopes, N., Freitas, C., Correia, O., Santos-Reis, M., Branquinho, C., Pinho, P., 2018. Green spaces are not all the same for the provision of air purification and climate regulation services: the case of urban parks. Environ. Res. 160, 306–313. https://doi.org/10.1016/j.envres.2017.10.006.

Voogt, J.A., Oke, T.R., 2003. Thermal remote sensing of urban climates. Remote Sens. Environ. 86 (3), 370–384. https://doi.org/10.1016/S0034-4257(03)00079-8.

Walsh, C.J., Leonard, A.W., Ladson, A.R., Fletcher, T.D., 2004. Urban Stormwater and the Ecology of Streams. Cooperative Research Centre for Freshwater Ecology. https://www.academia.edu/25484294/Urban_stormwater_and_the_ecology_of_streams.

Xu, Z., Zhao, G., 2016. Impact of urbanization on rainfall-runoff processes: case study in the Liangshui River Basin in Beijing, China. Proc. Int. Assoc. Hydrol. Sci. 373, 7–12. https://doi.org/10.5194/piahs-373-7-2016.

Zölch, T., Maderspacher, J., Wamsler, C., Pauleit, S., 2016. Using green infrastructure for urban climate-proofing: an evaluation of heat mitigation measures at the micro-scale. Urban For. Urban Green. 20, 305–316. https://doi.org/10.1016/j.ufug.2016.09.011.

SECTION D

# Political impacts

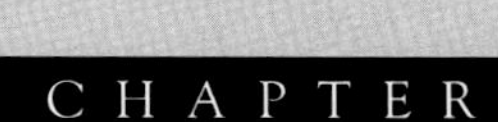

# CHAPTER 19

# Security implications of climate change: The climate-conflict nexus

*Elisabeth Lio Rosvold*[a,b]

[a]Department of Peace and Conflict Research, Uppsala University, Uppsala, Sweden [b]Department of Economic History and International Relations, Stockholm University, Stockholm, Sweden

OUTLINE



## 1 Introduction

The physical consequences of climate change that is projected for the coming years will affect and interact with a range of social processes in affected societies. The burden of climate change will not be shared equally among countries, many of the already disadvantaged countries will be hardest hit, and its impacts will likely be intensified by low coping capacities. Since the millennium, climate change has increasingly been "securitized" (Barnett and Neil Adger, 2007), with accelerating interest from policymakers and the scientific community alike. Particularly among the former group, the alarm has been sounded with frequent warnings that climate change will lead to more conflict and even new wars (prominent voices of



*The Impacts of Climate Change*
https://doi.org/10.1016/B978-0-12-822373-4.00015-X

concern being Ki-Moon, 2007; Obama, 2009). That an (already) changed climate will negatively impact social and ecological systems is clear from the leading body on the implications of climate change, the Intergovernmental Panel on Climate Change (IPCC). The IPCC synthesizes existing knowledge, and their most recent report (IPCC, 2014, see also their special report on 1.5°C warmings 2018) devoted an entire chapter to human security.

One aspect of human security that has attracted substantial attention is the potential effect of climate change on violent conflict. The wars in Syria and Darfur have been used to illustrate this, with arguments that droughts forced people out of work and increasing social tensions due to migration (e.g., Ki-Moon, 2007). The interest is also reflected in the academic community, and there has been a long debate among scholars seeking to determine whether and how climate change might lead to more violent conflict. However, the findings are mixed and we cannot conclude that there is a clear and sweeping effect of climate change on violent conflict. The Syrian case is illustrative of this disagreement. On the one hand, some hold that the 2007–10 drought that preceded the violent uprisings led to migration that can be linked to the violence (Ash and Obradovich, 2020; Kelley et al., 2015), while others argue that political and economic factors are far more important explanations (De Châtel, 2014; Fröhlich, 2016). That context matter is evident, as neighbors Jordan, Lebanon, and Cyprus were all affected by the same drought but did not see any subsequent conflict (Hendrix, 2017).

This chapter presents the emergence of the quantitative literature that over the past 20 years has investigated how climate variability (and disasters) have influenced conflict levels worldwide. In addition to a vast number of studies, there exists a myriad of review on the topic (see among others Adams et al., 2018; Brzoska, 2018; Dell et al., 2014; Koubi, 2019; Meierding, 2013; Sakaguchi et al., 2017; Theisen, 2017; Theisen et al., 2013). Consequently, the chapter does not claim to offer a complete nor systematic overview—for which the above contributions are excellent—but rather aims to provide a brief outline of a research program that while still being in its early phase has grown exponentially and received a lot of interest beyond the academic community.

## 2 Defining the concepts

### 2.1 Violent conflict

The types of conflicts that are investigated in the climate-conflict nexus are vast. Overarchingly conflict can be defined as a "process in which at least two social groups perceive their interests are contradictory and undertake [violent] actions to enforce or articulate these interests" (Ide et al., 2016, 286). In the literature, this includes (civil) war, farmer-herder conflicts, and communal conflict—which may or may not be lethal. In addition, different forms of collective violence such as protests and riots are sometimes investigated. Arguments concerning a direct influence of climate variables on psychological processes, for instance, that people become more angry/violent in hotter temperatures, and assessments of the effect of weather on crime rates, etc., can also be found.

Despite the variety, the vast majority are concerned with various forms of intrastate—within country—conflict (Sakaguchi et al., 2017). To distinguish between wars and conflict, it is standard to use causality thresholds of 1000 and 25 dead respectively, based on data from either the Uppsala Conflict Data Program (UCDP), the Armed Conflict Location and Event Data Project (ACLED), or the Social Conflict Analysis Database (SCAD). To determine whether conflict takes place or not, the most common indicator is the occurrence of a conflict event. Events can be used to determine the onset (the first occurrence of one particular conflict), but a more standard approach is to look at the *incidence*, indicating if a conflict was present or not in a given year. Nevertheless, the focus on ways in which a changed climate might further (new) conflict has experienced a decline over the past decade, giving way to investigations of how climate might impact the characteristics of ongoing conflict (Sakaguchi et al., 2017). Here the focus is rather on the number of people killed (conflict intensity), the location and targets of the violence, and the duration of conflicts.

## 2.2 Climate change versus climate variability

According to the World Meteorological Organization, climate encompasses the mean and the variability of weather variables such as temperature, precipitation, and wind for a given area over time, most typically 30 years (World Meteorological Organization, 2020). This means that climate *change* concerns statistically significant changes in long-term patterns, rather than changes in monthly or yearly weather patterns. Such short-term changes are typically referred to as climate (or weather) *variability* and denote deviations over a shorter period of time that goes beyond individual weather events (World Meteorological Organization, 2020).

The climate-conflict literature in large part concerns the latter of these two concepts, usually investigating short-term changes in weather (Koubi, 2019; Seter, 2016). Climate variabilities are part of the climate as a whole, and its causes can be both internal and external, encompassing both natural and anthropogenic changes (World Meteorological Organization, 2020). This means that while some of the changes in weather patterns might be due to climate change, it is difficult to determine when this is the case and when it is not. Nevertheless, an important reason why climate variability is still the focus in the empirical literature is the scarcity of suitable data. The time frame that would enable an investigation of the effects of climate change on something like collective violence should ideally span at least a century and few datasets on social variables (such as conflict) allow this. Consequently, past changes in the weather are used as proxies for future changes. As more frequent and more intense weather (both in terms of temperature and precipitation) are anticipated consequences of a changing climate, studying the impacts of weather extremes and short-term variability can hopefully tell us something about how social systems might be affected by future climate change.

The majority of the literature relies on meteorological data on precipitation, temperature, or a combination of the two in assessing the impacts of climate variability on conflict. Two standard data sources are the WMOs standardized precipitation index (SPI) and the standardized precipitation evapotranspiration index (SPEI) (Beguería et al., 2014; Vicente-Serrano et al., 2009). This data has the advantage of being independent of the (social/political/economic) situation on the ground and thus lends itself well to uncover causal relationships. Some studies also use other environmental factors that are affected by climate change, such as freshwater reservoirs and land cover from various sources.

Another type of climate impact that has become increasingly common to assess is that of climate-related disasters such as floods, storms, or droughts. While the social impacts of disasters are apparent, disaster measures are highly dependent (endogenous) on the situation where they hit. This hampers the ability to compare across contexts, and it is, therefore, more common to assess the influence of one particular disaster or disaster types within specific countries. Cross-national data do exist also for disasters, and the most used database is the Emergency Events Database (EM-DAT) maintained by the Centre for Epidemiology of Disasters at the Université Catholique de Louvain (Guha-Sapir et al., 2014). EM-DAT reports all disasters according to a set of criteria, and also includes disaster types that are not related to the climate, such as pandemics and earthquake. Another commonly used data source is the Dartmouth Flood Observatory (DFO) (Brakenridge, 2014), recording floods based on a combination of news sources and remote sensing.

# 3 The evolution of the climate-conflict nexus

## 3.1 Point of departure: Environmental security

The idea that the environment can negatively influence human security is not new, and the climate-conflict nexus is often traced back to the environmental security discussions that arose in the early 1990s. The classical work in this respect is Homer-Dixon's book Environment, Scarcity, and Violence (1999), where he proposes that through various social effects, environmental scarcity will lead to violent conflict. Following a Malthusian logic, the argument is that scarce resources, particularly in a context of inequitable distribution, will lead to fighting.

Even if it was not initially about climate, the environmental security approach became an important influence for the emergence of the climate-conflict literature. While being refuted for being highly complex (Gleditsch, 1998), deterministic and for removing the importance of social and political contexts (Koubi, 2019), its propositions have remained the point of departure for a substantial share of the later literature (Meierding, 2013)—and perhaps to blame for some of the disparities in the literature today.

## 3.2 First phase: A direct link between climate and conflict

Along with the increased focus on climate change as a security concern in its own right, the empirical literature directly assessing the relationship between climate variability and armed conflict started burgeoning in the early 2000s. This accelerated after a study by Miguel et al. (2004) predicted that negative economic growth had led to significant increases in civil conflict across 41 African countries between 1981 and 1991. Consequential for the development of climate-conflict nexus, the authors use negative rainfall growth as an instrument for economic growth. This means that—provided that the only way in which rainfall truly affects conflict is through its (derogative) effect on economic growth—the authors can estimate a causal link between economic conditions and armed conflict. The article sparked much debate as other scholars argued that the results were not robust when tested with different (better) specifications. For instance, Ciccone (2011; see also Jensen and Gleditsch, 2009) contended that rainfall deviations from the mean would be the correct specification, and employing this strategy, refuted a robust link between yearly rainfall growth and civil conflict.

Moving to a direct assessment of *climate* and conflict, Burke et al. (2009, 20670) found "strong historical linkages between civil war and temperature in Africa," an assessment that also included climate projections predicting future increases in conflict incidence. Also, this article received critique from several researchers for being overly deterministic on the suggested link, arguing that the cause was rather geopolitical with changes that co-occurred with warming temperatures (Sutton et al., 2010). Buhaug (2010b, and later 2010a) held that the results rely on a specific set of modeling choices and that changes to any one of these undermine the conclusions of Burke et al. Both of these accounts were subsequently refuted by Burke et al. (2010a,b).

Moving beyond organized armed conflicts, studies linking climate change to individual aggression found that high temperatures tend to increase aggression, and consequently increase interpersonal violence (assault, robbery, rape) (Anderson, 2001; Gamble and Hess, 2012; Mares, 2013; Mares and Moffett, 2016). Including this type of violence, Hsiang et al. (2013) performed a metaanalysis on existing evidence on climate and human conflict, concluding that "deviations from normal precipitation and mild temperatures systematically increase the risk of conflict, often substantially" (see also Hsiang and Burke, 2013). Grouping such widely different types of violence together and claiming that the effect of climate on violence is all-encompassing, the study received further criticism (Buhaug, 2014). Theisen et al. (2013) argued that if one makes the inference that what affects unorganized levels of violence also impacts organized, high-scale violence, climate change should be expected to increase the incidence of armed conflict. However, the degree to which these are caused by the same processes remains contested.

In this early period, the literature was diverse in terms of how both conflict and climate were operationalized. Raleigh and Urdal (2007) found that land degradation and water scarcities have a negligible effect on conflict risk, while Hendrix and Glaser (2007) investigated conflict risk by looking at both long-term and short-term climate impacts long-term trends by looking at land degradation and freshwater availability, and short-term trends by looking at annual rainfall. They found some support for the increased likelihood of conflict onset following a particularly large increase in rainfall. Since then, rainfall has become the most-used measure in the literature. Looking at different types of conflict, researchers proceeded to investigate links between rainfall anomalies and communal conflict (Fjelde and von Uexkull, 2012; Hendrix and Salehyan, 2012; Raleigh and Kniveton, 2012; Theisen, 2012). Here, both dry periods and abundant rainfall were found to correspond to the increased risk of conflict. O'Loughlin et al. (2012) found that wetter periods see more violent events across East Africa, contributing to the equivocal results of the literature. Looking at the conflict *between* countries, Gartzke (2012) did not find evidence for more wars in warmer temperatures. To investigate the potential consequences of drought, the combination of rainfall and temperature anomalies eventually became the preferred climate measure. Within this group, some found that there is a connection while others hold that it is weak at best (Couttenier and Soubeyran, 2014; Klomp and Bulte, 2013; Theisen et al., 2011).

Another type of impact that soon received attention was extreme weather events (i.e., disasters). While not taking a climate-perspective, Drury and Olson (1998) became an important starting point for this strand of the literature, finding that political violence increased with the severity of disasters in a range of low- and middle-income countries (in a somewhat related approach, Flores and Smith, 2012 found that leaders' positions are threatened following disasters). Looking specifically at civil conflict, Nel and Righarts (2008) found that disasters (also including earthquakes whose occurrence is not related to the climate) are associated with increased conflict, both in the long and medium-term. Omelicheva (2011) offers a more sobering account, finding that the effect of disasters on political instability disappears once the characteristics of the political regime in which the disasters occurred are taken into account. On the opposite side of the spectrum, disasters have also been presented as opportunities for more peaceful interactions. Kreutz (2012) found disasters to be associated with an increased likelihood that belligerents agree to ceasefires or initiate peace talks, while Slettebak (2012) found that the risk of conflict onset is lower in countries that are affected by climate-related disasters.

Whereas most of this early literature was concerned with historical (short-term) climate variability and its potential consequences, some studies have taken a wider perspective which better captures climate *change*. These studies found—perhaps contrary to the general expectation—that over the last millennium the prevalence of conflict has been higher in colder periods. In Eastern China, the increase in wars during cold phases is linked to loss of agricultural production (Zhang et al., 2007; Zhang et al., 2010; see also Lee et al., 2014). Looking at precipitation, Bai and Kung (2011) assessed the impact of rainfall shocks on Sino-nomadic conflicts in Han Chinese regions over 2000 years. They found that negative rainfall shocks—droughts—are positively associated with nomadic invasions of sedentary agricultural Han Chinese areas. Tol and Wagner (2010) repeated the process for Europe and found that between the years 1000 and 2000, cold weather coincided with violence (wars). While these studies all found some association between changes in climatic conditions and violence, they acknowledge that the link likely goes via socio-economic conditions but that this would be near impossible to test over the applied time-spans.

## 3.3 Testing contexts: A conditional influence of climate on conflict

Although the majority of the early studies do not claim that climate variabilities *directly* lead people to take up arms, the empirical analyses tend to test such a direct relationship. In most of this literature, the context in which a relationship is expected to take place is discussed, but it is only later that the empirical tests include them. Surveying the literature, Sakaguchi et al. (2017) refer to *mediating* factors comprising of economic, political, institutional, ethnocultural, and environmental conditions. Vesco and Buhaug (2020) refer to these as *conditional* pathways and identify eight conditional factors that have been tested in the literature. These are ethno-political exclusion, ethnic exclusion, socio-economic development, democracy, state capacity, population, urbanization, and mountainous terrain. The content of both classifications is more or less the same, and illustrate that changes in climatic conditions are assumed to influence the societal risk for conflict depending on the vulnerability of the affected societies.

The empirical evidence concerning contexts is, however, not particularly convergent across studies. Looking at the political system, Couttenier and Soubeyran (2014) found the risk of conflict following drought to be higher in less democratic countries, while Theisen et al. (2011) found no moderating effect of drought on conflict risk in democracies. On the opposite side of the spectrum, Bell and Keys (2016) found that droughts increased the vulnerability to the conflict in the contexts where we would expect the opposite, i.e., in democracies and states with higher state capacity, in countries with high levels of urbanization and countries without a recent history of conflict.

There is also disagreement in cases concerning socio-economic conditions. While Bell and Keys (2016) found that social and political inequalities do not make conflict more likely following droughts, Schleussner et al. (2016) found that in the majority of instances where drought (and heatwaves) coincided with conflict, the outbreaks were in ethnically fractionalized societies. Adverse climatic conditions (usually drought) taking place in the absence of (state or more traditional) institutions and the presence of rough terrain are found to consistently increase the risk of conflict (Couttenier and Soubeyran, 2014; Detges, 2016; Hendrix and Glaser, 2007; Meier et al., 2007; O'Loughlin et al., 2012).

These factors do not appear in isolation, and the interaction between political and economic factors, particularly that of political exclusion and agriculturally dependent economies appear to be relatively stable predictors of a negative effect of drought on conflict risk (von Uexkull, 2014; von Uexkull et al., 2016). Nevertheless, also for communal conflicts, other factors appear to be much more important than climatic factors (van Weezel, 2019).

The concept of climate change as a threat multiplier also falls within this approach as many of the conditions that would make an adverse climate impact likely to increase conflict risk, are also likely, in themselves, to lead to higher conflict risk. With some exceptions, the studies that take into account contextual factors do reveal that the risk of conflict following climate adversities (particularly droughts) are higher in countries that already experience conflict (Koubi et al., 2012).

## 3.4 A two-stage approach: Intermediary links between climate and conflict

In addition to contextual factors, the literature has advanced to also test the intermediate pathways, or mechanisms, that were proposed in the earlier literature. With the exception of some early works from Bergholt and Lujala (2012) and Koubi et al. (2012), who both assess the influence of climate on conflict via economic performance, most of the early studies did not empirically test the intermediate pathways they used to explain why/how changing climatic conditions would lead to more conflict. After 2014, however, two-stage approaches have become more common.

The main pathways that have been tested can be grouped into economic impacts, the resource situation, and migration—even if the line between these three might not always be clear cut. The economic pathway is by far the most studied (Sakaguchi et al., 2017), perhaps because of the already well-established link between climate change and economic hardships (Dell et al., 2014 provides an overview of this literature) on the one hand, and the well-established link between poverty and armed conflict (Collier and Hoeffler, 2004; Hegre and Sambanis, 2006) on the other. Among potential climate-induced economic impacts, income shocks, loss of livelihood, and food price shocks are the most commonly tested pathways (Vesco and Buhaug, 2020). Much of the early literature falls within this macro-perspective, but the unequivocal relationship established in the economic and conflict literature, respectively is not echoed when all three links are put together.

When it comes to food price or food production shocks, the expectations generally follow a Malthusian logic where climate impacts might cut food production (and consequently food supply), risking violent reactions. Taking a two-stage approach Raleigh et al. (2015) found that negative rainfall shocks increased the price of food across 113 African markets, which in turn saw increased levels of violence following price increases. This study included all types of political violence, and most studies have found a positive influence of food prices on protests and riots (Bellemare, 2015; Smith, 2014; see Rudolfsen, 2018 for a review of the literature on food insecurity and domestic instability). While food price shocks are seen to spur discontent among consumers, extremely wet and dry weather also adversely affect producers. Following economic theories of opportunity costs, a loss of livelihood means that the opportunity cost of rebellion decreases as farmers become out of work and are thus potentially cheaper to "employ" in a rebellion. Linking various climate impacts to the growing season of the main crops, several studies demonstrate a link between weather deviations and conflict incidence (Caruso et al., 2016; Harari and La Ferrara, 2018; Wischnath and Buhaug, 2014).

Other types of environmental resource (scarcities) and the large literature on water and conflict also falls into this category. Raleigh and Urdal (2007) look at both freshwater availability, soil capabilities, and human settlements, but found weak support for the notion that environmental scarcities induce conflict. However, unlike the immediate nature of the impacts explored concerning economic (and food) shocks, environmental degradation is a slow-moving process, making it difficult to empirically assess how climate change works through environmental change which then impacts conflict. In addition, linger time-spans means opportunities for adaptation efforts to mitigate negative consequences, making the attribution even more difficult. When it comes to water scarcities, cooperative behavior rather than conflict is increasingly found to be the case in situations of low water reserves (Ide and Detges, 2018).

The final intermediary link—which is increasingly receiving attention both among policymakers and scholars—is climate-induced migration. The proposed mechanism is that displaced people will create new or increase existing pressures in host regions. While it is clear that vast areas will likely become uninhabitable in the future due to climate change, the link between climate change and migration is still contested in the literature (Burrows and Kinney, 2016). The main reason for this has to do with attribution, which makes determining exactly who should constitute "climate migrants" challenging. Subsequently, linking then to the conflict has also proven difficult. Cattaneo and Bosetti (2017) did not find that migrant flows from climate-affected countries add any explanatory power to conflict in receiving areas. In reviewing the literature on climate migration and conflict, Brzoska and Fröhlich (2016) contend that the empirical (and the theoretical) basis for the assumptions is thin, but point to the absence of data on migration flows as an important reason for this (see also the review by Black et al., 2013).

Being easier to attribute to an actual event, the role of disaster-induced migration for conflict has received some attention. Assessing displacement following floods, Ghimire et al. (2015; see also Ghimire and Ferreira, 2016) found that flood-induced migration did exacerbate existing conflict, but that it did not have any impact on the risk of new conflicts. A potential explanation might be the temporary nature of migration after disasters (i.e., displacement) (Grace et al., 2018; Jülich, 2011; Lu et al., 2016), as people prefer to return to their homes as soon as it is feasible rather than to permanently relocate.

Of course, the above-mentioned processes rarely happen in isolation, and the combination of contextual and intermediary links have also been investigated. Examples of this are assessments of political and economic conditions in both affected areas and potential host communities (Reuveny, 2007; Xiao et al., 2013). Looking at so-called sons of the soil conflicts in India, Bhavnani and Lacina (2015) found that if migrants share the host population's political alignment, riots are less likely. Another example of an integrated approach is by Koubi et al. (2012), who found that climate-induced economic shocks had a stronger effect on conflict risk in autocratic regimes than in other regimes.

While contextual factors are becoming more important than merely being "controlled" for, social conditions are notoriously difficult to measure, particularly across contexts. When interacting with physical factors, comprehensive assessments become even more challenging. Notwithstanding, the search for causal influences across contexts continues. Some prominent studies have managed to integrate exposure, hazard, and adaptational capacities in assessing the vulnerabilities of affected societies to climate change and armed conflict—and their potential interactions (see Busby et al., 2013, 2014; Schilling et al., 2020). These provide important stepping stones for future work in the quest to capture interlinkages that are both complex and mutually reinforcing.

## 4 The way ahead: Past change predicting future uncertainties

Summing up, the now vast quantitative literature on climate change and conflict has not provided policymakers with unequivocal support for their gloomy predictions that more conflicts will manifest because of climate change. The failure to converge is often ascribed to the myriad of statistical methods, study areas, and data used (Koubi et al., 2012; Meierding, 2013; Theisen et al., 2013). While the majority of the literature concerns sub-Saharan Africa (Adams et al., 2018), a more pressing concern might be that the focus on a select few cases (where there has been conflict) also means that any detected link risks being overstated (Adams et al., 2018). Nevertheless, the research program has moved from a general, deterministic approach toward better theoretical and empirical specifications, allowing researchers to disentangle the contexts and intermediary pathways where climate change can be expected to influence violent conflict. Incorporating climate variables into forecasting models for conflict (such as Hegre et al., 2016) provides a promising lacuna for also expanding beyond the use of past variabilities as proxies for future change.

In addition to the burgeoning empirical literature, an impressive number of reviews of the climate-conflict nexus have emerged over the past decade. Every year since 2012 (with the exception of 2015), at least one review article has been published on the relationship between climate and armed conflict—excluding book chapters and commentaries on the topic (see for instance Abrahams, 2020; Ide and Scheffran, 2014; Mach et al., 2020; Selby, 2014; Buhaug, 2015, 2016; Theisen, 2017; Theisen et al., 2013). While they all concur that taken together, the evidence of a relationship between climate variability and conflict is mixed and rather weak, they do not take this to mean that a relationship does not exist, nor that it will not evolve in the future. A comprehensive analysis of expert opinions (so-called expert elicitation) was recently solicited in an attempt to bridge the disagreement (Mach et al., 2019, 196). Pointing to the fact that other societal, political, and economic factors by far exceed the role of climate as conflict drivers, the analysis concludes that

> (…) there is an agreement that climate variability and change shape the risk of organized armed conflict within countries. In conflicts to date, however, the role of climate is judged to be small compared to other drivers of conflict, and the mechanisms by which climate affects conflict are uncertain. As risks grow under future climate change, many more potential climate–conflict linkages become relevant and extend beyond historical experiences.

Moving forward, better specification of contexts and intermediary linkages, also on a theoretical level, and subsequent testing (of course relying on data availability, particularly on subnational levels) are the indisputable wishes for the research moving forward. Integrating knowledge from the qualitative literature appears imperative in this respect (van Baalen and Mobjörk, 2016); perhaps while leaving the theoretical assumptions from environmental security behind (Meierding, 2013). As the overlap between countries that have suffered protracted conflicts and those projected to be worst affected by climate adversaries is considerable, ongoing conflicts will be also be affected by climate change. A priority for scholars going forward should be better incorporation of characteristics such as the intensity and duration of the conflict (Scheffran, 2020). The empirical evidence concerning extreme events and disasters is more coherent than for climate variability, and there appears to be a positive (in the sense that they lead to an increase) influence of extreme events on violence levels (Brzoska, 2018; Scheffran, 2020). A fact that could well be a result of the tendency of these studies' to be more concerned with conflict dynamics than conflict onset.

## References

Abrahams, D., 2020. Conflict in abundance and peacebuilding in scarcity: challenges and opportunities in addressing climate change and conflict. World Dev. 132, 104998.

Adams, C., Ide, T., Barnett, J., Detges, A., 2018. Sampling bias in climate–conflict research. Nat. Clim. Chang. 8 (3), 200–203.

Anderson, C.A., 2001. Heat and violence. Curr. Dir. Psychol. Sci. 10 (1), 33–38.

Ash, K., Obradovich, N., 2020. Climatic stress, internal migration, and syrian civil war onset. J. Confl. Resolut. 64 (1), 3–31.

Bai, Y., Kung, J.K.-s., 2011. Climate shocks and sino-nomadic conflict. Rev. Econ. Stat. 93 (3), 970–981.

Barnett, J., Neil Adger, W., 2007. Climate change, human security and violent conflict. Polit. Geogr. 26 (6), 639–655.

Beguería, S., Vicente-Serrano, S.M., Reig, F., Latorre, B., 2014. Standardized precipitation evapotranspiration index (SPEI) revisited: parameter fitting, evapotranspiration models, tools, datasets and drought monitoring. Int. J. Climatol. 34 (10), 3001–3023.

Bell, C., Keys, P.W., 2016. Conditional relationships between drought and civil conflict in sub-saharan Africa. Foreign Policy Anal.. , orw002.

Bellemare, M.F., 2015. Rising food prices, food price volatility, and social unrest. Am. J. Agric. Econ. 97 (1), 1–21.

Bergholt, D., Lujala, P., 2012. Climate-related natural disasters, economic growth, and armed civil conflict. J. Peace Res. 49 (1), 147–162.

Bhavnani, R.R., Lacina, B., 2015. The effects of weather-induced migration on sons of the soil riots in India. World Polit. 67 (4), 760–794.

Black, R., et al., 2013. Migration, immobility and displacement outcomes following extreme events. Environ. Sci. Pol. 27, S32–S43.

Brakenridge, G.R., 2014. Global Active Archive of Large Flood Events. http://floodobservatory.colorado.edu/Archives/index.html.

Brzoska, M., 2018. Weather extremes, disasters, and collective violence: conditions, mechanisms, and disaster-related policies in recent research. Curr. Clim. Chan. Rep. 4 (4), 320–329.
Brzoska, M., Fröhlich, C., 2016. Climate change, migration and violent conflict: vulnerabilities, pathways and adaptation strategies. Migr. Dev. 5 (2), 190–210.
Buhaug, H., 2010a. Climate not to blame for African civil wars. Proc. Natl. Acad. Sci. U. S. A. 107 (38), 16477–16482.
Buhaug, H., 2010b. Reply to Burke et al.: bias and climate war research. Proc. Natl. Acad. Sci. U. S. A. 107 (51), E186–E187.
Buhaug, H., 2014. One effect to rule them all? A comment on climate and conflict. Clim. Chang. 127 (3–4), 391–397.
Buhaug, H., 2015. Climate-conflict research: some reflections on the way forward. Wiley Interdiscip. Rev. Clim. Chang. 6 (3), 269–275.
Buhaug, H., 2016. Climate change and conflict: taking stock. Peace Econ. Peace Sci. Public Policy 22 (4), 331–338.
Burke, M.B., et al., 2009. Warming increases the risk of civil war in Africa. Proc. Natl. Acad. Sci. U. S. A. 106 (49), 20670–20674.
Burke, M.B., et al., 2010a. Climate robustly linked to African civil war. Proc. Natl. Acad. Sci. U. S. A. 107 (51), E185.
Burke, M.B., et al., 2010b. Reply to Sutton et al.: relationship between temperature and conflict is robust. Proc. Natl. Acad. Sci. U. S. A. 107 (25), E103.
Burrows, K., Kinney, P.L., 2016. Exploring the climate change, migration and conflict Nexus. Int. J. Environ. Res. Public Health 13 (4), 443.
Busby, J.W., Smith, T.G., White, K.L., Strange, S.M., 2013. Climate change and insecurity mapping vulnerability in Africa. Int. Secur. 37 (4), 132–172.
Busby, J.W., Smith, T.G., Krishnan, N., 2014. Climate security vulnerability in Africa mapping 3.0. Polit. Geogr. 43, 51–67.
Caruso, R., Petrarca, I., Ricciuti, R., 2016. Climate change, rice crops, and violence: evidence from Indonesia. J. Peace Res. 53 (1), 66–83.
Cattaneo, C., Bosetti, V., 2017. Climate-induced international migration and conflicts. CESifo Econ. Stud. 63 (4), 500–528.
Ciccone, A., 2011. Economic shocks and civil conflict: a comment. Am. Econ. J. Appl. Econ. 3 (4), 215–227.
Collier, P., Hoeffler, A., 2004. Greed and grievance in civil war. Oxf. Econ. Pap. 56 (4), 563–595.
Couttenier, M., Soubeyran, R., 2014. Drought and civil war in Sub-Saharan Africa. Econ. J. 124 (575), 201–244.
De Châtel, F., 2014. The role of drought and climate change in the Syrian uprising: untangling the triggers of the revolution. Middle East. Stud. 50 (4), 521–535.
Dell, M., Jones, B.F., Olken, B.A., 2014. 19578 What Do We Learn From the Weather? The New Climate-Economy Literature. The National Bureau of Economic Research. http://www.nber.org/papers/w19578 June 1, 2016.
Detges, A., 2016. Local conditions of drought-related violence in Sub-Saharan Africa. J. Peace Res. 53 (5), 696–710.
Drury, A.C., Olson, R.S., 1998. Disasters and political unrest: an empirical investigation. J. Conting. Crisis Manag. 6 (3), 153–161.
Fjelde, H., von Uexkull, N., 2012. Climate triggers: rainfall anomalies, vulnerability and communal conflict in Sub-Saharan Africa. Polit. Geogr. 31 (7), 444–453.
Flores, A.Q., Smith, A., 2012. Leader survival and natural disasters. Br. J. Polit. Sci. 43 (4), 821–843.
Fröhlich, C.J., 2016. Climate migrants as protestors? Dispelling misconceptions about global environmental change in pre-revolutionary Syria. Contemp. Levant 1 (1), 38–50.
Gamble, J., Hess, J., 2012. Temperature and violent crime in Dallas, Texas: relationships and implications of climate change. West. J. Emerg. Med. 13 (3), 239–246.
Gartzke, E., 2012. Could climate change precipitate peace? J. Peace Res. 49 (1), 177–192.
Ghimire, R., Ferreira, S., 2016. Floods and armed conflict. Environ. Dev. Econ. 21 (1), 23–52.
Ghimire, R., Ferreira, S., Dorfman, J.H., 2015. Flood-induced displacement and civil conflict. World Dev. 66, 614–628.
Gleditsch, N.P., 1998. Armed conflict and the environment: a critique of the literature. J. Peace Res. 35 (3), 381–400.
Grace, K., Hertrich, V., Singare, D., Husak, G., 2018. Examining rural Sahelian out-migration in the context of climate change: an analysis of the linkages between rainfall and out-migration in two Malian villages from 1981 to 2009. World Dev. 109, 187–196.
Guha-Sapir, D., Below, R., Hoyois, P., 2014. EM-DAT: International Disaster Database. Brussels. Centre for Research on the Epidemiology of Disasters (CRED), Université Catholique de Louvain.https://www.emdat.be.
Harari, M., La Ferrara, E., 2018. Conflict, climate, and cells: a disaggregated analysis. Rev. Econ. Stat. 100 (4), 594–608.
Hegre, H., Sambanis, N., 2006. Sensitivity analysis of empirical results on civil war onset. J. Confl. Resolut. 50 (4), 508–535.

Hegre, H., et al., 2016. Forecasting civil conflict along the shared socioeconomic pathways. Environ. Res. Lett.. 11(5), 054002.
Hendrix, C.S., 2017. A comment on 'climate change and the Syrian civil war revisited'. Polit. Geogr. 60 (Supplement C), 251–252.
Hendrix, C.S., Glaser, S.M., 2007. Trends and triggers: climate, climate change and civil conflict in sub-Saharan Africa. Polit. Geogr. 26 (6), 695–715.
Hendrix, C.S., Salehyan, I., 2012. Climate change, rainfall, and social conflict in Africa. J. Peace Res. 49 (1), 35–50.
Homer-Dixon, T., 1999. Environment, Scarcity, and Violence. Princeton University Press.
Hsiang, S.M., Burke, M., 2013. Climate, conflict, and social stability: what does the evidence say? Clim. Chang. 123 (1), 39–55.
Hsiang, S.M., Burke, M., Miguel, E., 2013. Quantifying the influence of climate on human conflict. Science. 341(6151) http://www.sciencemag.org/content/341/6151/1235367.abstract.
Ide, T., Detges, A., 2018. International water cooperation and environmental Peacemaking. Glob. Environ. Polit. 18 (4), 63–84.
Ide, T., Scheffran, J., 2014. On climate, conflict and cumulation: suggestions for integrative cumulation of knowledge in the research on climate change and violent conflict. Glob. Chang. Peace Sec. 26 (3), 263–279.
Ide, T., Michael Link, P., Scheffran, J., Schilling, J., 2016. The climate-conflict nexus: pathways, regional links, and case studies. In: Handbook on Sustainability Transition and Sustainable Peace. Springer, pp. 285–304. https://www.springer.com/gp/book/9783319438825#aboutAuthors.
IPCC, 2014. Pachauri, R.K. et al., (Ed.), Climate Change 2014: Synthesis Report. Contribution of Working Groups I, II and III to the Fifth Assessment Report of the Intergovernmental Panel on Climate Change. IPCC.
IPCC, 2018. IPCC special report on global warming of 1.5°C. IPCC, Geneva, Switzerland.
Jensen, P.S., Gleditsch, K.S., 2009. Rain growth and civil war: the importance of location. Def. Peace Econ. 20 (5), 359–372.
Jülich, S., 2011. Drought triggered temporary migration in an East Indian Village. Int. Migr. 49 (s1), e189–e199.
Kelley, C.P., et al., 2015. Climate change in the fertile crescent and implications of the recent Syrian drought. Proc. Natl. Acad. Sci. U. S. A. 11 (112), 3241–3246.
Ki-Moon, B., 2007. A Climate Culprit in Darfur. The Washington Post.
Klomp, J., Bulte, E., 2013. Climate change, weather shocks, and violent conflict: a critical look at the evidence. Agric. Econ. 44 (s1), 63–78.
Koubi, V., 2019. Climate change and conflict. Annu. Rev. Polit. Sci. 22, 343–360.
Koubi, V., Bernauer, T., Kalbhenn, A., Spilker, G., 2012. Climate variability, economic growth, and civil conflict. J. Peace Res. 49 (1), 113–127.
Kreutz, J., 2012. From tremors to talks: do natural disasters produce ripe moments for resolving separatist conflicts? Int. Interact. 38 (4), 482–502.
Lee, J., et al., 2014. An experiment to model spatial diffusion process with nearest neighbor analysis and regression estimation. Int. J. Appl. Geospat. Res. (IJAGR) 5 (1), 1–15.
Lu, X., et al., 2016. Unveiling hidden migration and mobility patterns in climate stressed regions: a longitudinal study of six million anonymous mobile phone users in Bangladesh. Glob. Environ. Chang. 38, 1–7.
Mach, K.J., et al., 2019. Climate as a risk factor for armed conflict. Nature 571 (7764), 193–197.
Mach, K.J., et al., 2020. Directions for research on climate and conflict. Earth's Future. 8, e2020EF001532.
Mares, D., 2013. Climate change and levels of violence in socially disadvantaged neighborhood groups. J. Urban Health 90 (4), 768–783.
Mares, D.M., Moffett, K.W., 2016. Climate change and interpersonal violence: a 'global' estimate and regional inequities. Clim. Chang. 135 (2), 297–310.
Meier, P., Bond, D., Bond, J., 2007. Environmental influences on pastoral conflict in the horn of Africa. Polit. Geogr. 26 (6), 716–735.
Meierding, E., 2013. Climate change and conflict: avoiding small talk about the weather. Int. Stud. Rev. 15 (2), 185–203.
Miguel, E., Satyanath, S., Sergenti, E., 2004. Economic shocks and civil conflict: an instrumental variables approach. J. Polit. Econ. 112 (4), 725–753.
Nel, P., Righarts, M., 2008. Natural disasters and the risk of violent civil conflict. Int. Stud. Q. 52 (1), 159–185.
O'Loughlin, J., et al., 2012. Climate variability and conflict risk in East Africa, 1990-2009. Proc. Natl. Acad. Sci. U. S. A. 109 (45), 18344–18349.

Obama, B., 2009. It Will Do Little Good to Alleviate Poverty If You Can No Longer Harvest Your Crops or Find Drinkable Water. http://thinkprogress.org/climate/2009/09/22/204681/obama-un-speech-climate-change/.
Omelicheva, M.Y., 2011. Natural disasters: triggers of political instability? Int. Interact. 37 (4), 441–465.
Raleigh, C., Kniveton, D., 2012. Come rain or shine: an analysis of conflict and climate variability in East Africa. J. Peace Res. 49 (1), 51–64.
Raleigh, C., Urdal, H., 2007. Climate change, environmental degradation and armed conflict. Polit. Geogr. 26 (6), 674–694.
Raleigh, C., Choi, H.J., Kniveton, D., 2015. The devil is in the details: an investigation of the relationships between conflict, food price and climate across Africa. Glob. Environ. Chang. 32, 187–199.
Reuveny, R., 2007. Climate change-induced migration and violent conflict. Polit. Geogr. 26 (6), 656–673.
Rudolfsen, I., 2018. Food insecurity and domestic instability: a review of the literature. Terror. Polit. Violence 1–28.
Sakaguchi, K., Varughese, A., Auld, G., 2017. Climate wars? A systematic review of empirical analyses on the links between climate change and violent conflict. Int. Stud. Rev. 19 (4), 622–645.
Scheffran, J., 2020. Climate extremes and conflict dynamics. In: Climate Extremes and Their Implications for Impact and Risk Assessment. Elsevier, pp. 293–315. https://linkinghub.elsevier.com/retrieve/pii/B9780128148952000161 July 30, 2020.
Schilling, J., Hertig, E., Tramblay, Y., Scheffran, J., 2020. Climate change vulnerability, water resources and social implications in North Africa. Reg. Environ. Chang. 20 (1), 15.
Schleussner, C.-F., Donges, J.F., Donner, R.V., Schellnhuber, H.J., 2016. Armed-conflict risks enhanced by climate-related disasters in ethnically fractionalized countries. Proc. Natl. Acad. Sci. 113 (33), 9216–9221.
Selby, 2014. Positivist climate conflict research: a critique. Geopolitics 19 (4), 829–856.
Seter, H., 2016. Connecting climate variability and conflict: implications for empirical testing. Polit. Geogr. 53, 1–9.
Slettebak, R.T., 2012. Don't blame the weather! climate-related natural disasters and civil conflict. J. Peace Res. 49 (1), 163–176.
Smith, T.G., 2014. Feeding unrest: disentangling the causal relationship between food price shocks and sociopolitical conflict in urban Africa. J. Peace Res. 51 (6), 679–695.
Sutton, A.E., et al., 2010. Does warming increase the risk of civil war in Africa? Proc. Natl. Acad. Sci. 107 (25), E102.
Theisen, O.M., 2012. Climate clashes? Weather variability, land pressure, and organized violence in Kenya, 1989-2004. J. Peace Res. 49 (1), 81–96.
Theisen, O.M., 2017. Climate change and violence: insights from political science. Curr. Clim. Chang. Rep. 3 (4), 210–221.
Theisen, O.M., Holtermann, H., Buhaug, H., 2011. Climate wars? Assessing the claim that drought breeds conflict. Int. Secur. 36 (3), 79–106.
Theisen, O.M., Gleditsch, N.P., Buhaug, H., 2013. Is climate change a driver of armed conflict? Clim. Chang. 117 (3), 613–625.
Tol, R.S.J., Wagner, S., 2010. Climate change and violent conflict in Europe over the last millennium. Clim. Chang. 99 (1–2), 65–79.
van Baalen, S., Mobjörk, M., 2016. A Coming Anarchy? Pathways From Climate Change to Violent Conflict in East Africa. http://su.diva-portal.org/smash/get/diva2:928237/FULLTEXT01.pdf September 28, 2017.
van Weezel, S., 2019. On climate and conflict: precipitation decline and communal conflict in Ethiopia and Kenya. J. Peace Res. 56, 514–528.
Vesco, P., Buhaug, H., 2020. Climate change and environment. In: Hampson, F.O., Özerdem, A., Kent, J. (Eds.), Routledge Handbook of Peace, Security and Development. Routledgehttps://www.routledge.com/Routledge-Handbook-of-Peace-Security-and-Development/Hampson-Ozerdem-Kent/p/book/9780815397854?fbclid=IwAR1uQA0LJ1NnyBlM-RyeH0bEKRa04ScN6O_80sl1dS0UuPHQwb_CrGoi8IA.
Vicente-Serrano, S.M., Beguería, S., López-Moreno, J.I., 2009. A multiscalar drought index sensitive to global warming: the standardized precipitation evapotranspiration index. J. Clim. 23 (7), 1696–1718.
von Uexkull, N., 2014. Sustained drought, vulnerability and civil conflict in Sub-Saharan Africa. Polit. Geogr. 43, 16–26.
von Uexkull, N., Croicu, M., Fjelde, H., Buhaug, H., 2016. Civil conflict sensitivity to growing-season drought. Proc. Natl. Acad. Sci. U. S. A. 113 (40), 12391–12396.
Wischnath, G., Buhaug, H., 2014. Rice or riots: on food production and conflict severity across India. Polit. Geogr. 43, 6–15.

World Meteorological Organization, 2020. FAQs—Climate. https://public.wmo.int/en/about-us/FAQs/faqs-climate.

Xiao, L.B., Fang, X.Q., Ye, Y., 2013. Reclamation and revolt: social responses in eastern Inner Mongolia to flood/drought-induced refugees from the North China plain 1644–1911. J. Arid Environ. 88, 9–16.

Zhang, D.D., Zhang, J., Lee, H.F., He, Y.-q., 2007. Climate change and war frequency in eastern China over the last millennium. Hum. Ecol. 35 (4), 403–414.

Zhang, Z., et al., 2010. Periodic climate cooling enhanced natural disasters and wars in China during AD 10–1900. Proc. R. Soc. B Biol. Sci. 277 (1701), 3745–3753.

CHAPTER

20

# Climate change governance: Responding to an existential crisis

*Heike Schroeder*[a,c] *and Yuka Kobayashi*[b]

[a]School of International Development, University of East Anglia, Norwich, United Kingdom [b]Department of Politics and International Studies, SOAS, London, United Kingdom [c]Tyndall Centre for Climate Change Research, University of East Anglia, Norwich, United Kingdom

OUTLINE



## 1 Introduction

Current efforts of environmental policy, pollution control, and nature conservation do not adequately address the unfolding devastation of natural cycles and habitats at a global scale (Malhi et al., 2020). The sheer magnitude of soil depletion, climate change, and species collapse is driving home the message that humanity is standing at the brink of societal collapse. Given that they inherently link to all socioeconomic activity and underlying resources, they together make for an existential crisis to humanity. Some refer to such problems as "superwicked" given that they are highly urgent, uncertain, nonlinear, symptomatic of other problems, ever evolving and lacking a central authority (Levin et al., 2012), while others have declared climate and biodiversity emergencies (Gills and Morgan, 2020).


https://doi.org/10.1016/B978-0-12-822373-4.00006-9

Climate change has now been addressed internationally, nationally, and locally for over 30 years; yet global greenhouse gas emissions continue to rise year by year (Le Quéré et al., 2020). Deforestation rates also grow unabated, despite accelerated global efforts dating back some 15 years (IPCC, 2018). Key challenges in efforts to address these have been the sheer magnitude of national and vested interests that governments put above the wellbeing of current and future generations, as well as the prevalence of an environmentally destructive economic growth model. Thus, to steer ourselves away from this predicament of extinction, unprecedented collective and focused action across countries, organizations and individuals is urgently needed. We argue that, thus far, our approach to governance has failed us. By reviewing the climate change literature and its prevalent concepts and debates, we identify gaps we suggest need to be addressed to lead us beyond our current impasse. Our structure is as follows: We begin with reviewing the nature of governance in the context of climate change. Next, we examine the different levels of governance that are also subsumed under the concept of multilevel climate governance. We then discuss the key elements of governance, including climate-related actors and actor networks, their rulemaking systems and the rules emerging to address climate change. These terms are illustrated with examples from the international climate negotiations under the United Nations Framework Convention on Climate Change (UNFCCC) adopted in 1992, its 1997 Kyoto Protocol, and 2015 Paris Agreement, as well as climate governance arrangements at national, regional, provincial, and local levels. We end with concluding remarks and suggestions of key research gaps for addressing the governance of existential crises such as climate change.

## 2 What is governance in the context of climate change?

We define climate change governance here to be the interrelated and increasingly integrated system of formal and informal rules, rule-making systems, and actor networks at all levels of human society (from local to global) that are set up to steer societies toward preventing, mitigating, and adapting to climate change (Biermann et al., 2009). Governance thus opens up traditionally bounded intellectual and policy delivery systems to solve existential challenges such as climate change. The approach is to transcend national boundaries, link across levels of governance and enable conventional and nonconventional policy actors to play their parts in unearthing innovative forms of societal steering (see Fig. 1).

Despite some divergence on core elements (Biermann et al., 2010; Adger and Jordan, 2008), governance generally subsumes new forms of regulation that go beyond traditional hierarchical state activity and implies some form of self-regulation by political, private, or societal actors in addressing given problems as well as new forms of multilevel coordination or policy arrangements. The constellation of actors of a given governance arrangement could include a mix of governments, networks of experts, environmentalists, multinational corporations, government agencies, and intergovernmental bureaucracies (Newell et al., 2012; Di Gregorio et al., 2020).

Governance is particularly pertinent in the context of climate change given that it is an issue that cuts across virtually all human and industrial activity. Given the deep underlying causes and the nature of a superwicked problem, special attention is needed to improve institutional configurations, intersections, and instrument mixes (Cashore et al., 2010) at the

**FIG. 1** Governance and its basic components (based on the governance definition given in Biermann et al., 2009).

very least. Others (e.g., Schroeder et al., 2020; Gupta, 2016) argue that policy or governance learning, meaning a continuous response to feedback in a complex system, as acquisition of skills and knowledge is also needed, in particular with regard to addressing the existential climate crisis.

At the global level, the term "global governance" is used to describe processes of international and/or transnational politics (Young, 1999; Rosenau, 1995; Kanie and Haas, 2004). Global governance thus refers to a global system no longer exclusive to territorial states. Rather, it is defined as a more complex and multilayered global order characterized by increased institutional fragmentation, public-private cooperation, and transnational governance arrangements (Betsill and Bulkeley, 2004; Andonova et al., 2009). The idea of a global civil society—a new and distinct sphere of global collective life characterized by a plurality of members, voices and forms of social activity and activism—is central to this rethinking of global politics (Mitrani, 2013; Fisher et al., 2015). Others, in this context, have introduced the term "glocal" to denote the nature of the global as simultaneously local (e.g., Gupta, 2012), thus highlighting the multilevel nature of governance discussed below.

## 3 Multilevel governance of climate change

The concept of multilevel governance grew out of the study of the Europeanization process in the late 1980s and early 1990s (Benz and Eberlein, 1999), and the discussion still remains somewhat Eurocentric. It theorizes the reallocation of authority away from the central state to regional authorities (Hooghe and Marks, 2001), nonstate actors, and cross-boundary networks. The roots of multilevel governance are diverse and grow out of studies of federalism, decentralization, and European integration and its principle of subsidiarity, explained further below.

A multilevel governance approach highlights the dynamic interactions across levels of governance (Bulkeley and Betsill, 2005) that can take the form of, either or both, a more hierarchical model in which competencies are distributed rather than overlapping and a more fluid, issue-oriented alliance between levels of government and diverse actors, also referred to as a polycentric model (Hooghe and Marks, 2003). This difference is highlighted in Hooghe and Marks' (2001) two types characterization. Type I depicts the hierarchical approach, which focuses on the ways in which competences and authority are shared between different levels of government (e.g., renewable energy promotion through R&D, subsidies). Type II is a polycentric model in which multiple overlapping and interconnected horizontal spheres of authority are involved in governing particular issues (e.g., raising awareness on deforestation drivers through public and private initiatives). Authority has thus become dispersed across multiple territorial levels both or either vertically and/or diagonally and among a variety of private, public, or hybrid actors. While multilevel governance has this multidirectional and multidimensional nature, it fails thus far to capture the essence of the problem in the sense that it ignores the emotive nature of the issue it addresses.

Multilevel climate change governance can take a number of forms. The principle of *subsidiarity* ascribes governance to occur at the lowest level of governance possible, which is the urban level in many instances of EU climate governance (Bulkeley and Betsill, 2005). *Decentralization* is a redistribution or dispersion of functions, powers, people of things away from a central location or authority, which is common for forest governance in many countries (Krott et al., 2014). *Federalism* is a political system in which sovereignty is constitutionally divided between a central governing authority and constituent political units (e.g., states, provinces) (Blank, 2020). *Fragmentation* or polycentric governance refers to a state of multiple, partly overlapping centers of policymaking that characterizes the multiple arenas of global and regional climate change governance (Zelli and van Asselt, 2013). Lastly, adaptive *co-management* for climate change adaptation emphasizes pluralism and communication, shared decision-making and authority, linkages within and among levels of governance, actor autonomy and learning for adaptation in sharing the responsibility for a resource or public good (Baird et al., 2016).

Given the multilevel context, it can be argued that local decisions are never truly local. Instead, they are embedded in larger national and international structures (Steinberg, 2015). Each country has their own system of political and legal authority across its different levels of jurisdiction and thus different levels of autonomy for subnational entities. They are also subjected to varying access to resources and tax revenues as well as different policy instruments. This creates scalar dependence and governance capacity limitations at lower levels of governance, which leads to differences in the optimal approach to climate change governance across countries. As a result, urban climate change governance is more limited in the United States, where local authorities tend to have less financial autonomy compared to cities in Europe (Schroeder and Bulkeley, 2009).

## 4 Actors and actor networks

Actor networks may consist of a myriad of different actors, which, it should be noted, have varying degrees of agency. Agency refers to the ability of an actor to prescribe behavior and substantively participate in and/or set their own rules related to the interactions between humans and their natural environment (Biermann et al., 2009). Each actor also benefits from a unique set of resources or powers (Dingwerth and Pattberg, 2009).

*National governments* have exclusive access to the international climate change negotiating tables (Schroeder et al., 2012). As parties to the UNFCCC, they make up the Conference of the Parties (COPs), the UNFCCC's authoritative body. The *UNFCCC Secretariat* was set up by the parties to the UNFCCC as the coordinating body that oversees implementation of agreements, makes arrangements for meetings, coordinates across member countries, and compiles reports and data, among other tasks (Kuyper et al., 2018). *Environmental NGOs* and other constituencies under the UNFCCC[a] have in the past decades increased their influence over the international climate negotiations by developing and propagating creative policy solutions, constructing and disseminating knowledge, lobbying governments and campaigning for more climate awareness and action (Gough and Shackley, 2001; Betsill and Corell, 2001). Participants have to be accredited by the UNFCCC Secretariat, meaning only organizations that support the objectives of the UNFCCC and are not-for-profit can enter the official meeting venues. Thus, only not-for-profit *business associations* represent industry at these meetings. The businesses they represent will have done their lobbying of decision-makers prior (Vormedal, 2008). *Indigenous peoples* and *Youth*, for example, have increased their presence and voice at the UNFCCC meetings over the years (Schroeder, 2010; Wallbott, 2014; Thew, 2018; Kuyper et al., 2016). *Scientists* and *epistemic communities* have generated the ideas that create a basis for climate change governance (Schroeder and Lovell, 2012). The IPCC was created in 1988 to provide policymakers with regular, state-of-the-art assessments of the science of climate change, its implications and potential future risks as well as potential adaptation and mitigation options. Science has thus been given a strong institutional link into the negotiation, although there are accounts of IPCC recommendations being ignored or downplayed by certain countries, such as with the recent IPCC Special Report on 1.5°C (IPCC, 2018). Outside the UNFCCC, novel partnerships have emerged between *municipalities* and *local businesses* (Burch et al., 2013) and local governments and urban areas have experimented with climate change governance in innovative ways (Bulkeley et al., 2014; Bulkeley and Stripple, 2020).

In recognition of the key role nonparty stakeholders are playing in implementing the Paris Agreement, a number of institutions were set up to act as a clearing house for all the actions taken around the world. The Non-State Actor Zone for Climate Action (NAZCA Portal) and the Lima-Paris Action Agenda (LPAA) help highlight and synthesize nonstate climate actions (Morgan and Northrop, 2017). They confer legitimacy on their actions (Di Gregorio et al., 2020), which, alongside the orchestration efforts of the secretariat, are clearly necessary, but should not be seen as a substitute for meaningful, ambitious climate action by states (Allan, 2019). Thus, a multitude of actors are embedded into the complex social web of institutions under the UNFCCC, which create a forum for interaction, connections, and relationships to emerge, very much along the lines of "all hands on deck" (Hale, 2016). This social context feeds back into the structures of those institutions, which reflects the interconnected relationships between actors and institutions that embody the ideas and belief systems that surround the climate change issue.

[a] In the UNFCCC, recognized actor constituencies include Business and Industry NGOs, Environmental NGOs, Farmers and Agricultural NGOs, Indigenous Peoples Organizations, Local Government and Municipal Authorities, Research and Independent NGOs, Trade Union NGOs, Women and Gender and Youth NGOs.

## 5 Rule-making systems

Climate change governance is embedded in multiple and ever-evolving rule-making systems, particularly that of the UNFCCC, which by nature has a "framework" character. They include those set up by the UNFCCC and subsequent agreements as well as systems of rule-making that pertain to other levels of governance, such as the democratic or authoritarian systems of many countries and their climate active or inactive populace, organizations, and business sectors. They also include the specific stipulations that an international climate agreement has to go through to be ratified, such as the two-thirds Senate majority needed in the United States.

Rule-making systems are the institutions that set down what actors are eligible to take part in their governing processes and what types of rules, mechanisms, or measures can be adopted to reach a set goal. Because they are the overarching rules that define the parameters of the system and determine the lawful means of rule-making or decision-making, they are also referred to as "super rules" (Steinberg, 2015). Rule-making systems or institutions are thus the "rules of the game", that is, "clusters of rights, rules and decision-making procedures that give rise to social practices, assign roles to the participants in these practices, and guide interactions among occupants of these roles" (Young. 2008, xxii). Institutions can be statutory if written down and formally adopted, such as the UNFCCC adopted in 1992, or they can be customary if they have evolved gradually over time, such as belief systems, values and cultural contexts. Institutions are thus different from organizations, which are tangible, physical entities, usually with personnel, a budget and a location, such as the UNFCCC Secretariat as its operational arm.

Rule-making systems, or superrules, are much harder to create or change than simple rules. Yet, if there is a change, this changes the power dynamic among different actors or agents, as much discussed in the literature on the structure-agency debate (Okereke et al., 2009). One example of a recent change in the rule-making system under the UNFCCC has been the adoption of the Local Communities and Indigenous Peoples Platform through the Paris Agreement. This platform gives indigenous peoples representative the vehicle through which to allow for a greater representation of their world views and belief systems and engage with traditional ecological knowledge (TEK) for climate policy, and thus goes beyond previous tokenism and constraints on participation (Shawoo and Thornton, 2019; Belfer et al., 2019). While the impact from a platform as this may not be large, it still offers an opportunity for achieving a gradual shift away from a narrow Western-held worldview to one where multiple forms of knowledge formation are accepted as equal and complementary (Schroeder and González, 2019).

Because the UNFCCC's rules of procedure were never formally adopted, its decision-making procedures are based on the UN default of ruling by consensus. This means that it can take a long time for agreement to emerge within the UNFCCC and its various bodies. It also means that any agreement will tend to be a lowest common denominator outcome and include a lot of constructively ambiguous language to please all parties in some way. A "softer" rule-making system perhaps is the informal arena of side events and exhibits that parties and nonparty stakeholders can apply for to discuss climate change topics, exchange ideas, and/or promote their contributions, offerings, and solutions (Hjerpe and Linner, 2010).

## 6 Formal and informal rules

Rules have characteristics of being immaterial and invisible, yet influential, webs that penetrate the socioeconomic activity and determine relations of power among humans and their effects on the natural environment (Steinberg, 2015). Rules can be formal or statutory (codified or written down, legally operational) or informal or customary (socially accepted norms), and they can be enforced or just exist on paper (Young, 2008). There is a high variety of different rules, legislations, mechanisms, incentive schemes, etc., or modes of governance, to coerce or incentivize positive behavior within the defined social group. These are set through the various UNFCCC agreements and decisions as well as through other national, regional, transnational, local, or hybrid decision-making spaces. Law often moves from soft law to hard low. For example, a policy might be adopted first, which is a broad vision or goal or a statement on a long-term direction about an issue. It does not yet specify in detail the instruments or practices that might be used to implement it. Then, legislation is passed, which is a key instrument for implementing a policy by setting out rights and obligations and institutionalizing the rules. This can be done either through primary legislation enacted by parliament or secondary legislation in the form of regulations, decrees, ordinances, or by-laws. This then takes precedence over policy (Shaffer and Pollack, 2010).

*Formal rules and regulations* define traditional command and control (or sticks and carrots) approaches that are legislated through the legislative system of states and to some extent the formal provisions of international law or "hard" law. For example, this includes formal rules around emissions trading, set out in Article 17 of the Kyoto Protocol. It allows countries to trade and sell their excess capacity of greenhouse gas emissions according to rules set out in specific emission trading markets, such as the EU Emissions Trading Scheme (EU-ETS). The rules governing such transactions have often been criticized as too soft, allowing many heavily emitting industries to benefit from market distortions and design flaws, thus not achieving actual emission reductions through this scheme (Flachsland et al., 2020).

*Self-imposed rules and self-governance* refer to measures taken by the governing body itself, for example, the municipality, where a carbon emissions inventory is put together and subsequently operations are changed to reduce emissions. This could include changing municipal building light bulbs or the city's street light systems, greening municipal fleets or mandating or incentivizing walking or cycling to work if employees live within a certain distance from their workplace (Bulkeley et al., 2011; Schroeder and Bulkeley, 2009). It also includes the bottom-up approach of the Paris Agreement with regards to nationally determined contributions (NDCs). While the Kyoto Protocol required the implementation of legally binding emission reduction targets for developed countries, the Paris Agreement leaves the format and breadth of actions to mitigate and adapt to climate change up to each party (Kuyper et al., 2018).

*Informal rules of provision and enabling*, or "soft" law, refer to supplying information or delivering services to help others reduce their emissions, for the former, and offering advising, funding or training to increase the capacity of others to reduce their emissions, for the latter. For example, the NAZCA Portal mentioned above provides such a clearinghouse service. The various climate finance mechanisms, including the Green Climate Fund, enable payments to be made to developing countries to adapt to the adverse effects of climate change and to keep their forests standing (Corbera and Schroeder, 2011).

## 7 Conclusion

Climate change governance emerges from a complex and somewhat fluid web of actors, organizations, institutions, rules and goals operating and interacting across multiple levels of social organization. Governance is outcome oriented, offering a variety of approaches spanning the full spectrum of hard law to soft law and positive to negative incentives to guide social behavior and interaction. The course of three decades of international climate negotiations has contributed to the evolution of climate change governance through widening the realm and remit in a number of ways. First, the realm of actors and actor networks has widened from a more focused group of governments and business and environmental NGOs to today including diverse interest groups that are subsumed under now nine formal constituencies. Second, the architecture of the climate regime has broadened from a top-down approach of targets and timetables for developed countries under the Kyoto Protocol to a hybrid bottom-up and top-down approach under the Paris Agreement of voluntary national contributions and some form of formal review and ratcheting up of ambition over time, which also actively supports efforts made by nonparty stakeholders. Third, the diversity of approaches to governing have multiplied alongside the multiplication of actors in a manner of "all hands on deck" that is, the more, the better.

What is clear is that although climate change governance has grown remarkably in the ways summarized above, it has not in its current incarnation been ultimately successful at achieving the goal the UNFCCC set itself in 1992: to avoid dangerous anthropogenic interference with the global climate. It can be argued that this approach to governance focused on the collective, consensual, accountable, participatory, and adapting is not fit for purpose when it comes to superwicked, existential crises such as the climate crisis. For most people, there does not seem to be a willingness to engage or change, as it is just too difficult and emotionally concerning to contemplate, particularly if it means giving up much enjoyed activities and comforts. There is an underlying fear and despair of moral failure; the very notion of existential means that the "normal" patterns of governance simply cannot work.

Thus, the only way for an effective governance of climate change is for it to become embedded in social change and social values, which are alerted to grapple with existential crises in a progressive manner. To address this, we end by highlighting three key areas for future research. First, we recommend targeted engagement with learning through governance in order to more rapidly respond to our existential crisis by more systematically learning through trial and error and experimentation. Second, we recommend more appreciation for the trends in customary and soft realms, such as shifts in worldviews, belief systems, and values emerging out of new climate arenas and platforms. Taking these lessons and applying them toward shifting our mindset away from vested interests and the current destructive economic growth model to focusing on common goods for humankind that prioritizes human and planetary wellbeing is essential. Third, we recommend engaging with climate justice and care and reflecting this in climate governance in order to more authentically deal with the challenges and risks of the superwicked, existential problem of climate change.

## References

Adger, N.W., Jordan, A.J. (Eds.), 2008. Governing Sustainability. Cambridge University Press, Cambridge, UK.
Allan, J.I., 2019. Dangerous Incrementalism of the Paris Agreement. Glob. Environ. Polit. 19 (1), 4–11.

Andonova, L.B., Betsill, M.M., Bulkeley, H., 2009. Transnational climate governance. Glob. Environ. Polit. 9 (2), 52–73.

Baird, J., Plummert, R., Bodin, O., 2016. Collaborative governance for climate change adaptation in Canada: experimenting with adaptive co-management. Reg. Environ. Chang. 16, 747–758.

Belfer, E., Ford, J.D., Maillet, M., Araos, M., Flynn, M., 2019. Pursuing an indigenous platform: exploring opportunities and constraints for indigenous participation in the UNFCCC. Glob. Environ. Polit. 19 (1), 12–33.

Benz, A., Eberlein, B., 1999. The Europeanization of regional policies: patterns of multi-level governance. J. Eur. Publ. Policy 6, 329–348.

Betsill, M.M., Bulkeley, H., 2004. Transnational networks and global environmental governance: the cities for climate protection program. Int. Stud. Q. 48 (2), 471–493.

Betsill, M.M., Corell, E., 2001. NGO influence in international environmental negotiations: a framework for analysis. Glob. Environ. Polit. 1 (4), 65–85.

Biermann, F., Betsill, M., Gupta, J., Kanie, N., Lebel, L., Liverman, D., Schroeder, H., Siebenhuener, B., 2009. Earth System Governance: People, Places, and the Planet, Science and Implementation Plan of the Earth System Governance Project, IHDP Report 20. IHDP, Bonn.

Biermann, F., Betsill, M., Gupta, J., Kanie, N., Lebel, L., Liverman, D., Schroeder, H., Siebenhuener, B., 2010. Earth system governance: navigating the anthropocene. Int. Environ. Agreements 10 (4), 277–298.

Blank, Y., 2020. Federalism, subsidiarity, and the role of local governments in an age of global multilevel governance. Fordham Urban Law J. 37, 509–558.

Bulkeley, H., Betsill, M., 2005. Rethinking sustainable cities: multi-level governance and the 'urban' politics of climate change. Environ. Polit. 14, 42–63.

Bulkeley, H., Stripple, J., 2020. Climate smart city: new cultural political economies in the making in Malmö, Sweden. New Polit. Econ.. https://doi.org/10.1080/13563467.2020.1810219.

Bulkeley, H., Schroeder, H., Janda, K., Zhao, J., Armstrong, A., Chu, S., Ghosh, S., 2011. The role of institutions, governance and planning for mitigation and adaptation by cities. In: Hoornweg, D., Frire, M., Lee, M., Bhada, P., Yuen, B. (Eds.), Cities and Climate Change: Responding to an Urgent Agenda. The World Bank, Washington, DC, pp. 68–88.

Bulkeley, H.A., Castan Broto, V., Edwards, G.A.S., 2014. An Urban Politics of Climate Change: Experimentation and the Governing of Socio-Technical Transitions. Routledge, London.

Burch, S., Schroeder, H., Rayner, S., Wilson, J., 2013. Novel multisector networks and entrepreneurship: the role of small businesses in the multilevel governance of climate change. Eviron. Plann. C. Gov. Policy 31 (5), 822–840.

Cashore, B., Galloway, G., et al., 2010. Ability of institutions to address new challenges. In: Mery, G. (Ed.), Forests and Society: Responding to Global Drivers of Change. In: World Series, vol. 25. IUFRO, Vienna, pp. 441–486.

Corbera, E., Schroeder, H., 2011. Governing and implementing REDD+. Environ. Sci. Policy 14 (2), 89–100.

Di Gregorio, M., Massarella, K., Schroeder, H., Brockhaus, M., Pham, T.T., 2020. Building legitimacy in transnational climate change governance: evidence from a subnational government initiative from the global south. Glob. Environ. Chang. 64, 102126.

Dingwerth, K., Pattberg, P., 2009. Actors, arenas, and issues in global governance. In: Whitman, J. (Ed.), Palgrave Advances in Global Governance. Palgrave Advances. Palgrave Macmillan, London.

Fisher, D.R., Svendsen, E.S., Connolly, J., 2015. Urban Environmental Stewardship and Civic Engagement: How Planting Trees Strengthens the Roots of Democracy. Routledge Press, London.

Flachsland, C., Pahle, M., Burtraw, D., Edenhofer, O., Elkerbout, M., Fischer, C., Tietjen, O., Zetterberg, L., 2020. How to avoid history repeating itself: the case for an EU emissions trading system (EU ETS) price floor revisited. Clim. Pol. 20 (1), 133–142.

Gills, B., Morgan, J., 2020. Global climate emergency: after COP24, climate science, urgency, and the threat to humanity. Globalizations 17 (6), 885–902.

Gough, Shackley, 2001. The respectable politics of climate change: the epistemic communities and NGOs. Int. Aff. 77 (2), 329–346.

Gupta, J., 2012. Glocal forest and REDD+ governance: win–win or lose–lose? Curr. Opin. Environ. Sustain. 4 (6), 620–627.

Gupta, J., 2016. Climate change governance: history, future, and triple-loop learning? Wiley Interdiscip. Rev. Clim. Chang. 7 (2), 192–210.

Hale, T., 2016. "All hands on deck": the Paris agreement and nonstate climate action. Glob. Environ. Polit. 16 (3), 12–22.

Hjerpe, M., Linner, B.-O., 2010. Functions of COP side-events in climate-change governance. Clim. Pol. 10 (2), 167–180.

Hooghe, L., Marks, G., 2001. Types of multi-level governance. European integration online paper 5. http://eiop.or.at/eiop/texte/2001-2011a.htm.

Hooghe, L., Marks, G., 2003. Unraveling the central state, but how? Types of multilevel governance. Am. Polit. Sci. Rev. 97, 233–243.

IPCC, 2018. Masson-Delmotte, V., Zhai, P., Pörtner, H.-O., Roberts, D., Skea, J., Shukla, P.R., … Waterfield, T. (Eds.), Summary for Policymakers: Global Warming of 1.5°C. An IPCC Special Report on the Impacts of Global Warming of 1.5°C Above Pre-industrial Levels and Related Global Greenhouse Gas Emission Pathways, in the Context of Strengthening the Global Response to the Threat of Climate Change, Sustainable Development, and Efforts to Eradicate Poverty. World Meteorological Organization, Geneva, pp. 1–32.

Kanie, N., Haas, P.M. (Eds.), 2004. Emerging Forces in Environmental Governance. United Nations University Press, Tokyo.

Krott, M., Bader, A., Schusser, C., Devkota, R., Maryudi, A., Giessen, L., Aurenhammer, H., 2014. Actor-centred power: the driving force in decentralised community based forest governance. Forest Policy Econ. 49, 34–42.

Kuyper, J., Bäckstrand, K., Schroeder, H., 2016. Accountability of non-state actors in the UNFCCC: exit, voice, and loyalty. Rev. Policy Res. 34 (1), 88–109.

Kuyper, J.W., Schroeder, H., Linnér, B.-O., 2018. The evolution of the UNFCCC. Annu. Rev. Environ. Resour. 43, 343–368.

Le Quéré, C., Jackson, R.B., Jones, M.W., et al., 2020. Temporary reduction in daily global CO2 emissions during the COVID-19 forced confinement. Nat. Clim. Chang.. https://doi.org/10.1038/s41558-020-0797-x.

Levin, K., Cashore, B., Bernstein, S., Auld, G., 2012. Overcoming the tragedy of super wicked problems: constraining our future selves to ameliorate global climate change. Policy. Sci. 45 (2), 123–152.

Malhi, Y., Franklin, J., Seddon, N., Solan, M., Turner, M.G., Field, C.B., Knowlton, N., 2020. Climate change and ecosystems: threats, opportunities and solutions. Philos. Trans. R. Soc. B 375 (1794), 20190104.

Mitrani, M., 2013. Global civil society and international society: compete or complete? Alternatives 38 (2), 172–188.

Morgan, J., Northrop, E., 2017. Will the Paris Agreement accelerate the pace of change? WIREs Clim. Change 8, e471. https://doi.org/10.1002/wcc.471.

Newell, P., Pattberg, P., Schroeder, H., 2012. Multiactor governance and environment. Annu. Rev. Environ. Resour. 37, 365–387.

Okereke, C., Bulkeley, H., Schroeder, H., 2009. Conceptualizing climate governance beyond the international regime. Glob. Environ. Polit. 9 (1), 58–78.

Rosenau, J.N., 1995. Governance in the twenty-first century. Glob. Gov. 1 (1), 13–43.

Schroeder, H., Bulkeley, H., 2009. Global cities and the governance of climate change: what is the role of law in cities? Fordham Urban Law J. 36 (2), 313–359.

Schroeder, H., González, N.C., 2019. Bridging knowledge divides: the case of indigenous ontologies of territoriality and REDD+. Forest Policy Econ. 100, 198–206.

Schroeder, H., Lovell, H., 2012. The role of non-state actors and side events in the international climate negotiations. Clim. Pol. 12 (1), 23–37.

Schroeder, H., 2010. Agency in international climate negotiations: the case of indigenous peoples and avoided deforestation. Int. Environ. Agreem. Polit. Law Econ. 10 (4), 317–332.

Schroeder, H., Boykoff, M., Spiers, L., 2012. Equity and state representations in climate negotiations, commentary. Nat. Clim. Chang. 2 (12), 834–836.

Schroeder, H., Di Gregorio, M., Brockhaus, M., Pham, T.T., 2020. Policy learning in REDD+ donor countries: Norway, Germany and the UK. Glob. Environ. Chang. 63, 102106.

Shaffer, G.C., Pollack, M.A., 2010. Hard vs. soft law: alternatives, complements and antagonists in international governance. Minnesota Law Rev. 94, 706–799.

Shawoo, Z., Thornton, T., 2019. The UN local communities and indigenous peoples' platform: a traditional ecological knowledge-based evaluation. WIREs Clim. Change 10 (3), e575.

Steinberg, P., 2015. Who Rules the Earth? How Social Rules Shape Our Planet and Our Lives. Oxford University Press, Oxford.

Thew, H., 2018. Youth participation and agency in the United Nations framework convention on climate change. Int. Environ. Agreements 18, 369–389.

Vormedal, I., 2008. The influence of business and industry NGOs in the negotiation of the Kyoto mechanisms: the case of carbon capture and storage in the CDM. Glob. Environ. Polit. 8 (4), 36–65.

Wallbott, L., 2014. Indigenous peoples in UN REDD negotiations: "importing power" and lobbying for rights through discursive interplay management. Ecol. Soc.. 19(1).
Young, O.R., 1999. Governance in World Affairs. Cornell University Press, Ithaca, NY.
Young, O.R., 2008. Building regimes for socioecological systems: institutional diagnostics. In: Young, O.R., King, L.A., Schroeder, H. (Eds.), Institutions and Environmental Change: Principle Findings, Applications, and Research Frontiers. The MIT Press, Cambridge, MA.
Zelli, F., van Asselt, H., 2013. Introduction: the institutional fragmentation of global environmental governance: causes, consequences, and responses. Glob. Environ. Polit. 13 (3), 1–13.

# Further reading

Wamsler, C., Schäpke, N., Fraude, C., Stasiak, D., Bruhn, T., Lawrence, M., Schroeder, H., Mundaca, L., 2020. Enabling new mindsets and transformative skills for negotiating and activating climate action: lessons from the UNFCCC conferences of the parties. Environ. Sci. Pol. 112, 227–235.

CHAPTER

21

# Justice and climate change

*Steve Vanderheiden*

Department of Political Science, University of Colorado at Boulder, Colorado, CO, United States

OUTLINE



## 1 Introduction

Among the challenges posed by anthropogenic climate change are issues of inequality and injustice in its causes and consequences, making justice an important lens for viewing the problem. While all humans contribute toward it insofar as their activities generate greenhouse gas emissions or reduce the planetary capacity to safely absorb those emissions, they do so unequally, and this inequality can be in some cases be characterized as an injustice. Some states and persons emit at rates that are far above those of other persons or states, with many having become affluent through unlimited appropriations of the planet's sink capacity. In many cases, these emissions are also well beyond the threshold of sustainable per capita emissions, leading to concerns about injustice in the manner in which humans cause climate change. The same is true of climate impacts, which are expected to be disproportionately borne by the world's least advantaged, at both the individual and collective level, with poor countries and poor persons in all countries being more vulnerable to climate-related harm relative to their affluent counterparts. Given that those most responsible for causing climate change are also expected to be least vulnerable to its effects, and those least responsible for causing the problem are likely to suffer from it most, unmitigated climate change has widely

*The Impacts of Climate Change*
https://doi.org/10.1016/B978-0-12-822373-4.00001-X

and aptly been recognized as presenting a problem of distributive justice, or regarding how the benefits and burdens of collective activity are shared among those participating in it or affected by it.

Whether and how humans address anthropogenic climate change also raises issues of justice. Cooperative efforts to reduce the greenhouse emissions or land use changes that cause climate change (termed *mitigation*) entails costs that can either be assigned in accordance with justice principles or can be assigned in a manner that exacerbates existing inequalities. The same is true of *adaptation* measures, which seek to insulate persons or communities from the harmful impacts of climate change. Justice principles can help to determine which parties are entitled to adaptation assistance and which parties are obligated to fund or provide it, and in what amount. International climate finance mechanisms that fund adaptation efforts or provide compensation for victims for climate-related loss and damage have drawn scrutiny for the fairness of national contributions to them, as well as how they are used, with states that have thus far declined to contribute their fair shares incurring criticism for shirking their just burdens. Justice has thus become an important ideal for articulating the problem of unmitigated climate as well as guiding or evaluating various responses to it. In this chapter, I shall explore principles of justice as they have been applied to various issues in the human responses to climate change.

As a normative ideal, justice functions as a critical concept that specifies how institutions should be structured and how the benefits and burdens of cooperation ought to be shared as well as how problems of injustice are to be rectified. In its various forms, it can articulate and defend principles of distributive justice, which primarily concern allocations of benefits and burdens or other factors that influence the life chances of persons, as well as remedial justice, which focuses instead upon compensation or rectification of wrongs or injuries. Both are relevant to climate justice, which seeks both fair terms of cooperation in sharing the planetary sink (the main focus of mitigation) as well as a means by which those exposed to climate-related harm may be protected from or compensated for that harm (the focus of adaptation). In both cases, the lens of justice applies to societies or states, whether in an international context or within a given society or nation-state. Its focus is therefore not upon the conduct of the individual person, despite the common usage of the term often including an analysis of just persons or lives. For purposes of the exposition in this chapter, such questions of individual moral obligations to others in the context of climate change are confined to climate ethics (Gardiner, 2011), which is the subject of another chapter in this volume.

## 2 Justice and the climate treaty

As an analytical frame for evaluating climate change as a problem, along with the various efforts within and between nation-states to reduce its causes and minimize its effects, justice offers principled as well as practical advantages as a critical diagnostic criterion. Since responses to climate change involve the assignment of costly mitigation burdens to various state parties, the regime that organizes the international institutions and rules must justify the imposition of this coercion as legitimate. While it could ground this legitimacy in claims about procedural justice, for example, in reference to the inclusive and consensus-based processes of the United Nations Framework Convention on Climate Change (or UNFCCC), substantive justice principles that are included in the text of that landmark treaty (to be discussed further below) seek legitimacy in the justice of its overall objective of avoiding dangerous climate change as well as to the justice of its allocation of remedial burdens. As Rawls notes in his seminal treatise on distributive justice, underscoring this link between the legitimacy of coercive institutions and the justice of their ends and means, "justice is the first virtue of social institutions, as truth is of systems of thought" (Rawls, 1971, p. 1). International institutions like those charged with coordinating national climate change mitigation and adaptation efforts ought to serve defensible objectives such as securing or promoting justice, even if their coercive power of nation-state parties is minimal, because their overall aims are to reduce injustice and to do so in a manner that is itself just.

Incorporating justice into the design of institutions like the international climate regime also confers pragmatic benefits, especially where such institutions lack powers to enforce their terms beyond utilizing reputational accountability to shame those refusing to do their fair share (Keohane, 2006). Given the absence of any strong system of global governance capable of imposing regulatory burdens upon sovereign states without their consent, any cooperative effort to control greenhouse gas emissions is likely to garner greater support for and thus compliance with its terms if these are viewed as fair to all. As Athanasiou and Baer have noted of climate treaty proposals that unfairly limit development opportunities in large and fast-growing emitters like India and China by imposing excessive mitigation burdens on developing countries, which must participate in any successful international effort, such proposals "will not be accepted as fair and, finally, will not be accepted at all" (Athanasiou and Baer, 2002, p. 75). In a world where participation in such a cooperative effort cannot be compelled or coerced but must instead be the product of voluntary consent by participating parties, the fairness of the scheme may go a long way in determining its prospects for success, whereas fidelity to principle may be vital to conditions of legitimacy in an abstract sense, the shared perception that an arrangement is just is essential to it being regarded as such and may thus be vital to its political feasibility.

As noted above, climate change is widely regarded as involving issues of distributive justice in its known causes and expected effects, with its origin in the widely disparate per capita greenhouse gas emissions that form the backdrop of climate injustice noted above and its similarly inequitable expected impacts, which scientists predict "will fall disproportionately upon developing countries and poor persons within all countries, and thereby exacerbate inequities in health status and access to adequate food, clean water, and other resources" (Intergovernmental Panel on Climate Change (IPCC), 2011, §3.33). Since the global poor disproportionally reside in food insecure and drought-prone regions, expected changes in rainfall patterns and increases in the frequency and intensity of droughts are likely to primary affect these most vulnerable populations. Adaptive capacity to extreme weather events like storms and floods is likewise much lower among the world's most disadvantaged, relative to more affluent countries or affluent residents of the same countries, again increasing the vulnerability of the global poor to such impacts. Insofar as egalitarian justice principles seek to decouple poverty and vulnerability, such that being poor or otherwise socioeconomically disadvantaged would not also make a person or community more vulnerable to climate change, all of these forms of unequal vulnerability can be regarded as unjust (Vanderheiden, 2007, ch. 1).

Indeed, anthropogenic climate change functions as a global externality that transfers one unjust form of inequality in widely disparate per capita emissions patterns and generates another one in highly inequitable vulnerability. The injustice of unmitigated climate change along with the critical role for justice in shaping the international response to it have played prominent roles in political efforts to craft an international climate treaty. These are evidenced by the text of the UNFCCC, which calls upon the world's nation-state parties to "protect the climate system for the benefit of present and future generations of humankind," charging the international climate policy process with developing a treaty framework that assigns burdens to states "in accordance with their common but differentiated responsibilities and respective capabilities" (Principle 1). These ideals, which refer to principles of distributive and corrective justice in referencing equity and responsibility, require that climate change be addressed with specific attention to the global and intergenerational justice dimension noted in these treaty principles. The normative content of the treaty, which identifies these justice conceptions as guiding principles for the development of an international climate scheme, underscore the importance of justice in international politics and have provided important fodder to scholars of climate justice to develop these ideas in theory.

## 3 Mitigation, equity, and carbon budgets

The core objective of climate justice in the context of mitigation is to specify terms by which all persons and peoples can be assigned fair burdens of cooperation in reducing the human drivers of climate change. Since emissions reductions typically entail costs, this objective must be able to specify either how the abatement activities are assigned among various parties (e.g., in terms of national emissions targets) or how the costs associated with decarbonization are to be allocated. When it comes to mitigation efforts that involve the protection or enhancement of carbon sinks like forests (as through the Reducing Emissions from Deforestation and Forest Degradation program, or REDD), costs rather than the location of activities become the primary focus, as carbon abatement credits accrue to those parties carrying out the projects. Likewise with the international Clean Development Mechanism, which allows state parties to earn credit for emissions reduction projects funded and carried out in other countries. Both of these kinds of mitigation activities lend themselves to a burden-sharing framework for climate justice, as their focus is upon how to allocate the costs or burdens associated with mitigating climate change.

However, other mitigation activities can more cogently be analyzed through a normative framework that focuses upon how resources are shared, with the resource in question being the planet's ability to safely absorb greenhouse emissions without climate change (or sink capacity, which is an ecosystem service that depends on resources like forests). Here, the focus of climate justice is upon allocating carbon access among peoples and persons, with a two-stage process of first allocating allowable global emissions among nation-states and then allocating those within or among the persons or peoples that comprise it. As an international regime constructed for the purpose of coordinating the activities of nation-state parties, the UNFCCC climate regime only engages in this first stage of resource-sharing allocation, but similar principles to those used at the international level could apply to resource-sharing arrangements in the second stage, where nation-states implement their national commitments through domestic climate policies.

Within such a resource-sharing framework, carbon access can be intentionally rationed according to principles of distributive justice or the planet's sink capacity can remain open to all without restriction, as it has throughout most of human history. We now know that the latter would result in dangerous climate change, so the challenge of climate justice in mitigation is to specify fair assignments of carbon access, which can be set through a carbon budget. National emissions caps and the prescribed reductions that they entail can be readily analyzed in terms of distributive justice principles, with the first stage of this resource-sharing process focusing upon a fair international carbon budget, which rations carbon access among user groups and over time.

To meet a given global temperature target like the 2°C goal set by the Paris Agreement, a finite quantity of stored carbon can be emitted as carbon dioxide (from fossil fuel combustion) during the time period. Carbon budgets help to ensure that this finite quantity is not exceeded and that the burdens associated with meeting the target are assigned to all parties along a schedule by which necessary decarbonization occurs within the compliance period. If, as has been advocated, humanity can now release only 400 billion tons of carbon dioxide equivalent before phasing out carbon-based energy altogether (Shue, 2011), those 400 billion tons must first be budgeted over time (in terms of allowable emissions per year), with the costs of less aggressive decarbonization efforts in the near future falling upon those in the further future that are required as a result to undertake steeper reductions. With a schedule for phasing out emissions over time, each year's overall budget requires a second allocation, by which nation-state parties are assigned a carbon budget for the year. While climate justice scholars have largely focused upon these first two carbon budgeting decisions, national carbon budgets require subsequent internal allocations that raise climate justice questions of their own, where similar principles may apply to how the burdens of mitigation or benefits of carbon access are allocated within a country.

How can justice principles inform carbon budgeting or be used to evaluate proposals for the rationing of carbon access? International carbon rationing schemes based in some conception of equality have followed two basic approaches, with variations on each having been proposed along with hybrids that incorporate elements of both. The grandfathering approach, which was used in the setting of national carbon abatement targets under the 1997 Kyoto Protocol, seeks to assign roughly equal mitigation burdens regardless of the emission profile or other features of various nation-state parties. Under the protocol, Annex I developed country parties were tasked with reducing their emissions by an average of 5% of 1990 baseline levels, with some parties expected to make reductions somewhat beyond that level and others assigned somewhat less demanding targets. Assigning future carbon budgets on the basis of current emission levels, with roughly equal assigned reductions from current emission levels, the protocol did not ignore equity altogether but rather defined it in terms of equally shared burdens among developed countries, regardless of what their per capita emissions might have been at the treaty's outset.

While equal burdens offer one formulation of the demands of justice as applied to the problem of climate change mitigation, they are neither the only nor the most compelling one. Under a grandfather approach to declining overall carbon budgets, a county that emitted at twice the per capita rate as another at the start of the gradual phase-out of carbon access would be able to emit at twice the per capita rate throughout the compliance period. Percentage reductions would be equal between the two countries, but abatement costs would not necessarily be, given variation in such costs based on several factors. In countries that had already acted to harvest the "low-hanging fruit" of $CO_2$ abatement, for example, by modernizing building codes or upgrading old coal-fired power plants, additional reductions would be more expensive. For them, an equal reduction target would come at a higher cost since their cheaper decarbonization options would have already been exhausted. Under the Kyoto Protocol, for example, per capita emissions within the European Union were substantially lower than in the United States or Canada, given the prior modernization efforts that retired some of the EU's most polluting energy infrastructure, leaving costs per ton of further abatement higher than in North America.

In contrast to this equal burdens approach to incorporating justice principles into carbon budgets, an equal shares approach would conceive of justice as demanding that all parties enjoy roughly equal per capita carbon access, requiring bigger emitters to undertake significantly more costly mitigation measures than those with lower per capita emissions at the outset. Here, equity is understood in terms of the sharing of a resource rather than in the remedial burdens associated with decarbonization. Under such a resource-sharing approach, per capita carbon emissions would be assigned on a roughly equal basis, so this has been termed the equal shares approach. Unlike the equal burdens approach, which despite its political popularity enjoys little support among climate justice scholars, the equal per capita emissions entitlements that form the core of the equal shares approach enjoy wide support within the climate justice literature. On the other hand, the view commands little political support, especially among developed countries.

## 4 Adaptation and differentiated responsibility

Along with applications of justice principles to the design and evaluation of mitigation efforts, a primary concern for the development of remedial climate policy architecture involves the assignment of liability for the finance of mitigation and adaptation efforts (Paavola and Adger, 2006). Following the "common but differentiated responsibilities" (or CBDR) language of the UNFCCC, scholars have suggested that some version of a polluter-pays principle be used to allocate the burdens of climate finance for such purposes. Considerable variation exists among scholarly proposals for differentiating national responsibilities for climate change. Some would assign remedial liability in proportion to each country's full historical emissions (Shue, 1999), while others (often citing excusable ignorance about the harmful effects of climate change prior to then) base their assessments of responsibility only upon post-1990 (or some other recent baseline) emissions. Some assign liability for all emissions, while others assign it only to "luxury emissions" beyond what are necessary for survival. Some exempt developing countries from remedial liability while others do not. But all would in some way assign greater remedial liability to those countries that are by virtue of their higher emissions are viewed as more responsible for causing the problem.

Some also draw upon the "and respective capabilities" phrase that follows the CBDR principle in the UNFCCC principle to base remedial liability assignments in part on national wealth (typically measured in terms of per capita income or stages of economic development).

For example, Baer et al. (2008) develop a responsibility-capacity index that groups countries into tiers based on their stages of development and assigns remedial liability within each group according to each country's historical emissions but differentiates among those groupings. Since including capacity is typically urged in conjunction with an assessment of national responsibility (or, to follow the text of the UNFCCC, it is formulated as CBDR + RC), such approaches modify but do not reject the basic differentiation between rich countries with high current and historical emissions being required to should larger remedial burdens and poor ones with low current and historical emissions being assigned lower or no such burdens.

The justification for relying upon a polluter-pays principle involves remedial rather than distributive justice (Miller, 2009). By attaching remedial liability in proportion to a country's carbon emissions, activities causing climate that enrich or otherwise advantage one party while causing harm to another are directed to rectify that harm through some transfer from responsible party to victim. As with corrective justice accounts that require compensation for imposed injuries, the injustice arises with the injury caused by one party and borne by another, opening up an imbalance between them that can only be redressed by either removing the injury or somehow preventing it from harming (i.e., adaptation) or compensating victims for loss and damage suffered. Scholars have marshaled ethical theories about collective responsibility on behalf of justifications for versions of the polluter-pays principle, which is viewed as capturing the CBDR principle from the UNFCC. Accounts of corrective justice from legal theory are typically built on individualistic premises, where one person owes compensation to another for harming them, and so need some kind of account of collective causality and agency when applied to countries rather than persons. Suffice here to note that while some version of the polluter-pays principle enjoys wide support among climate justice scholars, conceptual issues such as those related to collective agency and causality have led some to seek other foundations for remedial liability.

Rather than grappling with objections to the use of a polluter-pays principle to assign national liability for climate change, some scholars advocate using a beneficiary-pays liability principle instead (Page, 2012). The rationale for such a principle is that parties have benefitted by many of the same actions that have led to climate change (early industrialization, fossil-based energy, etc.), and so now owe to maintain the climate system in proportion to their having used and benefitted from its open access condition, prior to access restrictions imposed to prevent it becoming a tragedy of the commons. Variations of beneficiary-based principles have been suggested, based on different ways of identifying and impugning beneficiaries, varied strategies for challenging national entitlements that have conventionally been attached to them, and different formulae for calculating the amount of benefit that various states have received from using an open access sink. For purposes here, these stand as alternative formulations of the burden-sharing problem at the core of climate justice and relying upon various but distinct conceptions of justice to provide their application and justification.

## 5 Climate change, justice, and human rights

Many of the social and economic impacts of climate change are expected to threaten human rights, so a popular way of expressing climate justice imperatives and mobilizing support for them has been through human rights law and discourse. Caney, for example, began by casting the injustice of climate change primarily in terms of cosmopolitan distributive justice principles (Caney, 2005), but later switched to a human rights frame (Caney, 2013) for climate injustice, partly because of what he sees as the latter's higher resonance within world politics. While largely aspirational in that few mechanisms exist for effectively promoting human rights or sanctioning those that violate them exist within international law and politics, human rights covenants like the Universal Declaration of Human Rights do have extant legal status within international law and have been given force through ratification by all those countries that are party to them. Insofar as climate change threatens to violate human rights on a wide scale—both in terms of individual rights to life, health, and subsistence as well as group rights to culture, self-determination, and territory—seeking to mobilize support for international efforts to mitigate climate change through a rights-based justice frame that better resonates with recalcitrant parties like the United States rather than one rooted in global distributive justice may yet prove to be strategically effective.

Whether the human rights frame for climate injustice will yield concrete results for the kind of aggressive action to mitigate climate change or shield vulnerable parties from its impacts remains to be seem. One case that sought to litigate US obstruction of progress on international climate treaty development suggests both the potential for such an approach as well as the limits of this potential. Bringing their challenge before the Inter-American Commission on Human Rights in 2005, the Inuit people of Arctic North America filed a human rights complaint against the United States for its domestic contributions to the problem and international obstructionism, claiming that human-caused changes to food sources and water flows threaten their human rights to culture, life, health, and subsistence. Although unsuccessful, this challenge represents a novel use of human rights law and politics in an effort to mobilize support for an international climate treaty, and in that such environment-mediated impacts could be used to identify correlative duties associated with human rights interests (Osofsky, 2006, p. 675).

A related climate justice concern has to do with how emissions caps or other mitigation burdens might limit the development opportunities (or, to use a more tendentious frame, infringe upon the development rights) of poor states. Since hard caps on emissions for developing states could impede their industrialization or limit their ability to benefit from it through increased access to energy and transport, advocates for such states have called for a "right to development" within climate treaty negotiations, with development recognized as another imperative that must be promoted alongside climate change mitigation (Vanderheiden, 2008). Such a principle is also recognized within the text of the UNFCCC, which declares that "responses to climate change should be coordinated with social and economic development in an integrated manner with a view to avoiding adverse impacts on the latter" and recognizes "the legitimate priority needs of developing countries for the achievement of sustained economic growth and the eradication of poverty." It was in recognition of this potential obstacle to development that the Kyoto Protocol exempted developing countries from binding emissions cap, creating what would come to be known as the "Annex I firewall," or view that those countries exempted from such binding caps in 1997 would remain so permanently, absent some mechanism by which countries like China could "graduate" from exempt into the category of countries held liable for climate change.

The right to development is not the only principle of international law that exists in some tension with climate justice imperatives. The principle of permanent sovereignty over natural resources, which was developed in response to practices of resource colonialism and grants to states the right to sell and control resources found within territorial borders, poses a potential obstacle to decarbonization efforts insofar as it grants to states autonomy over the development of their mineral resources, including coal and oil. Prohibitions upon developing some country's coal reserves in the interest of preventing carbon pollution could thus be viewed as limiting this aspect of its sovereignty (Armstrong, 2015). But the principle is not unlimited and may not be applicable to a transboundary pollutant like $CO_2$. As expressed in the UNFCCC, states may have "the sovereign right to exploit their own resources pursuant to their own environmental and developmental policies," but they also have "the responsibility to ensure that activities within their jurisdiction or control do not cause damage to the environment of other States or of areas beyond the limits of national jurisdiction." So understood, national emissions caps or other kinds of mitigation assignments would not directly impugn this principle, but situating climate justice imperatives alongside other normative claims like rights to subsistence or development illustrate the potential tensions as well as compatibilities that exist among these various objectives.

## 6 Conclusion: Putting climate justice into practice

Despite its popularity among scholars, egalitarian resource-sharing schemes like those calling for equal per capita national emissions entitlements are not seriously entertained within UNFCCC climate policy development processes. Since that system moved away from assigned national emissions targets and toward nationally determined contributions (NDCs) with the Paris Agreement, the CBDR principle has shifted from being an aspirational internal criterion that many urged to be built into an international carbon rationing scheme to an external criterion for assessing national NDCs for the purpose of encouraging greater ambition, but this does not in itself diminish the centrality of international equity or justice in the climate regime. Countries that fail to commit to their fair share of abatement burdens or fail to follow through on their pledged mitigation actions can be shamed for failing to discharge their responsibilities in the way that climate justice recommends. The same is true of national contributions to climate finance mechanisms like the Green Climate Fund, for which reputation accountability serves as the primary executive power to enforce adherence to UNFCCC principles (Vanderheiden, 2015). Compliance mechanisms like REDD as well as the certifications that they require in order to quantify the value of a sink enhancement project have likewise been subjected to scrutiny by climate justice scholars (Bachram, 2004), as have resource-sharing schemes like the European emissions trading system, as both depend on a climate justice framework for their justification.

Climate justice imperatives can also provide useful critical guidance for implementing national emission targets through internal state policy mechanisms like carbon taxes or domestic emissions trading schemes (ETS). As with international carbon rationing, the pricing of carbon can deter its use but also increases the costs of carbon-embedded goods and services, which imposes costs on downstream users bearing the costs. Absent some deliberate effort to assign those costs justly, they would be likely to fall disproportionally upon the poor, through increased home heating or transportation costs, in effect assigning remedial liability to those least able to pay and in many cases least responsible for causing climate change. In so doing, they would run afoul of principles that are based in equitable burden-sharing or would impose mitigation costs onto those less responsible for national carbon emissions—principles that were put in place to guide the allocation of international burdens. Although carbon pricing through a tax or ETS would in one sense reflect the polluter-pays principle, in that costs associated with carbon pricing would be borne in proportion to each person's carbon footprint, those costs would be regressive in raising energy and transport prices for the poor. For this reason, states implementing carbon pricing schemes may develop subsidies or offsets to blunt the impacts upon the poor, assigning remedial liability in a more progressive manner as a result.

When viewed as among other things a justice problem, and when constrained in available remedies to equitably share ecological goods and services assign remedial burdens justly, climate change challenges us to apply abstract theoretical principles to real-world problems. Alleviating the injustice of unmitigated climate change without exacerbating injustices among nation-states and peoples or between generations is not an easy task. Insofar as anthropogenic climate change violates human rights like those to health, subsistence, or territory while also violating principles of remedial or global distributive justice, either framework allows for the articulation of climate justice imperatives but each comes with its own strengths and shortcomings. While the latter is typically associated with more ambitious remedial measures be taken to reduce emissions and more equitable assign carbon budgets within and between states—as meeting the full demands of justice is often more demanding than merely refraining from violating the rights of the poor and disadvantaged—an approach that seeks merely to protect human rights would nonetheless do a lot to alleviate the primary injustice of climate change. Either way, theorizing the environmental problem as also involving a justice problem requires attention to and an appreciation for some of the most challenging but also vitally important dimensions of maintaining our shared planet.

## References

Armstrong, C., 2015. Against 'permanent sovereignty' over natural resources. Politics Philos. Econ. 14 (2), 129–151.

Athanasiou, T., Baer, P., 2002. Dead Heat: Global Justice and Global Warming. Seven Stories Press, New York.

Bachram, H., 2004. Climate fraud and carbon colonialism: the new trade in greenhouse gases. Capital. Nat. Social. 15 (4), 5–20.

Baer, P., Fieldman, G., Athanasiou, T., Kartha, S., 2008. Greenhouse development rights: towards an equitable framework for global climate policy. Camb. Rev. Int. Aff. 21 (4), 649–669.

Caney, S., 2005. Cosmopolitan justice, responsibility, and global climate change. Leiden J. Int. Law 18 (4), 747–775.

Caney, S., 2013. Human rights, human security, and climate change. In: Redclift, M.R., Grasso, M. (Eds.), Handbook on Climate Change and Human Security. Edward Elgar, Northampton, MA, pp. 402–422.

Gardiner, S.M., 2011. A Perfect Moral Storm: The Ethical Tragedy of Climate Change. Oxford University Press, New York, Oxford.

Intergovernmental Panel on Climate Change (IPCC), 2011. Climate Change 2001: Synthesis Report. Oxford University Press, New York.

Keohane, R.O., 2006. Accountability in world politics. Scand. Polit. Stud. 29 (2), 75–87.

Miller, D., 2009. Global justice and climate change: how should responsibilities be distributed? In: The Tanner Lectures on Human Values. vol. 28, pp. 119–156.

Osofsky, H.M., 2006. The Inuit petition as a bridge: beyond dialectics of climate change and indigenous peoples' rights. Am. Ind. Law Rev. 31 (2), 675–697.

Paavola, J., Adger, W.N., 2006. Fair adaptation to climate change. Ecol. Econ. 56, 594–609.

Page, E.A., 2012. Give it up for climate change: a defence of the beneficiary pays principle. Int. Theory 4 (2), 300–330.
Rawls, J., 1971. A Theory of Justice. Belknap Press, Cambridge, MA.
Shue, H., 1999. Global environment and international inequality. Int. Aff. 75 (3), 531–545.
Shue, H., 2011. Human rights, climate change, and the trillionth ton. In: Arnold, D.G. (Ed.), The Ethics of Global Climate Change. Cambridge University Press, New York, pp. 292–314.
Vanderheiden, S., 2007. Atmospheric Justice: A Political Theory of Climate Change. Oxford University Press, New York.
Vanderheiden, S., 2008. Climate change, environmental rights, and emissions shares. In: Vanderheiden (Ed.), Political Theory and Global Climate Change. The MIT Press, Cambridge, MA, pp. 43–66.
Vanderheiden, S., 2015. Justice and climate finance: differentiating responsibility in the Green Climate Fund. Int. Spect. 50 (1), 31–45.

CHAPTER

# 22

# Climate change and the law

*John F. McEldowney*

School of Law, University of Warwick, Coventry, United Kingdom

OUTLINE



## 1 Introduction

In this chapter, the question posed is what is the role, if any, for law and regulation in containing climate change and its impacts? Where is the law to be found and how effective it is likely to be in delivering what is expected? Let us start with the policy challenges. Climate change[a] necessitates reducing greenhouse gas emissions, as agreed at the Paris Climate Change Conference in 2015.[b] Climate change policy falls into two categories, mitigation, to reduce greenhouse gas emissions and adaptation, the means of adjusting to actual or expected climate change. In the United Kingdom, the main legal framework for climate

[a] House of Commons Library, *Debate Pack Climate Justice* Number CD 2020/0020 (3rd February 2020).

[b] See: *The Climate Issue*, *The Economist* 21st September 2019.


https://doi.org/10.1016/B978-0-12-822373-4.00018-5

impacts of change is the Climate Change Act 2008, landmark legislation that continues to shape and inform the UK's legal responses to climate change. Law and policy are intertwined in the 2008 Act through a long-term statutory framework for decarbonisation that sets targets for the United Kingdom to reduce its greenhouse gas emissions. Such policies are expected to operate across the entire UK economy. The 2008 Act establishes an independent body to oversee and advise the government, the Committee on Climate Change. Its reports are key to ensuring progress on policy-making and implementation. There are also a series of regular international climate change conferences where the United Kingdom participates and develop climate change policy (The Environment Agency, 2018).

Law and legal regulation are integrated with market-based economic instruments. Examples include emission trading, environmental taxes, and the setting of goals to achieve environmental standards and reductions in carbon emission. Invariably, law is called upon to adjudicate questions of fairness and justice, provide a forum for the resolution of conflict, and apply enforcement strategies. Many pressure groups use litigation to test the law or even fill gaps in policy-making. A popular movement for "environmental justice" is growing apace as many citizens fear the consequences of climate change. Legal regulation underpins a plethora of agencies and institutions used to monitor and regulate contributions to climate change. There are a range of standards, principles, and procedures in existence that are used to protect the environment, for example, to preserve fresh water and other natural resources and provide protection for fragile ecosystems, not exclusively, in the context of climate change. Examples include land contamination, the reduction in climate change gases, the protection of human health, and the monitoring of air contamination (McEldowney and McEldowney, 2011; Jewell and Steele, 1983; Fisher, 2012). COVID-19 underlines the fragility of human health with air pollution under investigation as a cause of many COVID-19 infection spikes within high density populations. The UK *Committee on Climate Change* has recommended rebuilding economic growth in the aftermath of COVID lockdowns through a cleaner, fairer and more sustainable economy.[c]

## 2 Climate change law and regulation

The sources of law are many and diverse. The UK Parliament (legislature) provides the major sources of law for the United Kingdom. Acts of Parliament (statutes) or statutory rules (statutory instruments) contain the main legal powers and responsibilities (McEldowney, 2018). In the case of climate change this may include the obligations, targets, and emission controls that are expected to be observed and applied. The devolved nations, Scotland, Wales, and Northern Ireland and also the London Assembly have subordinate law-making powers that apply within the territory of the devolved settlement. The protection of the environment

[c]UK Committee on Climate Change, *A Review of the Climate Change after COVID-19*, https://www.theccc.org.uk/publication/reducing-uk-emissions-2020-progress-report-to-parliament/ House of Commons Library, *Health and Social Care Key Issues and Sources* CMP 8887 (30th June 2020), House of Lords, Library Briefing *COVID-19 Fairer, Cleaner and More Sustainable Economy* debate 11th June 2020.

is a devolved matter, providing differences in approach and certainly funding. Law may also be found in the decisions of the courts, the outcome of litigation. The UK Supreme Court is the highest judicial authority and has jurisdiction over the devolved nations. Climate change litigation has risen in importance over the past 20 years with some notable decisions on air quality and its protection.[d]

Up until January 2020, the United Kingdom was a member of the European Union which provided the technical details of the bulk of environmental law. The EU will remain influential on environmental matters, but it is unclear to what extent EU law will be followed or applied in the United Kingdom after leaving the EU.[e] International law is also relevant and may provide an important means of influencing the legal system of each nation. The Paris Agreement, arrived in 2015, set a target that emissions "should be limited significantly well below 2.0 degrees centigrade" to inhibit global temperature rise. The target came from the efforts of the International Intergovernmental Panel on Climate Change (IPCC). The panel built on previous work undertaken from the United Nations Framework Convention on Climate Change at the Rio Conference (1992) and ratified by over 5O countries. A major breakthrough came with the Kyoto Protocol adopted in 1997 that set specific reduction targets for the main carbon gases.

Law has other important dimensions. It may provide a means of accountability and access to rights as well as means of allowing participation in decision-making through the consultation process. Law through regulation may provide enforcement powers and sanctions. In England, the Environment Agency has used various civil sanctions to provide pay-outs from Water Companies for sewage spills.[f] Law may also be useful for the purposes of environmental taxation (McEldowney and Salter, 2016) as well as the basis for emissions trading, a potential means of reducing greenhouse gas emissions.

Law and regulation are gradually being extended to new areas and concerns over climate change that require the law, not only, to be reactive, but also, proactive and flexible. There are many biodiversity influences of climate change that interact with conservation.[g] The UKs National Planning Policy Frameworks, for example, requires local authority plans to deliver local plans that produce net environmental gains in local developments and infrastructures,[h] Climate change is one of the largest drivers of biodiversity losses in the sea, land use, and the direct exploitation of wildlife including fishing. Mitigation and adaptation objectives of the Paris Agreement may be met through careful planning.[i] Regulatory systems contribute to forest restoration and the management of land is part of policy-making that is recognizable in

[d] *R (on the application of Client Earth) v Secretary of State for the Environment, Food and Rural Affairs* (2015) UKSC 28.

[e] House of Commons Library Briefing Paper: Commons Library analysis of the Environment Bill 2019–20 Number CBP 8824 (18th February 2020). The Environment Bill 2019–20 is pivotal to the future protection of the environment in the UK.

[f] *ENDS Report Focus on Civil Sanctions* 533 (August 2019) pp. 42–43.

[g] UK Parliament Climate Change—Biodiversity Interactions Number 617 (February 2020).

[h] UK Parliament POST NOTE Net Gain by Jonathan Wentworh (July 2018).

[i] House of Commons Library Analysis of the Environment Bill 2019–20 Briefing paper Number CBP 8824 (18th February 2020).

incentive schemes. The UK's Environmental Bill 2019–20 provides examples such as clause 90 and schedule 14 of the Bill amending the Town and Country Planning Act 1990 to include a new schedule to allow biodiversity gain as a responsibility on developers.

## 2.1 The Climate Change Act 2008

The Climate Change Act, described as the "flagship" piece of legislation (Bell et al., 2018), sets out the national emission targets of 80% by 2050. As we shall see this target has been further strengthened in 2019.[j] The Act contains a bundle of legal powers, obligations and economic instruments to achieve its aims. Powers are granted to the Secretary of State to achieve long-term goals of reducing greenhouse gas emissions. Linked to these powers is a system of carbon budgeting[k] intended to assist this process and constrain the amount of emissions. Setting the range of the carbon budget will be occur every 3 years in advance, guided by the Secretary of State and taking into account the limit of use of international carbon credits for each budgetary period. The setting of 5-yearly carbon budgets, 12 years in advance is expected to provide a means of establishing a way to achieve targeted savings of greenhouse gas emissions. The budgets are to encourage the clean use of energy and measures to inhibit greenhouse gas emissions. Such budgets operate at devolved levels through each administration setting their own targets to reduce emissions. The fourth carbon budget 2023–27 is implemented through the Carbon Budget Order 2011.The Fifth Carbon Budget 2028–32 is intended to drive forward measures to meet the Fourth Carbon Budget. It is possible to use bank lending powers to increase the budget from 1 year to the next. There are regular assessments of the progress being made and the adjustments needed.[l]

The Act established a Committee on Climate Change with annual reporting obligations and responses to Parliament in terms of a progress report. The new Committee is in the form of an independent advisory body with powers to advise central government as well as devolved governments. This covers both mitigation and adaptation strategies. The Committee advises on the overall strategy as well as the trading scheme. This is intended to provide the government with a wide range of policy options operating within an overall legal framework. There is some question about how targets will work[m] in terms of their enforceability and sanctions that arise for their breach. One interpretation is that there is an obligation to meet such targets and that this maybe enforced. This might be achieved through a declaratory judgment or a form of order requiring the government to take action and achieve the purchase of an emissions credit. It is far from clear that the courts will be prepared to go that far.

[j] There is equivalent legislation in Scotland, the Climate Change (Scotland) Act 2009.

[k] House of Commons Library, Briefing Paper, *UK Carbon Budgets* Number CBP7555 (9th July 2019).

[l] See the Energy Bill 2016.

[m] *R (Friends of the Earth) v Secretary of State for Business Enterprise and Regulatory Reform* [2010] ENV LR 11.

## 2.2 The Climate Change Act 2008 (2050) Target Amendment Order 2019 and the Climate Emergency

Environmental law works best when it is informed by sound science (McEldowney and McEldowney, 2011). Extreme weather conditions, across the world, highlight the crisis that confronts the planet. United in Science, a report coordinated by the World Meteorological Organisation, for the United Nations Climate Action Summit held in New York on September 22, 2019, provides details on the state of the world's climate and the increase in greenhouse gas emissions. There are a number of key findings such as the average global temperature for 2015–19 is the warmest of any equivalent period on record, estimated to be 1.1°C above preindustrial times; that global GHG emissions have grown at a rate of 1.6% per year from 2008 to 2017.The report explains that if the Nationally Determined Contributions (NDCs)[n] set by the Paris Agreement are not increased urgently and backed up by immediate action, exceeding the 1.5°C goal set by the Paris Agreement can no longer be avoided; the effects of warming so far are clear: Arctic summer sea-ice has declined at a rate of approximately 12% per decade while the amount of ice lost annually from the Antarctic ice sheet increased at least sixfold between 1979 and 2017; Summer 2019 experienced unprecedented wildfires in the Arctic region and there were multiple fires in the Amazon rainforest in 2019 (as well as fires in Australia in December 2019).

The consequences of such events led the United Kingdom in the Summer of 2019, in common with many Parliaments' throughout the world, to declare a climate emergency. In November 2019, the European Parliament adopted a similar declaration. The evidence for climate change is becoming more publicly acknowledged[o] in examples such as extreme weather conditions including the fires in Australia as politicians and the public begin to assess the risks and the costs of climate change.

Strengthening the Climate Change Act 2008 has come from legislation in 2019[p] implementing recommendations from the Committee on Climate Change[q] as well as the views given in 2018 by the IPCC Special Report. The findings came with a stark warning that limiting temperature increases to 1.5°C was possible, but required many rapid and far reaching changes. The steps toward such a goal are to be discussed at the next meeting in 2021 on progress on the Paris Agreement. The 2019 Order requires a target of 100% reduction of greenhouse gas emissions by 2050. This is a more robust goal than originally set by the 2008 Act. The use of the budgets has been partially successful with 2017 emissions down by 43% of the 1990 levels. However, this may be attributable to the decrease in use of coal for electricity generation and changes in road transport.[r]

[n]The Paris Agreement requires each party to the Agreement to prepare, communicate and maintain successive nationally delivered contributions that it intends to achieve in respect of domestic mitigation measures to reduce national emissions and adapt to the impacts of climate change.

[o]See: *The Economist* January 11th 2020 A Blaze that will keep on burning.

[p]House of Commons Library, Briefing Paper, *Net Zero in the UK* Number CBP8590 (16th December 2019).

[q]Committee on Climate Change, Report on Net Zero: the UK's contribution to global warming (2nd May 2019).

[r]House of Commons Library Briefing Papers UK Carbon Budgets CBP 7555 (9th Juley 2019).

Implementing a net zero policy by 2050 is demanding and difficult. The role of law and legal regulation is rather limited. Undoubtedly, legal arrangements facilitate the implementation of major policies, but the challenges require technical and scientific strategies for heavy industry, waste, land use, and agriculture as well as infrastructure, transport, and buildings. The five areas that set particular challenges[s] are the use of carbon capture and storage which has not yet started. The target that almost all buildings must be low-carbon by 2050, the overall supply of low-carbon electricity must quadruple and a fifth of the UK's agriculture must shift to support emissions reduction. In addition, industrial resource efficiency has to improve, linked to increasing use of reuse and recycling with materials lasting longer and less new material use.

The main effort is in attempting to meet the target through domestic action. However, the use of the international emissions trading market is a possibility and there is some potential to go beyond the target under the Order of 100% reduction. Various targets have been set as a means of meeting net zero, stimulated by the Climate Change Act 2008 which has acted as a catalyst for change. One concrete outcome was the setting up of Citizens' Assembly inviting people to become engaged in combatting climate change.

For example, translating legal targets into a form of direct action starts with attempting to remove $CO_2$ emissions from the atmosphere.[t] Planting tress provides a means of absorption of $CO_2$ from the atmosphere. Restoring peatlands and natural ecosystems through rewilding is also effective. Avoiding deforestation provides another opportunity by maintaining $CO_2$ that is already stored in plants and ensuring that the soil is safeguarded. More active means might be adopted such as supporting farmers by improving their agricultural practices. The recent Agriculture Bill 2017–19 provides for farmers to take steps to sequester carbon in return for payments. There are also newly emerging technologies which remove emissions from the atmosphere, one idea is to capture waste $CO_2$ from power stations and industrial processes and to encourage reductions in the production of chemicals, minerals, and the use of synthetic fuels.[u]

## 3 Climate change and agriculture

The agricultural sector (McEldowney, 2020) is particularly vulnerable to the impact of climate change (Foresight, 2011). It is also responsible for a large proportion of United Kingdom and global greenhouse gas emissions of around 10%. Predictions of the climate impact for the United Kingdom are hard to quantify, but projections are of warmer summers, and milder winters as well as more extreme weather events. Changes in rainfall amounts with greater rainfall intensity will impact on food supply, human habitation, and agricultural growth. Food security will be impacted as international weather patterns are relevant to a country that

[s]ENDS Report 532 (June 2019) pp. 20–23.

[t]House of Commons POST NOTE Greenhouse Gas Emissions Removal.

[u]House of Commons Briefing Papers POSTNOTE *Carbon Capture and Usage* (2nd November 2018). Also see UK Parliament POSTNOTE Bio-energy with Carbon Capture and Storage Number 618 (March 2018).

imports at least 40% of the food that it consumes. The legal requirements of the Climate Change Act 2008 and its enhancement through a net-zero target are particularly relevant to the agricultural sector. This has some stark messages for policy-making[v] underpinning the vulnerability of agriculture to weather and climatic conditions.[w] Climate change will affect the conditions necessary for growing fruit, vegetables, cereals, and livestock. The availability of water will affect the sustainability of farms and human settlements. Outside the UK temperature rise of 2°C or more will affect cereal crops in tropical areas, whereas in temperate areas there may not be so much of a difference. The global estimation is that 1°C rise in global temperatures will mean reductions in yields of wheat 6%, rice by 3.2%, and maize by 7.4%. This puts at risk millions of people in Africa, South Asia, and raises doubts about the sustainability of many populations. Global climate change is a likely catalyst for changes in agricultural production. This will lead to changes in the availability of fruit and vegetables. The significant impacts on agriculture underpin the importance of international collaboration in tackling climate change. The immediate impact will be the need to produce more food from land and emitting less greenhouse gas emissions.

Agricultural emissions set a difficult challenging. The main emission are nitrous oxide, methane, and carbon dioxide. Nitrous oxide arises from microorganisms in soils when fertilizers are applied to land and also from the deposition of ammonia. Methane is released through fermentation in the digestive systems of ruminant livestock, and the anaerobic breakdown of stored manure and slurry. It may also be caused by the decomposition of organic matter in flooded agriculture lands, including rice paddies. Carbon dioxide comes from land use change such as deforestation and draining water from peat and crop storage systems.[x]

The legal means to address such challenges involve the application from various adaptation and mitigation techniques that seek to reduce emissions, change demand for food production, and change agricultural practices. There are few workable options and the regulatory system has to be supportive of such measures. This is only possible through political consensus and robust policy-making.

## 4 Climate change and fisheries

Fisheries are an important source of food and healthy marine ecosystems are essential for the maintenance of food stocks. Marine ecosystems are exploited by marine fishing and aquaculture. They provide a staggering 11% of global average animal protein. Overfishing has resulted in declining catches and different forms of aquaculture outstrip global fish production. Furthermore, demand is rising and supply is diminishing. Fishing is a resource that is particularly vulnerable to climate change and greenhouse gas emissions. Fishing and fishery management struggle to address growing problems. In the United Kingdom, estimates vary

[v] House of Commons Briefing POSTNOTE *Climate Change and Agriculture* Number 600 (May 2019).

[w] See: European Parliament, *What Europe does for me: Conservation Farmers* EPRS EN-B95. (July 2018).

[x] Globally nitrous oxide is 38% from soils, methane is 11% from rice paddies and nitrous oxide and methane is 7% from manure management. The remaining 12% of Greenhouse Gas Emissions arise from biomass burning and an estimated 20% to 50% emissions come from deforestation expansion.

but in 2017 £1.53 billion was added to the UK economy with employment to 23,000 people in the industry. United Kingdom catches amount for about £980.1 million annually but UK consumption is largely import driven.[y]

## 5 Climate change and housing

Housing is responsible for 13% of emissions and more 22% if electricity is included. Climate change strategy[z] in the United Kingdom requires several legal targets, under the Climate Change Act 2008, to improve energy efficiency, upgrade all fuel poor homes and bring such homes within the higher level of Energy Performance Certification at Band C by 2030. Various strategies have been adopted including an October 2017 Clean Growth Strategy. Various policies are involved to bring these targets together. Specifically, there is an Energy Company Obligation for energy suppliers and installations especially for people with low income. There is the Green Deal offered as an energy saving scheme for the future, although this was stopped in 2015 because of low uptake. This has been replaced by the Future Homes Standard as part of Building Regulations, which is intended to improve installation. Energy policy is devolved to different nations giving the opportunity for diversity in meeting the various obligations.

The framework for net zero homes proved to be too ambitious and in 2016 the aim was canceled. However, as an alternative government adopted the Clean Growth Strategy in October 2017. This envisages improving energy efficiency through low carbon heating. There is an allied ambition to end the use of wood burning fires and greening the gas grid. Use of electricity is seen as the way forward. Significantly, there are plans to retrofit the wider housing stock, though it will be hard to estimate costs and effectiveness. Significantly, heating will move from gas or carbon burning to electricity. The Committee on Climate Change have concluded in their 2019 progress report that UK action to curb greenhouse gas emissions is lagging behind what is needed to meet legally binding targets.[aa] Despite some setbacks, any future strategy to limit greenhouse emissions will have to develop solutions for the efficient home heating and insulation.

## 6 Climate change and aviation

The aviation sector has a growing impact on climate change.[ab] Flights departing from the UK account for 7% of global greenhouse gas emissions. Recent studies have shown that 96%

[y] Houses of Parliament Climate Change and Fisheries POSTNOTE Number 604 (June 2019).

[z] House of Commons Library, Briefing Paper Housing and Net zero Number 8830 (22nd February 2020).

[aa] Committee on Climate Change 2019 *Progress Report*. Also see: House of Commons Library POSTNOTE Decarbonising the Gas Network (15th November 2017).

[ab] See: UK Parliament, *Climate Change and Aviation* POSTNOTE number 615 (February 2020). IPCC (1999) *Special Report on Aviation and the Global Atmosphere.*

of emissions are from international or long-haul passengers. This needs to be set in the context of passenger numbers tripling since 1990 and forecast to further increase by up to 49% between 2018 and 2050. If other sectors decrease emissions, this will leave aviation potentially as the largest contributor. Air pollution and noise are significant impacts of air travel. The United Kingdom is a major player in the aviation sector, through aircraft manufacture, fuel producers and navigation service providers as well as running airlines. The global picture is that projections about air travel is its likely to increase, possibly quadruple with the resultant emissions tripling by 2050. Limiting the aviation sector in terms of climate change impact is not easy for the United Kingdom. Various new technologies may be useful but new technology is unlikely to meet zero by the target date of 2050. Aviation emissions are managed by the International Civil Aviation Organisation (ICAO) which operates a Carbon Offsetting and Reduction Scheme for International Aviation as a global carbon offsetting scheme for airlines. It also sets standards for the maximum allowable fuel burn per kilometer of flight. The EU has its own reduction measures.

The UK target set at "net zero" emissions provides a legal framework to incentivize changes in the aviation sector. However, currently, UK emissions from international air travel are excluded and the Government has yet to indicate the policy choices going forward. The Committee on Climate Change has argued that aviation should be explicitly included in the UK's net zero emissions. Since 2018, the Committee argued that the way forward was to increase efforts to mitigate emissions from aviation. Technologies on offer include re-design to achieve more fuel-efficient engines, the adoption of low-carbon aviation fuels and the advancement of the use of electrical powered aircraft. Various hybrid aircraft engines are in development. Biofuels and elector-chemical fuels as well as the use of hydrogen may offer solutions.[ac] There are also various operational and airspace management strategies that change aircraft flight operations to conserve fuel and emit less pollution. Improving the efficiency of air routes and avoiding areas of high population where large vapor trails formed from condensation may be avoided and this may also contribute to savings in emissions. The legal target of "net zero" may also be met by managing demand for air travel. Air passengers and cargo might be studied to ensure that better transport options are available. Although there are plans to expand existing airports, questions about the growth in emissions from other sectors such as road transport linked to airports will need to be considered. A re-evaluation of the impact of a "net-zero" target will have a considerable incentive to re-think current transport policy-making. Legal frameworks include setting fiscal rules to encourage best practice and incentivize reductions in carbon usage. The cost of emissions may be incorporated into taxes on jet fuel, airline ticket prices and incentivize the development of alternative fuels. Jet fuels used in international flights are tax exempt under the 1944 Chicago Convention on International Civil Aviation and bilateral air service agreements. There maybe exceptions such as the Air Passenger Duty that charges passengers depending on flight distance and class of travel.[ad] Such taxes are regressive- they effect less-wealthy passengers more than well-off travelers. It is hard to ignore demand will be lessened by one country alone, but will require international co-operation. Proposals are also made to reduce the number of

[ac]See: UK Parliament POSTNOTE Low-Carbon Aviation Fuels number 616 (February 2020).

[ad]Sweden and France have used airline ticket taxation systems.

short-haul flights, however this may not be possible if there are no reliable alternative transport arrangements. It might be possible to link together different forms of transport to mitigate emissions.

The projected increase in demand for aircraft travel may make emissions offsetting the best option. There is a voluntary scheme starting in 2021 until 2027 as part of the ICAO's initiative but this stops short of setting ambitious targets and it is vulnerable to political intervention. Offsetting is a problem as it is hard to monitor and regulate and incentives through funding an offsetting scheme, may make it ineffective and it still permits aviation to grow. It might be desirable to co-ordinate any offsetting and to make international obligations effective through cross-country support. The EU operates the EU Emissions Trading Scheme (EU ETS) that has applied to all flights since 2012. It is an ambitious target of reducing emissions per kilometer flown by 75% by 2050. Currently, controlling UK aviation emissions awaits policy decisions and technological innovation. The legal framework will have to facilitate emission reduction in the future as explained in the recent Supreme Court decision taken by Friends of the Earth (see: [2020] UKSC 52).

## 7 Climate change and shipping

Shipping also contributes to climate change[ae] through the emission of greenhouse gases. Recognition of the challenges posed by shipping has been slow and responses erratic. However, the UK Government has drawn up a *Clean Maritime Plan* (Department for Transport, 2019). The development of carbon budgets makes it impossible to ignore such emissions. Shipping emissions (IAS) have to be estimated and the UK's share appropriately counted. The Committee on Climate Change has argued that such an inclusion is necessary. The Committee believes that it is feasible to include zero carbon for shipping through the use of alternative fuels such as zero-carbon hydrogen or ammonia. This is achievable through new ship-building and modernization programs. The role of the International Maritime Organisation is important and through international shipping agreements it might be to reduce emissions by international efforts. Maersk, one of the largest shipping companies has set its goal of reaching carbon neutrality by 2050.The incentives to include all IAS emissions, including shipping in common targets is that it will intensify efforts throughout the transport industry to meet targets. Shipping has also some non-$CO_2$ effects from the release of sulfur dioxide which causes local air pollution but has a cooling effect on the climate. There is a wide range of options available to reduce shipping emissions. Improving fuel efficiency and reducing water resistance of ships through effective coating of the hull, recovering waste heat and using different forms of propulsion are all possibilities. It is also possible to modify ship operations and reduce fuel use as well as employ software to plan and navigate the most efficient routes. The potential for alternative fuels, particularly the use of hydrogen is likely to be important. There is also the potential for the use of biofuels as there are other more effective alternatives. Finally, there are electrification options but this is limited at present. The UK government's *Clean Maritime Plan* (2019) also includes supply chains as part of the overall

[ae]See: House of Commons Environmental Audit Committee, *Reducing $CO_2$ and other Emissions from shipping* HC 528 Session 2007–8.

strategy. This includes refueling plans and co-ordinated approaches to developing port infrastructures. It is also possible to create a market in the sale of greenhouse gas emissions in shipping alongside other greener policies.[af]

## 8 Climate justice

Parallel to mainstream governmental institutional initiatives are a variety of informal and voluntary movements and pressure groups. Friends of the Earth, Greenpeace, Client Earth and various campaigns for climate justice are some of the examples of pressure groups. The Climate Justice Movement locates the protection of the climate in the development of human rights as central to their approach. The main argument is that the least developed countries, those that contributed the least to greenhouse gas emissions, will be most affected by climate change. It is a paradox that is hard to address. Climate justice suggests that environmental justice may be achieved through the equitable distribution of environmental risk, the recognition of diverse needs and the management of environmental policy through political processes. *The Mary Robinson Foundation* (Mary Robinson Foundation, 2020; Robinson, 2018) set up by the former President of Ireland, is an example of another such pressure group involved in human rights aspects of climate change. The Scottish government set up a Climate Justice Fund in 2012–17, but other parts of the United Kingdom have not followed suit.

## 9 Climate change litigation

One response to climate change involves various forms of litigation.[ag] Climate change litigation has become more significant over the last decade and is particularly prevalent in the United States and Australia (Baxi, 1986; Kyritsis, 2012; The Institute of Race Relations, 2017; Peel and Osofsky, 2015). It has gained some acceptance in the United Kingdom. Many see (Robinson, 2018) the role of the courts as a means of interest groups and individuals accessing enforceable outcomes. Generally, climate change litigation focuses on the reduction of greenhouse gas emissions, and providing a means to encourage adaptation. It may also assist regulatory bodies in clarifying legal obligations as well as stimulate a change in the norms and practices of business. A recent survey of climate change litigation, carried out by the United Nations, found that it has increased to a historically high level:

> As of March 2017 climate change cases had been filed in 24 countries (25 if one counts the European Union) with 654 cases filed in the US and over 230 cases filed in all other countries combined.

Many more cases have been filed in Australia than any other non-US country. The United Kingdom and the EU, individually have seen about half as many as Australia. There are a number of aspects that motivate litigation such as holding government to account for their

[af] Letter from Lord Deben, Chair of the Committee for Climate Change, to the Secretary of State for Transport, *Net Zero and the approach to international aviation and shipping emissions* (24th September 2019).

[ag] House of Commons Library: Briefing Paper, *Brexit and air quality* Number CBP8195 (21st May 2019).

policy making and commitments; making a link between the use of resources and climate change to improve resilience; establishing particular emissions that cause an adverse climate change impact and applying a broadly defined liability to meet the demands set by climate change; and finally an emerging conceptual framework of public trust applied to climate change. This suggest that the good of the community may become a basis for the litigation of the state when the latter is said to be in breach of its duties.

Many reasons for the increased use of climate change litigation can be advanced such as a changing attitudes to climate change in the midst of variable and unpredictable severe weather events. More than any other phenomenon, this has helped create a link between climate change and adverse effects such as droughts or flooding. The persuasive advocacy of some pressure groups and NGOs has also helped put the spotlight on climate change issues.

It is estimated that the number of national climate change laws and policies has increased remarkably as standards over the whole area of environmental protection have increased. Water, air and land are comprehensively protected in legislation that sets standards and operates a polluter pays principle and applies the precautionary principle.[ah] One survey in 2012 found that over 177 countries had laws that regulated the right to a clean or healthy environment. Included in the study were numerous court cases and decisions that needed to be taken into account for their influence in protecting the environment. There are instances of changes in the politics of the environment that lead to constitutional change. In *Greenpeace v Norway* in 2014 there was a change to the basis of the Canadian Charter of Rights to include environmental issues. More subtly the election victory of Mr. Leghari in Pakistan came about through an environmental pledge to enact new guidance to protect the environment. The Paris Agreement with its expectation of only a 2 degrees centigrade rise in global temperatures has enabled litigation to be taken that centers around this expectation as a means of examining government information and company systems. Reliable data is likely to help as does improved understanding of the causes of carbon emissions. The ability to track pollution country by country is very important accompanied by the scientific determination of the main companies that are polluting. Being able to draw on this data provides a foundation for seeking liability in court. It makes campaigning more effective and targeted and also raises the ability of customers to make choices that may lead to companies having to complying with a demand for greener credentials.

It is relatively early days in the use of litigation to influence policy making. Long running court hearings alleging carbon emissions have contributed to global warming and resulted in large scale natural disasters have proved difficult to sustain and have not resulted in large-scale damages. Legal battles in the United States from 2005 to 2012 failed to achieve damages over claims that Hurricane Katrina was caused by 34 carbon emitters. Procedural obstacles such as showing standing in the case or that any injury could be traced back to the defendants proved impossible to overcome. This may not always be the case. As scientific data improves attributing climate impacts is likely to be easier, supporting the use of systems of liability. It is

[ah] Article 191(2) TFEU makes clear that Union policy on the environment shall aim at a high level of protection taking into account the diversity of situations in the various regions of the Union. It shall be based on the precautionary principle and on the principles that preventative action should be taken, that environmental damage should as a priority be rectified at source and the polluter should pay. Such principles are part of EU law.

certainly the case that the insurance industry is able to make climate change impact a basis for calculating insurance rates.

The role of the International Panel on Climate Change and its scientific data has become critical as a source of information that may be useful for litigation strategies. Various cases taken by Greenpeace in Norway have been possible because of better data. Innovation abounds, an example is the argument that rising sea levels caused by carbon emissions gives rise to liability because of trespass to land. One model in use is based on the litigation over cigarettes, with agreement that energy firms are aware of the harm caused by their action and should be liable. Defensive positions taken by companies include the delineation of country liability in Treaties rather than the companies themselves. This is an attempt to provide immunity from any liability based on the use of oil and gas resources by companies. Limitations exist on the ability of courts to liberally interpret their constitutional arrangements. Often the courts will see that regulators are primarily responsible for regulating the environment, removing the courts' role to enforce environmental law. Time will tell if litigation strategies are going to be effective and create change.

*The Economist*,[ai] a leading UK weekly periodical has noted that the potential for environmental litigation is beginning to be transformed by the work of scientists in tracking the carbon emissions from America and the European Union that raise the frequency of devastating heatwaves in Argentina by at least one third. The use of scientific estimates of probability and risk align with the methodology of courts when assessing liability. Cases involving workers protection and liability are often based on risk assessment that incorporates the extent of hazard and likelihood of exposure. In one high profile case in the United States *Massachusetts v Environmental Protection Agency*,[aj] the challenge was to the role of the Environment Protection Agency in a claim that it abused its powers in failing to regulate greenhouse gas emissions appropriately. Such claims excite possibilities that there is a new era whereby environmental liability will influence the path of climate change. A pattern is emerging showing that climate change litigation is increasing. In Pakistan, the Lahore High Court[ak] found that the National government's delay in the implementation of the country's climate change framework, violated the fundamental rights of citizens. In another case, the Hague District Court[al] was critical of the Dutch government's failure to adopt tougher measures to reduce greenhouse gas emissions. Both cases proceeded on the basis that variations of a rights analysis might support setting standards and applying them through environmental protection strategies.

Despite limitations, the rise in environmental litigation[am] is likely to continue. Concerned citizens[an] will seek redress as the impact on their lives of environmental harm associated with

[ai] *The Economist* 4th November 2017 pp. 67–68.

[aj] 549 US 497 (2007).

[ak] *Asgar Leghari v Federation of Pakistan* W.P. No 25501/2015.

[al] *Urgenda Foundation v The State of Netherlands* C/09/456689/hA ZA 13-1396.

[am] *R(Greenpeace) v Secretary of State for Trade and Industry* [2007] EWHC 311 the application of the Aarhus Convention on consultation should be applied to the Government's plans for nuclear electricity production – the Convention is often seen as advisory but it was used as a standard setting measure.

[an] *Ex parte Greenpeace* (No.2) [2000] Env LR 221 The High Court held that the Habitats Directive should have been applied beyond territorial waters partly based on EU Directive.

climate warming increases. Flooding due to increased extreme climate events and air pollution are two examples where there are tangible impacts on individuals.[ao] Client Earth has managed to successfully litigate in the United Kingdom over excessive air pollution that contravenes the EU Directive on Air Quality.[ap] The case is an ongoing example of how litigation may monitor, adapt and change government policy making.[aq] Various studies on air pollution in London and capital cities show that air pollution has reached sufficiently high levels to result in a detriment to public health. A recent study at King's College[ar] showed that air pollution could amount to a "mortality" burden of the equivalent of 9500 people per year. The World Health Organisation is clear that there is an established link between air pollution and the risk of disease and premature death. Legal limits have been breached and the risk to health that is evidenced from medical opinion is well established.[as] European Legislation in 2008, the ambient air quality directive 2008/50/EC set legal limits for major air pollutants such as nitrogen dioxide and various particulates. In England the directive was enacted under the Air Quality Standards Regulations 2010. Various standards in the United Kingdom were evaluated and it was discovered that the United Kingdom did not meet the standards set for nitrogen oxide or particulate emissions. Studies, using various monitoring stations was carried out and in 40 out of the 43 zones the United Kingdom was in breach of the regulations. In 2013 the UK Supreme Court heard a case brought by a pressure group,[at] Client Earth, established that the United Kingdom was in breach of Article 13 of the Air Standards Regulations and requested guidance from the Court of Justice on the appropriate guidance to Member States. This led to a second judgment, requiring the Government to draw up further plans to meet EU rules by the end of 2015.[au] The time has passed and during 2017 there have been further returns to court to ask the court to ensure that policy making will be in alignment with the requirements of the Air Standards Regulations. In addition, recently, a High Court

[ao] The High Court, in July 2009, found that there was a credible link between these birth defects and the exposure of the mothers during pregnancy to contaminants from the clean-up of the Corby Steelworks.

The link between the closure of the steel works in 1980 and the birth defects arose because of the removal of large amounts of toxic material containing cadmium, chromium, nickel, dioxins and polycyclic aromatic hydrocarbons from the site. The method of removing the contaminated soil was by road, partly through a housing estate and had not been subject to an appropriate risk assessment. Mr. Justice Akenhead held that Corby Council had been "extensively negligent" and that the link between the birth defects and the contaminants "could realistically have caused the types of birth defects of which the complaint has been made".

[ap] *Client Earth litigation v. Defra* [2015] UKSC.28.

[aq] The continued failure of the UK government since 2010 to tackle nitrogen dioxide emissions under the EU Directive 2008/50/EC- these regulations are in UK law through the Air Quality Regulations 2010.

Limits were set under the Directive for the "promotion of human health" setting certain limits for nitrogen dioxide which should not be exceeded. Most of our major cities including London breach the limits for nitrogen dioxide. Governments are given time to put in place plans to meet the requirements set by the Directive. The UK created strategic plans but these did not meet the requirements of the Directive and were unlikely to do so.

[ar] House of Lords, *Air Quality and Health in the UK* LIF 2015/024 (18th November 2015).

[as] House of Lords, *Air Quality and Health in the UK* LIF 2015/024 (18th November 2015).

[at] *R (Client Earth) v Secretary of State for DEFRA R (Client Earth) v Secretary of State for DEFRA*[2013] UKSC 25.

[au] *R (Client Earth) v Secretary of State for DEFRA* [2015] UKSC 28.

judge has supported a planning inspector's decision to dismiss the council's housing development in Kent due to its likely impact on air quality. This is an unprecedented[av] example of the role of courts through judicial review procedures and access to justice on influencing environmental standards. Client Earth has continued to press the Government to take action in terms of 45 local authorities in England that have illegal levels of pollution, as well as action to be taken in Wales. The aim of the latest litigation is to require the Government to supplement existing plans and enhance their effectiveness.[aw] It is clear that not all litigation addresses climate change directly, but it will lead to improvements in environmental standards with indirect effects on climate change.

Perhaps the most significant case on climate change was decided in the Court of Appeal in February 2020.[ax] The case involved the Airports National Policy Statement (ANPS) prepared by the UK government for the South East of England in 2018. The policy is part of the regulatory arrangements under section 5(1) of the Planning Act 2008. Challenges are permitted under section 13(1) of the Planning Act. The challenges taken by a number of pressure groups and the Mayor of London relates to the proposed expansion of Heathrow Airport by building a third runway. The challenges included the application of the Habitats Directive and also the related Strategic Environmental Directive. Both challenges failed, however the challenge on the basis that the government's commitment to addressing climate change as a result of the Paris Agreement was accepted. It was held that the ANPS had failed to take into account the Paris Agreement in mitigating and adapting to climate change. The Court of Appeal concluded that ANPS had not complied with the legal requirements set by Parliament. The Paris Agreement ought to have been fully taken into account. A declaration was granted by the court. The main reason was that the UK's commitment to reduce carbon emissions and litigate climate change had not been taken into consideration. The government may be able to re-visit the issue and amend the ANPS to make it compatible with the Paris Agreement. However, the government decided not to appeal the case. It seems that Heathrow Airport may seek leave to appeal which they did and succeeded in reversing the Court of Appeal decision in the Supreme Court. The Supreme Court concluded that the Government had given adequate consideration to climate change emissions and sustainability. The case suggests that courts in the future might be less willing to review policy making decisions made by central government

Many view the development of climate change litigation as engaging human rights, a link that was recognized in 2008 by the UN Human Rights Council. Adopting a human rights approach is certainly a promising way forward but rights and the environment have for many years promised much only to lead to disappointment (Cox, 2012). There are possibilities that rights litigation may be successful, for example through, indigenous peoples making claims regarding their communities. Such claims, however, do not always succeed and in some cases litigation may not be productive or much good. More important will be the opportunity to broaden the development of climate change litigation to provide effective remedies based on risk and the application of the precautionary principle. This avoids the pattern of human

[av] High Court makes first planning ruling over air quality ENDS (13th November 2017).

[aw] Client Earth 4th December 2017.

[ax] *R(Friends of the Earth) v Secretary of State for Transport and Others* [2020] EWCA Civ. 213.

rights cases to be mired in domestic systems where rights are often used as an impediment to change, given the relatively few environmental cases that are heard.

Even the most optimistic admit that the regulatory systems relevant to climate change are relatively loosely drawn with little directly enforceable principles. Litigation is often premised on past mistakes and retrospective analysis. Prospective litigation based around predicting or assessing the future will always have limitations in directing policy making choices.

## 10 Conclusions

Law, through legislation and regulation provides an important means of taking forward policy choices and providing an enabling platform for implementation of strategies. Law is both a catalyst for change as well as a facilitator of policies and may aid debate, discussion and public awareness. In many important and significant ways, law can also reward good behavior as well as penalize bad practices and infringement of the rules. Law may provide a mechanism to advance economic instruments such as taxation. The question is how effective is law likely to be in addressing the problem of climate change? The answer will depend on a number of uncertainties in the law at present. The UK's legal and regulatory arrangements for climate change are intertwined with the UK's membership of the European Union which ceased at the end of January 2020. The United Kingdom is currently in the transition stage of leaving the EU leading to a final agreement at the end of December 2020. It is difficult to assess the impact of leaving the European Union will have on the UK's climate change strategy. Any drop in economic activity will have a consequence for carbon emissions. However, the extent of future regulatory and legal alignment with the EU and the policy making options to be adopted after the transition arrangements end in 2020 is far from clear. The Fisheries Act 2020 and the Environmental Bill 2020 will be influential in setting the legal arrangements for important aspects of climate change. The recently agreed EU-UK Trade and Cooperation Agreement December 24th 2020 includes UK and EU agreements on emission standards and Climate change strategies (see: [2020] UKSC 52). The aftermath of the COVID-19 pandemic may stimulate economic revival through a greener and fairer economy that addresses climate change.

## References

Baxi, (Ed.), 1986. Inconvenient Forum and Convenient Catastrophy: The Bhopal Case. Tripathi, Bombay.
Bell, S., et al., 2018. Environmental Law, nineth ed. Oxford University Press, Oxford, pp. 561–562.
Cox, R., 2012. Revolution Justified. Stichting Planet Prosperity.
Department for Transport, 2019. Clean Maritime Plan.
Fisher, E., 2012. The rise of transnational environmental law and the expertise of environmental lawyers. Transl. Environ. Law 1, 43.
Foresight, 2011. International Dimensions of Climate Change. The Government Office for Science, London.
Jewell, T., Steele, J., 1983. Law in environmental decision-making. In: Steele, J., Jewell, T. (Eds.), Law in Environmental Decision-Making: National, European and International Perspectives. Oxford University Press, p. 3.
Kyritsis, D., 2012. Constitutional review in representative democracy. Oxf. J. Leg. Stud. 32 (2), 297–324.
Mary Robinson Foundation, 2020. Principles of Climate Justice.

McEldowney, J.F., 2018. Public Law, fourth ed. Sweet and Maxwell, London.
McEldowney, J., 2020. EU Briefing Parliament, EU Agricultural Policy and Climate Change. March 2020. European Parliament.
McEldowney, J., McEldowney, S., 2011. Science and environmental law: collaboration across the double helix. Environ. Law Rev. 13 (3), 169.
McEldowney, J.F., Salter, D., 2016. Environmental taxation in the UK: the climate change levy and policy making. Denn. Law J. 28, 37–65.
Peel, J., Osofsky, H., 2015. Climate Change Litigation. Cambridge University Press, Cambridge.
Robinson, M., 2018. Climate Justice. Bloomsbury, London.
The Environment Agency, 2018. Climate Change Impacts and Adaptation. The Environment Agency, London.
The Institute of Race Relations, 2017. Humanitarianism: The Unacceptable Face of Solidarity. The Institute of Race Relations, London.

CHAPTER

# 23

# The ethics of measuring climate change impacts

*Kian Mintz-Woo*[a,b]

[a]University Center for Human Values and Princeton School of Public and International Affairs, Princeton University, Princeton, NJ, United States [b]Department of Philosophy and Environmental Research Institute, University College Cork, Cork, Ireland

OUTLINE



## 1 Introduction

On the one hand, estimating the economic costs of climate impacts involves physical, social, and moral uncertainties, making estimation difficult and sensitive to assumptions. On the other hand, without estimates of those costs to guide policy, we would be adrift with respect to optimal policy choice. In particular, if trying to generate the appropriate balance of costs and benefits from marginal emissions, we need some estimates of the effects that those emissions will cause.



https://doi.org/10.1016/B978-0-12-822373-4.00023-9

There are ways around requiring estimates of these economic costs. Instead of the more comprehensive *cost-benefit analysis* (CBA), we could adopt *cost-effectiveness analysis* (CEA), where we adopt some exogenously given intended target (e.g., from a political decision-maker), and the question is how to most inexpensively reach that target. For instance, we might think that it is important to get to net-zero greenhouse gas emissions by 2050, and the question is just how to get there with least cost (CEA) instead of whether, all things considered, that is the optimal or appropriate target (CBA) (Kaufman et al., 2020).

However, this is insufficient for more ambitious intentions such as if the goal is to internalize the climate externality from emissions (Mintz-Woo and Leroux, forthcoming). For instance, if we wish to introduce optimal carbon taxes, this will require estimates of the economic costs of these climate impacts. Carbon taxes are influential, and also interact with other conditions and circumstances (e.g., including the coronavirus disease 2019 (COVID-19) context (Mintz-Woo et al., 2021)).

The primary economic concept for these costs is the *social cost of carbon* (SCC), which is an estimate of the cost to society of a marginal ton of $CO_2$ (i.e., the SCC is usually expressed in terms of USD/t$CO_2$) (Fleurbaey et al., 2019).

One of the key tools used to estimate the SCC are *integrated assessment models* (IAMs) which combine simple climate modules with simple economic modules. For concreteness, this chapter will focus on models similar to *DICE-2016R2* (DICE stands for "Dynamic Integrated Climate-Economy") models (Nordhaus, 2018a) but, as the discussion here is more methodological than quantitative, they apply to many similar models (Nordhaus, 2018b). These models are sets of equations which try to represent the social and physical implications of perturbing the climate system. However, they involve both simplifications and assumptions which require moral consideration and evaluation. Providing these is the purpose of this chapter.

To start with, we can set aside the physical components of these models. They are relatively simple: an atmospheric stock of $CO_2$ drives temperatures and $CO_2$ decays as it settles through ocean layers (Dietz and Stern, 2015). The issues of interest here arise in the factual and moral assumptions. Section 2 discusses several key factual assumptions pertaining to the economics. Here, "factual assumptions" mean economic phenomena that are in principle observable without moral evaluation. Section 3 discusses the moral assumptions which have to do with how the states of the model are valued, where this can be subject to, inter alia, how wide the scope of evaluation is (e.g., does it consider regional/national or international impacts) or how future impacts are to be traded off against current ones. Section 4 concludes with an overview of the moral importance of considering modeling assumptions.

## 2 Factual assumptions

This section addresses several key structural economic assumptions, identifying them as factual (or nonmoral) because they are in principle observable given sufficient macroeconomic or population information. The difference between considerations in this and the next section is sometimes expressed as the difference between positive and normative economics, or between descriptive and prescriptive economics, where the former is meant to be based on observed values and the latter involves moral judgment (Arrow et al., 1996; Manne, 1995). For instance, we can observe that the mean member of a population saves at a particular rate which implies a certain level of discounting or care about long-term consumption behavior, but for prescriptive or normative purposes that does not settle the normative question of how society *should* or *ought* to invest or save for the future (Broome, 1994, 2008).

The assumptions in this section are not normative or prescriptive in this sense; they are in principle observable in an unproblematic way. That is not to say that they are *easy* to observe. It may be that it will take a long time either to analyze or to collect the relevant information. For instance, with damage functions about responsiveness to climate change we may have to make decisions before we have observed the relevant harms and before we are able to base the damage functional forms on observed data. However, it is still factual in the sense adverted here because afterwards if we had sufficient data, we could in principle observe the impact of various mean surface temperatures on consumption and production—and we could do so without making moral judgments.

The factual assumptions discussed here are as follows:

1. Consumption—its measurement and homogeneity.
2. Equality of consumption—the distribution of consumption and ways of relaxing this assumption.
3. Endogenous versus exogenous growth—the total factor productivity growth and climate impacts on it.
4. Damage functions—the functional form of various damage functions.

## 2.1 Consumption

One key structural assumption in most models is that all costs and benefits to the public at large are expressed in terms of consumption equivalents. Whatever goods and services are being demanded, there is a homogenous good which we call *consumption* (usually represented $c$) that it can be converted into. This is necessary because climate impacts are not directly expressed in currency; the SCCs require the assumption that those impacts can be expressed in monetary equivalents. In practice, consumption can be measured or estimated by using a financial currency like dollars which are fungible and can be used to purchase goods of various types.

Some environmental theorists reject this assumption. They point out that some of the systems affected by climate change, like wetlands and tropical forests, provide *ecosystem services*, or contributions of the natural world to goods people value. These ecosystem services, they worry, are either (i) difficult to price in markets, and thus to convert into consumption via financial currency, especially when they are taken to have intrinsic value; or (ii) difficult to exchange for other forms of consumption in principle. (Similar concerns have been raised with respect to pricing outcomes in terms of mortality and morbidity.)

In terms of the first difficulty, pricing goods, it is correct that ecosystem services are rarely priced in markets, and if reflected in market prices, usually indirectly. However, familiar valuation methods from economics can be used even for environmental and ecosystem services, ranging from adjusted market prices (where market prices are adjusted for distortions); production function methods (where the ecosystem service is taken to be an input to a production function and valued according to its effect on output); revealed preference methods (where some choices are observed and valuations inferred from the selections made); and stated preference methods (where choices are posited and valuers indicate their preferences in surveys) (Atkinson et al., 2014).

In terms of the second difficulty, substitutability, it is true that there are limited substitution options for some ecosystem services. However, this is standardly accommodated by positing rapidly increasing marginal value as the services are reduced or degraded (Atkinson et al., 2014). However, some believe this is not enough. They distinguish between weak and strong sustainability, where *weak sustainability* indicates that we can in principle substitute between consumption and natural capital providing ecosystem services and *strong sustainability* indicates that the damages to ecosystems and services are in principle not substitutable with respect to other types of consumption [e.g., Norton distinguishes between "natural capital" and "human built capital" (Norton, 2005)]. Doing so requires more than simply thinking some ecosystem services are very valuable; it requires the stronger claim that without some (minimal) level of ecosystem services, there can be no meaningful type of well-being, regardless of the level of human built capital. Intuitively, there is no level of human consumption that can make up for degraded natural capital. An influential economic analysis which distinguish between human built capital and environmental capital finds that this distinction can greatly increase the SCC (Sterner and Persson, 2008).

## 2.2 Equal per capita consumption

Even granting that we can measure the climate impacts in terms of a single homogenous good, which we call consumption, many IAMs further simplify by assuming that the good is spread evenly, or that we are considering a representative median agent, or that the world is represented by a single homogenous region. All of these ways of simplifying avoid the actual heterogeneity of consumption. Taking a representative agent which effectively ignores this heterogeneity may appear to be a worrying structural assumption.

There is an important reason that equal per capita consumption is assumed. First, each agent has a utility function which translates consumption to utility. If consumption can be costlessly shared and utility functions are identical (both idealizing assumptions), then the most preferable distribution of consumption is a uniform distribution. Section 3.3 expands on these utility functions, which are key moral assumptions in the model.

Given these identical agents and uniform consumption distributions, it is more tractable computationally to have a single region with a single agent. However, this assumption can and has been relaxed. This assumption can be relaxed by taking an outcome which is equivalently valued to the uniform distribution, or by making the model itself more complex.

The first option is straightforward. For any given heterogeneous consumption in a population, we can find the hypothetical level of perfectly equally distributed consumption that would match it in value. This hypothetical level is the "equally distributed equivalent" (EDE) (e.g., Fleurbaey et al., 2019). In this manner, working with a homogenous population can be thought of as the proxy for all of the heterogeneous distributions taken to be equivalent by the objective function which represents the goals of the model (sometimes a social welfare or value function).

The second option is to explicitly model regional or spatial inequality. First, the DICE model, which uses a single world region, also has a "regional" counterpart called RICE (Nordhaus and Yang, 1996). Second, we can add greater regional heterogeneity by having various income levels within regions, yielding a "nested inequality" counterparts of DICE called NICE (Budolfson et al., 2017; Dennig et al., 2015). The NICE authors find that if we are averse to inequality, that points both *against* sending resources to the (presumably richer) future and *for* sending resources to those worse-off in the future who, on heterogeneous models, may be poor like the present poor even when mean consumption has risen in the future. But the extent of these effects depends on whether the climate damages are proportional to wealth or more or less than proportional. All are plausible; it could be that having greater wealth means that you have more and more valuable capital at risk to various extreme events or it could be that having greater wealth means having greater adaptive and protective capacity so that more capital is shielded from climate impacts.

The key point is that while many climate IAMs are highly simplified, the structural assumptions that yield their simplifications can be relaxed with more data and computing power. With more heterogeneous populations, we can have more complex policy analysis but also more realistic outcomes to evaluate.

## 2.3 Endogenous versus exogenous growth

In IAMs like DICE, growth of consumption is given by a combination of exogenous factors that exponentially increase total production. These factors are the number of producers, or labor, and an (assumed) increasing productivity for any given level of labor and capital. A key assumption in the standard models is that the growth is exogenous, meaning that its increase is mostly independent of the other parameters involved in production. (Mostly, because damage in a year can lead to losses in future productivity by, for instance, reducing the amount of savings. But this is a relatively indirect mechanism for climate impacts to harm growth prospects.)

In response to this, Dietz and Stern (2015) suggest an extension of DICE which incorporates endogenous growth by expanding the ways that climate impacts could harm growth. First, they suggest that it could either reduce the capital stock, via mechanisms like loss of built environment in the face of coastal incursion, or could reduce the productivity of capital, via mechanisms like poorer infrastructure performance in changing environmental conditions. Second, they suggest that climate impacts could reduce the total factor productivity. In other words, given the same capital and labor inputs in subsequent periods, you have less production. For instance, in a warming climate, people may be uncomfortable and move more slowly leading to less productive economic activity.

Stern (2015) pointed out that modeling growth as endogenous in climate context makes the downside risk much greater. If there are ways that climate could damage the trajectory of economic activity, that is much worse than having annual climate costs, since a single hit to productivity permanently damages growth, whereas the annual climate costs are each temporary.

## 2.4 Damage functions

One of the most difficult structural assumptions involve the functional form and calibration of damage functions. Almost by definition, climate changed impacts will be unlike those previously observed so extrapolation is highly uncertain. Some economists like Robert Pindyck claim that the damage functions are so indefensible empirically or theoretically that they should be viewed as little more than crude guesses (Pindyck, 2013, 2017). This is exacerbated by the limited incentives for improving damage functions in IAMs, since these improvements are often assumed to be insufficiently novel or important to justify publication. Indeed, for related reasons, Nordhaus designated the damage functions "placeholders subject to further research" (Stern, 2013).

The standard reduced form damage function in DICE is quadratic:

$$D_t = 1 - \frac{1}{1 + \pi_1 T_t + \pi_2 T_t^2}$$

where $D_t$ is the damage at time $t$, and $T_t$ is the global mean temperature at time $t$. The parameters $\pi_1$ and $\pi_2$ are estimated by fitting the damage function to various expected costs at particular temperatures. However, since observed temperatures have not varied outside of a range of 2°C for millennia, this fitting is highly speculative. Furthermore, even if points at low temperatures are correct, they underdetermine the functional form beyond those temperatures.

Recent alternative ways of estimating the effects of temperature rise have drawn on large datasets of *local* climate anomalies to distinguish potential effects on human behavior and macroeconomic variables. These new empirical techniques might be more successful than the discussions of the functional forms of damages. For instance, we can run statistical analyses on large-*n* samples (Burke et al., 2015; Carleton and Hsiang, 2016; Hsiang et al., 2017). By comparing longitudinally (comparisons within regions) as opposed to cross-sectionally (comparisons across regions), it is easier to avoid cultural and other factors that might lead to spurious correlations. Roughly speaking, the idea is that by looking at the same regions over longer historical periods, we have a large number of data points which we can analyze in terms of the deviation from historical norms and index to social outcomes of interest (whether social outcomes like violent conflict or macroeconomic factors like productivity). They find various relationships, but several of them are roughly u-shaped, in that deviations both colder and warmer from the mean correlate with increases in the social outcomes of interest, for example. Increased conflict both when it is colder and warmer than average. Of course, even these new methods run up against the predictive limitations of data when future weather is far outside the norm; in these cases, there are fewer historical cases and more uncertainty. Assuming the climate continues to warm, these limits on observation matter.

These ways of estimating the damages from climate impacts help address the factual assumptions involved in damage functions. While more drastic changes in climate will outrun the observed weather events, it is certainly the case that it has remained difficult to improve and calibrate these damage functions.

Having introduced some important factual or structural assumptions, we can now turn to the moral assumptions.

## 3 Moral assumptions

This section discusses moral assumptions and introduces arguments for various choices. The primary intentions are to give a sense of the variety of moral assumptions that are taken for granted in SCC estimates and to introduce some alternatives to the standard assumptions.

## 3.1 Utilitarianism versus other moral theories

The most prominent moral assumption in calculating the cost of climate impacts is the moral theory adopted. DICE is implicitly committed to *(total) discounted utilitarianism*, a particular member of the family of theories called *consequentialist*. Total discounted utilitarianism means that evaluation of actions is conducted solely in terms of outcomes (consequentialism), the value of those outcomes is governed by the sum of utility (where utility is an increasing function of consumption) (total), all the utilities at a time are valued equally (utilitarianism), but future consumption or utility is discounted to generate a net present value (discounted). All of these assumptions can be relaxed. For our purposes, we can be agnostic between various different interpretations of utility, but the classical interpretations are pleasure and satisfaction of preferences, with the latter having been especially influential in economics.

A moral theory that does not evaluate actions purely in terms of outcomes is *deontological*. For these theories, some actions are ruled out as morally impermissible because, for instance, they violate rights or side constraints. That means that an action could have very positive consequences overall but might involve sacrificing or killing an innocent; deontological theories will rule out this type of action as immoral or impermissible regardless of the consequences (or perhaps the consequences are weighted very lightly or are only relevant in terms of tie-breaking). In the context of climate economics, if we are considering actions that might be side constraints which cannot be violated, then IAMs and similar tools are unnecessary. We could model this by increasing the social cost of such actions to very high finite values or infinite values, but these options introduce further complexities having to do with comparing risk with large finite or infinite disvalues (Jackson and Smith, 2006). Alternatively, we can have a two-step procedure where an act is first checked for its permissibility and then subjected to CBA.

Setting such deontological side constraints to the side, we have *consequentialist* theories which evaluate actions purely in terms of outcomes, whether actual or expected. The most common class of consequentialist theories evaluate actions purely by considering outcomes in terms of utility. If the utilities at a time are unweighted in the evaluation, then we have *utilitarianism*. If they are weighted, then we have one of several variations on utilitarianism. Before considering utilitarianism, we can consider two particular ways of weighting utilities which are of especial interest in climate economics. The first because it is commonly done and the second because it yields an interesting class of moral theories. The first is *Negishi weighting*, where subpopulations are weighted in such a way that the weights are proportional to the inverse of the marginal utility meaning that the subpopulational distribution of consumption is optimal, regardless of the shape or pattern of distribution (e.g., Nordhaus, 2008). The ostensive purpose of this is to remove moral evaluation of extant distributions, yielding some type of value freedom (Fleurbaey et al., 2019), but of course Negishi weighting introduces its own value judgments: current distributions, whatever they are, are taken to be morally justifiable (at least from the optimizing utilitarian perspective).

The second variation for weighting utility is where those with lower utility are weighed more heavily. This yields some form of *prioritarianism*, since (moral) priority is given to those who have less utility. In *deterministic* contexts (where the outcome is known or correctly believed conditional on actions chosen), this can be difficult to distinguish from utilitarianism, since utilitarians already give priority to those with less consumption (which we can take to indicate financial means), so prioritarianism just indicates that there is another weighting on top of the utilitarian one. In these contexts, there is an additional moral question: not only do we need to determine the elasticity of marginal utility of consumption (discussed in Section 3.3), but we also need to determine the degree of priority of those at a lower level of utility (Adler and Treich, 2015; Adler et al., 2017).

If we do not weight the utility, then all utility is valued equally (e.g., regardless of who it accrues to). This yields *utilitarianism*. Note that valuing all utility equally does not imply—and under normal assumptions rejects—the claim that all *consumption* is valued equally. That is because it is assumed that one additional unit of consumption generates greater utility when it is given to those with lower initial consumption. This chapter takes the degree of this effect to be a moral issue about how much we are averse to consumption inequality in society, and is discussed in Section 3.3.

However, even if we take all the simultaneous (intratemporal) utility to be equally valuable, it does not follow that we weight future (intertemporal) utility to be equally valuable. If we do take it all equally, we have *total (undiscounted) utilitarianism*. If we weight future utility less when evaluating outcomes, then we have *total discounted utilitarianism*. The next section discusses reasons for both of these positions.

## 3.2 Time preference/discounting

One of the most highly debated moral assumptions in IAMs arises out of the discounting problem, that is, the issue of how (future) costs and benefits are "discounted" to make them comparable in present terms [overviews include (Davidson, 2015; Gollier, 2012; Greaves, 2017a; Mintz-Woo, forthcoming), the last of which this section draws heavily on]. This issue is also discussed in terms of "time preference."

One important point to recognize is that economists tend to talk about discounting in impure, or consumption terms, since they are concerned with more observable and measurable units, whereas philosophers tend to talk about discounting in pure, or utility/welfare terms, since they are concerned with the morally relevant units and consumption is not intrinsically valuable. This has led to some confusion in ethical discussions, as discounting in pure and impure terms are not directly comparable. This is especially challenging in contexts where "the discount rate" is ambiguously invoked.

The key to time preference in standard IAMs is the Ramsey Rule (Dasgupta, 2008; Goulder and Williams, 2012). The way IAMs approach trade-offs over time is governed by the parameters that are used by the Ramsey Rule; these parameters, discussed in this and the next section, are of very significant importance in determining the urgency of addressing climate change because climate change policies mostly involve up-front costs and delayed benefits. The Ramsey Rule involves discounting in terms of the consumption of individuals. It governs the optimal consumption path given some common assumptions—and is technically an approximation. It can be stated in a very compact form which relates the social discount rate $r$ on the left-hand side to what is sometimes called the "social rate of time preference" on the right-hand side.

The social rate of time preference is the sum of the pure rate of time preference (the Greek letter $\delta$, "delta") and a term which is the product of the growth rate of consumption $g$ and the elasticity of marginal utility of consumption (the Greek letter $\eta$, "eta"), which is discussed at greater length in the following Section 3.3:

$$r = \delta + (\eta \times g)$$

Intuitively, discounting of consumption depends on both pure discounting, that is, factors which are not particular to consumption but time itself (or things correlated with time) ($\delta$) and the diminishing marginal value of consumption as consumption grows ($\eta \times g$). It may also be helpful to think of pure discounting as discounting applied in the pure units of utility or welfare, that is, as a utility discount rate. To give a sense of potential values, $\delta$ is usually taken to be between 0 and 0.015 (i.e., 0%–1.5% per year); $\eta$ is usually taken to be anywhere between 0.5 and 4; and $g$ is usually taken to be between 0.01 and 0.03 (i.e., 1%–3% per year).

There are two morally interesting questions regarding discounting. The first is what arguments can be adduced for a positive $\delta$ and whether any of these are convincing; the second is what the role of $\eta$ is and how it should be determined. The first of these is discussed in this section; the second in the following section. Here, the growth rate of consumption $g$ can be set aside, because its value is factual or nonmoral in the sense that if we have sufficient macroeconomic data about consumption patterns, we do not need to make moral judgments to indicate at what rate consumption grows.

While the vast majority of moral philosophers think that pure time preference is immoral [indeed, the creator of the optimal savings approach advocated a zero rate of pure time preference Ramsey (1928)], there are notable dissenters. The common view among philosophers is that the IAMs are utilitarian (or at least consequentialist) and utilitarianism is committed to impartial increase in utility. Impartiality requires treating utility at different times equally so pure discounting is immoral (Broome, 1994; Greaves, 2017a). Here, I will discuss two reasons for questioning the common view.

First, we might think that our connections to those further away are weaker, and Arrow argues that it is reasonable to consider them less heavily since we have "agent-centered prerogatives" to weight them less (Arrow, 1999), drawing on (Scheffler, 1994). This is a *permission*; we are allowed to count them less according to this view since we are able to prioritize our own connections and projects. Arrow claims that this arises because the cost to our own connections and projects would be overwhelming or overdemanding if we did not have a positive pure time preference.

Second, we might distinguish between the present value of the future from the eventual value (Heath, 2017) and emphasize that there are important differences between the future as viewed from the present and at that point. As indicated before, positive values of $\delta$ can be justified for considerations that grow with time. In this vein, one important difference between the present and the future is that the future is *uncertain* relative to the present and, insofar as that uncertainty grows in a predictable manner, that gives a principled reason for discounting in pure terms as a way of introducing that uncertainty (Mintz-Woo, 2019).

## 3.3 Elasticity of marginal utility of consumption

The next moral assumption involves the conversion of consumption into something morally relevant like welfare or utility (which we can take as synonymous). The parameter that governs this conversion, the *elasticity of marginal utility of consumption*, tells us how utility increases in percentage terms given a percent increase in consumption. This is a key parameter characterizing *individual utility functions*.

It is standard to assume that marginal consumption, or equal small additions of consumption, yield decreasing utility (i.e., the function has a positive first derivative or is increasing, but a negative second derivative or increases at a diminishing rate). In IAMs, this parameter can reflect multiple considerations, all of which have moral importance, but we will focus on two that are especially relevant. The first is inequality aversion, where the evaluator is averse to a given unit of consumption being distributed to someone who has greater consumption compared to less consumption. The second is risk aversion, where the evaluator is averse to a given unit of consumption being distributed to an outcome with greater consumption compared to an outcome with less consumption.

While we can observe individual's risk aversion and inequality aversion within their own lives, this assumption is moral because there is no moral reason to privilege observed risk aversion and inequality aversion. A slightly different way of expressing the same idea is that observing how people trade-off between various goods in their own lives gives us no guidance about how to trade-off between various goods across different people (Dasgupta, 2008).

## 3.4 Population ethics

One branch of formal ethics involves the comparison of various populations (Greaves, 2017b). While this is called population ethics, or population axiology, the moral issues are not about optimal population size or number of children, but the comparison of population outcomes. These outcomes may vary in size but also in which individuals are taken to be identical (i.e., the same individual) across the outcomes. Intuitively, there is an important moral difference when outcomes involve substituting a person with another who has greater utility (say) compared to if someone improves the utility of one person from one outcome to another.

In the context of the SCC, we can lay out two versions of utilitarianism depending on different population ethics assumptions and see that the change can have significant effects. The standard form of utilitarianism assumed in IAMs is *total utilitarianism* (or "*totalism*"), where the evaluation of an outcome is the sum total of utility across all agents in that outcome. The most influential alternative is *average utilitarianism* (or "*averagism*"), where the evaluation of an outcome is the average (mean) utility across all agents in that outcome.

While there are many alternative theories in population ethics (Greaves, 2017b), the distinction between total and average utilitarianism is already highly significant. Scovronick et al. (2017) found that the choice between these theories can have as much impact on the SCC as different positions on the discounting problem. More precisely, with a larger population, increases in temperature are relatively worse under total utilitarianism than average utilitarianism, since climate impacts lower the average more slowly than they affect the total value of the outcome, especially when the population is large in expectation.

While this comparison between two theories in population ethics gives us some understanding of the moral assumptions, there are many more theories that could be adopted with a wide array of potential impacts on estimates of the SCC.

## 3.5 Scope of morality

Philosophers often use the term "scope" to indicate the range of morally considerable beings. This is a morally important assumption that is easy to overlook. This section will consider whether the scope should be regional/national ("domestic") or global; the next section will consider whether the scope should be broadened beyond human beings to other sentient or other beings. This section draws heavily on work done in Mintz-Woo (2018).

In March 2017, the Trump administration issued an Executive Order (EO) on Promoting Energy Independence and Economic Growth. Among other things, the EO limits the SCC to only the expected domestic impacts, as opposed to the global impacts.

While the choice between domestic or global SCCs may appear arcane, here are two reasons to think otherwise. First, the actual impact of this rule change is significant. Since greenhouse gases are long-lived and disperse freely in the atmosphere, the effects of that marginal ton fall on many different countries in a highly varied manner; restricting the effects to any single country ignores international impacts, resulting to a lower SCC. Second, it acts as synecdoche for the broader orientation of the Trump administration. The question that the administration poses is whether nationals should care about foreigners.

The morally interesting puzzle is why, as the Obama administration and others contended, carbon regulatory analysis should appeal to (at least some) global costs and benefits, but not analyses of other targets of regulation (Gayer and Viscusi, 2016). Other potential targets include, but are not limited to, terrorism, illegal drugs, and infectious diseases (Gayer and Viscusi, 2017). This is on its face surprising; what could be so special about the SCC when typical US regulatory impact analyses only consider domestic impacts?

This section offers the following response to this moral puzzle, based on reinforcing norms in the face of commons tragedies with observable responses. The relevant fact about carbon is that its negative effects are predominantly global and not domestic. In this sense, considering only domestic impacts would greatly bias or distort estimates of the SCC.

We can see setting domestic and global SCCs as a form of defecting or cooperating, since each time a country considers only domestic impacts, it will underreduce relative to the social optimum and since the optimal outcome for all is that every country considers global impacts. However, climate change is different from other commons problems in two especially important aspects.

First, it is iterated, meaning that instead of a one-off decision, countries can respond to observed decisions of others. Second, those decisions are observable, meaning that we can see whether countries adopt global or domestic SCCs.

In fact, countries which have instituted SCCs have in almost all cases introduced ones with values that are greater than they would be if they included only domestic impacts (Howard and Schwartz, 2017, Appendix B).

In this context, a moral argument in favor of continuing to adopt global SCCs lies in the minimal principle that, when facing iterated commons problems where other members are known to be cooperating, in order to generate and reinforce cooperative norms, one also ought to cooperate. This is minimal, since it does not require that one cooperate first. It is also weaker than moral claims made in the literature that, in light of financial capacity and historical emissions, countries like the United States morally ought to *lead* with respect to climate action (Maltais, 2014; Shue, 2011). If we accept this minimal principle, then the SCC should reflect global climate impacts.

### 3.6 Anthropocentrism

The next moral scope assumption is the restriction to human beings' utility. Existing SCC estimates rarely include nonhuman impacts, but in principle we might adopt several views that would support their inclusion.

It is worth distinguishing three categories of moral worth as discussed by environmental ethicists (Jamieson, 2009; Pepper, 2018). We start with *anthropocentrism*, the view that only human beings are morally considerable, as this is the default assumption in these types of economic contexts.

Singer forcefully argued that anthropocentrism was too narrow, since we can take experiences or preferences to be the morally relevant objects, and human beings are not the only beings with experiences or preferences (Singer, 2011). This justifies *sentientism*, the view that any beings with the relevant experiential states are morally considerable. Singer's version of sentientism is the view that all interests matter equally, and only and all beings that can experience pain or pleasure thereby have interests.

Goodpaster argued that Singer's view was itself too narrow, since, he suggested experiences are not morally considerable by themselves, since sentience is an adaptive capacity of living organisms to improve their fitness (Goodpaster, 1978). This suggests, Goodpaster claims, that sentience cannot be of moral importance itself as it is there to support something else, which presumably is of moral importance. Goodpaster concluded that the only nonarbitrary designation is every living being, whether sentient or not (e.g., including both flora and fauna).

Both extensions beyond human beings for a SCC would involve radical changes to estimating SCCs. However, drawing on the discussion on ecosystem service valuation, there are multiple standard ways of coming to value goods not priced in markets (Atkinson et al., 2014). Each of these would require some assumptions about what is taken to be of (equal) moral importance. However, most of these methods appeal to human estimations of those values. For instance, contingent valuation of ecosystems requires people to estimate the worth *to them* of various types of ecosystems and their services. Trying to get beyond human valuation is more challenging. For instance, adopting Singer's consideration of equal interests would require some way of estimating and comparing the interests of nonhumans while adopting Goodpaster's view would require all living beings to be comparable in some way. One way, for instance, that this could be attempted is via lost biomass where the relevant unit is living matter. While this moral assumption is still in the process of being discussed (McShane, 2018; Sebo, forthcoming), recognizing this as a moral assumption is the first step to incorporating it into the SCC.

## 4 Conclusion

When we are presented with the economic costs of a climate policy, a figure given in financial terms can appear value-free. The purpose of this chapter was to draw attention to the variety of assumptions that are needed to generate these estimates. This is not to say that these estimates should be rejected or not included in our policy evaluations. Instead, it is to say that the choices that we make in terms of both moral and nonmoral assumptions can have great impact on the choices of polices that society adopts. For that reason, we should consider our assumptions carefully and adopt them carefully with an eye toward the long term.

## References

Adler, M.D., Treich, N., 2015. Prioritarianism and climate change. Environ. Resour. Econ. 62 (2), 279–308. https://doi.org/10.1007/s10640-015-9960-7.

Adler, M.D., Anthoff, D., Bosetti, V., Garner, G., Keller, K., Treich, N., 2017. Priority for the worse-off and the social cost of carbon. Nat. Clim. Chang. 7 (6), 443–449. https://doi.org/10.1007/s10584-010-9967-6.

Arrow, K., 1999. Discounting, morality, and gaming. In: Portnoy, P.R., Weyant, J.P. (Eds.), Discounting and Intergenerational Equity. Resources for the Future, New York, pp. 12–22.

Arrow, K.J., Cline, W., Maler, K., Munasinghe, M., Squiteri, R., Stiglitz, J., 1996. Intertemporal equity, discounting, and economic efficiency. In: Intergovernmental Panel on Climate Change, (Ed.), Climate Change Second Assessment Report. Cambridge University Press, Cambridge, UK, pp. 125–144.

Atkinson, G., Bateman, I.J., Mourato, S., 2014. Valuing ecosystem services and biodiversity. In: Helm, D., Hepburn, C. (Eds.), Nature in the Balance: The Economics of Biodiversity. Oxford University Press, Oxford, pp. 101–125.

Broome, J., 1994. Discounting the future. Philos. Public Aff. 23 (2), 128–156. https://doi.org/10.1111/j.1088-4963.1994.tb00008.x.

Broome, J., 2008. The ethics of climate change. Sci. Am.. 298 (6)https://doi.org/10.1038/scientificamerican0608-96 96–100, 102.

Budolfson, M.B., Dennig, F., Fleurbaey, M., Siebert, A., Socolow, R.H., 2017. The comparative importance for optimal climate policy of discounting, inequalities and catastrophes. Clim. Chang. 15 (3–4), 481–494. https://doi.org/10.1007/s10584-017-2094-x.

Burke, M., Hsiang, S.M., Miguel, E., 2015. Global non-linear effect of temperature on economic production. Nature 527 (7577), 235–239. https://doi.org/10.1038/nature15725.

Carleton, T.A., Hsiang, S.M., 2016. Social and economic impacts of climate. Science 353 (6304), aad9837. https://doi.org/10.1126/science.aad9837.

Dasgupta, P., 2008. Discounting climate change. J. Risk Uncertain. 37 (2), 141–169. https://doi.org/10.1007/s11166-008-9049-6.

Davidson, M.D., 2015. Climate change and the ethics of discounting. Wiley Interdiscip. Rev. Clim. Chang. 6 (4), 401–412. https://doi.org/10.1002/wcc.347.

Dennig, F., Budolfson, M.B., Fleurbaey, M., Siebert, A., Socolow, R.H., 2015. Inequality, climate impacts on the future poor, and carbon prices. Proc. Natl. Acad. Sci. USA 112 (52), 15827–15832. https://doi.org/10.1073/pnas.1513967112.

Dietz, S., Stern, N., 2015. Endogenous growth, convexity of damage and climate risk: how Nordhaus' framework supports deep cuts in carbon emissions. Econ. J. 125 (583), 574–620. https://doi.org/10.1111/ecoj.12188.

Fleurbaey, M., Ferranna, M., Budolfson, M.B., Dennig, F., Mintz-Woo, K., Socolow, R.H., et al., 2019. The social cost of carbon: valuing inequality, risk, and population for climate policy. Monist 102 (1), 84–109. https://doi.org/10.1093/monist/ony023.

Gayer, T., Viscusi, W.K., 2016. Determining the proper scope of climate change policy benefits in U.S. regulatory analyses: domestic versus global approaches. Rev. Environ. Econ. Policy 10 (2), 245–263. https://doi.org/10.1093/reep/rew002.

Gayer, T., Viscusi, W.K., 2017. Letter—the social cost of carbon: maintaining the integrity of economic analysis—a response to Revesz et al. (2017). Rev. Environ. Econ. Policy 11 (1), 174–175. https://doi.org/10.1093/reep/rew021.

Gollier, C., 2012. The debate on discounting: reconciling positivists and ethicists. Chic. J. Int. Law 13 (2), 549–564. http://heinonline.org/hol-cgi-bin/get_pdf.cgi?handle=hein.journals/cjil13§ion=26.

Goodpaster, K.E., 1978. On being morally considerable. J. Philos. 75 (6), 308–325. https://doi.org/10.2307/2025709.

Goulder, L.H., Williams, R.C., 2012. The choice of discount rate for climate change policy evaluation. Clim. Chang. Econ. 3 (4), 1250024. https://doi.org/10.1142/S2010007812500248.

Greaves, H., 2017a. Discounting for public policy: a survey. Econ. Philos. 33 (3), 391–439. https://doi.org/10.1017/S0266267117000062.

Greaves, H., 2017b. Population axiology. Philos Compass. 12(11)https://doi.org/10.1111/phc3.12442.
Heath, J., 2017. Climate ethics: justifying a positive social time preference. J Moral Philos 14 (4), 435–462. https://doi.org/10.1163/17455243-46810051.
Howard, P.H., Schwartz, J.A., 2017. Think global: international reciprocity as justification for a global social cost of carbon. Columbia J. Environ. Law 42 (Symposium Issue), 203–294.
Hsiang, S.M., Kopp, R., Jina, A., Rising, J., Delgado, M., Mohan, S., et al., 2017. Estimating economic damage from climate change in the United States. Science 356 (6345), 1362–1369. https://doi.org/10.1126/science.aal4369.
Jackson, F., Smith, M., 2006. Absolutist moral theories and uncertainty. J. Philos. 103 (6), 267–283. http://www.jstor.org/stable/20619943.
Jamieson, D., 2009. The value of nature. In: Ethics and the Environment. Cambridge University Press, Cambridge, pp. 145–180. https://doi.org/10.1017/CBO9780511806186.007.
Kaufman, N., Barron, A.R., Krawczyk, W., Marsters, P., McJeon, H., 2020. A near-term to net zero alternative to the social cost of carbon for setting carbon prices. Nat. Clim. Chang. 1–10. https://doi.org/10.1038/s41558-020-0880-3.
Maltais, A., 2014. Failing international climate politics and the fairness of going first. Polit. Stud. 62 (3), 618–633. https://doi.org/10.1080/14747730500377156.
Manne, A.S., 1995. The rate of time preference. Energy Policy 23 (4/5), 391–394. https://doi.org/10.1016/0301-4215(95)90163-2.
McShane, K., 2018. Why animal welfare is not biodiversity, ecosystem services, or human welfare. Les Ateliers De L'éthique 13 (1), 43–64. https://doi.org/10.7202/1055117ar.
Mintz-Woo, K., 2018. Two moral arguments for a global social cost of carbon. Ethics Policy Environ. 21 (1), 60–63. https://doi.org/10.1080/21550085.2018.1448038.
Mintz-Woo, K., 2019. Principled utility discounting under risk. Moral Philos. Polit. 6 (1), 89–112. https://doi.org/10.1515/mopp-2018-0060.
Mintz-Woo, K., forthcoming. A philosopher's guide to discounting. In: Budolfson, M., McPherson, T., Plunkett, D. (Eds.), Philosophy and Climate Change. Oxford University Press, Oxford
Mintz-Woo, K., Leroux, J., forthcoming. What do climate change winners owe, and to whom?. Econ. Philos.
Mintz-Woo, K, Dennig, F., Liu, H., Schinko, T., 2021. Carbon pricing and COVID-19. Clim. Pol.. https://doi.org/10.1080/14693062.2020.1831432 In press.
Nordhaus, W., 2008. A Question of Balance. Yale University Press, New Haven.
Nordhaus, W.D., 2018a. Projections and uncertainties about climate change in an era of minimal climate policies. Am. Econ. J. Econ. Pol. 10 (3), 333–360. https://doi.org/10.1257/pol.20170046.
Nordhaus, W.D., 2018b. Evolution of modeling of the economics of global warming: changes in the DICE model, 1992–2017. Clim. Chang. 148 (4), 623–640. https://doi.org/10.1007/s10584-018-2218-y.
Nordhaus, W.D., Yang, Z., 1996. A regional dynamic general-equilibrium model of alternative climate-change strategies. Am. Econ. Rev. 86 (4), 741–765. https://www.jstor.org/stable/2118303.
Norton, B.G., 2005. Sustainability: A Philosophy of Adaptive Ecosystem Management. University of Chicago Press, Chicago.
Pepper, A., 2018. Delimiting justice: animal, vegetable, ecosystem? Les Ateliers De L'éthique 13 (1), 210–230. https://doi.org/10.7202/1055125ar.
Pindyck, R.S., 2013. Climate change policy: what do the models tell us? J. Econ. Lit. 51 (3), 860–872. https://doi.org/10.1257/jel.51.3.860.
Pindyck, R.S., 2017. The use and misuse of models for climate policy. Rev. Environ. Econ. Policy 11 (1), 100–114. https://doi.org/10.1093/reep/rew012.
Ramsey, F.P., 1928. A mathematical theory of saving. Econ. J. 38 (152), 543–559. https://doi.org/10.1111/ecoj.12229.
Scheffler, S., 1994. The Rejection of Consequentialism: A Philosophical Investigation of the Considerations Underlying Rival Moral Conceptions. Oxford University Press, Oxford.
Scovronick, N., Budolfson, M.B., Dennig, F., Fleurbaey, M., Siebert, A., Socolow, R.H., et al., 2017. Impact of population growth and population ethics on climate change mitigation policy. Proc. Natl. Acad. Sci. USA 114 (46), 12338–12343. https://doi.org/10.1073/pnas.1618308114.
Sebo, J., forthcoming. Animals and climate change. In: Budolfson, M., McPherson, T., Plunkett, D. (Eds.), Philosophy and Climate Change. Oxford University Press, Oxford.

Shue, H., 2011. Face reality? After you!—a call for leadership on climate change. Ethics Int. Aff. 25 (1), 17–26. https://doi.org/10.1017/S0892679410000055.
Singer, P., 2011. Practical Ethics, third ed. Cambridge University Press, Cambridge.
Stern, N., 2013. The structure of economic modeling of the potential impacts of climate change: grafting gross underestimation of risk onto already narrow science models. J. Econ. Lit. 51 (3), 838–859. https://doi.org/10.1257/jel.51.3.838.
Stern, N., 2015. Why Are We Waiting? MIT Press, Cambridge, MA.
Sterner, T., Persson, U.M., 2008. An even sterner review: introducing relative prices into the discounting debate. Rev. Environ. Econ. Policy 2 (1), 61–76. https://doi.org/10.1093/reep/rem024.

CHAPTER

# 24

# Climate change and refugees


*John F. McEldowney*[a] *and Julie L. Drolet*[b]

[a]School of Law, University of Warwick, Coventry, United Kingdom [b]Faculty of Social Work, University of Calgary, Edmonton, AB, Canada


OUTLINE



## 1 Climate refugees and the law

As we shall see, national and international legal responses to climate refugees are limited, ineffective, and are frequently unable to deal with the problem. The legal terminology in the area of climate refugees is underdeveloped. This is partly because of the absence of a clear definition of "climate refugee" and the inadequacy of the 1951 Refugee Convention drafted in the aftermath of the second world war and only applicable to people who have a well-grounded fear of prosecution because of their race, religion nationality membership of a particular group, or political opinion and are unable to be protected within their home country. This leaves the category of climate refugee without a legal status that fully recognizes their needs and vulnerability. The best that seems to have been achieved is recognition of climate refugee, through global compacts on safe, orderly, and regular migration, which fall short of what is needed.




https://doi.org/10.1016/B978-0-12-822373-4.00010-0

On 19th September 2016, the UN General Assembly in the New York *Declaration for Refugees and Migrants* called for various global compacts one for refugees and the other for "other migrants." This was further endorsed in New York in December 2018 and on 10th September 2018 in Marrakech. Monitoring of the climate refugee phenomenon is undertaken by the UN Internal Displacement Monitoring Council and the Norwegian Refugee Council. Much of the monitoring is linked to extreme weather events and rising sea levels.

In 2018, the UN's *Global compact on safe, orderly, and regular migration* under Objective 2 recognized climate migrants because of natural disasters and climate change. This is a limited achievement as there are no discernible protections that flow from such recognition. The defect in the 1951 *Convention on the Status of the Refugee* leaves people displaced because of the consequences of environment disasters, lacking any precise legal protection. Environmental degradation and subsequent displacement is not the same as "persecution," a necessary qualification to come within the 1951 Convention. This leaves many climate refugees who leave their own country facing considerable obstacles in terms of rights and protection of their vulnerability. The problem of internal migration within a country because of climate change[a] is also an important consideration, and may have profound consequences for the economics and social cohesion of a country.

The United Nations (UN) 2030 Agenda for Sustainable Development includes a process of review of migration and a better disaggregation of the causes of the migration including aspects of climate change. There are many recent instances such as the 200,000 Bangladeshi's made homeless each year because of flooding. The 2030 Agenda for Sustainable Development and its guiding principle, "Leave no one behind," brought together members of the international community who have set the goal of improving the living conditions of poor and marginalized groups, including refugees and migrants (Koch and Khunt, 2020).

## 2 Climate refugees and the European Union

The European Union's approach to climate refugees is unsatisfactory and disappointing. The term "climate refugee" is not legally recognized in EU law and it is not easy to find the meaning of the term through any interpretation of existing legal provisions. Despite this major shortcoming in the legal recognition of climate refugees, the EU does recognize a causal link between refugees and climate change.[b] However, despite this, there is little progress toward formal legal recognition. In 2009, the Stockholm Programme included the EU's internal security and the need to address climate change and resulting migratory flows, climate change to be addressed was included as linked to the likely increase in migratory flows. This was followed in 2011–13 by an EU strategy including a 179 Euro million budget, to include cooperation over climate change and migration in third countries.

The 2011 study made by the Directorate General for Internal Policies of the EU (Citizens' Rights and Constitutional Affairs) on Climate Refugees: Legal and Policy Responses to

[a] World Bank, Groundswell: Preparing for Internal Climate Migration Washington, 2018.

[b] See EU Commission President Jean-Claude Juncker speech (2015), Notably the EU Parliament identified migration and refugees as well as climate change on the list of *Ten Issues to Watch in 2020* (EPRS, January 2020), PE 646.

environmentally induced migration made some significant conclusions. The study concluded that reforming the Geneva Convention through an extension or amendment to include climate change refugees was not feasible. The concern was that this might devalue the current protection given to refugees. The absence of political will and a necessary consensus made it unlikely that there could be a new framework applicable to the environment or climate change displacement. It also canvassed the suggestion that various forms of temporary protection might be advanced according displaced people with some protection in their phase of displacement and resettlement. There is also an option of broadening the Guiding Principles on Internal Displacement, but this was thought to lack any credible legal force.

In April 2013, the EU Commission published its own document on climate refugees. This was focused mainly on population movements between developed countries rather than on the plight of refugees. It rejected the idea that it was necessary to create a category of refugee that linked climate change to refugee status. It is hard to identify the main reasons for the inaction. While some countries have supported the general idea of creating a new category of climate refugee, this was not generally accepted. The Director General for Migration and Home Affairs suggested a status of "permanently forced migration," the idea being to deal with the issue of refugees but not to make a link with climate change. The contention that climate change is the cause of refugees being forced to leave their own country brings to the fore contentious issues that play into the hands of climate change skeptics, who deny climate change is taking place.

There are also some legal arguments and difficulties about the addition of a new category of climate refugee. The first is that it might be unnecessary as the EU has already accepted migration patterns that apply to EU citizens moving within Member States. EU citizens are already protected because of free movement of people and also the application of the EU Charter of Human Rights. This leaves the challenge of those outside the EU where refugees are already a contentious issue and might be best left to be resolved in international law. The EU Commission's Directorate General for Environment in its publication Migration in Response to Environmental Change recognized the pattern of refugees from within the EU as well as from outside. In terms of EU citizens, the Commission's 2016 working document, in line with the Paris Agreement recognizes that Goal 10 of the Paris Agreement addresses already accepts the irregular pattern of migration caused by the impacts of climate change. The 2014–17 Road Map for the Joint Africa-EU strategy makes clear the need for cooperation with the African nations and the recognition of protecting migrants' human rights.

This falls short of the need, identified by a 2011 study on climate refugees, of providing asylum legislation, which provides examples from their country because of the rapidly changing nature of climate change. Providing asylum legislation is a sensible proposal that would advance asylum seeking to incorporate climate refugees. The use of the Temporary Protection Directive (TPD) might apply to any influx of migrants. It is mainly a financial and political response rather than defining enforceable legal rights and obligations. The TPD arrangements were originally intended to aid migrants from armed conflict, it might be possible to fit the Directive to meet climate refugees. However, a report by the Commission in 2015 on A European Agenda on Migration failed to mention climate change. Another Commission report in 2016 into the working of the TPD made no mention of climate change or natural disasters.

Despite this lack of progress and failure in respect of climate refugees, there are some encouraging signs. The April 2018 European Parliament resolution on Progress on UN Global Compacts for safe, orderly, and regular migration and on refugees put a focus on the various diverse drivers of irregular migration and forced displacement including extreme poverty, climate change, or natural disasters.

# 3 The United Nation's International Organization for Migration

The term climate refugee is relatively recent, dating back to the 1985 period when the United Nations Environment Programme identified the phenomenon. The UN's *International Organization for Migration* (IOM) is an expert organization that identifies emerging migration issues and brings visibility to the challenges. Its first publication in 1992 made the link between the environment and migration. Since then the IOM has monitored and updated its data. Establishing clear legal categories of climate refugees are very challenging. The IOM's analysis is that an individual's rights and the obligation of states are engaged by the climate refugees. There is broad acceptance that the 1951 Refugee Convention is inadequate to the problem, but principles of humanitarian law and of environmental law are relevant. There is concern that because of the highly politically sensitive issue of climate change it is difficult to adopt a satisfactory definition. There is not internationally accepted legal definition and no legal instrument that addresses the problem. In the case of internal migration, the 1998 UN Guiding principle on Internal Displacement may apply to persons who are avoiding natural or human-made disasters and who have NOT crossed an internationally recognized state border. This leaves a wide margin for interpretation and states are free to interpret their obligations.

The IOM prefers "soft law" solutions that encourage informal mechanisms building on rights based approaches rather than formal legal sanctions. Their conclusions are that the difficulties of legal conceptualization make it best to focus on relevant rights tailored to the environmental situation including the consequences for individuals. This is an approach that reflects the problems that the legal system operates under. The influential *World Migration Report* 2020 devotes a chapter to "human mobility and adaptation to environmental change." The report accepts that "anthropogenic climate change is expected to increasingly affect migration and other forms of people moving to manage those changing risks."[c] Nonlegal solutions are proposed including adaptation and change as well as mitigation strategies. Many examples are cited in the report including the drylands of the Senegal, the mountains of the Himalayas and Central Asia and the coastal areas of the Pacific Islands as well as the urban areas of Kenya. Many consequences of climate change are mentioned including moving for employment opportunities as well as changes in land fertility and in the fishing industry. Small islands are particularly visible as examples of the challenges that climate change presents. In common with failures in law and legal regulation, it seems that policy-making lacks impetus to create a common framework for the future. A small start has been made through the United Nations Framework Climate Change Convention and Global Compact for Safe,

[c]World Migration Report, 2020, Washington 2020. p. 253 paragraph 2.

Orderly, and Regular Migration. The Task Force established after COP21 at Paris will be further considered in the upcoming meeting in 2021.

## 4 Forced displacement

According to UNHCR, 100 million people were forced to flee their homes in the past decade, and forced displacement has almost doubled since 2010, from 41 million to 79.5 million (UNHCR, 2020). It is estimated that 80% of the world's displaced people are in countries or territories affected by acute food insecurity and malnutrition, of which many countries are facing climate and other disaster risk (UNHCR, 2020). The reality of forced displacement is widespread and long term. As discussed earlier, most people displaced across an international border solely by the effects of climate change or disaster will not fall within the definition of a refugee under international law even though they may still be in need of international protection, on a temporary or longer term basis (UNHCR, 2019b).

## 5 Environmental migration

The concept of environmental migration is controversial, largely because of the difficulty in measuring the extent to which environmental factors compel people to move (Drolet et al., 2014). "Environmentally induced migrants" is a term used to describe persons on the move in response to immediate life-threatening events or because the environment has deteriorated their livelihoods so much that they can no longer support themselves (Warner et al., 2008). With the impacts of climate change and climate-induced environmental disasters such as floods and wildfires, Alston et al. (2019) discuss the trauma of suddenly having to decide how to move forward following catastrophic events. Individuals, families, and communities are often torn apart by loss of life, homes, infrastructure, and destroyed livelihoods. Individuals and families are forced to relocate in search of new livelihoods, housing, schools and health services, and to seek income to be generated elsewhere (Alston et al., 2019).

Betts (2013) developed the concept of "survival migration" to highlight the situation of people whose own countries are unable or unwilling to ensure their most fundamental human rights (yet fall outside the framework of the refugee protection regime) by not focusing on the underlying cause of movement—whether persecution, conflict, or environment. "Climate refugees" fall outside the dominant interpretation of a refugee under the international framework created after the Second World War (Betts, 2013).

While migration is a coping mechanism used throughout history by societies as a means of coping with climate variability, the right to migration at a time when there are increasing inequities in international labor flow practice is contested (Adger et al., 2009). In some parts of the world, particularly in developing countries, migration may be a limited option. This explains the new emphasis on enhancing resilience and supportive adaptive capacity in order to build on existing coping strategies in international developments in climate policy (Adger et al., 2009).

## 6 Planned relocation

"Climigration" is a term used to describe the planned relocation of entire communities to new locations further from harm, and may be the only feasible solution and best adaption response to climate change in some cases (Matthews, 2019). Low lying coastal communities are vulnerable to sea-level rise, inundations of tides, increased intensity of storm surges and coastal erosion, and extreme weather events such as hurricanes or cyclones contribute to migration. The loss of coastal land to rising seas and storm surges forced residents of Isle de Jean Charles in Louisiana in the United States to undergo federally sanctioned resettlement when the state bought land to develop a new community (Louisiana Office of Community Development, 2020).

Similarly, several communities in Fiji have undergone relocation because of the effects of climate change. Planned Relocation Guidelines were first proposed at the National Climate Change Summit held in Labasa in 2012, recognizing that planned relocation within Fiji represents an option of last resort, but is expected to become a more common response to climate-related events in the future (Republic of Fiji, 2018). "Relocation is the voluntary, planned, and coordinated movement of climate-displaced persons within States to suitable locations, away from risk-prone areas, where they can enjoy the full spectrum of rights including housing, land, and property rights and all other livelihood and related rights" (Republic of Fiji, 2018, p. 6). Over time, the cumulative effects of climate change in affected communities render traditional places of living uninhabitable, especially when compounded by preexisting pressure such as overcrowding, unemployment, poor infrastructure, pollution and environmental fragility (Republic of Fiji, 2018).

Massive population movements are occurring within countries and across borders because of climate-induced disasters. Global population growth is exacerbating this movement from rural to urban areas and across borders, particularly from developing to developed nations (DESA, 2019). The world population is expected to reach 9.7 billion by 2050 (DESA, 2019). A growing population is placing increased pressure on urban cities and fuelling hyperurbanization (Dominelli, 2020).

On January 20, 2020, the United Nations Human Rights Committee ruled that climate migrants cannot be returned to countries where their lives may be threatened by the climate crisis (Kaduuli, 2020). The judgment opens the door to protection claims for those threatened by the effects of climate change and the legal obligations of countries under international law. Action is needed to protect the human rights of those most impacted from climate change.

## 7 Climate change and conflict

Climate change and disasters can exacerbate threats that force people to flee within their country or across international borders (UNHCR, 2019a). Increasing extreme weather events because of climate change are contributing factors to conflict in some regions of the world. Environmental degradation because of drought and desertification, erratic rainfall, land degradation, and population growth can lead to shortages in food, fresh water and energy, contributing to political instability, insecurity, conflict, and violence. For example, in Syria, the drought that occurred between 2006 and 2010 created multiple stresses that contributed to the ongoing conflict since 2011. Political instability, chronic poverty and inequality, and climate change influence displacement within and across countries. Many people who fled environmental disasters in countries such as Somalia, Iraq, and Yemen had previously been displaced by conflict. The loss and grief associated with forced displacement can be unsettling, traumatic, and contribute to socioeconomic challenges and mental health concerns. In practice and research, traumatic events include disasters, war, forced migration and displacement, forced separations of children from their parents, abuse and neglect, torture, accidents and injuries, health crises and assaults to emotional, physical, social and spiritual well-being (Harms, 2015).

## 8 Canada's immigration and refugee policy

Canada is among the refugee resettlement countries that has signed international agreements to protect refugee rights (Kaduuli, 2020; Murray, 2010). Yet there is currently no specific provision for climate change migrants in Canada's Immigration Refugee Protection Act. While the federal government has responsed to environmental disasters such as the 2010 earthquake in Haiti by temporarily suspending removal orders, resettlement options may need to be considered to plan for climate migration in the near future. Permanent residency could be granted on humanitarian and compassionate grounds. Climate migrants are not refugees, but they deserve the same protections as convention refugees (Kaduuli, 2020).

Canada admits approximately 300,000 immigrants of all classes per year, including family class, economic immigrants, and refugees. While Canada has a reputation for being open to immigrants as a diverse and welcoming society, very little action is being taken to address forced migration because of the impact of climate change. Migration is best managed by developing a well-planned, comprehensive, and cooperative approach that facilitates the movement of people both within and beyond borders (Dickson et al., 2014). Canadian immigration policies need to be reformed to create a new category of immigration status for climate migrants in order to address migration dynamics. Canada needs to contribute to international dialog on managing migrants who are displaced by diverse factors including climate change, which is necessary for climate justice.

## 9 Climate justice

The climate crisis involves profound issues of political, economic, and social justice, which cannot be resolved without equally profound changes in the political, economic, and social systems that are causing the crisis (Angus, 2010). Climate justice recognizes that identity is shaped by cultural, social, political, and environmental conditions. Increasingly, social workers and human service practitioners are being called upon to promote community capacity building in response to social, economic, and environmental challenges that contribute to displacement and migration (Drolet et al., 2014). It is critical to explore climate change adaptation strategies that address the challenges experienced by climate migrants and climate "refugees" in their country of origin, relocation, and potential resettlement. Many of the most marginalized and vulnerable populations are forced to flee because of intersecting factors that are intensified by the climate crisis.

As discussed in this chapter, displacements induced by climate change may be temporary or may require permanent resettlement. Industrialized countries have disproportionately benefitted from the combustion of fossil fuels, and have moral obligations to provide humanitarian assistance and to take actions in solidarity with other parts of the world who have contributed least to climate change yet will disproportionately feel its impacts (Dickson et al., 2014).

## 10 Conclusions

Law is often described as goal-oriented setting standards, principles, and procedures that seek to address environmental problems, including climate change. Law and regulation assists in setting priorities and advancing fairness and justice. Such concepts and ideals seem remote to the needs of climate refugees. The complex task of obtaining an acceptable legal definition of climate change refugee is a major challenge. The language of a "refugee" is opposed by many organizations on the grounds that refugee status, itself contested by many countries, may be further complicated and further confused by public debate and controversy surrounding the existence of climate change. Making the link between refugees and climate change is thought to be counterproductive. Alternative terminology such as "environmentally/climate displaced person" seems inelegant and may not seem to be appropriate. Environment migrant is widely used but falls outside the remit of the 1951 Refugee Convention. The World Migration Report 2020 recognizes the three elements that define and frame human mobility in the context of climate change namely securitization, protection and adaptation and climate risk management. The report seeks to argue for better integration of migration into global climate policy-making. This would place human mobility and the deterioration of the environment as part of environmental policy and at the heart of addressing climate change.

## References

Adger, W.N., Huq, S., Brown, K., Conway, D., Hulme, M., 2009. Adaptation to climate change in the developing world. In: Schipper, E.L.F., Burton, I. (Eds.), The Earthscan Reader on Adapation to Climate Change. Earthscan, London, pp. 295–311.

Alston, M., Hazeleger, T., Hargreaves, D. (Eds.), 2019. Social Work and Disasters: A Handbook for Practice. Routledge, Oxon & New York.

Angus, I., 2010. Introduction. In: Angus, I., Rebick's, J. (Eds.), The Global Fight for Climate Justice. Fernwood Publishing, Winnipeg, MB, pp. 11–14.

Betts, A., 2013. Survival Migration: Failed Governance and the Crisis of Displacement. Cornell University Press, Ithaca, NY.

DESA (UN Department of Economic and Social Affairs), 2019 June 17. Growing at a Slower Pace, World Population Is Expected to Reach 9.7 Billion in 2050 and Could Peak at Nearly 11 Billion Around 2100. https://www.un.org/development/desa/en/news/population/world-population-prospects-2019.html.

Dickson, S., Webber, S., Takara, T.K., 2014 November. Preparing BC for Climate Migration. Canadian Centre for Policy Alternatives BC Office, Vancouver, BC. https://www.policyalternatives.ca/sites/default/files/uploads/publications/BC%20Office/2014/11/ccpa-bc_ClimateMigration_web.pdf.

Dominelli, L., 2020. Community development in greening cities. In: Todd, S., Drolet, J. (Eds.), Community Practice and Social Development in Social Work. Springer Nature, Signapore, pp. 189–202.

Drolet, J., Sampson, T., Prashanthi, D., Richard, L., 2014. Social work and environmentally induced displacement: a commentary. Refuge 29 (2), 55–62.

Harms, L., 2015. Understanding Trauma and Resilience. Palgrave Macmillan, London.

Kaduuli, S., 2020 February 5. Canada has a moral obligation to accept climate migrants. Policy Options Politiques. Retrieved from: https://policyoptions.irpp.org/magazines/february-2020/canada-has-a-moral-obligation-to-accept-climate-migrants/.

Koch, A., Khunt, J., 2020 July. Migration and the 2030 Agenda: Making Everyone Count. https://www.swp-berlin.org/en/publication/migration-and-the-2030-agenda-making-everyone-count/.

Louisiana Office of Community Development, 2020 June 9. Resettlement of Isle de Jean Charles—Background and Overview. http://isledejeancharles.la.gov/sites/default/files/public/IDJC-Background-and-Overview-6-20_web.pdf.

Matthews, T., 2019 September 15. 'Climigration': When Communities Must Move Because of Climate Change. The Conversation. https://theconversation.com/climigration-when-communities-must-move-because-of-climate-change-122529.

Murray, S., 2010. Environmental migrants and Canada's refugee policy. Refuge 27 (1), 89–102.

Republic of Fiji, Ministry of Economy, 2018. Fiji: Planned Relocation Guidelines: A Framework to Undertaken Climate Change Related Relocation. https://www.refworld.org/docid/5c3c92204.html.

UNHCR, 2019a. Global Trends Forced Displacement in 2019. https://www.unhcr.org/globaltrends2019/.

UNHCR, 2019b. Key Messages and Commitments on Climate Change and Disaster Displacement. https://www.unhcr.org/protection/environment/5e01e3857/key-messages-for-cop25.html.

UNHCR, 2020 June 18. 1 Per Cent of Humanity Displaced: UNHCR Global Trends Report. https://www.unhcr.org/news/press/2020/6/5ee9db2e4/1-cent-humanity-displaced-unhcr-global-trends-report.html.

Warner, K., Afifi, T., Dun, O., Stal, M., Schmidl, S., 2008. Human Security, Climate Change and Environmentally Induced Migration. United Nations University. http://www.iesp.de/fileadmin/user_upload/pdf/_QOXBVUDYWX-6102009181115-IUBILTNZOG_.pdf.

# Index

Note: Page numbers followed by *f* indicate figures, *t* indicate tables, and *np* indicate footnotes.

## B



## C

## D

## K



## L



## M



## N

Printed in the United States
by Baker & Taylor Publisher Services